INTRODUCTION TO FINITE ELEMENT VIBRATION ANALYSIS, SECOND EDITION

There are many books on finite element methods but few give more than a brief description of their application to structural vibration analysis. This book presents an introduction to the mathematical basis of finite element analysis as applied to vibrating systems. Finite element analysis is a technique that is very important in modelling the response of structures to dynamic loads. Although this book assumes no previous knowledge of finite element methods, those who do have knowledge will still find the book to be useful. It can be utilised by aeronautical, civil, mechanical and structural engineers as well as naval architects. This second edition includes information on the many developments that have taken place over the last 20 years. Existing chapters have been expanded, where necessary, and three new chapters have been included that discuss the vibration of shells and multilayered elements and provide an introduction to the hierarchical finite element method.

Maurice Petyt is an Emeritus Professor of Structural Dynamics, Institute of Sound and Vibration Research, University of Southampton. He has also held appointments as a Research Professor at George Washington University and Visiting Professor at the University d'Aix-Marseille II, France, and the National University of Singapore. He is a Charted Mathematician, a Fellow of the Institute of Mathematics and Its Applications, a Fellow of the International Institute of Acoustics and Vibration, and a Fellow of the Institute of Acoustics. Formerly European Editor and Editor-in-Chief, he is now Editor Emeritus of the *Journal of Sound and Vibration.*

Introduction to Finite Element Vibration Analysis

Second Edition

Maurice Petyt
University of Southampton

CAMBRIDGE
UNIVERSITY PRESS

CAMBRIDGE
UNIVERSITY PRESS

32 Avenue of the Americas, New York NY 10013-2473, USA

Cambridge University Press is part of the University of Cambridge.

It furthers the University's mission by disseminating knowledge in the pursuit of education, learning and research at the highest international levels of excellence.

www.cambridge.org
Information on this title: www.cambridge.org/9781107507357

First edition published 1990
Second edition published 2010
First paperback edition 2015

A catalogue record for this publication is available from the British Library

Library of Congress Cataloguing in Publication data

Petyt, M.
Introduction to finite element vibration analysis / Maurice Petyt. – 2nd ed.
p. cm.
Includes bibliographical references and index.
ISBN 978-0-521-19160-9
1. Vibration. 2. Finite element method. I. Title.
TA356.P47 2010
624.1'76–dc22 2010029494

ISBN 978-0-521-19160-9 Hardback
ISBN 978-1-107-50735-7 Paperback

Contents

Preface *page* xi
Notation xv

1 **Formulation of the Equations of Motion** 1
1.1 Dynamic Equilibrium 1
1.2 Principle of Virtual Displacements 3
1.3 Hamilton's Principle 4
1.4 Lagrange's Equations 8
1.5 Equations of Motion for a System with Constraints 12
Problems 15

2 **Element Energy Functions** 19
2.1 Axial Element 20
2.2 Torque Element 21
2.3 Beam Bending Element 23
2.4 Deep Beam Bending Element 25
2.5 Membrane Element 26
2.6 Thin Plate Bending Element 28
2.7 Thick Plate Bending Element 30
2.8 Three-Dimensional Solid 31
2.9 Axisymmetric Solid 33
2.10 The Dissipation Function 35
2.11 Equations of Motion and Boundary Conditions 36
Problems 40

3 **Introduction to the Finite Element Displacement Method** 45
3.1 Rayleigh–Ritz Method 45
3.2 Finite Element Displacement Method 53
3.3 Axial Vibration of Rods 56
3.4 Torsional Vibration of Shafts 70
3.5 Bending Vibration of Beams 72
3.6 Vibration of Plane Frameworks 77

3.7 Vibration of Three-Dimensional Frameworks 84
3.8 Techniques for Increasing the Accuracy of Elements 92
3.9 Shear Deformation and Rotary Inertia Effects 95
3.10 Numerical Integration 101
3.11 Other Considerations for Beams 112
Problems 115

4 In-plane Vibration of Plates 119

4.1 Linear Triangular Element 120
4.2 Linear Rectangular Element 127
4.3 Linear Quadrilateral Element 132
4.4 Area Coordinates for Triangles 137
4.5 Linear Triangle in Area Coordinates 138
4.6 Increasing the Accuracy of Elements 139
Problems 144

5 Vibration of Solids 148

5.1 Axisymmetric Solids 148
5.2 Applied Loading 149
5.3 Displacements 152
5.4 Reduced Energy Expressions 152
5.5 Linear Triangular Element 153
5.6 Core Elements 160
5.7 Arbitrary Shaped Solids 163
5.8 Rectangular Hexahedron 165
5.9 Isoparametric Hexahedron 170
5.10 Right Pentahedron 174
5.11 Volume Coordinates for Tetrahedra 178
5.12 Tetrahedron Element 180
5.13 Increasing the Accuracy of Elements 182
Problems 190

6 Flexural Vibration of Plates 192

6.1 Thin Rectangular Element (non-conforming) 193
6.2 Thin Rectangular Element (conforming) 204
6.3 Thick Rectangular Element 209
6.4 Thin Triangular Element (non-conforming) 215
6.5 Thin Triangular Element (conforming) 222
6.5.1 Cartesian Coordinates 222
6.5.2 Area Coordinates 228
6.6 Thick Triangular Element 234
6.7 Other Plate Bending Elements 237
Problems 245

7 Vibration of Stiffened Plates and Folded Plate Structures 248

7.1 Stiffened Plates I 248

7.2 Stiffened Plates II 252
7.3 Folded Plates I 257
7.4 Folded Plates II 259
7.5 Folded Plates III 262
Problems 265

8 **Vibration of Shells** 266
8.1 Thin Shell Elements 266
8.2 Thick Shell Elements 268
8.2.1 Middle Surface Shell Element 269
8.3 Thick Axisymmetric Shell Elements 277
8.3.1 Middle Surface Axisymmetric Shell Element 277
Problems 283

9 **Vibration of Laminated Plates and Shells** 284
9.1 Laminated Plate Elements 285
9.1.1 Classical Laminated Plate Theory 285
9.1.2 First-Order Shear Deformation Plate Theory 288
9.1.3 Third-Order Shear Deformation Plate Theory 290
9.2 Laminated Shell Elements 295
9.3 Sandwich Plate and Shell Elements 295

10 **Hierarchical Finite Element Method** 297
10.1 Polynomial Functions 297
10.1.1 Axial Vibration of Rods 298
10.1.2 Bending Vibration of Beams 300
10.1.3 Flexural Vibration of Plates 303
10.1.4 Vibration of Shells 306
10.2 Trigonometric Functions 307
10.2.1 Axial Vibration of Rods 307
10.2.2 Bending Vibration of Beams 309
10.2.3 In-plane Vibration of Plates 310
10.2.4 Flexural Vibration of Plates 311
10.2.5 Vibration of Shells 314

11 **Analysis of Free Vibration** 316
11.1 Some Preliminaries 316
11.1.1 Orthogonality of Eigenvectors 322
11.1.2 Transformation to Standard Form 322
11.2 Sturm Sequences 326
11.3 Orthogonal Transformation of a Matrix 334
11.4 Eigenproblem Solution Methods 334
11.5 Reducing the Number of Degrees of Freedom 336
11.5.1 Making Use of Symmetry 336
11.5.2 Rotationally Periodic Structures 338

11.5.3 Elimination of Unwanted Degrees of Freedom 343
11.5.4 Component Mode Synthesis 349
11.5.4.1 Fixed Interface Method 349
11.5.4.2 Free Interface Method 352
Problems 355

12 **Forced Response I** 357
12.1 Modal Analysis 357
12.2 Representation of Damping 358
12.2.1 Structural Damping 358
12.2.2 Viscous Damping 359
12.3 Harmonic Response 361
12.3.1 Modal Analysis 361
12.3.2 Direct Analysis 370
12.4 Response to Periodic Excitation 377
12.5 Transient Response 381
12.5.1 Modal Analysis 381
12.5.1.1 Central Difference Method 384
12.5.1.2 The Houbolt Method 389
12.5.1.3 The Newmark Method 394
12.5.1.4 The Wilson θ Method 400
12.5.2 Direct Analysis 402
12.5.2.1 Central Difference Method 403
12.5.2.2 The Houbolt Method 408
12.5.2.3 The Newmark Method 409
12.5.2.4 The Wilson θ Method 409
12.5.3 Selecting a Time Step 410
12.5.4 Additional Methods 411
Problems 411

13 **Forced Response II** 413
13.1 Response to Random Excitation 413
13.1.1 Representation of the Excitation 413
13.1.2 Response of a Single Degree of Freedom System 423
13.1.3 Direct Response of a Multi-Degree of Freedom System 426
13.1.4 Modal Response of a Multi-Degree of Freedom System 430
13.2 Truncation of the Modal Solution 431
13.2.1 Mode Acceleration Method 435
13.2.2 Residual Flexibility 437
13.3 Ritz Vector Analysis 438
13.4 Response to Imposed Displacements 440
13.4.1 Direct Response 440
13.4.2 Modal Response 442
13.5 Reducing the Number of Degrees of Freedom 443
13.5.1 Making Use of Symmetry 443
13.5.2 Rotationally Periodic Structures 444

13.5.3 Elimination of Unwanted Degrees of Freedom 446
13.5.4 Component Mode Synthesis 446
Problems 448

14 **Computer Analysis Techniques** 449
14.1 Program Format 450
14.1.1 Pre-processing 451
14.1.2 Solution Phase 456
14.1.3 Post-processing 457
14.2 Modelling 458
14.3 Using Commercial Codes 463

APPENDIX 1: Equations of Motion of Multi-Degree of Freedom Systems 467

APPENDIX 2: Transformation of Strain Components 471

Answers to Problems 473
Bibliography 477
References 479
Index 497

Preface

There are many books on finite element methods but very few give more than a brief description of their application to structural vibration analysis. I have given lecture courses on this topic to undergraduates, postgraduates and those seeking post experience training for many years. Being unable to recommend a single suitable text led me to write this book.

The book assumes no previous knowledge of finite element methods. However, those with a knowledge of static finite element analysis will find a very large proportion of the book useful. It is written in such a way that it can be used by Aeronautical, Civil, Mechanical and Structural Engineers as well as Naval Architects. References are given to applications in these fields.

The text has been written in modular style. This will facilitate its use for courses of varying length and level. A prior knowledge of strength of materials and fundamentals of vibration is assumed. Mathematically, there is a need to be able to differentiate and integrate polynomials and trigonometric functions. Algebraic manipulation is used extensively but only an elementary knowledge of vector methods is required. A knowledge of matrix analysis is essential. The reader should be able to add, subtract, multiply, transpose, differentiate and integrate matrices. Methods of solving linear equations and the existence of a matrix inverse is a prerequisite, and the evaluation of determinants is also required.

This second edition includes information on the many developments that have taken place over the last 20 years. Existing chapters have been expanded where necessary and three new chapters included.

Chapter 1 deals with methods of formulating the equations of motion of a dynamical system. A number of methods are introduced. The advantages and disadvantages of each are discussed and recommendations made. The treatment is simple for ease of understanding, with more advanced aspects being treated in an appendix. The simplest methods derive the equations of motion from the expressions for kinetic and strain energy and the virtual work done by externally applied loads. Expressions for these are derived for various structural elements in Chapter 2.

The response of practical structures cannot be obtained using analytical techniques due to their complexity. This difficulty is overcome by seeking approximate solutions. Chapter 3 begins by describing the technique known as the Rayleigh–Ritz method. The finite element displacement method is then introduced as a generalised Rayleigh–Ritz method. The principal features of the method are introduced by considering rods, shafts, beams and frameworks. In this chapter specific element matrices are evaluated explicitly. However, many of the elements presented in later chapters can only be evaluated using numerical integration techniques. In preparation for this, numerical integration in one dimension is introduced. The extension to two and three dimensions is presented where required. Section 3.11 has been expanded to contain additional information on open and closed thin-walled section beams and curved beams.

In Chapter 4, various membrane elements are derived. These can be used for analysing flat plate structures which vibrate in their plane. Chapter 5 deals with the vibration of solids using both axisymmetric and three-dimensional elements. Details of a 10-node tetrahedron are given. Chapter 6 indicates the difficulties encountered in the development of accurate plate-bending elements. This has led to a large number of elements being developed in attempting to overcome these problems. Details of the IMDKT and MITC4 elements have been added. Chapter 7 describes methods of analysing the vibrations of stiffened plates and folded plate structures. This involves combining the framework, membrane and plate-bending elements described in previous chapters. The problems which arise and how to overcome them are described. This includes further information on how to calculate the real stiffness and inertia coefficients for the drilling degrees of freedom.

There follow three new chapters. The first is on the vibration of shells. This includes thin, thick and axisymmetric elements. The next deals with multi-layered elements. Details are given for plate elements using classical, first-order and third-order shear deformation theory. References are given for shell elements and sandwich plate and shell elements. The final chapter of this trio, Chapter 10, consists of an introduction to the hierarchical finite element method. Beam, plate and shell elements are presented using orthogonal polynomials and trigonometric and mixed functions.

Chapters 11–13 (formerly 8–10) present methods of solving the equations of motion. Chapter 11 considers the equations for free vibration of an undamped structure. These take the form of a linear eigenproblem. The methods of solution to be found in the major finite element systems are discussed, the mathematical details having been excluded. The presentation is designed to give the finite element user an appreciation of the methods. Program developers will need to consult the references given. Methods of reducing the number of degrees of freedom are presented. These consist of making use of symmetry, the analysis of rotationally periodic structures, Guyan reduction and component mode synthesis.

Methods of predicting the response of structures to harmonic, periodic, transient and random loads are described in Chapters 12 and 13. Both direct and modal analysis techniques are presented. Methods of representing damping are discussed. This now includes frequency-dependent damping. Also modal analysis has been expanded to include information on classical and non-classical damping. The prediction of the response to transient loads involves the use of step-by-step integration methods. The stability and accuracy of such methods are discussed. Chapter 13 now includes a section on Ritz vector analysis. However, the sections on fatigue and failure and response spectrum methods have been omitted.

The final chapter on computer analysis techniques assumes that the reader intends to use a commercial program. Those wishing to write programs are referred to suitable texts.

Chapters 3 to 10 present details of the simpler elements. Reference to more advanced techniques are given at the end of each chapter. Each has its own extensive list of references. Throughout the book numerical examples are presented to illustrate the accuracy of the methods described. At the end of several chapters a number of problems are presented to give the reader practice using the techniques described. Many of these can be solved by hand. Those requiring the use of an existing finite element program are indicated. Readers who do not have one available are referred to a suitable one in Chapter 14.

In preparing such a text it is very difficult to acknowledge all the help given to the author. First and foremost I am indebted to the finite element community who have undertaken research and development that have led to the techniques described. Without their publications, many of which are listed, the task would have been all the greater. I should like to thank all my past research students and those of my colleagues who have stimulated my interest in finite element techniques, also, all the students who have taken the courses on which this book is based.

I am indebted to Maureen Mew, whose excellent typing skills speeded up the process of converting my handwritten notes into the final typescript. I should also like to thank Deborah Chase, Marilyn Cramer and Chris Jones for converting my drawings into reproducible form.

Notation

The following is a list of principal symbols used. Those which have local meaning only and may have different meanings in different contexts are defined when used.

Mathematical Symbols

[ ]	A rectangular or square matrix
⌈ ⌋	A diagonal matrix
⌊ ⌋	A row matrix
{ }	A column matrix
| |	Matrix determinant
$[\ ]^{\mathrm{T}}$	Matrix transpose
$[\ ]^{-1}$	Matrix inverse
$[\ ]^{-\mathrm{T}}$	Inverse transpose: $[\ ]^{-\mathrm{T}} \equiv ([\ ]^{-1})^{\mathrm{T}} \equiv ([\ ]^{\mathrm{T}})^{-1}$
$[\ ]^{\mathrm{H}}$	Complex conjugate of transposed matrix

Latin Symbols

A	Area
$[\mathbf{B}]$	Strain-displacement matrix
$[\mathbf{C}]$	Structural damping matrix (Global)
D	Dissipation function
$[\mathbf{D}]$	Matrix of material constants
E	Young's modulus
$\{\mathbf{f}\}$	Equivalent nodal forces
G	Shear modulus
h	Plate thickness
I	Second moment of area of beam cross-section
$[\mathbf{I}]$	Unit matrix
J	Torsion constant
$[\mathbf{J}]$	Jacobian matrix
k	Spring stiffness
$[\mathbf{k}]$	Element stiffness matrix
$[\mathbf{K}]$	Structural stiffness matrix (Global)

$[\mathbf{m}]$	Element inertia matrix
$[\mathbf{M}]$	Structural inertia matrix (Global)
$\lfloor \mathbf{N} \rfloor$, $[\mathbf{N}]$	Matrix of assumed displacement functions
$\{\mathbf{q}\}$	Modal coordinates
$\{\mathbf{Q}\}$	Modal forces
r, θ, z	Cylindrical coordinates
t	Time
T	Kinetic energy
u, v, w	Components of displacement
$\{\mathbf{u}\}$	Column matrix of nodal displacements
U	Strain energy
$\boldsymbol{V}$	Volume
$\boldsymbol{W}$	Work done by applied forces
x, y, z	Local Cartesian coordinates
X, Y, Z	Global Cartesian coordinates

Greek Symbols

$[\boldsymbol{\alpha}]$	Receptance matrix
γ	Damping ratio
Δ	Increment operator
δ	Virtual operator
$\{\boldsymbol{\varepsilon}\}$	Strain components
$\theta_x, \theta_y, \theta_z$	Rotations about Cartesian axes
κ	Shear factor
λ	Eigenvalue
ν	Poisson's ratio
ξ, η, ζ	Isoparametric coordinates
ρ	Mass per unit volume
$\{\sigma\}$	Stress components
$\boldsymbol{\phi}$	Eigenvector (mode shape)
$\boldsymbol{\Phi}$	Modal matrix
ω	Circular frequency in radians per second

1 Formulation of the Equations of Motion

The first step in the analysis of any structural vibration problem is the formulation of the equations of motion. It is an important part of the exercise, since the success of the analysis is dependent upon the equations of motion being formulated correctly. This process will be less prone to errors if a routine procedure for formulating the equations can be established. In this chapter a number of methods will be presented and discussed.

1.1 Dynamic Equilibrium

The equations of motion of any dynamic system can be written down using Newton's second law of motion, which states that 'the rate of change of momentum of a mass is equal to the force acting on it'.

Consider a mass, m, which is displaced a distance $u(t)$ when acted upon by a force $f(t)$, both being functions of time, t, as shown in Figure 1.1, then Newton's second law of motion gives

$$\frac{\mathrm{d}}{\mathrm{d}t}\left(m\frac{\mathrm{d}u}{\mathrm{d}t}\right) = f(t) \tag{1.1}$$

For constant m, which will be assumed throughout this book, equation (1.1) reduces to

$$m\frac{\mathrm{d}^2u}{\mathrm{d}t^2} = f \tag{1.2}$$

or

$$m\ddot{u} = f \tag{1.3}$$

where dots denote differentiation with respect to time.

Equation (1.3) can be rewritten in the form

$$f - m\ddot{u} = 0 \tag{1.4}$$

If the term $-m\ddot{u}$ is now regarded as a force, then equation (1.4) represents an equation of equilibrium, that is, the sum of the forces acting on the mass is equal to zero. The introduction of this fictitious force, which is referred to as an inertia force, of magnitude $m\ddot{u}$, acting in the opposite direction to the acceleration, $\ddot{u}$, allows an

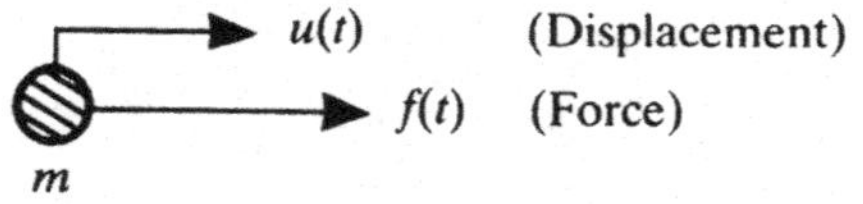

Figure 1.1. Motion of a single mass.

equation of dynamic equilibrium to be formulated using the concepts of static equilibrium. This equation of dynamic equilibrium, when rearranged, gives the equation of motion of the system. This concept is known as d'Alembert's principle.

EXAMPLE 1.1 Derive the equation of motion of the single mass, spring, damper system shown in Figure 1.2(a).

The forces acting on the mass consist of the externally applied force f, a restoring force ku due to the spring, a damping force $c\dot{u}$ due to the viscous damper and a fictitious inertia force $m\ddot{u}$. All act in the directions shown in Figure 1.2(b). For equilibrium

$$-m\ddot{u} - c\dot{u} - ku + f = 0 \tag{1.5}$$

Rearranging, gives the equation of motion

$$m\ddot{u} + c\dot{u} + ku = f \tag{1.6}$$

The above concepts can be extended to multi-degree of freedom systems. Consider a system of N masses. The equations of dynamic equilibrium are obtained by equating the sums of the forces and moments on each mass of the system to zero. This gives

$$\vec{f}_j - \frac{\mathrm{d}}{\mathrm{d}t}(m_j \dot{\vec{u}}_j) = 0 \qquad j = 1, 2, \ldots, N \tag{1.7}$$

and

$$\vec{L}_j - \frac{\mathrm{d}}{\mathrm{d}t}(\vec{J}_j) = 0 \qquad j = 1, 2, \ldots, N \tag{1.8}$$

In these equations $\vec{u}_j$ is the displacement of the mass m_j, $\vec{f}_j$ is the sum of the applied forces, $\vec{J}_j$ is the angular momentum, and $\vec{L}_j$ is the sum of the applied moments. If the vectors $\vec{u}_j$ do not represent independent motions, equations (1.7) and (1.8) must be modified by constraints of the form

$$g_j(\vec{u}_1, \vec{u}_2, \ldots, \vec{u}_N) = 0 \qquad j = 1, 2, \ldots, m \tag{1.9}$$

where m is the number of constraints. This aspect is discussed in Section 1.5.

EXAMPLE 1.2 Derive the equations of motion of the system shown in Figure 1.3.

The mass m_1 has two forces acting on it due to the extension of the two springs joining it to the masses m_2 and m_3.

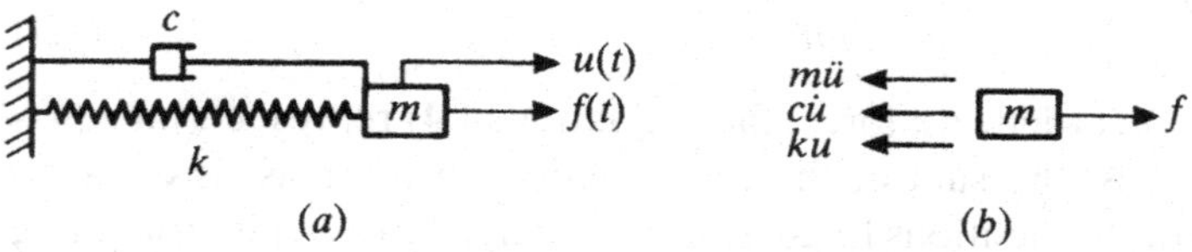

Figure 1.2. Single mass, spring, damper system.

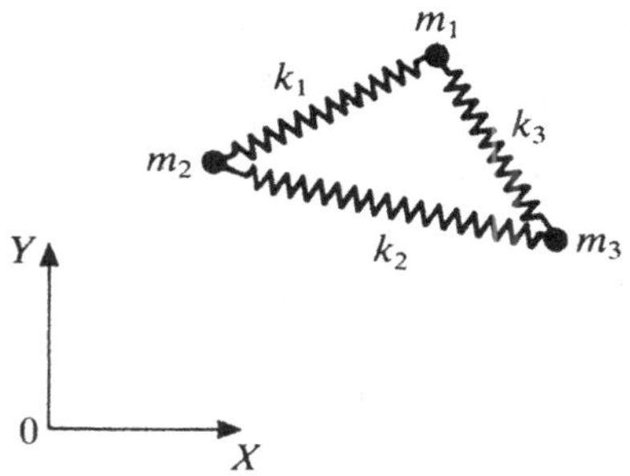

Figure 1.3. Multi-mass, spring system.

If the position vectors of m_1 and m_2 are $\vec{V}1$ and $\vec{V}2$ respectively, then the unit vector $\vec{n}_1$, along the line 2–1 is

$$\vec{n}_1 = \frac{1}{L_1}(\vec{V}1 - \vec{V}2) \tag{1.10}$$

where

$$L_1 = \text{abs}(\vec{V}1 - \vec{V}2)$$

If the displacements of m_1 and m_2 are denoted by $\vec{U}_1$ and $\vec{U}_2$ then the extension, e_1, of the spring joining m_1 and m_2 is given by the scalar product

$$e_1 = (\vec{U}_1 - \vec{U}_2) \cdot \vec{n}_1 \tag{1.11}$$

If the stiffness of the spring is k_1, then the force, f_1, acting on the mass m_1 in the direction $\vec{n}_1$ is

$$f_1 = -k_1 e_1 = k_1(\vec{U}_2 - \vec{U}_1) \cdot \vec{n}_1 \tag{1.12}$$

Similarly, the force, f_3, acting on the mass m_1 in the direction $\vec{n}_3$ is

$$f_3 = k_3(\vec{U}_3 - \vec{U}_1) \cdot \vec{n}_3 \tag{1.13}$$

where

$$\vec{n}_3 = \frac{1}{L_3}(\vec{V}1 - \vec{V}3) \tag{1.14}$$

and

$$L_3 = \text{abs}(\vec{V}1 - \vec{V}3).$$

The equation of dynamic equilibrium for m_1 is therefore

$$f_1\vec{n}_1 + f_3\vec{n}_3 - m_1\ddot{\vec{U}}_1 = 0 \tag{1.15}$$

When the components of each of the vectors are substituted in this equation, two scalar equations will be obtained. These can then be rearranged, in the manner shown in Example 1.1, to give the equations of motion of the mass m_1. The equations of motion of the masses m_2 and m_3 are obtained in a similar way.

1.2 Principle of Virtual Displacements

If the structure to be analysed is a complex one, then the vectoral addition of all the forces acting at each mass point is difficult. This difficulty may be overcome by first using d'Alembert's principle and then the principle of virtual displacements. By this means the equations of dynamic equilibrium and hence the equations of motion, are formulated indirectly.

The principle of virtual displacements states that 'if a system, which is in equilibrium under the action of a set of forces, is subjected to a virtual displacement, then the total work done by the forces will be zero'. In this context, a virtual displacement is a physically possible one, that is, any displacement which is compatible with the system constraints.

EXAMPLE 1.3 Use the principle of virtual displacements to derive the equation of motion of the system shown in Figure 1.2.

Figure 1.2(*b*) shows the forces acting after the application of d'Alembert's principle. If the system is given a virtual displacement δu, then the principle of virtual displacements gives

$$-m\ddot{u}\delta u - c\dot{u}\delta u - ku\delta u + f\delta u = 0 \tag{1.16}$$

Rearranging gives

$$(-m\ddot{u} - c\dot{u} - ku + f)\,\delta u = 0 \tag{1.17}$$

Since δu is arbitrary and non-zero, then

$$m\ddot{u} + c\dot{u} + ku = f \tag{1.18}$$

The advantage of this approach is that the virtual work contributions are scalar quantities which can be added algebraically.

For a multi-degree of freedom system, the principle of virtual work gives

$$\sum_{j=1}^{N}\left(\vec{f}_j - \frac{\mathrm{d}}{\mathrm{d}t}(m_j\dot{\vec{u}}_j)\right)\cdot\delta\vec{u}_j + \sum_{j=1}^{N}\left(\vec{L}_j - \frac{\mathrm{d}}{\mathrm{d}t}(\vec{J}_j)\right)\cdot\delta\vec{\theta}_j = 0 \tag{1.19}$$

where the $\delta\vec{u}_j$ are virtual displacements and the $\delta\vec{\theta}_j$ virtual rotations. Since each of these is arbitrary, equations (1.7) and (1.8) must hold.

1.3 Hamilton's Principle

Although the principle of virtual displacements overcomes the problem of vectorial addition of forces, virtual work itself is calculated from the scalar product of two vectors, one representing a force and one a virtual displacement. This disadvantage can be largely overcome by using Hamilton's principle to determine the equations of motion.

Consider a mass, m, which is acted upon by a force, f_T, causing a displacement, u, as shown in Figure 1.4. f_T represents the sum of all the applied forces, both conservative and non-conservative.

The work done by a conservative force in moving a mass from one point to another depends only on the position of the two points and is independent of the

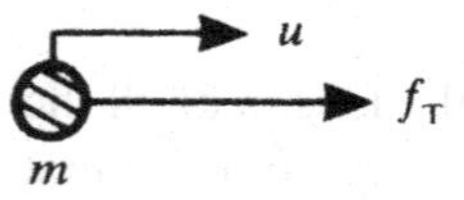

Figure 1.4. Motion of a single mass.

Figure 1.5. Path taken by a mass.

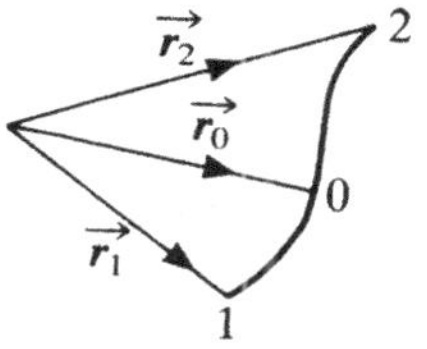

path taken between them. The work done by non-conservative forces does depend upon the path taken between the two points. Non-conservative forces are energy dissipating forces such as friction forces, or forces imparting energy to the system such as external forces.

The work done by a conservative force can be obtained from the change in potential energy. The potential energy $V(\vec{r})$ associated with position $\vec{r}$ is defined as the work done by a conservative force $\vec{f}$ in moving a mass from position $\vec{r}$ to a reference position $\vec{r}_0$. That is

$$V(\vec{r}) = \int_{\vec{r}}^{\vec{r}_0} \vec{f} \cdot \mathrm{d}\vec{r} \tag{1.20}$$

The work done by a conservative force $\vec{f}$ in moving a mass from position $\vec{r}_1$ to position $\vec{r}_2$, as shown in Figure 1.5, is

$$\begin{aligned} W &= \int_{\vec{r}_1}^{\vec{r}_2} \vec{f} \cdot \mathrm{d}\vec{r} \\ &= \int_{\vec{r}_1}^{\vec{r}_0} \vec{f} \cdot \mathrm{d}\vec{r} - \int_{\vec{r}_2}^{\vec{r}_0} \vec{f} \cdot \mathrm{d}\vec{r} \\ &= -\{V(\vec{r}_2) - V(\vec{r}_1)\} \end{aligned} \tag{1.21}$$

Since the force is a conservative one, the work done is independent of the path, and so in Figure 1.5 the path has been chosen to pass through the reference point 0.

Equation (1.21) states that the work done by a conservative force is minus the change in potential energy. In differential form this is

$$\delta W = -\delta V \tag{1.22}$$

The type of potential energy which will be considered in this book is the elastic potential energy, or strain energy U.

Consider a linear elastic spring of stiffness, k, which is stretched by an amount u. Then the force, f, in the spring in the direction of u is

$$f = -ku \tag{1.23}$$

and the potential energy

$$U = \int_u^0 f\,\mathrm{d}u = -\int_u^0 ku\,\mathrm{d}u = \frac{1}{2}ku^2 \tag{1.24}$$

Applying the principle of virtual displacements to the system in Figure 1.4 gives

$$f_{\mathrm{T}}\delta u - m\ddot{u}\delta u = 0 \tag{1.25}$$

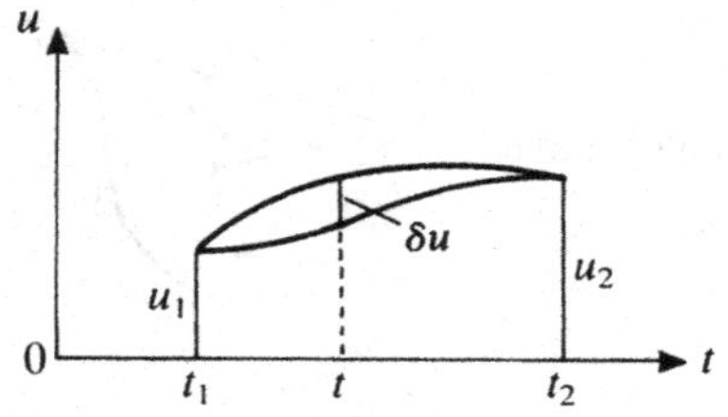

Figure 1.6. Variation in the motion of a mass.

where δu is a virtual displacement.

$$\text{Now } f_T \delta u = \delta W = \text{work done by the forces} \tag{1.26}$$

and

$$m\ddot{u}\delta u = m\frac{\mathrm{d}}{\mathrm{d}t}(\dot{u}\delta u) - m\dot{u}\delta\dot{u} \tag{1.27}$$

where it has been assumed that

$$\frac{\mathrm{d}}{\mathrm{d}t}(\delta u) = \delta\left(\frac{\mathrm{d}u}{\mathrm{d}t}\right) = \delta\dot{u}$$

Equation (1.27) can be further modified as follows

$$\begin{aligned} m\ddot{u}\delta u &= m\frac{\mathrm{d}}{\mathrm{d}t}(\dot{u}\delta u) - \delta\left(\frac{1}{2}m\dot{u}^2\right) \\ &= m\frac{\mathrm{d}}{\mathrm{d}t}(\dot{u}\delta u) - \delta T \end{aligned} \tag{1.28}$$

where

$$T = \tfrac{1}{2}m\dot{u}^2 \tag{1.29}$$

represents the kinetic energy of the system.

Substituting equations (1.26) and (1.28) into equation (1.25) gives

$$\delta W - m\frac{\mathrm{d}}{\mathrm{d}t}(\dot{u}\delta u) + \delta T = 0$$

or, on rearranging

$$\delta T + \delta W = m\frac{\mathrm{d}}{\mathrm{d}t}(\dot{u}\delta u) \tag{1.30}$$

If the position of the mass is known at two instants of time t_1 and t_2, then its motion during this interval of time can be represented by a curve, as shown in Figure 1.6. A slightly different curve or path is obtained if, at any instant, a small variation in position δu is allowed with no associated change in time; that is $\delta t = 0$ (Figure 1.6). The stipulation is made, however, that at times t_1 and t_2 the two paths coincide, that is

$$\delta u = 0 \qquad \text{at } t = t_1 \text{ and } t = t_2 \tag{1.31}$$

The problem is to choose the true path from u_1 to u_2 from all the possible ones.

Multiplying equation (1.30) by $\mathrm{d}t$ and integrating between t_1 and t_2 gives

$$\begin{aligned}\int_{t_1}^{t_2} (\delta T + \delta W)\,\mathrm{d}t &= \int_{t_1}^{t_2} m\frac{\mathrm{d}}{\mathrm{d}t}(\dot{u}\delta u)\,\mathrm{d}t \\ &= [m\dot{u}\delta u]_{t_1}^{t_2} = 0\end{aligned} \tag{1.32}$$

by virtue of equation (1.31). Equation (1.32), therefore, states that

$$\int_{t_1}^{t_2} (\delta T + \delta W)\,\mathrm{d}t = 0 \tag{1.33}$$

Separating the forces into conservative and non-conservative forces, gives

$$\delta W = \delta W_{\mathrm{c}} + \delta W_{\mathrm{nc}} \tag{1.34}$$

Using equation (1.22), namely,

$$\delta W_{\mathrm{c}} = -\delta V \tag{1.35}$$

equation (1.34) becomes

$$\delta W = -\delta V + \delta W_{\mathrm{nc}} \tag{1.36}$$

Substituting equation (1.36) into equation (1.33) gives

$$\int_{t_1}^{t_2} (\delta T - \delta V + \delta W_{\mathrm{nc}})\,\mathrm{d}t = 0 \tag{1.37}$$

or

$$\int_{t_1}^{t_2} (\delta(T - V) + \delta W_{\mathrm{nc}})\,\mathrm{d}t = 0 \tag{1.38}$$

Note that equation (1.37) cannot be written in the form

$$\int_{t_1}^{t_2} \delta(T - V + W_{\mathrm{nc}})\,\mathrm{d}t = 0 \tag{1.39}$$

since a work function W_{nc} does not exist for non-conservative forces. However, the virtual work can always be calculated. Equation (1.38) is the mathematical statement of Hamilton's principle. For a conservative system $\delta W_{\mathrm{nc}} = 0$. In this case equation (1.38) shows that the integral of $(T - V)$ along the true time path is stationary. It can be shown, for the applications considered in this book, that the stationary value of the integral is a minimum.

The application of this principle leads directly to the equations of motion for any system. It can be applied to both discrete, multi-degree of freedom systems (as shown in Appendix 1) and continuous systems (as illustrated in Section 2.11). The advantage of this formulation is that it uses scalar energy quantities. Vector quantities may only be required in calculating the work done by the non-conservative forces. As previously stated, the only potential energy of interest in this book is elastic strain energy U. The form of Hamilton's principle to be used is therefore

$$\int_{t_1}^{t_2} (\delta(T - U) + \delta W_{\mathrm{nc}})\,\mathrm{d}t = 0 \tag{1.40}$$

EXAMPLE 1.4 Use Hamilton's principle to derive the equations of motion of the system shown in Figure 1.2.

For this system

$$\begin{aligned} T &= \tfrac{1}{2}m\dot{u}^2 \\ U &= \tfrac{1}{2}ku^2 \\ \delta W_{\mathrm{nc}} &= f\delta u - c\dot{u}\delta u \end{aligned} \tag{1.41}$$

Substituting into equation (1.40) gives

$$\int_{t_1}^{t_2} \delta\left(\frac{1}{2}m\dot{u}^2 - \frac{1}{2}ku^2\right) \mathrm{d}t + \int_{t_1}^{t_2} (f\delta u - c\dot{u}\delta u)\,\mathrm{d}t = 0 \tag{1.42}$$

that is

$$\int_{t_1}^{t_2} (m\dot{u}\delta\dot{u} - ku\delta u + f\delta u - c\dot{u}\delta u)\,\mathrm{d}t = 0 \tag{1.43}$$

Now

$$\delta\dot{u} = \delta\left(\frac{\mathrm{d}u}{\mathrm{d}t}\right) = \frac{\mathrm{d}}{\mathrm{d}t}(\delta u)$$

Hence integrating the first term by parts gives

$$\begin{aligned} \int_{t_1}^{t_2} m\dot{u}\delta\dot{u}\,\mathrm{d}t &= [m\dot{u}\delta u]_{t_1}^{t_2} - \int_{t_1}^{t_2} m\ddot{u}\delta u\,\mathrm{d}t \\ &= -\int_{t_1}^{t_2} m\ddot{u}\delta u\,\mathrm{d}t \end{aligned} \tag{1.44}$$

by virtue of equation (1.31).

Substituting equation (1.44) into equation (1.43) gives

$$\int_{t_1}^{t_2} (-m\ddot{u} - c\dot{u} - ku + f)\delta u\,\mathrm{d}t = 0 \tag{1.45}$$

Since δu is arbitrary, equation (1.45) is satisfied only if

$$m\ddot{u} + c\dot{u} + ku = f \tag{1.46}$$

1.4 Lagrange's Equations

When Hamilton's principle is applied to discrete systems it can be expressed in a more convenient form. To illustrate this, consider the system shown in Figure 1.2. The kinetic and strain energies are given by

$$T = \tfrac{1}{2}m\dot{u}^2 = T(\dot{u}), \qquad U = \tfrac{1}{2}ku^2 = U(u) \tag{1.47}$$

and the virtual work done by the non-conservative forces is

$$\delta W_{\mathrm{nc}} = (f - c\dot{u})\delta u \tag{1.48}$$

Equation (1.40) therefore becomes

$$\int_{t_1}^{t_2} \left(\frac{\partial T}{\partial \dot{u}}\delta\dot{u} - \frac{\partial U}{\partial u}\delta u + (f - c\dot{u})\delta u\right) \mathrm{d}t = 0 \tag{1.49}$$

Integrating the first term by parts gives

$$\begin{aligned}\int_{t_1}^{t_2} \frac{\partial T}{\partial \dot{u}} \delta \dot{u} \, \mathrm{d}t &= \left[\frac{\partial T}{\partial \dot{u}} \delta u \right]_{t_1}^{t_2} - \int_{t_1}^{t_2} \frac{\mathrm{d}}{\mathrm{d}t} \left(\frac{\partial T}{\partial \dot{u}} \right) \delta u \, \mathrm{d}t \\ &= - \int_{t_1}^{t_2} \frac{\mathrm{d}}{\mathrm{d}t} \left(\frac{\partial T}{\partial \dot{u}} \right) \delta u \, \mathrm{d}t \end{aligned} \tag{1.50}$$

as a consequence of using equation (1.31).

Substituting equation (1.50) into equation (1.49) gives

$$\int_{t_1}^{t_2} \left\{ -\frac{\mathrm{d}}{\mathrm{d}t} \left(\frac{\partial T}{\partial \dot{u}} \right) - \frac{\partial U}{\partial u} + f - c\dot{u} \right\} \delta u \, \mathrm{d}t = 0 \tag{1.51}$$

Since δu is arbitrary, then

$$\frac{\mathrm{d}}{\mathrm{d}t} \left(\frac{\partial T}{\partial \dot{u}} \right) + \frac{\partial U}{\partial u} + c\dot{u} = f \tag{1.52}$$

Introducing a dissipation function, D, which is defined by

$$D = \tfrac{1}{2} c\dot{u}^2 \tag{1.53}$$

the damping force is given by

$$c\dot{u} = \frac{\partial D}{\partial \dot{u}} \tag{1.54}$$

The dissipation function represents the instantaneous rate of energy dissipation which is given by

$$\tfrac{1}{2} \times \text{damping force} \times \text{rate of extension of damper}$$

Substituting the relationship (1.54) into equation (1.52) gives

$$\frac{\mathrm{d}}{\mathrm{d}t} \left(\frac{\partial T}{\partial \dot{u}} \right) + \frac{\partial D}{\partial \dot{u}} + \frac{\partial U}{\partial u} = f \tag{1.55}$$

Equation (1.55) is Lagrange's equation for a single degree of freedom system. Substituting equations (1.47) and (1.53) into equation (1.55) gives

$$m\ddot{u} + c\dot{u} + ku = f \tag{1.56}$$

which is the equation of motion of the system. It can be seen that the term $(\mathrm{d}/\mathrm{d}t)(\partial T/\partial \dot{u})$ gives the inertia force and $\partial U/\partial u$ the restoring force due to the spring.

In the case of a multi-degree of freedom system, the deformation of which is described by n independent displacements $q_1, q_2, \ldots, q_n$, then the kinetic energy is a function of the velocities $\dot{q}_j (j = 1, 2, \ldots, n)$ only and the strain energy a function of the displacements $q_j (j = 1, 2, \ldots, n)$ only, that is

$$\begin{aligned} T &= T(\dot{q}_1, \dot{q}_2, \ldots, \dot{q}_n) \\ U &= U(q_1, q_2, \ldots, q_n) \end{aligned} \tag{1.57}$$

Similarly, the dissipation function is a function of the velocities $\dot{q}_j$, that is

$$D = D(\dot{q}_1, \dot{q}_2, \ldots, \dot{q}_n) \tag{1.58}$$

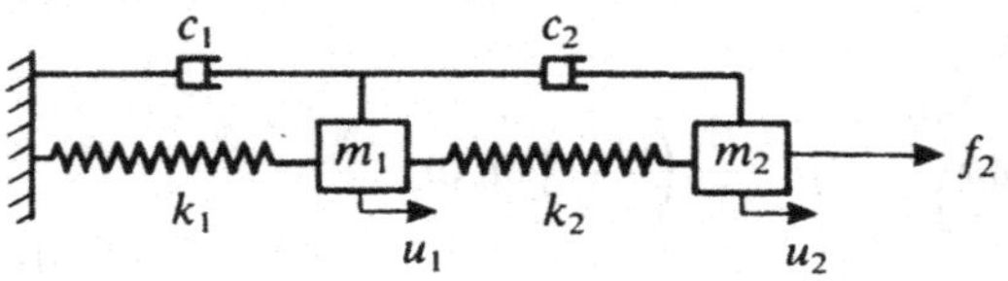

Figure 1.7. Two degree of freedom mass, spring, damper system.

Also, the work done by the non-conservative forces can be written in the form (see Appendix 1)

$$\delta W_{\text{nc}} = \sum_{j=1}^{n} \left(Q_j - \frac{\partial D}{\partial \dot{q}_j} \right) \delta q_j \tag{1.59}$$

where the Q_j are generalised forces.

Lagrange's equations now take the form

$$\frac{\text{d}}{\text{d}t}\left(\frac{\partial T}{\partial \dot{q}_j}\right) + \frac{\partial D}{\partial \dot{q}_j} + \frac{\partial U}{\partial q_j} = Q_j, \qquad j = 1, 2, \ldots, n \tag{1.60}$$

These equations are derived in Appendix 1.

EXAMPLE 1.5 Use Lagrange's equations to derive the equations of motion of the system shown in Figure 1.7

The kinetic energy is given by

$$T = \tfrac{1}{2} m_1 \dot{u}_1^2 + \tfrac{1}{2} m_2 \dot{u}_2^2 \tag{1.61}$$

the dissipation function by

$$\begin{aligned} D &= \tfrac{1}{2} c_1 \dot{u}_1^2 + \tfrac{1}{2} c_2 (\dot{u}_2 - \dot{u}_1)^2 \\ &= \tfrac{1}{2}(c_1 + c_2)\dot{u}_1^2 - c_2 \dot{u}_1 \dot{u}_2 + \tfrac{1}{2} c_2 \dot{u}_2^2 \end{aligned} \tag{1.62}$$

and the strain energy by

$$\begin{aligned} U &= \tfrac{1}{2} k_1 u_1^2 + \tfrac{1}{2} k_2 (u_2 - u_1)^2 \\ &= \tfrac{1}{2}(k_1 + k_2) u_1^2 - k_2 u_1 u_2 + \tfrac{1}{2} k_2 u_2^2 \end{aligned} \tag{1.63}$$

The virtual work done by the applied force is

$$\delta W = f_2 \delta u_2 \tag{1.64}$$

Applying Lagrange's equations (1.60) gives

$$\begin{aligned} m_1 \ddot{u}_1 + (c_1 + c_2)\dot{u}_1 - c_2 \dot{u}_2 + (k_1 + k_2) u_1 - k_2 u_2 &= 0 \\ m_2 \ddot{u}_2 - c_2 \dot{u}_1 + c_2 \dot{u}_2 - k_2 u_1 + k_2 u_2 &= f_2 \end{aligned} \tag{1.65}$$

The procedure can be made even more systematic, and therefore less prone to errors, by using matrix notation. The kinetic energy, dissipation function and strain energy can all be written in the following forms

$$\begin{aligned} T &= \tfrac{1}{2}\{\dot{\mathbf{q}}\}^{\text{T}}[\mathbf{M}]\{\dot{\mathbf{q}}\} \\ D &= \tfrac{1}{2}\{\dot{\mathbf{q}}\}^{\text{T}}[\mathbf{C}]\{\dot{\mathbf{q}}\} \\ U &= \tfrac{1}{2}\{\mathbf{q}\}^{\text{T}}[\mathbf{K}]\{\mathbf{q}\} \end{aligned} \tag{1.66}$$

where

$\{\mathbf{q}\}$ = column matrix of system displacements
$\{\dot{\mathbf{q}}\}$ = column matrix of system velocities
$[\mathbf{M}]$ = square symmetric matrix of inertia coefficients
$[\mathbf{C}]$ = square symmetric matrix of damping coefficients
$[\mathbf{K}]$ = square symmetric matrix of stiffness coefficients

Using equations (1.66), the separate terms in Lagrange's equations become

$$\left\{\frac{\mathrm{d}}{\mathrm{d}t}\left(\frac{\partial T}{\partial \dot{q}}\right)\right\} = [\mathbf{M}]\{\ddot{\mathbf{q}}\}$$
$$\left\{\frac{\partial D}{\partial \dot{q}}\right\} = [\mathbf{C}]\{\dot{\mathbf{q}}\} \qquad (1.67)$$
$$\left\{\frac{\partial U}{\partial q}\right\} = [\mathbf{K}]\{\mathbf{q}\}$$

Lagrange's equations (1.60) therefore yield the following equations of motion in matrix form

$$[\mathbf{M}]\{\ddot{\mathbf{q}}\} + [\mathbf{C}]\{\dot{\mathbf{q}}\} + [\mathbf{K}]\{\mathbf{q}\} = \{\mathbf{Q}\} \qquad (1.68)$$

Equations (1.66) and (1.68) show that it is only necessary to obtain the energy expressions in matrix form in order to determine the matrix coefficients in the equations of motion.

EXAMPLE 1.6 Determine the equations of motion of the system in Figure 1.7 in matrix form.

In matrix form the energy expressions are as follows

$$T = \tfrac{1}{2}m_1\dot{u}_1^2 + \tfrac{1}{2}m_2\dot{u}_2^2$$
$$= \frac{1}{2}\begin{bmatrix}\dot{u}_1\\ \dot{u}_2\end{bmatrix}^{\mathrm{T}}\begin{bmatrix}m_1 & 0\\ 0 & m_2\end{bmatrix}\begin{bmatrix}\dot{u}_1\\ \dot{u}_2\end{bmatrix} \qquad (1.69)$$

giving

$$[\mathbf{M}] = \begin{bmatrix}m_1 & 0\\ 0 & m_2\end{bmatrix} \qquad (1.70)$$

$$D = \tfrac{1}{2}(c_1 + c_2)\dot{u}_1^2 - c_2\dot{u}_1\dot{u}_2 + \tfrac{1}{2}c_2\dot{u}_2^2$$
$$= \frac{1}{2}\begin{bmatrix}\dot{u}_1\\ \dot{u}_2\end{bmatrix}^{\mathrm{T}}\begin{bmatrix}(c_1 + c_2) & -c_2\\ -c_2 & c_2\end{bmatrix}\begin{bmatrix}\dot{u}_1\\ \dot{u}_2\end{bmatrix} \qquad (1.71)$$

giving

$$[\mathbf{C}] = \begin{bmatrix}(c_1 + c_2) & -c_2\\ -c_2 & c_2\end{bmatrix} \qquad (1.72)$$

$$U = \tfrac{1}{2}(k_1 + k_2)u_1^2 - k_2u_1u_2 + \tfrac{1}{2}k_2u_2^2$$
$$= \frac{1}{2}\begin{bmatrix}u_1\\ u_2\end{bmatrix}^{\mathrm{T}}\begin{bmatrix}(k_1 + k_2) & -k_2\\ -k_2 & k_2\end{bmatrix}\begin{bmatrix}u_1\\ u_2\end{bmatrix} \qquad (1.73)$$

giving

$$[\mathbf{K}] = \begin{bmatrix} (k_1 + k_2) & -k_2 \\ -k_2 & k_2 \end{bmatrix} \tag{1.74}$$

Also

$$\{\mathbf{q}\} = \begin{bmatrix} u_1 \\ u_2 \end{bmatrix}, \qquad \{\mathbf{Q}\} = \begin{bmatrix} 0 \\ f_2 \end{bmatrix} \tag{1.75}$$

The equations of motion are therefore

$$\begin{aligned} \begin{bmatrix} m_1 & 0 \\ 0 & m_2 \end{bmatrix} \begin{bmatrix} \ddot{u}_1 \\ \ddot{u}_2 \end{bmatrix} + \begin{bmatrix} (c_1 + c_2) & -c_2 \\ -c_2 & c_2 \end{bmatrix} \begin{bmatrix} \dot{u}_1 \\ \dot{u}_2 \end{bmatrix} \\ + \begin{bmatrix} (k_1 + k_2) & -k_2 \\ -k_2 & k_2 \end{bmatrix} \begin{bmatrix} u_1 \\ u_2 \end{bmatrix} = \begin{bmatrix} 0 \\ f_2 \end{bmatrix} \end{aligned} \tag{1.76}$$

Note that equations (1.65) and (1.76) are identical.

The inertia, damping and stiffness matrices are all symmetric matrices. In addition, the inertia matrix is positive definite and the stiffness matrix either positive definite or positive semi-definite.

A positive definite matrix is one whose elements are the coefficients of a positive definite quadratic form. The kinetic energy is represented by a positive definite quadratic form (equation (1.66)) since $T > 0$ for all possible values of $\{\dot{\mathbf{q}}\} \neq 0$. If the structural system is supported, then the strain energy is also represented by a positive definite quadratic form since $U > 0$ for all possible values of $\{\mathbf{q}\} \neq 0$. An unsupported structure is capable of rigid body motion without distortion. In this case $U = 0$ for some $\{\mathbf{q}\} \neq 0$. Therefore, for such a structure $U \geq 0$ for all $\{\mathbf{q}\} \neq 0$. In this case, the quadratic form for U is said to be positive semi-definite.

1.5 Equations of Motion for a System with Constraints

Sometimes it is easier to express the energy functions in terms of a set of displacements q_j ($j = 1, 2, \ldots, n$) which are not independent. In this case there will be a set of m constrains of the form

$$g_j(q_1, q_2, \ldots, q_n) = 0, \qquad j = 1, 2, \ldots, m \tag{1.77}$$

In this text only linear constraint equations will be considered. Equations (1.77) can therefore be written in the following matrix form

$$[\mathbf{G}]\{\mathbf{q}\} = 0 \tag{1.78}$$

where $[\mathbf{G}]$ is a matrix of coefficients having m rows and n columns.

The column matrix $\{\mathbf{q}\}$ can be partitioned into a set of $(n - m)$ independent displacements $\{\mathbf{q}_1\}$ and a set of m dependent displacements $\{\mathbf{q}_2\}$. Partitioning the matrix $[\mathbf{G}]$ in a compatible manner, equation (1.78) can be written in the form

$$[\mathbf{G}_1 \quad \mathbf{G}_2] \begin{bmatrix} \mathbf{q}_1 \\ \mathbf{q}_2 \end{bmatrix} = 0 \tag{1.79}$$

The sub-matrices $[\mathbf{G}_1]$ and $[\mathbf{G}_2]$ will have $(n - m)$ and m columns respectively. Expanding equation (1.79) gives

$$[\mathbf{G}_1]\{\mathbf{q}_1\} + [\mathbf{G}_2]\{\mathbf{q}_2\} = 0 \tag{1.80}$$

Since $[\mathbf{G}_2]$ is a square matrix it can be inverted to give $[\mathbf{G}_2]^{-1}$. Premultiplying equation (1.80) by $[\mathbf{G}_2]^{-1}$ and rearranging gives

$$\{\mathbf{q}_2\} = -[\mathbf{G}_2]^{-1}[\mathbf{G}_1]\{\mathbf{q}_1\} \tag{1.81}$$

This equation expresses the relationship between the dependent displacements and the independent ones. Combining the two sets of displacements gives

$$\{\mathbf{q}\} = \begin{bmatrix} \mathbf{q}_1 \\ \mathbf{q}_2 \end{bmatrix} = \begin{bmatrix} \mathbf{I} \\ -\mathbf{G}_2^{-1}\mathbf{G}_1 \end{bmatrix} \{\mathbf{q}_1\} = [\mathbf{T}_\mathrm{G}]\{\mathbf{q}_1\} \tag{1.82}$$

where $\mathbf{I}$ is the unit matrix.

The transformation expressed in equation (1.82) is now substituted into the energy function (1.66) to give

$$\begin{aligned} T &= \tfrac{1}{2}\{\dot{\mathbf{q}}_1\}^\mathrm{T}[\bar{\mathbf{M}}]\{\dot{\mathbf{q}}_1\} \\ D &= \tfrac{1}{2}\{\dot{\mathbf{q}}_1\}^\mathrm{T}[\bar{\mathbf{C}}]\{\dot{\mathbf{q}}_1\} \\ U &= \tfrac{1}{2}\{\mathbf{q}_1\}^\mathrm{T}[\bar{\mathbf{K}}]\{\mathbf{q}_1\} \end{aligned} \tag{1.83}$$

where

$$\begin{aligned} [\bar{\mathbf{M}}] &= [\mathbf{T}_G]^\mathrm{T}[\mathbf{M}][\mathbf{T}_G] \\ [\bar{\mathbf{C}}] &= [\mathbf{T}_G]^\mathrm{T}[\mathbf{C}][\mathbf{T}_G] \\ [\bar{\mathbf{K}}] &= [\mathbf{T}_G]^\mathrm{T}[\mathbf{K}][\mathbf{T}_G] \end{aligned} \tag{1.84}$$

The virtual work done by the applied forces also becomes, on being transformed

$$\begin{aligned} \delta W &= \sum_{j=1}^{n} Q_j \delta q_j = \{\delta \mathbf{q}\}^\mathrm{T}\{\mathbf{Q}\} \\ &= \{\delta \mathbf{q}_1\}^\mathrm{T}[\mathbf{T}_G]^\mathrm{T}\{\mathbf{Q}\} = \{\delta \mathbf{q}_1\}^\mathrm{T}\{\bar{\mathbf{Q}}\} \end{aligned} \tag{1.85}$$

where

$$\{\bar{\mathbf{Q}}\} = [\mathbf{T}_G]^\mathrm{T}\{\mathbf{Q}\} \tag{1.86}$$

The energy and virtual work functions are now expressed in terms of independent displacements and so may be substituted into Lagrange's equations (1.60) to give the equations of motion

$$[\bar{\mathbf{M}}]\{\ddot{\mathbf{q}}_1\} + [\bar{\mathbf{C}}]\{\dot{\mathbf{q}}_1\} + [\bar{\mathbf{K}}]\{\mathbf{q}_1\} = \{\bar{\mathbf{Q}}\} \tag{1.87}$$

EXAMPLE 1.7 Derive the equations of motion of the torsional system shown in Figure 1.8, which consists of three rigid gears with torsional inertias I_1, I_2 and I_3 and two light shafts having torsional stiffnesses k_1 and k_2.

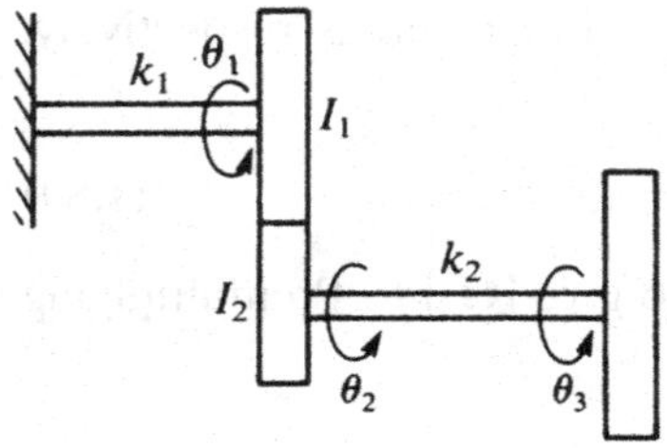

Figure 1.8. Branched, gear system.

For the system shown in Figure 1.8 it is easier to express the energy functions in terms of the angular displacements of the three gears θ_1, θ_2 and θ_3. These are

$$T = \frac{1}{2}\lfloor \dot{\theta}_1 \quad \dot{\theta}_2 \quad \dot{\theta}_3 \rfloor \begin{bmatrix} I_1 & 0 & 0 \\ 0 & I_2 & 0 \\ 0 & 0 & I_3 \end{bmatrix} \begin{bmatrix} \dot{\theta}_1 \\ \dot{\theta}_2 \\ \dot{\theta}_3 \end{bmatrix}$$

$$U = \frac{1}{2}\lfloor \theta_1 \quad \theta_2 \quad \theta_3 \rfloor \begin{bmatrix} k_1 & 0 & 0 \\ 0 & k_2 & -k_2 \\ 0 & -k_2 & k_2 \end{bmatrix} \begin{bmatrix} \theta_1 \\ \theta_2 \\ \theta_3 \end{bmatrix} \tag{1.88}$$

$$D = 0, \qquad \delta W = 0$$

If the teeth of gears I_1 and I_2 remain in contact, then θ_1 and θ_2 must satisfy the constraint relation

$$r_1\theta_1 = -r_2\theta_2 \tag{1.89}$$

where r_1 and r_2 are the radii of the two gears. This can be written in the simpler form

$$\theta_1 = -n\theta_2 \tag{1.90}$$

where $n = r_2/r_1$.

Since there are three coordinates and one constraint relation, only two coordinates are independent. Choosing θ_2 and θ_3 as the independent coordinates, the transformation equation (1.82) becomes

$$\begin{bmatrix} \theta_1 \\ \theta_2 \\ \theta_3 \end{bmatrix} = \begin{bmatrix} -n & 0 \\ 1 & 0 \\ 0 & 1 \end{bmatrix} \begin{bmatrix} \theta_2 \\ \theta_3 \end{bmatrix} \tag{1.91}$$

Substituting equation (1.91) into equations (1.88) gives

$$T = \frac{1}{2}\lfloor \dot{\theta}_2 \quad \dot{\theta}_3 \rfloor \begin{bmatrix} (n^2 I_1 + I_2) & 0 \\ 0 & I_3 \end{bmatrix} \begin{bmatrix} \dot{\theta}_2 \\ \dot{\theta}_3 \end{bmatrix}$$

$$U = \frac{1}{2}\lfloor \theta_2 \quad \theta_3 \rfloor \begin{bmatrix} (n^2 k_1 + k_2) & -k_2 \\ -k_2 & k_2 \end{bmatrix} \begin{bmatrix} \theta_2 \\ \theta_3 \end{bmatrix} \tag{1.92}$$

The equations of motion are obtained by substituting equations (1.92) into Lagrange's equation (1.60). This results in

$$\begin{bmatrix} (n^2 I_1 + I_2) & 0 \\ 0 & I_3 \end{bmatrix} \begin{bmatrix} \ddot{\theta}_2 \\ \ddot{\theta}_3 \end{bmatrix} + \begin{bmatrix} (n^2 k_1 + k_2) & -k_2 \\ -k_2 & k_2 \end{bmatrix} \begin{bmatrix} \theta_2 \\ \theta_3 \end{bmatrix} = 0 \tag{1.93}$$

Problems

1.1 Three masses m_1, m_2 and m_3 are supported by four springs of stiffness k_1, k_2, k_3 and k_4 respectively as shown in Figure P1.1. The masses are subjected to dynamic forces f_1, f_2 and f_3 along the axis of the springs as shown. Derive the equations of motion in terms of the displacements u_1, u_2 and u_3 of the masses along the axis of the springs.

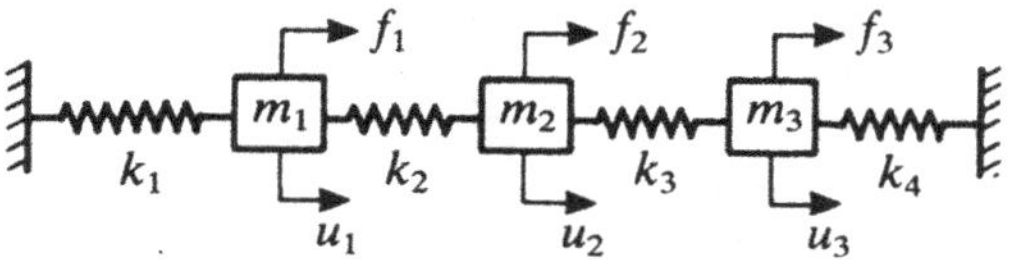

Figure P1.1

1.2 A delicate package of mass m_1 is to be transported on a land vehicle. In order to protect it from damage it is mounted on springs of stiffness k_1 inside a container of mass m_2. The container is, in turn, mounted on a rigid platform by means of springs of stiffness k_2, as shown in Figure P1.2. The platform is then securely mounted on the transportation vehicle. Derive the equations of motion relative to the vehicle assuming that the motion takes place only in the vertical direction.

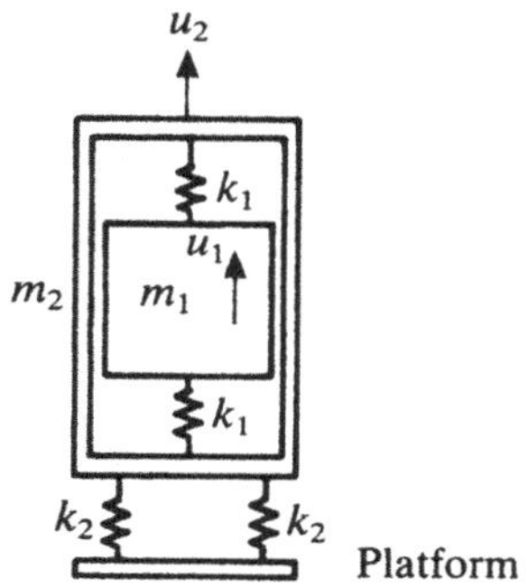

Figure P1.2

1.3 Figure P1.3(b) shows a simplified model of the drop hammer in Figure P1.3(a) after the tup has fallen. The foundation m_1 is supported by a spring k_1 and viscous damper c_1 representing the ground. m_2 represents the mass of the anvil, sow, tup and part of the columns. The stiffness k_2 and damper c_2 represent the isolators between the anvil and the foundation. m_3 represents the mass of the headgear and the remainder of the columns. The stiffness and damping of the columns is k_3 and c_3 respectively. Each mass is constrained to move vertically. Derive the equations of free vibration.

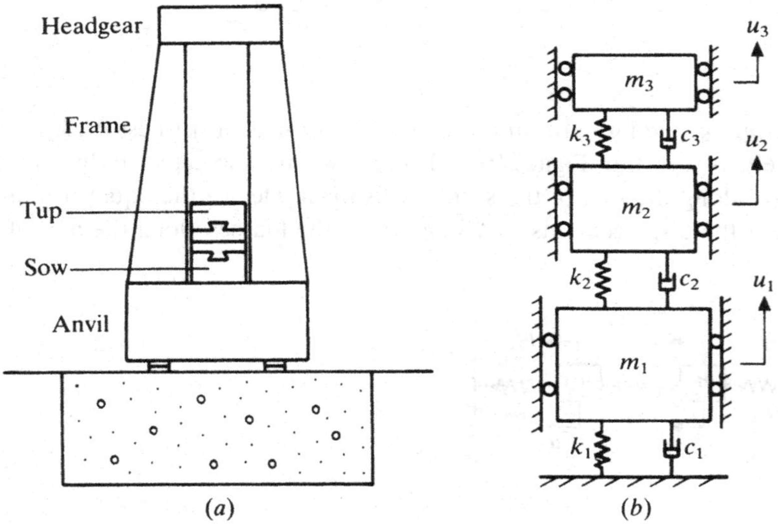

Figure P1.3

1.4 An automobile body is represented by a rigid body of mass m and pitching moment of inertia I_p about an axis through the centre of gravity (CG), as shown in Figure P1.4. The front suspension has stiffness k_1 and damping c_1 and the rear one k_2 and c_2. Derive the equations for motion in the fore and aft plane.

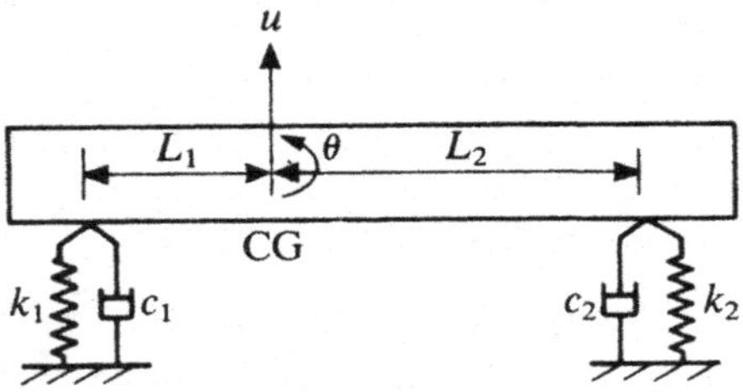

Figure P1.4

1.5 A concrete block is mounted on four isolators of stiffness k and damping c at its four corners, as shown in Figure P1.5. After a vibration test rig has been mounted on the top surface the distances of the centre of gravity (CG) from the front and back face in the x-direction are a_1 and a_2, and the distances from the side faces are b_1 and b_2 in the y-direction. The total mass is m and the moments of inertia about the axes through the centre of gravity are I_x and I_y. Derive the equations of free vibration, assuming that the mounts move only in the z-direction.

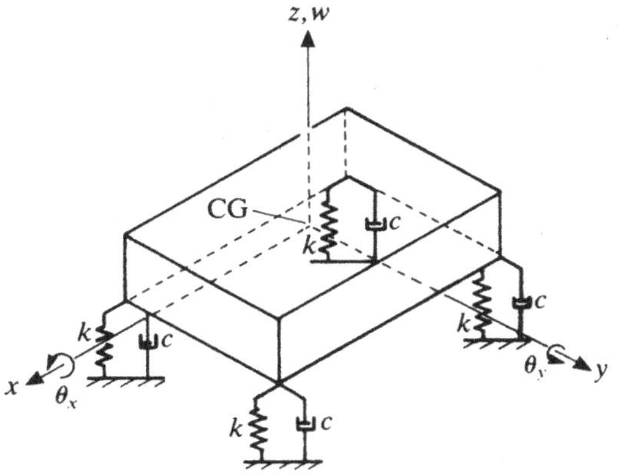

Figure P1.5

1.6 A simplified model of a three-storey building consists of three horizontal rigid floors of mass m_1, m_2 and m_3 supported by massless, elastic columns, as shown in Figure P1.6. The stiffnesses k_1, k_2 and k_3 indicate the horizontal stiffness of all columns in a storey. The building is subjected to a distributed dynamic pressure load along one wall. This load can be represented by three equivalent point forces f_1, f_2 and f_3 acting at the floor levels as shown. Derive the equations of motion of the model assuming that the floors move only horizontally.

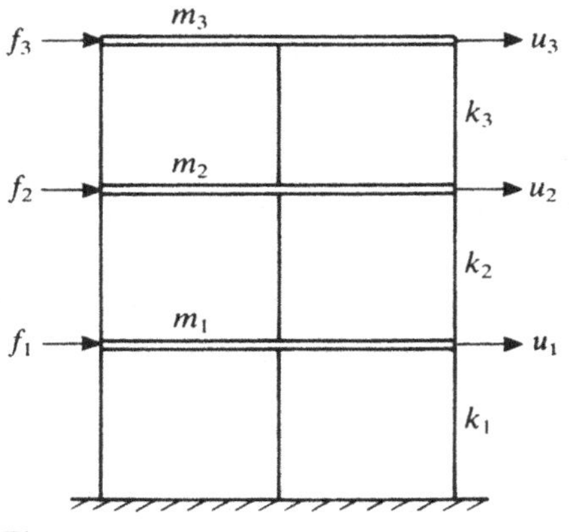

Figure P1.6

1.7 The torsional system shown in Figure P1.7 consists of three rigid gears of radii R_1, R_2 and R_2 which have moments of inertia I_1, I_2 and I_2 and three light shafts having torsional stiffensses k_1, k_2 and k_2. Each shaft is fixed at one end as shown. Derive the equation of free vibration of the system.

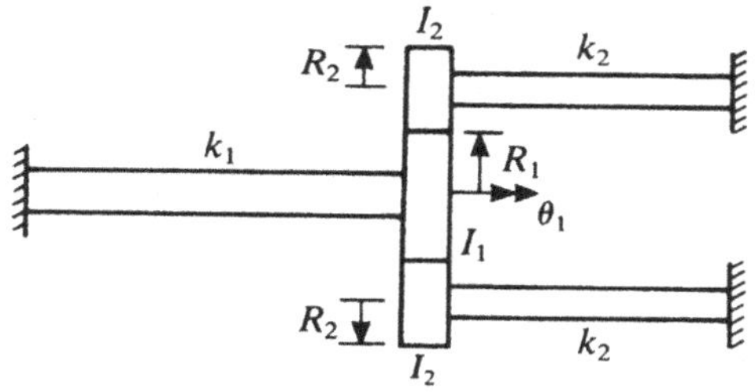

Figure P1.7

1.8 The branched-geared system shown in Figure P1.8 consists of three rigid discs having moments of inertia I_1, I_5 and I_6 and three rigid gears of radii R_1, $R_1/2$ and $R_1/3$ having moments of inertia I_2, I_3 and I_4 which are connected by three light-shafts having torsional stiffnesses k_1, k_2 and k_3. Derive the equations of free vibration.

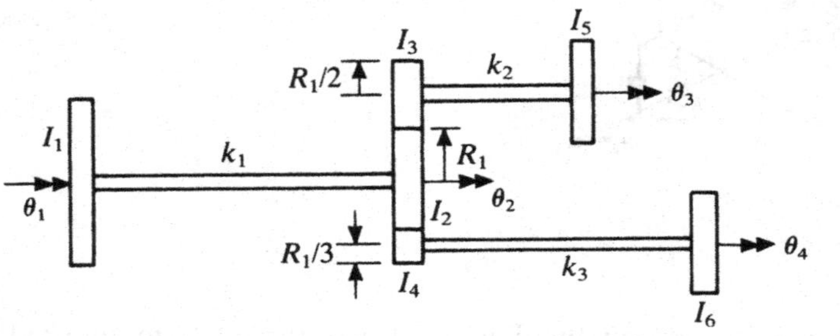

Figure P1.8

2 Element Energy Functions

In Chapter 1 it is shown that the equations of motion of any structural system can be obtained from the energy functions of the system. These energy functions consist of the strain and kinetic energies, the dissipation function and the virtual work done by the applied loads. In this chapter, the energy functions are derived for various structural elements.

The derivations are based upon the linear theory of elasticity. This means that both the stress-strain and the strain-displacement relations are linear. The state of stress in a three-dimensional elastic body is defined by the stress components (which are referred to Cartesian axes x, y, z).

$$[\boldsymbol{\sigma}] = \begin{bmatrix} \sigma_x & \tau_{xy} & \tau_{xz} \\ \tau_{yx} & \sigma_y & \tau_{yz} \\ \tau_{zx} & \tau_{zy} & \sigma_z \end{bmatrix} \tag{2.1}$$

with

$$\tau_{yx} = \tau_{xy}, \qquad \tau_{zx} = \tau_{xz}, \qquad \tau_{zy} = \tau_{yz} \tag{2.2}$$

In relating the stresses to the strains, anisotropic, orthotropic and isotropic materials will be considered, except in the case of one-dimensional elements (axial, torque and beam) when only the isotropic case is treated.

The state of strain in an elastic body is defined by the strain components

$$[\boldsymbol{\varepsilon}] = \begin{bmatrix} \varepsilon_x & \gamma_{xy} & \gamma_{xz} \\ \gamma_{xy} & \varepsilon_y & \gamma_{yz} \\ \gamma_{xz} & \gamma_{yz} & \varepsilon_z \end{bmatrix} \tag{2.3}$$

If the displacement components in the directions of the axes are denoted by (u, v, w), then the strain-displacement relations are

$$\begin{aligned} &\varepsilon_x = \frac{\partial u}{\partial x}, \qquad \varepsilon_y = \frac{\partial v}{\partial y}, \qquad \varepsilon_z = \frac{\partial w}{\partial z} \\ &\gamma_{xy} = \frac{\partial u}{\partial y} + \frac{\partial v}{\partial x}, \qquad \gamma_{xz} = \frac{\partial u}{\partial z} + \frac{\partial w}{\partial x}, \qquad \gamma_{yz} = \frac{\partial v}{\partial z} + \frac{\partial w}{\partial y} \end{aligned} \tag{2.4}$$

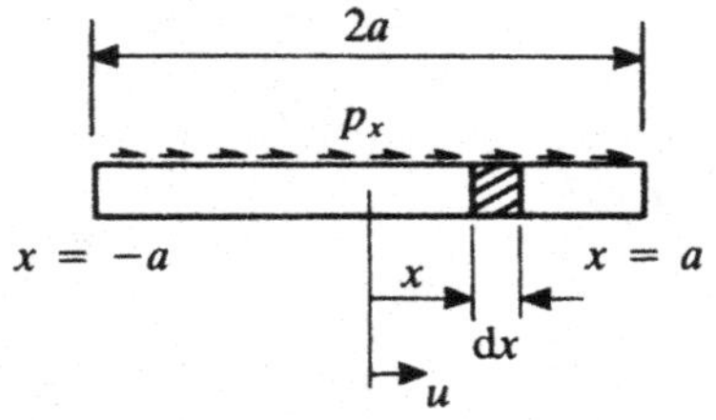

Figure 2.1. Axial element in the local coordinate system.

It should be noted that all displacement components are time dependent. In considering the spatial variation of these quantities explicit mention of time dependence is omitted.

2.1 Axial Element

An axial element of constant cross-sectional area A and length $2a$ is shown in Figure 2.1.

For wavelengths which are greater than ten times the cross-sectional dimensions of the element, it can be assumed that each plane cross-section remains plane during the motion. Also, all the stress components are negligible except for the axial component, σ_x, which is uniform over each cross-section.

The axial force on one of the faces of the increment, dx, in Figure 2.1 is, therefore, $\sigma_x A$. The extension of the element is $\varepsilon_x \, dx$, where ε_x is the axial strain component. The work done on the element, dW, is therefore

$$\begin{aligned} dW &= \tfrac{1}{2}\sigma_x A \cdot \varepsilon_x \, dx \\ &= \tfrac{1}{2}\sigma_x \varepsilon_x A \, dx \end{aligned} \tag{2.5}$$

This work will be stored as strain energy dU, and so

$$dU = \tfrac{1}{2}\sigma_x \varepsilon_x A \, dx \tag{2.6}$$

The total strain energy in the complete element is therefore

$$U = \frac{1}{2}\int_{-a}^{+a} \sigma_x \varepsilon_x A \, dx \tag{2.7}$$

Assuming a linear stress-strain relationship, the direct stress, σ_x, is given by

$$\sigma_x = E\varepsilon_x \tag{2.8}$$

where E is Young's modulus for the material.

Substituting equation (2.8) into equation (2.7) gives

$$U = \frac{1}{2}\int_{-a}^{+a} EA\varepsilon_x^2 \, dx \tag{2.9}$$

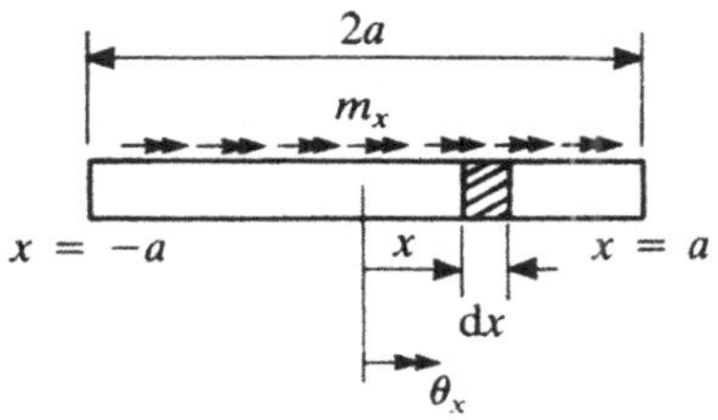

Figure 2.2. Torque element in the local coordinate system.

From equations (2.4), the axial strain component can be expressed in terms of the axial displacement, $u(x)$, by means of the relation

$$\varepsilon_x = \frac{\partial u}{\partial x} \tag{2.10}$$

Substituting equation (2.10) into equation (2.9), the strain energy becomes

$$U = \frac{1}{2}\int_{-a}^{+a} EA\left(\frac{\partial u}{\partial x}\right)^2 \mathrm{d}x \tag{2.11}$$

The kinetic energy of a small increment, dx, is $\frac{1}{2}\dot{u}^2\rho A\,\mathrm{d}x$, where ρ is the mass per unit volume of the material. The kinetic energy of the complete element is therefore

$$T = \frac{1}{2}\int_{-a}^{+a} \rho A\dot{u}^2\,\mathrm{d}x \tag{2.12}$$

If there is an applied load of magnitude p_x per unit length, as shown in Figure 2.1, then the force on the increment dx is $p_x\,\mathrm{d}x$, and the work done in a virtual displacement δu is $\delta u \cdot p_x\,\mathrm{d}x$. The virtual work for the complete element is therefore

$$\delta W = \int_{-a}^{+a} p_x\delta u\,\mathrm{d}x \tag{2.13}$$

2.2 Torque Element

A torque element of constant cross-sectional area A and length $2a$ is shown in Figure 2.2.

The element is assumed to undergo twisting deformations about the x-axis only. The rotation at position x is denoted by θ_x.

For wavelengths which are greater than ten times the cross-sectional dimensions of the element, the Saint-Venant theory of torsion can be used. This theory assumes that the deformation of the twisted element consists of (a) rotations of the cross-sections about the axis and (b) warping of the cross-sections. Taking axes y and z perpendicular to the x-axis, the displacements u, v, w in the directions of the x, y, z axes are given by

$$\begin{aligned} u(x, y, z) &= \frac{\partial\theta_x(x)}{\partial x}\psi(y, z) \\ v(x, y, z) &= -\theta_x(x)z \\ w(x, y, z) &= \theta_x(x)y \end{aligned} \tag{2.14}$$

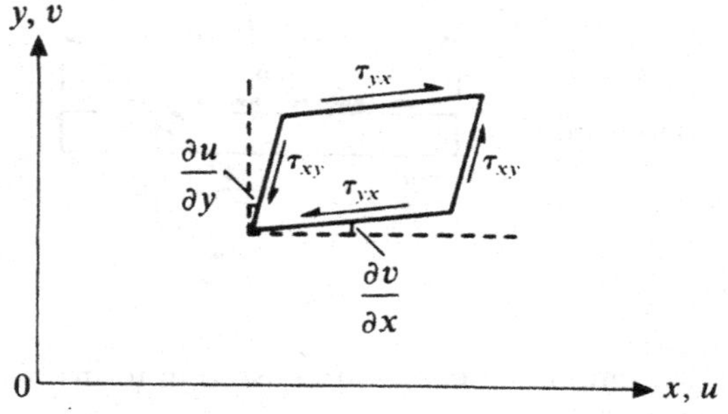

Figure 2.3. Stresses and distortion in the xy-plane.

where $\psi(y, z)$ is a function which represents the warping of a cross-section. With these displacements, the components of strain are

$$
\begin{aligned}
\varepsilon_y &= \varepsilon_z = \gamma_{yz} = 0 \\
\varepsilon_x &= \frac{\partial^2 \theta}{\partial x^2}\psi \\
\gamma_{xy} &= \frac{\partial \theta_x}{\partial x}\left(\frac{\partial \psi}{\partial y} - z\right) \\
\gamma_{xz} &= \frac{\partial \theta_x}{\partial x}\left(\frac{\partial \psi}{\partial z} + y\right)
\end{aligned}
\tag{2.15}
$$

In most cases the axial strain component, ε_x, is negligible. (An exception to this rule, thin-walled, open-section beams, is discussed in Section 3.11.) Therefore, the only strain components to be considered are the shear strains γ_{xy} and γ_{xz}. The corresponding stresses are given by

$$
\tau_{xy} = G\gamma_{xy}, \qquad \tau_{xz} = G\gamma_{xz} \tag{2.16}
$$

where G denotes the shear modulus of the material.

With reference to Figure 2.3, it can be seen that the shear stresses τ_{xy} form a couple of magnitude $\tau_{xy}\,\mathrm{d}y\,\mathrm{d}z\,\mathrm{d}x$ on an element of thickness $\mathrm{d}z$. This couple causes a rotation $\partial v/\partial x$. Similarly, the stresses τ_{yx} form a couple of magnitude $\tau_{yx}\,\mathrm{d}x\,\mathrm{d}z\,\mathrm{d}y$ which causes a rotation $\partial u/\partial y$. Since $\tau_{yx} = \tau_{xy}$, the total work done on the element is

$$
\begin{aligned}
\mathrm{d}W &= \frac{1}{2}\tau_{xy}\left(\frac{\partial v}{\partial x} + \frac{\partial u}{\partial y}\right)\mathrm{d}x\,\mathrm{d}y\,\mathrm{d}z \\
&= \tfrac{1}{2}\tau_{xy}\gamma_{xz}\,\mathrm{d}A\,\mathrm{d}x
\end{aligned}
\tag{2.17}
$$

where $\mathrm{d}A$ is an element of area of the cross-section.

The work done by the stresses τ_{xz} is, therefore,

$$
\mathrm{d}W = \tfrac{1}{2}\tau_{xz}\gamma_{xz}\,\mathrm{d}A\,\mathrm{d}x \tag{2.18}
$$

Since the strain energy is equal to the total work done, then

$$
U = \frac{1}{2}\int_{-a}^{+a}\int_A (\tau_{xy}\gamma_{xy} + \tau_{xz}\gamma_{xz})\,\mathrm{d}A\,\mathrm{d}x \tag{2.19}
$$

Figure 2.4. Straight beam element in local coordinate system.

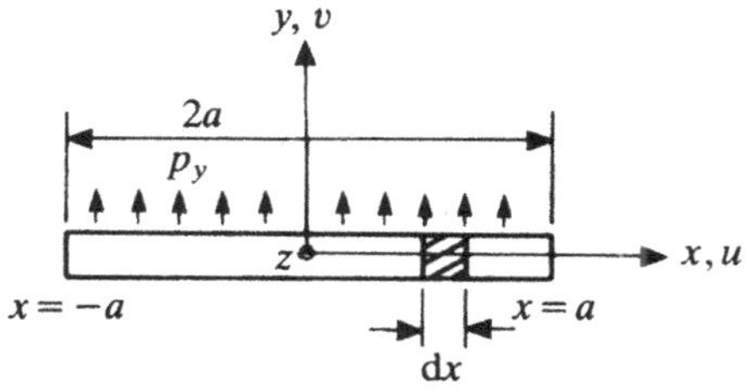

Substituting equations (2.15) and (2.16) into equation (2.19) gives

$$U = \frac{1}{2}\int_{-a}^{+a} GJ\left(\frac{\partial\theta_x}{\partial x}\right)^2 \mathrm{d}x \tag{2.20}$$

where

$$J = \int_A \left\{\left(\frac{\partial\psi}{\partial y} - z\right)^2 + \left(\frac{\partial\psi}{\partial z} + y\right)^2\right\} \mathrm{d}A \tag{2.21}$$

is the torsion constant of the cross-section. In the case of a circular shaft $\psi = 0$ and J reduces to the polar moment of area of the cross-section, I_x. For other solid cross-sectional shapes J is given by the approximate expression

$$J \simeq 0.025A^4/I_x$$

This equation should not, however, be used for elongated sections. For a more complete discussion of this topic see, for example, reference [2.1].

In deriving an expression for the kinetic energy of the element, the longitudinal displacement due to warping of the cross-sections can be neglected. The kinetic energy of a small increment, dx, of the element is, therefore, $\frac{1}{2}\dot{\theta}_x^2\,\rho I_x\,\mathrm{d}x$, where I_x is the second moment of area of the cross-section about the x-axis. The kinetic energy of the complete element is therefore

$$T = \frac{1}{2}\int_{-a}^{+a} \rho I_x \dot{\theta}_x^2 \,\mathrm{d}x \tag{2.22}$$

If there is a twisting moment of magnitude m_x per unit length, as shown in Figure 2.2, then the torque on the increment, dx, is $m_x\,\mathrm{d}x$, and the work done in a virtual displacement $\delta\theta_x$, is $\delta\theta_x\,m_x\,\mathrm{d}x$. The virtual work for the complete element is therefore

$$\delta W = \int_{-a}^{+a} m_x \delta\theta_x \,\mathrm{d}x \tag{2.23}$$

2.3 Beam Bending Element

In deriving the energy functions for a beam bending element it is assumed that the vibration occurs in one of the principal planes of the beam. The beam, which is of length $2a$ and has a constant cross-sectional area A, is shown in Figure 2.4. The xy-plane is the principal plane in which the beam is vibrating and the x-axis coincides with the centroidal axis.

For wavelengths which are greater than ten times the cross-sectional dimensions of the element, the elementary theory of bending can be used. This theory assumes that the stress components σ_y, σ_z, τ_{yz} and τ_{xz} are zero. It also assumes that plane sections which are normal to the undeformed centroidal axis remain plane after bending and are normal to the deformed axis. With this assumption, the axial displacement, u, at a distance y from the centroidal axis is

$$u(x, y) = -y\frac{\partial v}{\partial x} \tag{2.24}$$

where $v = v(x)$ is the displacement of the centroidal axis in the y-direction at position x. The strain components ε_x and γ_{xy} are therefore

$$\begin{aligned}\varepsilon_x &= \frac{\partial u}{\partial x} = -y\frac{\partial^2 v}{\partial x^2} \\ \gamma_{xy} &= \frac{\partial u}{\partial y} + \frac{\partial v}{\partial x} = 0\end{aligned} \tag{2.25}$$

The strain energy stored in the element is therefore given by

$$U = \frac{1}{2}\int_V \sigma_x \varepsilon_x \, dv \tag{2.26}$$

The normal stress is given by

$$\sigma_x = E\varepsilon_x \tag{2.27}$$

and so equation (2.26) becomes

$$U = \frac{1}{2}\int_V E\varepsilon_x^2 \, dV \tag{2.28}$$

Substituting the first of equations (2.25) into equation (2.28) gives, since $dV = dA \cdot dx$

$$U = \frac{1}{2}\int_{-a}^{+a} EI_z \left(\frac{\partial^2 v}{\partial x^2}\right)^2 dx \tag{2.29}$$

where

$$I_z = \int_A y^2 \, dA \tag{2.30}$$

is the second moment of area of the cross-section about the z-axis.

The stress-strain relations

$$\tau_{xy} = G\gamma_{xy}$$

together with equations (2.25) suggest that τ_{xy} is zero. In fact this component is non-zero, as can be shown by considering equilibrium. The resulting shear force, Q, in the y-direction is given by equation (2.131).

The kinetic energy of a small increment, dx, is $\frac{1}{2}v^2 \rho A \, dx$. The kinetic energy of the complete element is therefore

$$T = \frac{1}{2}\int_{-a}^{+a} \rho A \dot{v}^2 \, dx \tag{2.31}$$

If there is a distributed load of magnitude p_y per unit length, as shown in Figure 2.4, then the force on the increment, dx, is $p_y \, dx$, and the work done in a virtual displacement δv is $\delta v\, p_y\, \mathrm{d}x$. The virtual work for the complete element is therefore

$$\delta W = \int_{-a}^{+a} p_y\, \delta v\, \mathrm{d}x \tag{2.32}$$

2.4 Deep Beam Bending Element

Flexural wave speeds are much lower than the speed of either longitudinal or torsional waves. Therefore, flexural wavelengths which are less than ten times the cross-sectional dimensions of the beam will occur at much lower frequencies. This situation occurs when analysing deep beams at low frequencies and slender beams at higher frequencies. In these cases, deformation due to transverse shear and kinetic energy due to the rotation of the cross-section become important.

In developing energy expressions which include both shear deformation and rotary inertia, the assumption that plane sections which are normal to the undeformed centroidal axis remain plane after bending, will be retained. However, it will no longer be assumed that these sections remain normal to the deformed axis. Consequently, the axial displacement, u, at a distance y from the centroidal axis is now written as

$$u(x, y) = -y\theta_z(x) \tag{2.33}$$

where $\theta_z(x)$ is the rotation of the cross-section at position x.

The strain components ε_x and γ_{xy} in this case are given by

$$\begin{aligned} \varepsilon_x &= \frac{\partial u}{\partial x} = -y\frac{\partial \theta_z}{\partial x} \\ \gamma_{xy} &= \frac{\partial u}{\partial y} + \frac{\partial v}{\partial x} = -\theta_z + \frac{\partial v}{\partial x} \end{aligned} \tag{2.34}$$

Thus the strain energy stored in the element is the sum of the energies due to bending and shear deformation; which is given by

$$U = \frac{1}{2}\int_V \sigma_x \varepsilon_x\, \mathrm{d}V + \frac{1}{2}\int_V \tau_{xy}\, \gamma_{xy}\, \mathrm{d}V \tag{2.35}$$

As in the previous section, the normal stress is given by

$$\sigma_x = E\varepsilon_x \tag{2.36}$$

The shear stresses, τ_{xy}, corresponding to a given shear force, vary over the cross-section. It follows that the corresponding shear strains will also vary over the cross-section. In assuming that plane sections remain plane, the variation of strain over the cross-section has been neglected. This variation can be accounted for by introducing a numerical factor κ, which depends upon the shape of the cross-section, such that

$$\tau_{xy} = \kappa G \gamma_{xy} \tag{2.37}$$

where τ_{xy} is the average shear stress. The values of κ for various cross-sectional shapes are given in reference [2.2].

Substituting equations (2.36) and (2.37) into (2.35) gives

$$U = \frac{1}{2}\int_V E\varepsilon_x^2\, \mathrm{d}V + \frac{1}{2}\int_V \kappa G \gamma_{xy}^2\, \mathrm{d}V \tag{2.38}$$

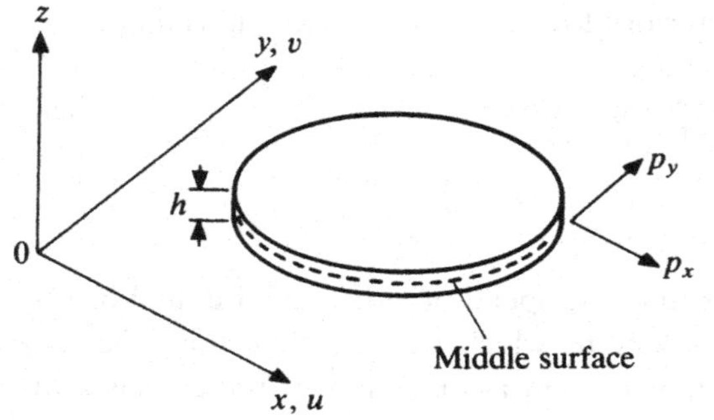

Figure 2.5. Membrane element lying in xy-plane.

The substitution of equations (2.34) into equation (2.38) and the use of $\mathrm{d}V = \mathrm{d}A \cdot \mathrm{d}x$ results in

$$U = \frac{1}{2}\int_{-a}^{+a} EI_z\left(\frac{\partial \theta_z}{\partial x}\right)^2 \mathrm{d}x + \frac{1}{2}\int_{-a}^{+a} \kappa AG\left(\frac{\partial v}{\partial x} - \theta_z\right)^2 \mathrm{d}x \tag{2.39}$$

The kinetic energy of the beam consists of kinetic energy of translation and kinetic energy of rotation which is expressed as

$$T = \frac{1}{2}\int_{-a}^{+a} \rho A\dot{v}^2\, \mathrm{d}x + \frac{1}{2}\int_{-a}^{+a} \rho I_z\dot{\theta}_z^2\, \mathrm{d}x \tag{2.40}$$

The virtual work of the distributed loading is given by equation (2.32).

2.5 Membrane Element

Figure 2.5 shows a thin plate of constant thickness, h, which is subject to distributed boundary loads. These loads are applied in directions which are parallel to the middle plane of the plate and are uniformly distributed through the thickness. There are no forces acting on the surfaces $z = \pm h/2$ and so the stress components σ_z, τ_{zx}, τ_{zy} are zero on these surfaces. Under the above stated conditions it is reasonable to assume that these stresses are negligibly small everywhere within the plate. The state of stress is then defined by the components σ_x, σ_y, τ_{xy}, which are assumed to be independent of z. Such a state is called 'plane stress' and the element a 'membrane element'.

The strain energy stored in the element is given by

$$U = \frac{1}{2}\int_V (\sigma_x\varepsilon_x + \sigma_y\varepsilon_y + \tau_{xy}\gamma_{xy})\, \mathrm{d}V \tag{2.41}$$

which can be expressed in the following matrix form

$$U = \frac{1}{2}\int_V \{\boldsymbol{\sigma}\}^{\mathrm{T}}\{\boldsymbol{\varepsilon}\}\, \mathrm{d}V \tag{2.42}$$

where

$$\{\boldsymbol{\sigma}\}^{\mathrm{T}} = \lfloor \sigma_x \quad \sigma_y \quad \tau_{xy} \rfloor$$

and

$$\{\boldsymbol{\varepsilon}\}^{\mathrm{T}} = \lfloor \varepsilon_x \quad \varepsilon_y \quad \gamma_{xy} \rfloor \tag{2.43}$$

The stress-strain relationships take the form

$$\{\boldsymbol{\sigma}\} = [\mathbf{D}]\{\boldsymbol{\varepsilon}\} \tag{2.44}$$

where, for an anisotropic material, which has different material properties in each direction

$$[\mathbf{D}] = \begin{bmatrix} d_{11} & d_{12} & d_{13} \\ & d_{22} & d_{23} \\ \text{Sym} & & d_{33} \end{bmatrix} \tag{2.45}$$

The coefficients d_{ij} $(i, j = 1, 2, 3)$ are material constants.

If the material is orthotropic it will have two lines of symmetry. Taking these lines as coordinate axes $\bar{x}$, $\bar{y}$, the matrix of material constants takes the form

$$[\mathbf{D}^*] = \begin{bmatrix} E'_{\bar{x}} & E'_{\bar{x}}\nu_{\bar{x}\bar{y}} & 0 \\ & E'_{\bar{y}} & 0 \\ \text{Sym} & & G_{\bar{x}\bar{y}} \end{bmatrix} \tag{2.46}$$

where

$$E'_{\bar{x}} = \frac{E_{\bar{x}}}{(1 - \nu_{\bar{x}\bar{y}}\nu_{\bar{y}\bar{x}})}, \qquad E'_{\bar{y}} = \frac{E_{\bar{y}}}{(1 - \nu_{\bar{x}\bar{y}}\nu_{\bar{y}\bar{x}})} \tag{2.47}$$

$E_{\bar{x}}$ = modulus of elasticity in the $\bar{x}$-direction
$E_{\bar{y}}$ = modulus of elasticity in the $\bar{y}$-direction
$\nu_{\bar{x}\bar{y}}$ = strain in the $\bar{x}$-direction due to a unit strain in the $\bar{y}$-direction
$\nu_{\bar{y}\bar{x}}$ = strain in the $\bar{y}$-direction due to a unit strain in the $\bar{x}$-direction
$G_{\bar{x}\bar{y}}$ = shear modulus with respect to $\bar{x}$-, $\bar{y}$-directions.

These constants are related as follows

$$E_{\bar{x}}\nu_{\bar{x}\bar{y}} = E_{\bar{y}}\nu_{\bar{y}\bar{x}}, \qquad E'_{\bar{x}}\nu_{\bar{x}\bar{y}} = E'_{\bar{y}}\nu_{\bar{y}\bar{x}} \tag{2.48}$$

In general, the material axes will be inclined at some angle β to the geometric axes. By considering the relationship between strains related to both material and geometric axes, it can be shown that the matrix of material constants, referred to geometric axes, is given by

$$[\mathbf{D}] = [\mathbf{R}^*]^{\mathrm{T}}[\mathbf{D}^*][\mathbf{R}^*] \tag{2.49}$$

where the transformation matrix $[\mathbf{R}^*]$ is given by

$$[\mathbf{R}^*] = \begin{bmatrix} \cos^2\beta & \sin^2\beta & \frac{1}{2}\sin 2\beta \\ \sin^2\beta & \cos^2\beta & -\frac{1}{2}\sin 2\beta \\ -\sin 2\beta & \sin 2\beta & \cos 2\beta \end{bmatrix} \tag{2.50}$$

For isotropic materials the elastic properties are the same in all directions. The matrix of material constants therefore reduces to

$$[\mathbf{D}] = \begin{bmatrix} E' & E'\nu & 0 \\ & E' & 0 \\ \text{Sym} & & G \end{bmatrix} \tag{2.51}$$

where

$$E' = \frac{E}{(1-\nu^2)}, \qquad G = \frac{E}{2(1+\nu)} \tag{2.52}$$

E = Young's modulus
ν = Poisson's ratio

Substituting equation (2.44) into the energy expression (2.42) gives

$$U = \frac{1}{2}\int_V \{\boldsymbol{\varepsilon}\}^{\mathrm{T}} [\mathbf{D}] \{\boldsymbol{\varepsilon}\} \, \mathrm{d}V \tag{2.53}$$

Since the stresses $\{\boldsymbol{\sigma}\}$ are assumed independent of z, then the strains $\{\boldsymbol{\varepsilon}\}$ and the displacement components, u, v, will also be independent of z. Integrating (2.53) with respect to z gives

$$U = \frac{1}{2}\int_A h \{\boldsymbol{\varepsilon}\}^{\mathrm{T}} [\mathbf{D}] \{\boldsymbol{\varepsilon}\} \, \mathrm{d}A \tag{2.54}$$

where A is the area of the middle surface. In expression (2.54) the strains are expressed in terms of the displacements as follows

$$\{\boldsymbol{\varepsilon}\} = \begin{bmatrix} \partial u/\partial x \\ \partial v/\partial y \\ \partial u/\partial y + \partial v/\partial x \end{bmatrix} \tag{2.55}$$

The kinetic energy of the membrane element is given by

$$T = \frac{1}{2}\int_A \rho h (\dot{u}^2 + \dot{v}^2) \, \mathrm{d}A \tag{2.56}$$

If p_x, p_y are the components of the applied boundary forces per unit arc length of the boundary, then the virtual work is

$$\delta W = \int_s (p_x \delta u + p_y \delta v) \, \mathrm{d}s \tag{2.57}$$

where s denotes the boundary of the element.

2.6 Thin Plate Bending Element

Figure 2.6 shows a thin plate of constant thickness, h, which is subject to distributed surface loads. These loads are normal to the middle surface, which is the plane $z = 0$.

In deriving the energy functions for a thin plate, it is assumed that the direct stress in the transverse direction, σ_z, is zero. Also, normals to the middle surface of the undeformed plate remain straight and normal to the middle surface during deformation and are inextensible. Thus, the displacements parallel to the undeformed middle surface are given by

$$\begin{aligned} u(x, y, z) &= -z\frac{\partial w}{\partial x} \\ v(x, y, z) &= -z\frac{\partial w}{\partial y} \end{aligned} \tag{2.58}$$

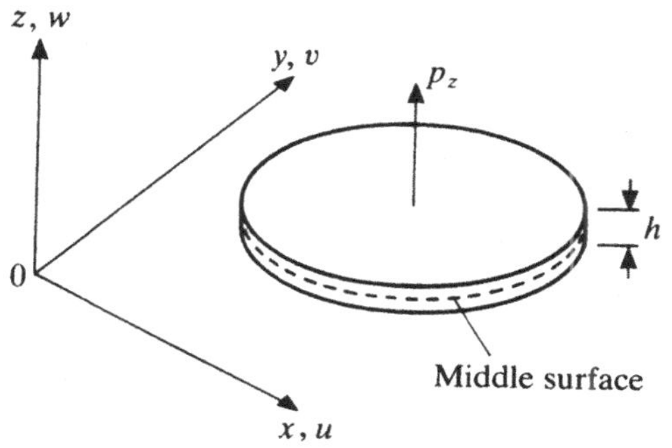

Figure 2.6. Plate bending element.

where $w(x, y)$ denotes the displacement of the middle surface in the z-direction. The components of strain are given as follows:

$$\varepsilon_x = \frac{\partial u}{\partial x} = -z\frac{\partial^2 w}{\partial x^2}$$
$$\varepsilon_y = \frac{\partial v}{\partial y} = -z\frac{\partial^2 w}{\partial y^2} \tag{2.59}$$

$$\gamma_{xy} = \frac{\partial u}{\partial y} + \frac{\partial v}{\partial x} = -z2\frac{\partial^2 w}{\partial x\,\partial y}$$
$$\gamma_{xz} = \frac{\partial u}{\partial z} + \frac{\partial w}{\partial x} = 0, \qquad \gamma_{yz} = \frac{\partial v}{\partial z} + \frac{\partial w}{\partial y} = 0 \tag{2.60}$$

Since σ_z, γ_{xz} and γ_{yz} are all zero, then the strain energy stored in the element is given by

$$U = \frac{1}{2}\int_V (\sigma_x\varepsilon_x + \sigma_y\varepsilon_y + \tau_{xy}\gamma_{xy})\,\mathrm{d}V \tag{2.61}$$

which is identical to equation (2.41) for a membrane element. This equation can be expressed in the matrix form

$$U = \frac{1}{2}\int_V \{\boldsymbol{\sigma}\}^{\mathrm{T}}\{\boldsymbol{\varepsilon}\}\,\mathrm{d}V \tag{2.62}$$

where $\{\boldsymbol{\sigma}\}$ and $\{\boldsymbol{\varepsilon}\}$ are defined by (2.43).

Since $\sigma_z = 0$ the stress–strain relations take the form

$$\{\boldsymbol{\sigma}\} = [\mathbf{D}]\{\boldsymbol{\varepsilon}\} \tag{2.63}$$

where $[\mathbf{D}]$ is defined by equation (2.45).

Substituting (2.63) into (2.62) gives

$$U = \frac{1}{2}\int_V \{\boldsymbol{\varepsilon}\}^{\mathrm{T}}[\mathbf{D}]\{\boldsymbol{\varepsilon}\}\,\mathrm{d}V \tag{2.64}$$

Using equations (2.59), the strain matrix can be written in the form

$$\{\boldsymbol{\varepsilon}\} = -z\{\boldsymbol{\chi}\} \tag{2.65}$$

where

$$\{\boldsymbol{\chi}\} = \begin{bmatrix} \partial^2 w/\partial x^2 \\ \partial^2 w/\partial y^2 \\ 2\partial^2 w/\partial x\,\partial y \end{bmatrix} \tag{2.66}$$

Substituting equation (2.65) into equation (2.64) and integrating with respect to z gives

$$U = \frac{1}{2}\int_A \frac{h^3}{12}\{\boldsymbol{\chi}\}^{\mathrm{T}}[\mathbf{D}]\{\boldsymbol{\chi}\}\,\mathrm{d}A \tag{2.67}$$

The kinetic energy of the plate is given by

$$T = \frac{1}{2}\int_A \rho h\dot{w}^2\,\mathrm{d}A \tag{2.68}$$

and the virtual work of the transverse loading

$$\delta W = \int_A p_z \delta w\,\mathrm{d}A \tag{2.69}$$

2.7 Thick Plate Bending Element

As in the case of deep beams, when the wavelengths are less than ten times the plate thickness, shear deformation and rotary inertia effects must be included.

When shear deformation is important, it cannot be assumed that normals to the middle surface remain normal to it. In this case the displacements parallel to the middle surface are given by

$$\begin{aligned} u(x, y, z) &= z\theta_y(x, y) \\ v(x, y, z) &= -z\theta_x(x, y) \end{aligned} \tag{2.70}$$

where θ_x, θ_y are the rotations about the x- and y-axes of lines originally normal to the middle plane before deformation. The in-plane strains are now given by

$$\{\boldsymbol{\varepsilon}\} = -z\{\boldsymbol{\chi}\} \tag{2.71}$$

where

$$\{\boldsymbol{\chi}\} = \begin{bmatrix} -\partial\theta_y/\partial x \\ \partial\theta_x/\partial y \\ \partial\theta_x/\partial x - \partial\theta_y/\partial y \end{bmatrix} \tag{2.72}$$

The relationships between the transverse shear strains γ_{xz}, γ_{yz} and the displacements are

$$\gamma_{xz} = \frac{\partial u}{\partial z} + \frac{\partial w}{\partial x}, \qquad \gamma_{yz} = \frac{\partial v}{\sigma z} + \frac{\partial w}{\partial y} \tag{2.73}$$

Substituting for the displacements from (2.70) gives

$$\{\boldsymbol{\gamma}\} = \begin{bmatrix} \gamma_{xz} \\ \gamma_{yz} \end{bmatrix} = \begin{bmatrix} \theta_y + \dfrac{\partial w}{\partial x} \\ -\theta_x + \dfrac{\partial w}{\partial y} \end{bmatrix} \tag{2.74}$$

Note that when the transverse shears are negligible, equation (2.74) gives $\theta_y = -\partial w/\partial x$ and $\theta_x = \partial w/\partial y$. With these relationships it can be seen that equation (2.72) reduces to equation (2.66).

The strain energy stored in the element is the sum of the energies due to bending and shear deformation, which is given by

$$U = \frac{1}{2}\int_V \{\boldsymbol{\varepsilon}\}^{\mathrm{T}}[\mathbf{D}]\{\boldsymbol{\varepsilon}\}\,\mathrm{d}V + \frac{1}{2}\int_V \{\boldsymbol{\tau}\}^{\mathrm{T}}\{\boldsymbol{\gamma}\}\,\mathrm{d}V \tag{2.75}$$

provided the plane $z = 0$ is a plane of symmetry for the material. The relationship between the average shear stresses, $\{\boldsymbol{\tau}\}$, and the shear strains, as given by (2.74) is,

$$\{\boldsymbol{\tau}\} = \begin{bmatrix}\tau_{xz}\\ \tau_{yz}\end{bmatrix} = \kappa\,[\mathbf{D}^{\mathrm{s}}]\{\boldsymbol{\gamma}\} \tag{2.76}$$

where κ is a constant which is introduced to account for the variation of the shear stresses and strains through the thickness. It is usual to take κ to be either $\pi^2/12$ or $5/6$. If the material is orthotropic, then

$$[\mathbf{D}^{\mathrm{s}}] = \begin{bmatrix}\cos\beta & -\sin\beta\\ \sin\beta & \cos\beta\end{bmatrix}\begin{bmatrix}G_{\bar{x}\bar{z}} & 0\\ 0 & G_{\bar{y}\bar{z}}\end{bmatrix}\begin{bmatrix}\cos\beta & \sin\beta\\ -\sin\beta & \cos\beta\end{bmatrix} \tag{2.77}$$

If the material is isotropic, then

$$[\mathbf{D}^{\mathrm{s}}] = \begin{bmatrix}G & 0\\ 0 & G\end{bmatrix} = \frac{E}{2(1+\nu)}\begin{bmatrix}1 & 0\\ 0 & 1\end{bmatrix} \tag{2.78}$$

Substituting equations (2.71) and (2.76) into equation (2.75) and integrating through the thickness gives

$$U = \frac{1}{2}\int_A \frac{h^3}{12}\{\boldsymbol{\chi}\}^{\mathrm{T}}[\mathbf{D}]\{\boldsymbol{\chi}\}\,\mathrm{d}A + \frac{1}{2}\int_A \kappa h\{\boldsymbol{\gamma}\}^{\mathrm{T}}[\mathbf{D}^{\mathrm{s}}]\{\boldsymbol{\gamma}\}\,\mathrm{d}A \tag{2.79}$$

The kinetic energy of the plate is given by

$$T = \frac{1}{2}\int_V \rho(\dot{u}^2 + \dot{v}^2 + \dot{w}^2)\,\mathrm{d}V \tag{2.80}$$

Substituting for u, v from equation (2.70) and integrating with respect to z after putting $\mathrm{d}V = \mathrm{d}z\cdot\mathrm{d}A$ gives

$$T = \frac{1}{2}\int_A \rho\left(h\dot{w}^2 + \frac{h^3}{12}\dot{\theta}_x^2 + \frac{h^3}{12}\dot{\theta}_y^2\right)\mathrm{d}A \tag{2.81}$$

The virtual work of the transverse loading is again given by equation (2.69).

2.8 Three-Dimensional Solid

Consider a three-dimensional solid of volume V which is enclosed by a surface S as shown in Figure 2.7. The state of stress and strain at a point is defined by the six independent components given in expressions (2.1) and (2.3). The strain energy is

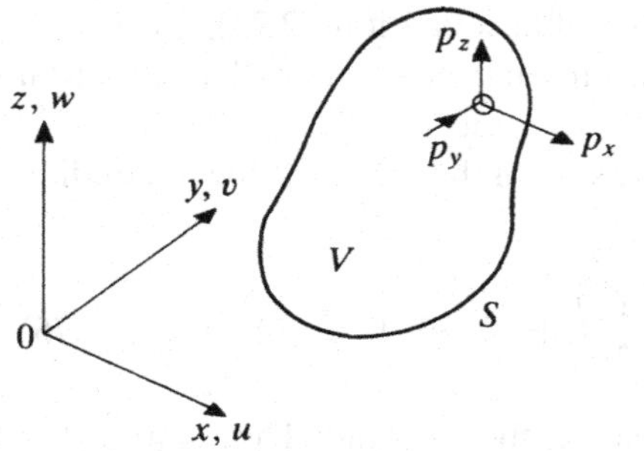

Figure 2.7. Three-dimensional solid.

therefore given by

$$U = \frac{1}{2}\int_V (\sigma_x\varepsilon_x + \sigma_y\varepsilon_y + \sigma_z\varepsilon_z + \tau_{xy}\gamma_{xy} + \tau_{xz}\gamma_{xz} + \tau_{yz}\gamma_{yz})\, \mathrm{d}V \tag{2.82}$$

which can be expressed in matrix form as

$$U = \frac{1}{2}\int_V \{\boldsymbol{\sigma}\}^{\mathrm{T}} \{\boldsymbol{\varepsilon}\}\, \mathrm{d}V \tag{2.83}$$

where

$$\begin{aligned} \{\boldsymbol{\sigma}\}^{\mathrm{T}} &= \lfloor \sigma_x \quad \sigma_y \quad \sigma_z \quad \tau_{xy} \quad \tau_{xz} \quad \tau_{yz} \rfloor \\ \{\boldsymbol{\varepsilon}\}^{\mathrm{T}} &= \lfloor \varepsilon_x \quad \varepsilon_y \quad \varepsilon_z \quad \gamma_{xy} \quad \gamma_{xz} \quad \gamma_{yz} \rfloor \end{aligned} \tag{2.84}$$

The stress–strain relationships take the form

$$\{\boldsymbol{\sigma}\} = [\mathbf{D}]\{\boldsymbol{\varepsilon}\} \tag{2.85}$$

where $[\mathbf{D}]$ is a symmetric matrix. For an anisotropic material it contains 21 independent material constants. In the case of an isotropic material it is

$$[\mathbf{D}] = \frac{E}{(1+\nu)(1-2\nu)} \times \begin{bmatrix} (1-\nu) & \nu & \nu & 0 & 0 & 0 \\ & (1-\nu) & \nu & 0 & 0 & 0 \\ & & (1-\nu) & 0 & 0 & 0 \\ & & & \frac{1}{2}(1-2\nu) & 0 & 0 \\ & & & & \frac{1}{2}(1-2\nu) & 0 \\ \text{Sym} & & & & & \frac{1}{2}(1-2\nu) \end{bmatrix} \tag{2.86}$$

where

E = Young's modulus, ν = Poisson's ratio

Substituting equation (2.85) into equation (2.83) gives

$$U = \frac{1}{2}\int_V \{\boldsymbol{\varepsilon}\}^{\mathrm{T}} [\mathbf{D}]\{\boldsymbol{\varepsilon}\}\, \mathrm{d}V \tag{2.87}$$

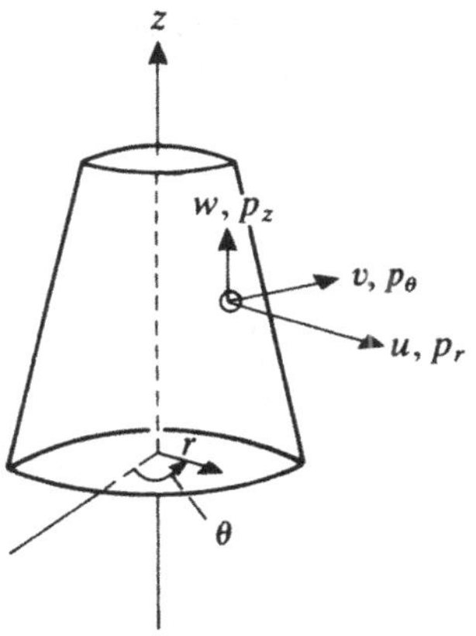

Figure 2.8. Axisymmetric solid.

The strain-displacement relations to be used in equation (2.87) are given by (2.4), namely

$$\{\boldsymbol{\varepsilon}\} = \begin{bmatrix} \dfrac{\partial u}{\partial x} \\ \dfrac{\partial v}{\partial y} \\ \dfrac{\partial w}{\partial z} \\ \dfrac{\partial u}{\partial y} + \dfrac{\partial v}{\partial x} \\ \dfrac{\partial u}{\partial z} + \dfrac{\partial w}{\partial x} \\ \dfrac{\partial v}{\partial z} + \dfrac{\partial w}{\partial y} \end{bmatrix} \tag{2.88}$$

The kinetic energy is given by

$$T = \frac{1}{2}\int_V \rho(\dot{u}^2 + \dot{v}^2 + \dot{w}^2)\,\mathrm{d}V \tag{2.89}$$

If p_x, p_y, p_z are the components of the applied surface forces per unit area, then the virtual work is

$$\delta W = \int_S (p_x\delta u + p_y\delta v + p_z\delta w)\,\mathrm{d}S \tag{2.90}$$

2.9 Axisymmetric Solid

In the previous section the solid body considered had a general shape. If the shape of the body can be generated by rotating a plane area through a full revolution about an axis lying in the plane of the area, it is called a solid of revolution or an axisymmetric solid. In this case it is more convenient to use cylindrical polar coordinates r, θ, z than Cartesian coordinates x, y, z.

An axisymmetric solid together with its axes, displacements and loadings is shown in Figure 2.8.

The strain energy is again given by equation (2.87) where the elasticity matrix $[\mathbf{D}]$ is given by equation (2.86) for an isotropic material. That is

$$U = \frac{1}{2}\int_V \{\boldsymbol{\varepsilon}\}^{\mathrm{T}}[\mathbf{D}]\{\boldsymbol{\varepsilon}\}\,\mathrm{d}V \tag{2.91}$$

On this occasion the strain components are referred to cylindrical polar coordinates, namely

$$\{\boldsymbol{\varepsilon}\} = \begin{bmatrix} \varepsilon_r \\ \varepsilon_\theta \\ \varepsilon_z \\ \gamma_{zr} \\ \gamma_{r\theta} \\ \gamma_{\theta z} \end{bmatrix} \tag{2.92}$$

The strain-displacement relations are

$$\{\boldsymbol{\varepsilon}\} = \begin{bmatrix} \dfrac{\partial u}{\partial r} \\ \dfrac{u}{r} + \dfrac{1}{r}\dfrac{\partial v}{\partial \theta} \\ \dfrac{\partial w}{\partial z} \\ \dfrac{\partial u}{\partial z} + \dfrac{\partial w}{\partial r} \\ \dfrac{1}{r}\dfrac{\partial u}{\partial \theta} + \dfrac{\partial v}{\partial r} - \dfrac{v}{r} \\ \dfrac{\partial v}{\partial z} + \dfrac{1}{r}\dfrac{\partial w}{\partial \theta} \end{bmatrix} \tag{2.93}$$

where u, v, w are the displacements in the r, θ, z directions.

The kinetic energy is given by

$$T = \frac{1}{2}\int_V \rho(\dot{u}^2 + \dot{v}^2 + \dot{w}^2)\,\mathrm{d}V \tag{2.94}$$

In both equations (2.91) and (2.94) the element of volume is given by

$$\mathrm{d}V = r\,\mathrm{d}r\,\mathrm{d}\theta\,\mathrm{d}z \tag{2.95}$$

If p_r, p_θ, p_z are the components of the applied surface forces per unit area, then the virtual work is

$$\delta W = \int_S (p_r\delta u + p_\theta\delta v + p_z\delta w)\,\mathrm{d}S \tag{2.96}$$

The element of surface area is given by

$$\mathrm{d}S = r_{\mathrm{S}}\,\mathrm{d}\theta\,\mathrm{d}s \tag{2.97}$$

where $\mathrm{d}s$ is an increment of arc length on the surface in the r, z plane and r_{S} is the value of r on the surface.

2.10 The Dissipation Function

The dissipation function has not been derived for each of the structural elements considered in the previous sections since damping is not necessarily an inherent property of the vibrating structure. Damping forces depend not only on the structure itself, but also on the surrounding medium. Structural damping is caused by internal friction within the material and at joints between components. Viscous damping occurs when a structure is moving in air or a fluid.

Generally, the formulation of mathematical expressions for the damping forces in a structure is quite complicated. For this reason, simplified models have been developed which, in many cases, have been found to be adequate. Of these models, the viscous damping force leads to the simplest mathematical treatment. This type of force is introduced in Chapter 1. Because of its simplicity, damping forces of a complicated nature are very often replaced by equivalent viscous damping forces.

The energy dissipated in one cycle of oscillation by a viscous damping force is proportional to the frequency of oscillation and the square of the amplitude of vibration. This can be shown by considering a single degree of freedom system. Denoting the displacement by u, then for harmonic motion

$$u = |u| \cos(\omega t - \alpha) \tag{2.98}$$

where $|u|$ is the amplitude of vibration, ω the frequency and α the phase difference between excitation and response. Since the viscous damping force is $-c\dot{u}$, then the energy dissipated per cycle is

$$\begin{aligned} W_{\mathrm{d}} &= \int c\dot{u}\,\mathrm{d}u \\ &= \int_0^{2\pi/\omega} c\dot{u}^2\,\mathrm{d}t \\ &= \int_0^{2\pi/\omega} c\omega^2 |u|^2 \sin^2(\omega t - \alpha)\,\mathrm{d}t \\ &= \pi c\omega |u|^2 \end{aligned} \tag{2.99}$$

Experimental investigations have indicated that for most structural metals, the energy dissipated per cycle is independent of frequency, over a wide frequency range, and proportional to the square of the amplitude of vibration. Structural damping can, therefore, be represented by a frequency dependent damping coefficient

$$c(\omega) = \frac{h}{\omega} \tag{2.100}$$

In this case

$$\text{damping force} = -\frac{h}{\omega}\dot{u} = -\mathrm{i}hu \tag{2.101}$$

for harmonic motion. Thus, it can be seen that the damping force is in antiphase to velocity and proportional to displacement.

The equation of motion of a single degree of freedom system, consisting of a mass m, a spring of stiffness k, a structural damping force $-\mathrm{i}hu$ and an applied force f, is

$$m\ddot{u} + ihu + ku = f \tag{2.102}$$

This can be rewritten in the form

$$m\ddot{u} + (k + ih)\,u = f \tag{2.103}$$

The quantity $(k + ih)$ is called a 'complex stiffness', which is usually written in the form

$$k + ih = k(1 + i\eta) \tag{2.104}$$

where η is the 'loss factor' of the system. A physical interpretation of the loss factor can be obtained as follows. From equations (2.99) and (2.100), the energy dissipated per cycle for a structurally damped system is

$$\begin{aligned} W_{\mathrm{d}} &= \pi\eta k\,|u|^2 \\ &= 2\pi\eta \cdot \tfrac{1}{2}k\,|u|^2 \\ &= 2\pi\eta \cdot U_{\mathrm{m}} \end{aligned} \tag{2.105}$$

where

$$U_{\mathrm{m}} = \tfrac{1}{2}k|u|^2 \tag{2.106}$$

is the maximum strain energy stored. Therefore, from equation (2.105)

$$\eta = \frac{1}{2\pi}\frac{W_{\mathrm{d}}}{U_{\mathrm{m}}} = \frac{1}{2\pi}\frac{\text{energy dissipated per cycle}}{\text{maximum strain energy}} \tag{2.107}$$

Strictly speaking, this method of representing structural damping should only be used for frequency domain analysis where the excitation is harmonic.

Because of the difficulties which were referred to above, it is useful to use simplified damping models for the complete structure, rather than for individual structural elements. The techniques used for representing both viscous and structural damping type forces are presented in Chapter 9.

2.11 Equations of Motion and Boundary Conditions

In this section two examples are given of how the equations of motion and boundary conditions of continuous structural elements can be formulated using Hamilton's principle. In Chapter 1 the mathematical statement of Hamilton's principle is shown to be (see equation (1.40))

$$\int_{t_1}^{t_2} (\delta(T - U) + \delta W_{\mathrm{nc}})\,\mathrm{d}t = 0 \tag{2.108}$$

where T is the kinetic energy, U the strain energy and δW_{nc} the virtual work done by the non-conservative forces. Expressions for these quantities are derived in the preceding sections of this chapter. For the present exercise the damping forces will be assumed to be zero.

EXAMPLE 2.1 Derive the equation of motion and boundary conditions for an axial element.

From equations (2.11), (2.12) and (2.13) the energy expressions are

$$T = \frac{1}{2}\int_{-a}^{+a} \rho A \dot{u}^2 \, dx \tag{2.109}$$

$$U = \frac{1}{2}\int_{-a}^{+a} EA \left(\frac{\partial u}{\partial x}\right)^2 dx \tag{2.110}$$

$$\delta W_{nc} = \int_{-a}^{+a} p_x \delta u \, dx \tag{2.111}$$

Substituting equations (2.109), (2.110) and (2.111) into equation (2.108) gives

$$\int_{t_1}^{t_2} \left\{ \int_{-a}^{+a} \rho A \dot{u} \delta \dot{u} \, dx - \int_{-a}^{+a} EA \frac{\partial u}{\partial x} \delta \left(\frac{\partial u}{\partial x}\right) dx + \int_{-a}^{+a} p_x \delta u \, dx \right\} dt = 0 \tag{2.112}$$

Assuming that the operators δ and $\partial(\,)/\partial t$ as well as δ and $\partial(\,)/\partial x$ are commutative, and also that integrations with respect to t and x are interchangeable, the first and second terms can be integrated by parts as follows.

Integrating the first term of equation (2.112) with respect to t gives

$$\begin{aligned} \int_{t_1}^{t_2} \rho A \dot{u} \delta \dot{u} \, dt &= \int_{t_1}^{t_2} \rho A \dot{u} \frac{\partial}{\partial t} (\delta u) \, dt \\ &= [\rho A \dot{u} \delta u]_{t_1}^{t_2} - \int_{t_1}^{t_2} \rho A \ddot{u} \delta u \, dt \\ &= -\int_{t_1}^{t_2} \rho A \ddot{u} \delta u \, dt \end{aligned} \tag{2.113}$$

since δu vanishes at $t = t_1$ and $t = t_2$.

Integrating the second term of equation (2.112) with respect to x gives

$$\begin{aligned} \int_{-a}^{+a} EA \frac{\partial u}{\partial x} \delta \left(\frac{\partial u}{\partial x}\right) dx &= \int_{-a}^{+a} EA \frac{\partial u}{\partial x} \frac{\partial}{\partial x} (\delta u) \, dx \\ &= \left[EA \frac{\partial u}{\partial x} \delta u \right]_{-a}^{+a} - \int_{-a}^{+a} EA \frac{\partial^2 u}{\partial x^2} \delta u \, dx \end{aligned} \tag{2.114}$$

Introducing equations (2.113) and (2.114) into equation (2.112) gives

$$\int_{t_1}^{t_2} \left\{ \int_{-a}^{+a} \left(EA \frac{\partial^2 u}{\partial x^2} - \rho A \ddot{u} + p_x \right) \delta u \, dx - \left[EA \frac{\partial u}{\partial x} \delta u \right]_{-a}^{+a} \right\} dt = 0 \tag{2.115}$$

Since δu is arbitrary within both the space and time intervals, then

$$EA \frac{\partial^2 u}{\partial x^2} - \rho A \frac{\partial^2 u}{\partial t^2} + p_x = 0 \tag{2.116}$$

throughout the region $-a \le x \le +a$, and either

$$EA \frac{\partial u}{\partial x} = 0 \quad \text{or} \quad \delta u = 0 \tag{2.117}$$

at $x = -a$ and $x = +a$.

Equation (2.116) is the equation of motion of the element whilst equations (2.117) constitute a statement of the boundary conditions. These conditions can be interpreted physically as follows. $\delta u = 0$ implies that $u = 0$ and therefore the boundary is fixed. Using equations (2.8) and (2.10), it can be seen that $EA\,\partial u/\partial x$ represents the total force on a cross-section. Its vanishing implies that the boundary is free.

EXAMPLE 2.2 Derive the equation of motion and boundary conditions for a beam bending element.

From equations (2.29), (2.31) and (2.32), the energy expressions are

$$T = \frac{1}{2}\int_{-a}^{+a} \rho A \dot{v}^2 \, \mathrm{d}x \tag{2.118}$$

$$U = \frac{1}{2}\int_{-a}^{+a} EI_z \left(\frac{\partial^2 v}{\partial x^2}\right)^2 \mathrm{d}x \tag{2.119}$$

$$\delta W_{\mathrm{nc}} = \int_{-a}^{+a} p_y \delta v \, \mathrm{d}x \tag{2.120}$$

Substituting equations (2.118), (2.119) and (2.120) into equation (2.108) gives

$$\int_{t_1}^{t_2} \left[\int_{-a}^{+a} \rho A \dot{v} \delta \dot{v} \, \mathrm{d}x - \int_{-a}^{+a} EI_z \frac{\partial^2 v}{\partial x^2} \delta \left(\frac{\partial^2 v}{\partial x^2}\right) \mathrm{d}x + \int_{-a}^{+a} p_y \delta v \, \mathrm{d}x \right] \mathrm{d}t = 0 \tag{2.121}$$

The first term is integrated by parts with respect to t as follows

$$\begin{aligned} \int_{t_1}^{t_2} \rho A \dot{v} \delta \dot{v} \, \mathrm{d}t &= \int_{t_1}^{t_2} \rho A \dot{v} \frac{\partial}{\partial t}(\delta v) \, \mathrm{d}t \\ &= [\rho A \dot{v} \delta v]_{t_1}^{t_2} - \int_{t_1}^{t_2} \rho A \ddot{v} \delta v \, \mathrm{d}t \\ &= -\int_{t_1}^{t_2} \rho A \ddot{v} \delta v \, \mathrm{d}t \end{aligned} \tag{2.122}$$

since $\delta v = 0$ when $t = t_1$ and $t = t_2$.

The second term in equation (2.121) is integrated by parts with respect to x

$$\begin{aligned} \int_{-a}^{+a} EI_z \frac{\partial^2 v}{\partial x^2} \delta \left(\frac{\partial^2 v}{\partial x^2}\right) \mathrm{d}x &= \int_{-a}^{+a} EI_z \frac{\partial^2 v}{\partial x^2} \frac{\partial^2}{\partial x^2}(\delta v) \, \mathrm{d}x \\ &= \left[EI_z \frac{\partial^2 v}{\partial x^2} \frac{\partial}{\partial x}(\delta v) \right]_{-a}^{+a} \\ &\quad - \int_{-a}^{+a} EI_z \frac{\partial^3 v}{\partial x^3} \frac{\partial}{\partial x}(\delta v) \, \mathrm{d}x \\ &= \left[EI_z \frac{\partial^2 v}{\partial x^2} \delta \left(\frac{\partial v}{\partial x}\right) - EI_z \frac{\partial^3 v}{\partial x^3} \delta v \right]_{-a}^{+a} \\ &\quad + \int_{-a}^{+a} EI_z \frac{\partial^4 v}{\partial x^4} \delta v \, \mathrm{d}x \end{aligned} \tag{2.123}$$

Substituting equations (2.122) and (2.123) into equation (2.121) gives

$$\int_{t_1}^{t_2}\left\{\int_{-a}^{+a}\left(-EI_z\frac{\partial^4 v}{\partial x^4}-\rho A\ddot{v}+p_y\right)\delta v\,\mathrm{d}x \right. \\ \left. -\left[EI_z\frac{\partial^2 v}{\partial x^2}\delta\left(\frac{\partial v}{\partial x}\right)-EI_z\frac{\partial^3 v}{\partial x^3}\delta v\right]_{-a}^{+a}\right\}\mathrm{d}t=0 \tag{2.124}$$

Since δv is arbitrary, then

$$-EI_z\frac{\partial^2 v}{\partial x^4}-\rho A\frac{\partial^2 v}{\partial t^2}+p_y=0 \tag{2.125}$$

throughout $-a\le x\le +a$. In addition, either

$$EI_z\frac{\partial^2 v}{\partial x^2}=0 \qquad \text{or} \qquad \frac{\partial v}{\partial x}=0$$

and either

$$EI_z=\frac{\partial^3 v}{\partial x^3}=0 \qquad \text{or} \qquad v=0 \tag{2.126}$$

at $x=-a$ and $x=+a$

Equations (2.125) and (2.126) represent the equations of motion and boundary conditions respectively. Note that in this case two conditions are required at each boundary. The conditions $v=0$ and $\partial v/\partial x=0$ represent the vanishing of the displacement and slope respectively. To interpret the other two, consider the distribution of stress over a cross-section given by equations (2.27) and (2.25), that is

$$\sigma_x=-Ey\frac{\partial^2 v}{\partial x^2} \tag{2.127}$$

The moment of this distribution about the z-axis is

$$M_z=-\int_A\sigma_x y\,\mathrm{d}A \\ =\int_A Ey^2\frac{\partial^2 v}{\partial x^2}\mathrm{d}A=EI_z\frac{\partial^2 v}{\partial x^2} \tag{2.128}$$

The vanishing of $EI_z\,\partial^2 v/\partial x^2$ therefore represents the vanishing of the bending moment. Using equation (2.128) the other term in (2.126) is

$$EI_z\frac{\partial^3 v}{\partial x^3}=\frac{\partial M_z}{\partial x} \tag{2.129}$$

For moment equilibrium of an element of length dx of the beam, as shown in Figure 2.9

$$-M_z+\left(M_z+\frac{\partial M_z}{\partial x}\mathrm{d}x\right)+Q\frac{1}{2}\,\mathrm{d}x+\left(Q+\frac{\partial Q}{\partial x}\,\mathrm{d}x\right)\frac{1}{2}\,\mathrm{d}x=0 \tag{2.130}$$

The last term can be neglected in comparison with the others, and so equation (2.130) simplies to

$$\frac{\partial M_z}{\partial x}+Q=0. \tag{2.131}$$

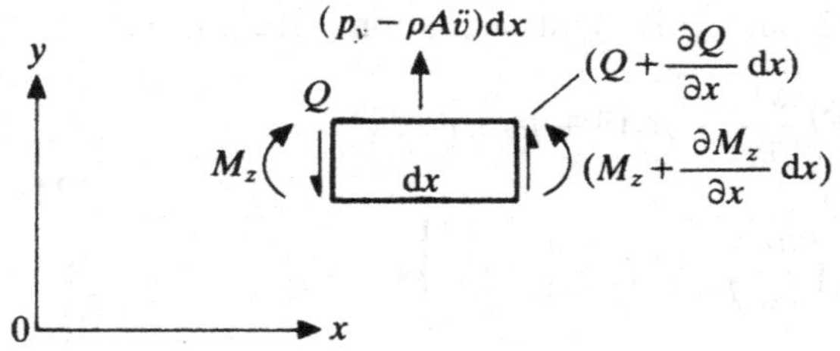

Figure 2.9. Forces acting on an element of a beam.

Equations (2.129) and (2.131) together give

$$EI_z \frac{\partial^3 v}{\partial x^3} = -Q \tag{2.132}$$

The vanishing of $EI_z \partial^3 v / \partial x^3$ therefore indicates the vanishing of the shear force.

The boundary conditions for a beam are therefore as follows:

$$\begin{aligned}
&\text{Simply supported} \quad v = 0, \qquad M_z = EI_z \frac{\partial^2 v}{\partial x^2} = 0 \\
&\text{Clamped} \quad v = 0, \qquad \frac{\partial v}{\partial x} = 0 \\
&\text{Free} \quad Q = -EI_z \frac{\partial^3 v}{\partial x^3} = 0, \qquad M_z = EI_z \frac{\partial^2 v}{\partial x^2} = 0
\end{aligned} \tag{2.133}$$

In the above two examples it can be seen that the boundary conditions consist of two types. One type consists of the vanishing of displacements or rotations. This type is called a 'geometric' boundary condition. The other type, which consists of the vanishing of forces or moments, is referred to as a 'natural' boundary condition. In general, it will be found that if the mathematical statement of Hamilton's principle involves derivatives of order p, then boundary conditions which involve derivatives up to and including those of order $(p - 1)$ will be geometric boundary conditions, and those involving derivatives of order p, $(p + 1), \ldots, (2p - 1)$ will be natural boundary conditions.

Problems

2.1 Discuss the changes which should be made to the derivation of the energy expressions for an axial element in Section 2.1 if the cross-sectional area varies along its length.

Show that in this case the expressions for the kinetic and strain energies become

$$T = \frac{1}{2} \int_{-a}^{+a} \rho A(x) \dot{u}^2 \, dx, \qquad U = \frac{1}{2} \int_{-a}^{+a} EA(x) \left(\frac{\partial u}{\partial x} \right)^2 dx$$

where $A(x)$ is the cross-sectional area at position x.

2.2 Assuming that the Saint–Venant theory of torsion still holds and that warping in the axial direction is negligible, show that the expressions for the kinetic and strain energy of a torque element of variable cross-section are

$$T = \frac{1}{2}\int_{-a}^{+a} \rho I_x(x)\dot{\theta}_x^2 \, dx, \qquad U = \frac{1}{2}\int_{-a}^{+a} GJ(x)\left(\frac{\partial \theta_x}{\partial x}\right)^2 dx$$

where $I_x(x)$ and $J(x)$ are the second moment of area about the x-axis and the torsion constant at position x.

2.3 Show that when shear deformation and rotary inertia effects are neglected, the expressions for kinetic and strain energy of a beam bending element of variable cross-section are

$$T = \frac{1}{2}\int_{-a}^{+a} \rho A(x)\dot{v}^2 \, dx, \qquad U = \frac{1}{2}\int_{-a}^{+a} EI_z(x)\left(\frac{\partial^2 v}{\partial x^2}\right)^2 dx$$

where $A(x)$ and $I_z(x)$ are the area and second moment of area about the z-axis of the cross-section at position x.

2.4 Show that if shear deformation and rotary inertia effects are included, the expressions for the kinetic and strain energy of a beam bending element of variable cross-section are

$$T = \frac{1}{2}\int_{-a}^{+a} \rho A(x)\dot{v}^2 \, dx + \frac{1}{2}\int_{-a}^{+a} \rho I_z(x)\dot{\theta}_z^2 \, dx$$

$$U = \frac{1}{2}\int_{-a}^{+a} EI_z(x)\left(\frac{\partial \theta_z}{\partial x}\right)^2 dx + \frac{1}{2}\int_{-a}^{+a} \kappa A(x) G\left(\frac{\partial v}{\partial x} - \theta_z\right)^2 dx$$

where $A(x)$ and $I_z(x)$ are as defined in Problem 2.3.

2.5 Show that the cross-sectional area, $A(\xi)$, and the second moment of area of the cross-section, $I_z(\xi)$, of the linearly tapered beam element shown in Figure P2.5(a) can be expressed in the form

$$A(\xi) = A(0)\{1 + a_1\xi + a_2\xi^2\}$$

$$I_z(\xi) = I_z(0)\{1 + b_1\xi + b_2\xi^2 + b_3\xi^3 + b_4\xi^4\}$$

Also show that for

(1) a rectangular cross-section (Figure P.25(b)), the coefficients are given by

$$a_1 = -\left(\frac{1-B}{1+B}\right) - \left(\frac{1-D}{1+D}\right) \qquad a_2 = \left(\frac{1-B}{1+B}\right)\left(\frac{1-D}{1+D}\right)$$

$$b_1 = -\left(\frac{1-B}{1+B}\right) - 3\left(\frac{1-D}{1+D}\right)$$

$$b_2 = 3\left(\frac{1-B}{1+B}\right)\left(\frac{1-D}{1+D}\right) + 3\left(\frac{1-D}{1+D}\right)^2$$

$$b_3 = -3\left(\frac{1-B}{1+B}\right)\left(\frac{1-D}{1+D}\right)^2 - \left(\frac{1-D}{1+D}\right)^3$$

$$b_4 = \left(\frac{1-B}{1+B}\right)\left(\frac{1-D}{1+D}\right)^3$$

with $B = b(+1)/b(-1)$ and $D = d(+1)/d(-1)$, and

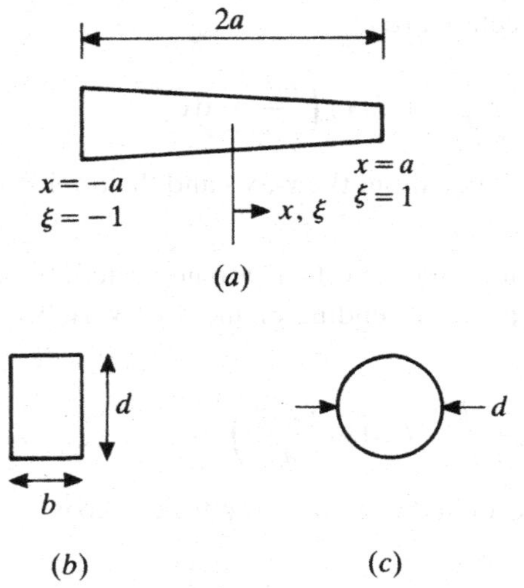

Figure P2.5

(2) for a circular cross-section (Figure P2.5(*c*)) the coefficients are as above with *B* replaced by *D*.

2.6 Use Hamilton's principle to show that the equation of motion and boundary conditions for a uniform rod with a mass at one end, which is vibrating axially (Figure P2.6), are

$$EA\frac{\partial^2 u}{\partial x^2} - \rho A\frac{\partial^2 u}{\partial t^2} + p_x = 0$$

Either

$$EA\frac{\partial u}{\partial x} = 0 \qquad \text{or} \qquad u = 0 \text{ at } x = -a$$

and either

$$EA\frac{\partial u}{\partial x} + M\ddot{u} = 0 \qquad \text{or} \qquad u = 0 \text{ at } x = +a$$

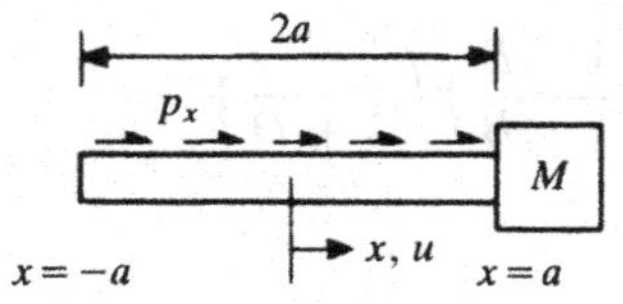

Figure P2.6

2.7 Use Hamilton's principle to show that the equations of motion of a deep uniform beam element are

$$\kappa AG\frac{\partial^2 v}{\partial x^2} - \kappa AG\frac{\partial \theta_z}{\partial x} - \rho A\ddot{v} + p_y = 0$$

$$\kappa AG\frac{\partial v}{\partial x} + EI_z\frac{\partial^2 \theta_z}{\partial x^2} - \kappa AG\theta_z - \rho I_z\ddot{\theta}_z = 0$$

and the boundary conditions are either

$$v = 0 \qquad \text{or} \qquad \kappa AG\left(\frac{\partial v}{\partial x} - \theta_z\right) = 0$$

and either

$$\theta_z = 0 \qquad \text{or} \qquad EI_z\frac{\partial \theta_z}{\partial x} = 0.$$

2.8 Use Hamilton's principle to show that the equation of motion for axial vibration of a tapered rod is

$$\frac{\partial}{\partial x}\left(EA(x)\frac{\partial u}{\partial x}\right) - \rho A(x)\ddot{u} + p_x = 0$$

and the boundary conditions are either

$$v = 0 \qquad \text{or} \qquad EA(x)\partial u/\partial x = 0$$

2.9 Use Hamilton's principle to show that the equation of motion for flexural vibrations of a slender tapered beam is

$$\frac{\partial^2}{\partial x^2}\left\{EI_z(x)\frac{\partial^2 v}{\partial x^2}\right\} + \rho A(x)\ddot{v} - p_y = 0$$

and the boundary conditions are either

$$v = 0 \qquad \text{or} \qquad \frac{\partial}{\partial x}\left\{EI_z(x)\frac{\partial^2 v}{\partial x^2}\right\} = 0$$

and either

$$\frac{\partial v}{\partial x} = 0 \qquad \text{or} \qquad EI_z(x)\frac{\partial^2 v}{\partial x^2} = 0$$

2.10 Use Hamilton's principle to show that the equations of motion for flexural vibrations of a deep tapered beam are

$$\frac{\partial}{\partial x}\left(\kappa A(x)G\left(\frac{\partial v}{\partial x} - \theta_z\right)\right) - \rho A(x)\ddot{v} + p_y = 0$$

$$\kappa A(x)G\left(\frac{\partial v}{\partial x} - \theta_z\right) + \frac{\partial}{\partial x}\left(EI_z(x)\frac{\partial \theta_z}{\partial x}\right) - \rho I_z(x)\ddot{\theta}_z = 0$$

and the boundary conditions are either

$$v = 0 \qquad \text{or} \qquad \kappa A(x)G\left(\frac{\partial v}{\partial x} - \theta_z\right) = 0$$

and either

$$\theta_z = 0 \qquad \text{or} \qquad EI_z(x)\frac{\partial \theta_z}{\partial x} = 0$$

2.11 Show that the expressions for the kinetic and strain energy of a membrane element of variable thickness are

$$T = \frac{1}{2}\int_A \rho h(\dot{u}^2 + \dot{v}^2)\mathrm{d}A, \qquad U = \frac{1}{2}\int_A h\{\boldsymbol{\varepsilon}\}^{\mathrm{T}}[\mathbf{D}]\{\boldsymbol{\varepsilon}\}\,\mathrm{d}A$$

where $h = h(x, y)$, the other notation being defined in Section 2.5, provided the plate is sufficiently thin.

2.12 Show that the expressions for the kinetic and strain energy of a thin plate bending element of variable thickness are

$$T = \frac{1}{2}\int_A \rho h \dot{w}\,\mathrm{d}A, \qquad U = \frac{1}{2}\int_A \frac{h^3}{12}\{\boldsymbol{\chi}\}^{\mathrm{T}}[\mathbf{D}]\{\boldsymbol{\chi}\}\,\mathrm{d}A$$

where $h = h(x, y)$, the other notation being defined in Section 2.6.

3 Introduction to the Finite Element Displacement Method

The response of simple structures, such as uniform axial, torque and beam elements, may be obtained by solving the differential equations of motion together with the appropriate boundary conditions, as derived in Section 2.11. In many practical situations either the geometrical or material properties vary, or it may be that the shape of the boundaries cannot be described in terms of known functions. Also, practical structures consist of an assemblage of components of different types, namely, beams, plates, shells and solids. In these situations it is impossible to obtain analytical solutions to the equations of motion which satisfy the boundary conditions. This difficulty is overcome by seeking approximate solutions which satisfy Hamilton's principle (see Section 1.3).

There are a number of techniques available for determining approximate solutions to Hamilton's principle. One of the most widely used procedures is the Rayleigh–Ritz method, which is described in the next section. A generalisation of the Rayleigh–Ritz method, known as the finite element displacement method, is then introduced. The principlal features of this method are described by considering rods, shafts, beams and frameworks.

3.1 Rayleigh–Ritz Method

The Rayleigh–Ritz method is first described with reference to the problem of determining the axial motion of the rod shown in Figure 3.1.

Hamilton's principle (Section 1.3) requires that

$$\int_{t_2}^{t_1} (\delta(T - U) + \delta W)\, dt = 0 \tag{3.1}$$

From Section 2.1 the energy functions are

$$\begin{aligned} T &= \frac{1}{2}\int_0^L \rho A \dot{u}^2\, dx \\ U &= \frac{1}{2}\int_0^L EA\left(\frac{\partial u}{\partial x}\right)^2 dx \\ \delta W &= F\delta u(L) \end{aligned} \tag{3.2}$$

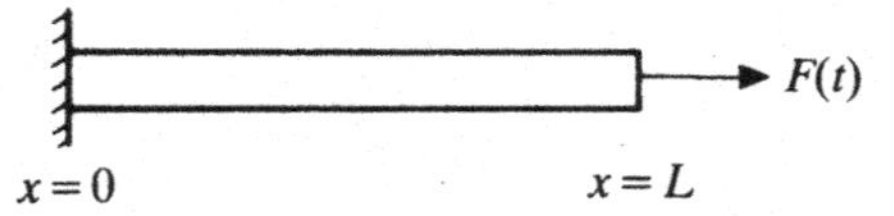

Figure 3.1. Rod subject to an axial force.

Since Hamilton's principle is derived using the principle of virtual displacements, then the solution $u(x, t)$ which is required is the one which satisfies both (3.1) and the geometric boundary condition

$$u(0) = 0 \tag{3.3}$$

Satisfaction of Hamilton's principle will ensure that both the equation of motion and the natural boundary condition at $x = L$ will be satisfied (see Section 2.11).

The Rayleigh–Ritz method approximates the solution with a finite expansion of the form

$$u^n(x, t) = \sum_{j=1}^{n} \phi_j(x) q_j^n(t) \tag{3.4}$$

where the $q_j^n(t)$ are unknown functions of time, t, and the $\phi_j(x)$ are prescribed functions of x, which are linearly independent. A set of functions are said to be linearly independent if

$$\sum_{j=1}^{n} \alpha_j \phi_j(x) = 0 \qquad \text{for all } x \tag{3.5}$$

implies that

$$\alpha_j = 0 \qquad \text{for } j = 1, 2, \ldots, n \tag{3.6}$$

Each of the functions $\phi_j(x)$ must satisfy the geometric boundary condition (3.3) in order to ensure that the solution, as given by equation (3.4), satisfies this condition. Therefore

$$\phi_j(0) = 0 \qquad j = 1, 2, \ldots, n \tag{3.7}$$

Since the strain energy expression (3.2) involves the first derivative of u with respect to x, then each of the functions $\phi_j(x)$ should have a finite derivative. This implies that these functions must be continuous.

A continuous deformable body, such as the rod considered here, consists of an infinity of material points, and therefore, it has infinitely many degrees of freedom. By assuming that the motion is given by the expression (3.4), the continuous system has been reduced to a system with a finite number of degrees of freedom. This has been achieved by applying the constraints

$$q_{n+1}^n = q_{n+2}^n = \cdots = 0 \tag{3.8}$$

The expression (3.4) is substituted into equation (3.1) and the $q_j^n(t)$ found. Since the system has been reduced to one with a finite number of degrees of freedom, then the application of Hamilton's principle leads to Lagrange's equations (Sections 1.4 and A1.2). These give, in matrix form (Section 1.4)

$$[\mathbf{m}]\{\ddot{\mathbf{q}}^n\} + [\mathbf{K}]\{\mathbf{q}^n\} = \{\mathbf{Q}^n\} \tag{3.9}$$

where

$$\{\mathbf{q}^n\}^{\mathrm{T}} = \lfloor q_1^n \quad q_2^n \quad \cdots \quad q_n^n \rfloor \tag{3.10}$$

The inertia and stiffness matrices are determined by substituting (3.4) into the kinetic and strain energy expressions (3.2) respectively. The elements of these matrices are given by

$$\begin{aligned} M_{jk} &= \int_0^L \rho A \phi_j(x)\phi_k(x)\,\mathrm{d}x \\ K_{jk} &= \int_0^L EA\phi_j'(x)\phi_k'(x)\,\mathrm{d}x \end{aligned} \tag{3.11}$$

where primes denote differentiation with respect to x. The generalised forces Q_j^n are obtained by calculating the virtual work done by the applied load $F(t)$. From (3.2) and (3.4)

$$\begin{aligned} \delta W = F(t)\delta u(L) &= F(t)\sum_{j=1}^{n} \phi_j(L)\delta q_j^n(t) \\ &= \sum_{j=1}^{n} Q_j^n \delta q_j^n \end{aligned} \tag{3.12}$$

This gives

$$Q_j^n = \phi_j(L)F(t) \tag{3.13}$$

Equation (3.9) is solved for the $\{\mathbf{q}^n\}$, which are then substituted into (3.4) to give an approximate solution for $u(x, t)$. Methods of solving equation (3.9) are described in Chapters 12 and 13.

If the integrals in (3.1) involve derivatives up to order p, then the functions $\phi_j(x)$ of equation (3.4) must satisfy the following criteria in order to ensure convergence of the solution.

(1) Be linearly independent.
(2) Be continuous and have continuous derivatives up to order $(p-1)$. In this book only the cases $p = 1$ and 2 will be considered.
(3) Satisfy the geometric boundary conditions. These involve derivatives up to order $(p-1)$ (see Section 2.11).
(4) Form a complete series.

A series of functions is said to be complete if the 'mean square error' vanishes in the limit, that is

$$\lim_{n\to\infty} \int_0^L \left(u - \sum_{j=1}^{n} \phi_j q_j^n \right) \mathrm{d}x = 0 \tag{3.14}$$

Polynomials (i.e., $1, x, x^2, \ldots$), trigonometric functions, Legendre, Tchebycheff and Jacobi or hypergeometric polynomials are all series of functions which are complete. An approximate solution which satisfies (3.14) is said to 'converge in the mean'.

In order to assess the convergence of the method, solutions are obtained using the sequence of functions $u^1, u^2, u^3, \ldots, u^n$. This sequence is called a minimising sequence. Using a minimising sequence ensures monotonic convergence of the solution. Using functions $\phi_j(x)$, which form a complete series, ensures monotonic convergence to the true solution.

The proof of convergence of the Rayleigh–Ritz method is based upon the proof of convergence of the expansion of an arbitrary function by means of an infinite series of linearly independent functions. If polynomials are used, then use can be made of Weierstrass's Approximation Theorem which states that: 'Any function which is continuous in the interval $a \leq x \leq b$ may be approximated uniformly by polynomials in this interval.' This theorem [3.1] asserts the possibility of uniform convergence rather than just convergence in the mean. Since the functions are required to have continuous derivatives up to order $(p - 1)$, then all derivatives up to this order will converge uniformly.

These statements can be extended to functions of more than one variable. Further details are given in references [3.2–3.5]. It should be noted that in using the Rayleigh–Ritz method the equations of motion and natural boundary conditions will only be satisfied approximately.

Another problem of interest in vibration analysis is that of determining the natural frequencies and modes of free vibration of a structure. In this case $\delta W = 0$ in (3.1). The value of the integral of $(T - U)$, I^n, obtained by substituting (3.4) into it, will be greater than the true minimum because of the application of the constraints (3.8). Using the sequence of functions $u^1, u^2, u^3, \ldots, u^n$, it follows that

$$I^1 \geq I^2 \geq I^3 \geq \cdots \geq I^n \tag{3.15}$$

since the inclusion of more terms in (3.4) is equivalent to relaxing successive constraints. In this equation (3.9) reduces to

$$[\mathbf{M}]\{\ddot{\mathbf{q}}^n\} + [\mathbf{K}]\{\mathbf{q}^n\} = 0 \tag{3.16}$$

Since the motion is harmonic then

$$\{\mathbf{q}^n(t)\} = \{\mathbf{A}^n\} \sin \omega t \tag{3.17}$$

where the amplitudes $\{\mathbf{A}^n\}$ are independent of time and ω is the frequency of vibration. Substituting (3.17) into (3.16) gives

$$[\mathbf{K} - \omega^2\mathbf{M}]\{\mathbf{A}^n\} = 0 \tag{3.18}$$

Equation (3.18) represents a set of n linear homogeneous equations in the unknowns $A_1^n, A_2^n, \ldots, A_n^n$. The condition that these equations should have a non-zero solution is that the determinant of coefficients should vanish, that is

$$\det[\mathbf{K} - \omega^2\mathbf{M}] = |\mathbf{K} - \omega^2\mathbf{M}| = 0 \tag{3.19}$$

Equation (3.19) can be expanded to give a polynomial of degree n in ω^2. This polynomial equation will have n roots $\omega_1^2, \omega_2^2, \ldots, \omega_n^2$. Such roots are called 'eigenvalues'. Since $[\mathbf{M}]$ is positive definite, and $[\mathbf{K}]$ is either positive definite or positive semi-definite (see Section 1.4), the eigenvalues are all real and either positive or zero [3.6]. However, they are not necessarily all different from one another. The quantities $\omega_1, \omega_2, \ldots, \omega_n$, which are also real and either positive or zero, are approximate values of the first n natural frequencies of the system. Moreover, these approximate values will be greater than the true frequencies of the system [3.7].

Corresponding to each eigenvalue ω^2, there exists a unique solution (to within an arbitrary constant) to equation (3.18) for $\{\mathbf{A}^n\}$. These solutions are known as 'eigenvectors'. When combined with the prescribed functions $\phi_j(x)$ they define the shapes of the modes of vibration in an approximate sense. The approximate shape of a mode of vibration is given by (see equation (3.4)):

$$u^n(x) = \sum_{j=1}^{n} \phi_j(x) A_j^n \tag{3.20}$$

The solution of equation (3.18) is known as an 'eigenproblem'. Numerical methods of determining the solutions of eigenproblems, as defined by this equation, are presented in Chapter 11. These solutions, as indicated above, give approximate solutions for the natural frequencies and modes of free vibration. Convergence to the true frequencies and mode shapes is obtained as the number of terms in the approximating expression (3.4) is increased. This statement is illustrated by means of the examples below.

EXAMPLE 3.1 Use the Rayleigh–Ritz method to estimate the lower frequencies and mode shapes of the clamped-free rod shown in Figure 3.1. Compare the results with the exact solution.

For free vibration, the equation of motion of the rod is (see equation (2.116))

$$EA\frac{\partial^2 u}{\partial x^2} - \rho A\frac{\partial^2 u}{\partial t^2} = 0 \tag{3.21}$$

Assuming harmonic motion

$$u(x, t) = \psi(x) \sin \omega t \tag{3.22}$$

Substituting (3.22) into equation (3.21) gives

$$\frac{\mathrm{d}^2\psi}{\mathrm{d}x^2} + \omega^2 \left(\frac{\rho}{E}\right) \psi = 0 \tag{3.23}$$

The boundary conditions are (see Section 2.11)

$$u(0, t) = 0, \qquad \frac{\partial u(L, t)}{\partial x} = 0 \tag{3.24}$$

Substituting (3.22) into the boundary conditions (3.24) gives

$$\psi(0) = 0, \qquad \frac{\mathrm{d}\psi(L)}{\mathrm{d}x} = 0 \tag{3.25}$$

The solutions of equation (3.23) subject to the boundary conditions (3.25) are

$$\begin{aligned} \omega_r &= \frac{(2r-1)\pi}{2}\left(\frac{E}{\rho L^2}\right)^{1/2}, \\ \psi_r(x) &= \sin(2r-1)\frac{\pi x}{2L} \qquad r = 1, 2, \ldots \end{aligned} \tag{3.26}$$

To obtain an approximate solution using the Rayleigh–Ritz method, assume the prescribed functions in (3.4) to be

$$\phi_j(x) = x^j \tag{3.27}$$

Note that each of these satisfy the geometric boundary condition $\phi_j(0) = 0$.

The elements of the stiffness and inertia matrices in equation (3.18) are, from equations (3.11)

$$
\begin{aligned}
K_{jk} &= \int_0^L EAj \cdot k \cdot x^{j+k-2}\,dx = \frac{jk}{(j+k-1)} EAL^{j+k-1} \\
M_{jk} &= \int_0^L \rho A x^{j+k}\,dx = \frac{1}{(j+k+1)} \rho AL^{j+k+1}
\end{aligned}
\tag{3.28}
$$

One term solution

Using only one term in the series (3.4), equation (3.18) reduces to

$$\left(EAL - \omega^2 \frac{\rho AL^3}{3}\right) A_1^1 = 0$$

the solution of which gives $\omega_1 = 1.732(E/\rho L^2)^{1/2}$.

Two term solution

Increasing the number of terms to two in the series (3.4), gives the following equation:

$$\left[EA \begin{bmatrix} L & L^2 \\ L^2 & 4L^3/3 \end{bmatrix} - \omega^2 \rho A \begin{bmatrix} L^3/3 & L^4/4 \\ L^4/4 & L^5/5 \end{bmatrix}\right] \begin{bmatrix} A_1^2 \\ A_2^2 \end{bmatrix} = 0$$

Letting $\omega^2 \rho L^2/E = \lambda$, the above equation simplifies to

$$\begin{bmatrix} (1-\lambda/3) & (1-\lambda/4) \\ (1-\lambda/4) & (4/3-\lambda/5) \end{bmatrix} \begin{bmatrix} A_1^2 \\ A_2^2 L \end{bmatrix} = 0$$

This equation has a non-zero solution provided

$$\begin{vmatrix} (1-\lambda/3) & (1-\lambda/4) \\ (1-\lambda/4) & (4/3-\lambda/5) \end{vmatrix} = 0$$

Expanding gives

$$\frac{\lambda^2}{240} - \frac{13\lambda}{90} + \frac{1}{3} = 0$$

The two roots of this equation are

$$\lambda = 2.486 \quad \text{and} \quad 32.18$$

and the natural frequencies of the system are

$$\omega_1 = \lambda_1^{1/2}\left(\frac{E}{\rho L^2}\right)^{1/2} = 1.577\left(\frac{E}{\rho L^2}\right)^{1/2}$$

and

$$\omega_2 = \lambda_2^{1/2}\left(\frac{E}{\rho L^2}\right)^{1/2} = 5.673\left(\frac{E}{\rho L^2}\right)^{1/2}$$

Table 3.1. *Comparison of approximate frequencies with exact solution for a rod*

	R–R solutions		
Mode	1 term	2 term	Exact solution
1	1.732	1.577	1.571
2	–	5.673	4.712

From the homogeneous equations

$$A_2^2 = -\frac{(1-\lambda/3)}{(1-\lambda/4)\,L}A_1^2$$

When

$$\lambda = 2.486, \qquad A_2^2 = -0.4527\frac{A_1^2}{L}$$

$$\lambda = 32.18, \qquad A_2^2 = -1.3806\frac{A_1^2}{L}$$

The modes of vibration are therefore given by

$$u = A_1^2 L\left\{\frac{x}{L} - 0.4527\left(\frac{x}{L}\right)^2\right\}$$

and

$$u = A_1^2 L\left\{\frac{x}{L} - 1.3806\left(\frac{x}{L}\right)^2\right\}$$

The approximate values of $\omega(\rho L^2/E)^{1/2}$ are compared with the exact values in Table 3.1. As postulated, the approximate frequencies are greater than the exact ones and approach the exact ones as the number of terms is increased.

The approximate mode shapes for the two term solution are compared with the exact mode shapes in Figure 3.2. The differences between the approximate and exact shapes for the first mode are too small to show up on the scale used.

EXAMPLE 3.2 Use the Rayleigh–Ritz method to estimate the lower frequencies of the cantilever beam shown in Figure 3.3. Compare the results with the exact solution.

From Section 2.11 the equation of motion of the beam is

$$EI_z\frac{\partial^4 v}{\partial x^4} + \rho A\frac{\partial^2 v}{\partial t^2} = 0 \tag{3.29}$$

and the boundary conditions are

$$\begin{aligned} &v(0,t) = 0, && \frac{\partial v}{\partial x}(0,t) = 0 \\ &\frac{\partial^2 v}{\partial x^2}(L,t) = 0, && \frac{\partial^3 v}{\partial x^3}(L,t) = 0 \end{aligned} \tag{3.30}$$

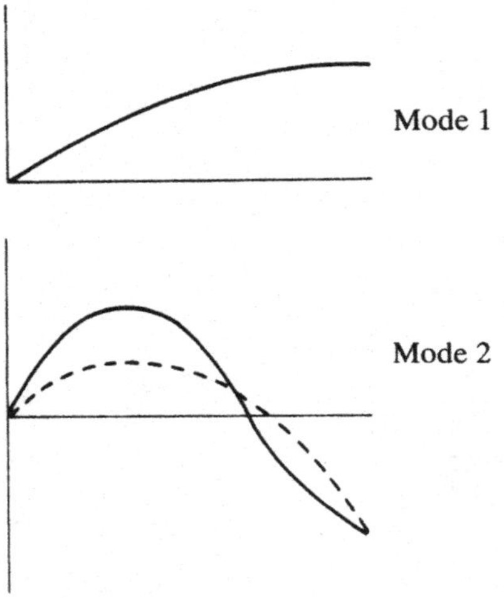

Figure 3.2. Axial modes of vibration of a rod. —— Exact; - - - approximate (R–R).

The solutions of equation (3.29) subject to the boundary conditions (3.30) are given by [3.8]:

$$v_r(x, t) = \psi_r(x) \sin \omega_r t$$

where

$$\omega_r = (\beta_r L)^2 \left(\frac{EI_z}{\rho A L^4} \right)^{1/2} \tag{3.31}$$

$$\psi_r(x) = \{\cosh \beta_r x - \cos \beta_r x - \eta_r (\sinh \beta_r x - \sin \beta_r x)\} \tag{3.32}$$

and

$$\eta_r = \frac{\cos \beta_r L + \cosh \beta_r L}{\sin \beta_r L + \sinh \beta_r L}$$

From Section 2.3 the energy functions are

$$\begin{aligned} T &= \frac{1}{2} \int_0^L \rho A \dot{v}^2 \, dx \\ U &= \frac{1}{2} \int_0^L EI_z \left(\frac{\partial^2 v}{\partial x^2} \right)^2 dx \end{aligned} \tag{3.33}$$

To obtain an approximate solution using the Rayleigh–Ritz method, assume an expansion of the form

$$v^n(x, t) = \sum_{j=1}^{n} \phi_j(x) A_j^n \sin \omega t \tag{3.34}$$

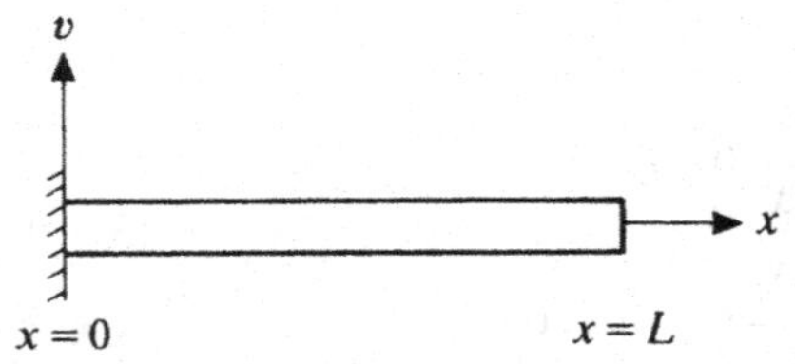

Figure 3.3. Cantilever beam.

Table 3.2. *Comparison of approximate frequencies with exact solution for a beam*

Mode	Approximate solutions		Exact solution
	1 term	2 term	
1	4.472	3.533	3.516
2	–	34.807	22.035

where

$$\phi_j(x) = x^{j+1}$$

Each of the functions $\phi_j(x)$ satisfy the geometric boundary conditions at $x = 0$, that is

$$\phi_j(0) = 0, \qquad \frac{d\phi_j}{dx}(0) = 0$$

Substituting (3.34) into the energy expressions (3.33) gives the elements of the inertia and stiffness matrices in equation (3.18), namely

$$\begin{aligned} M_{jk} &= \int_0^L \rho A x^{j+k+2}\,dx = \frac{1}{(j+k+3)}\rho A L^{j+k+3} \\ K_{jk} &= \int_0^L EI_z (j+1)j(k+1)k x^{j+k-2}\,dx \\ &= \frac{(j+1)j(k+1)k}{(j+k-1)} EI_z\, L^{j+k-1} \end{aligned} \tag{3.35}$$

The approximate values of $\omega(\rho A L^4/EI_z)^{1/2}$ are compared with the exact solutions in Table 3.2 for various values of n.

3.2 Finite Element Displacement Method

When analysing either structures of complex shape or built-up structures, difficulties arise in constructing a set of prescribed functions which satisfy the geometric boundary conditions. These difficulties can be overcome by using the Finite Element Displacement Method. This method provides an automatic procedure for constructing the approximating functions in the Rayleigh–Ritz method.

The prescribed functions are constructed in the following manner:

(1) Select a set of reference or 'node' points on the structure.
(2) Associate with each node point a given number of degrees of freedom (displacement, slope, etc.).
(3) Construct a set of functions such that each one gives a unit value for one degree of freedom and zero values for all the others.

This procedure is illustrated for the axial motion of a rod in Figure 3.4 and the bending vibration of a beam in Figure 3.5.

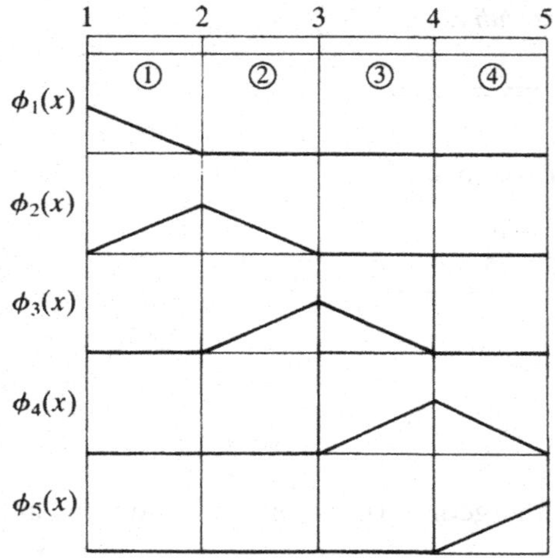

Figure 3.4. Prescribed functions for a rod.

In Figure 3.4 five node points have been selected at equal intervals. The region between each pair of adjacent nodes is referred to as an 'element'. It is shown in the previous section that only the prescribed functions themselves need be continuous for a rod. (This implies that the first derivative, which appears in the strain energy expression, can be discontinuous.) Therefore, the axial displacement, u, is the only degree of freedom required at each node point. In the figure, five prescribed functions are illustrated. They have been constructed by giving each node point in turn a unit axial displacement, whilst maintaining zero displacement at all other nodes. If these functions were to be used to analyse a clamped-free rod, then the first function,

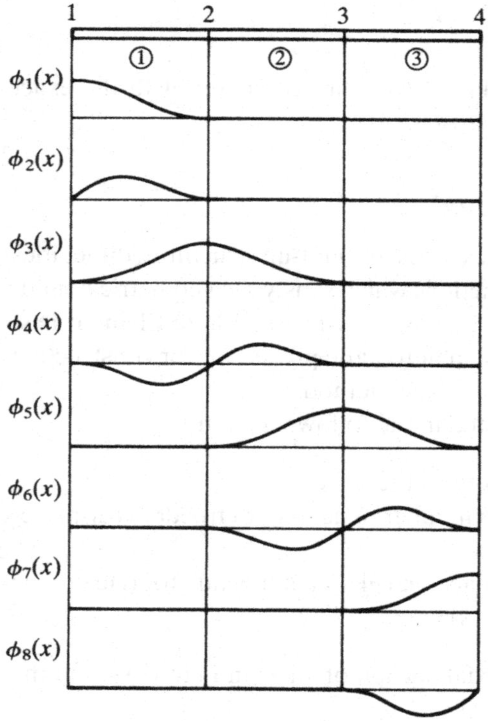

Figure 3.5. Prescribed functions for a beam.

$\phi_1(x)$, would be omitted, since it does not satisfy the geometric boundary condition at $x = 0$. For a clamped-clamped rod both $\phi_1(x)$ and $\phi_5(x)$ would be omitted.

In Figure 3.5, four node points have been selected at equal intervals. Thus the beam has been divided into three elements. The highest derivative appearing in the energy expressions for a beam is the second (see equations (3.33)). Therefore, the Rayleigh–Ritz procedure requires the prescribed functions and their first derivative to be continuous. Hence, it will be necessary to take v and $\partial v/\partial x$ as degrees of freedom at each node. In the figure the odd numbered prescribed functions have been constructed by giving each node point in turn a unit lateral displacement, whilst maintaining zero displacement at all other nodes. At the same time the rotations are kept zero at all nodes. The even numbered prescribed functions are constructed by giving each node in turn a unit rotation, whilst the rotations at all other nodes are kept zero. In addition, the displacements at all nodes are zero. Again, the geometric boundary conditions are satisfied by omitting the appropriate functions. For example, the functions $\phi_1(x)$ and $\phi_2(x)$ are omitted when analysing a cantilever beam.

Referring back to Figure 3.4, it can be seen that the variation of axial displacement over each element is zero except for two cases, the number being equal to the number of nodes (2) multiplied by the number of degrees of freedom at each node (1) for a single element. These two displacement variations are identical for each element. In the same way, each element of the beam in Figure 3.5 deforms in only four of the prescribed functions, being equal to the number of nodes (2) multiplied by the number of degrees of freedom at each node (2). Again the displacement variations for each element are identical. Because of this feature, it is simpler to evaluate the energy expressions for each element and then add the contributions from the elements together. This technique is illustrated in the following sections where explicit expressions for the prescribed functions over a single element are derived. These functions are referred to as 'element displacement functions'. In some texts the term 'shape function' is used, but it will not be used here.

In order to satisfy the convergence criteria of the Rayleigh–Ritz method, the element displacement functions should satisfy the following conditions:

(1) Be linearly independent.
(2) Be continuous and have continuous derivatives up to order $(p - 1)$ both within the element and across element boundaries. An element which satisfies this condition is referred to as a 'conforming' element.
(3) If polynomial functions are used, then they must be complete polynomials of at least degree p. If any terms of degree greater than p are used, they need not be complete. (A complete polynomial of degree n in m variables has $(n+m)!/n!m!$ independent terms.) However, the rate of convergence is governed by the order of completeness of the polynomial. The element displacement functions need not be polynomials as shown in Chapter 10.
(4) Satisfy the geometric boundary conditions.

In the Rayleigh–Ritz method, convergence is obtained as the number of prescribed functions is increased. To increase the number of prescribed functions in the finite element method, the number of node points, and therefore the number of elements, is increased. A complete discussion of the convergence of the finite element method is given in reference [3.9].

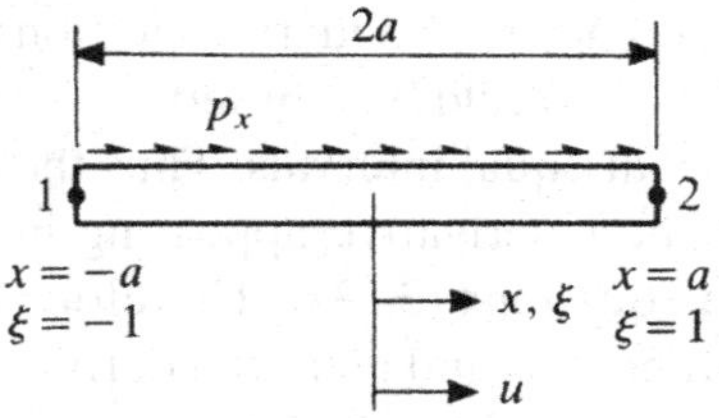

Figure 3.6. Geometry of a single axial element.

3.3 Axial Vibration of Rods

There are a number of ways of determining the displacement functions of a single element. The most common of these are as follows:

(1) By inspection.
(2) Assume a polynomial function having the appropriate number of terms. Then evaluate it and, if necessary, its derivatives at the nodes to obtain the coefficients in terms of the nodal degrees of freedom.
(3) Solve the equations of static equilibrium to determine the deformation of the element due to prescribed boundary displacements.

In practice the most appropriate method is used for each type of element. All three procedures are now illustrated using an axial element.

It is shown in Figure 3.4 that the deformation of an axial element is given by the combination of two linear functions. Using the non-dimensional coordinate $\xi = x/a$, defined in Figure 3.6, it is easily seen that the displacement variation for such an element is given by

$$u = \tfrac{1}{2}(1-\xi)u_1 + \tfrac{1}{2}(1+\xi)u_2 \tag{3.36}$$

where u_1, u_2 are the axial displacements of nodes 1 and 2.

Alternatively, since the element has 2 nodes and 1 degree of freedom at each node, the displacement variation can be represented by a polynomial having 2 constants, namely

$$u = \alpha_1 + \alpha_2\xi \tag{3.37}$$

Note that the highest derivative which occurs in the energy expressions is the first (see equations (2.11) to (2.13)), and so a polynomial of at least degree one must be used to satisfy the convergence criteria.

Evaluating (3.37) at $\xi = \mp 1$ gives

$$u_1 = \alpha_1 - \alpha_2, \qquad u_2 = \alpha_1 + \alpha_2 \tag{3.38}$$

Solving for α_1 and α_2 gives

$$\alpha_1 = \tfrac{1}{2}(u_1 + u_2), \qquad \alpha_2 = \tfrac{1}{2}(u_2 - u_1) \tag{3.39}$$

Substituting (3.39) into (3.37) gives

$$\begin{aligned} u &= \tfrac{1}{2}(u_1 + u_2) + \tfrac{1}{2}(u_2 - u_1)\xi \\ &= \tfrac{1}{2}(1-\xi)u_1 + \tfrac{1}{2}(1+\xi)u_2 \end{aligned} \tag{3.40}$$

An expression which is identical to (3.36) has therefore been obtained.

The equation of static equilibrium for the element can be deduced from equation (2.116) to be

$$\frac{\mathrm{d}^2 u}{\mathrm{d}x^2} = 0 \tag{3.41}$$

(It is assumed that there is no distributed loading, only end forces necessary to sustain prescribed displacements.) Changing to the non-dimensional coordinate ξ gives

$$\frac{\mathrm{d}^2 u}{\mathrm{d}\xi^2} = 0 \tag{3.42}$$

The general solution of this equation is

$$u = \alpha_1 + \alpha_2 \xi. \tag{3.43}$$

The constants of integration α_1, α_2 are found from the boundary conditions

$$u(-1) = u_1, \qquad u(+1) = u_2 \tag{3.44}$$

as before, to give expression (3.40).

The static displacement of an element is used in preference to the dynamic displacement because of its simplicity. An expression for the dynamic displacement can be obtained by solving the equation of motion (see equation (2.116))

$$\frac{\partial^2 u}{\partial x^2} - \frac{1}{c^2}\frac{\partial^2 u}{\partial t^2} = 0 \tag{3.45}$$

where $c^2 = E/\rho$, subject to the boundary conditions

$$u(-a, t) = u_1(t), \qquad u(+a, t) = u_2(t) \tag{3.46}$$

Changing to the non-dimensional coordinate ξ, equations (3.45) and (3.46) become

$$\frac{\partial^2 u}{\partial \xi^2} - \left(\frac{a}{c}\right)^2 \frac{\partial^2 u}{\partial t^2} = 0 \tag{3.47}$$

and

$$u(-1, t) = u_1(t), \qquad u(+1, t) = u_2(t) \tag{3.48}$$

The solution to equations (3.47) and (3.48) is [3.10]

$$\begin{aligned} u(\xi, t) = {} & \frac{c}{a}\sum_{n=1}^{\infty}(-1)^{n+1}\sin\left\{\frac{n\pi}{2}(1-\xi)\right\}\int_0^t u_1(\tau)\sin\left\{\frac{n\pi c}{2a}(t-\tau)\right\}\mathrm{d}\tau \\ & + \frac{c}{a}\sum_{n=1}^{\infty}(-1)^{n+1}\sin\left\{\frac{n\pi}{2}(1+\xi)\right\}\int_0^t u_2(\tau)\sin\left\{\frac{n\pi c}{2a}(t-\tau)\right\}\mathrm{d}\tau \end{aligned} \tag{3.49}$$

From this expression it can be seen that the displacement at ξ depends upon the past history of the displacements at $\xi = \mp 1$.

The penalty to be paid in using the static deformation of an element, in obtaining an approximate solution for the dynamic response, is an increase in the number of elements required for a given accuracy of solution. However, this is more than offset by the simplicity of the mathematical analysis it provides.

The expressions (3.36) and (3.40) can be written in the alternative form

$$u = N_1(\xi)u_1 + N_2(\xi)u_2 \tag{3.50}$$

where

$$N_j(\xi) = \tfrac{1}{2}(1 + \xi_j\xi) \tag{3.51}$$

In (3.51) ξ_j represents the coordinate of node point j. Figure 3.6 indicates that $\xi_1 = -1$ and $\xi_2 = +1$. The expression (3.50) can be rewritten in matrix form as follows

$$u = \lfloor N_1(\xi) \quad N_2(\xi) \rfloor \begin{bmatrix} u_1 \\ u_2 \end{bmatrix} = \lfloor \mathbf{N}(\xi) \rfloor \{\mathbf{u}\}_e \tag{3.52}$$

The energy expressions for the single element shown in Figure 3.6 are from section 2.1

$$T_e = \frac{1}{2}\int_{-a}^{+a} \rho A \dot{u}^2 \, \mathrm{d}x \tag{3.53}$$

$$U_e = \frac{1}{2}\int_{-a}^{+a} EA\left(\frac{\partial u}{\partial x}\right)^2 \mathrm{d}x \tag{3.54}$$

$$\delta W_e = \int_{-a}^{+a} p_x \delta u \, \mathrm{d}x \tag{3.55}$$

Substituting the displacement expression (3.52) into the kinetic energy (3.53) gives

$$\begin{aligned} T_e &= \frac{1}{2}\int_{-a}^{+a} \rho A \dot{u}^2 \, \mathrm{d}x \\ &= \frac{1}{2}\int_{-1}^{+1} \rho A \dot{u}^2 a \, \mathrm{d}\xi \\ &= \frac{1}{2}\{\dot{\mathbf{u}}\}_e^{\mathrm{T}} \rho A a \int_{-1}^{+1} \lfloor \mathbf{N}(\xi) \rfloor^{\mathrm{T}} \lfloor \mathbf{N}(\xi) \rfloor \, \mathrm{d}\xi \, \{\dot{\mathbf{u}}\}_e \end{aligned} \tag{3.56}$$

Therefore, the kinetic energy can be expressed in the form

$$T_e = \tfrac{1}{2}\{\dot{\mathbf{u}}\}_e^{\mathrm{T}} [\mathbf{m}]_e \{\dot{\mathbf{u}}\}_e \tag{3.57}$$

where

$$[\mathbf{m}]_e = \rho A a \int_{-1}^{+1} \lfloor \mathbf{N}(\xi) \rfloor^{\mathrm{T}} \lfloor \mathbf{N}(\xi) \rfloor \, \mathrm{d}\xi \tag{3.58}$$

which is referred to as the 'element inertia matrix'. Substituting for the functions $N_j(\xi)$ from (3.51) and integrating gives

$$\begin{aligned} [\mathbf{m}]_e &= \rho A a \int_{-1}^{+1} \begin{bmatrix} \frac{1}{2}(1-\xi) \\ \frac{1}{2}(1+\xi) \end{bmatrix} \lfloor \tfrac{1}{2}(1-\xi) \quad \tfrac{1}{2}(1+\xi) \rfloor \, \mathrm{d}\xi \\ &= \rho A a \begin{bmatrix} \frac{2}{3} & \frac{1}{3} \\ \frac{1}{3} & \frac{2}{3} \end{bmatrix} \end{aligned} \tag{3.59}$$

Note that the sum of the terms in the matrix is $\rho A2a$, which is the total mass of the element.

Substituting the displacement expression (3.52) into the strain energy (3.54) gives

$$\begin{aligned} U_e &= \frac{1}{2}\int_{-a}^{+a} EA\left(\frac{\partial u}{\partial x}\right)^2 \mathrm{d}x \\ &= \frac{1}{2}\int_{-1}^{+1} EA\frac{1}{a^2}\left(\frac{\partial u}{\partial \xi}\right)^2 a\,\mathrm{d}\xi \\ &= \frac{1}{2}\{\mathbf{u}\}_e^{\mathrm{T}}\frac{EA}{a}\int_{-1}^{+1} \lfloor \mathbf{N}'(\xi)\rfloor^{\mathrm{T}}\lfloor \mathbf{N}'(\xi)\rfloor\,\mathrm{d}\xi\,\{\mathbf{u}\}_e \end{aligned} \tag{3.60}$$

where $\lfloor \mathbf{N}'(\xi)\rfloor = \lfloor \partial\mathbf{N}(\xi)/\partial\xi\rfloor$. Therefore, the strain energy can be expressed in the form

$$U_e = \tfrac{1}{2}\{\mathbf{u}\}_e^{\mathrm{T}}\,[\mathbf{k}]_e\,\{\mathbf{u}\}_e \tag{3.61}$$

where

$$[\mathbf{k}]_e = \frac{EA}{a}\int_{-1}^{+1} \lfloor N'(\xi)\rfloor^{\mathrm{T}}\,\lfloor \mathbf{N}'(\xi)\rfloor\,\mathrm{d}\xi \tag{3.62}$$

is the 'element stiffness matrix'. Substituting the functions $N_j(\xi)$ from (3.51) gives

$$\begin{aligned} [\mathbf{k}]_e &= \frac{EA}{a}\int_{-1}^{+1} \begin{bmatrix} -\frac{1}{2} \\ +\frac{1}{2} \end{bmatrix} \lfloor -\tfrac{1}{2} \quad +\tfrac{1}{2}\rfloor\,\mathrm{d}\xi \\ &= \frac{EA}{a}\begin{bmatrix} \frac{1}{2} & -\frac{1}{2} \\ -\frac{1}{2} & \frac{1}{2} \end{bmatrix} \end{aligned} \tag{3.63}$$

Note that the sum of the terms in each row of this matrix is zero. This indicates that when the element moves as a rigid body, the elastic restoring forces are zero.

The virtual work done by the distributed forces is from (3.55),

$$\begin{aligned} \delta W_e &= \int_{-a}^{+a} p_x \delta u\,\mathrm{d}x \\ &= \int_{-1}^{+1} p_x \delta u\, a\,\mathrm{d}\xi \\ &= \{\delta\mathbf{u}\}_e^{\mathrm{T}} a\int_{-1}^{+1} p_x \lfloor \mathbf{N}(\xi)\rfloor^{\mathrm{T}}\,\mathrm{d}\xi \end{aligned} \tag{3.64}$$

This can be expressed in the form

$$\delta W_e = \{\delta\mathbf{u}\}_e^{\mathrm{T}}\{\mathbf{f}\}_e \tag{3.65}$$

where

$$\{\mathbf{f}\}_e = a\int_{-1}^{+1} p_x \lfloor \mathbf{N}(\xi)\rfloor^{\mathrm{T}}\,\mathrm{d}\xi \tag{3.66}$$

is the 'element load matrix'. Substituting for the functions $N_j(\xi)$ and assuming p_x to have the constant value p_x^e over the element gives

$$\{\mathbf{f}\}_e = p_x^e a \int_{-1}^{+1} \begin{bmatrix} \frac{1}{2}(1-\xi) \\ \frac{1}{2}(1+\xi) \end{bmatrix} \mathrm{d}\xi$$
$$= p_x^e a \begin{bmatrix} 1 \\ 1 \end{bmatrix} \tag{3.67}$$

The energy expressions for a complete rod are obtained by adding together the energies for all the individual elements. Before carrying this out it is necessary to relate the degrees of freedom of a single element, $\{\mathbf{u}\}_e$, to the set of degrees of freedom for the complete rod, $\{\mathbf{u}\}$. For the rod shown in Figure 3.4 this is

$$\{\mathbf{u}\}^{\mathrm{T}} = \lfloor u_1 \quad u_2 \quad u_3 \quad u_4 \quad u_5 \rfloor \tag{3.68}$$

For element e the relationship is

$$\{\mathbf{u}\}_e = [\mathbf{a}]_e \{\mathbf{u}\} \tag{3.69}$$

The transformation matrices $[\mathbf{a}]_e$ for the four elements are as follows

$$[\mathbf{a}]_1 = \begin{bmatrix} 1 & 0 & 0 & 0 & 0 \\ 0 & 1 & 0 & 0 & 0 \end{bmatrix}$$
$$[\mathbf{a}]_2 = \begin{bmatrix} 0 & 1 & 0 & 0 & 0 \\ 0 & 0 & 1 & 0 & 0 \end{bmatrix}$$
$$[\mathbf{a}]_3 = \begin{bmatrix} 0 & 0 & 1 & 0 & 0 \\ 0 & 0 & 0 & 1 & 0 \end{bmatrix} \tag{3.70}$$
$$[\mathbf{a}]_4 = \begin{bmatrix} 0 & 0 & 0 & 1 & 0 \\ 0 & 0 & 0 & 0 & 1 \end{bmatrix}$$

Substituting the transformation (3.69) into (3.57), (3.61) and (3.65) and summing over all elements gives

$$T = \frac{1}{2}\{\dot{\mathbf{u}}\}^{\mathrm{T}} \sum_{e=1}^{4} [\mathbf{a}]_e^{\mathrm{T}} [\mathbf{m}]_e [\mathbf{a}]_e \{\dot{\mathbf{u}}\}$$
$$= \tfrac{1}{2}\{\dot{\mathbf{u}}\}^{\mathrm{T}}[\mathbf{M}]\{\dot{\mathbf{u}}\} \tag{3.71}$$

$$U = \frac{1}{2}\{\mathbf{u}\}^{\mathrm{T}} \sum_{e=1}^{4} [\mathbf{a}]_e^{\mathrm{T}} [\mathbf{k}]_e [\mathbf{a}]_e \{\mathbf{u}\}$$
$$= \tfrac{1}{2}\{\mathbf{u}\}^{\mathrm{T}} [\mathbf{K}] \{\mathbf{u}\} \tag{3.72}$$

and

$$\delta W = \{\delta\mathbf{u}\}^{\mathrm{T}} \sum_{e=1}^{4} [\mathbf{a}]_e^{\mathrm{T}} \{\mathbf{f}\}_e$$
$$= \{\delta u\}^{\mathrm{T}}\{\mathbf{f}\} \tag{3.73}$$

[M], **[K]** and **{f}** are the inertia, stiffness and load matrices for the complete rod. It will be left to the reader to verify the following results

$$[\mathbf{M}] = \frac{\rho A a}{3}\begin{bmatrix} 2 & 1 & & & \\ 1 & 4 & 1 & & \\ & 1 & 4 & 1 & \\ & & 1 & 4 & 1 \\ & & & 1 & 2 \end{bmatrix} \tag{3.74}$$

$$[\mathbf{K}] = \frac{EA}{2a}\begin{bmatrix} 1 & -1 & & & \\ -1 & 2 & -1 & & \\ & -1 & 2 & -1 & \\ & & -1 & 2 & -1 \\ & & & -1 & 1 \end{bmatrix} \tag{3.75}$$

$$\{\mathbf{f}\} = a\begin{bmatrix} p_x^1 \\ p_x^1 + p_x^2 \\ p_x^2 + p_x^3 \\ p_x^3 + p_x^4 \\ p_x^4 \end{bmatrix} \tag{3.76}$$

The matrix product $[\mathbf{a}]_e^T[\mathbf{m}]_e[\mathbf{a}]_e$ in (3.71) effectively locates the positions in **[M]** to which the elements of $[\mathbf{m}]_e$ have to be added. In practice it is unnecessary to form $[\mathbf{a}]_e$ and carry out the matrix multiplication. The information required can be obtained from the element node numbers. Element number e has nodes e and $(e+1)$. Therefore, the two rows and columns of the element inertia matrix (equation (3.59)) are added into the rows and columns e and $(e+1)$ of the inertia matrix for the complete rod. This procedure is known as the 'assembly process'. This procedure also applies to the stiffness matrix. In the case of the load matrix, the two rows of the element load matrix (equation (3.67)) are added into rows e and $(e+1)$ of the load matrix for the complete rod.

The next step in the analysis is to ensure that the geometric boundary conditions are satisfied. As it stands, the analysis refers to a free-free rod. If the rod is now clamped at node 1, then the nodal displacement u_1 is zero. This condition can be introduced by omitting u_1 from the set of degrees of freedom for the complete rod, equation (3.68), and at the same time omitting the first row and column from the inertia and stiffness matrices for the complete rod, equations (3.74) and (3.75), and also the first row of the load matrix (3.76).

The energy expressions (3.71) to (3.73) are now substituted into Lagrange's equations (Sections 1.4 and A1.2) which give the equations of motion

$$[\mathbf{M}]\{\ddot{\mathbf{u}}\} + [\mathbf{K}]\{\mathbf{u}\} = \{\mathbf{f}\} \tag{3.77}$$

Methods of solving these equations are described in Chapters 12 and 13.

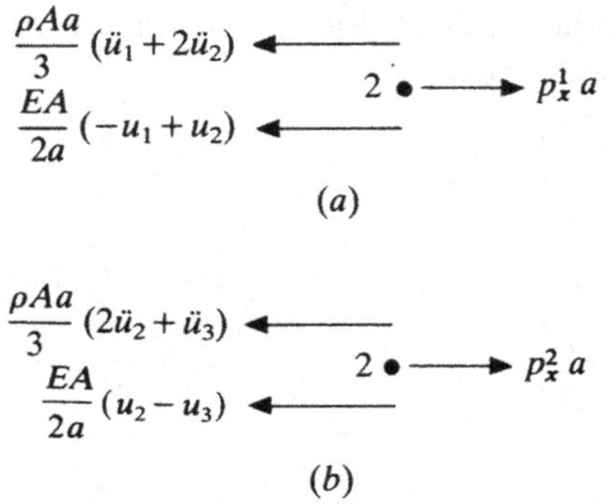

Figure 3.7. Forces acting at node 2.

The physical significance of equations (3.77) will be illustrated by considering node 2. The equation of motion of node 2 is the second of the equations in (3.77), namely

$$\frac{\rho Aa}{3}(\ddot{u}_1 + 4\ddot{u}_2 + \ddot{u}_3) + \frac{EA}{2a}(-u_1 + 2u_2 - u_3) = a\left(p_x^1 + p_x^2\right) \tag{3.78}$$

The forces acting at node 2 are illustrated in Figure 3.7. The set (*a*) arise from element 1 and the set (*b*) from element 2. These forces are obtained by substituting the element energy expressions (3.57), (3.61) and (3.65) into Lagrange's equations. Equilibrium of the forces shown in Figure 3.7 gives equation (3.78).

EXAMPLE 3.3 Use the finite element displacement method to estimate the lower frequencies and mode shapes of the clamped-free rod shown in Figure 3.1. Compare the results with the exact solution.

One element solution

The kinetic and strain energies of an element of length L are, from equations (3.57), (3.59), (3.61) and (3.63)

$$T = \frac{1}{2}\lfloor \dot{u}_1 \quad \dot{u}_2 \rfloor \frac{\rho AL}{6}\begin{bmatrix} 2 & 1 \\ 1 & 2 \end{bmatrix}\begin{bmatrix} \dot{u}_1 \\ \dot{u}_2 \end{bmatrix}$$

$$U = \frac{1}{2}\lfloor u_1 \quad u_2 \rfloor \frac{EA}{L}\begin{bmatrix} 1 & -1 \\ -1 & 1 \end{bmatrix}\begin{bmatrix} u_1 \\ u_2 \end{bmatrix}$$

Imposing the condition that $u_1 = 0$ and substituting into Lagrange's equations gives the equation of free vibration

$$\left[\frac{EA}{L} \cdot 1 - \omega^2 \frac{\rho AL}{6} \cdot 2\right] A_2 = 0$$

where ω is the frequency of vibration. The solution of this equation is

$$\omega_1 = 1.732\left(\frac{E}{\rho L^2}\right)^{1/2}$$

Two element solution

If the rod is now divided into two elements of length $L/2$, then the equation of motion becomes, from equations (3.74) and (3.75)

$$\left[\frac{2EA}{L}\begin{bmatrix} 2 & -1 \\ -1 & 1 \end{bmatrix} - \omega^2 \frac{\rho AL}{12}\begin{bmatrix} 4 & 1 \\ 1 & 2 \end{bmatrix}\right]\begin{bmatrix} A_2 \\ A_3 \end{bmatrix} = 0$$

Letting $(\omega^2 \rho L^2/24E) = \lambda$, the above equation simplifies to

$$\begin{bmatrix} (2-4\lambda) & -(1+\lambda) \\ -(1+\lambda) & (1-2\lambda) \end{bmatrix}\begin{bmatrix} A_2 \\ A_3 \end{bmatrix} = 0$$

This equation has a non-zero solution provided

$$\begin{vmatrix} (2-4\lambda) & -(1+\lambda) \\ -(1+\lambda) & (1-2\lambda) \end{vmatrix} = 0$$

Expanding gives

$$7\lambda^2 - 10\lambda + 1 = 0$$

The two roots of this equation are

$$\lambda = 0.108 \quad \text{and} \quad 1.320$$

and the natural frequencies of the system are

$$\omega_1 = (24\lambda_1)^{1/2}\left(\frac{E}{\rho L^2}\right)^{1/2} = 1.610\left(\frac{E}{\rho L^2}\right)^{1/2}$$

and

$$\omega_2 = (24\lambda_2)^{1/2}\left(\frac{E}{\rho L^2}\right)^{1/2} = 5.628\left(\frac{E}{\rho L^2}\right)^{1/2}$$

From the homogeneous equations

$$A_2 = \frac{(1+\lambda)}{(2-4\lambda)}A_3$$

When

$$\lambda = 0.108, \qquad A_2 = 0.707\,A_3$$
$$\lambda = 1.320, \qquad A_2 = -0.707\,A_3$$

The modes of vibration are therefore given by

$$\begin{bmatrix} A_1 \\ A_2 \\ A_3 \end{bmatrix} = \begin{bmatrix} 0 \\ 0.707 \\ 1.0 \end{bmatrix} A_3 \quad \text{and} \quad \begin{bmatrix} A_1 \\ A_2 \\ A_3 \end{bmatrix} = \begin{bmatrix} 0 \\ -0.707 \\ 1.0 \end{bmatrix} A_3$$

The approximate values of $\omega(\rho L^2/E)^{1/2}$ are compared with the exact values (see Example 3.1) in Table 3.3. The approximate frequencies are greater than the exact ones and approach the exact ones as the number of elements is increased.

Table 3.3. *Comparison of approximate frequencies with exact solution for a rod*

Mode	FEM solutions		Exact solution
	1 Element	2 Elements	
1	1.732	1.610	1.571
2	–	5.628	4.712

The approximate mode shapes obtained using two elements are compared with the exact mode shapes in Figure 3.8.

If the analysis is repeated using three and four elements it will be necessary to use one of the methods described in Chapter 11 to solve the resulting eigenproblem. Comparing the results obtained with the exact frequencies gives the percentage errors indicated in Figure 3.9.

The application of the finite element method to non-uniform structures adds no more complications, as illustrated by the following example.

EXAMPLE 3.4 Determine the inertia and stiffness matrices for the non-uniform rod shown in Figure 3.10.

From equation (3.59) the inertia matrices for elements 1 and 2 are

$$[\mathbf{m}]_1 = \rho AL\begin{bmatrix} \frac{2}{3} & \frac{1}{3} \\ \frac{1}{3} & \frac{2}{3} \end{bmatrix}, \qquad [\mathbf{m}]_2 = \rho AL\begin{bmatrix} \frac{1}{3} & \frac{1}{6} \\ \frac{1}{6} & \frac{1}{3} \end{bmatrix}$$

Element 3 is a lumped mass and its kinetic energy is

$$T_3 = \tfrac{1}{2}M\dot{u}_3^2$$

The inertia matrix is therefore

$$[\mathbf{m}]_3 = M \qquad \text{(a scalar)}$$

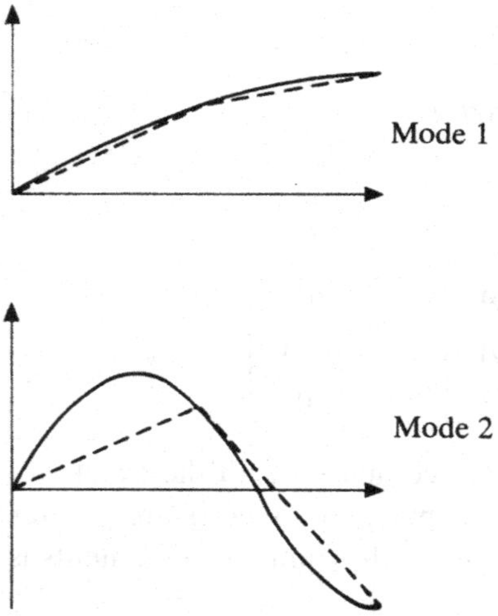

Figure 3.8. Axial modes of vibration of a rod. —— Exact; - - - approximate (FEM).

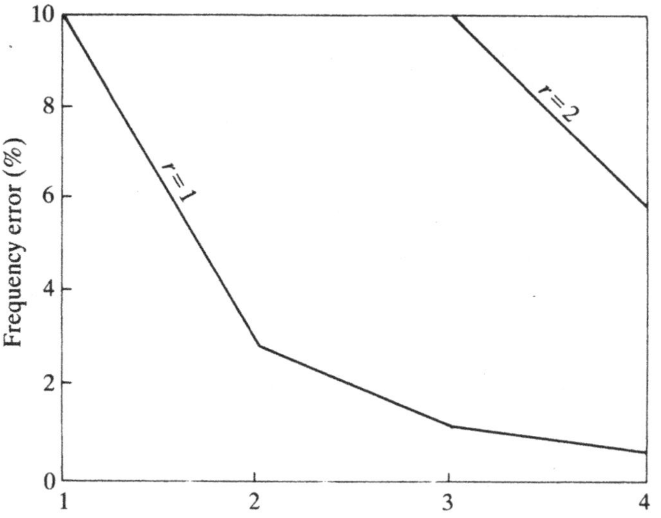

Figure 3.9. Axial vibration of a clamped-free rod.

Assembling these three element matrices and applying the condition $u_1 = 0$ gives

$$[\mathbf{M}] = \begin{bmatrix} (2\rho AL/3 + \rho AL/3) & \rho AL/6 \\ \rho AL/6 & (\rho AL/3 + M) \end{bmatrix}$$

$$= \begin{bmatrix} \rho AL & \rho AL/6 \\ \rho AL/6 & (\rho AL/3 + M) \end{bmatrix}$$

From equation (3.63) the stiffness matrices for elements 1 and 2 are

$$[\mathbf{k}]_1 = \frac{EA}{L}\begin{bmatrix} 2 & -2 \\ -2 & 2 \end{bmatrix}, \qquad [\mathbf{k}]_2 = \frac{EA}{L}\begin{bmatrix} 1 & -1 \\ -1 & 1 \end{bmatrix}$$

Element 3 is considered to be a rigid mass and so does not contribute to the stiffness matrix of the complete structure. Assembling the two stiffness matrices and applying the condition $u_1 = 0$ gives

$$[\mathbf{K}] = \frac{EA}{L}\begin{bmatrix} 2+1 & -1 \\ -1 & 1 \end{bmatrix}$$

$$= \frac{EA}{L}\begin{bmatrix} 3 & -1 \\ -1 & 1 \end{bmatrix}$$

Figure 3.10. Non-uniform rod.

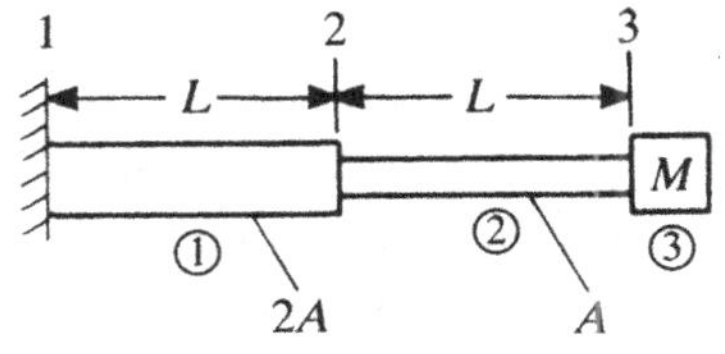

To determine the stress distribution when using the finite element method, the distribution of stress is determined for each element in turn.

From equations (2.8) and (2.10) the axial component of stress, σ_x, is given by

$$\sigma_x = E\frac{\partial u}{\partial x} \tag{3.79}$$

For element number e the displacement variation is given by equation (3.52), namely

$$u = \lfloor \mathbf{N}(\xi) \rfloor \{\mathbf{u}\}_e \tag{3.80}$$

Substituting (3.80) into (3.79) gives

$$\sigma_x = \frac{E}{a}\lfloor \partial \mathbf{N}(\xi)/\partial \xi \rfloor \{\mathbf{u}\}_e = \lfloor \mathbf{s} \rfloor_e \{\mathbf{u}\}_e \tag{3.81}$$

where $\lfloor \mathbf{s} \rfloor_e$ is the 'stress matrix' for the element. Substituting for the functions $N_j(\xi)$ from equation (3.51) gives

$$\lfloor \mathbf{s} \rfloor_e = \frac{E}{a}\lfloor -\tfrac{1}{2} \quad +\tfrac{1}{2} \rfloor \tag{3.82}$$

Thus the stress is constant within each element.

EXAMPLE 3.5 Use the two element solution for a clamped-free rod given in Example 3.3 to determine the stress distribution when the rod is vibrating in its lowest frequency mode. Compare the results with the exact solution.

For element 1

$$\begin{aligned}\sigma_x &= \frac{E}{L}\lfloor -2 \quad +2 \rfloor \begin{bmatrix} 0 \\ 0.707 \end{bmatrix} A_3 \\ &= 1.414\frac{E}{L}A_3\end{aligned}$$

For element 2

$$\begin{aligned}\sigma_x &= \frac{E}{L}\lfloor -2 \quad +2 \rfloor \begin{bmatrix} 0.707 \\ 1.0 \end{bmatrix} A_3 \\ &= 0.586\frac{E}{L}A_3\end{aligned}$$

From Example 3.1 the exact solution for the displacement is given by

$$\psi_1(x) = \sin\frac{\pi x}{2L}$$

The axial stress is therefore

$$\sigma_x = E\frac{\partial \psi_1}{\partial x} = \frac{\pi}{2}\frac{E}{L}\cos\frac{\pi x}{2L}$$

The distributions of $\sigma_x L/E$ given by the two methods are shown in Figure 3.11. Using the finite element method the stress distribution is constant within each element, as noted before. In addition, it can be seen that the stress distribution is discontinuous at the junction between elements. This is a feature

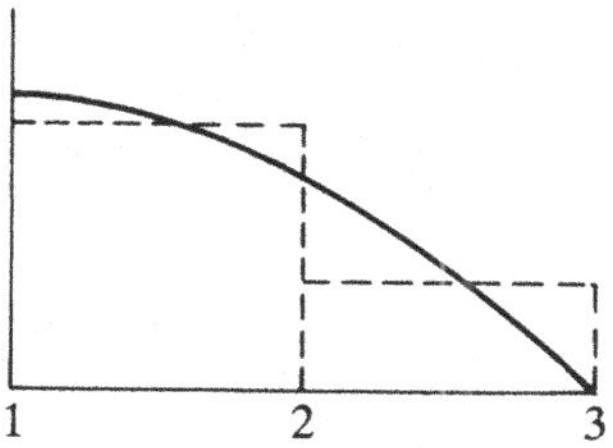

Figure 3.11. Distribution of axial stress for the first mode of vibration of a clamped-free rod. —— Exact; - - - approximate (FEM).

of the finite element displacement method, since the convergence criteria require that the displacement functions be continuous and have continuous derivatives up to order $(p-1)$ across element boundaries. The strain components, and hence the stresses, involve derivatives up to order p which can be discontinuous.

The reason why the predicted stresses in the above example are poor, even when the predicted displacements are good, is because the equations of motion and natural boundary conditions are only approximately satisfied.

This can be seen by considering the equilibrium of part of an element as shown in Figure 3.12. The introduction of an inertia force of magnitude $\rho A\ddot{u}$ per unit length in the opposite direction to the acceleration allows the concepts of static equilibrium to be used (see Section 1.1). For equilibrium

$$A\sigma_x(x) = A\sigma_x(-a) + \int_{-a}^{x} \rho A\ddot{u}\,\mathrm{d}x \tag{3.83}$$

Dividing by A, converting to the ξ coordinate and assuming harmonic motion gives

$$\sigma_x(\xi) = \sigma_x(-1) - \rho\omega^2 a \int_{-1}^{\xi} u(\xi)\,\mathrm{d}\xi \tag{3.84}$$

Since the displacement $u(\xi)$ has been assumed to vary linearly over the element, then from equation (3.84) the stress $\sigma_x(\xi)$ will vary quadratically. However, equations (3.81) and (3.82) give only constant stress.

An improved estimate of the stress distribution can be obtained using equation (3.84). Substituting for $u(\xi)$ from (3.50) and (3.51) gives

$$\begin{aligned}\sigma_x(\xi) &= \sigma_x(-1) - \rho\omega^2 a \int_{-1}^{\xi} \lfloor \tfrac{1}{2}(1-\xi) \quad \tfrac{1}{2}(1+\xi) \rfloor\,\mathrm{d}\xi\,\{\mathbf{u}\}_e \\ &= \sigma_x(-1) - \rho\omega^2 a \lfloor \tfrac{1}{2}\left(\xi - \tfrac{1}{2}\xi^2 + \tfrac{3}{2}\right) \quad \tfrac{1}{2}\left(\xi + \tfrac{1}{2}\xi^2 + \tfrac{1}{2}\right) \rfloor \{\mathbf{u}\}_e\end{aligned} \tag{3.85}$$

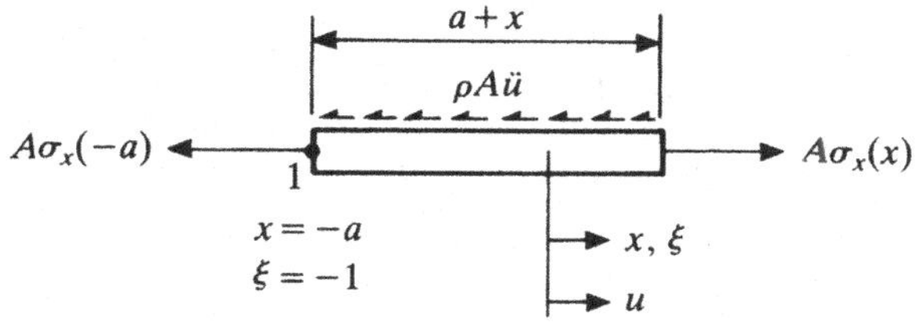

Figure 3.12. Equilibrium of part of an axial element.

The stress at the first node can be calculated from the total force acting on the element. This force is obtained by substituting the element energy expressions into Lagrange's equations. This process gives (compare Figure 3.7(b))

$$\sigma_x(-1) = -\frac{\text{Force at node 1}}{\text{Area}}$$
$$= \frac{E}{a}\lfloor -\tfrac{1}{2} \quad \tfrac{1}{2} \rfloor \{\mathbf{u}\}_e + \rho\omega^2 a \lfloor \tfrac{2}{3} \quad \tfrac{1}{3} \rfloor \{\mathbf{u}\}_e \tag{3.86}$$

Using the same method, the stress at the second node is

$$\sigma_x(+1) = \frac{\text{Force at node 2}}{\text{Area}}$$
$$= \frac{E}{a}\lfloor -\tfrac{1}{2} \quad \tfrac{1}{2} \rfloor \{\mathbf{u}\}_e - \rho\omega^2 a \lfloor \tfrac{2}{3} \quad \tfrac{1}{3} \rfloor \{\mathbf{u}\}_e \tag{3.87}$$

This expression can also be derived by evaluating equation (3.85) at $\xi = +1$ and using (3.86).

EXAMPLE 3.6 Use equations (3.85) and (3.86) to determine the stress distribution for the clamped-free rod given in Example 3.3 when it is vibrating in its lowest frequency mode. Compare the results with the exact solution.

For element 1

$$\sigma_x(-1) = \frac{4E}{L}\lfloor -\tfrac{1}{2} \quad \tfrac{1}{2} \rfloor \begin{bmatrix} 0 \\ 0.707 \end{bmatrix} A_3 + \rho\omega^2 \frac{L}{4} \lfloor \tfrac{2}{3} \quad \tfrac{1}{3} \rfloor \begin{bmatrix} 0 \\ 0.707 \end{bmatrix} A_3$$

Since

$$\lambda = \rho\omega^2 L^2/24E = 0.108, \qquad \rho\omega^2 L/4 = 0.648E/L$$

Therefore

$$\sigma_x(-1) = 1.567\frac{E}{L}A_3$$

$$\sigma_x(0) = \sigma_x(-1) - \rho\omega^2 \frac{L}{4} \lfloor \tfrac{3}{4} \quad \tfrac{1}{4} \rfloor \begin{bmatrix} 0 \\ 0.707 \end{bmatrix} A_3$$
$$= 1.452\frac{E}{L}A_3$$

$$\sigma_x(+1) = \frac{4E}{L}\lfloor -\tfrac{1}{2} \quad \tfrac{1}{2} \rfloor \begin{bmatrix} 0 \\ 0.707 \end{bmatrix} A_3 - \rho\omega^2 \frac{L}{4} \lfloor \tfrac{1}{3} \quad \tfrac{2}{3} \rfloor \begin{bmatrix} 0 \\ 0.707 \end{bmatrix} A_3$$
$$= 1.108\frac{E}{L}A_3$$

Table 3.4. *Comparison of approximate and exact stresses in the fundamental mode of a rod*

	Stresses	
x/L	FEM	Exact
0.0	1.567	1.571
0.25	1.452	1.451
0.5	1.108	1.111
0.75	0.605	0.601
1.0	0.0	0.0

For element 2

$$\sigma_x(-1) = \frac{4E}{L}\lfloor -\tfrac{1}{2} \quad \tfrac{1}{2} \rfloor \begin{bmatrix} 0.707 \\ 1.0 \end{bmatrix} A_3 + \rho\omega^2 \frac{L}{4} \lfloor \tfrac{2}{3} \quad \tfrac{1}{3} \rfloor \begin{bmatrix} 0.707 \\ 1.0 \end{bmatrix} A_3$$

$$= 1.108 \frac{E}{L} A_3$$

$$\sigma_x(0) = \sigma_x(-1) - \rho\omega^2 \frac{L}{4} \lfloor \tfrac{3}{4} \quad \tfrac{1}{4} \rfloor \begin{bmatrix} 0.707 \\ 1.0 \end{bmatrix} A_3$$

$$= 0.605 \frac{E}{L} A_3$$

$$\sigma_x(+1) = \frac{4E}{L}\lfloor -\tfrac{1}{2} \quad \tfrac{1}{2} \rfloor \begin{bmatrix} 0.707 \\ 1.0 \end{bmatrix} A_3 - \rho\omega^2 \frac{L}{4} \lfloor \tfrac{1}{3} \quad \tfrac{2}{3} \rfloor \begin{bmatrix} 0.707 \\ 1.0 \end{bmatrix} A_3$$

$$= 0.0$$

Note that $\sigma_x(-1)$ for element 2 is equal to $\sigma_x(+1)$ for element 1. This is because the stresses at the node points have been calculated from the nodal forces which are in equilibrium.

The approximate values of $(\sigma_x L/E)$, corresponding to a unit displacement at the tip, are compared with the exact values (see Example 3.5) in Table 3.4. The modified procedure for calculating stresses has produced accurate estimates at all positions of the rod.

If an element is subjected to an applied load of magnitude p_x per unit length, as well as the inertia force of magnitude $\rho A\ddot{u}$ per unit length (see Figure 3.12), then for equilibrium

$$A\sigma_x(x) = A\sigma_x(-a) + \int_{-a}^{x} (\rho A\ddot{u} - p_x)\,\mathrm{d}x \tag{3.88}$$

Dividing by A, and converting to the ξ coordinate gives

$$\sigma_x(\xi) = \sigma_x(-1) + \rho a \int_{-a}^{\xi} \ddot{u}(\xi)\,\mathrm{d}\xi - (a/A) \int_{-1}^{\xi} p_x(\xi)\,\mathrm{d}\xi \tag{3.89}$$

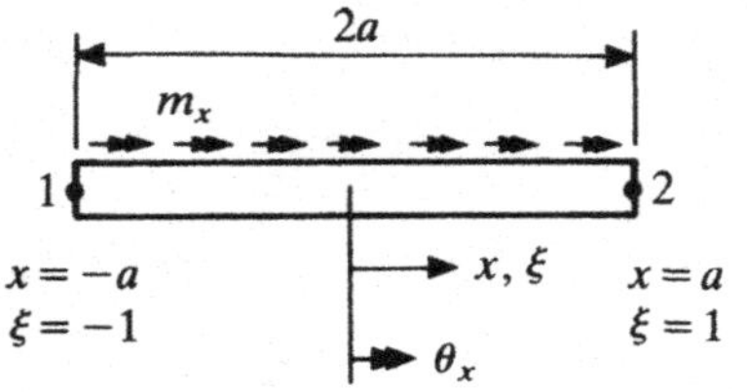

Figure 3.13. Geometry of a single torque element.

It should be remembered that all the force, displacement and stress quantities in (3.89) are time dependent also. Substituting for $u(\xi)$ from (3.50) and (3.51) and assuming p_x to have the constant value p_x^e over the element gives

$$\sigma_x(\xi) = \sigma_x(-1) + \rho a \lfloor \tfrac{1}{2}(\xi - \tfrac{1}{2}\xi^2 + \tfrac{3}{2}) \quad \tfrac{1}{2}(\xi + \tfrac{1}{2}\xi^2 + \tfrac{1}{2}) \rfloor \{\ddot{\mathbf{u}}\}_e - (p_x^e a/A)(\xi + 1) \tag{3.90}$$

The stresses at the two nodes of the element are again calculated from the total forces acting, which gives

$$A\begin{bmatrix} -\sigma_x(-1) \\ \sigma_x(+1) \end{bmatrix} = \frac{EA}{2a}\begin{bmatrix} 1 & -1 \\ -1 & 1 \end{bmatrix}\{\mathbf{u}\}_e + \frac{\rho A a}{3}\begin{bmatrix} 2 & 1 \\ 1 & 2 \end{bmatrix}\{\ddot{\mathbf{u}}\}_e - p_x^e a\begin{bmatrix} 1 \\ 1 \end{bmatrix} \tag{3.91}$$

The values of the nodal displacements $\{\mathbf{u}\}_e$ are obtained from the solution of equation (3.77).

3.4 Torsional Vibration of Shafts

The energy expressions for the single torque element shown in Figure 3.13 are from Section 2.2

$$T_e = \frac{1}{2}\int_{-a}^{+a} \rho I_x \dot{\theta}_x^2 \, dx \tag{3.92}$$

$$U_e = \frac{1}{2}\int_{-a}^{+a} GJ \left(\frac{\partial \theta_x}{\partial x}\right)^2 dx \tag{3.93}$$

$$\delta W_e = \int_{-a}^{+a} m_x \delta\theta_x \, dx \tag{3.94}$$

The highest derivative appearing in these expressions is the first and so only the rotation about the x-axis (i.e., the twist), θ_x, need be continuous. Therefore, θ_x is the only degree of freedom required for each node point. This means that the variation of θ_x with x is the same as the variation of u with x for an axial element, that is, linear. The element displacement function is therefore

$$\theta_x = \lfloor N_1(\xi) \quad N_2(\xi) \rfloor \begin{bmatrix} \theta_{x1} \\ \theta_{x2} \end{bmatrix} = \lfloor \mathbf{N}(\xi) \rfloor \{\boldsymbol{\theta}\}_e \tag{3.95}$$

where

$$N_j(\xi) = \tfrac{1}{2}(1 + \xi_j \xi) \tag{3.96}$$

with $\xi_1 = -1$ and $\xi_2 = +1$.

Note that

$$\frac{\partial \theta_x}{\partial x} = \frac{1}{a}\begin{bmatrix} -\frac{1}{2} & +\frac{1}{2} \end{bmatrix} \begin{bmatrix} \theta_{x1} \\ \theta_{x2} \end{bmatrix} \tag{3.97}$$

which is constant within the element.

Substituting the functions (3.95) to (3.97) into the energy expressions (3.92) to (3.94) gives the following results

$$T_e = \tfrac{1}{2}\{\dot{\boldsymbol{\theta}}\}_e^{\mathrm{T}}[\mathbf{m}]_e\{\dot{\boldsymbol{\theta}}\}_e \tag{3.98}$$

$$U_e = \tfrac{1}{2}\{\boldsymbol{\theta}\}_e^{\mathrm{T}}[\mathbf{k}]_e\{\boldsymbol{\theta}\}_e \tag{3.99}$$

$$\delta W_e = \{\delta\boldsymbol{\theta}\}_e^{\mathrm{T}}\{\mathbf{f}\}_e \tag{3.100}$$

where

$$[\mathbf{m}]_e = \rho I_x a \begin{bmatrix} \frac{2}{3} & \frac{1}{3} \\ \frac{1}{3} & \frac{2}{3} \end{bmatrix} \tag{3.101}$$

$$[\mathbf{k}]_e = \frac{GJ}{a}\begin{bmatrix} \frac{1}{2} & -\frac{1}{2} \\ -\frac{1}{2} & \frac{1}{2} \end{bmatrix} \tag{3.102}$$

$$\{\mathbf{f}\}_e = m_x^e a \begin{bmatrix} 1 \\ 1 \end{bmatrix} \tag{3.103}$$

where m_x^e is the constant value of m_x for the element.

These results are also obtainable from those for an axial rod by replacing pA, EA, p_x by ρI_x, GJ, m_x respectively. The assembly and application of geometric boundary conditions is exactly the same as for the axial vibration of rods.

The shear stresses are shown in Section 2.2 to be given by

$$\begin{bmatrix} \tau_{xy} \\ \tau_{xz} \end{bmatrix} = G \begin{bmatrix} \partial\psi/\partial y - z \\ \partial\psi/\partial z + y \end{bmatrix} \frac{\partial \theta_x}{\partial x} \tag{3.104}$$

where $\psi(y, z)$ is the warping function. Methods of determining this are given in reference [3.11].

The total moment about the x-axis, M_x, is

$$M_x = \int_A (y\tau_{xz} - z\tau_{xy})\,\mathrm{d}A \tag{3.105}$$

Substituting for τ_{xy}, τ_{xz} from (3.104) into (3.105) gives

$$M_x = G\frac{\partial \theta_x}{\partial x}\int_A \left\{ y\left(\frac{\partial \psi}{\partial z} + y\right) - z\left(\frac{\partial \psi}{\partial y} - z\right)\right\} \mathrm{d}A \tag{3.106}$$

It can be shown that the integral in this expression is equivalent to the expression for J as given by equation (2.21) [3.12]. Equation (3.106) therefore becomes

$$M_x = GJ\frac{\partial \theta_x}{\partial x} \tag{3.107}$$

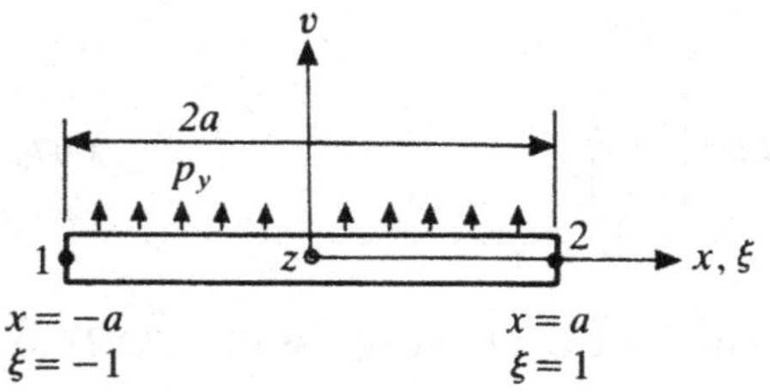

Figure 3.14. Geometry of a single beam element.

Substituting for $\partial\theta_x/\partial x$ in equation (3.104) from (3.107) gives

$$\begin{bmatrix} \tau_{xy} \\ \tau_{xz} \end{bmatrix} = \begin{bmatrix} \partial\psi/\partial y - z \\ \partial\psi/\partial z + y \end{bmatrix} \frac{M_x}{J} \tag{3.108}$$

The twisting moments at the nodes are obtained by substituting the element energy expressions (3.98) to (3.100) into Lagrange's equations. This gives

$$\begin{bmatrix} M_x(-1) \\ M_x(+1) \end{bmatrix} = \frac{GJ}{2a}\begin{bmatrix} 1 & -1 \\ -1 & 1 \end{bmatrix}\{\boldsymbol{\theta}\}_e + \frac{\rho I_x a}{3}\begin{bmatrix} 2 & 1 \\ 1 & 2 \end{bmatrix}\{\ddot{\boldsymbol{\theta}}\}_e - m_x^e a\begin{bmatrix} 1 \\ 1 \end{bmatrix} \tag{3.109}$$

The twisting moment at any section can be obtained by considering equilibrium of part of the element between -1 and ξ.

$$M_x(\xi) = -M_x(-1) + \rho I_x a \int_{-1}^{\xi} \ddot{\theta}_x(\xi)\,\mathrm{d}\xi - a\int_{-1}^{\xi} m_x(\xi)\,\mathrm{d}\xi \tag{3.110}$$

Substituting for θ_x from (3.95) and assuming m_x is constant gives

$$M_x(\xi) = -M_x(-1) + \rho I_x a \lfloor \tfrac{1}{2}(\xi - \tfrac{1}{2}\xi^2 + \tfrac{3}{2}) \quad \tfrac{1}{2}(\xi + \tfrac{1}{2}\xi^2 + \tfrac{1}{2}) \rfloor \{\ddot{\boldsymbol{\theta}}\}_e - m_x^e a(\xi + 1) \tag{3.111}$$

The shear stresses are given by equations (3.108) and (3.111) combined.

3.5 Bending Vibration of Beams

It is shown in Section 3.2 that it is necessary to take v and $\partial v/\partial x$ as degrees of freedom at each node of a beam element. Therefore the element shown in Figure 3.14, which has two nodes, has a total of four degrees of freedom. The displacement function can thus be represented by a polynomial having four constants, namely

$$v = \alpha_1 + \alpha_2\xi + \alpha_3\xi^2 + \alpha_4\xi^3 \tag{3.112}$$

This expression can be written in the following matrix form

$$v = \lfloor 1 \quad \xi \quad \xi^2 \quad \xi^3 \rfloor \begin{bmatrix} \alpha_1 \\ \alpha_2 \\ \alpha_3 \\ \alpha_4 \end{bmatrix} \tag{3.113}$$

or

$$v = \lfloor \mathbf{P}(\xi) \rfloor \{\boldsymbol{\alpha}\} \tag{3.114}$$

Differentiating (3.112) gives

$$a\theta_z = a\frac{\partial v}{\partial x} = \frac{\partial v}{\partial \xi} = \alpha_2 + 2\alpha_3\xi + 3\alpha_4\xi^2 \tag{3.115}$$

Evaluating (3.112) and (3.115) at $\xi = \mp 1$ gives

$$\begin{bmatrix} v_1 \\ a\theta_{z1} \\ v_2 \\ a\theta_{z2} \end{bmatrix} = \begin{bmatrix} 1 & -1 & 1 & -1 \\ 0 & 1 & -2 & 3 \\ 1 & 1 & 1 & 1 \\ 0 & 1 & 2 & 3 \end{bmatrix} \begin{bmatrix} \alpha_1 \\ \alpha_2 \\ \alpha_3 \\ \alpha_4 \end{bmatrix} \tag{3.116}$$

or

$$\{\bar{\mathbf{v}}\}_e = [\mathbf{A}]_e \{\boldsymbol{\alpha}\}. \tag{3.117}$$

Solving for $\{\boldsymbol{\alpha}\}$ gives

$$\{\boldsymbol{\alpha}\} = [\mathbf{A}]_e^{-1} \{\bar{\mathbf{v}}\}_e \tag{3.118}$$

where

$$[\mathbf{A}]_e^{-1} = \frac{1}{4}\begin{bmatrix} 2 & 1 & 2 & -1 \\ -3 & -1 & 3 & -1 \\ 0 & -1 & 0 & 1 \\ 1 & 1 & -1 & 1 \end{bmatrix} \tag{3.119}$$

Equation (3.118) can be written in the alternative form

$$\{\boldsymbol{\alpha}\} = [\mathbf{C}]_e \{\mathbf{v}\}_e \tag{3.120}$$

where

$$\{\mathbf{v}\}_e^{\mathrm{T}} = \lfloor v_1 \quad \theta_{z1} \quad v_2 \quad \theta_{z2} \rfloor \tag{3.121}$$

and

$$[\mathbf{C}]_e = \frac{1}{4}\begin{bmatrix} 2 & a & 2 & -a \\ -3 & -a & 3 & -a \\ 0 & -a & 0 & a \\ 1 & a & -1 & a \end{bmatrix} \tag{3.122}$$

Substituting (3.120) into (3.114) gives

$$v = \lfloor \mathbf{P}(\xi) \rfloor [\mathbf{C}]_e \{\mathbf{v}\}_e \tag{3.123}$$

This can be expressed in the form

$$v = \lfloor \mathbf{N}(\xi) \rfloor \{\mathbf{v}\}_e \tag{3.124}$$

where

$$\lfloor \mathbf{N}(\xi) \rfloor = \lfloor N_1(\xi) \quad aN_2(\xi) \quad N_3(\xi) \quad aN_4(\xi) \rfloor \tag{3.125}$$

The displacement functions in (3.125) are given by

$$\begin{aligned} N_1(\xi) &= \tfrac{1}{4}(2 - 3\xi + \xi^3) \\ N_2(\xi) &= \tfrac{1}{4}(1 - \xi - \xi^2 + \xi^3) \\ N_3(\xi) &= \tfrac{1}{4}(2 + 3\xi - \xi^3) \\ N_4(\xi) &= \tfrac{1}{4}(-1 - \xi + \xi^2 + \xi^3) \end{aligned} \tag{3.126}$$

The energy expressions for the single element shown in Figure 3.14 are from Section 2.3

$$T_e = \frac{1}{2}\int_{-a}^{+a} \rho A \dot{v}^2 \, \mathrm{d}x \tag{3.127}$$

$$U_e = \frac{1}{2}\int_{-a}^{+a} EI_z \left(\frac{\partial^2 v}{\partial x^2}\right)^2 \mathrm{d}x \tag{3.128}$$

$$\delta W_e = \int_{-a}^{+a} p_y \delta v \, \mathrm{d}x \tag{3.129}$$

Substituting the displacement expression (3.124) into the kinetic energy (3.127) gives

$$\begin{aligned} T_e &= \frac{1}{2}\int_{-a}^{+a} \rho A \dot{v}^2 \, \mathrm{d}x = \frac{1}{2}\int_{-1}^{+1} \rho A \dot{v}^2 a \, \mathrm{d}\xi \\ &= \frac{1}{2}\{\dot{\mathbf{v}}\}_e^{\mathrm{T}} \rho A a \int_{-1}^{+1} \lfloor \mathbf{N}(\xi) \rfloor^{\mathrm{T}} \lfloor \mathbf{N}(\xi) \rfloor \, \mathrm{d}\xi \, \{\dot{\mathbf{v}}\}_e \end{aligned} \tag{3.130}$$

Therefore the element inertia matrix is given by

$$[\mathbf{m}]_e = \rho A a \int_{-1}^{+1} \lfloor \mathbf{N}(\xi) \rfloor^{\mathrm{T}} \lfloor \mathbf{N}(\xi) \rfloor \, \mathrm{d}\xi \tag{3.131}$$

Substituting for the functions $N_j(\xi)$ from (3.126) and integrating gives

$$[\mathbf{m}]_e = \frac{\rho A a}{105} \begin{bmatrix} 78 & 22a & 27 & -13a \\ 22a & 8a^2 & 13a & -6a^2 \\ 27 & 13a & 78 & -22a \\ -13a & -6a^2 & -22a & 8a^2 \end{bmatrix} \tag{3.132}$$

In deriving this result, it is simpler to use the expression (3.123) for the displacement v. This approach requires the integral $\int_{-1}^{+1} \lfloor \mathbf{P}(\xi) \rfloor^{\mathrm{T}} \lfloor \mathbf{P}(\xi) \rfloor \, \mathrm{d}\xi$ to be evaluated, which is much simpler than the expression (3.131).

Substituting the displacement expression (3.124) into the strain energy (3.128) gives

$$\begin{aligned} U_e &= \frac{1}{2}\int_{-a}^{+a} EI_z \left(\frac{\partial^2 v}{\partial x^2}\right)^2 \mathrm{d}x = \frac{1}{2}\int_{-1}^{+1} EI_z \frac{1}{a^4}\left(\frac{\partial^2 v}{\partial \xi^2}\right)^2 a \, \mathrm{d}\xi \\ &= \frac{1}{2}\{\mathbf{v}\}_e^{\mathrm{T}} \frac{EI_z}{a^3} \int_{-1}^{+1} \lfloor \mathbf{N}''(\xi) \rfloor^{\mathrm{T}} \lfloor \mathbf{N}''(\xi) \rfloor \, \mathrm{d}\xi \, \{\mathbf{v}\}_e \end{aligned} \tag{3.133}$$

The element stiffness matrix is therefore

$$[\mathbf{k}]_e = \frac{EI_z}{a^3}\int_{-1}^{+1} \lfloor \mathbf{N}''(\xi)\rfloor^{\mathrm{T}} \lfloor \mathbf{N}''(\xi)\rfloor \,\mathrm{d}\xi \tag{3.134}$$

Substituting for the functions $N_j(\xi)$ from (3.126) and integrating gives

$$[\mathbf{k}]_e = \frac{EI_z}{2a^3}\begin{bmatrix} 3 & 3a & -3 & 3a \\ 3a & 4a^2 & -3a & 2a^2 \\ -3 & -3a & 3 & -3a \\ 3a & 2a^2 & -3a & 4a^2 \end{bmatrix} \tag{3.135}$$

The virtual work done by the distributed forces becomes, after substituting (3.124) into (3.129)

$$\begin{aligned} \delta W_e &= \int_{-a}^{+a} p_y \delta v \,\mathrm{d}x = \int_{-1}^{+1} p_y \delta v \, a \,\mathrm{d}\xi \\ &= \{\delta \mathbf{v}\}_e^{\mathrm{T}} a \int_{-1}^{+1} p_y \lfloor \mathbf{N}(\xi)\rfloor^{\mathrm{T}} \mathrm{d}\xi \end{aligned} \tag{3.136}$$

The element load matrix is therefore

$$\{\mathbf{f}\}_e = a\int_{-1}^{+1} p_y \lfloor \mathbf{N}(\xi)\rfloor^{\mathrm{T}} \,\mathrm{d}\xi \tag{3.137}$$

Substituting for the functions $N_j(\xi)$ from (3.126) and assuming p_y to have the constant value p_y^e over the element gives

$$\{\mathbf{f}\}_e = p_y^e \frac{a}{3}\begin{bmatrix} 3 \\ a \\ 3 \\ -a \end{bmatrix} \tag{3.138}$$

The assembly process for a beam element is similar to that of an axial element. For element e with nodes e and $(e+1)$, the four rows and columns of the inertia matrix (3.132) are added into rows and columns $(2e-1)$ to $(2e+2)$ of the inertia matrix for the complete beam. The stiffness matrix is treated in the same way. The four rows of the element load matrix are added into rows $(2e-1)$ to $(2e+2)$ of the assembled load matrix.

EXAMPLE 3.7 Use the finite element displacement method to estimate the lower frequencies of the cantilever beam shown in Figure 3.3. Compare the results with the exact solution.

One element solution

The kinetic and strain energies of a beam of length, L, which is represented by a single element, are given by the expressions (3.130) to (3.135) with $a = L/2$. Imposing the conditions that $v_1 = \theta_{z1} = 0$ and substituting into Lagrange's equations gives the equation of free vibration

$$\left[\frac{EI_z}{L^3}\begin{bmatrix} 12 & -6L \\ -6L & 4L^2 \end{bmatrix} - \omega^2 \frac{\rho AL}{210}\begin{bmatrix} 78 & -11L \\ -11L & 2L^2 \end{bmatrix}\right]\begin{bmatrix} v_2 \\ \theta_{z2} \end{bmatrix} = 0$$

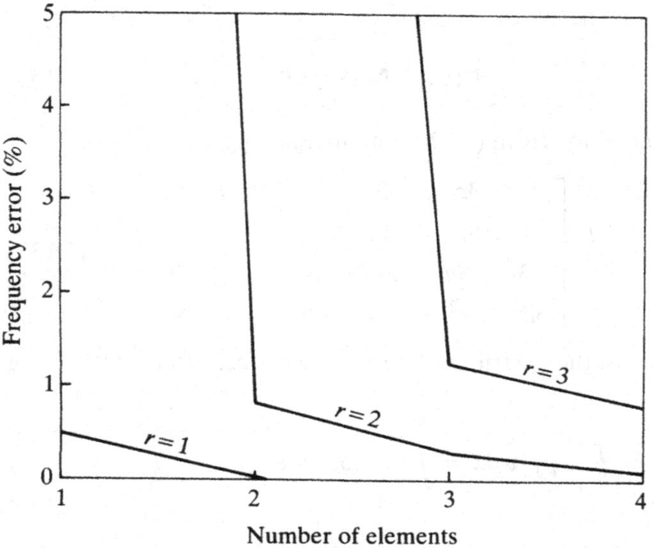

Figure 3.15. Flexural vibration of a cantilever beam.

Letting $(\omega^2 \rho A L^4 / 210 E I_z) = \lambda$, this equation simplifies to

$$\begin{bmatrix} (12 - 78\lambda) & (-6 + 11\lambda) \\ (-6 + 11\lambda) & (4 - 2\lambda) \end{bmatrix} \begin{bmatrix} v_2 \\ L\theta_{z2} \end{bmatrix} = 0$$

This equation has a non-zero solution provided

$$\begin{vmatrix} (12 - 78\lambda) & (-6 + 11\lambda) \\ (-6 + 11\lambda) & (4 - 2\lambda) \end{vmatrix} = 0$$

Expanding gives

$$35\lambda^2 - 204\lambda + 12 = 0$$

The two roots of this equation are

$$\lambda = 0.0594295 \quad \text{and} \quad 5.76914$$

and the natural frequencies of the system are

$$\omega_1 = (210\lambda_1)^{1/2} \left(\frac{EI_z}{\rho A L^4} \right)^{1/2} = 3.533 \left(\frac{EI_z}{\rho A L^4} \right)^{1/2}$$

and

$$\omega_2 = (210\lambda_2)^{1/2} \left(\frac{EI_z}{\rho A L^4} \right)^{1/2} = 34.807 \left(\frac{EI_z}{\rho A L^4} \right)^{1/2}$$

Table 3.2 shows that the values of the coefficient for these two frequencies should be 3.516 and 22.035 respectively. The errors produced by a one element solution are therefore 0.48 and 58% respectively.

Repeating the analysis using two, three and four elements gives the errors shown in Figure 3.15 when compared with the exact solution. Notice that the convergence in this case is better than that obtained for the rod (Figure 3.9).

This is in keeping with the observation made in reference [3.13] that the convergence of the Rayleigh–Ritz method is improved if the order of the derivatives in the energy expressions is higher. Results for a variety of boundary conditions are presented in reference [3.14].

The shear force and bending moment at the two nodes are obtained by substituting the element energy expressions (3.130) to (3.138) into Lagrange's equations. This gives

$$\begin{bmatrix} Q(-1) \\ M_z(-1) \\ Q(+1) \\ M_z(+1) \end{bmatrix} = [\mathbf{k}]_e\{\mathbf{v}\}_e + [\mathbf{m}]_e\{\ddot{\mathbf{v}}\}_e - \{\mathbf{f}\}_e \tag{3.139}$$

where $[\mathbf{k}]_e$, $[\mathbf{m}]_e$, $\{\mathbf{f}\}_e$ and $\{\mathbf{v}\}_e$ are defined by equations (3.135), (3.132), (3.138) and (3.121) respectively.

The shear force and bending moment at any section can be obtained by considering equilibrium of the part of the element between -1 and ξ. This gives

$$Q(\xi) = Q(-1) + \rho A a \int_{-1}^{\xi} \ddot{v}(\xi_1)\,\mathrm{d}\xi_1 - a \int_{-1}^{\xi} p_y(\xi_1)\,\mathrm{d}\xi_1 \tag{3.140}$$

$$\begin{aligned} M(\xi) = M(-1) - Q(-1)a(1+\xi) - \rho A a^2 \int_{-1}^{\xi} \ddot{v}(\xi_1)(\xi - \xi_1)\,\mathrm{d}\xi_1 \\ + a^2 \int_{-1}^{\xi} p_y(\xi_1)(\xi - \xi_1)\,\mathrm{d}\xi_1 \end{aligned} \tag{3.141}$$

The integrals are evaluated after substituting for v from (3.124). Some applications of this method can be found in reference [3.15].

The distribution of the direct stress component, σ_x, over a cross-section can be calculated using a combination of equations (2.126) and (2.127), namely

$$\sigma_x = -yM_z/I_z \tag{3.142}$$

The method of determining the distribution of shear stress depends upon the shape of the cross-section [3.16].

3.6 Vibration of Plane Frameworks

Consider a plane framework, such as the one shown in Figure 3.16, which is vibrating in its own plane. It can be seen that the framework consists of members which are inclined to one another at various angles. When applying the finite element method to such a structure, the following procedure is used:

(1) Divide each member into the appropriate number of elements.
(2) Derive the energy expressions for each element in terms of nodal degrees of freedom relative to a 'local' set of axes.
(3) Transform the energy expressions for each element into expressions involving nodal degrees of freedom relative to a common set of 'global' axes.
(4) Add the energies of the elements together.

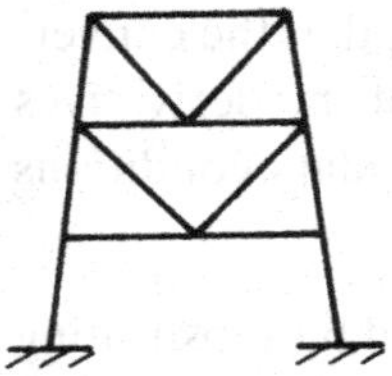

Figure 3.16. Example of a plane framework.

Figure 3.17 shows a typical element together with its local axes x and y which are inclined to the global axes X and Y. The local axis of x lies along the centroidal axis which joins nodes 1 and 2. The local y-axis is perpendicular to the x-axis and passes through the mid-point of the line joining 1 and 2.

Each member of a plane framework is capable of both axial and bending deformations. Therefore the energy functions for an element are a combination of the energy functions derived in Sections 2.1 and 2.3. These are, in terms of local coordinates, as follows:

$$\begin{aligned} T_e &= \frac{1}{2}\int_{-a}^{+a} \rho A(\dot{u}^2 + \dot{v}^2)\,\mathrm{d}x \\ U_e &= \frac{1}{2}\int_{-a}^{+a} EA\left(\frac{\partial u}{\partial x}\right)^2 \mathrm{d}x + \frac{1}{2}\int_{-a}^{+a} EI_z\left(\frac{\partial^2 v}{\partial x^2}\right)^2 \mathrm{d}x \qquad (3.143) \\ \delta W_e &= \int_{-a}^{+a} p_x \delta u\,\mathrm{d}x + \int_{-a}^{+a} p_y \delta v\,\mathrm{d}x \end{aligned}$$

In these expressions u, v are the displacement components of the centroid of a cross-section and p_x, p_y are the components of the load per unit length, both relative to the local axes x and y.

Since the axial and bending deformations are uncoupled, they can be treated separately as in Sections 3.3 and 3.5. The displacement functions can therefore be taken to be the ones defined by equations (3.52) and (3.124), which are

$$\begin{aligned} u &= \lfloor \mathbf{N}_u(\xi) \rfloor \{\mathbf{u}\}_e \\ v &= \lfloor \mathbf{N}_v(\xi) \rfloor \{\mathbf{v}\}_e \end{aligned} \qquad (3.144)$$

The subscripts u and v are introduced here to differentiate between axial and lateral displacements.

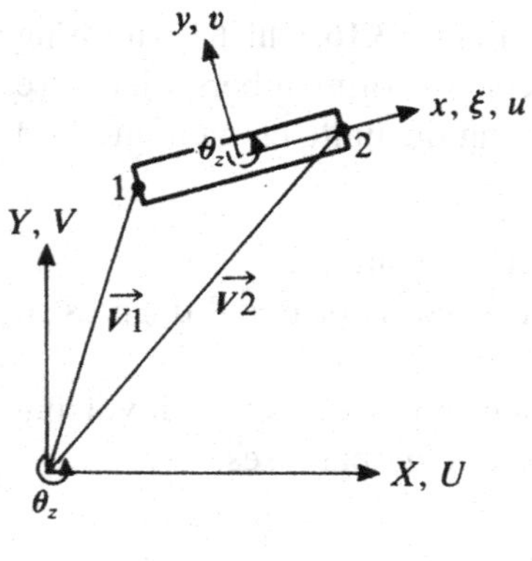

Figure 3.17. Geometry of a plane framework element.

Substituting equations (3.144) into the kinetic energy expression in (3.143) gives, on integration, the sum of the kinetic energies given in Sections 3.3 and 3.5, namely

$$T_e = \frac{1}{2}\begin{bmatrix}\dot{u}_1\\ \dot{u}_2\end{bmatrix}^{\mathrm{T}} \frac{\rho Aa}{3}\begin{bmatrix}2 & 1\\ 1 & 2\end{bmatrix}\begin{bmatrix}\dot{u}_1\\ \dot{u}_2\end{bmatrix}$$

$$+\frac{1}{2}\begin{bmatrix}\dot{v}_1\\ \dot{\theta}_{z1}\\ \dot{v}_2\\ \dot{\theta}_{z2}\end{bmatrix}^{\mathrm{T}} \frac{\rho Aa}{105}\begin{bmatrix}78 & 22a & 27 & -13a\\ 22a & 8a^2 & 13a & -6a^2\\ 27 & 13a & 78 & -22a\\ -13a & -6a^2 & -22a & 8a^2\end{bmatrix}\begin{bmatrix}\dot{v}_1\\ \dot{\theta}_{z1}\\ \dot{v}_2\\ \dot{\theta}_{z2}\end{bmatrix} \quad (3.145)$$

This expression can also be written in the more compact form

$$T_e = \frac{1}{2}\begin{bmatrix}\dot{u}_1\\ \dot{v}_1\\ \dot{\theta}_{z1}\\ \dot{u}_2\\ \dot{v}_2\\ \dot{\theta}_{z2}\end{bmatrix}^{\mathrm{T}} \frac{\rho Aa}{105}\begin{bmatrix}70 & 0 & 0 & 35 & 0 & 0\\ 0 & 78 & 22a & 0 & 27 & -13a\\ 0 & 22a & 8a^2 & 0 & 13a & -6a^2\\ 35 & 0 & 0 & 70 & 0 & 0\\ 0 & 27 & 13a & 0 & 78 & -22a\\ 0 & -13a & -6a^2 & 0 & -22a & 8a^2\end{bmatrix}\begin{bmatrix}\dot{u}_1\\ \dot{v}_1\\ \dot{\theta}_{z1}\\ \dot{u}_2\\ \dot{v}_2\\ \dot{\theta}_{z2}\end{bmatrix}$$

$$= \tfrac{1}{2}\{\dot{\mathbf{u}}\}_e^{\mathrm{T}}[\bar{\mathbf{m}}]_e\{\dot{\mathbf{u}}\}_e \quad (3.146)$$

Substituting the displacement functions (3.144) into the strain energy function (3.143) gives

$$U_e = \tfrac{1}{2}\{\mathbf{u}\}_e{}^{\mathrm{T}}\left[\bar{\mathbf{k}}\right]_e\{\mathbf{u}\}_e \quad (3.147)$$

where

$$\left[\bar{\mathbf{k}}\right]_e = \frac{EI_z}{2a^3}\begin{bmatrix}(a/r_z)^2 & 0 & 0 & -(a/r_z)^2 & 0 & 0\\ 0 & 3 & 3a & 0 & -3 & 3a\\ 0 & 3a & 4a^2 & 0 & -3a & 2a^2\\ -(a/r_z)^2 & 0 & 0 & (a/r_z)^2 & 0 & 0\\ 0 & -3 & 0 & 0 & 3 & -3a\\ 0 & 3a & 2a^2 & 0 & -3a & 4a^2\end{bmatrix} \quad (3.148)$$

and $r_z^2 = I_z/A$. (Note, r_z represents the radius of gyration of the cross-section about the z-axis.)

Similarly, the virtual work done by the applied loads is

$$\delta W_e = \{\mathbf{u}\}_e^{\mathrm{T}}\{\bar{\mathbf{f}}\}_e \quad (3.149)$$

If the applied loads are constant, then the load matrix is

$$\{\bar{\mathbf{f}}\}_e = \begin{bmatrix} p_x^e \\ p_y^e \\ ap_y^e/3 \\ p_x^e \\ p_y^e \\ -ap_y^e/3 \end{bmatrix} \tag{3.150}$$

The next step is to transform the energy expressions (3.146), (3.147) and (3.149) into expressions involving nodal degrees of freedom relative to the global axes.

The vector displacement $\vec{\boldsymbol{u}}$ of a single node is given by

$$\vec{\boldsymbol{u}} = u\hat{\boldsymbol{x}} + v\hat{\boldsymbol{y}} \tag{3.151}$$

relative to the local axes, where $\hat{\boldsymbol{x}}$ and $\hat{\boldsymbol{y}}$ are unit vectors in the x- and y-directions. Relative to global axes, this same vector displacement is

$$\vec{\boldsymbol{u}} = U\hat{\boldsymbol{X}} + V\hat{\boldsymbol{Y}} \tag{3.152}$$

where $\hat{\boldsymbol{X}}$, $\hat{\boldsymbol{Y}}$ and U, V are unit vectors and displacement components in the direction of the global axes X and Y (see Figure 3.17).

Taking the scalar product of equations (3.151) and (3.152) with $\hat{\boldsymbol{x}}$ and $\hat{\boldsymbol{y}}$ respectively gives

$$\begin{aligned} \hat{\boldsymbol{x}} \cdot \vec{\boldsymbol{u}} = u &= \hat{\boldsymbol{x}} \cdot \hat{\boldsymbol{X}}U + \hat{\boldsymbol{x}} \cdot \hat{\boldsymbol{Y}}V \\ &= \cos(x, X)U + \cos(x, Y)V \end{aligned} \tag{3.153}$$

and

$$\begin{aligned} \hat{\boldsymbol{y}} \cdot \vec{\boldsymbol{u}} = v &= \hat{\boldsymbol{y}} \cdot \hat{\boldsymbol{X}}U + \hat{\boldsymbol{y}} \cdot \hat{\boldsymbol{Y}}V \\ &= \cos(y, X)U + \cos(y, Y)V \end{aligned} \tag{3.154}$$

where cos (x, X) denotes the cosine of the angle between $\hat{\boldsymbol{x}}$ and $\hat{\boldsymbol{X}}$, etc. Since the local z-axis is parallel to the global Z-axis, then

$$\hat{\boldsymbol{z}} = \hat{\boldsymbol{Z}} \quad \text{and} \quad \theta_z = \theta_Z \tag{3.155}$$

Combining equations (3.153) to (3.155) together in matrix form gives

$$\begin{aligned} \begin{bmatrix} u \\ v \\ \theta_z \end{bmatrix} &= \begin{bmatrix} \cos(x, X) & \cos(x, Y) & 0 \\ \cos(y, X) & \cos(y, Y) & 0 \\ 0 & 0 & 1 \end{bmatrix} \begin{bmatrix} U \\ V \\ \theta_Z \end{bmatrix} \\ &= [\mathbf{L}_2] \begin{bmatrix} U \\ V \\ \theta_Z \end{bmatrix} \end{aligned} \tag{3.156}$$

where $[\mathbf{L}_2]$ is a direction cosine array.

The degrees of freedom at both the nodes of the element can therefore be transformed from local to global axes by means of the relation

$$\begin{bmatrix} u_1 \\ v_1 \\ \theta_{z1} \\ u_2 \\ v_2 \\ \theta_{z2} \end{bmatrix} = \begin{bmatrix} [\mathbf{L}_2] & [\mathbf{0}] \\ & \\ [\mathbf{0}] & [\mathbf{L}_2] \end{bmatrix} \begin{bmatrix} U_1 \\ V_1 \\ \theta_{z1} \\ U_2 \\ V_2 \\ \theta_{z2} \end{bmatrix} \qquad (3.157)$$

or

$$\{\mathbf{u}\}_e = [\mathbf{R}]_e \{\mathbf{U}\}_e \qquad (3.158)$$

Substituting (3.158) into the energy expressions (3.146), (3.147) and (3.149) gives

$$\begin{aligned} T_e &= \tfrac{1}{2}\{\dot{\mathbf{U}}\}_{\mathbf{e}}^{\mathrm{T}}[\mathbf{m}]_{\mathbf{e}}\{\dot{\mathbf{U}}\}_{\mathbf{e}} \\ U_e &= \tfrac{1}{2}\{\mathbf{U}\}_e^{\mathrm{T}}[\mathbf{k}]_e\{\mathbf{U}\}_e \\ \delta W_e &= \{\mathbf{U}\}_e^{\mathrm{T}}\{\mathbf{f}\}_e \end{aligned} \qquad (3.159)$$

where

$$\begin{aligned} [\mathbf{m}]_e &= [\mathbf{R}]_e^{\mathrm{T}}[\bar{\mathbf{m}}]_e[\mathbf{R}]_e \\ [\mathbf{k}]_e &= [\mathbf{R}]_e^{\mathrm{T}}[\bar{\mathbf{k}}]_e[\mathbf{R}]_e \\ \{\mathbf{f}\}_e &= [\mathbf{R}]_e^{\mathrm{T}}\{\bar{\mathbf{f}}\}_e \end{aligned} \qquad (3.160)$$

In order to evaluate the expressions (3.160) it is necessary to calculate the element length, $2a$, and the elements of the direction cosine array, $[\mathbf{L}_2]$, from the global coordinates of the nodes 1 and 2. The position vectors of nodes 1 and 2 are (see Figure 3.17)

$$\vec{\boldsymbol{V}}1 = X_1\hat{\boldsymbol{X}} + Y_1\hat{\boldsymbol{Y}}$$

and

$$\vec{\boldsymbol{V}}2 = X_2\hat{\boldsymbol{X}} + Y_2\hat{\boldsymbol{Y}}. \qquad (3.161)$$

The length of the element is equal to the magnitude of the vector $(\vec{\boldsymbol{V}}2 - \vec{\boldsymbol{V}}1)$, and so

$$2a = |\vec{\boldsymbol{V}}2 - \vec{\boldsymbol{V}}1| \qquad (3.162)$$

Substituting (3.161) into (3.162) gives

$$\begin{aligned} 2a &= \{(X_2 - X_1)^2 + (Y_2 - Y_1)^2\}^{1/2} \\ &= \left\{X_{21}^2 + Y_{21}^2\right\}^{1/2} \end{aligned} \qquad (3.163)$$

where

$$X_{21} = X_2 - X_1, \qquad Y_{21} = Y_2 - Y_1 \qquad (3.164)$$

Now

$$\hat{\boldsymbol{x}} = \frac{(\vec{\boldsymbol{V}}2 - \vec{\boldsymbol{V}}1)}{|\vec{\boldsymbol{V}}2 - \vec{\boldsymbol{V}}1|} = \frac{X_{21}}{2a}\hat{\boldsymbol{X}} + \frac{Y_{21}}{2a}\hat{\boldsymbol{Y}} \qquad (3.165)$$

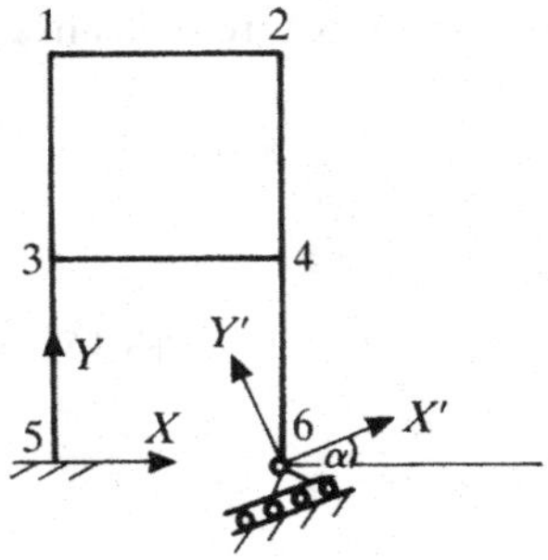

Figure 3.18. Example of a plane framework.

Therefore

$$\hat{\boldsymbol{x}} \cdot \hat{\boldsymbol{X}} = X_{21}/2a, \qquad \hat{\boldsymbol{x}} \cdot \hat{\boldsymbol{Y}} = Y_{21}/2a$$

Since $\hat{\boldsymbol{y}}$ is perpendicular to $\hat{\boldsymbol{x}}$, then

$$\hat{\boldsymbol{y}} = \hat{\boldsymbol{Z}} \wedge \hat{\boldsymbol{x}} = -\frac{Y_{21}}{2a}\hat{\boldsymbol{X}} + \frac{X_{21}}{2a}\hat{\boldsymbol{Y}} \tag{3.166}$$

where $(\wedge)$ denotes a vector product. Thus

$$\hat{\boldsymbol{y}} \cdot \hat{\boldsymbol{X}} = -Y_{21}/2a, \qquad \hat{\boldsymbol{y}} \cdot \hat{\boldsymbol{Y}} = X_{21}/2a \tag{3.167}$$

The direction cosine array, which is defined in (3.156) is therefore

$$[\mathbf{L}_2] = \begin{bmatrix} X_{21}/2a & Y_{21}/2a & 0 \\ -Y_{21}/2a & X_{21}/2a & 0 \\ 0 & 0 & 1 \end{bmatrix} \tag{3.168}$$

The assembly process for an element of a plane framework is slightly different from that used for the previous elements. The reason for this is that the nodes at the two ends of an element do not always have consecutive numbers as shown by the example in Figure 3.18. The general rule in this case is that for an element with node numbers n_1 and n_2, then columns 1 to 3 and 4 to 6 of the element matrices are added into columns $(3n_1 - 2)$ to $3n_1$ and $(3n_2 - 2)$ to $3n_2$ of the matrices for the complete framework respectively. At the same time, rows 1 to 3 and 4 to 6 are added into rows $(3n_1 - 2)$ to $3n_1$ and $(3n_2 - 2)$ to $3n_2$ respectively. To illustrate this the positions of the terms representing the framework element 1–3 in Figure 3.18 are indicated in Figure 3.19.

The example shown in Figure 3.18 will now be used to illustrate two types of constraint which are frequently encountered. These consist of linear relationships between the degrees of freedom, either at a single node or at two or more nodes.

First of all consider node 6 which is supported by an inclined roller. The condition to be applied here is that the displacement in the Y'-direction is zero, that is

$$(U_6\hat{\boldsymbol{X}} + V_6\hat{\boldsymbol{Y}}) \cdot \hat{\boldsymbol{Y}}' = 0 \tag{3.169}$$

Now

$$\hat{\boldsymbol{Y}}' = -\sin\alpha\hat{\boldsymbol{X}} + \cos\alpha\hat{\boldsymbol{Y}} \tag{3.170}$$

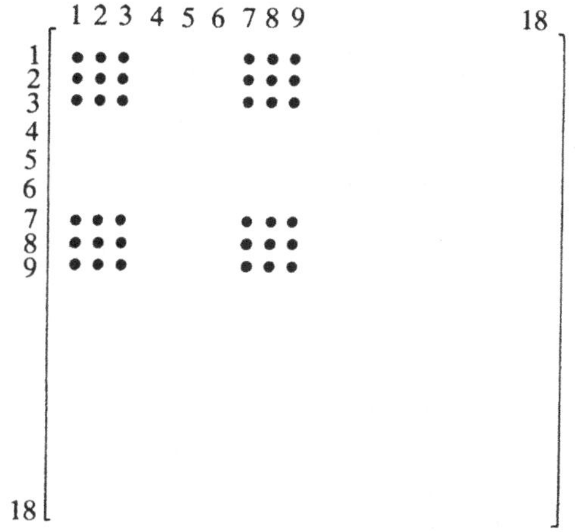

Figure 3.19. Illustration of the assembly process.

Substituting (3.170) into (3.169) gives

$$-\sin\alpha U_6 + \cos\alpha\, V_6 = 0 \tag{3.171}$$

This is a linear relationship between the degrees of freedom at a single node. It may be used to eliminate U_6 from the energy expression using the method described in Section 1.5.

Reference [3.17] shows that when the members of a framework are slender, the axial deformation of each member can be neglected. This fact can be used to reduce the number of degrees of freedom in the following manner.

Since node 5 is clamped, then

$$U_5 = 0, \qquad V_5 = 0, \qquad \theta_{Z5} = 0 \tag{3.172}$$

The axial deformation of members 1–3 and 3–5 can be neglected by imposing the conditions

$$V_1 = 0, \qquad V_3 = 0 \tag{3.173}$$

The axial deformation of the remaining members can be neglected by applying the constraints

$$\begin{aligned} U_1 - U_2 &= 0, \qquad U_3 - U_4 = 0 \\ V_2 - V_6 &= 0, \qquad V_4 - V_6 = 0 \end{aligned} \tag{3.174}$$

The constraints (3.172) and (3.173) are applied in the manner described in Section 3.3. Equations (3.174) are linear relationships between degrees of freedom at two nodes. They may be used to eliminate four degrees of freedom, say U_1, U_3, V_2 and V_4, from the energy expressions using the method described in Section 1.5.

EXAMPLE 3.8 Calculate the first five antisymmetric frequencies and modes of the two-dimensional, steel framework shown in Figure 3.20. Compare the results with the analytical solution [3.18]. Take $E = 206.84$ GN/m^2 and $\rho = 7.83 \times 10^3$ kg/m^3.

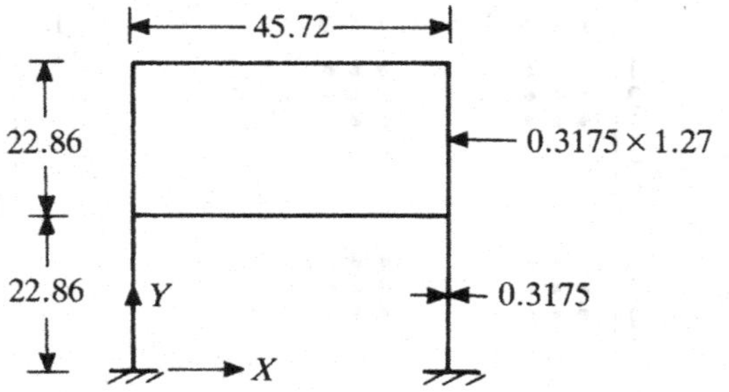

Figure 3.20. Geometry of a two-dimensional framework. Dimensions are in centimetres.

Since the framework has one axis of symmetry, the antisymmetric and symmetric modes can be calculated separately by idealising half the structure and applying appropriate boundary conditions on the axis of symmetry (see Chapter 8). The results presented in reference [3.14] indicate that modes which involve the individual vertical members and half the horizontal members deforming in not more than one complete flexural wave, can be represented adequately by three elements per member. Figure 3.21, therefore, indicates an adequate idealisation for the modes to be predicted.

There are three degrees of freedom at each node, namely, linear displacements U and V in the X- and Y-directions and a rotation θ_z about the Z-axis, which is orthogonal to X and Y. Since node 1 is fully fixed, all three degrees of freedom there are constrained to be zero. The antisymmetric modes are obtained by setting the V displacement to zero at nodes 9 and 13.

Reference [3.17] indicates that the framework is slender and so the axial deformation of each member can be neglected. The axial deformation of the vertical members can be eliminated by setting the V displacement to zero at nodes 2, 3, 4, 6, 8 and 10. For the horizontal members, the U displacement at nodes 5, 7 and 9 are set equal to the U displacement at node 4 and the U displacement at nodes 11, 12 and 13 are set equal to the U displacement at node 10.

The predicted frequencies are compared with the analytical ones in Table 3.5 and the corresponding mode shapes are shown in Figure 3.22. The predicted frequencies are in very close agreement.

3.7 Vibration of Three-Dimensional Frameworks

The procedure for analysing a three-dimensional framework is the same as the one described in Section 3.6 for a plane framework. Figure 3.23 shows a typical element

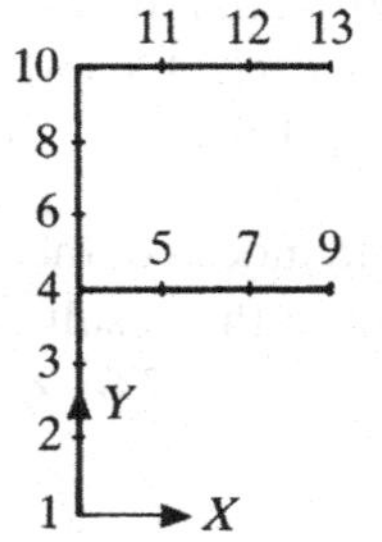

Figure 3.21. Idealisation of half the framework.

Table 3.5. *Comparison of predicted frequencies* (Hz) *with exact ones for the antisymmetric modes of a framework*

Mode	FEM	Analytical [3.18]	% Difference
1	15.14	15.14	0.00
2	53.32	53.32	0.00
3	155.48	155.31	0.11
4	186.51	186.23	0.15
5	270.85	270.07	0.29

together with its local axes x, y and z and global axes X, Y and Z. The local x-axis lies along the centroidal axis which joins nodes 1 and 2. The local y- and z-axes coincide with the principal axes of the cross-section of the element.

In this section it will be assumed that the shear centre of a cross-section coincides with the centroid. This assumption is restrictive only when both properties are important in the same problem. The modifications required when the shear centre does not coincide with centroid is discussed in Section 3.11.

Each member of a three-dimensional framework is capable of axial deformation, bending in two principal planes, and torsion about its axis. The energy functions are, therefore, a combination of the energy functions derived in Sections 2.1

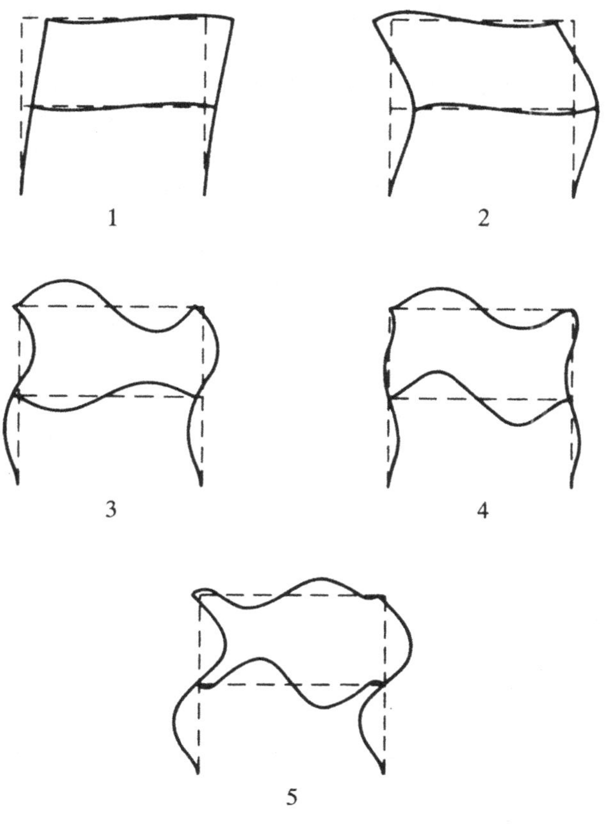

Figure 3.22. Antisymmetric mode shapes of a framework.

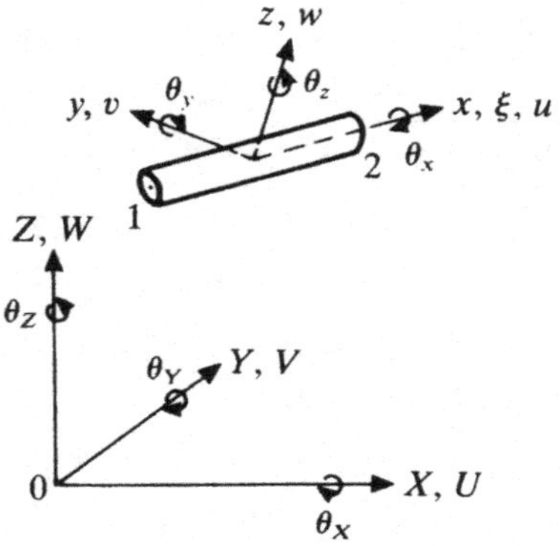

Figure 3.23. Geometry of a three-dimensional framework element.

to 2.3. These are, in terms of local coordinates, as follows:

$$T_e = \frac{1}{2}\int_{-a}^{+a} \rho A(\dot{u}^2 + \dot{v}^2 + \dot{w}^2)\,\mathrm{d}x + \frac{1}{2}\int_{-a}^{+a} \rho I_x \dot{\theta}_x^2\,\mathrm{d}x$$

$$U_e = \frac{1}{2}\int_{-a}^{+a} EA\left(\frac{\partial u}{\partial x}\right)^2 \mathrm{d}x + \frac{1}{2}\int_{-a}^{+a}\left\{EI_z\left(\frac{\partial^2 v}{\partial x^2}\right) + EI_y\left(\frac{\partial^2 w}{\partial x^2}\right)^2\right\}\mathrm{d}x$$

$$+\frac{1}{2}\int_{-a}^{+a} GJ\left(\frac{\partial \theta_x}{\partial x}\right)^2 \mathrm{d}x \qquad (3.175)$$

$$\delta W_e = \int_{-a}^{+a} p_x \delta u\,\mathrm{d}x + \int_{-a}^{+a} (p_y\delta v + p_z\delta w)\,\mathrm{d}x + \int_{-a}^{+a} m_x \delta\theta_x\,\mathrm{d}x$$

In these expressions u, v and w are the displacement components of the centroid of the cross-section relative to the local axes, x, y and z, and θ_x the rotation of the cross-section about the local x-axis. Also p_x, p_y and p_z are the components of the load per unit length relative to the local axes and m_x the twisting moment per unit length about the local x-axis.

Since the axial, bending and twisting deformations are uncoupled, they can be treated separately as in Sections 3.3 to 3.5. The displacement functions can therefore be taken to be the ones defined by equations (3.52), (3.124) and (3.95), which are

$$\begin{aligned} u &= \lfloor \mathbf{N}_u(\xi)\rfloor\{\mathbf{u}\}_e \\ v &= \lfloor \mathbf{N}_v(\xi)\rfloor\{\mathbf{v}\}_e \\ w &= \lfloor \mathbf{N}_w(\xi)\rfloor\{\mathbf{w}\}_e \\ \theta_x &= \lfloor N_x(\xi)\rfloor\{\boldsymbol{\theta}_x\}_e \end{aligned} \qquad (3.176)$$

Note that

$$\lfloor \mathbf{N}_x(\xi)\rfloor = \lfloor \mathbf{N}_u(\xi)\rfloor$$

and

$$\lfloor \mathbf{N}_w(\xi)\rfloor = \lfloor N_1(\xi) \quad -aN_2(\xi) \quad N_3(\xi) \quad -aN_4(\xi)\rfloor \qquad (3.177)$$

where the functions $N_1(\xi)$ to $N_4(\xi)$ are defined by equations (3.126). The change in signs from the expression for $\lfloor \mathbf{N}_v(\xi) \rfloor$ (see (3.124)) is because $\theta_y = -\partial w/\partial x$, whilst $\theta_z = +\partial v/\partial x$.

Substituting equations (3.176) into the kinetic energy expression in (3.175) gives, after integration:

$$T_e = \tfrac{1}{2}\{\dot{\mathbf{u}}\}_e^{\mathrm{T}} [\bar{\mathbf{m}}]_e \{\dot{\mathbf{u}}\}_e \tag{3.178}$$

where

$$\{\mathbf{u}\}_e^{\mathrm{T}} = \lfloor u_1 \quad v_1 \quad w_1 \quad \theta_{x1} \quad \theta_{y1} \quad \theta_{z1} \quad u_2 \quad v_2 \quad w_2 \quad \theta_{x2} \quad \theta_{y2} \quad \theta_{z2} \rfloor \tag{3.179}$$

and

$$[\bar{\mathbf{m}}]_e = \begin{bmatrix} \bar{\mathbf{m}}_{11} & \bar{\mathbf{m}}_{12} \\ \bar{\mathbf{m}}_{12}^{\mathrm{T}} & \bar{\mathbf{m}}_{22} \end{bmatrix} \tag{3.180}$$

where

$$\bar{\mathbf{m}}_{11} = \frac{\rho A a}{105} \begin{bmatrix} 70 & 0 & 0 & 0 & 0 & 0 \\ 0 & 78 & 0 & 0 & 0 & 22a \\ 0 & 0 & 78 & 0 & -22a & 0 \\ 0 & 0 & 0 & 70r_x^2 & 0 & 0 \\ 0 & 0 & -22a & 0 & 8a^2 & 0 \\ 0 & 22a & 0 & 0 & 0 & 8a^2 \end{bmatrix} \tag{3.181}$$

$$\bar{\mathbf{m}}_{12} = \frac{\rho A a}{105} \begin{bmatrix} 35 & 0 & 0 & 0 & 0 & 0 \\ 0 & 27 & 0 & 0 & 0 & -13a \\ 0 & 0 & 27 & 0 & 13a & 0 \\ 0 & 0 & 0 & 35r_x^2 & 0 & 0 \\ 0 & 0 & -13a & 0 & -6a^2 & 0 \\ 0 & 13a & 0 & 0 & 0 & -6a^2 \end{bmatrix} \tag{3.182}$$

$$\bar{\mathbf{m}}_{22} = \frac{\rho A a}{105} \begin{bmatrix} 70 & 0 & 0 & 0 & 0 & 0 \\ 0 & 78 & 0 & 0 & 0 & -22a \\ 0 & 0 & 78 & 0 & 22a & 0 \\ 0 & 0 & 0 & 70r_x^2 & 0 & 0 \\ 0 & 0 & 22a & 0 & 8a^2 & 0 \\ 0 & -22a & 0 & 0 & 0 & 8a^2 \end{bmatrix} \tag{3.183}$$

and $r_x^2 = I_x/A$.

Similarly, substituting equations (3.176) into the strain energy expression in (3.175) gives

$$U_e = \tfrac{1}{2}\{\mathbf{u}\}_e^{\mathrm{T}} [\bar{\mathbf{k}}]_e \{\mathbf{u}\}_e \tag{3.184}$$

where

$$[\bar{\mathbf{k}}]_e = \begin{bmatrix} \bar{\mathbf{k}}_{11} & \bar{\mathbf{k}}_{12} \\ \bar{\mathbf{k}}_{12}^{\mathrm{T}} & \bar{\mathbf{k}}_{22} \end{bmatrix} \tag{3.185}$$

and

$$\bar{\mathbf{k}}_{11} = \frac{AE}{8a^3}\begin{bmatrix} 4a^2 & 0 & 0 & 0 & 0 & 0 \\ 0 & 12r_z^2 & 0 & 0 & 0 & 12ar_z^2 \\ 0 & 0 & 12r_y^2 & 0 & -12ar_y^2 & 0 \\ 0 & 0 & 0 & 2a^2r_J^2/(1+\nu) & 0 & 0 \\ 0 & 0 & -12ar_y^2 & 0 & 16a^2r_y^2 & 0 \\ 0 & 12ar_z^2 & 0 & 0 & 0 & 16a^2r_z^2 \end{bmatrix} \tag{3.186}$$

$$\bar{\mathbf{k}}_{12} = \frac{AE}{8a^3}\begin{bmatrix} -4a^2 & 0 & 0 & 0 & 0 & 0 \\ 0 & -12r_z^2 & 0 & 0 & 0 & 12ar_z^2 \\ 0 & 0 & -12r_y^2 & 0 & -12ar_y^2 & 0 \\ 0 & 0 & 0 & -2a^2r_J^2/(1+\nu) & 0 & 0 \\ 0 & 0 & 12ar_y^2 & 0 & 8a^2r_y^2 & 0 \\ 0 & -12ar_z^2 & 0 & 0 & 0 & 8a^2r_z^2 \end{bmatrix} \tag{3.187}$$

$$\bar{\mathbf{k}}_{22} = \frac{AE}{8a^3}\begin{bmatrix} 4a^2 & 0 & 0 & 0 & 0 & 0 \\ 0 & 12r_z^2 & 0 & 0 & 0 & -12ar_z^2 \\ 0 & 0 & 12r_y^2 & 0 & 12ar_y^2 & 0 \\ 0 & 0 & 0 & 2a^2r_J^2/(1+\nu) & 0 & 0 \\ 0 & 0 & 12ar_y^2 & 0 & 16a^2r_y^2 & 0 \\ 0 & -12ar_z^2 & 0 & 0 & 0 & 16a^2r_z^2 \end{bmatrix} \tag{3.188}$$

In the above $r_y^2 = I_y/A, r_z^2 = I_z/A$ and $r_J^2 = J/A$.

The work done by the applied loads is

$$\delta W_e = \{\mathbf{u}\}_e^{\mathrm{T}}\{\bar{\mathbf{f}}\}_e \tag{3.189}$$

If the applied loads are constant, then the load matrix is given by

$$\{\bar{\mathbf{f}}\}_e^{\mathrm{T}} = a\lfloor p_x^e \;\; p_y^e \;\; p_z^e \;\; m_x^e \;\; -p_z^e a/3 \;\; p_y^e a/3 \quad p_x^e \;\; p_y^e \;\; p_z^e \;\; m_x^e \;\; p_z^e a/3 \;\; -p_y^e a/3\rfloor \tag{3.190}$$

The energy expressions (3.178), (3.184) and (3.189) are now transformed into expressions involving nodal degrees of freedom relative to the global axes.

The vector displacement $\vec{u}$ of a single node is given by

$$\vec{u} = U\hat{X} + V\hat{Y} + W\hat{Z} \tag{3.191}$$

relative to the global axes, where $\hat{\mathbf{X}}$, $\hat{\mathbf{Y}}$ and $\hat{\mathbf{Z}}$ are unit vectors along the X, Y and Z-axes. The components of $\vec{\boldsymbol{u}}$ relative to the local axes are given by

$$\begin{aligned} u &= \hat{\boldsymbol{x}} \cdot \vec{\boldsymbol{u}} = \hat{\boldsymbol{x}} \cdot \hat{\mathbf{X}}U + \hat{\boldsymbol{x}} \cdot \hat{\mathbf{Y}}V + \hat{\boldsymbol{x}} \cdot \hat{\mathbf{Z}}W \\ &= \cos(x, X)U + \cos(x, Y)V + \cos(x, Z)W \end{aligned} \tag{3.192}$$

$$\begin{aligned} v &= \hat{\boldsymbol{y}} \cdot \vec{\boldsymbol{u}} = \hat{\boldsymbol{y}} \cdot \hat{\mathbf{X}}U + \hat{\boldsymbol{y}} \cdot \hat{\mathbf{Y}}V + \hat{\boldsymbol{y}} \cdot \hat{\mathbf{Z}}W \\ &= \cos(y, X)U + \cos(y, Y)V + \cos(y, Z)W \end{aligned} \tag{3.193}$$

$$\begin{aligned} w &= \hat{\boldsymbol{z}} \cdot \vec{\boldsymbol{u}} = \hat{\boldsymbol{z}} \cdot \hat{\mathbf{X}}U + \hat{\boldsymbol{z}} \cdot \hat{\mathbf{Y}}V + \hat{\boldsymbol{z}} \cdot \hat{\mathbf{Z}}W \\ &= \cos(z, X)U + \cos(z, Y)V + \cos(z, Z)W \end{aligned} \tag{3.194}$$

Equations (3.192) to (3.194) can be combined in the following matrix form

$$\begin{aligned} \begin{bmatrix} u \\ v \\ w \end{bmatrix} &= \begin{bmatrix} \cos(x, X) & \cos(x, Y) & \cos(x, Z) \\ \cos(y, X) & \cos(y, Y) & \cos(y, Z) \\ \cos(z, X) & \cos(z, Y) & \cos(z, Z) \end{bmatrix} \begin{bmatrix} U \\ V \\ W \end{bmatrix} \\ &= [\mathbf{L}_3] \begin{bmatrix} U \\ V \\ W \end{bmatrix} \end{aligned} \tag{3.195}$$

It can be shown in a similar manner that

$$\begin{bmatrix} \theta_x \\ \theta_y \\ \theta_z \end{bmatrix} = [\mathbf{L}_3] \begin{bmatrix} \theta_X \\ \theta_Y \\ \theta_Z \end{bmatrix} \tag{3.196}$$

The degrees of freedom at all the nodes of the element can therefore be transformed from local to global axes by means of the relation

$$\{\mathbf{u}\}_e = [\mathbf{R}]_e \{\mathbf{U}\}_e \tag{3.197}$$

where

$$\{\mathbf{u}\}_e^{\mathrm{T}} = \lfloor u_1 \quad v_1 \quad w_1 \quad \theta_{x1} \quad \theta_{y1} \quad \theta_{z1} \quad u_2 \quad v_2 \quad w_2 \quad \theta_{x2} \quad \theta_{y2} \quad \theta_{z2} \rfloor \tag{3.198}$$

$$\{\mathbf{U}\}_e^{\mathrm{T}} = \lfloor U_1 \quad V_1 \quad W_1 \quad \theta_{X1} \quad \theta_{Y1} \quad \theta_{Z1} \quad U_2 \quad V_2 \quad W_2 \quad \theta_{X1} \quad \theta_{Y2} \quad \theta_{Z2} \rfloor \tag{3.199}$$

$$[\mathbf{R}]_e = \begin{bmatrix} [\mathbf{L}_3] & & & \\ & [\mathbf{L}_3] & & \\ & & [\mathbf{L}_3] & \\ & & & [\mathbf{L}_3] \end{bmatrix} \tag{3.200}$$

Substituting the transformation (3.197) into the energy expressions (3.178), (3.184) and (3.189) gives expressions of the same form as (3.159) and (3.160).

In order to evaluate these expressions it is necessary to calculate the element length, $2a$, and the elements of the direction cosine array $[\mathbf{L}_3]$. The position vectors of nodes 1 and 2 are

$$\begin{aligned} \vec{V}1 &= X_1\hat{\mathbf{X}} + Y_1\hat{\mathbf{Y}} + Z_1\hat{\mathbf{Z}} \\ \vec{V}2 &= X_2\hat{\mathbf{X}} + Y_2\hat{\mathbf{Y}} + Z_2\hat{\mathbf{Z}} \end{aligned} \tag{3.201}$$

The length of the element is given by

$$2a = |\vec{V}2 - \vec{V}1| = \{X_{21}^2 + Y_{21}^2 + Z_{21}^2\}^{1/2} \tag{3.202}$$

where

$$X_{21} = X_2 - X_1, \qquad Y_{21} = Y_2 - Y_1, \qquad Z_{21} = Z_2 - Z_1 \tag{3.203}$$

Now

$$\hat{\boldsymbol{x}} = \frac{(\vec{V}2 - \vec{V}1)}{|\vec{V}2 - \vec{V}1|} = \frac{X_{21}}{2a}\hat{\boldsymbol{X}} + \frac{Y_{21}}{2a}\hat{\boldsymbol{Y}} + \frac{Z_{21}}{2a}\hat{\boldsymbol{Z}} \tag{3.204}$$

Therefore

$$\hat{\boldsymbol{x}} \cdot \hat{\boldsymbol{X}} = \frac{X_{21}}{2a}, \qquad \hat{\boldsymbol{x}} \cdot \hat{\boldsymbol{Y}} = \frac{Y_{21}}{2a}, \qquad \hat{\boldsymbol{x}} \cdot \hat{\boldsymbol{Z}} = \frac{Z_{21}}{2a} \tag{3.205}$$

The orientation of the local yz-axes can be defined by specifying the position of any convenient point in the local xy-plane but not on the x-axis. This point will be referred to as node 3 and its position indicated by the vector $\vec{V}3$. Since the local z-axis is perpendicular to both $(\vec{V}2 - \vec{V}1)$ and $(\vec{V}3 - \vec{V}1)$, then

$$\hat{\boldsymbol{z}} = \frac{(\vec{V}2 - \vec{V}1) \wedge (\vec{V}3 - \vec{V}1)}{|(\vec{V}2 - \vec{V}1) \wedge (\vec{V}3 - \vec{V}1)|} \tag{3.206}$$

Now the position of node 3 is defined by

$$\vec{V}3 = X_3\hat{\boldsymbol{X}} + Y_3\hat{\boldsymbol{Y}} + Z_3\hat{\boldsymbol{Z}} \tag{3.207}$$

Substituting the expressions (3.201) and (3.207) into (3.206) gives

$$\hat{\boldsymbol{z}} = \frac{1}{2A_{123}}\{(Y_{21}Z_{31} - Y_{31}Z_{21})\hat{\boldsymbol{X}} + (Z_{21}X_{31} - Z_{31}X_{21})\hat{\boldsymbol{Y}} + (X_{21}Y_{31} - X_{31}Y_{21})\hat{\boldsymbol{Z}}\} \tag{3.208}$$

where

$$2A_{123} = \{(Y_{21}Z_{31} - Y_{31}Y_{21})^2 + (Z_{21}X_{31} - Z_{31}X_{21})^2 + (X_{21}Y_{31} - X_{31}Y_{21})^2\}^{1/2} \tag{3.209}$$

In equations (3.208) and (3.209) the following notation has been used

$$\begin{aligned} X_{ij} &= X_i - X_j, & Y_{ij} &= Y_i - Y_j, \\ Z_{ij} &= Z_i - Z_j, & i, j &= 1, 2, 3 \end{aligned}$$

Using (3.208) it can be seen that

$$\begin{aligned} \hat{\boldsymbol{z}} \cdot \hat{\boldsymbol{X}} &= \frac{1}{2A_{123}}(Y_{21}Z_{31} - Y_{31}Z_{21}) \\ \hat{\boldsymbol{z}} \cdot \hat{\boldsymbol{Y}} &= \frac{1}{2A_{123}}(Z_{21}X_{31} - Z_{31}X_{21}) \\ \hat{\boldsymbol{z}} \cdot \hat{\boldsymbol{Z}} &= \frac{1}{2A_{123}}(X_{21}Y_{31} - X_{31}Y_{21}) \end{aligned} \tag{3.210}$$

Since $\hat{\boldsymbol{y}}$ is perpendicular to both $\hat{\boldsymbol{x}}$ and $\hat{\boldsymbol{z}}$ then

$$\hat{\boldsymbol{y}} = \hat{\boldsymbol{z}} \wedge \hat{\boldsymbol{x}} \tag{3.211}$$

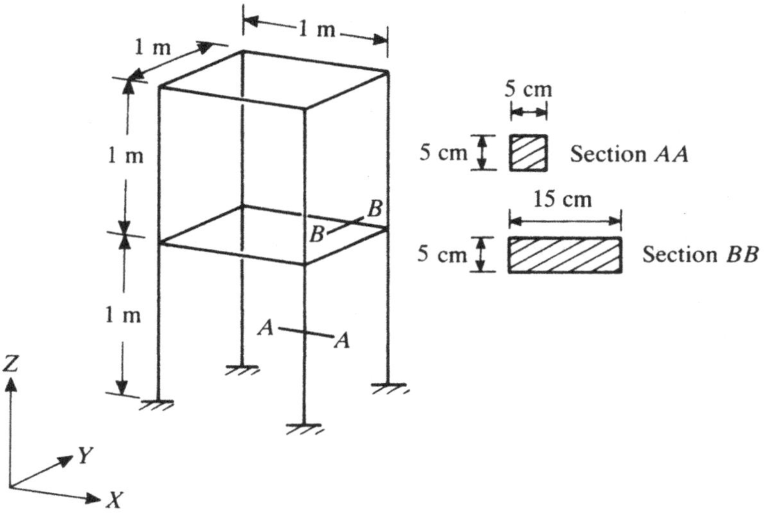

Figure 3.24. Geometry of a three-dimensional framework.

This relationship gives

$$\begin{aligned}\cos(y, X) &= \cos(z, Y)\cos(x, Z) - \cos(z, Z)\cos(x, Y)\\ \cos(y, Y) &= \cos(z, Z)\cos(x, X) - \cos(z, X)\cos(x, Z)\\ \cos(y, Z) &= \cos(z, X)\cos(x, Y) - \cos(z, Y)\cos(x, X)\end{aligned} \tag{3.212}$$

The right hand sides of these expressions can be evaluated using (3.205) and (3.210).

The assembly process for an element of a three-dimensional frame is similar to that for an element of a plane framework. For an element with node numbers n_1 and n_2, then columns 1 to 6 and 7 to 12 of the element matrices are added into columns $(6n_1 - 5)$ to $6n_1$ and $(6n_2 - 5)$ to $6n_2$ respectively. An identical rule applies to the rows.

EXAMPLE 3.9 Calculate the frequencies and shapes of the first two swaying modes of the three-dimensional, steel framework shown in Figure 3.24. All vertical members are square and all horizontal members are rectangular, as shown. Take $E = 219.9\,\text{GN/m}^2$ and $\rho = 7.9 \times 10^3\ \text{kg/m}^3$.

Since the framework has two planes of symmetry, the swaying modes can be calculated by idealising one quarter of the structure and applying symmetric boundary conditions on one plane of symmetry and antisymmetric conditions on the other plane. Figure 3.25 shows an idealisation of one quarter of the framework using two elements per member.

There are six degrees of freedom at each node namely, linear displacements U, V and W in the X, Y and Z-directions and rotations θ_X, θ_Y and θ_Z about the same axes. Since node 1 is fully fixed, all six degrees of freedom there are constrained to be zero. Motion which is symmetrical about the XZ-plane is obtained by setting V, θ_X and θ_Z to zero at nodes 3 and 7. Antisymmetric motion about the YZ-plane is obtained by setting V, W and θ_X to zero at nodes 5 and 9.

The predicted frequencies and mode shapes are shown in Figure 3.26.

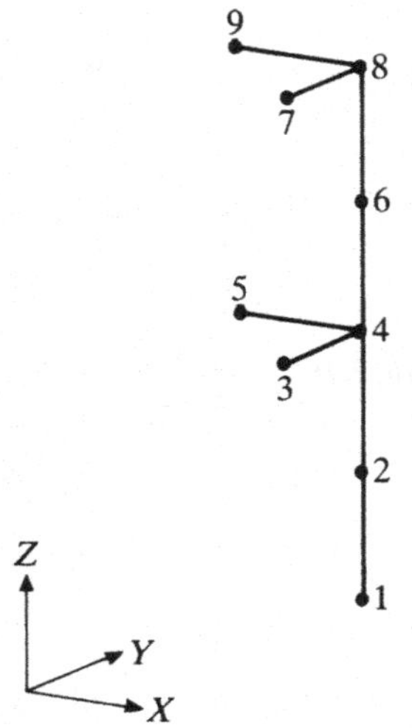

Figure 3.25. Idealisation of one quarter of the framework.

3.8 Techniques for Increasing the Accuracy of Elements

The accuracy of the solution of a given problem can be increased by either increasing the number of elements, as demonstrated in previous sections, or by increasing the order of the polynomial representation of the displacements within each element. This latter course of action results in an increased number of degrees of freedom for an element. The additional degrees of freedom can be located either at existing nodes or at additional node points.

To illustrate these procedures, consider the axial element shown in Figure 3.6. Taking u and $\partial u/\partial x$ as degrees of freedom at the two nodes, the displacement function can be expressed in the form

$$u = N_1(\xi)u_1 + aN_2(\xi)\frac{\partial u_1}{\partial x} + N_3(\xi)u_2 + aN_4(\xi)\frac{\partial u_2}{\partial x} \tag{3.213}$$

where the functions $N_1(\xi)$ to $N_4(\xi)$ are defined by (3.126). Since the element has four degrees of freedom, the displacement is approximated by a cubic polynomial.

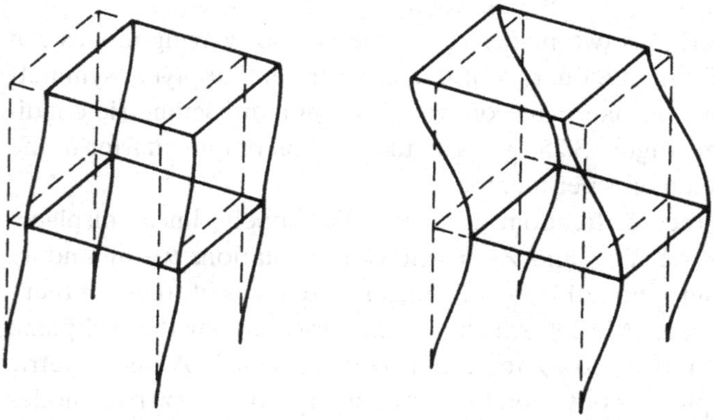

Figure 3.26. Swaying modes of a framework.

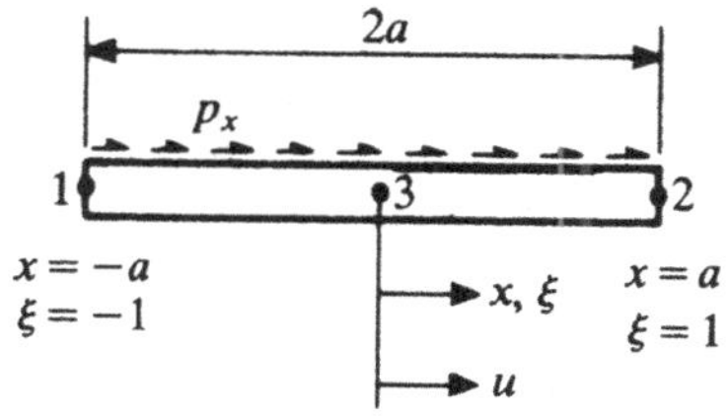

Figure 3.27. Geometry of a three node axial element.

With this displacement function it is possible to satisfy the natural boundary condition $\partial u/\partial x = 0$ at a free end as well as the geometric boundary condition $u = 0$ at a fixed end. This will result in increased accuracy since all the boundary conditions will be satisfied exactly and only the equations of motion will be satisfied approximately. However, the function does have the disadvantage that when adjacent elements have different cross-sectional areas or elastic properties, then continuity of $\partial u/\partial x$ cannot be enforced at nodes. This follows from the fact that $EA\partial u/\partial x$ must be continuous to satisfy equilibrium.

In Figure 3.27 the number of nodes for an axial element have been increased to three. Taking u as the only degree of freedom at each node gives a total of three degrees of freedom for the element. This means that the axial displacement can be represented by a quadratic function, which can be written in the form

$$u = N_1(\xi)u_1 + N_2(\xi)u_2 + N_3(\xi)u_3 \tag{3.214}$$

where

$$N_j(\xi) = \tfrac{1}{2}\xi_j\xi(1 + \xi_j\xi), \qquad j = 1, 2 \quad \text{and} \quad N_3(\xi) = (1 - \xi^2) \tag{3.215}$$

These functions are illustrated in Figure 3.28.

Substituting the displacement function (3.214) into the energy expressions (3.53), (3.54) and (3.55) gives the following element matrices

$$[\mathbf{m}]_e = \frac{\rho Aa}{15}\begin{bmatrix} 4 & -1 & 2 \\ -1 & 4 & 2 \\ 2 & 2 & 16 \end{bmatrix} \tag{3.216}$$

$$[\mathbf{k}]_e = \frac{EA}{6a}\begin{bmatrix} 7 & 1 & -8 \\ 1 & 7 & -8 \\ -8 & -8 & 16 \end{bmatrix} \tag{3.217}$$

$$\{\mathbf{f}\}_e = \frac{p_x^e a}{3}\begin{bmatrix} 1 \\ 1 \\ 4 \end{bmatrix} \tag{3.218}$$

In evaluating the element load matrix it has been assumed that p_x^e is constant over the element.

The accuracy obtained when using the quadratic element to predict the first two natural frequencies of a clamped-free rod is shown in Figure 3.29. The figure also shows the results obtained with a linear element. It can be seen that six linear elements are required to predict the lowest frequency with a better accuracy than the estimate obtained with one quadratic element.

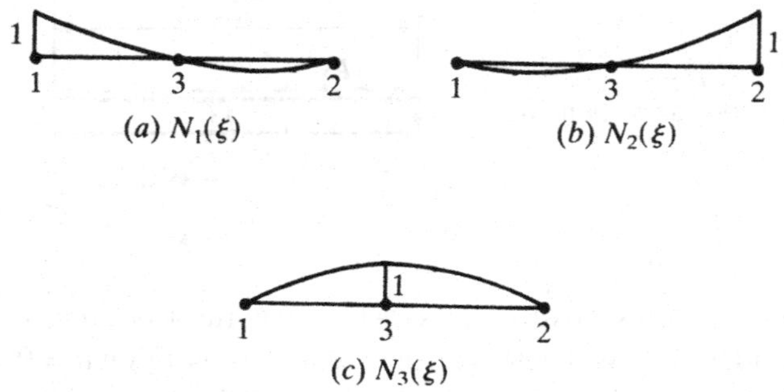

Figure 3.28. Quadratic displacement functions for an axial element.

Since the displacement u is the only degree of freedom at the nodes, then non-uniform rods, such as the one illustrated in Figure 3.10, can be analysed with the quadratic, three-node element without any complications.

This technique of increasing the number of nodes in an element can also be used for analysing shafts in torsion. Results for clamped-free shafts using 2, 3 and 4 node elements are given in reference [3.19]. The bending vibration of slender, clamped-free beams are also analysed using a three node element. The degrees of freedom at each node are v and θ_z. Using only one element, the errors for the first three frequencies are 0.0, 0.56 and 2.68%. The corresponding errors obtained when using two elements with two nodes each are 0.04, 0.85 and 21.82%. Note that this increase in accuracy, obtained with the three node element, has been achieved without any increase in the total number of degrees of freedom.

The displacement functions for elements requiring continuity of the dependent variables, but not their derivatives, can be constructed using Lagrange interpolation

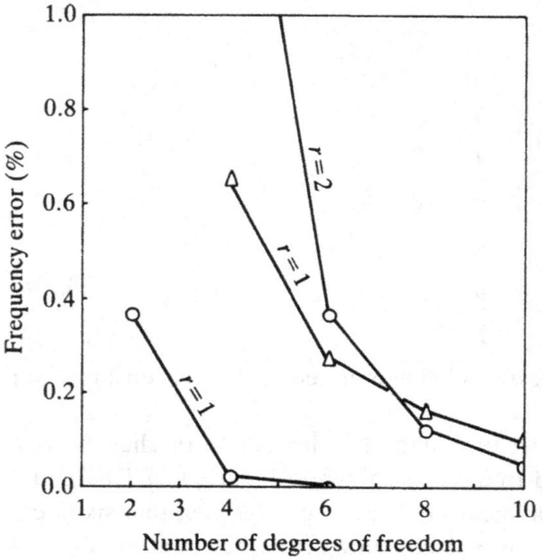

Figure 3.29. Axial vibration of clamped-free rod. Δ——Δ Linear element; ○——○ quadratic element.

functions [3.20, 3.21]. The mth order Lagrange interpolation function, $l_j^m(\xi)$, is defined as:

$$l_j^m(\xi) = \frac{(\xi - \xi_1)\cdots(\xi - \xi_{j-1})(\xi - \xi_{j+1})\cdots(\xi - \xi_{m+1})}{(\xi_j - \xi_1)\cdots(\xi_j - \xi_{j-1})(\xi_j - \xi_{j+1})\cdots(\xi_j - \xi_{m+1})} \tag{3.219}$$

This function has the following properties:

$$l_j^m(\xi_k) = \begin{cases} 1 & \text{for } k = j \\ 0 & \text{for } k \neq j \end{cases} \tag{3.220}$$

For an element with $(m+1)$ nodes, whose coordinates are $\xi_1, \xi_2 \ldots, \xi_{m+1}$, the function $l_j^m(\xi)$ has the same properties as the displacement function corresponding to node j.

To illustrate this procedure, consider the axial element with two nodes ($\xi_1 = -1$, $\xi_2 = +1$) as shown in Figure 3.6.

$$\begin{aligned} l_1^1(\xi) &= \frac{(\xi - \xi_2)}{(\xi_1 - \xi_2)} = \frac{(\xi - 1)}{(-2)} = \frac{1}{2}(1 - \xi) \\ l_2^1(\xi) &= \frac{(\xi - \xi_1)}{(\xi_2 - \xi_1)} = \frac{(\xi + 1)}{2} = \frac{1}{2}(1 + \xi) \end{aligned} \tag{3.221}$$

These agree with the functions defined in (3.51).

For the axial element with three nodes ($\xi_1 = -1$, $\xi_2 = +1$, $\xi_3 = 0$), as shown in Figure 3.27, the interpolation functions become:

$$\begin{aligned} l_1^2(\xi) &= \frac{(\xi - \xi_2)(\xi - \xi_3)}{(\xi_1 - \xi_2)(\xi_1 - \xi_3)} = \frac{(\xi - 1)\xi}{(-2)(-1)} = -\frac{1}{2}\xi(1 - \xi) \\ l_2^2(\xi) &= \frac{(\xi - \xi_1)(\xi - \xi_3)}{(\xi_2 - \xi_1)(\xi_2 - \xi_3)} = \frac{(\xi + 1)}{2(1)} = \frac{1}{2}\xi(1 + \xi) \\ l_3^2(\xi) &= \frac{(\xi - \xi_1)(\xi - \xi_2)}{(\xi_3 - \xi_1)(\xi_3 - \xi_2)} = \frac{(\xi + 1)(\xi - 1)}{1(-1)} = (1 - \xi^2) \end{aligned} \tag{3.222}$$

These expressions are identical to the ones defined in (3.215).

The displacement functions for elements requiring continuity of both the dependent variables and their first derivative can be constructed using Hermitian, osculating interpolation functions. Details are given in references [3.20, 3.21].

3.9 Shear Deformation and Rotary Inertia Effects

Section 2.4 indicates that shear deformation and rotary inertia effects become important when analysing deep beams at low frequencies or slender beams at high

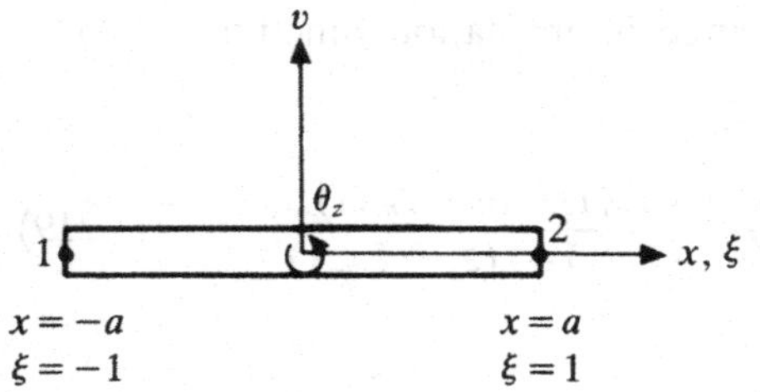

Figure 3.30. Geometry of a single beam element.

frequencies. In this case the appropriate energy expressions for the element shown in Figure 3.30 are

$$T_e = \frac{1}{2}\int_{-a}^{+a} \rho A \dot{v}^2 \,\mathrm{d}x + \frac{1}{2}\int_{-a}^{+a} \rho I_z \dot{\theta}_z^2 \,\mathrm{d}x \tag{3.223}$$

$$U_e = \frac{1}{2}\int_{-a}^{+a} EI_z \left(\frac{\partial \theta_z}{\partial x}\right)^2 \mathrm{d}x + \frac{1}{2}\int_{-a}^{+a} \kappa AG \left(\frac{\partial v}{\partial x} - \theta_z\right)^2 \mathrm{d}x \tag{3.224}$$

$$\delta W_e = \int_{-a}^{+a} p_y \delta v \,\mathrm{d}x \tag{3.225}$$

In this section two methods of deriving suitable displacement functions are presented. The first one involves solving the equations of static equilibrium. The second method uses assumed polynomials with the correct number of terms.

The highest derivative, of both v and θ_z, appearing in the energy expressions is the first. Therefore, v and θ_z are the only degrees of freedom required at the node points.

The equations of static equilibrium for a beam, including shear deformation effects, are

$$\kappa AG \frac{\mathrm{d}^2 v}{\mathrm{d}x^2} - \kappa AG \frac{\mathrm{d}\theta_z}{\mathrm{d}x} = 0 \tag{3.226}$$

$$\kappa AG \frac{\mathrm{d}v}{\mathrm{d}x} + EI_z \frac{\mathrm{d}^2 \theta_z}{\mathrm{d}x^2} - \kappa AG \theta_z = 0 \tag{3.227}$$

These can be derived using the method illustrated in Section 2.11, assuming no time variation.

Eliminating θ_z and v in turn gives

$$\frac{\mathrm{d}^4 v}{\mathrm{d}x^4} = 0 \quad \text{and} \quad \frac{\mathrm{d}^3 \theta_z}{\mathrm{d}x^3} = 0 \tag{3.228}$$

respectively. Changing to the ξ coordinate ($\xi = x/a$) yields

$$\frac{\mathrm{d}^4 v}{\mathrm{d}\xi^4} = 0 \quad \text{and} \quad \frac{\mathrm{d}^3 \theta_z}{\mathrm{d}\xi^3} = 0 \tag{3.229}$$

The general solutions of these two equations are

$$v = a_1 + a_2\xi + a_3\xi^2 + a_4\xi^3 \tag{3.230}$$

$$\theta_z = b_1 + b_2\xi + b_3\xi^2 \tag{3.231}$$

The seven constants of integration are not independent since the solutions (3.230) and (3.231) must also satisfy equation (3.227), which represents moment equilibrium. This gives rise to the following relationships:

$$b_1 = \frac{1}{a}a_2 + \frac{6\beta}{a}a_4, \qquad b_2 = \frac{2}{a}a_3, \qquad b_3 = \frac{3}{a}a_4 \tag{3.232}$$

where

$$\beta = \frac{EI_z}{\kappa AGa^2} \tag{3.233}$$

This leaves only four independent constants which can be determined by evaluating (3.230) and (3.231) at $\xi = \pm 1$. The resulting displacement functions are:

$$\begin{aligned} v &= \lfloor N_1(\xi) \quad aN_2(\xi) \quad N_3(\xi) \quad aN_4(\xi) \rfloor \{\mathbf{v}\}_e \\ \theta_z &= \left[\frac{1}{a}N_5(\xi) \quad N_6(\xi) \quad \frac{1}{a}N_7(\xi) \quad N_8(\xi)\right] \{\mathbf{v}\}_e \end{aligned} \tag{3.234}$$

where

$$\{\mathbf{v}\}_e^{\mathrm{T}} = \lfloor v_1 \quad \theta_{z1} \quad v_2 \quad \theta_{z2} \rfloor \tag{3.235}$$

and

$$\begin{aligned}
N_1(\xi) &= \frac{1}{4(1+3\beta)}\{2 + 6\beta - 3(1+2\beta)\xi + \xi^3\} \\
N_2(\xi) &= \frac{1}{4(1+3\beta)}\{1 + 3\beta - \xi - (1+3\beta)\xi^2 + \xi^3\} \\
N_3(\xi) &= \frac{1}{4(1+3\beta)}\{2 + 6\beta + 3(1+2\beta)\xi - \xi^3\} \\
N_4(\xi) &= \frac{1}{4(1+3\beta)}\{-(1+3\beta) - \xi + (1+3\beta)\xi^2 + \xi^3\} \\
N_5(\xi) &= \frac{1}{4(1+3\beta)}(-3 + 3\xi^2) \\
N_6(\xi) &= \frac{1}{4(1+3\beta)}\{-1 + 6\beta - (2+6\beta)\xi + 3\xi^2\} \\
N_7(\xi) &= \frac{1}{4(1+3\beta)}(3 - 3\xi^2) \\
N_8(\xi) &= \frac{1}{4(1+3\beta)}\{-1 + 6\beta + (2+6\beta)\xi + 3\xi^2\}
\end{aligned} \tag{3.236}$$

Note that in the case of a slender beam, when $\beta = 0$, the functions $N_1(\xi)$ to $N_4(\xi)$ reduce to the functions given in (3.126) and $N_5(\xi)$ to $N_8(\xi)$ are such that $\theta_z = \partial v/\partial x$, as required.

Substituting the displacement functions (3.234) into the energy expressions (3.223) to (3.225) gives the following element matrices.

$$[\mathbf{m}]_e = \frac{\rho A a}{210(1+3\beta)^2}\begin{bmatrix} m_1 & & & \text{Sym} \\ m_2 & m_5 & & \\ m_3 & -m_4 & m_1 & \\ m_4 & m_6 & -m_2 & m_5 \end{bmatrix} + \frac{\rho I_z}{30a(1+3\beta)^2}\begin{bmatrix} m_7 & & & \text{Sym} \\ m_8 & m_9 & & \\ -m_7 & -m_8 & m_7 & \\ m_8 & m_{10} & -m_8 & m_9 \end{bmatrix} \tag{3.237}$$

where

$$\begin{aligned} m_1 &= 156 + 882\beta + 1260\beta^2 \\ m_2 &= (44 + 231\beta + 315\beta^2)a \\ m_3 &= 54 + 378\beta + 630\beta^2 \\ m_4 &= (-26 - 189\beta - 315\beta^2)a \\ m_5 &= (16 + 84\beta + 126\beta^2)a^2 \\ m_6 &= (-12 - 84\beta - 126\beta^2)a^2 \\ m_7 &= 18 \\ m_8 &= (3 - 45\beta)a \\ m_9 &= (8 + 30\beta + 180\beta^2)a^2 \\ m_{10} &= (-2 - 30\beta + 90\beta^2)a^2 \end{aligned} \tag{3.238}$$

$$[\mathbf{k}]_e = \frac{EI_z}{2a^3(1+3\beta)}\begin{bmatrix} 3 & & & \text{Sym} \\ 3a & (4+3\beta)a^2 & & \\ -3 & -3a & 3 & \\ 3a & (2-3\beta)a^2 & -3a & (4+3\beta)a^2 \end{bmatrix} \tag{3.239}$$

and

$$\{\mathbf{f}\}_e = p_y^e \frac{a}{3}\begin{bmatrix} 3 \\ a \\ 3 \\ -a \end{bmatrix} \tag{3.240}$$

for a constant value of p_y over the element.

There are a number of ways of deriving the element matrices. A survey of the various methods, together with an indication of their equivalence is given in reference [3.22]. More recent derivations of the element matrices are described in references [3.23, 3.24].

Note that (3.240) is identical to the corresponding expression for a slender beam, (3.138). Also, when $\beta = 0$, the matrices (3.237) and (3.239) reduce to the ones given in (3.132) and (3.135) respectively.

EXAMPLE 3.10 Use the finite element displacement method to estimate the lower frequencies of a cantilever beam having the following properties:

$$r_z/L = 0.08, \qquad \kappa = 2/3, \qquad E/G = 8/3$$

Compare the results with the exact solution.

For a deep beam, the equations of free vibration can be derived using the technique illustrated in Section 2.11 (see Problem 2.7). These are:

$$\begin{aligned} &\kappa AG\frac{\partial^2 v}{\partial x^2} - \rho A\frac{\partial^2 v}{\partial t^2} - \kappa AG\frac{\partial \theta_z}{\partial x} = 0 \\ &\kappa AG\frac{\partial v}{\partial x} + EI_z\frac{\partial^2 \theta_z}{\partial x^2} - \kappa AG\theta_z - \rho I_z\frac{\partial^2 \theta}{\partial t^2} = 0 \end{aligned} \tag{3.241}$$

and the boundary conditions for a cantilever are

$$\begin{aligned} &v(0,t) = 0, \qquad \theta_z(0,t) = 0 \\ &\frac{\partial \theta_z}{\partial x}(L,t) = 0, \qquad \frac{\partial v}{\partial x}(L,t) - \theta_z(L,t) = 0 \end{aligned} \tag{3.242}$$

The solutions of equations (3.241) subject to the boundary conditions (2.242) are given in references [3.25, 3.26].

One element solution

Representing a beam, of length L, by one element means that $a = L/2$ and

$$\beta = \frac{E}{\kappa G}\left(\frac{r_z}{a}\right)^2 = \frac{4}{\kappa}\frac{E}{G}\left(\frac{r_z}{L}\right)^2 = 0.1024$$

Imposing the conditions that $v_1 = \theta_{z1} = 0$, the equations of motion are obtained using equations (3.238) and (3.239). This gives

$$\left[\frac{EI_z}{L_3}\frac{4}{1.3072}\begin{bmatrix} 3 & -1.5L \\ -1.5L & 1.0768L^2 \end{bmatrix}\right.$$

$$\left.-\omega^2\frac{\rho AL}{420(1.3072)^2}\begin{bmatrix} 262.7545 & -35.3346L \\ -35.3346L & -7.0613L \end{bmatrix}\right]\begin{bmatrix} v_2 \\ \theta_{z2} \end{bmatrix} = 0$$

Letting $\omega^2\rho AL^4/2196.096EI_z = \lambda$, this equation simplifies to

$$\begin{bmatrix} (3 - 262.7545\lambda) & -(1.5 - 35.3346\lambda) \\ -(1.5 - 35.3346\lambda) & (1.0768 - 7.0613\lambda) \end{bmatrix}\begin{bmatrix} v_2 \\ L\theta_{z2} \end{bmatrix} = 0$$

The eigenvalues of this equation are given by the roots of the equation

$$606.8544\lambda^2 - 198.1141\lambda + 0.9804 = 0$$

which are

$$\lambda = 0.0050260 \quad \text{and} \quad 0.3214346$$

The natural frequencies are

$$\omega_1 = (2196.096\lambda_1)^{1/2}\left(\frac{EI_z}{\rho AL^4}\right)^{1/2} = 3.322\left(\frac{EI_z}{\rho AL^4}\right)^{1/2}$$

Table 3.6. *Effect of changing the slenderness ratio on the accuracy of the frequencies of a cantilever beam*

	$r_z/L = 0.08$		$r_z/L = 0.02$	
Mode	$\lambda^{1/2}$ Exact	% Error	$\lambda^{1/2}$ Exact	% Error
1	3.284	0.12	3.500	0.02
2	15.488	2.47	21.353	0.59
3	34.301	8.80	57.474	2.37

and

$$\omega_2 = (2196.096\lambda_2)^{1/2}\left(\frac{EI_z}{\rho AL^4}\right)^{1/2} = 26.569\left(\frac{EI_z}{\rho AL^4}\right)^{1/2}$$

The exact values of the coefficient for these two frequencies are 3.284 and 15.488. The errors produced by a one element solution are therefore 1.16 and 71.54%.

Repeating the analysis using two, three and four elements gives the errors shown in Figure 3.31 when compared with the exact solution.

The effect of changing the slenderness ratio, r_z/L, on the accuracy of the solutions is indicated in Table 3.6. The results have been obtained using three elements and $\lambda = \rho AL^4\omega^2/EI_z$. The values of κ and E/G remain unchanged.

It can be seen that the accuracy of the predicted frequencies increases as the slenderness ratio decreases. Equation (3.233) shows that as r_z/L decreases β decreases also.

In order to increase the accuracy of the element, reference [3.27] represents the lateral displacement by a quintic polynomial and the cross-sectional rotation by a

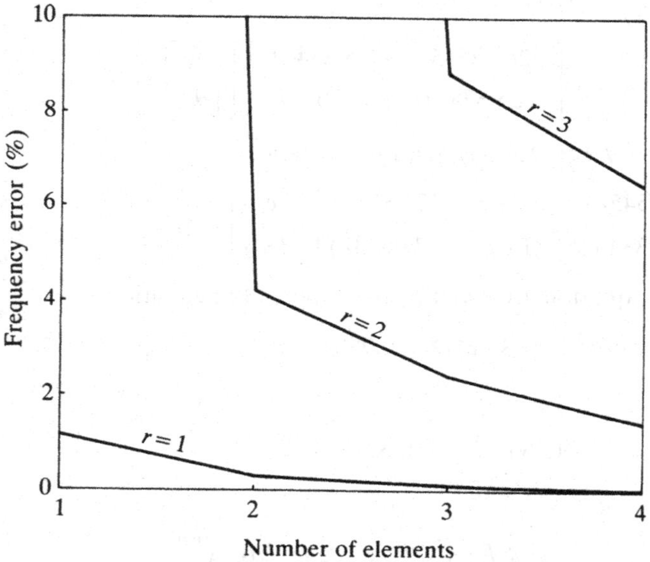

Figure 3.31. Flexural vibration of a deep cantilever beam.

quartic as follows:

$$v = a_1 + a_2\xi + a_3\xi^2 + a_4\xi^3 + a_5\xi^4 + a_6\xi^5 \tag{3.243}$$

$$\theta_z = b_1 + b_2\xi + b_3\xi^2 + b_4\xi^3 + b_5\xi^4 \tag{3.244}$$

The eleven coefficients are not independent since the expressions (3.243) and (3.244) are required to satisfy the equation of static, moment equilibrium (3.227). This gives

$$\begin{aligned}
b_1 &= \frac{1}{a}a_2 + \frac{6\beta}{a}a_4 + \frac{120\beta^2}{a}a_6 \\
b_2 &= \frac{2}{a}a_3 + \frac{24\beta}{a}a_5 \\
b_3 &= \frac{3}{a}a_4 + \frac{60\beta}{a}a_6 \\
b_4 &= \frac{4}{a}a_5 \\
b_5 &= \frac{5}{a}a_6
\end{aligned} \tag{3.245}$$

Only six of the constants are independent. They are determined by evaluating (3.243) and (3.244) at $\xi = \mp 1, 0$. The element has, therefore, three nodes with two degrees of freedom at each node.

Using only one element to represent a cantilever beam, for which $r_z/L = 0.08$, the errors for the first two modes are less than 2%. This element is much more accurate than the previous one which produces an error of 71.54% for the second frequency.

When $\beta = 0$, the element reduces to the three node, slender beam element of reference [3.19]. In this case the errors for the first two modes of a cantilever represented by one element are less than 1%.

A simpler technique, which is used extensively for thick plates and shells, is to use independent functions for v and θ_z. The number of terms in each function is therefore equal to the number of nodes, since v and θ_z are the only degrees of freedom at each node. Reference [3.19] presents a four node deep beam element. Both v and θ_z are represented by cubic polynomials. The element displacement functions can easily be derived using the Lagrange interpolation functions presented in Section 3.8. A one element solution for a cantilever with $r_z/L = 0.05$ gives errors of 0.17% and 18.15% for the first two modes. Although not as accurate as the previous element, in spite of an increased number of degrees of freedom, convergence is very rapid with an increase in the number of elements. The main disadvantage with this approach is that the elements of the stiffness matrix increase as β decreases and a slender beam element is not obtained when $\beta = 0$. This phenomenon is known as shear locking. Methods of overcoming this are presented in the next section.

3.10 Numerical Integration

Exact integration of the expressions for the inertia, stiffness and load matrices is often tedious, time consuming and prone to human error. In some instances, as can

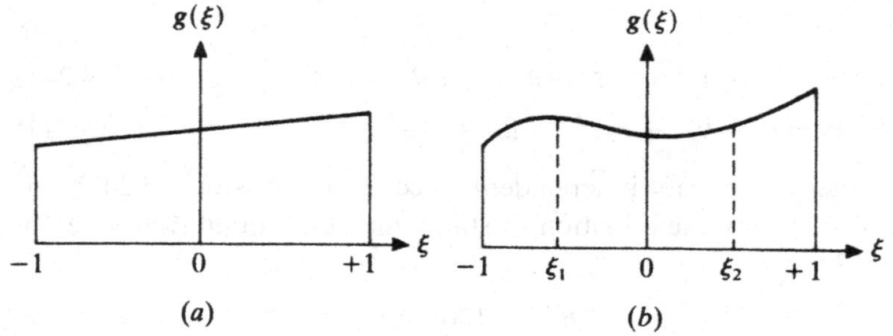

Figure 3.32. Numerical integration of (*a*) a linear function, (*b*) a cubic function.

be seen in the following chapters, it is impossible to carry out the integration exactly. These difficulties are overcome by using numerical integration. There are a number of techniques available, but only the Gauss–Legendre method [3.20, 3.21] will be discussed here.

The integral of a function $g(\xi)$ can be evaluated using the formula

$$\int_{-1}^{+1} g(\xi)\,d\xi = \sum_{j=1}^{n} H_j g(\xi_j) \tag{3.246}$$

where the H_j are weight coefficients and the ξ_j are sampling points. If the positions of the sampling points are located so as to achieve the best accuracy, then it is easy to see that a polynomial of degree $(2n - 1)$ will be integrated exactly by a suitable choice of n sampling points and n weight coefficients.

Consider first a linear function as shown in Figure 3.32(*a*). This is to be integrated by means of the formula

$$\int_{-1}^{+1} g(\xi)\,d\xi = H_1 g(\xi_1) \tag{3.247}$$

The function can be represented by

$$g(\xi) = a_1 + a_2\xi \tag{3.248}$$

The exact integral of this is

$$\int_{-1}^{+1} g(\xi)\,d\xi = 2a_1 \tag{3.249}$$

Therefore, it can be seen that taking $\xi_1 = 0$ and $H_1 = 2$ in (3.247) will give the exact value of the integral.

Now consider a cubic function as shown in Figure 3.32(*b*). This is to be integrated using $n = 2$ in equation (3.246).

A linear function, $G(\xi)$, can be constructed, using Lagrange interpolation functions, to coincide with $g(\xi)$ at the sampling points ξ_1 and ξ_2, namely,

$$G(\xi) = \frac{(\xi_2 - \xi)}{(\xi_2 - \xi_1)} g(\xi_1) + \frac{(\xi - \xi_1)}{(\xi_2 - \xi_1)} g(\xi_2) \tag{3.250}$$

The function $g(\xi)$ can therefore be expressed in the form

$$g(\xi) = G(\xi) + p_2(\xi)(b_1 + b_2\xi) \tag{3.251}$$

where $p_2(\xi)$ is a quadratic polynomial whose roots are ξ_1 and ξ_2. The integrals of the functions $g(\xi)$ and $G(\xi)$ will be equal if

$$\int_{-1}^{+1} p_2(\xi)(b_1 + b_2\xi)\, d\xi = 0 \tag{3.252}$$

This equation can be used to determine $p_2(\xi)$ and hence the positions of the sampling points ξ_1 and ξ_2. These will be independent of b_1 and b_2 if

$$\int_{-1}^{+1} p_2(\xi)\, d\xi = 0 \quad \text{and} \quad \int_{-1}^{+1} p_2(\xi)\xi\, d\xi = 0 \tag{3.253}$$

Letting

$$p_2(\xi) = a_1 + a_2\xi + \xi^2$$

and substituting into (3.253) gives

$$2a_1 + \tfrac{2}{3} = 0, \qquad \tfrac{2}{3}a_2 = 0,$$

giving

$$p_2(\xi) = \xi^2 - \tfrac{1}{3} \tag{3.254}$$

The roots of $p_2(\xi) = 0$ are therefore $\xi_1 = -1/3^{1/2}$, $\xi_2 = +1/3^{1/2}$. The weight coefficients are given by

$$H_1 = \int_{-1}^{+1} \frac{(\xi_2 - \xi)}{(\xi_2 - \xi_1)}\, d\xi = \frac{2\xi_2}{(\xi_2 - \xi_1)} = 1$$

and

$$H_2 = \int_{-1}^{+1} \frac{(\xi - \xi_1)}{(\xi_2 - \xi_1)}\, d\xi = \frac{-2\xi_2}{(\xi_2 - \xi_1)} = 1 \tag{3.255}$$

The validity of the above result can be checked by considering the function:

$$g(\xi) = c_1 + c_2\xi + c_3\xi^2 + c_4\xi^3 \tag{3.256}$$

The exact integral of this function is

$$\begin{aligned} \int_{-1}^{+1} g(\xi) d\xi &= \left[c_1\xi + \tfrac{1}{2}c_2\xi^2 + \tfrac{1}{3}c_3\xi^3 + \tfrac{1}{4}c_4\xi^4\right]_{-1}^{+1} \\ &= 2c_1 + \tfrac{2}{3}c_3 \end{aligned} \tag{3.257}$$

Table 3.7. *Integration points and weight coefficients for the Gauss integration formula*

n	$\mp\xi_j$	H_j
1	0	2
2	$1/3^{1/2}$	1
3	$(0.6)^{1/2}$	5/9
	0	8/9
4	$\left[\frac{3+(4.8)^{1/2}}{7}\right]^{1/2}$	$\left[\frac{1}{2}-\frac{(30)^{1/2}}{36}\right]$
	$\left[\frac{3-(4.8)^{1/2}}{7}\right]^{1/2}$	$\left[\frac{1}{2}+\frac{(30)^{1/2}}{36}\right]$

Using the formula (3.246) the value of the integral is

$$\begin{aligned}\int_{-1}^{+1} g(\xi)\,\mathrm{d}\xi &= \left(c_1 - c_2\frac{1}{3^{1/2}} + c_3\frac{1}{3} - c_4\frac{1}{3(3^{1/2})}\right)\\ &\quad + \left(c_1 + c_2\frac{1}{3^{1/2}} + c_3\frac{1}{3} + c_4\frac{1}{3(3^{1/2})}\right)\\ &= 2c_1 + \tfrac{2}{3}c_3.\end{aligned} \tag{3.258}$$

The above approach can be used for other values of n. In general, n sampling points are given by the roots of the Legendre polynomial of degree n, $P_n(\xi)$, which can be generated by means of the relation

$$P_{j+1}(\xi) = \frac{1}{(j+1)}\{(2j+1)\xi P_j(\xi) - jP_{j-1}(\xi)\} \qquad (j = 1, 2, \ldots, n-1) \tag{3.259}$$

with

$$P_0(\xi) = 1, \qquad P_1(\xi) = \xi. \tag{3.260}$$

The corresponding weight coefficients are given by the integrals of the n Lagrange interpolation functions of order $(n-1)$. Table 3.7 gives the positions of the sampling points, which are usually referred to as integration points, and the corresponding weight coefficients for $n = 1$ to 4. The integration points are positioned symmetrically about $\xi = 0$ and so only the numerical values are given.

EXAMPLE 3.11 Use numerical integration to derive the element matrices for the two node axial element described in Section 3.3.

The inertia matrix is given by equation (3.59), namely

$$[\mathbf{m}]_e = \rho A a \int_{-1}^{+1} \begin{bmatrix} \frac{1}{2}(1-\xi) \\ \frac{1}{2}(1+\xi) \end{bmatrix} \begin{bmatrix} \frac{1}{2}(1-\xi) & \frac{1}{2}(1+\xi) \end{bmatrix} \mathrm{d}\xi \tag{3.261}$$

The integrand is a quadratic polynomial and so it is necessary to use two integration points. These are positioned at $\xi_1 = -1/3^{1/2}$, $\xi_2 = +1/3^{1/2}$ both with a weight coefficient of 1. Therefore

$$[\mathbf{m}]_e = \rho A a \begin{bmatrix} (1+1/3^{1/2})/2 \\ (1-1/3^{1/2})/2 \end{bmatrix} [(1+1/3^{1/2})/2 \quad (1-1/3^{1/2})/2]$$

$$+ \rho A a \begin{bmatrix} (1-1/3^{1/2})/2 \\ (1+1/3^{1/2})/2 \end{bmatrix} [(1-1/3^{1/2})/2 \quad (1+1/3^{1/2})/2] \quad (3.262)$$

$$= \rho A a \begin{bmatrix} \frac{2}{3} & \frac{1}{3} \\ \frac{1}{3} & \frac{2}{3} \end{bmatrix}$$

The stiffness matrix is given by equation (3.63), namely

$$[\mathbf{k}]_e = \frac{EA}{a} \int_{-1}^{+1} \begin{bmatrix} -\frac{1}{2} \\ +\frac{1}{2} \end{bmatrix} [-\tfrac{1}{2} \quad +\tfrac{1}{2}] \, d\xi \quad (3.263)$$

The integrand is constant and so only one integration point is necessary. This is located at $\xi_1 = 0$ and has a weight coefficient of 2. Therefore

$$[\mathbf{k}]_e = \frac{EA}{a} 2 \begin{bmatrix} -\frac{1}{2} \\ +\frac{1}{2} \end{bmatrix} [-\tfrac{1}{2} \quad +\tfrac{1}{2}]$$

$$= \frac{EA}{a} \begin{bmatrix} \frac{1}{2} & -\frac{1}{2} \\ -\frac{1}{2} & \frac{1}{2} \end{bmatrix} \quad (3.264)$$

The load matrix is given by equation (3.67), namely

$$\{\mathbf{f}\}_e = p_x^e a \int_{-1}^{+1} \begin{bmatrix} \frac{1}{2}(1-\xi) \\ \frac{1}{2}(1+\xi) \end{bmatrix} d\xi \quad (3.265)$$

In this case the integrand is a linear function which can also be integrated using one integration point. This gives

$$\{\mathbf{f}\}_e = p_x^e a 2 \begin{bmatrix} \frac{1}{2} \\ \frac{1}{2} \end{bmatrix} = p_x^e a \begin{bmatrix} 1 \\ 1 \end{bmatrix} \quad (3.266)$$

The use of too many integration points does not affect the result. For example, if the stiffness matrix in the example above is evaluated using two integration points instead of one, then

$$[\mathbf{k}]_e = \frac{EA}{a} \begin{bmatrix} -\frac{1}{2} \\ +\frac{1}{2} \end{bmatrix} [-\tfrac{1}{2} \quad \tfrac{1}{2}] + \frac{EA}{a} \begin{bmatrix} -\frac{1}{2} \\ +\frac{1}{2} \end{bmatrix} [-\tfrac{1}{2} \quad \tfrac{1}{2}]$$

$$= \frac{EA}{a} \begin{bmatrix} \frac{1}{2} & -\frac{1}{2} \\ -\frac{1}{2} & \frac{1}{2} \end{bmatrix} \quad (3.267)$$

as before. However, the use of too few integration points should be avoided. To illustrate this, consider the stiffness matrix for a slender beam element as given by equation (3.134). The row matrix $\lfloor \mathrm{N}''(\xi) \rfloor$ is linear and so two integration points are required to evaluate the stiffness matrix exactly. If, however, only one integration point was used then the resulting stiffness matrix would be

$$[\mathbf{k}]_e = \frac{EI_z}{a^3}2\begin{bmatrix} 0 \\ -a/2 \\ 0 \\ +a/2 \end{bmatrix}[0 \quad -a/2 \quad 0 \quad +a/2]$$
$$= \frac{EI_z}{2a}\begin{bmatrix} 0 & 0 & 0 & 0 \\ 0 & 1 & 0 & -1 \\ 0 & 0 & 0 & 0 \\ 0 & -1 & 0 & 1 \end{bmatrix} \tag{3.268}$$

This clearly does not agree with (3.135).

In general, when using numerical integration, the correct result will not be known beforehand. However, a check can be made on the resulting matrix to ensure that a sufficient number of integration points have been used. The stiffness matrix (3.268) should be positive semi-definite. This means that its eigenvalues should be either positive or zero. The number of zero eigenvalues should be equal to the number of rigid body displacements the element is capable of performing. In this case the number is two, a translation and a rotation. The eigenvalues, $\bar{\lambda}$, of the matrix (3.268) are given by the roots of the equation

$$\begin{bmatrix} -\bar{\lambda} & 0 & 0 & 0 \\ 0 & (1-\bar{\lambda}) & 0 & -1 \\ 0 & 0 & -\bar{\lambda} & 0 \\ 0 & -1 & 0 & (1-\bar{\lambda}) \end{bmatrix} = 0 \tag{3.269}$$

where $\bar{\lambda} = 2a\lambda/EI_z$. Expanding the determinant gives

$$\bar{\lambda}^3(\bar{\lambda} - 2) = 0 \tag{3.270}$$

Therefore, the stiffness matrix (3.268) has three zero eigenvalues and one positive eigenvalue, which indicates that it is incorrect.

Element inertia matrices should be positive definite. Hence, all their eigenvalues should be positive.

Section 3.3 demonstrates the fact that the finite element displacement method predicts the nodal displacements accurately, whilst the element stress distributions, which are calculated from the derivatives of the element displacement functions, are less accurate and discontinuous between elements. However, at certain points of the element, the stresses are more accurate than at any other point.

If the predicted stress distribution, $\tilde{\sigma}$, is given by a polynomial of order m, then it can be considered to be a least squares approximation to a more accurate distribution, σ, of order $(m + 1)$. Now any polynomial can be expressed as a linear combination of Legendre polynomials, and so

$$\sigma = \sum_{j=0}^{m+1} a_j P_j(\xi) \tag{3.271}$$

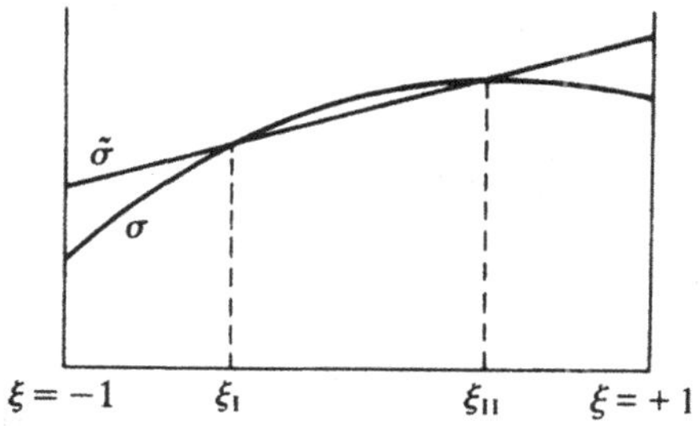

Figure 3.33. Linear approximation to a parabolic stress distribution.

Expressing the distribution, $\tilde{\sigma}$, in the form

$$\tilde{\sigma} = \sum_{j=0}^{m} b_j P_j(\xi) \tag{3.272}$$

the coefficients b_j are found from the condition that the integral

$$I = \int_{-a}^{+a} \left\{ \sigma - \sum_{j=0}^{m} b_j P_j(\xi) \right\}^2 \mathrm{d}\xi \tag{3.273}$$

is a minimum, that is

$$\frac{\partial I}{\partial b_j} = 0 \qquad j = 0, 1, \ldots, m \tag{3.274}$$

This gives

$$b_j = \frac{(2j+1)}{2} \int_{-1}^{+1} \sigma P_j(\xi)\, \mathrm{d}\xi \tag{3.275}$$

since

$$\int_{-1}^{+1} P_j(\xi) P_k(\xi)\, \mathrm{d}\xi = 0 \qquad \text{for } k \neq j \tag{3.276}$$

and

$$\int_{-1}^{+1} \{P_j(\xi)\}^2\, \mathrm{d}\xi = \frac{2}{(2j+1)} \tag{3.277}$$

Substituting (3.271) into (3.275) and using (3.276) and (3.277) gives

$$b_j = a_j \qquad j = 0, 1, \ldots, m \tag{3.278}$$

Therefore

$$\tilde{\sigma} = \sum_{j=0}^{m} a_j P_j(\xi) \tag{3.279}$$

Expressions (3.271) and (3.279) have the same value whenever

$$P_{m+1}(\xi) = 0 \tag{3.280}$$

This equation has $(m + 1)$ real roots. At these $(m + 1)$ points the accuracy of the distribution, $\tilde{\sigma}$, is higher than at all other points. This is illustrated for the case $m = 1$ in Figure 3.33.

When $m = 1$ equation (3.280) becomes

$$P_2(\xi) = \tfrac{1}{2}(3\xi^2 - 1) = 0 \tag{3.281}$$

and so in Figure 3.33

$$\xi_{\mathrm{I}} = -\frac{1}{3^{1/2}}, \qquad \xi_{\mathrm{II}} = +\frac{1}{3^{1/2}} \tag{3.282}$$

These points are the same as the integration points used in Gauss–Legendre integration. Thus an approximate stress distribution of order m has higher accuracy at $(m + 1)$ Gauss–Legendre integration points.

The axial element described in Section 3.3 gives a constant stress distribution (see Figure 3.11) and so $m = 0$ and the most accurate stress occurs at $\xi = 0$. The slender beam element of Section 3.5 produces a linear variation of stress which gives the best results at the points $\xi = \mp 1/3^{1/2}$.

The above statements strictly apply to uniform elements. In the case of non-uniform elements, if the size of the element is sufficiently small, then the variation of the element properties will be slight and the above technique can again be used. Further discussions of this technique are presented in references [3.28–3.31].

In Section 3.9 it is indicated that if independent functions are used for v and θ_z when developing a deep beam element, then problems arise if the element is used to analyse a slender beam problem. This is because the elements of the stiffness matrix increase without limit as the beam becomes more slender. This problem can be overcome by using numerical integration techniques.

The strain energy of a deep beam element (Section 2.4) is

$$U_e = \frac{1}{2}\int_{-a}^{+a} EI_z\left(\frac{\partial \theta_z}{\partial x}\right)^2 \mathrm{d}x + \frac{1}{2}\int_{-a}^{+a} \kappa AG\left(\frac{\partial v}{\partial x} - \theta_z\right)^2 \mathrm{d}x \tag{3.283}$$

The first term is the strain energy due to bending and the second term the strain energy due to shear.

Consider an element with two nodes, as shown in Figure 3.30, with v and θ_z as degrees of freedom at each node. If independent functions are used for v and θ_z, then each one will be linear, that is

$$v = \lfloor N_1(\xi) \quad N_2(\xi) \rfloor \begin{bmatrix} v_1 \\ v_2 \end{bmatrix}, \qquad \theta_z = \lfloor N_1(\xi) \quad N_2(\xi) \rfloor \begin{bmatrix} \theta_{z1} \\ \theta_{z2} \end{bmatrix} \tag{3.284}$$

where

$$N_1(\xi) = \tfrac{1}{2}(1 - \xi), \qquad N_2(\xi) = \tfrac{1}{2}(1 + \xi) \tag{3.285}$$

Therefore

$$\frac{\partial \theta_z}{\partial x} = \lfloor 0 \quad -1/2a \quad 0 \quad 1/2a \rfloor \{\mathbf{v}\}_e \tag{3.286}$$

and

$$\frac{\partial v}{\partial x} - \theta_z = \lfloor -1/2a \quad -(1-\xi)/2 \quad 1/2a \quad -(1+\xi)/2 \rfloor \{\mathbf{v}\}_e \tag{3.287}$$

where

$$\{\mathbf{v}\}_e^{\mathrm{T}} = \lfloor v_1 \quad \theta_{z1} \quad v_2 \quad \theta_{z2} \rfloor \tag{3.288}$$

These equations indicate that the bending strain is constant and the shear strain varies linearly. On substitution into the strain energy expression (3.283), the integrals can be evaluated exactly by using one integration point for the bending strain energy and two integration points for the shear strain energy. However, this procedure produces an element which is too stiff as already noted. The reason for this is that the presence of the linear term in the expression for shear strain places too much emphasis on the shear strain energy in comparison with the energy due to bending. This effect increases if the beam is slender, which is just the opposite of the true situation since the shear strain in a slender beam is negligible. To overcome this difficulty the linear shear strain distribution is replaced by a constant one using a least squares fit procedure. These two distributions coincide at the single integration point $\xi = 0$, and so the constant value of shear strain is obtained by evaluating the linear variation at this point. Since the shear strain is now constant, the integral for the shear strain energy can be evaluated using one integration point. Thus a better element can be obtained by evaluating both integrals in (3.283) by means of one integration point.

Substituting the displacement functions (3.284) into the strain energy expression (3.283) gives [3.32]

$$U_e = \tfrac{1}{2}\{\mathbf{v}\}_e^{\mathrm{T}}[\mathbf{k}]_e\{\mathbf{v}\}_e \tag{3.289}$$

where

$$[\mathbf{k}]_e = [\mathbf{k}]_{\mathrm{b}} + [\mathbf{k}]_{\mathrm{s}} \tag{3.290}$$

Using one integration point gives the exact value of $[\mathbf{k}]_{\mathrm{b}}$, namely

$$[\mathbf{k}]_{\mathrm{b}} = \frac{EI_z}{2a^3}\begin{bmatrix} 0 & 0 & 0 & 0 \\ 0 & a^2 & 0 & -a^2 \\ 0 & 0 & 0 & 0 \\ 0 & -a^2 & 0 & a^2 \end{bmatrix} \tag{3.291}$$

Exact integration of the shear strain energy gives

$$[\mathbf{k}]_{\mathrm{s}} = \frac{EI_z}{2a^3\beta}\begin{bmatrix} 1 & a & -1 & a \\ a & 4a^2/3 & -a & 2a^2/3 \\ -1 & -a & 1 & -a \\ a & 2a^2/3 & -a & 4a^2/3 \end{bmatrix} \tag{3.292}$$

where $\beta = EI_z/\kappa AGa^2$.

Using one integration point produces the following result

$$[\mathbf{k}]_{\mathrm{s}} = \frac{EI_z}{2a^3\beta}\begin{bmatrix} 1 & a & -1 & a \\ a & a^2 & -a & a^2 \\ -1 & -a & 1 & -a \\ a & a^2 & -a & a^2 \end{bmatrix} \tag{3.293}$$

Combining these results shows that exact integration produces the following stiffness matrix

$$[\mathbf{k}]_e = \frac{EI_z}{2a^3(3\beta)}\begin{bmatrix} 3 & & & \text{Sym} \\ 3a & (4+3\beta)a^2 & & \\ -3 & -3a & 3 & \\ 3a & (2-3\beta)a^2 & -3a & (4+3\beta)a^2 \end{bmatrix} \qquad (3.294)$$

and that reduced integration (using one point) produces the following result

$$[\mathbf{k}]_e = \frac{EI_z}{2a^3(3\beta)}\begin{bmatrix} 3 & & & \text{Sym} \\ 3a & (3+3\beta)a^2 & & \\ -3 & -3a & 3 & \\ 3a & (3-3\beta)a^2 & -3a & (3+3\beta)a^2 \end{bmatrix} \qquad (3.295)$$

Both these results should be compared with the one given by equation (3.239).

EXAMPLE 3.12 Calculate the tip displacement of a cantilever beam subject to a static tip load. Use both one and two point integration and compare the results with the exact solution for both a deep and a slender beam having the following properties:

$$\frac{r_z}{L} = \frac{1}{8(3)^{1/2}}, \qquad \kappa = \frac{5}{6}, \qquad \frac{E}{G} = \frac{8}{3} \quad \text{and} \quad \frac{8}{3}\times 10^{-5}$$

The exact solution can be obtained by using one element and the stiffness matrix given by equation (3.239). In the case of static analysis the inertia and damping forces are zero and so the application of Lagrange's equations results in the following equation

$$\frac{EI_z}{2a^3(1+3\beta)}\begin{bmatrix} 3 & -3a \\ -3a & (4+3\beta)a^2 \end{bmatrix}\begin{bmatrix} v \\ \theta_z \end{bmatrix} = \begin{bmatrix} P \\ 0 \end{bmatrix}$$

where P is the tip load. Solving for v gives

$$v_{\text{ex}} = \frac{2a^3}{3EI_z}(4+3\beta)P.$$

Using one element and the stiffness matrices given by equations (3.294) and (3.295) results in the following solutions

$$v_2 = \frac{3\beta}{(1+3\beta)}v_{\text{ex}}, \qquad v_1 = \frac{(3+3\beta)}{(4+3\beta)}v_{\text{ex}}$$

Representing the beam by one element means that $a = L/2$ and so

$$\beta = \frac{EI_z}{\kappa AGa^2} = \frac{E}{\kappa G}\left(\frac{r_z}{a}\right)^2 = \frac{4}{\kappa}\frac{E}{G}\left(\frac{r_z}{L}\right)^2$$

$$= \tfrac{1}{15}(\text{deep beam}) \quad \text{or} \quad \frac{10^{-5}}{15}(\text{slender beam}).$$

Taking $\beta = 1/15$ gives

$$\frac{v_2}{v_{\text{ex}}} = 0.1667 \qquad \frac{v_1}{v_{\text{ex}}} = 0.7619$$

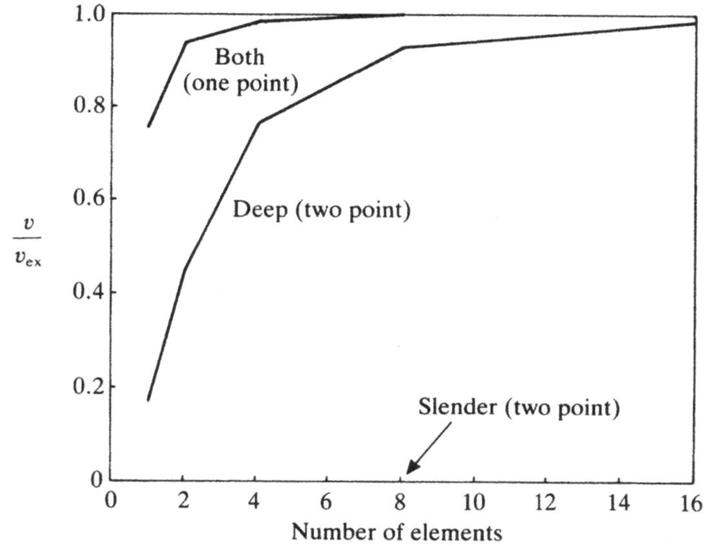

Figure 3.34. Static deflection of cantilevers using both 1 and 2 point integration [3.32].

Taking $\beta = 10^{-5}/15$ gives

$$\frac{v_2}{v_{\text{ex}}} = 0.2 \times 10^{-5} \qquad \frac{v_1}{v_{\text{ex}}} = 0.750$$

It can be seen that the results which have been obtained using a single integration point are more accurate than those obtained with two.

Repeating the analysis for an increasing number of elements gives the results shown in Figure 3.34. It can be seen that exact, two point integration gives reasonable results for a deep beam, but for a slender beam the displacements approach zero. When reduced, one point integration is used the results are greatly improved in both cases.

The kinetic energy of a deep beam element (Section 2.4) is

$$T_e = \frac{1}{2}\int_{-a}^{+a} \rho A \dot{v}^2 \, \mathrm{d}x + \frac{1}{2}\int_{-a}^{+a} \rho I_z \dot{\theta}_z^2 \, \mathrm{d}x \tag{3.296}$$

Substituting the displacement functions (3.284) into the kinetic energy expression (3.296) gives [3.33]

$$T_e = \tfrac{1}{2}\{\dot{\mathbf{v}}\}_e^{\mathrm{T}}[\mathbf{m}]_e\{\dot{\mathbf{v}}\}_e \tag{3.297}$$

where $\{\mathbf{v}\}_e$ is defined by (3.288) and

$$[\mathbf{m}]_e = \frac{\rho A a}{3}\begin{bmatrix} 2 & 0 & 1 & 0 \\ 0 & 2r_z^2 & 0 & r_z^2 \\ 1 & 0 & 2 & 0 \\ 0 & r_z^2 & 0 & 2r_z^2 \end{bmatrix} \tag{3.298}$$

where $r_z^2 = I_z/A$.

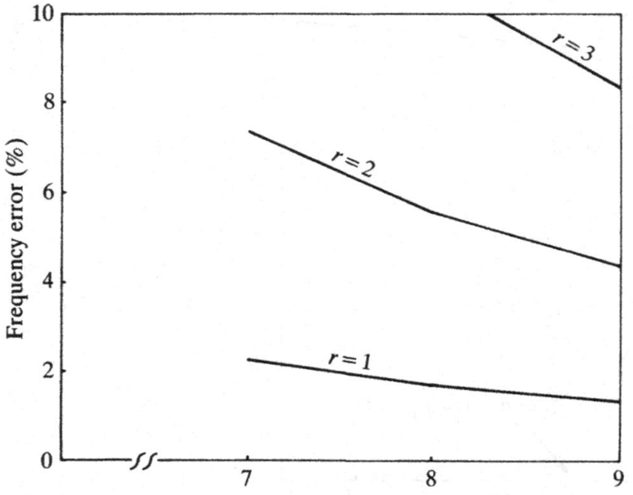

Figure 3.35. Flexural vibration of a deep hinged-hinged beam [3.33].

EXAMPLE 3.13 Use the element matrices (3.295) and (3.298) to estimate the lower frequencies of a hinged-hinged beam having the following properties:

$$r_z/L = 0.08, \qquad \kappa = 0.85, \qquad E/G = 2.6$$

Compare the results with the exact solution.

The equations of motion for a deep beam are (3.241). The boundary conditions for a hinged end are:

$$v(x,t) = 0, \qquad \frac{\partial \theta_z}{\partial x}(x,t) = 0 \tag{3.299}$$

The solutions of equations (3.241) subject to the boundary conditions (3.299) at $x = 0$ and $x = L$ are given in references [3.25, 3.26].

Finite element solutions using 7, 8 and 9 elements are quoted in reference [3.33]. A comparison between the two sets of results is given in Figure 3.35. Various other techniques have been suggested for avoiding shear locking. These are surveyed in reference [3.34].

3.11 Other Considerations for Beams

When analysing frameworks, it has been assumed so far that each individual element has a constant cross-sectional area, its node points lie on its centroidal axis, the shear centre of a cross-section coincides with the centroid, and cross-sections are free to warp without restraint during torsion. In particular applications these assumptions may be too restrictive and some of the following features may have to be included in the analysis:

(1) Node points offset from the centroidal axis
(2) Shear centre offset from the centroid
(3) Warping restraint
(4) Variable cross-section
(5) Twist
(6) Curvature

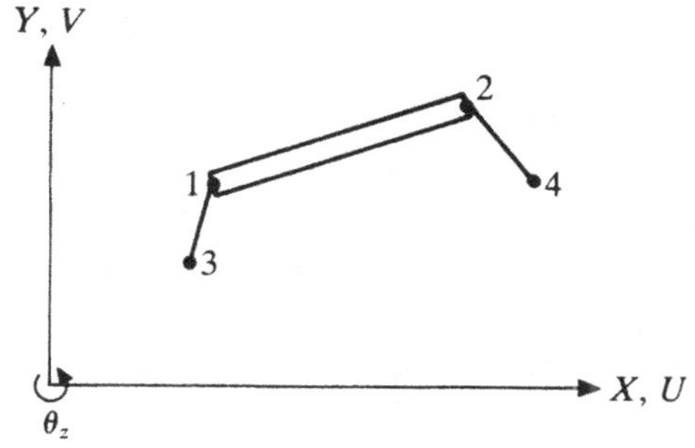

Figure 3.36. Plane framework element with offset nodes.

In many practical applications the centroids of beam members do not coincide at joints. This is the case when two or more members are joined by means of a rigid gusset plate. One way of dealing with this is to assume the various beams meeting at the joint to have different node points. The rigid constraint provided by the gusset plate can then be introduced by means of linear relationships between the degrees of freedom at the various node points as discussed in Section 3.6. An alternative way is to transform the beam element into one with nodes which are off-set from the centroidal axis.

Consider first the case of a plane framework element as shown in Figure 3.36. The element matrices referred to node points 1 and 2 on the centroidal axis are derived in Section 3.6. A transformation is applied to these matrices so that the resulting element matrices refer to the offset nodes 3 and 4. This transformation is obtained by expressing the displacement components at node 1 in terms of the displacement components at node 3. A similar relationship will hold between the displacement components at nodes 2 and 4. From geometrical considerations

$$\begin{aligned} U_1 &= U_3 - (Y_1 - Y_3)\theta_{Z3} \\ V_1 &= V_3 + (X_1 - X_3)\theta_{Z3} \\ \theta_{Z1} &= \theta_{Z3} \end{aligned} \tag{3.300}$$

or, in matrix form

$$\begin{bmatrix} U \\ V \\ \theta_Z \end{bmatrix}_1 = \begin{bmatrix} 0 & 0 & -Y_{13} \\ 0 & 1 & X_{13} \\ 0 & 0 & 1 \end{bmatrix} \begin{bmatrix} U \\ V \\ \theta_Z \end{bmatrix}_3 \tag{3.301}$$

where $Y_{13} = Y_1 - Y_3$ and $X_{13} = X_1 - X_3$. Similarly, for node 2

$$\begin{bmatrix} U \\ V \\ \theta_Z \end{bmatrix}_2 = \begin{bmatrix} 1 & 0 & -Y_{24} \\ 0 & 1 & X_{24} \\ 0 & 0 & 1 \end{bmatrix} \begin{bmatrix} U \\ V \\ \theta_Z \end{bmatrix}_4 \tag{3.302}$$

The inertia, stiffness and load matrices referred to nodes 3 and 4 are $[\mathbf{T}]_e^{\mathrm{T}}[\mathbf{m}]_e[\mathbf{T}]_e$, $[\mathbf{T}]_e^{\mathrm{T}}[\mathbf{k}]_e\,[\mathbf{T}]_e$ and $[\mathbf{T}]_e^{\mathrm{T}}\{\mathbf{f}\}_e$ respectively, where

$$[\mathbf{T}]_e = \begin{bmatrix} 1 & 0 & -Y_{13} & & & \\ 0 & 1 & X_{13} & & [\mathbf{0}] & \\ 0 & 0 & 1 & & & \\ & & & 1 & 0 & -Y_{24} \\ & [\mathbf{0}] & & 0 & 1 & X_{24} \\ & & & 0 & 0 & 1 \end{bmatrix} \tag{3.303}$$

This procedure is not recommended when beams are used as plate stiffeners [3.35]. The techniques to be used in this situation are presented in Chapter 7.

In general, the shear centre of a thin-walled, open section beam does not coincide with the centroid. Also, cross-sections that are plane before deformation, warp in the axial direction during torsion, so that they are no longer plane. In addition, the cross-sections of closed-section beams distort in their plane, a phenomenon which does not occur with open-sections [3.36]. Thus, it is convenient to develop an element which includes all these features. An open-section beam element is described in reference [3.37]. Although rotary inertia is included, shear deformation is not taken into account. References [3.38–3.40] derive elements which can be used for both open- and closed-sections based upon the assumption that the cross-sections are rigid. A method of treating the in-plane distortion of closed-section beams is presented in reference [3.41].

Three-dimensional beam elements with variable cross-sections can be obtained by treating A, I_x, I_z and J as functions of x in the energy expressions (3.53), (3.54), (3.92), (3.93), (3.127) and (3.128). This technique has been used to develop a beam bending element in references [3.42, 3.43].

Reference [3.44] derives several twisted beam elements and compares their performance. An alternative derivation is given in reference [3.45]. An element which is both tapered and twisted has been developed in reference [3.46]. Such elements are used to analyse the vibration of turbine, compressor and helicopter rotor blades, and aircraft propellers.

There is a large number of references dealing with curved beam elements. Typical examples are references [3.47–3.52] which discuss the following aspects:

(1) In- and out-of-plane vibration
(2) Constant and variable curvature
(3) Thin and thick arches
(4) Shallow and deep arches
(5) Constant and variable cross-sections
(6) Thin-walled cross-sections

Note that an arch is thin or thick according to whether the slenderness ratio $R/h \gtrless 40$ where h is the thickness and R the radius of curvature. Also, the arch is shallow or deep depending on whether the subtended angle $\alpha \lesseqgtr 40°$.

Curved components can be represented by an assemblage of straight beam elements, but many more elements are required in order to obtain acceptable accuracy. This can be illustrated by considering the in-plane vibrations of a circular ring. A circular ring has two perpendicular axes of symmetry. Therefore, it is necessary to represent only one-quarter of the ring by an assemblage of elements of the type described in Section 3.6. Comparing the frequencies obtained with the exact frequencies gives the percentage errors shown in Figure 3.37. In this figure n indicates the number of waves around the complete ring. The axial and bending actions of a curved beam are coupled. These actions are uncoupled with a straight beam element. When analysing curved beams using straight beam elements, the necessary coupling is obtained only at the node points. This necessitates the use of more elements for acceptable accuracy.

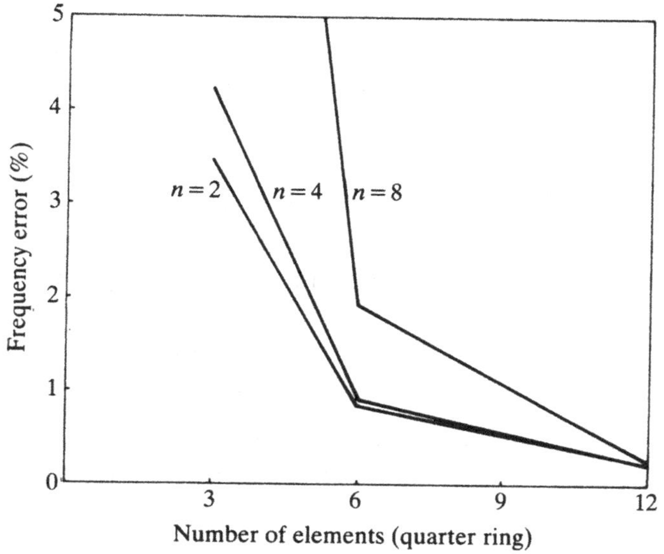

Figure 3.37. In-plane vibration of a circular ring [3.53].

Problems

Note: Problems 3.12–3.15 require the use of a digital computer.

3.1 Use the functions

$$\phi_1(x) = (L^2 - x^2), \qquad \phi_2(x) = (L^2 - x^2)x$$

where $-L \leq x \leq +L$, in a Rayleigh–Ritz analysis to estimate the two lowest frequencies of axial vibration of a uniform rod which is fixed at both ends. Compare these values with the analytical solution

$$\omega_1 = \frac{\pi}{2}\left(\frac{E}{\rho L^2}\right)^{1/2} \quad \text{and} \quad \omega_2 = \pi\left(\frac{E}{\rho L^2}\right)^{1/2}$$

3.2 Use the functions

$$\phi_1(x) = x, \qquad \phi_2(x) = x^2$$

where $0 \leq x \leq L$, in a Rayleigh–Ritz analysis to estimate the two lowest frequencies of axial vibration of a non-uniform rod which is fixed at $x = 0$ and free at $x = L$, given that the cross-sectional area at position x is $A(x) = A_0(1 - 0.2x/L)$ where A_0 is the area at $x = 0$. Compare these values with the analytical solution

$$\omega_1 = 1.642\left(\frac{E}{\rho L^2}\right)^{1/2} \quad \text{and} \quad \omega_2 = 4.737\left(\frac{E}{\rho L^2}\right)^{1/2}$$

3.3 Use the functions

$$\phi_1(x) = (L^2 - x^2), \qquad \phi_2 = (L^2 - x^2)x$$

where $-L \leq x \leq +L$, in a Rayleigh–Ritz analysis to estimate the two lowest frequencies for flexural vibration of a uniform beam which is simply supported at both ends. Compare these values with the analytical solution

$$\omega_1 = \left(\frac{\pi}{2}\right)^2 \left(\frac{EI_z}{\rho A L^4}\right)^{1/2} \quad \text{and} \quad \omega_2 = \pi^2 \left(\frac{EI_z}{\rho A L^4}\right)^{1/2}$$

3.4 Use the functions

$$\phi_1(x) = (L^2 - x^2)^2, \qquad \phi_2(x) = (L^2 - x^2)^2 x$$

where $-L \leq x \leq +L$, in a Rayleigh–Ritz analysis to estimate the two lowest frequencies for flexural vibration of a uniform beam which is clamped at both ends. Compare these values with the analytical solution

$$\omega_1 = 5.593 \left(\frac{EI_z}{\rho A L^4}\right)^{1/2} \quad \text{and} \quad \omega_2 = 15.482 \left(\frac{EI_z}{\rho A L^4}\right)^{1/2}$$

3.5 A tall chimney of height L is fixed at $x = 0$ and free at $x = L$. The cross-sectional area, $A(x)$, and second moment of area of the cross-section, $I_z(x)$, at position x are given by

$$A(\xi) = A_0(1 - 1.4\xi + 0.48\xi^2)$$
$$I_z(\xi) = I_0(1 - 2.6\xi + 2.52\xi^2 - 1.08\xi^3 + 0.1728\xi^4)$$

where $\xi = x/L$ and A_0, I_0 are the area and second moment of area at $x = 0$. Use the functions

$$\phi_1(x) = x^2, \qquad \phi_2(x) = x^3$$

in a Rayleigh–Ritz analysis to estimate the two lowest frequencies for flexural vibration, neglecting shear deformation and rotary inertia. Compare these values with the analytical solution

$$\omega_1 = 5.828 \left(\frac{EI_0}{\rho A_0 L^4}\right)^{1/2} \quad \text{and} \quad \omega_2 = 20.393 \left(\frac{EI_0}{\rho A_0 L^4}\right)^{1/2}$$

3.6 Analyse Problem 3.1 using three axial finite elements having two nodes each.

3.7 Estimate the lowest frequency of axial vibration of a uniform rod of length $2L$ which is free at both ends using one axial finite element. Compare this value with the analytical solution $\omega_1 = (\pi/2)(E/\rho L^2)^{1/2}$ .What does the second solution of the equations of motion represent?

3.8 Analyse Problem 3.3 using one beam bending finite element having two nodes.

3.9 Analyse Problem 3.4 using two beam bending finite elements having two nodes each.

3.10 Show that the cubic polynomial (3.112) satisfies the equation of static equilibrium for a uniform beam subject to forces and moments at its ends.

3.11 Use the inertia and stiffness matrices (3.132) and (3.135) to calculate the kinetic and strain energy of a beam element when it undergoes (i) a rigid body translation, v, and (ii) a rigid body rotation, θ, about its centre of mass. What can be deduced from the results? Check the answers by direct calculation.

3.12 A one-dimensional model of a stringer stiffened panel, as used in aircraft construction, consists of a uniform beam on equally spaced simple supports, as shown in Figure P3.12. Calculate the first five lowest natural frequencies and mode shapes using 20 beam elements. Take $A = 2.438 \times 10^{-5}$ m^2, $I_z = 3.019 \times 10^{-12}$ m^4, $E = 68.9 \times 10^9$ N/m^2 and $\rho = 2720$ kg/m^3. Compare the frequencies with the analytical values 102.2, 113.4, 141.8, 178.6, 214.4 Hz.

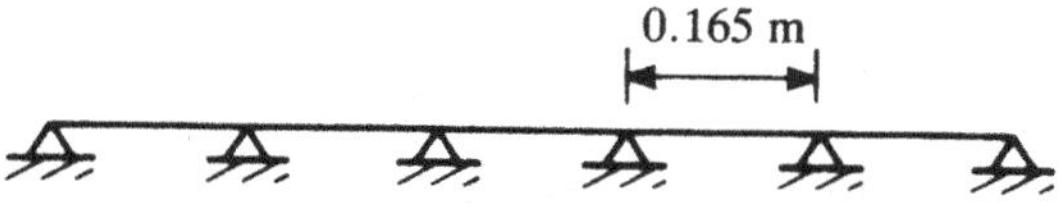

Figure P3.12

3.13 Figure P3.13 shows a two-dimensional framework which consists of two identical inclined members which are rigidly joined together. The other two ends are fully fixed. Their cross-sectional area is 24×10^{-4} m^2 and the second moment of area of the cross-section is 48×10^{-8} m^4. Take $E = 206 \times 10^9$ N/m^2 and $\rho = 7830$ kg/m^3. Calculate the four lowest frequencies and mode shapes for in-plane vibration using 8 elements. Compare the frequencies with the analytical values 88.9, 128.6, 286.9, 350.9 Hz.

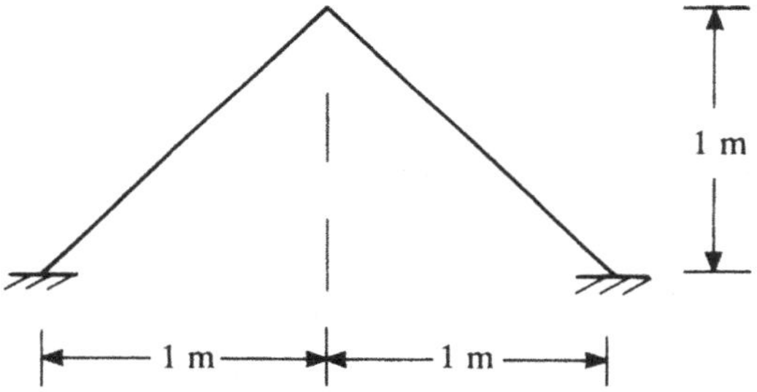

Figure P3.13

3.14 Figure P3.14 shows a three-bay portal frame which consists of 7 identical members having an area of 6.048×10^{-5} m^2 and a second moment of area of 1.143×10^{-10} m^4. E is 207×10^9 N/m^2 and ρ is 7786 kg/m^3. The vertical members are fully fixed at their lower end. Calculate the four lowest frequencies and mode shapes using 3 elements for every member. Compare these frequencies with the analytical values 138.3, 575, 663, 812 Hz.

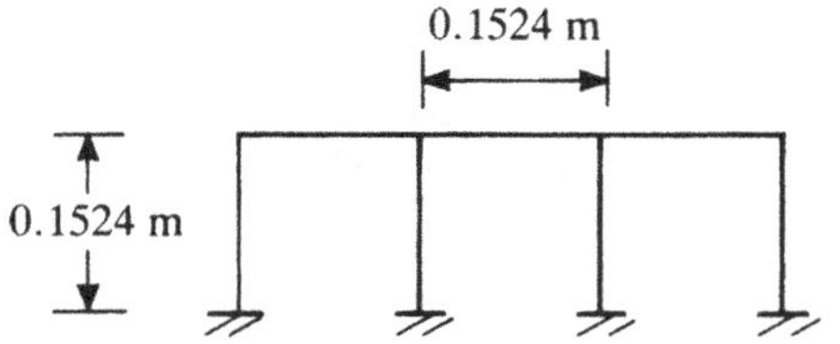

Figure P3.14

3.15 Calculate the frequencies and mode shapes of the first two modes which are antisymmetrical with respect to both the YZ-, ZX-planes for the three-dimensional framework of Example 3.9.

3.16 Calculate the eigenvector corresponding to the first mode of vibration of a fixed-free rod of length L using one three-node axial element. Use this eigenvector to determine the distribution of direct stress in the element. Calculate the stress at the two Gauss integration points $\xi = \mp 1/3^{1/2}$. Compare the result with the exact solution given in Section 3.3.

3.17 Show that the displacement functions for a four node axial element are

$$N_i(\xi) = \tfrac{1}{16}(1 + \xi_i \xi)(9\xi^2 - 1) \qquad \text{for } \xi_i = \mp 1$$

$$N_i(\xi) = \tfrac{9}{16}(1 - \xi^2)(1 + 9\xi_i \xi) \qquad \text{for } \xi_i = \mp 1/3$$

3.18 Verify that a quintic polynomial can be integrated exactly using three Gauss integration points.

3.19 Show that the inertia and stiffness matrices of a tapered axial element of length $2a$, having two nodes and whose cross-sectional area is given by

$$A(\xi) = A(0)(1 + a_1\xi + a_2\xi^2) \qquad -1 \le \xi \le +1$$

are

$$[\mathbf{m}] = \frac{\rho A(0)a}{15}\begin{bmatrix} (10 - 5a_1 + 4a_2) & (5 + a_2) \\ (5 + a_2) & (10 + 5a_1 + 4a_2) \end{bmatrix}$$

$$[\mathbf{k}] = \frac{EA(0)}{6a}\begin{bmatrix} (3 + a_2) & -(3 + a_2) \\ -(3 + a_2) & (3 + a_2) \end{bmatrix}$$

3.20 Analyse Problem 3.2 using two tapered axial elements with two nodes each.

4 In-plane Vibration of Plates

Flat plate structures which vibrate in their plane, such as shear wall buildings, can be analysed by dividing the plate up into an assemblage of two-dimensional finite elements, called membrane elements. The most common shapes of element used are triangles, rectangles and quadrilaterals. These elements can also be used to analyse the low frequency vibrations of complex shell-type structures such as aircraft and ships. In these cases the membrane action of the walls of the structures are more predominant than the bending action.

In Chapter 3 it is shown that in order to satisfy the convergence criteria, the element displacement functions should be derived from complete polynomials. In one dimension the polynomial terms are 1, x, x^2, $x^3, \ldots$, etc. Complete polynomials in two variables, x and y, can be generated using Pascal's triangle, as shown in Figure 4.1. Node points are normally situated at the vertices of the element, although additional ones are sometimes situated along the sides of the element in order to increase accuracy. (This technique is analogous to having additional node points along the length of a one-dimensional element, as described in Section 3.8.) When two adjacent elements are joined together, they are attached at their node points. The nodal degrees of freedom and element displacement functions should be chosen to ensure that the elements are conforming, that is, the displacement functions and their derivatives up to order $(p-1)$, are continuous at every point on the common boundary (see Section 3.2). In some cases it is not possible to achieve the necessary continuity using complete polynomials [4.1, 4.2]. This is overcome by using some additional terms of higher degree. When selecting these terms care should be taken to ensure that the displacement pattern is independent of the direction of the coordinate axes. This property is known as geometric invariance. For the two-dimensional case, the additional terms should be chosen in pairs, one from either side of the axis of symmetry in Figure 4.1. As an example, consider the derivation of a quadratic model with eight terms. Selecting all the constant, linear and quadratic terms plus the x^2y and xy^2 terms, produces a function which is quadratic in x along $y =$ constant and quadratic in y along $x =$ constant. Thus the deformation pattern will be the same whatever the orientation of the axes. This would not be true if the terms x^3 and x^2y had been selected. In this case the function is cubic in x along $y =$ constant and quadratic in y along $x =$ constant. Therefore, the deformation

				1					Constant
			x		y				Linear
		x^2		xy		y^2			Quadratic
	x^3		x^2y		xy^2		y^3		Cubic
x^4		x^3y		x^2y^2		xy^3		y^4	Quartic

Axis of symmetry

Figure 4.1. Complete polynomials in two variables.

pattern depends upon the orientation of the axes. Note that complete polynomials are invariant.

The energy expressions for a membrane element are, from Section 2.5

$$T_e = \frac{1}{2}\int_{A_e} \rho h \left(\dot{u}^2 + \dot{v}^2\right) \mathrm{d}A \tag{4.1}$$

$$U_e = \frac{1}{2}\int_{A_e} h \{\boldsymbol{\varepsilon}\}^{\mathrm{T}} [\mathbf{D}] \{\boldsymbol{\varepsilon}\} \,\mathrm{d}A \tag{4.2}$$

with

$$\{\boldsymbol{\varepsilon}\} = \begin{bmatrix} \partial u/\partial x \\ \partial v/\partial y \\ \partial u/\partial y + \partial v/\partial x \end{bmatrix} \tag{4.3}$$

$[\mathbf{D}]$ is a matrix of material constants which is defined by (2.45), (2.49) or (2.51) depending upon whether the material is anisotropic, orthotropic or isotropic. Also

$$\delta W_e = \int_{S_e} (p_x \delta u + p_y \delta v) \,\mathrm{d}s \tag{4.4}$$

The highest derivative appearing in these expressions is the first. Hence, it is only necessary to take u and v as degrees of freedom at each node to ensure continuity. Also, complete polynomials of at least degree 1 should be used (see Section 3.2).

4.1 Linear Triangular Element

The simplest way of idealising a flat plate of irregular shape is to use an assemblage of triangular elements. Figure 4.2 shows a triangular element with three node points, one at each vertex. There are two degrees of freedom at each node, namely, the components of displacement, u and v, in the directions of the x- and y-axes respectively. Each component can, therefore, be represented by polynomials having three terms each. Figure 4.1 shows that a complete linear function, which is the

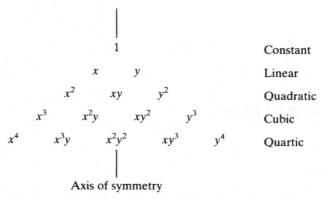

Figure 4.2. Geometry of a triangular element.

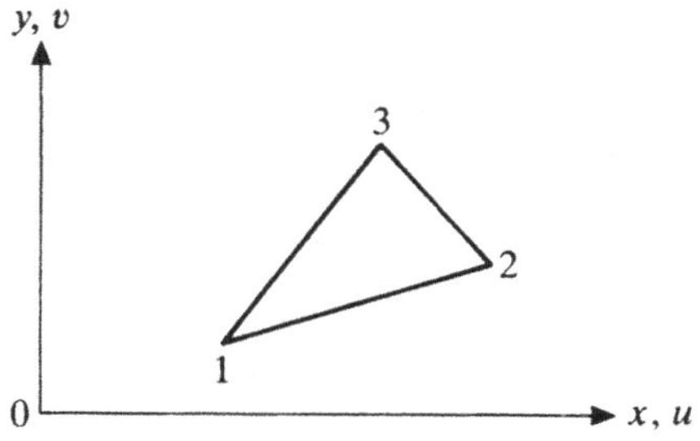

minimum requirement, has three terms. Therefore, the displacement variation can be represented by

$$\begin{aligned} u &= \alpha_1 + \alpha_2 x + \alpha_3 y \\ v &= \alpha_4 + \alpha_5 x + \alpha_6 y \end{aligned} \tag{4.5}$$

Since the functions for u and v are of the same form, only one need be considered in detail. Evaluating the expression for u at the three nodes gives

$$\begin{bmatrix} u_1 \\ u_2 \\ u_3 \end{bmatrix} = [\mathbf{A}] \begin{bmatrix} \alpha_1 \\ \alpha_2 \\ \alpha_3 \end{bmatrix} \tag{4.6}$$

where u_1, u_2, u_3 are the x-components of the displacement at the three nodes and

$$[\mathbf{A}] = \begin{bmatrix} 1 & x_1 & y_1 \\ 1 & x_2 & y_2 \\ 1 & x_3 & y_3 \end{bmatrix} \tag{4.7}$$

(x_i, y_i) are the coordinates of node $i (i = 1, 2, 3)$. Solving equation (4.6) gives

$$\begin{bmatrix} \alpha_1 \\ \alpha_2 \\ \alpha_3 \end{bmatrix} = [\mathbf{A}]^{-1} \begin{bmatrix} u_1 \\ u_2 \\ u_3 \end{bmatrix} \tag{4.8}$$

where

$$[\mathbf{A}]^{-1} = \frac{1}{2A} \begin{bmatrix} A_1^0 & A_2^0 & A_3^0 \\ a_1 & a_2 & a_3 \\ b_1 & b_2 & b_3 \end{bmatrix} \tag{4.9}$$

in which

$$\begin{aligned} A_i^0 &= x_j y_l - x_l y_j \\ a_i &= y_j - y_l \\ b_i &= x_l - x_j \end{aligned} \tag{4.10}$$

with the other coefficients obtained by a cyclic permutation of the subscripts in the order i, j, l. Also, A is the area of the triangle, which is given by

$$A = \tfrac{1}{2} \det [\mathbf{A}] = \tfrac{1}{2}(A_1^0 + A_2^0 + A_3^0) = \tfrac{1}{2}(a_1 b_2 - a_2 b_1) \tag{4.11}$$

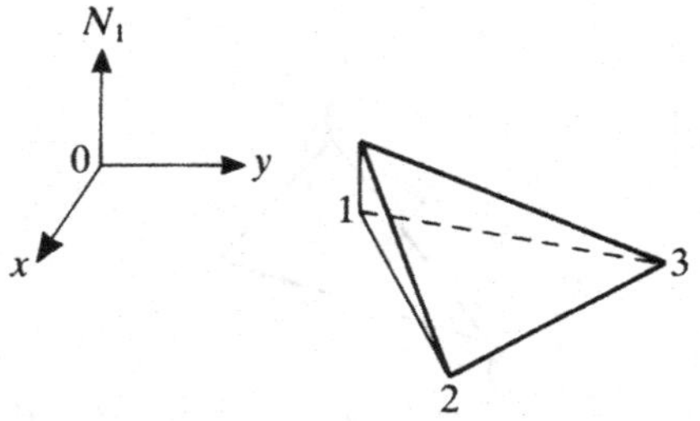

Figure 4.3. Displacement function N_1.

Substituting (4.8) into the expression for u, (4.5), gives

$$u = \lfloor \mathbf{N} \rfloor \begin{bmatrix} u_1 \\ u_2 \\ u_3 \end{bmatrix} \tag{4.12}$$

where

$$\lfloor \mathbf{N} \rfloor = \lfloor N_1 \quad N_2 \quad N_3 \rfloor = \lfloor 1 \quad x \quad y \rfloor [\mathbf{A}]^{-1} \tag{4.13}$$

hence

$$N_i = \frac{1}{2A} \left(A_i^0 + a_i x + b_i y \right) \tag{4.14}$$

Similarly

$$v = \lfloor \mathbf{N} \rfloor \begin{bmatrix} v_1 \\ v_2 \\ v_3 \end{bmatrix} \tag{4.15}$$

where v_1, v_2, v_3 are the y-components of the displacement at the three nodes.

Combining (4.12) and (4.15) gives

$$\begin{bmatrix} u \\ v \end{bmatrix} = \begin{bmatrix} N_1 & 0 & N_2 & 0 & N_3 & 0 \\ 0 & N_1 & 0 & N_2 & 0 & N_3 \end{bmatrix} \begin{bmatrix} u_1 \\ v_1 \\ u_2 \\ v_2 \\ u_3 \\ v_3 \end{bmatrix} \tag{4.16}$$

The function $N_i(x, y)$ varies linearly, has a unit value at node i and zero values at the other two nodes. These features are illustrated in Figure 4.3 for the function N_1.

If the displacement functions (4.16) are evaluated at a point on the side joining nodes 2 and 3, then N_1 will be zero. This means that the components of displacement will be uniquely determined by the values of the displacements at nodes 2 and 3, the nodes at the the two ends of the side being considered, and the position of the point on the side. Also, the variation of displacement along the side 2–3 is linear. Therefore, if the element is attached to another triangular element at nodes 2 and 3, then no gaps will occur between the two elements when the nodes are displaced, as illustrated in Figure 4.4. This means that the displacements are continuous along the common side, as required.

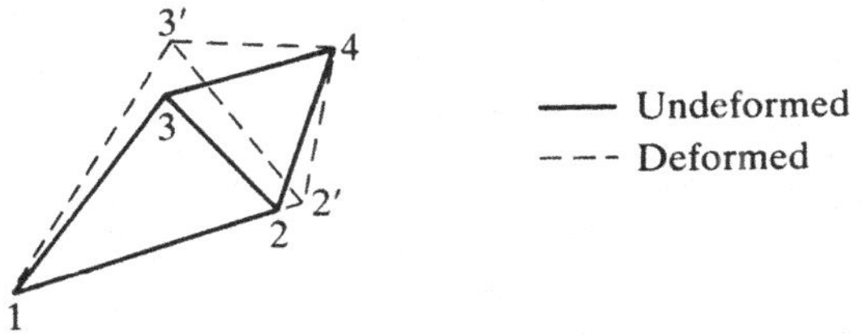

Figure 4.4. Continuity of displacements between adjacent elements.

The expression (4.16) is now written in the form

$$\begin{bmatrix} u \\ v \end{bmatrix} = [\mathbf{N}]\,\{\mathbf{u}\}_e \tag{4.17}$$

for convenience. Substituting (4.17) into (4.1) gives

$$T_e = \tfrac{1}{2}\{\dot{\mathbf{u}}\}_e^{\mathrm{T}}[\mathbf{m}]_e\{\dot{\mathbf{u}}\}_e \tag{4.18}$$

where

$$[\mathbf{m}]_e = \int_{A_e} \rho[\mathbf{N}]^{\mathrm{T}}[\mathbf{N}]\,\mathrm{d}A \tag{4.19}$$

is the element inertia matrix. Using (4.14) and (4.16) in (4.19) gives the following result:

$$[\mathbf{m}]_e = \frac{\rho h A}{12}\begin{bmatrix} 2 & 0 & 1 & 0 & 1 & 0 \\ 0 & 2 & 0 & 1 & 0 & 1 \\ 1 & 0 & 2 & 0 & 1 & 0 \\ 0 & 1 & 0 & 2 & 0 & 1 \\ 1 & 0 & 1 & 0 & 2 & 0 \\ 0 & 1 & 0 & 1 & 0 & 2 \end{bmatrix} \tag{4.20}$$

The details of this calculation are not given as they are rather tedious. A more elegant formulation is given in Section 4.5.

Substituting (4.17) into (4.3) and (4.2) gives

$$U_e = \tfrac{1}{2}\{\mathbf{u}\}_e^{\mathrm{T}}[\mathbf{k}]_e\{\mathbf{u}\}_e \tag{4.21}$$

where

$$[\mathbf{k}]_e = \int_{A_e} h[\mathbf{B}]^{\mathrm{T}}[\mathbf{D}][\mathbf{B}]\,\mathrm{d}A \tag{4.22}$$

is the element stiffness matrix, and

$$[\mathbf{B}] = \begin{bmatrix} \partial/\partial x & 0 \\ 0 & \partial/\partial y \\ \partial/\partial y & \partial/\partial x \end{bmatrix}[\mathbf{N}] \tag{4.23}$$

is the element 'strain matrix'. Using (4.14) and (4.16) gives

$$[\mathbf{B}] = \frac{1}{2A}\begin{bmatrix} a_1 & 0 & a_2 & 0 & a_3 & 0 \\ 0 & b_1 & 0 & b_2 & 0 & b_3 \\ b_1 & a_1 & b_2 & a_2 & b_3 & a_3 \end{bmatrix} \tag{4.24}$$

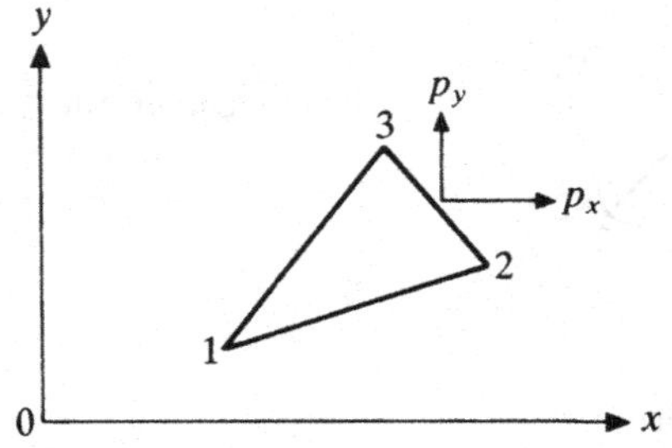

Figure 4.5. A triangular element subject to boundary loading.

Therefore, in this case, $[\mathbf{B}]$ is a constant matrix. Thus (4.22) becomes

$$[\mathbf{k}]_e = hA\,[\mathbf{B}]^{\mathrm{T}}\,[\mathbf{D}]\,[\mathbf{B}] \tag{4.25}$$

A constant $[\mathbf{B}]$ matrix implies that the strain components are constant within the element. Because of this, the element is sometimes referred to as a 'constant strain triangle'.

As an example of the calculation of equivalent nodal forces, consider the case of a distributed load applied to the side 2–3, as illustrated in Figure 4.5. This load has components (p_x, p_y) per unit length in the x- and y-directions. Equation (4.4) can be written in the form

$$\delta W_e = \int_{s_e} \begin{bmatrix} \delta u \\ \delta v \end{bmatrix}^{\mathrm{T}} \begin{bmatrix} p_x \\ p_y \end{bmatrix} \mathrm{d}s \tag{4.26}$$

Substituting (4.17) into (4.26) gives

$$\delta W_e = \{\delta\mathbf{u}\}_e^{\mathrm{T}} \{\mathbf{f}\}e \tag{4.27}$$

where

$$\{\mathbf{f}\}_e = \int_{s_e} [\mathbf{N}]_{2-3}^{\mathrm{T}} \begin{bmatrix} p_x \\ p_y \end{bmatrix} \mathrm{d}s \tag{4.28}$$

is the element equivalent nodal force matrix. Assuming p_x and p_y to be constant, (4.28) gives

$$\{\mathbf{f}\}_e = \frac{1}{2} l_{2-3} \begin{bmatrix} 0 \\ 0 \\ p_x \\ p_y \\ p_x \\ p_y \end{bmatrix} \tag{4.29}$$

where $l_{2\text{-}3}$ is the length of the side 2–3 which can be calculated from

$$l_{2-3} = \{(x_3 - x_2)^2 + (y_3 - y_2)^2\}^{1/2} \tag{4.30}$$

Since the components of the total force on side 2–3 are $p_x l_{2\text{-}3}$ and $p_y l_{2\text{-}3}$, it can be seen that (4.29) represents the application of half the total force at each node on the side.

The assembly process for a triangular, membrane element is as follows. If the element has node numbers n_1, n_2 and n_3, then columns 1 and 2, 3 and 4, 5 and 6 of

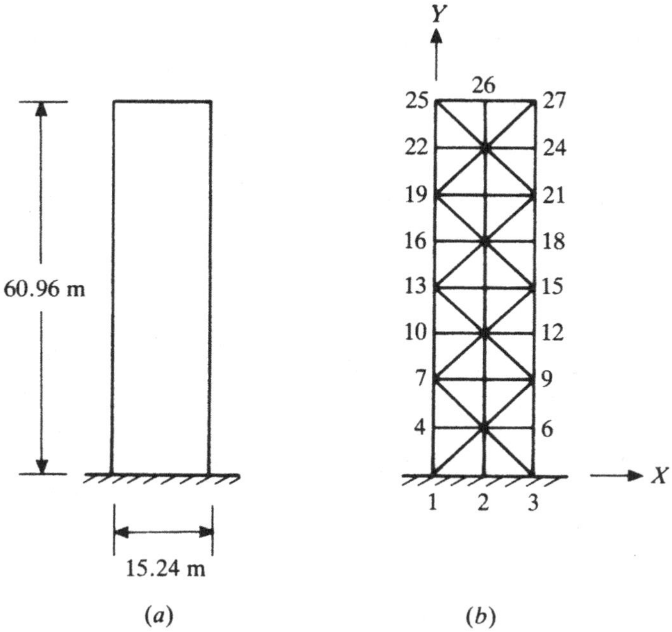

Figure 4.6. Geometry of a cantilever shear wall.

the element matrices are added into columns $(2n_1 - 1)$ and $2n_1$, $(2n_2 - 1)$ and $2n_2$, $(2n_3 - 1)$ and $2n_3$ respectively. An identical rule applies to the rows.

The stress components within the element are given by (2.44), namely

$$\{\boldsymbol{\sigma}\} = [\mathbf{D}]\{\boldsymbol{\varepsilon}\} \tag{4.31}$$

Using (4.3) and (4.17) to evaluate the strain components gives

$$\{\boldsymbol{\sigma}\} = [\mathbf{D}][\mathbf{B}]\{\mathbf{u}\}_e \tag{4.32}$$

where $[\mathbf{B}]$ is given by (4.24). The stresses are, therefore, constant within the element. It is usual to assign these constant values to the centroid of the element. It is shown in Section 3.10 that predicted stresses are more accurate at the integration points. Numerical integration of a constant function over a triangle requires one integration point, at the centroid (see Section 5.5).

EXAMPLE 4.1 Calculate the first five natural frequencies and modes for in-plane vibration of the cantilever shear wall shown in Figure 4.6(*a*). Compare the results with an analytical solution obtained by treating the wall as a deep beam [4.3]. Take $E = 34.474 \times 10^9$ N/m^2, $\nu = 0.11$, $\rho = 568.7$ kg/m^3 and the thickness to be 0.2286 m. Use the idealisation shown in Figure 4.6(*b*).

There are two degrees of freedom at each node, namely, linear displacements U and v in the X- and Y-directions. Since nodes 1, 2 and 3 are fully fixed, both degrees of freedom there, are constrained to be zero.

The predicted frequencies are compared with the analytically derived ones in Table 4.1. The corresponding mode shapes are shown in Figure 4.7. The agreement between the two sets of frequencies is not particularly good for the flexural modes 1, 2 and 4, the differences being greater than 20%. This result

Table 4.1. *Comparison of predicted frequencies (Hz) using triangular element with analytical beam frequencies*

Mode	FEM	Analytical [4.3]	% Difference
1	6.392	4.973	28.53
2	32.207	26.391	22.04
3	32.010	31.944	0.21
4	74.843	62.066	20.59
5	96.900	95.832	1.11

is obtained in spite of using 32 elements in the idealisation. Satisfactory results can only be obtained by increasing the number of elements. This would suggest that the linear triangular element does not provide an efficient method of solution. On the other hand, the frequencies of the longitudinal modes 3 and 5 agree quite closely.

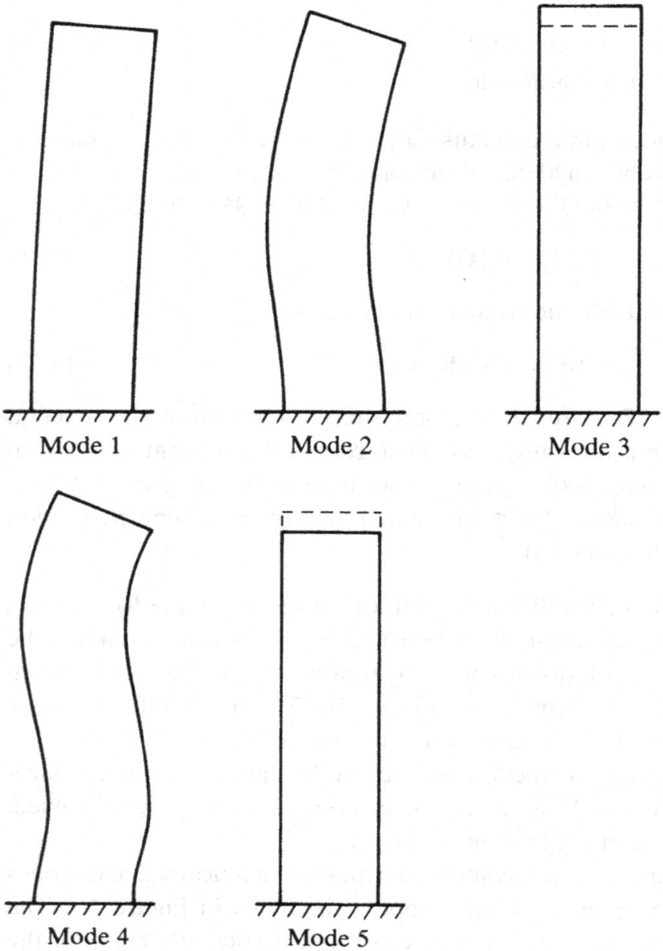

Figure 4.7. Mode shapes of cantilever shear wall.

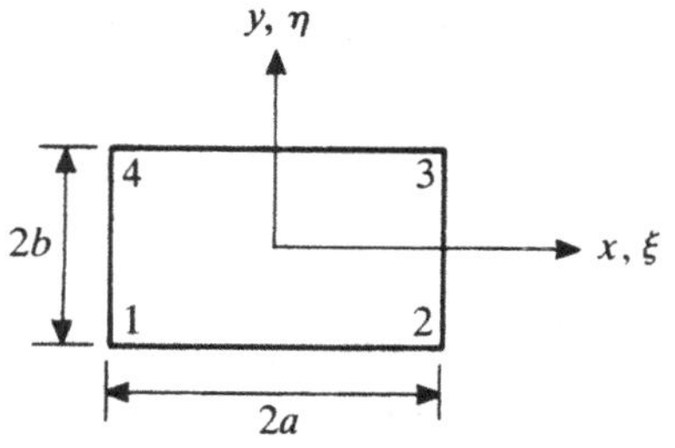

Figure 4.8. Geometry of a rectangular element $\xi = x/a$, $\eta = y/b$.

4.2 Linear Rectangular Element

The analysis procedure can be made more efficient by using rectangular elements as far as possible and filling in with triangular elements where the shape of the boundary makes it necessary. Figure 4.8 shows a rectangular element with four node points, one at each corner. There are two degrees of freedom at each node, namely, the components of displacement, u and v, in the directions of the x- and y-axes respectively. Each component can, therefore, be represented by polynomials having four terms each. Figure 4.1 shows that a complete linear function has three terms. Therefore, it is necessary to choose one of the three quadratic terms. The xy term is chosen in order to ensure that the displacement functions are invariant. This choice results in the displacement variation being linear along lines $x =$ constant and $y =$ constant. The displacements can, therefore, be represented by

$$\begin{aligned} u &= \alpha_1 + \alpha_2 x + \alpha_3 y + \alpha_4 xy \\ v &= \alpha_5 + \alpha_6 x + \alpha_7 y + \alpha_8 xy \end{aligned} \tag{4.33}$$

The coefficients α_1 to α_8 can be expressed in terms of the components of displacement at the four nodes by evaluating (4.33) at the four node points and solving the resulting equations, as demonstrated in the last section. However, in this case it is simpler to write down the displacement functions by inspection. The displacement functions are required in a similar form to (4.12) and (4.15), that is

$$u = \sum_{j=1}^{4} N_j u_j, \qquad v = \sum_{j=1}^{4} N_j v_j \tag{4.34}$$

The function N_j is required to have a unit value at node j and zero values at the other three nodes. Noting that the expressions (4.33) can also be written in the form of a product of two linear functions, namely $(\beta_1 + \beta_2 x)$ and $(\beta_3 + \beta_4 y)$, then it can easily be seen that the four functions, N_j, can be obtained by taking products of the linear displacement functions derived for axial vibrations of rods in Section 3.3. This gives

$$\begin{aligned} N_1 &= \tfrac{1}{4}(1 - \xi)(1 - \eta) \\ N_2 &= \tfrac{1}{4}(1 + \xi)(1 - \eta) \\ N_3 &= \tfrac{1}{4}(1 + \xi)(1 + \eta) \\ N_4 &= \tfrac{1}{4}(1 - \xi)(1 + \eta) \end{aligned} \tag{4.35}$$

Denoting the non-dimensional coordinates of node j by (ξ_j, η_j), then the function N_j can be written in the form

$$N_j = \tfrac{1}{4}(1 + \xi_j \xi)(1 + \eta_j \eta) \tag{4.36}$$

Like the linear triangle, the displacements vary linearly along each side of the element and are uniquely determined by the values of the displacements at the nodes at the ends of the side. Therefore, the displacements will be continuous between elements.

The expressions (4.34) can be written in the combined form (4.17), where

$$\{\mathbf{u}\}_e^{\mathrm{T}} = \lfloor u_1 \quad v_1 \quad u_2 \quad v_2 \quad u_3 \quad v_3 \quad u_4 \quad v_4 \rfloor \tag{4.37}$$

and

$$[\mathbf{N}] = \begin{bmatrix} N_1 & 0 & N_2 & 0 & N_3 & 0 & N_4 & 0 \\ 0 & N_1 & 0 & N_2 & 0 & N_3 & 0 & N_4 \end{bmatrix} \tag{4.38}$$

The inertia matrix is again given by (4.19). A typical element of this matrix is

$$\begin{aligned} &\rho hab \int_{-1}^{+1} \int_{-1}^{+1} N_i N_j \, \mathrm{d}\xi \, \mathrm{d}\eta \\ &= \frac{\rho hab}{16} \int_{-1}^{+1} (1 + \xi_i \xi)(1 + \xi_j \xi) \, \mathrm{d}\xi \int_{-1}^{+1} (1 + \eta_i \eta)(1 + \eta_j \eta) \, \mathrm{d}\eta \end{aligned} \tag{4.39}$$

The first integral has the following value

$$\int_{-1}^{+1} (1 + \xi_i \xi)(1 + \xi_j \xi) \, \mathrm{d}\xi = 2\left(1 + \tfrac{1}{3}\xi_i \xi_j\right) \tag{4.40}$$

The second integral can be evaluated in a similar manner. Using these results gives the following inertial matrix:

$$[\mathbf{m}]_e = \frac{\rho hab}{9} \begin{bmatrix} 4 & 0 & 2 & 0 & 1 & 0 & 2 & 0 \\ 0 & 4 & 0 & 2 & 0 & 1 & 0 & 2 \\ 2 & 0 & 4 & 0 & 2 & 0 & 1 & 0 \\ 0 & 2 & 0 & 4 & 0 & 2 & 0 & 1 \\ 1 & 0 & 2 & 0 & 4 & 0 & 2 & 0 \\ 0 & 1 & 0 & 2 & 0 & 4 & 0 & 2 \\ 2 & 0 & 1 & 0 & 2 & 0 & 4 & 0 \\ 0 & 2 & 0 & 1 & 0 & 2 & 0 & 4 \end{bmatrix} \tag{4.41}$$

The stiffness matrix is given by (4.22). Using (4.23), (4.36) and (4.38) gives the strain matrix

$$[\mathbf{B}] = \frac{1}{4}\begin{bmatrix} -\frac{(1-\eta)}{a} & 0 & \frac{(1-\eta)}{a} & 0 & \frac{(1+\eta)}{a} & 0 & -\frac{(1+\eta)}{a} & 0 \\ 0 & -\frac{(1-\xi)}{b} & 0 & -\frac{(1+\xi)}{b}) & 0 & \frac{(1+\xi)}{b} & 0 & \frac{(1-\xi)}{b} \\ -\frac{(1-\xi)}{b} & -\frac{(1-\eta)}{a} & -\frac{(1+\xi)}{b} & \frac{(1-\eta)}{a} & \frac{(1+\xi)}{b} & \frac{(1+\eta)}{a} & \frac{(1-\xi)}{b} & -\frac{(1+\eta)}{a)} \end{bmatrix} \tag{4.42}$$

Substituting (4.42) into (4.22) and integrating will give the element stiffness matrix. However, this is a tedious process. It is far simpler to use numerical integration. In terms of (ξ, η) coordinates (4.22) becomes

$$[\mathbf{k}]_e = \int_{-1}^{+1}\int_{-1}^{+1} abh[\mathbf{B}]^{\mathrm{T}}[\mathbf{D}][\mathbf{B}]\,\mathrm{d}\xi\,\mathrm{d}\eta \tag{4.43}$$

Section 3.10 describes how to integrate a function in one dimension between −1 and +1 using Gauss–Legendre integration. This method can be extended to square regions as follows. Consider the integral

$$I = \int_{-1}^{+1}\int_{-1}^{+1} g(\xi, \eta)\,\mathrm{d}\xi\,\mathrm{d}\eta \tag{4.44}$$

First evaluate $\int_{-1}^{+1} g(\xi, \eta)\mathrm{d}\xi$ keeping η constant. Using (3.246) gives

$$\int_{-1}^{+1} g(\xi, \eta)\,\mathrm{d}\xi = \sum_{i=1}^{n} H_i g(\xi_i, \eta) = \psi(\eta) \tag{4.45}$$

where H_i are the weight coefficients, ξ_i the sampling points and n the number of sampling points. Next evaluate $\int_{-1}^{+1} \psi(\eta)\,\mathrm{d}\eta$. Again (3.246) gives

$$\int_{-1}^{+1} \psi(\eta)\,\mathrm{d}\eta = \sum_{j=1}^{m} H_j \psi(\eta_j) \tag{4.46}$$

Combining (4.45) and (4.46) gives

$$I = \sum_{i=1}^{n}\sum_{j=1}^{m} H_i H_j g(\xi_i, \eta_j) \tag{4.47}$$

The number of integration points n and m in the ξ and η directions will depend upon the degree of the function $g(\xi, \eta)$ in ξ and η. For example, if $g(\xi, \eta)$ is cubic in both ξ and η, then $n = m = 2$. There are, therefore, a total of four integration points. Table 3.7 and equation (4.47) indicate that the integration points will be at $(\mp 1/3^{1/2}, \mp 1/3^{1/2})$ and all the weight coefficients are 1. The positions of the integration points are illustrated in Figure 4.9.

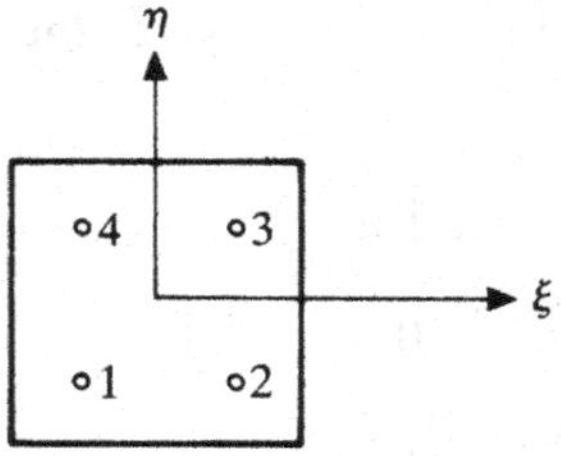

Figure 4.9. Integration points for $n = m = 2$ in a square region.

Substituting (4.42) into (4.43) indicates the presence of terms which are quadratic in either ξ or η. Therefore, it will be necessary to take $n = m = 2$ when using (4.47). However, in some situations, this procedure leads to unrepresentative properties. This can be illustrated by subjecting the element to nodal displacements which are consistent with pure bending in the ξ-direction, that is

$$u_1 = u_3 = u, \qquad u_2 = u_4 = -u$$
$$v_1 = v_2 = v_3 = v_4 = 0 \tag{4.48}$$

The deformation of the element, when subject to these displacements, is shown in Figure 4.10(*a*). The components of strain are given by

$$\{\boldsymbol{\varepsilon}\} = [\mathbf{B}]\{\mathbf{u}\}_e \tag{4.49}$$

where $[\mathbf{B}]$ is given by (4.42) and $\{\mathbf{u}\}_e$ is defined by (4.37). Substituting (4.48) into (4.49) gives the following distribution of strain

$$\varepsilon_x = \frac{u}{a}\eta, \qquad \varepsilon_y = 0, \qquad \gamma_{xy} = \frac{u}{b}\xi \tag{4.50}$$

The exact deformation, according to slender beam theory, is shown in Figure 4.10(*b*) and the corresponding strain distribution is

$$\varepsilon_x = \frac{u}{a}\eta, \qquad \varepsilon_y = 0, \qquad \gamma_{xy} = 0 \tag{4.51}$$

The element, therefore, predicts the direct strains correctly. However, there exists a parasitic shear strain which is only correct along $\xi = 0$. Similarly, by considering pure bending in the η-direction, it can be shown that there exists a parasitic shear which is only correct along $\eta = 0$. In general, the element will predict the correct shear strain at the point $\xi = 0$, $\eta = 0$, only. Therefore, an improvement can be

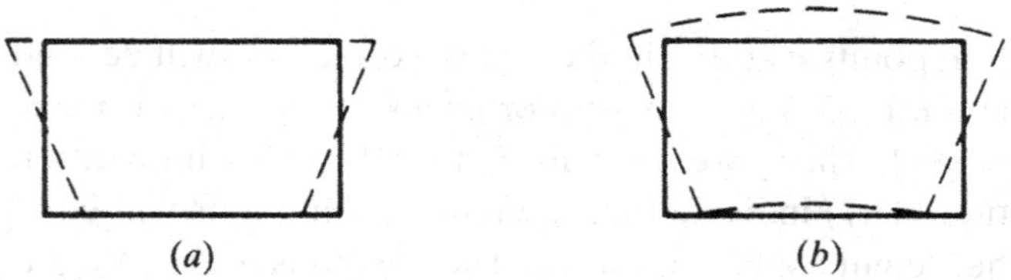

Figure 4.10. Rectangular element subject to pure bending: (*a*) Deformation given by (4.38), (*b*) exact deformation.

obtained by evaluating the shear strain at the point (0, 0) in (4.42). This gives

$$[\mathbf{B}] = \frac{1}{4}\begin{bmatrix} -\frac{(1-\eta)}{a} & 0 & \frac{(1-\eta)}{a} & 0 & \frac{(1+\eta)}{a} & 0 & -\frac{(1+\eta)}{a} & 0 \\ 0 & -\frac{(1-\xi)}{b} & 0 & -\frac{(1+\xi)}{b} & 0 & \frac{(1+\xi)}{b} & 0 & \frac{(1-\xi)}{b} \\ -\frac{1}{b} & -\frac{1}{a} & -\frac{1}{b} & \frac{1}{a} & \frac{1}{b} & \frac{1}{a} & \frac{1}{b} & -\frac{1}{a} \end{bmatrix} \tag{4.52}$$

Using (4.52) the integral in (4.43) is evaluated using a (2×2) array of integration points as illustrated in Figure 4.9. The stiffness matrix is, therefore, given by

$$[\mathbf{K}]_e = \sum_{j=1}^{4} abh\,[\mathbf{B}(\xi_j, \eta_j)]^{\mathrm{T}}\,[\mathbf{D}]\,[\mathbf{B}(\xi_j, \eta_j)] \tag{4.53}$$

where (ξ_j, η_j) are the coordinates of the jth integration point. Applying this procedure results in a non-conforming element. In this case, eigenvalues do not necessarily converge monotonically from above, as previously illustrated.

Section 3.10 indicates that for one-dimensional elements the stresses are more accurate at the integration points. Reference [4.4] shows that in the present case the best position is at the single point (0, 0). The stresses at this point are obtained by evaluating (4.52) there and then substituting the result into (4.32).

The equivalent nodal forces can be obtained in the way described in the last section. For example, if there is a distributed load of magnitude p_x per unit length along side 2–3 in Figure 4.8, then the equivalent nodal force matrix is

$$\{\mathbf{f}\}_e = bp_x \begin{bmatrix} 0 \\ 0 \\ 1 \\ 0 \\ 1 \\ 0 \\ 0 \\ 0 \end{bmatrix} \tag{4.54}$$

assuming p_x to be constant. Equation (4.54) represents the application of half the total force at each node on the side.

EXAMPLE 4.2 Calculate the first five frequencies of the shear wall shown in Figure 4.6(*a*) using the idealisation shown in Figure 4.11.

As in Example 4.1, there are two degrees of freedom at each node. These two degrees of freedom are constrained to be zero at nodes 1, 2 and 3.

The predicted frequencies are compared with the analytically derived ones in Table 4.2. The mode shapes are the same as those given in Figure 4.7. These results have been obtained using the strain matrix (4.42).

Table 4.2. *Comparison of predicted frequencies (Hz) using rectangular elements and analytical beam frequencies*

Mode	FEM	Analytical [4.3]	% Difference
1	5.250	4.973	5.57
2	27.991	26.391	6.06
3	32.016	31.994	0.23
4	67.518	62.066	8.78
5	97.250	95.832	1.48

The idealisations in Figures 4.6(*b*) and 4.11 have the same number of node points. The solution in Example 4.1 and the present one, therefore, have the same number of degrees of freedom, that is 48. Comparing the results in Tables 4.1 and 4.2, it can be seen that the use of linear rectangular elements as opposed to linear triangles, leads to greater accuracy.

This time the frequencies of the flexural modes 1, 2 and 4 agree to within 9% of the analytical results. Again the frequencies of the longitudinal modes 3 and 5 are much more accurate than the flexural ones.

Examples 4.1 and 4.2 between them, illustrate the fact that it is better to use rectangular elements as far as possible and only use triangular ones where the shape of the structure requires them.

4.3 Linear Quadrilateral Element

The rectangular element presented in the previous section will be much more versatile if it can be transformed into a quadrilateral element, as shown in Figure 4.12(*a*). Any point (ξ, η) within the square element, shown in Figure 4.12(*b*), can be mapped on to a point (x, y) within the quadrilateral element, shown in Figure 4.12(*a*), by means of the relationships

$$x = \sum_{j=1}^{4} N_j(\xi, \eta) x_j, \qquad y = \sum_{j=1}^{4} N_j(\xi, \eta) y_j \tag{4.55}$$

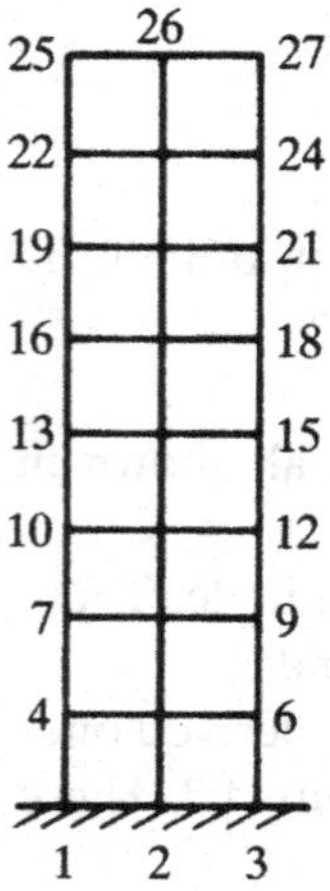

Figure 4.11. Idealisation of a cantilever shear wall.

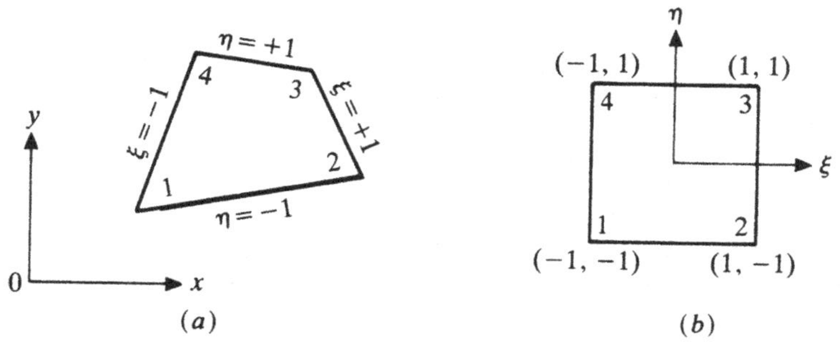

Figure 4.12. Geometry of a quadrilateral element: (*a*) physical coordinates, (*b*) isoparametric coordinates.

where (x_j, y_j) are the coordinates of node point j and the functions $N_j(\xi, \eta)$ are defined by (4.36).

Note that putting $\xi = 1, \eta = -1$ in (4.55) gives $x = x_2$ and $y = y_2$. Similarly $\xi = 1$, $\eta = 1$ gives $x = x_3$ and $y = y_3$. Also, putting $\xi = 1$ into (4.55) gives

$$\begin{aligned} x &= \tfrac{1}{2}(1-\eta)x_2 + \tfrac{1}{2}(1+\eta)x_3 \\ y &= \tfrac{1}{2}(1-\eta)y_2 + \tfrac{1}{2}(1+\eta)y_3 \end{aligned} \tag{4.56}$$

Rearranging (4.56) gives

$$\begin{aligned} x &= \tfrac{1}{2}(x_2 + x_3) + \tfrac{1}{2}\eta(x_3 - x_2) \\ y &= \tfrac{1}{2}(y_2 + y_3) + \tfrac{1}{2}\eta(y_3 - y_2) \end{aligned} \tag{4.57}$$

Eliminating η gives

$$y - \tfrac{1}{2}(y_2 + y_3) = \frac{(y_3 - y_2)}{(x_3 - x_2)}\left\{x - \tfrac{1}{2}(x_2 + x_3)\right\} \tag{4.58}$$

which is the equation of a straight line through the points (x_2, y_2) and (x_3, y_3). Thus, the side 2–3 in Figure 4.12(*b*) maps onto the side 2–3 in Figure 4.12(*a*). Similarly for the other three sides.

The variations in displacement can be expressed in (ξ, η) coordinates using (4.34). Notice the similarity between (4.34) and (4.55). The same functions have been used to define both the geometry of the element and its deformation. Because of this, the element is referred to as an 'isoparametric element'.

The position and displacement of any point, η, on the side 2–3 in Figure 4.12(*a*), is uniquely determined from the coordinates and displacements of nodes 2 and 3 using (4.55) and (4.34). Two adjacent elements, having the side 2–3 as a common side, will have the same nodes and the same variation with η along 2–3. Therefore, the displacements will be continuous between elements.

The inertia matrix is again given by equation (4.19), that is

$$[\mathbf{m}]_e = \int_{A_e} \rho h \, [\mathbf{N}]^{\mathrm{T}} [\mathbf{N}] \, \mathrm{d}A \tag{4.59}$$

where $[\mathbf{N}]$ is defined by (4.38). Because the N_i in (4.38) are expressed in (ξ, η) coordinates, it is simpler to evaluate (4.59) if the integral is transformed from (x, y) coordinates to (ξ, η) coordinates. In general, the lines ξ = constant and η = constant in the x-, y-plane will not be orthogonal. The vectors

$$\begin{aligned} \mathrm{d}\vec{\xi} &= \left(\frac{\partial x}{\partial \xi}, \frac{\partial y}{\partial \xi}\right) \mathrm{d}\xi \\ \mathrm{d}\vec{\eta} &= \left(\frac{\partial x}{\partial \eta}, \frac{\partial y}{\partial \eta}\right) \mathrm{d}\eta \end{aligned} \tag{4.60}$$

are directed along the lines $\eta = $ constant and $\xi = $ constant respectively. The element of area in (ξ, η) coordinates is given by the modulus of their vector product, that is

$$\begin{aligned} \mathrm{d}A &= \left|\mathrm{d}\vec{\xi} \wedge \mathrm{d}\vec{\eta}\right| \\ &= \left(\frac{\partial x}{\partial \xi}\frac{\partial y}{\partial \eta} - \frac{\partial y}{\partial \xi}\frac{\partial x}{\partial \eta}\right) \mathrm{d}\xi \, \mathrm{d}\eta \end{aligned} \tag{4.61}$$

This can be written in the form

$$\mathrm{d}A = \det[\mathbf{J}] \, \mathrm{d}\xi \, \mathrm{d}\eta \tag{4.62}$$

where

$$[\mathbf{J}] = \begin{bmatrix} \dfrac{\partial x}{\partial \xi} & \dfrac{\partial y}{\partial \xi} \\ \dfrac{\partial x}{\partial \eta} & \dfrac{\partial y}{\partial \eta} \end{bmatrix} \tag{4.63}$$

is known as the Jacobian matrix.

Substituting (4.55) into (4.63) gives

$$[\mathbf{J}] = \begin{bmatrix} \dfrac{\partial N_1}{\partial \xi} & \dfrac{\partial N_2}{\partial \xi} & \dfrac{\partial N_3}{\partial \xi} & \dfrac{\partial N_4}{\partial \xi} \\ \dfrac{\partial N_1}{\partial \eta} & \dfrac{\partial N_2}{\partial \eta} & \dfrac{\partial N_3}{\partial \eta} & \dfrac{\partial N_4}{\partial \eta} \end{bmatrix} \begin{bmatrix} x_1 & y_1 \\ x_2 & y_2 \\ x_3 & y_3 \\ x_4 & y_4 \end{bmatrix} \tag{4.64}$$

The expression for the inertia matrix, (4.59), now becomes

$$[\mathbf{m}]_e = \int_{-1}^{+1} \int_{-1}^{+1} \rho h \, [\mathbf{N}]^{\mathrm{T}} [\mathbf{N}] \det[\mathbf{J}] \, \mathrm{d}\xi \, \mathrm{d}\eta \tag{4.65}$$

The integral in (4.65) can be evaluated using numerical integration, as described in the previous section. In order to determine the number of integration points required, it is necessary to determine the order of the function $\det[\mathbf{J}]$. Substituting (4.36) into (4.64) gives

$$[\mathbf{J}] = \frac{1}{4} \begin{bmatrix} (e_1 + e_2\eta) & (f_1 + f_2\eta) \\ (e_3 + e_2\xi) & (f_3 + f_2\xi) \end{bmatrix} \tag{4.66}$$

where

$$\begin{aligned} e_1 &= (-x_1 + x_2 + x_3 - x_4) \\ e_2 &= (x_1 - x_2 + x_3 - x_4) \\ e_3 &= (-x_1 - x_2 + x_3 + x_4) \\ f_1 &= (-y_1 + y_2 + y_3 - y_4) \\ f_2 &= (y_1 - y_2 + y_3 - y_4) \\ f_3 &= (-y_1 - y_2 + y_3 + y_4) \end{aligned} \tag{4.67}$$

Therefore

$$\det[\mathbf{J}] = \tfrac{1}{16}(c_1 + c_2\xi + c_3\eta) \tag{4.68}$$

where

$$\begin{aligned} c_1 &= e_1 f_3 - e_3 f_1 \\ c_2 &= e_1 f_2 - e_2 f_1 \\ c_3 &= e_2 f_3 - e_3 f_2 \end{aligned} \tag{4.69}$$

In general, det [**J**] is a linear function of (ξ, η). However, in the case of either a rectangle or parallelogram, $c_2 = c_3 = 0$, and so det [**J**] is a constant.

[**N**] is a bi-linear function of (ξ, η) and so $[\mathbf{N}]^{\mathrm{T}}[\mathbf{N}]$ is a bi-quadratic function. This means that $[\mathbf{N}]^{\mathrm{T}}[\mathbf{N}]\det[\mathbf{J}]$ is either a bi-quadratic or bi-cubic function. In either case (4.65) can be evaluated using a (2×2) array of integration points.

The stiffness matrix is given by (4.22), that is

$$[\mathbf{k}]_e = \int_{A_e} h\,[\mathbf{B}]^{\mathrm{T}}[\mathbf{D}][\mathbf{B}]\,\mathrm{d}A \tag{4.70}$$

where [**B**] is defined by (4.23). Transforming to (ξ, η) coordinates using (4.62) gives

$$[\mathbf{k}]_e = \int_{-1}^{+1}\int_{-1}^{+1} h\,[\mathbf{B}]^{\mathrm{T}}[\mathbf{D}][\mathbf{B}]\,\det[\mathbf{J}]\,\mathrm{d}\xi\,\mathrm{d}\eta \tag{4.71}$$

In evaluating the matrix [**B**], it is necessary to express the differentials in the (ξ, η) coordinates, since the matrix [**N**] is expressed in terms of these.

Noting that

$$\frac{\partial}{\partial\xi} = \frac{\partial x}{\partial\xi}\frac{\partial}{\partial x} + \frac{\partial y}{\partial\xi}\frac{\partial}{\partial y}$$

and

$$\frac{\partial}{\partial\eta} = \frac{\partial x}{\partial\eta}\frac{\partial}{\partial x} + \frac{\partial y}{\partial\eta}\frac{\partial}{\partial y} \tag{4.72}$$

can be expressed in the matrix form

$$\begin{bmatrix} \dfrac{\partial}{\partial\xi} \\[2mm] \dfrac{\partial}{\partial\eta} \end{bmatrix} = [\mathbf{J}] \begin{bmatrix} \dfrac{\partial}{\partial x} \\[2mm] \dfrac{\partial}{\partial y} \end{bmatrix} \tag{4.73}$$

where [**J**] is the Jacobian matrix defined in (4.63), then

$$\begin{bmatrix} \dfrac{\partial}{\partial x} \\[2mm] \dfrac{\partial}{\partial y} \end{bmatrix} = [\mathbf{J}]^{-1} \begin{bmatrix} \dfrac{\partial}{\partial\xi} \\[2mm] \dfrac{\partial}{\partial\eta} \end{bmatrix} \tag{4.74}$$

The elements of [**B**] require $\partial N_j/\partial_x$ and $\partial N_j/\partial y$ to be evaluated for $j = 1, 2, 3, 4$. These can be calculated using the expression

$$\begin{bmatrix} \dfrac{\partial N_1}{\partial x} & \dfrac{\partial N_2}{\partial x} & \dfrac{\partial N_3}{\partial x} & \dfrac{\partial N_4}{\partial x} \\[2mm] \dfrac{\partial N_1}{\partial y} & \dfrac{\partial N_2}{\partial y} & \dfrac{\partial N_3}{\partial y} & \dfrac{\partial N_4}{\partial y} \end{bmatrix} = [\mathbf{J}]^{-1} \begin{bmatrix} \dfrac{\partial N_1}{\partial\xi} & \dfrac{\partial N_2}{\partial\xi} & \dfrac{\partial N_3}{\partial\xi} & \dfrac{\partial N_4}{\partial\xi} \\[2mm] \dfrac{\partial N_1}{\partial\eta} & \dfrac{\partial N_2}{\partial\eta} & \dfrac{\partial N_3}{\partial\eta} & \dfrac{\partial N_4}{\partial\eta} \end{bmatrix} \tag{4.75}$$

Notice that the second matrix on the right hand side of (4.75) is also required in (4.64) for evaluating **[J]**.

Expressions (4.66), (4.68) and (4.75) show that the elements of **[B]** are obtained by dividing a bi-linear function of ξ and η by a linear function. Therefore, the elements of $[\mathbf{B}]^{\mathrm{T}}[\mathbf{D}][\mathbf{B}]\,\det[\mathbf{J}]$ are bi-quadratic functions divided by a linear function. This means that **[k]** cannot be evaluated exactly using numerical integration.

From practical considerations it is best to use as few integration points as is possible without causing numerical difficulties. A smaller number of integrating points results in lower computational cost. Also, a lower order rule tends to counteract the over-stiff behaviour associated with the displacement method. A lower limit on the number of integration points can be obtained by observing that as the mesh is refined, the state of strain within an element approaches a constant. In this case the stiffness matrix, equation (4.71), becomes

$$[\mathbf{k}]_e \simeq [\mathbf{B}]^{\mathrm{T}}[\mathbf{D}][\mathbf{B}] \int_{-1}^{+1}\int_{-1}^{+1} h \det[\mathbf{J}]\,\mathrm{d}\xi\,\mathrm{d}\eta \tag{4.76}$$

The integral in (4.76) represents the volume of the element. Therefore, the minimum number of integration points, is the number required to evaluate exactly the volume of the element. Taking the thickness, h, to be constant and noting that det **[J]** is linear, indicates that the volume can be evaluated exactly using one integration point. However, in the present case, one integration point is unacceptable since it gives rise to zero-energy deformation modes. These are modes of deformation which give rise to zero strain energy. This will be the case if one of these modes gives zero strain at the integration point. The existence of these modes is indicated by the stiffness matrix having more zero eigenvalues than rigid body modes (see Section 3.10). Experience has shown that the best order of integration is a (2×2) array of points. As in the case of the rectangular element, Section 4.2, the shear strain should be evaluated at $\xi = 0$, $\eta = 0$ when forming the matrix **[B]** in (4.71).

The equivalent nodal forces are obtained in the same way as in previous sections. For example, if there is a distributed load having components (p_x, p_y) per unit length along side 2–3 in Figure 4.12(a), then the equivalent nodal force matrix is

$$\{\mathbf{f}\}_e = \tfrac{1}{2} l_{2\text{–}3} \begin{bmatrix} 0 \\ 0 \\ p_x \\ p_y \\ p_x \\ P_y \\ 0 \\ 0 \end{bmatrix} \tag{4.77}$$

if p_x and p_y are constants. $l_{2\text{–}3}$ is the length of the side 2–3.

The best position to evaluate the stresses is at the single point $\xi = 0$, $\eta = 0$ as indicated in Section 4.2 for the rectangle. These are obtained by evaluating the matrix **[B]** there, using (4.23), (4.38) and (4.75), and then substituting into (4.32).

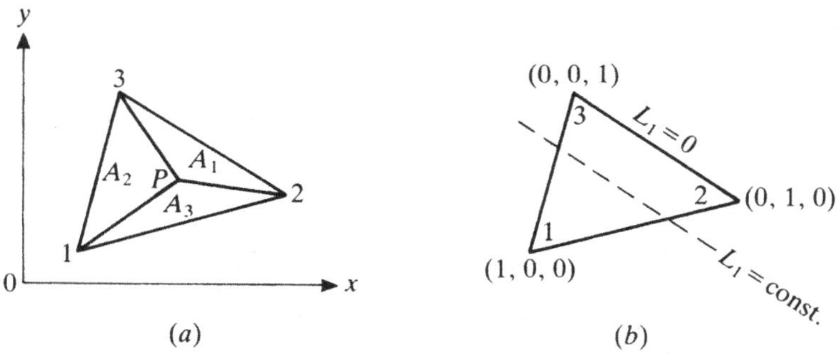

Figure 4.13. Definition of area coordinates for a triangle.

4.4 Area Coordinates for Triangles

In dealing with triangular regions it is advantagous to use area coordinates. Referring to Figure 4.13(*a*), the area coordinates (L_1, L_2, L_3) of the point P are defined as

$$L_1 = A_1/A, \qquad L_2 = A_2/A, \qquad L_3 = A_3/A \tag{4.78}$$

where A_1, A_2, A_3 denote the areas of the sub-triangles indicated and A is the area of the complete triangle. Since

$$A_1 + A_2 + A_3 = A \tag{4.79}$$

then the three area coordinates are related by the expression

$$L_1 + L_2 + L_3 = 1 \tag{4.80}$$

It can easily be seen that the coordinates of the three vertices are (1, 0, 0), (0, 1, 0) and (0, 0, 1) respectively.

Area coordinates can also be interpreted as ratios of lengths. For example

$$L_1 = \frac{\text{distance from } P \text{ to side 2–3}}{\text{distance from 1 to side 2–3}} \tag{4.81}$$

This definition indicates that the line $L_1 =$ constant is parallel to the side 2–3 whose equation is $L_1 = 0$, as shown in Figure 4.13(*b*).

Cartesian and area coordinates are related by

$$\begin{aligned} x &= x_1 L_1 + x_2 L_2 + x_3 L_3 \\ y &= y_1 L_1 + y_2 L_2 + y_3 L_3 \end{aligned} \tag{4.82}$$

where (x_i, y_i) are the Cartesian coordinates of vertex i. Combining (4.80) and (4.82) gives

$$\begin{bmatrix} 1 \\ x \\ y \end{bmatrix} = \begin{bmatrix} 1 & 1 & 1 \\ x_1 & x_2 & x_3 \\ y_1 & y_2 & y_3 \end{bmatrix} \begin{bmatrix} L_1 \\ L_2 \\ L_3 \end{bmatrix} = [\mathbf{A}]^{\mathrm{T}} \begin{bmatrix} L_1 \\ L_2 \\ L_3 \end{bmatrix} \tag{4.83}$$

where $[\mathbf{A}]$ is defined in (4.7). Inverting this relationship gives

$$\begin{bmatrix} L_1 \\ L_2 \\ L_3 \end{bmatrix} = [\mathbf{A}]^{-\mathbf{T}} \begin{bmatrix} 1 \\ x \\ y \end{bmatrix} \tag{4.84}$$

Using (4.9) in (4.84) shows that

$$L_i = \frac{1}{2A}\left(A_i^0 + a_i x + b_i y\right) \qquad i = 1, 2, 3 \tag{4.85}$$

where A_i^0, a_i, b_i are defined in (4.10).

4.5 Linear Triangle in Area Coordinates

The linear triangle is presented in Section 4.1 using Cartesian coordinates. It is shown there that the inertia matrix is given by

$$[\mathbf{m}]_e = \int_{A_e} \rho h [\mathbf{N}]^{\mathrm{T}} [\mathbf{N}] \, \mathrm{d}A \tag{4.86}$$

where

$$[\mathbf{N}] = \begin{bmatrix} N_1 & 0 & N_2 & 0 & N_3 & 0 \\ 0 & N_1 & 0 & N_2 & 0 & N_3 \end{bmatrix} \tag{4.87}$$

and

$$N_i = \frac{1}{2A}\left(A_i^0 + a_i x + b_i y\right) \tag{4.88}$$

Comparing (4.88) with (4.85) shows that in terms of area coordinates

$$N_i = L_i \tag{4.89}$$

This means that a typical element of the inertia matrix is of the form $\rho h \int_A L_i L_j \, \mathrm{d}A$. Integrals of this form can be evaluated using the following formula [4.5]

$$\int_A L_1^m L_2^n L_3^p \, \mathrm{d}A = \frac{m!n!p!}{(m+n+p+2)!} 2A \tag{4.90}$$

where ! is the factorial sign. Remember that $0! = 1$. Therefore

$$\int_A L_i L_j \, \mathrm{d}A = \frac{1!1!0!}{4!} 2A = \frac{A}{12} \tag{4.91}$$

and

$$\int_A L_i^2 \, \mathrm{d}A = \frac{2!0!0!}{4!} 2A = \frac{A}{6} \tag{4.92}$$

Substituting (4.91) and (4.92) into (4.86) gives (4.20).

The stiffness matrix is given by

$$[\mathbf{k}]_e = \int_{A_e} h [\mathbf{B}]^{\mathrm{T}} [\mathbf{D}] [\mathbf{B}] \, \mathrm{d}A \tag{4.93}$$

where

$$[\mathbf{B}] = \begin{bmatrix} \partial/\partial x & 0 \\ 0 & \partial/\partial y \\ \partial/\partial y & \partial/\partial x \end{bmatrix} [\mathbf{N}] \tag{4.94}$$

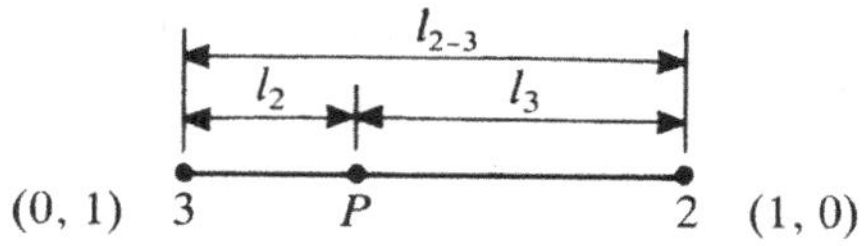

Figure 4.14. One-dimensional natural coordinates.

Functions expressed in area coordinates can be differentiated with respect to Cartesian coordinates as follows

$$\frac{\partial}{\partial x} = \sum_{j=1}^{3} \frac{\partial L_j}{\partial x} \frac{\partial}{\partial L_j}, \qquad \frac{\partial}{\partial y} = \sum_{j=1}^{3} \frac{\partial L_j}{\partial y} \frac{\partial}{\partial L_j} \tag{4.95}$$

Using (4.85) these become

$$\frac{\partial}{\partial x} = \frac{1}{2A} \sum_{j=1}^{3} a_j \frac{\partial}{\partial L_j}, \qquad \frac{\partial}{\partial y} = \frac{1}{2A} \sum_{j=1}^{3} b_j \frac{\partial}{\partial L_j} \tag{4.96}$$

(4.89) and (4.96) together show that

$$\frac{\partial N_i}{\partial x} = \frac{a_i}{2A}, \qquad \frac{\partial N_i}{\partial y} = \frac{b_i}{2A} \tag{4.97}$$

Substituting (4.97) into (4.94) gives the strain matrix (4.24) which is constant. Thus, again, the evaluation of (4.93) is a trivial matter. The result is given by (4.25).

When the displacement functions are evaluated along one side of the triangle, in order to calculate the equivalent nodal force matrix, the area coordinates become one-dimensional natural coordinates. This is illustrated for side 2–3 in Figure 4.14.

The natural coordinates of P are defined as

$$L_2 = l_2/l_{2-3}, \qquad L_3 = l_3/l_{2-3} \tag{4.98}$$

with

$$L_2 + L_3 = 1 \tag{4.99}$$

Evaluation of (4.28) involves integrals of the form $\int_s L_i \, ds$. These can be evaluated using the formula [4.5]

$$\int_s L_2^n L_3^p \, ds = \frac{n!p!}{(n+p+1)!} l_{2-3} \tag{4.100}$$

Thus

$$\int_s L_i \, ds = \frac{1!0!}{2!} l_{2-3} = \frac{l_{2-3}}{2} \tag{4.101}$$

This gives the result (4.29).

4.6 Increasing the Accuracy of Elements

In Section 3.8 it is shown that the accuracy of rod and beam elements can be increased by increasing the order of the polynomial representation of the displacements within the element. This results in an increased number of degrees of freedom which may be either at existing nodes or at additional nodes. These ideas can be extended to membrane elements.

The accuracy of the linear rectangular element shown in Figure 4.8 can be increased by assuming that the displacements are given by [4.6]

$$u = \sum_{j=1}^{4} N_j u_j + \alpha_1(1 - \xi^2) + \alpha_2(1 - \eta^2)$$
$$v = \sum_{j=1}^{4} N_j v_j + \alpha_3(1 - \xi^2) + \alpha_4(1 - \eta^2) \tag{4.102}$$

where the functions $N_j(\xi, \eta)$ are defined by (4.35). Comparing (4.102) with (4.34), it can be seen that the previously used bi-linear functions have been augmented by two quadratic functions each. This will ensure that in pure bending situations the deformations are more like Figure 4.10(b) than Figure 4.10(a).

The parameters α_1, to α_4 are not nodal displacements. However, since the functions $(1 - \xi^2)$ and $(1 - \eta^2)$ are zero at the four node points in Figure 4.8, α_1 to α_4 can be considered to be generalised coordinates associated with the interior of the element.

On substituting (4.102) into (4.23) and (4.22) a stiffness matrix of order (12×12) will be obtained. In order to reduce this to an (8×8) matrix, the element strain energy is minimised with respect to the additional degrees of freedom α_1 to α_4. Writing the strain energy in partitioned form as follows

$$U_e = \tfrac{1}{2}\lfloor \{\mathbf{u}\}_e^{\mathrm{T}} \{\boldsymbol{\alpha}\}^{\mathrm{T}} \rfloor \begin{bmatrix} [k_{uu}] & [k_{u\alpha}] \\ [k_{\alpha u}] & [k_{\alpha\alpha}] \end{bmatrix} \begin{bmatrix} \{\mathbf{u}\}_e \\ \{\boldsymbol{\alpha}\} \end{bmatrix} \tag{4.103}$$

where $\{\mathbf{u}\}_e$ is defined as in (4.37) and $\{\boldsymbol{\alpha}\}^{\mathrm{T}} = \lfloor \alpha_1 \alpha_2 \alpha_3 \alpha_4 \rfloor$, then $\partial U_e / \partial\{\boldsymbol{\alpha}\} = 0$ gives

$$[\mathbf{k}]_{\alpha u}\{\mathbf{u}\}_e + [\mathbf{k}]_{\alpha\alpha}\{\boldsymbol{\alpha}\} = 0 \tag{4.104}$$

Solving for $\{\boldsymbol{\alpha}\}$ gives

$$\{\boldsymbol{\alpha}\} = -[\mathbf{k}]_{\alpha\alpha}^{-1}[\mathbf{k}]_{\alpha u}\{\mathbf{u}\}_e \tag{4.105}$$

Introducing (4.105) into (4.103) shows that the strain energy can be expressed in terms of the following (8×8) stiffness matrix

$$[\mathbf{k}] = [\mathbf{k}]_{uu} - [\mathbf{k}]_{u\alpha}\,[\mathbf{k}]_{\alpha\alpha}^{-1}\,[\mathbf{k}]_{\alpha u} \tag{4.106}$$

This procedure is known as static condensation. Since the additional degrees of freedom are not node point degrees of freedom, displacements will not be continuous between elements and so the element will be a non-conforming one.

The displacement functions (4.102) are only used to calculate the element stiffness matrix. The inertia matrix is evaluated using (4.34) as described in Section 4.2.

If the shear wall shown in Figure 4.11 is analysed using this modified stiffness matrix, then the results obtained are as shown in Table 4.3. Comparing these with those given in Table 4.2, which were obtained using the unmodified stiffness matrix, shows that the accuracy of the flexural modes has increased substantially.

An alternative way of increasing the accuracy of a rectangular element is to introduce additional node points at the mid points of the edges as shown in Figure 4.15. The additional node points 5 to 8 have two degrees of freedom each,

Table 4.3. *Comparison of predicted frequencies (Hz) using rectangular elements with extra displacement functions and analytical beam frequencies*

Mode	FEM	Analytical [4.3]	% Difference
1	4.984	4.973	0.22
2	26.882	26.391	1.86
3	32.014	31.944	0.22
4	65.376	62.066	5.33
5	97.234	95.832	1.46

namely the components of displacement u and v, just like nodes 1 to 4. The displacements are given by [4.7]

$$u = \sum_{j=1}^{8} N_j(\xi, \eta)\, u_j, \qquad v = \sum_{j=1}^{8} N_j(\xi, \eta) v_j \tag{4.107}$$

where

$$N_j(\xi, \eta) = \tfrac{1}{4}(1 + \xi_j \xi)(1 + \eta_j \eta)(\xi_j \xi + \eta_j \eta - 1) \tag{4.108}$$

for nodes 1 to 4

$$N_j(\xi, \eta) = \tfrac{1}{2}(1 - \xi^2)(1 + \eta_j \eta) \tag{4.109}$$

for nodes 5 and 7, and

$$N_j(\xi, \eta) = \frac{1}{2}(1 + \xi_j \xi)(1 - \eta^2) \tag{4.110}$$

for nodes 6 and 8, where (ξ_j, η_j) are the coordinates of node j.

The stiffness matrix should be evaluated using either (2×2) or (3×3) array of Gauss integration points depending upon the application [4.6]. The inertia matrix can be evaluated exactly using a (3×3) mesh of integration points.

If the shear wall of Example 4.2 is analysed using the idealisation shown in Figure 4.16, then the results obtained using a (3×3) array of integration points for both stiffness and inertia matrices are as shown in Table 4.4.

Comparing Tables 4.4 and 4.3 shows that an improvement in both the flexural and longitudinal mode frequencies has been obtained. In addition the total number of degrees of freedom has reduced from 48 to 40.

The element shown in Figure 4.15 can be transformed into a straight sided quadrilateral, as shown in Figure 4.17, using the relationships (4.55). Such an

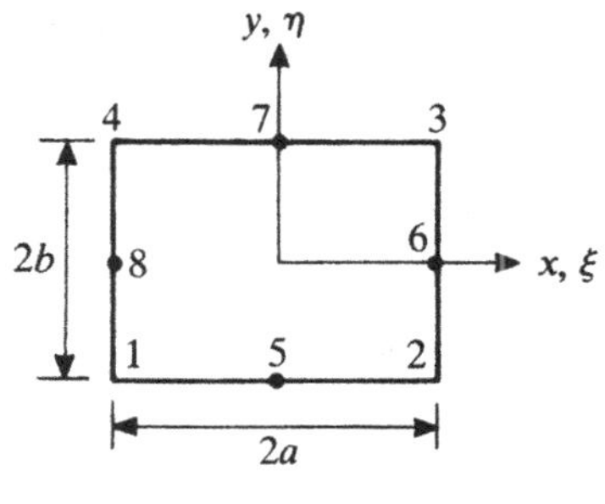

Figure 4.15. Geometry of an eight node rectangular element $\xi = x/a$, $\eta = y/b$.

Table 4.4. *Comparison of predicted frequencies (Hz) using rectangular 8-node elements and analytical beam frequencies*

Mode	FEM	Analytical [4.3]	% Difference
1	4.986	4.973	0.26
2	26.327	26.391	−0.24
3	31.964	31.944	0.06
4	62.648	62.066	0.94
5	95.955	95.832	0.13

element is referred to as a 'sub-parametric' element. In this case the inertia matrix can be evaluated exactly using a (3×3) array of Gauss integration points. It is also recommended [4.8] that a (3×3) array be used for the stiffness matrix.

The accuracy of the triangular element described in Sections 4.1 and 4.5 can also be increased by introducing additional node points at the mid-points of the sides as shown in Figure 4.18. In this case the components of displacement are given by [4.7]

$$u = \sum_{j=1}^{6} N_j u_j, \qquad v = \sum_{j=1}^{6} N_j v_j \tag{4.111}$$

where

$$N_j = (2L_j - 1)\, L_j \qquad \text{for } j = 1, 2, 3$$

and

$$N_4 = 4L_1 L_2, \qquad N_5 = 4L_2 L_3, \qquad N_6 = 4L_3 L_1 \tag{4.112}$$

The inertia and stiffness matrices can be evaluated in the way described in Section 4.5.

If the above elements are used to idealise a membrane with a curved boundary (e.g., the boundary of a cut-out), then the curved boundary will be replaced by a polygonal one. The number of straight line segments can be increased until a desired geometrical accuracy is obtained. However, an accurate geometry does not

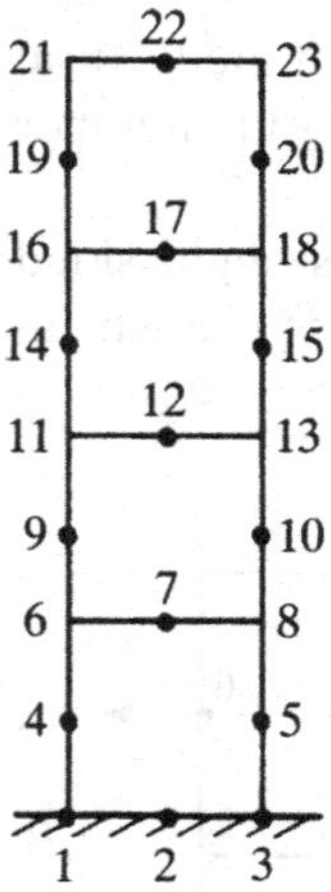

Figure 4.16. Idealisation of a cantilever shear wall.

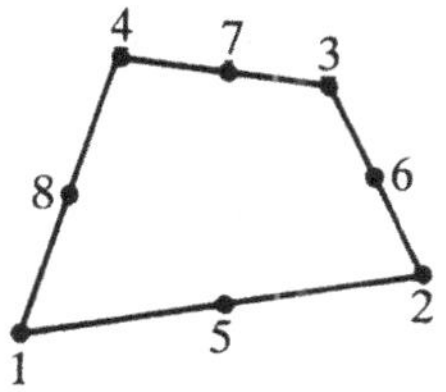

Figure 4.17. Geometry of a higher order quadrilateral element.

necessarily indicate an accurate numerical solution, as discussed in reference [4.9] and the many references cited. In such a situation, it is better to use elements with curved sides near the curved boundary and straight sided elements elsewhere.

The eight node rectangle (Figure 4.15) can be transformed into a quadrilateral with curved sides, as shown in Figure 4.19(*a*), using the relationships

$$x = \sum_{j=1}^{8} N_j(\xi, \eta)x_j, \qquad y = \sum_{j=1}^{8} N_j(\xi, \eta)y_j \tag{4.113}$$

where the functions $N_j(\xi, \eta)$ are defined by (4.108) to (4.110). Hence the element is an isoparametric one. In this case the determinant of the Jacobian, det $[\mathbf{J}]$, is cubic [4.8]. Therefore, the inertia matrix can be integrated exactly using a (4×4) array of integration points. The stiffness matrix should be integrated using either a (3×3) or a (4×4) array.

The six node triangle (Figure 4.18) can be transformed into one with curved sides, as shown in Figure 4.19(*b*), using the relationships

$$x = \sum_{j=1}^{6} N_j x_j, \qquad y = \sum_{j=1}^{6} N_j y_j \tag{4.114}$$

where the functions N_j are defined by (4.112).

The inertia and stiffness matrices can be evaluated using (4.65) and (4.71) in Section 4.3 if the (ξ, η) coordinates are defined by

$$L_1 = \xi, \qquad L_2 = \eta, \qquad L_3 = 1 - \xi - \eta \tag{4.115}$$

This gives

$$\frac{\partial}{\partial \xi} = \frac{\partial}{\partial L_1} - \frac{\partial}{\partial L_3}, \qquad \frac{\partial}{\partial \eta} = \frac{\partial}{\partial L_2} - \frac{\partial}{\partial L_3} \tag{4.116}$$

The expressions (4.65) and (4.71) are again evaluated using numerical integration. Details of some numerical integration schemes for triangles are given in Section 5.5. The displacement functions are quadratic and so is det $[\mathbf{J}]$. The inertia matrix can, therefore, be evaluated exactly using twelve integration points. At least three integration points should be used for the stiffness matrix in order to evaluate the volume

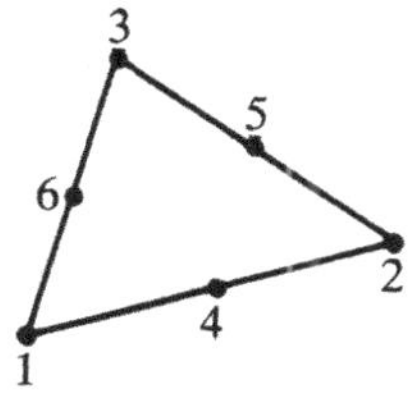

Figure 4.18. Geometry of a higher order triangular element.

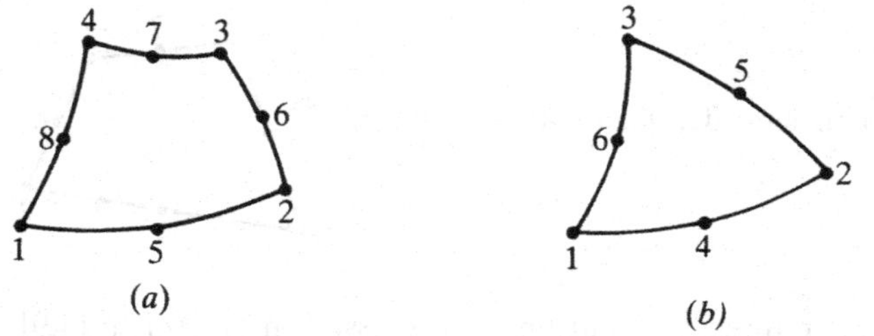

Figure 4.19. Geometry of elements with curved boundaries: (*a*) quadrilateral, (*b*) triangle.

of the element correctly. It may be that in some situations more may be necessary to obtain good accuracy. More details regarding these elements can be found in references [4.7] and [4.10].

Problems

Note: Problem 4.12 requires the use of a digital computer.

4.1 Explain why the sum of the elements in the first, third and fifth rows and columns of the inertia matrix of the constant strain triangle (4.20), is equal to the mass of the element.

4.2 Show that the sum of the first, third and fifth columns of the stiffness matrix of the constant strain triangle (4.25) is zero. Explain this result.

4.3 Show that when the strain matrix (4.52) is used the stiffness matrix of the linear rectangle is

$$[\mathbf{k}]_e = \frac{Eh}{24(1-\nu^2)} \begin{bmatrix} \mathbf{k}_{11} & \mathbf{k}_{21}^{\mathrm{T}} \\ \mathbf{k}_{21} & \mathbf{k}_{22} \end{bmatrix}$$

where

$$\mathbf{k}_{11} = \begin{bmatrix} 8\beta + 3\alpha\,(1-\nu) & & & \text{Sym} \\ 3\,(1+\nu) & 8\alpha + 3\beta\,(1-\nu) & & \\ -8\beta + 3\alpha\,(1-\nu) & -3\,(3\nu-1) & 8\beta + 3\alpha\,(1-\nu) & \\ 3\,(3\nu-1) & 4\alpha - 3\beta\,(1-\nu) & -3\,(1+\nu) & 8\alpha + 3\beta\,(1-\nu) \end{bmatrix}$$

where $\alpha = a/b$, $\beta = b/a$.

$$\mathbf{k}_{21} = \begin{bmatrix} -4\beta - 3\alpha(1-\nu) & & & \text{Sym} \\ -3(1+\nu) & -4\alpha - 3\beta(1-\nu) & & \\ 4\beta - 3\alpha(1-\nu) & 3(3\nu-1) & -4\beta - 3\alpha(1-\nu) & \\ -3(3\nu-1) & -8\alpha + 3\beta(1-\nu) & 3(1+\nu) & -4\alpha - 3\beta(1-\nu) \end{bmatrix}$$

$\mathbf{k}_{22} = \mathbf{k}_{11}$

4.4 An alternative proceudre for eliminating parasitic shear in a rectangular element, Figure P4.4(*a*), is to divide it into four overlapping triangles as shown in Figure P4.4(*b*) [4.11]. Each triangle is treated as a linear triangle (Section 4.1) and a weighted average shear strain defined by

$$\bar{\gamma}_{xy} = \frac{\sum_{i=1}^{4} A_i \gamma_i}{\sum_{i=1}^{4} A_i}$$

where A_i, γ_i are the area and shear strain of an individual triangle.

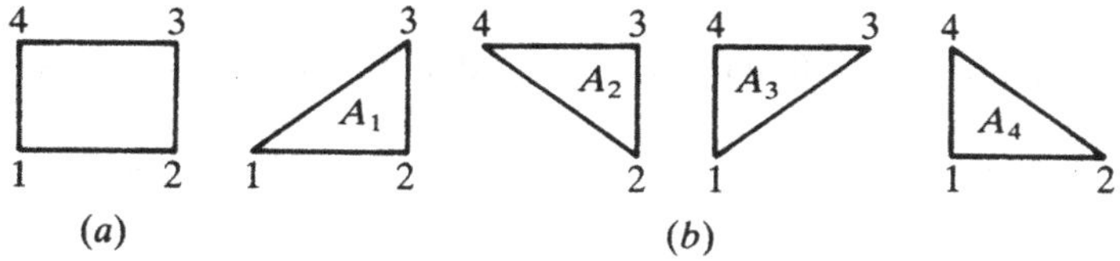

Figure P4.4

Show that this procedure gives the same shear strain as the third row of (4.52).

4.5 Show that rigid body motions do not cause any strain in the linear quadrilateral element (Section 4.3).

4.6 Show that the inertia matrix of the six node triangle is

$$[\mathbf{M}]_e = \begin{bmatrix} \mathbf{m}_{11} & \mathbf{m}_{12} \\ \mathbf{m}_{12}^{\mathrm{T}} & \mathbf{m}_{22} \end{bmatrix}$$

where

$$\mathbf{m}_{11} = \frac{\rho A h}{180} \begin{bmatrix} 6 & 0 & -1 & 0 & -1 & 0 \\ & 6 & 0 & -1 & 0 & -1 \\ & & 6 & 0 & -1 & 0 \\ \text{Sym} & & & 6 & 0 & -1 \\ & & & & 6 & 0 \\ & & & & & 6 \end{bmatrix}$$

$$\mathbf{m}_{12} = \frac{\rho A h}{180} \begin{bmatrix} 0 & 0 & -4 & 0 & 0 & 0 \\ 0 & 0 & 0 & -4 & 0 & 0 \\ 0 & 0 & 0 & 0 & -4 & 0 \\ 0 & 0 & 0 & 0 & 0 & -4 \\ -4 & 0 & 0 & 0 & 0 & 0 \\ 0 & -4 & 0 & 0 & 0 & 0 \end{bmatrix}$$

$$\mathbf{m}_{22} = \frac{\rho A h}{180} \begin{bmatrix} 32 & 0 & 16 & 0 & 16 & 0 \\ & 32 & 0 & 16 & 0 & 16 \\ & & 32 & 0 & 16 & 0 \\ \text{Sym} & & & 32 & 0 & 16 \\ & & & & 32 & 0 \\ & & & & & 32 \end{bmatrix}$$

4.7 Show that the stiffness matrix of an isotropic linear triangular element whose thickness varies linearly is

$$[\mathbf{k}]_e = \bar{h}A\,[\mathbf{B}]^{\mathrm{T}}\,[\mathbf{D}]\,[\mathbf{B}]$$

where $[\mathbf{B}]$ and $[\mathbf{D}]$ are given by (4.24) and (2.51) respectively, A is the area of the triangle and $\bar{h}$ the mean thickness $\frac{1}{3}(h_1 + h_2 + h_3)$, where h_1, h_2 and h_3 are the nodal thicknesses.

4.8 Show that the inertia matrix of a linear triangular element whose thickness varies linearly is

$$[\mathbf{m}]_e = \frac{\rho\bar{h}A}{60}\begin{bmatrix} (6+4\alpha_1) & & & & & \\ 0 & (6+4\alpha_1) & & \text{Sym} & & \\ (6-\alpha_3) & 0 & (6+4\alpha_2) & & & \\ 0 & (6-\alpha_3) & 0 & (6+4\alpha_2) & & \\ (6-\alpha_2) & 0 & (6-\alpha_1) & 0 & (6+4\alpha_3) & \\ 0 & (6-\alpha_2) & 0 & (6-\alpha_1) & 0 & (6+4\alpha_3) \end{bmatrix}$$

where ρ is the density, A the area, $\bar{h}$ the mean thickness and $\alpha_j = h_j/\bar{h}$ for $j = 1, 2, 3$.

4.9 If the thickness variation of a linear rectangular element is given by

$$h(\xi, \eta) = \sum_{j=1}^{4} N_j h_j$$

where the N_j are defined by (4.36) and the h_j are the nodal values of thickness, how many Gauss integration points are required to evaluate the inertia and stiffness matrices?

4.10 If the thickness variation of a linear quadrilateral is the same as the rectangular element in Problem 4.9, how many Gauss integration points are required to evaluate the inertia matrix exactly? How many Gauss integration points are required to integrate the volume exactly?

4.11 Show that the sum of the first, third, fifth and seventh columns of the stiffness matrix of the linear rectangular element given in Problem 4.3 is zero. Explain this result.

4.12 Figure P4.12 shows the cross-section of an earth dam which is assumed to be rigidly fixed at its base. Assume that a plane strain condition exists, that E is 5.605×10^8 N/m^2, ν is 0.45 and ρ is 2082 kg/m^3. Calculate the four lowest frequencies and mode shapes using linear triangles. Compare the frequencies with the

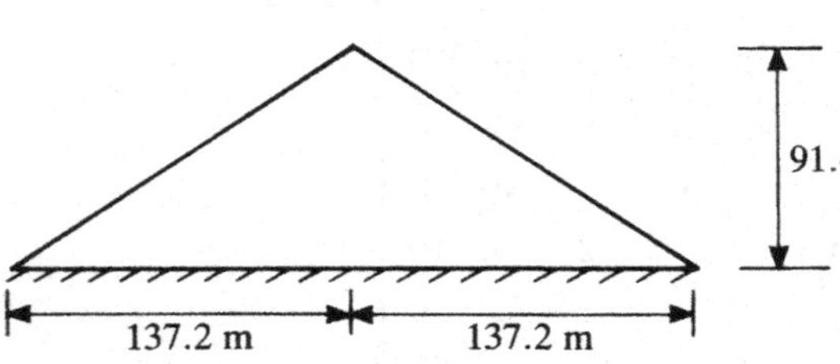

Figure P4.12

values obtained using 100 elements [4.12] 1.227 (A), 1.993 (S), 2.324 (A), 3.073 (A) Hz, where A and S indicate antisymmetric and symmetric modes respectively.

Note: A plane strain element can be derived from a plane stress element by replacing E by $E/(1-\nu^2)$ and ν by $\nu/(1-\nu)$.

4.13 Derive the equivalent nodal forces corresponding to a distributed load p_x per unit length in the x-direction along the side 2–3 in Figure 4.15.

5 Vibration of Solids

Solid type structures, such as machinery components, can be analysed using three-dimensional finite elements. These can be either tetrahedral, pentahedral or hexahedral in shape. However, if the structure is axisymmetric, the three-dimensional analysis can be replaced by a sequence of two-dimensional problems by expanding the loading and displacements as Fourier series in the circumferential coordinate. These two-dimensional problems can be solved by techniques similar to those described in Chapter 4.

5.1 Axisymmetric Solids

An axisymmetric solid can be generated by rotating a plane area, Figure 5.1(*a*), through a full revolution about the z-axis, which lies in the plane of the area. The resulting solid is shown in Figure 5.1(*b*).

The energy expressions for an axisymmetric solid are, from Section 2.9

$$T_e = \frac{1}{2}\int_V \rho(\dot{u}^2 + \dot{v}^2 + \dot{w}^2)\,\mathrm{d}V \tag{5.1}$$

$$U_e = \frac{1}{2}\int_V (\{\boldsymbol{\varepsilon}_1\}^{\mathrm{T}}[\mathbf{D}_1]\{\boldsymbol{\varepsilon}_1\} + \{\boldsymbol{\varepsilon}_2\}^{\mathrm{T}}[\mathbf{D}_2]\{\boldsymbol{\varepsilon}_2\})\,\mathrm{d}V \tag{5.2}$$

with

$$\{\boldsymbol{\varepsilon}_1\} = \begin{bmatrix} \dfrac{\partial u}{\partial r} \\ \dfrac{u}{r} + \dfrac{1}{r}\dfrac{\partial v}{\partial \theta} \\ \dfrac{\partial w}{\partial z} \\ \dfrac{\partial u}{\partial z} + \dfrac{\partial w}{\partial r} \end{bmatrix}, \qquad \{\boldsymbol{\varepsilon}_2\} = \begin{bmatrix} \dfrac{1}{r}\dfrac{\partial u}{\partial \theta} + \dfrac{\partial v}{\partial r} - \dfrac{v}{r} \\ \dfrac{\partial v}{\partial z} + \dfrac{1}{r}\dfrac{\partial w}{\partial \theta} \end{bmatrix} \tag{5.3}$$

$$[\mathbf{D}_1] = \frac{E}{(1+\nu)(1-2\nu)}\begin{bmatrix} (1-\nu) & \nu & \nu & 0 \\ \nu & (1-\nu) & \nu & 0 \\ \nu & \nu & (1-\nu) & 0 \\ 0 & 0 & 0 & \frac{1}{2}(1-2\nu) \end{bmatrix} \tag{5.4}$$

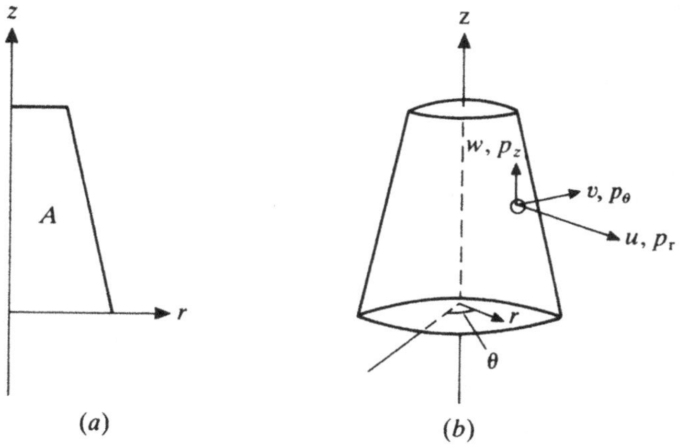

Figure 5.1. Solid of revolution.

and

$$[\mathbf{D}_2] = \frac{E}{2(1+\nu)}\begin{bmatrix}1 & 0\\ 0 & 1\end{bmatrix} \tag{5.5}$$

Also

$$\delta W = \int_S (p_r\delta u + p_\theta\delta v + p_z\delta w)\,\mathrm{d}S \tag{5.6}$$

5.2 Applied Loading

Even though the solid is axisymmetric, the applied loading need not be. If this is the case, then the loading distribution is represented by means of a Fourier series in the angular coordinate θ. This means that the components of the applied load can be expressed in the following form:

$$\begin{aligned}\begin{bmatrix}p_r\\ p_z\end{bmatrix} &= \sum_{n=0}^{\infty}\begin{bmatrix}p_{rn}\\ p_{zn}\end{bmatrix}\cos n\theta + \sum_{n=1}^{\infty}\begin{bmatrix}\bar{p}_{rn}\\ \bar{p}_{zn}\end{bmatrix}\sin n\theta\\ p_\theta &= \sum_{n=1}^{\infty} p_{\theta n}\sin n\theta - \sum_{n=0}^{\infty}\bar{p}_{\theta n}\cos n\theta\end{aligned} \tag{5.7}$$

The terms without a bar represent loading which is symmetric about $\theta = 0$ and those with a bar represent loading which is antisymmetric about $\theta = 0$. The terms (p_{r0}, p_{z0}) represent axisymmetric loading and $\bar{p}_{\theta 0}$ represents pure torsion.

Using the following relationships

$$\begin{aligned}&\int_{-\pi}^{+\pi}\sin m\theta\,\sin n\theta\,\mathrm{d}\theta = \begin{cases}\pi & m = n \neq 0\\ 0 & m \neq n,\ m = n = 0\end{cases}\\ &\int_{-\pi}^{+\pi}\cos m\theta\,\cos n\theta\,\mathrm{d}\theta = \begin{cases}2\pi & m = n = 0\\ \pi & m = n \neq 0\\ 0 & m \neq 0\end{cases}\\ &\int_{-\pi}^{+\pi}\sin m\theta\,\cos n\theta\,\mathrm{d}\theta = 0 \qquad \text{all } m \text{ and } n\end{aligned} \tag{5.8}$$

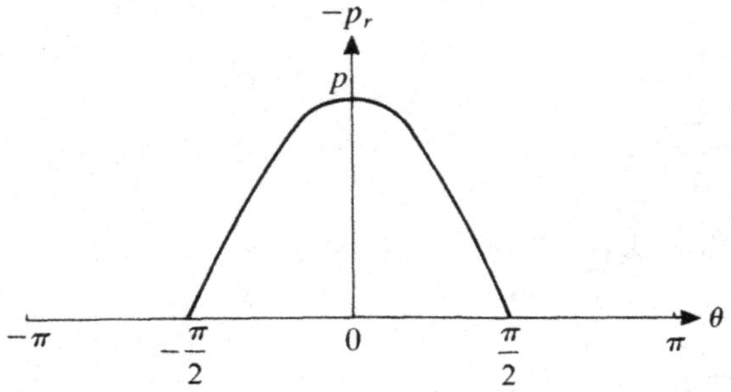

Figure 5.2. Pressure loading distribution.

It can be shown that the coefficients in (5.7) are given by

$$
\begin{aligned}
\begin{bmatrix} p_{r0} \\ p_{z0} \end{bmatrix} &= \frac{1}{2\pi}\int_{-\pi}^{+\pi}\begin{bmatrix} p_r \\ p_z \end{bmatrix} \mathrm{d}\theta \\
\begin{bmatrix} p_{rn} \\ p_{zn} \end{bmatrix} &= \frac{1}{\pi}\int_{-\pi}^{+\pi}\begin{bmatrix} p_r \\ p_z \end{bmatrix} \cos n\theta \,\mathrm{d}\theta \qquad n \neq 0 \\
\begin{bmatrix} \bar{p}_{rn} \\ \bar{p}_{zn} \end{bmatrix} &= \frac{1}{\pi}\int_{-\pi}^{+\pi}\begin{bmatrix} p_r \\ p_z \end{bmatrix} \sin n\theta \,\mathrm{d}\theta \qquad n \neq 0 \\
p_{\theta n} &= \frac{1}{\pi}\int_{-\pi}^{+\pi} p_\theta \sin n\theta \,\mathrm{d}\theta \qquad n \neq 0 \\
\bar{p}_{\theta 0} &= -\frac{1}{2\pi}\int_{-\pi}^{+\pi} p_\theta \,\mathrm{d}\theta \\
\bar{p}_{\theta n} &= -\frac{1}{\pi}\int_{-\pi}^{+\pi} p_\theta \cos n\theta \,\mathrm{d}\theta \qquad n \neq 0
\end{aligned}
\tag{5.9}
$$

EXAMPLE 5.1 Find the Fourier series representation of the loading

$$
p_r = \begin{cases} -p\cos\theta & -\dfrac{\pi}{2} < \theta < +\dfrac{\pi}{2} \\ 0 & -\pi < \theta < -\dfrac{\pi}{2}, \dfrac{\pi}{2} < \theta < \pi \end{cases}
$$

which is illustrated in Figure 5.2.

The relationships (5.9) give

$$
p_{r0} = -\frac{1}{2\pi}\int_{-\pi/2}^{+\pi/2} p\cos\theta \,\mathrm{d}\theta = -\frac{p}{\pi}
$$

$$
p_{r1} = -\frac{1}{\pi}\int_{-\pi/2}^{+\pi/2} p\cos^2\theta \,\mathrm{d}\theta = -\frac{p}{2}
$$

$$
p_{rn} = -\frac{1}{\pi}\int_{-\pi/2}^{+\pi/2} p\cos\theta\cos n\theta \,\mathrm{d}\theta \qquad \text{for } n > 1
$$

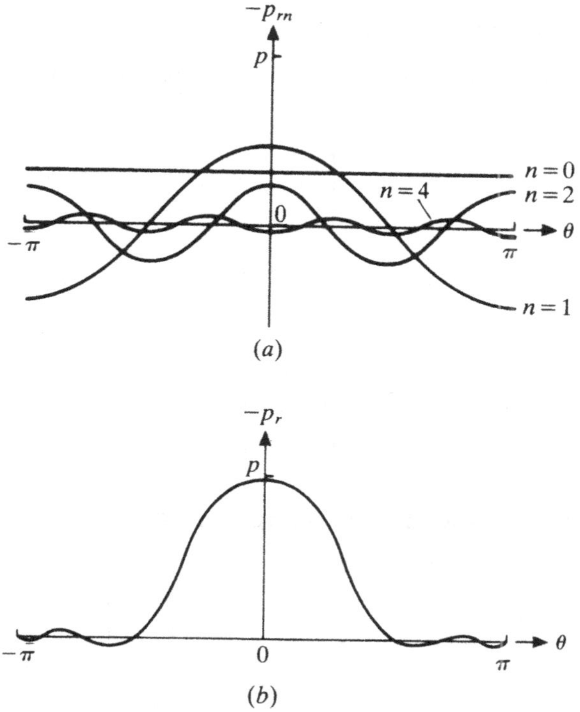

Figure 5.3. (*a*) Harmonic components; (*b*) Fourier series representation.

Now

$$\begin{aligned}
&\int_{-\pi/2}^{+\pi/2} \cos\theta \cos n\theta \, d\theta \\
&= \frac{1}{2}\int_{-\pi/2}^{+\pi/2} \{\cos(n+1)\theta + \cos(n-1)\theta\}\, d\theta \\
&= \left\{\frac{\sin(n+1)\pi/2}{(n+1)} + \frac{\sin(n-1)\pi/2}{(n-1)}\right\} \\
&= -\frac{2\cos(n\pi/2)}{(n^2-1)}
\end{aligned}$$

Therefore

$$p_{rn} = \begin{cases} 0 & n \text{ odd} \\ \dfrac{2p(-1)^{n/2}}{\pi(n^2-1)} & n \text{ even} \end{cases}$$

The Fourier series expansion is, therefore

$$p_r = -\frac{p}{\pi} - \frac{p}{2}\cos\theta - \frac{2p}{3\pi}\cos 2\theta + \frac{2p}{15\pi}\cos 4\theta \cdots$$

The harmonic components are illustrated in Figure 5.3(*a*) and the Fourier series representation in Figure 5.3(*b*).

5.3 Displacements

The displacement components can be represented in a similar manner to the applied loading, that is

$$\begin{bmatrix} u \\ w \end{bmatrix} = \sum_{n=0}^{\infty} \begin{bmatrix} u_n \\ w_n \end{bmatrix} \cos n\theta - \sum_{n=1}^{\infty} \begin{bmatrix} \bar{u}_n \\ \bar{w}_n \end{bmatrix} \sin n\theta$$
$$v = \sum_{n=1}^{\infty} v_n \sin n\theta - \sum_{n=0}^{\infty} \bar{v}_n \cos n\theta \tag{5.10}$$

where the unbarred and barred terms represent motion which is symmetric and anti-symmetric about $\theta = 0$. The reason for using a negative sign in the expression for v is explained below equation (5.12).

5.4 Reduced Energy Expressions

The expressions (5.7) and (5.10) are substituted into (5.1), (5.2) and (5.6) and integrated with respect to θ between the limits $-\pi$ and $+\pi$. Using the relations (5.8) reduces the energy expressions to sums of the energies corresponding to each harmonic component. That is, there is no coupling between them. Thus, the motion corresponding to each harmonic component can be determined separately. The energy expressions for a single harmonic component are, for $n \neq 0$.

$$T = \frac{1}{2}\pi \int_A \rho(\dot{u}_n^2 + \dot{v}_n^2 + \dot{w}_n^2) r \, dA$$
$$U = \frac{1}{2}\pi \int_A [\{\boldsymbol{\varepsilon}_1\}_n^{\mathrm{T}} [\mathbf{D}_1] \{\boldsymbol{\varepsilon}_1\}_n + \{\boldsymbol{\varepsilon}_2\}_n^{\mathrm{T}} [\mathbf{D}_2] \{\boldsymbol{\varepsilon}_2\}_n] r \, dA \tag{5.11}$$
$$\delta W = \pi \int_s (p_{rn}\delta u_n + p_{\theta n}\delta v_n + p_{zn}\delta w_n) r_s \, ds$$

where

$$\{\boldsymbol{\varepsilon}_1\}_n = \begin{bmatrix} \dfrac{\partial u_n}{\partial r} \\ \dfrac{u_n}{r} + \dfrac{n v_n}{r} \\ \dfrac{\partial w_n}{\partial z} \\ \dfrac{\partial u_n}{\partial z} + \dfrac{\partial w_n}{\partial r} \end{bmatrix}, \qquad \{\boldsymbol{\varepsilon}_2\}_n = \begin{bmatrix} \dfrac{-n u_n}{r} + \dfrac{\partial v_n}{\partial r} - \dfrac{v_n}{r} \\ \dfrac{\partial v_n}{\partial z} - \dfrac{n w_n}{r} \end{bmatrix} \tag{5.12}$$

r_S is the value of r on the surface S, ds is an element of arc along the generator curve of the surface, and A is the area of the generator plane (see Figure 5.1).

When considering the barred terms in (5.10) the strain matrix $\{\boldsymbol{\varepsilon}_1\}_n$ remains unchanged and $\{\boldsymbol{\varepsilon}_2\}_n$ changes sign. This means that the energy expressions (5.11) are the same for both symmetric and antisymmetric motion. (This is the reason for taking the negative sign in the expression for v in (5.10).) The case $n = 0$ needs to be considered separately.

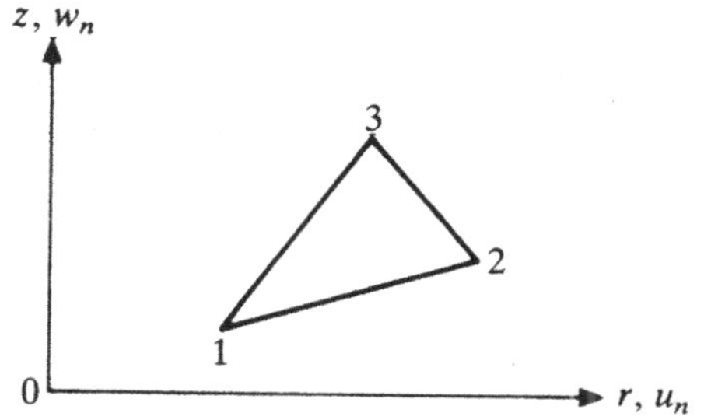

Figure 5.4. Geometry of a triangular element.

For axisymmetric motion

$$
\begin{aligned}
T &= \frac{1}{2}(2\pi)\int_A \rho(\dot{u}_0^2 + \dot{w}_0^2)r\,\mathrm{d}A \\
U &= \frac{1}{2}(2\pi)\int_A \{\boldsymbol{\varepsilon}_1\}_0^{\mathrm{T}}[\mathbf{D}_1]\{\boldsymbol{\varepsilon}_1\}_0\, r\,\mathrm{d}A \\
\delta W &= 2\pi\int_s (p_{r0}\delta u_0 + p_{z0}\delta w_0)r_s\,\mathrm{d}s
\end{aligned} \tag{5.13}
$$

For antisymmetric motion (pure torsion)

$$
\begin{aligned}
T &= \frac{1}{2}(2\pi)\int_A \rho\dot{\bar{v}}_0^2 r\,\mathrm{d}A \\
U &= \frac{1}{2}(2\pi)\int_A \{\boldsymbol{\varepsilon}_2\}_0^{\mathrm{T}}[\mathbf{D}_2]\{\boldsymbol{\varepsilon}_2\}_0\, r\,\mathrm{d}A \\
\delta W &= 2\pi\int_s \bar{p}_{\theta 0}\delta\bar{v}_0 r_s\,\mathrm{d}s
\end{aligned} \tag{5.14}
$$

The motion corresponding to a single harmonic component is obtained using a two-dimensional finite element idealisation. Any of the element shapes and associated displacement functions used for membrane analysis in Chapter 4 can be used. The procedure is illustrated in the next section.

5.5 Linear Triangular Element

Figure 5.4 shows a triangular element with three node points, one at each vertex. There are three degrees of freedom at each node, namely, u_n, v_n and w_n corresponding to the harmonic component n. u_n and w_n are in the directions of the r- and z-axes as indicated. v_n is in the θ-direction which is perpendicular to both r and z.

Following the procedure presented in Section 4.5 for a linear membrane triangle, the displacement functions can be expressed in the form

$$
\begin{aligned}
u_n &= \sum_{j=1}^{3} N_j u_{nj} \\
v_n &= \sum_{j=1}^{3} N_j v_{nj} \\
w_n &= \sum_{j=1}^{3} N_j w_{nj}
\end{aligned} \tag{5.15}
$$

where

$$N_j = L_j \tag{5.16}$$

(Remember that the L_j are area coordinates for the triangle), and (u_{nj}, v_{nj}, w_{nj}) are the displacement components at node j.

The expression (5.15) can be written in the combined form

$$\begin{bmatrix} u_n \\ v_n \\ w_n \end{bmatrix} = [\mathbf{N}]\,\{\mathbf{u}_n\}_e \tag{5.17}$$

with

$$\{\mathbf{u}_n\}_e^{\mathrm{T}} = \lfloor u_{n1} \quad v_{n1} \quad w_{n1} \quad u_{n2} \quad v_{n2} \quad w_{n2} \quad u_{n3} \quad v_{n3} \quad w_{n3} \rfloor \tag{5.18}$$

and

$$[\mathbf{N}] = \begin{bmatrix} N_1 & 0 & 0 & N_2 & 0 & 0 & N_3 & 0 & 0 \\ 0 & N_1 & 0 & 0 & N_2 & 0 & 0 & N_3 & 0 \\ 0 & 0 & N_1 & 0 & 0 & N_2 & 0 & 0 & N_3 \end{bmatrix} \tag{5.19}$$

Considering the case $n \neq 0$, the energy expressions are obtained by substituting (5.17) into the expressions (5.11) and integrating over the area of the element.

The kinetic energy is given by

$$T_e = \tfrac{1}{2}\,\{\dot{\mathbf{u}}_n\}_e^{\mathrm{T}}\,[\mathbf{m}]_e\,\{\dot{\mathbf{u}}\}_e \tag{5.20}$$

where

$$[\mathbf{m}]_e = \pi\rho \int_A [\mathbf{N}]^{\mathrm{T}}[\mathbf{N}]\,r\,\mathrm{d}A \tag{5.21}$$

and A now denotes the area of the element.

A typical element of the inertia matrix is of the form $\pi\rho \int_A L_i L_j r\,\mathrm{d}A$. By analogy with (4.82), r can be expressed as

$$r = r_1 L_1 + r_2 L_2 + r_3 L_3 \tag{5.22}$$

where r_j is the value of r at node j. Therefore

$$\pi\rho \int_A L_i L_j\,\mathrm{d}A = \pi\rho \int_A L_i L_j (r_1 L_1 + r_2 L_2 + r_3 L_3)\,\mathrm{d}A$$

Integrals of the form $\int_A L_i L_j L_l\,\mathrm{d}A$ can be evaluated using (4.90). After evaluating these integrals, the inertia matrix is given by

$$[\mathbf{m}]_e = \frac{\pi\rho A}{60} \begin{bmatrix} I_{11}[\mathbf{I}] & I_{12}[\mathbf{I}] & I_{13}[\mathbf{I}] \\ I_{12}[\mathbf{I}] & I_{22}[\mathbf{I}] & I_{23}[\mathbf{I}] \\ I_{13}[\mathbf{I}] & I_{23}[\mathbf{I}] & I_{33}[\mathbf{I}] \end{bmatrix} \tag{5.23}$$

where $[\mathbf{I}]$ is a unit matrix of order 3 and

$$\begin{aligned} I_{11} &= 2(3r_1 + r_2 + r_3) \\ I_{12} &= (2r_1 + 2r_2 + r_3) \\ I_{13} &= (2r_1 + r_2 + 2r_3) \\ I_{22} &= 2(r_1 + 3r_2 + r_3) \\ I_{23} &= (r_1 + 2r_2 + 2r_3) \\ I_{33} &= 2(r_1 + r_2 + 3r_3) \end{aligned} \tag{5.24}$$

Note that the inertia matrix is independent of n.

The strain energy is given by

$$U_e = \tfrac{1}{2}\{\mathbf{u}_n\}_e^{\mathrm{T}}[\mathbf{k}]_e\{\mathbf{u}_n\}_e \tag{5.25}$$

where

$$[\mathbf{k}]_e = \pi\int_A [\mathbf{B}_1]^{\mathrm{T}}[\mathbf{D}_1][\mathbf{B}_1]r\,\mathrm{d}A + \pi\int_A [\mathbf{B}_2]^{\mathrm{T}}[\mathbf{D}_2][\mathbf{B}_2]r\,\mathrm{d}A \tag{5.26}$$

is the element stiffness matrix.

The strain matrices $[\mathbf{B}_1]$ and $[\mathbf{B}_2]$ are of the form

$$\begin{aligned}[]
[\mathbf{B}_1] &= [\mathbf{B}_{11} \quad \mathbf{B}_{12} \quad \mathbf{B}_{13}]\\
[\mathbf{B}_2] &= [\mathbf{B}_{21} \quad \mathbf{B}_{22} \quad \mathbf{B}_{23}]
\end{aligned} \tag{5.27}$$

where

$$\mathbf{B}_{1i} = \begin{bmatrix} \partial N_i/\partial r & 0 & 0 \\ N_i/r & nN_i/r & 0 \\ 0 & 0 & \partial N_i/\partial z \\ \partial N_i/\partial z & 0 & \partial N_i/\partial r \end{bmatrix} \tag{5.28}$$

and

$$\mathbf{B}_{2i} = \begin{bmatrix} -nN_i/r & (\partial N_i/\partial r - N_i/r) & 0 \\ 0 & \partial N_i/\partial z & -nN_i/r \end{bmatrix} \tag{5.29}$$

Now from (5.16)

$$N_i = L_i \tag{5.30}$$

Following the techniques presented in Section (4.5), differentiation with respect to r and z can be carried out as follows:

$$\frac{\partial}{\partial r} = \sum_{j=1}^{3} \frac{\partial L_j}{\partial r}\frac{\partial}{\partial L_j}, \qquad \frac{\partial}{\partial z} = \sum_{j=1}^{3} \frac{\partial L_j}{\partial z}\frac{\partial}{\partial L_j} \tag{5.31}$$

By analogy with (4.85)

$$L_j = \frac{1}{2A}(A_j^0 + a_j r + b_j z) \tag{5.32}$$

where A is the area of the triangle and

$$\begin{aligned}
A_j^0 &= r_l z_i - r_i z_l\\
a_j &= z_l - z_i\\
b_j &= r_i - r_l
\end{aligned} \tag{5.33}$$

with j, l, i being a cyclic permutation of 1, 2, 3.

Relations (5.30) to (5.32) together give

$$\frac{\partial N_i}{\partial r} = \frac{a_i}{2a}, \qquad \frac{\partial N_i}{\partial z} = \frac{b_i}{2a} \tag{5.34}$$

Substituting (5.34) into (5.28) and (5.29) gives

$$\mathbf{B}_{1i} = \frac{1}{2A}\begin{bmatrix} a_i & 0 & 0 \\ 2AL_i/r & 2AL_i/r & 0 \\ 0 & 0 & b_i \\ b_i & 0 & a_i \end{bmatrix} \tag{5.35}$$

and

$$\mathbf{B}_{2i} = \frac{1}{2A}\begin{bmatrix} -2AL_i/r & (a_i - 2AL_i/r) & 0 \\ 0 & b_i & -2AL_i/r \end{bmatrix} \tag{5.36}$$

In order to obtain the stiffness matrix, as given by (5.26), it is necessary to evaluate integrals of the form

$$\int_A r\,\mathrm{d}A, \qquad \int_A L_i\,\mathrm{d}A, \qquad \int_A \frac{L_i L_j}{r}\,\mathrm{d}A$$

Using (5.22)

$$\int_A r\,\mathrm{d}A = \int_A (r_1 L_1 + r_2 L_2 + r_3 L_3)\,\mathrm{d}A \tag{5.37}$$

Each term can be evaluated using (4.90). This gives

$$\int_A L_i\,\mathrm{d}A = \frac{A}{3} \tag{5.38}$$

Therefore

$$\int_A r\,\mathrm{d}A = \frac{A}{3}(r_1 + r_2 + r_3) \tag{5.39}$$

The integral $\int_A (L_i L_j/r)\,\mathrm{d}A$ can be evaluated explicitly in terms of the nodal coordinates. However, this involves a considerable amount of algebraic manipulation. It is more convenient to evaluate it numerically.

Details of numerical integration for rectangular regions are given in Section 4.2. This can be extended to triangular regions as follows. Consider the integral

$$I = \int_A g(L_1, L_2, L_3)\,\mathrm{d}A \tag{5.40}$$

This can be evaluated numerically using the following expression

$$I = A\sum_{j=1}^{N} H_j g(\xi_j, \eta_j, \zeta_j) \tag{5.41}$$

where (ξ_j, η_j, ζ_j) are the area coordinates of the N integration points and H_j the corresponding weight coefficients.

A large number of integration schemes have been proposed, most of which have their basis in reference [5.1]. Some typical schemes are given in Table 5.1. The formula number indicates the highest degree of polynomial which will be integrated exactly. The positions of the integration points are illustrated in Figure 5.5. Further schemes are tabulated in reference [5.2].

The integral $\int_A (L_i L_j/r)\,\mathrm{d}A$ cannot be evaluated exactly using numerical integration, since the integrand is a polynomial of degree two divided by a polynomial

Table 5.1. *Integration points and weight coefficients for triangles*

Formula no.	j	ξ_j	η_j	ζ_j	H_i
1	1	$\frac{1}{3}$	$\frac{1}{3}$	$\frac{1}{3}$	1
2	1	$\frac{2}{3}$	$\frac{1}{6}$	$\frac{1}{6}$	$\frac{1}{3}$
	2	$\frac{1}{6}$	$\frac{2}{3}$	$\frac{1}{6}$	$\frac{1}{3}$
	3	$\frac{1}{6}$	$\frac{1}{6}$	$\frac{2}{3}$	$\frac{1}{3}$
3	1	$\frac{1}{3}$	$\frac{1}{3}$	$\frac{1}{3}$	$-\frac{27}{48}$
	2	$\frac{3}{5}$	$\frac{1}{5}$	$\frac{1}{5}$	$\frac{25}{48}$
	3	$\frac{1}{5}$	$\frac{3}{5}$	$\frac{1}{5}$	$\frac{25}{48}$
	4	$\frac{1}{5}$	$\frac{1}{5}$	$\frac{3}{5}$	$\frac{25}{48}$
4	1	0.816 847 57	0.091 576 21	0.091 576 21	0.109 951 74
	2	0.091 576 21	0.816 847 57	0.091 576 21	0.109 951 74
	3	0.091 576 21	0.091 576 21	0.816 847 57	0.109 951 74
	4	0.108 103 02	0.445 948 49	0.445 948 49	0.223 381 59
	5	0.445 948 49	0.108 103 02	0.445 948 49	0.223 381 59
	6	0.445 948 49	0.445 948 49	0.108 303 02	0.223 381 59
5	1	0.333 333 33	0.333 333 33	0.333 333 33	0.225 000 00
	2	0.797 426 99	0.101 286 51	0.101 286 51	0.125 939 18
	3	0.101 286 51	0.797 426 99	0.101 286 51	0.125 939 18
	4	0.101 286 51	0.101 286 51	0.797 426 99	0.125 939 18
	5	0.059 715 87	0.470 142 06	0.470 142 06	0.132 394 15
	6	0.470 142 06	0.059 715 87	0.470 142 06	0.132 394 15
	7	0.407 142 06	0.470 142 06	0.059 715 87	0.132 394 15

of degree one. The minimum number of integration points required depends upon the size of the element and its distance from the z-axis. Small elements situated far away from the axis can be evaluated accurately with as few as three integration points, whilst large elements close to the axis require many more.

The integrals $\int_A L_i \,\mathrm{d}A$ and $\int_A r \,\mathrm{d}A$ can both be integrated exactly using three integration points. Therefore, the most convenient way of computing the stiffness matrix is to evaluate the complete matrix using numerical integration. The stiffness matrix (5.26) is, therefore, given by

$$\begin{aligned}[\mathbf{k}]_e = \pi A \sum_{j=1}^{N} H_j r_j ([\mathbf{B}_1(\xi_j, \eta_j, \zeta_j)]^{\mathrm{T}}[\mathbf{D}_1][\mathbf{B}_1(\xi_j, \eta_j, \zeta_j)] \\ + [\mathbf{B}_2(\xi_j, \eta_j, \zeta_j)]^{\mathrm{T}}[\mathbf{D}_2][\mathbf{B}_2(\xi_j, \eta_j, \zeta_j)])\end{aligned} \tag{5.42}$$

The stiffness matrix can also be expressed in the form

$$[\mathbf{k}]_e = [\mathbf{k}_1]_e + n[\mathbf{k}_2]_e + n^2[\mathbf{k}_3]_e \tag{5.43}$$

where $[\mathbf{k}_1]_e$, $[\mathbf{k}_2]_e$ and $[\mathbf{k}_3]_e$ are independent of n. It is therefore, better to evaluate these matrices and then obtain the stiffness matrices for various values of n from (5.43).

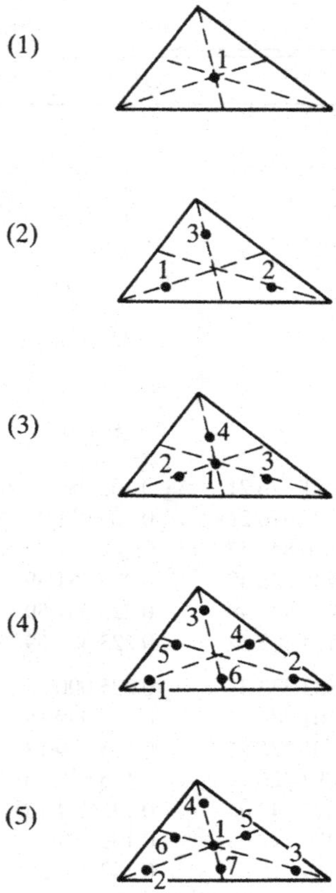

Figure 5.5. Integration points for triangles.

The equivalent nodal forces corresponding to a distributed load along the side 2–3 of the element can be calculated as follows. Substituting (5.17) into (5.11) gives

$$\delta W_e = \{\delta \mathbf{u}_n\}_e^{\mathrm{T}} \{\mathbf{f}\}_e \tag{5.44}$$

where

$$\{\mathbf{f}\}_e = \pi \int_{s_e} [\mathbf{N}]_{2-3}^{\mathrm{T}} \begin{bmatrix} p_{rn} \\ p_{\theta n} \\ p_{zn} \end{bmatrix} r_s \, \mathrm{d}s \tag{5.45}$$

Now along the side 2–3, $L_1 = 0$ and so

$$[\mathbf{N}]_{2-3}^{\mathrm{T}} = \begin{bmatrix} 0 & 0 & 0 \\ 0 & 0 & 0 \\ 0 & 0 & 0 \\ L_2 & 0 & 0 \\ 0 & L_2 & 0 \\ 0 & 0 & L_2 \\ L_3 & 0 & 0 \\ 0 & L_3 & 0 \\ 0 & 0 & L_3 \end{bmatrix} \tag{5.46}$$

Also, using (5.22)

$$r_s = r_2 L_2 + r_3 L_3 \tag{5.47}$$

Assuming p_{rn}, $p_{\theta n}$, p_{zn} to be constant on 2–3, integrals of the form

$$\int_{s_e} L_i^2 \, \mathrm{d}s, \qquad \int_{s_e} L_i L_j \, \mathrm{d}s$$

have to be evaluated. Noting that L_2 and L_3 are now one-dimensional natural coordinates, these integrals can be evaluated using (4.100). This gives

$$\int_{s_e} L_i L_j \, \mathrm{d}s = \begin{cases} l_{2\text{–}3}/3 & \text{for } j = i \\ l_{2\text{–}3}/6 & \text{for } j \neq i \end{cases} \tag{5.48}$$

where $l_{2\text{-}3}$ is the length of the side 2–3.

The equivalent nodal forces are, therefore,

$$\{\mathbf{f}\}_e = \frac{\pi l_{2-3}}{6} \begin{bmatrix} 0 \\ 0 \\ 0 \\ (2r_2 + r_3)p_{rn} \\ (2r_2 + r_3)p_{\theta n} \\ (2r_2 + r_3)p_{zn} \\ (r_2 + 2r_3)p_{rn} \\ (r_2 + 2r_3)p_{\theta n} \\ (r_2 + 2r_3)p_{zn} \end{bmatrix} \tag{5.49}$$

The inertia, stiffness and equivalent nodal force matrices corresponding to $n = 0$ are obtained by substituting (5.17) into (5.13) and (5.14). For axisymmetric motion the inertia matrix is of the form (5.23) with an overall factor of $\pi \rho A/30$ and $[\mathbf{I}]$ a unit matrix of order 2. The stiffness matrix is given by

$$[\mathbf{k}]_e = 2\pi A \sum_{j=1}^{N} H_j r_j [\mathbf{B}_1(\xi_j, \eta_j, \zeta_j)]^{\mathrm{T}} [\mathbf{D}_1][\mathbf{B}_1(\xi_j, \eta_j, \zeta_j)] \tag{5.50}$$

where $[\mathbf{B}_1]$ is defined by (5.27) with

$$\mathbf{B}_{1i} = \frac{1}{2A} \begin{bmatrix} a_i & 0 \\ 2AL_i/r & 0 \\ 0 & b_i \\ b_i & a_i \end{bmatrix} \tag{5.51}$$

The equivalent nodal forces corresponding to a constant load on side 2–3 is

$$\{\mathbf{f}\}_e = \frac{\pi l_{2-3}}{3} \begin{bmatrix} 0 \\ 0 \\ (2r_2 + r_3)p_{rn} \\ (2r_2 + r_3)p_{zn} \\ (r_2 + 2r_3)p_{rn} \\ (r_2 + 2r_3)p_{\theta n} \end{bmatrix} \tag{5.52}$$

In this case the nodal degrees of freedom are (u_0, w_0).

For antisymmetric motion there is only one nodal degree of freedom $\bar{v}_0$. The inertia matrix is

$$\{\mathbf{m}\}_e = \frac{\pi \rho A}{30} \begin{bmatrix} I_{11} & I_{12} & I_{13} \\ I_{12} & I_{22} & I_{23} \\ I_{13} & I_{23} & I_{33} \end{bmatrix} \tag{5.53}$$

with I_{ij} as defined in (5.24). The stiffness matrix is given by

$$[\mathbf{k}]_e = 2\pi A \sum_{j=1}^{N} H_j r_j [\mathbf{B}_2(\xi_j, \eta_j, \zeta_j)]^{\mathrm{T}} [\mathbf{D}_2][\mathbf{B}_2(\xi_j, \eta_j, \zeta_j)] \tag{5.54}$$

where $[\mathbf{B}_2]$ is defined by (5.27) with

$$[\mathbf{B}_{2i}] = \frac{1}{2A} \begin{bmatrix} (a_i - 2AL_i/r) \\ b_i \end{bmatrix} \tag{5.55}$$

The equivalent nodal forces corresponding to a constant load on side 2–3 is

$$\{\mathbf{f}\}_e = \frac{\pi l_{2-3}}{3} \begin{bmatrix} 0 \\ (2r_2 + r_3) p_{\theta n} \\ (r_2 + 2r_3) p_{\theta n} \end{bmatrix} \tag{5.56}$$

The stresses within the element are given by

$$\{\boldsymbol{\sigma}_1\} = \begin{bmatrix} \sigma_r \\ \sigma_\theta \\ \sigma_z \\ \tau_{zr} \end{bmatrix} = [\mathbf{D}_1]\{\boldsymbol{\varepsilon}_1\} \tag{5.57}$$

and

$$\{\boldsymbol{\sigma}_2\} = \begin{bmatrix} \tau_{r\theta} \\ \tau_{\theta z} \end{bmatrix} = [\mathbf{D}_2]\{\boldsymbol{\varepsilon}_2\} \tag{5.58}$$

Substituting for $\{\boldsymbol{\varepsilon}_1\}$ and $\{\boldsymbol{\varepsilon}_2\}$ gives

$$\begin{aligned} \{\boldsymbol{\sigma}_1\} &= [\mathbf{D}_1][\mathbf{B}_1]\{\mathbf{u}_n\}_e \\ \{\boldsymbol{\sigma}_2\} &= [\mathbf{D}_2][\mathbf{B}_2]\{\mathbf{u}_n\}_e \end{aligned} \tag{5.59}$$

It is usual to evaluate the stresses at the centroid of the element $(\frac{1}{3}, \frac{1}{3}, \frac{1}{3})$ as in the case of the triangular membrane element (Section 4.1). At this point $r = \frac{1}{3}(r_1 + r_2 + r_3)$.

5.6 Core Elements

If either one or two nodes of the element lie on the z-axis, as shown in Figure 5.6, then the integrand $L_i L_j / r$ is singular when $r = 0$. Several suggestions for overcoming this problem have been made. One method is to place the nodes slightly off the axis. Another is to use a numerical integration scheme which does not have integration points on the sides of the triangle. (Note that all the schemes presented in Figure 5.5 satisfy this requirement.) Although these techniques have been used with

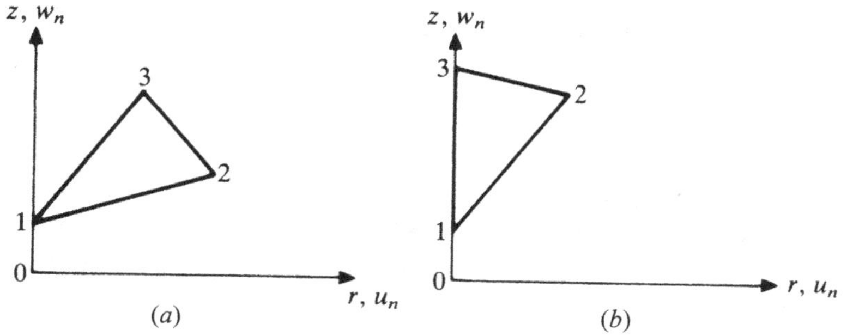

Figure 5.6. A triangular core element with (a) one and (b) two nodes on the z-axis.

some success for static analysis, they are not entirely satisfactory for dynamic analysis [5.3–5.5].

The correct procedure is to investigate the conditions that the strains be finite when $r = 0$. The expressions (5.12) indicate the following requirements. For $n = 0$

$$\begin{aligned} u_0 &= 0 \quad \text{for axisymmetric motion} \\ \bar{v}_0 &= 0 \quad \text{for antisymmetric motion} \end{aligned} \tag{5.60}$$

For $n \neq 0$

$$u_n + nv_n = 0 \tag{5.61a}$$

$$nu_n + v_n = 0 \tag{5.61b}$$

$$w_n = 0 \tag{5.61c}$$

Eliminating v_n from (5.61a, b) gives

$$(n^2 - 1)u_n = 0 \tag{5.62a}$$

Similarly, eliminating u_n from (5.61a, b) gives

$$(n^2 - 1)v_n = 0 \tag{5.62b}$$

Equations (5.62a, b) indicate that, provided $n \neq 1$, then the requirements are

$$u_n = 0, \qquad v_n = 0 \tag{5.63}$$

Also from (5.61c)

$$w_n = 0 \tag{5.64}$$

When $n = 1$, the requirements are

$$u_1 + v_1 = 0, \qquad w_1 = 0 \tag{5.65}$$

Therefore, for $n = 0$ and $n > 1$ the problem is overcome by applying the constraints (5.60) and (5.63), (5.64) respectively to the degrees of freedom at the nodes on the z-axis. In these cases special core elements of reduced order are valuated. When node 1 is on the axis, the element degrees of freedom are

$$\lfloor w_{01} \quad u_{02} \quad w_{02} \quad u_{03} \quad w_{03} \rfloor \qquad \text{for } n = 0, \text{ symmetric}$$

$$\lfloor \bar{v}_{02} \quad \bar{v}_{03} \rfloor \qquad \text{for } n = 0, \text{ antisymmetric}$$

and

$$\lfloor u_{n2} \quad v_{n2} \quad w_{n2} \quad u_{n3} \quad v_{n3} \quad w_{n3} \rfloor \qquad \text{for } n > 1$$

In each case the inertia matrix is obtained by deleting the rows and columns corresponding to the omitted degrees of freedom. The stiffness matrices are obtained by deleting the columns of $\mathbf{B}_{11}$ and $\mathbf{B}_{21}$ corresponding to the omitted degrees of freedom. Finally, the elements of the nodal force matrix which correspond to the omitted degrees of freedom are deleted.

When $n = 1$ the element matrices are derived using displacement functions which satisfy conditions (5.65). These can be written in the form (5.17) with

$$\{\mathbf{u}_1\}_e^{\mathrm{T}} = \lfloor u_{11} \quad u_{12} \quad v_{12} \quad w_{12} \quad u_{13} \quad v_{13} \quad w_{13} \rfloor \tag{5.66}$$

and

$$[\mathbf{N}] = \begin{bmatrix} L_1 & L_2 & 0 & 0 & L_3 & 0 & 0 \\ -L_1 & 0 & L_2 & 0 & 0 & L_3 & 0 \\ 0 & 0 & 0 & L_2 & 0 & 0 & L_3 \end{bmatrix} \tag{5.67}$$

Substituting (5.67) into (5.21) and integrating gives the following inertia matrix

$$[\mathbf{m}]_e = \frac{\pi \rho A}{60} \begin{bmatrix} 2I_{11} & I_{12} \quad -I_{12} \quad 0 & I_{13} \quad -I_{13} \quad 0 \\ I_{12} & & \\ -I_{12} & I_{22}[\mathbf{I}] & I_{23}[\mathbf{I}] \\ 0 & & \\ I_{13} & & \\ -I_{13} & I_{23}[\mathbf{I}] & I_{33}[\mathbf{I}] \\ 0 & & \end{bmatrix} \tag{5.68}$$

where $[\mathbf{I}]$ is a unit matrix of order 3 and the I_{ij} are as defined in (5.24).

Substituting (5.17) and (5.67) into (5.12) gives

$$\mathbf{B}_{11} = \frac{1}{2A} \begin{bmatrix} a_1 \\ 0 \\ 0 \\ b_1 \end{bmatrix} \tag{5.69}$$

and

$$\mathbf{B}_{21} = \frac{1}{2A} \begin{bmatrix} -a_1 \\ -b_1 \end{bmatrix} \tag{5.70}$$

$\mathbf{B}_{1i}$ and $\mathbf{B}_{2i}$ with $i = 2, 3$ are defined by (5.35) and (5.36).

The stiffness matrix is again given by (5.42) with the new definitions of $[\mathbf{B}_1]$ and $[\mathbf{B}_2]$.

Table 5.2. *Comparison of predicted frequencies (kHz) of circular disc with exact frequencies*

Mode	FEM [5.8]	Exact	% Error
F	0.2822	0.2727	3.5
F	1.0855	1.0432	4.1
E	1.1703	1.1650	0.45
F	2.1181	1.9646	7.8
E	3.0077	3.0183	−0.35

F, Flexural modes; E, Extensional modes.

The equivalent nodal force matrix becomes

$$\{\mathbf{f}\}_e = \frac{\pi l_{2-3}}{6} \begin{bmatrix} 0 \\ (2r_2 + r_3)p_{r1} \\ (2r_2 + r_3)p_{\theta 1} \\ (2r_2 + r_3)p_{z1} \\ (r_2 + 2r_3)p_{r1} \\ (r_2 + 2r_3)p_{\theta 1} \\ (r_2 + 2r_3)p_{z1} \end{bmatrix} \tag{5.71}$$

EXAMPLE 5.2 Calculate the first five axisymmetric ($n = 0$) frequencies of a circular disc of radius 1.27 m and thickness 0.254 m. Compare the results with the exact solution [5.6, 5.7]. Take $\nu = 0.31$ and $E/\rho = 18.96$ Nm/kg.

Figure 5.7 shows an idealisation of the disc. There are 65 nodes and 98 elements. Since $n = 0$ and the motion is symmetric, then the condition $u_0 = 0$ is applied at the three nodes on the axis of symmetry.

The predicted frequencies are compared with the exact ones in Table 5.2. The accuracy of the frequencies predicted by the finite element method is good. It can be seen that the accuracy of the frequencies of the extensional modes are better than the accuracy of the flexural modes. This is typical of low order elements.

5.7 Arbitrary Shaped Solids

Non-axisymmetric solids of arbitrary shape are analysed by dividing them up into an assemblage of three-dimensional finite elements. The most common shapes of element used are tetrahedral, pentahedral and hexahedral.

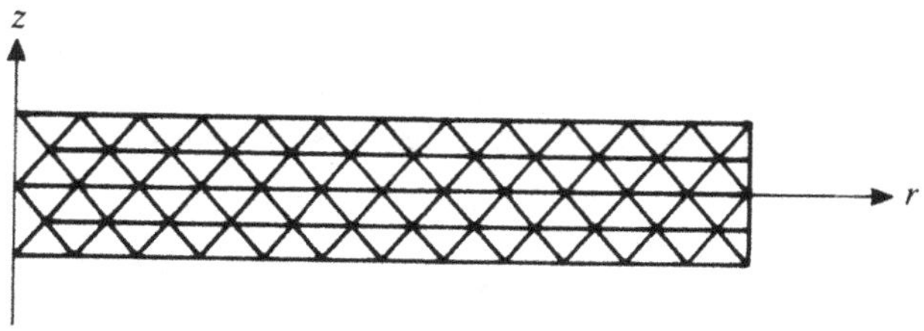

Figure 5.7. Idealisation of a disc into axisymmetric triangular elements.

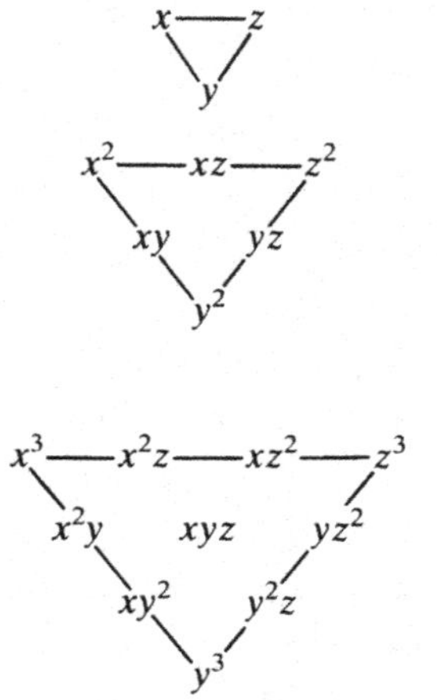

Figure 5.8. Complete polynomials in three variables.

In Chapter 4 it is shown that complete polynomials in two variables can be generated using Pascal's triangle. The displacement functions of three-dimensional solid elements are derived using polynomials in the three variables x, y and z, the Cartesian coordinates of a point within the element.

The three-dimensional analogue of Pascal's triangle is a tetrahedron, as shown in Figure 5.8.

The energy expressions for a solid element are, from Section 2.8

$$T_e = \frac{1}{2}\int_{V_e} \rho(\dot{u}^2 + \dot{v}^2 + \dot{w}^2)\,\mathrm{d}V \tag{5.72}$$

$$U_e = \frac{1}{2}\int_{V_e} \{\boldsymbol{\varepsilon}\}^{\mathrm{T}}\,[\mathbf{D}]\,\{\boldsymbol{\varepsilon}\}\,\mathrm{d}V \tag{5.73}$$

with

$$\{\boldsymbol{\varepsilon}\} = \begin{bmatrix} \dfrac{\partial u}{\partial x} \\ \dfrac{\partial v}{\partial y} \\ \dfrac{\partial w}{\partial z} \\ \dfrac{\partial u}{\partial y} + \dfrac{\partial v}{\partial x} \\ \dfrac{\partial u}{\partial z} + \dfrac{\partial w}{\partial x} \\ \dfrac{\partial v}{\partial z} + \dfrac{\partial w}{\partial y} \end{bmatrix} \tag{5.74}$$

$[\mathbf{D}]$ is a matrix of material constants which is defined by (2.85) for an isotropic material. Also

$$\delta W_e = \int_{S_e} (p_x\delta u + p_y\delta v + p_z\delta w)\,\mathrm{d}S \tag{5.75}$$

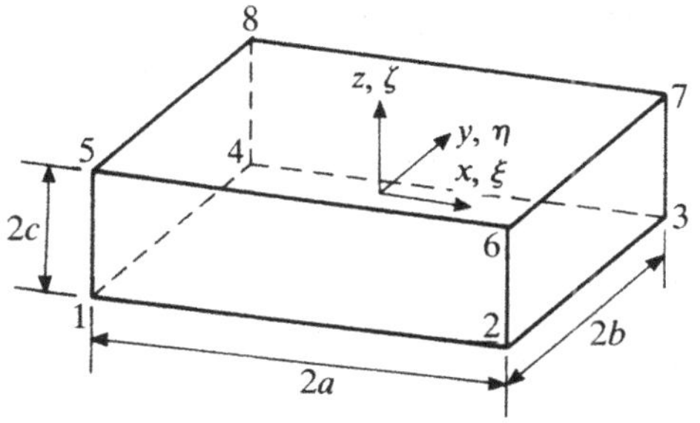

Figure 5.9. Geometry of a rectangular hexahedron element, $\xi = x/a$, $\eta = y/b$, $\zeta = z/c$.

The highest derivative appearing in these expressions is the first. Hence, it is only necessary to take u, v and w as degrees of freedom at each node to ensure continuity. Also, complete polynomials of at least degree 1 should be used.

5.8 Rectangular Hexahedron

Figure 5.9 shows a rectangular hexahedron element with eight node points, one at each corner. There are three degrees of freedom at each node, namely, the components of displacement u, v and w in the directions of the x-, y- and z-axes respectively. Each component can, therefore, be represented by polynomials having eight terms each. Figure 5.8 shows that a complete linear function has four terms. Therefore, it is necessary to choose four more terms of higher degree. Three quadratic terms xy, yz, xz and one cubic term xyz are chosen in order to ensure geometric invariance.

The displacements can, therefore, be represented by expressions of the form

$$u = \alpha_1 + \alpha_2 x + \alpha_3 y + \alpha_4 z + \alpha_5 xy + \alpha_6 yz + \alpha_7 xz + \alpha_8 xyz \tag{5.76}$$

The coefficients α_1 to α_8 can be expressed in terms of the values of the u-component of displacement at the eight node points by evaluating (5.76) at the nodes and solving the resulting equations. However, it is much simpler to write down the displacement functions by inspection in a similar manner to that used for a rectangular membrane element in Section 4.2. The displacement functions are required in the form

$$u = \sum_{j=1}^{8} N_j u_j, \qquad v = \sum_{j=1}^{8} N_j v_j, \qquad w = \sum_{j=1}^{8} N_j w_j \tag{5.77}$$

where the functions N_j are required to have a unit value at node j and zero values at the other seven nodes. Noting that the expression (5.76) can be expressed as a product of three linear functions in x, y and z respectively, it can easily be seen that the three-dimensional version of (4.36) is

$$N_j = \tfrac{1}{8}(1 + \xi_j \xi)(1 + \eta_j \eta)(1 + \zeta_j \zeta) \tag{5.78}$$

where (ξ_j, η_j, ζ_j) are the coordinates of node j.

On each face the variation of displacement is bilinear and uniquely determined by the values at the four node points. There will, therefore, be continuity of displacements between adjacent elements.

The expressions (5.77) can be written in the combined form

$$\begin{bmatrix} u \\ v \\ w \end{bmatrix} = [\mathbf{N}]\{\mathbf{u}\}_e \tag{5.79}$$

where

$$\{\mathbf{u}\}_e^{\mathrm{T}} = \lfloor u_1 \quad v_1 \quad w_1 \quad \cdots \quad u_8 \quad v_8 \quad w_8 \rfloor \tag{5.80}$$

and

$$[\mathbf{N}] = \begin{bmatrix} N_1 & 0 & 0 & \cdots & N_8 & 0 & 0 \\ 0 & N_1 & 0 & \cdots & 0 & N_8 & 0 \\ 0 & 0 & N_1 & \cdots & 0 & 0 & N_8 \end{bmatrix} \tag{5.81}$$

Substituting (5.79) into (5.72) gives

$$T_e = \tfrac{1}{2}\{\dot{\mathbf{u}}\}_e^{\mathrm{T}}[\mathbf{m}]_e\{\dot{\mathbf{u}}\}_e \tag{5.82}$$

where

$$[\mathbf{m}]_e = \int_{V_e} \rho[\mathbf{N}]^{\mathrm{T}}[\mathbf{N}]\,\mathrm{d}V \tag{5.83}$$

is the element inertia matrix.

A typical element of this matrix is

$$\begin{aligned} \rho abc \int_{-1}^{+1}\int_{-1}^{+1}\int_{-1}^{+1} N_i N_j \,\mathrm{d}\xi\,\mathrm{d}\eta\,\mathrm{d}\zeta &= \frac{\rho abc}{64}\int_{-1}^{+1}(1+\xi_i\xi)(1+\xi_j\xi)\,\mathrm{d}\xi \int_{-1}^{+1}(1+\eta_i\eta)(1+\eta_j\eta)\,\mathrm{d}\eta \\ &\quad \times \int_{-1}^{+1}(1+\zeta_i\zeta)(1+\zeta_j\zeta)\,\mathrm{d}\zeta \\ &= \frac{\rho abc}{8}\left(1+\tfrac{1}{3}\xi_i\xi_j\right)\left(1+\tfrac{1}{3}\eta_i\eta_j\right)\left(1+\tfrac{1}{3}\zeta_i\zeta_j\right) \end{aligned} \tag{5.84}$$

Using this result gives the following inertia matrix:

$$[\mathbf{m}]_e = \begin{bmatrix} \mathbf{m}_1 & \mathbf{m}_2 \\ \mathbf{m}_2 & \mathbf{m}_1 \end{bmatrix} \tag{5.85}$$

with

$$\mathbf{m}_1 = 2\mathbf{m}_2 \tag{5.86}$$

and

$$\mathbf{m}_2 = \frac{\rho abc}{27}\begin{bmatrix} 4 & 0 & 0 & 2 & 0 & 0 & 1 & 0 & 0 & 2 & 0 & 0 \\ 0 & 4 & 0 & 0 & 2 & 0 & 0 & 1 & 0 & 0 & 2 & 0 \\ 0 & 0 & 4 & 0 & 0 & 2 & 0 & 0 & 1 & 0 & 0 & 2 \\ 2 & 0 & 0 & 4 & 0 & 0 & 2 & 0 & 0 & 1 & 0 & 0 \\ 0 & 2 & 0 & 0 & 4 & 0 & 0 & 2 & 0 & 0 & 1 & 0 \\ 0 & 0 & 2 & 0 & 0 & 4 & 0 & 0 & 2 & 0 & 0 & 1 \\ 1 & 0 & 0 & 2 & 0 & 0 & 4 & 0 & 0 & 2 & 0 & 0 \\ 0 & 1 & 0 & 0 & 2 & 0 & 0 & 4 & 0 & 0 & 2 & 0 \\ 0 & 0 & 1 & 0 & 0 & 2 & 0 & 0 & 4 & 0 & 0 & 2 \\ 2 & 0 & 0 & 1 & 0 & 0 & 2 & 0 & 0 & 4 & 0 & 0 \\ 0 & 2 & 0 & 0 & 1 & 0 & 0 & 2 & 0 & 0 & 4 & 0 \\ 0 & 0 & 2 & 0 & 0 & 1 & 0 & 0 & 2 & 0 & 0 & 4 \end{bmatrix} \tag{5.87}$$

Substituting (5.79) into (5.74) and (5.73) gives

$$U_e = \tfrac{1}{2}\{\mathbf{u}\}_e^{\mathrm{T}}[\mathbf{k}]_e\{\mathbf{u}\}_e \tag{5.88}$$

where

$$[\mathbf{k}]_e = \int_{V_e} [\mathbf{B}]^{\mathrm{T}}[\mathbf{D}][\mathbf{B}]\,\mathrm{d}V \tag{5.89}$$

is the element stiffness matrix. The strain matrix $[\mathbf{B}]$ is of the form

$$[\mathbf{B}] = [\mathbf{B}_1 \quad \cdots \quad \mathbf{B}_8] \tag{5.90}$$

where

$$\mathbf{B}_i = \begin{bmatrix} \dfrac{\partial N_i}{\partial x} & 0 & 0 \\ 0 & \dfrac{\partial N_i}{\partial y} & 0 \\ 0 & 0 & \dfrac{\partial N_i}{\partial z} \\ \dfrac{\partial N_i}{\partial y} & \dfrac{\partial N_i}{\partial x} & 0 \\ \dfrac{\partial N_i}{\partial z} & 0 & \dfrac{\partial N_i}{\partial x} \\ 0 & \dfrac{\partial N_i}{\partial z} & \dfrac{\partial N_i}{\partial y} \end{bmatrix} \tag{5.91}$$

Using (5.78) gives

$$\begin{aligned} \frac{\partial N_i}{\partial x} &= \frac{1}{a}\frac{\partial N_i}{\partial \xi} = \frac{\xi_i}{8a}(1+\eta_i\eta)(1+\zeta_i\zeta) \\ \frac{\partial N_i}{\partial y} &= \frac{1}{b}\frac{\partial N_i}{\partial \eta} = \frac{\eta_i}{8b}(1+\xi_i\xi)(1+\zeta_i\zeta) \\ \frac{\partial N_i}{\partial z} &= \frac{1}{c}\frac{\partial N_i}{\partial \zeta} = \frac{\zeta_i}{8c}(1+\xi_i\xi)(1+\eta_i\eta) \end{aligned} \tag{5.92}$$

Substituting (5.90) to (5.92) into (5.89) and integrating will give the element stiffness matrix. As this is a tedious process it is simpler to use numerical integration. In terms of (ξ, η, ζ) coordinates (5.89) becomes

$$[\mathbf{k}]_e = \int_{-1}^{+1}\int_{-1}^{+1}\int_{-1}^{+1} abc[\mathbf{B}]^{\mathrm{T}}[\mathbf{D}][\mathbf{B}]\,\mathrm{d}\xi\,\mathrm{d}\eta\,\mathrm{d}\zeta \tag{5.93}$$

Sections (3.10) and (4.2) describe how to integrate functions in one and two dimensions using Gauss–Legendre integration. Extending this to three dimensions, it can be shown that the integral

$$I = \int_{-1}^{+1}\int_{-1}^{+1}\int_{-1}^{+1} g(\xi, \eta, \zeta)\,\mathrm{d}\xi\,\mathrm{d}\eta\,\mathrm{d}\zeta \tag{5.94}$$

can be evaluated using the formula

$$I = \sum_{i=1}^{n}\sum_{j=1}^{m}\sum_{l=1}^{p} H_i H_j H_l g(\xi_i, \eta_j, \zeta_l) \tag{5.95}$$

where n, m and p are the number of integration points in the ξ, η, ζ directions, H_i H_j and H_l are the weight coefficients and ξ_i, η_j, ζ_l the integration points as given in Table 3.6.

The integrand (5.93) contains terms which are quadratic in either ξ, η or ζ. Therefore the exact value of the integral can be obtained using $n = m = p = 2$. However, as in the case of the rectangular membrane, this procedure leads to unrepresentative properties, especially if the element undergoes bending deformation. In this case it is better to evaluate the shear strains at $\xi = \eta = \zeta = 0$.

The equivalent nodal forces due to a distributed load over the face $\xi = 1$ are obtained by substituting (5.79) into (5.75). This gives

$$\delta W_e = \{\delta\mathbf{u}\}_e^{\mathrm{T}}\{\mathbf{f}\}_e \tag{5.96}$$

where

$$\{\mathbf{f}\}_e = \int_{-1}^{+1}\int_{-1}^{+1} [\mathbf{N}]_{\xi=1}^{\mathrm{T}} \begin{bmatrix} p_x \\ p_y \\ p_z \end{bmatrix} bc\,\mathrm{d}\eta\,\mathrm{d}\zeta \tag{5.97}$$

When $\xi = 1$

$$N_i = \begin{cases} 0 & \text{for } i = 1, 4, 5, 8 \\ \frac{1}{4}(1 + \eta_i\eta)(1 + \zeta_i\zeta) & \text{for } i = 2, 3, 6, 7 \end{cases} \tag{5.98}$$

Assuming that p_x, p_y and p_z are constant, the equivalent nodal forces, as given by (5.97) are

$$\{\mathbf{f}\}_e = \begin{bmatrix} \mathbf{f}_1 \\ \vdots \\ \mathbf{f}_8 \end{bmatrix} \tag{5.99}$$

where

$$\{\mathbf{f}_i\} = 0 \qquad \text{for } i = 1, 4, 5, 8$$

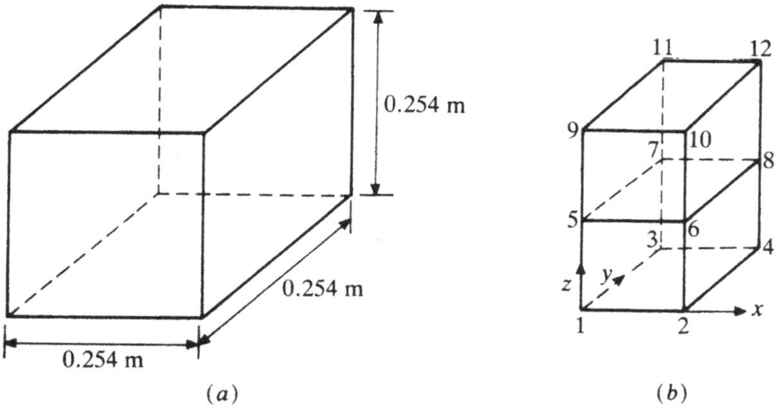

Figure 5.10. Geometry of a solid cube.

and

$$\{\mathbf{f}_i\} = bc \begin{bmatrix} p_x \\ p_y \\ p_z \end{bmatrix} \qquad \text{for } i = 2, 3, 6, 7 \tag{5.100}$$

Thus one quarter of the total force is applied at each node of the face of application.

The stresses within the element are given by (2.84), namely

$$\{\boldsymbol{\sigma}\} = [\mathbf{D}]\{\boldsymbol{\varepsilon}\} \tag{5.101}$$

where

$$\{\boldsymbol{\sigma}\}^{\mathrm{T}} = \lfloor \sigma_x \quad \sigma_y \quad \sigma_z \quad \tau_{xy} \quad \tau_{xy} \quad \tau_{yz} \rfloor \tag{5.102}$$

Substituting for $\{\boldsymbol{\varepsilon}\}$ gives

$$\{\boldsymbol{\sigma}\} = [\mathbf{D}][\mathbf{B}]\{\mathbf{u}\}_e \tag{5.103}$$

The best position to evaluate the stresses is at the point $\xi = 0,\ \eta = 0,\ \zeta = 0$.

EXAMPLE 5.3 Calculate the first four natural frequencies and modes of the cube shown in Figure 5.10(a) which is fixed at its base. Compare the results with the exact solution [5.9]. Take $E = 68.95 \times 10^9$ N/m^2, $\nu = 0.3$ and $\rho = 2560$ kg/m^3.

Since the cube has two planes of symmetry, the modes can be obtained by idealising one quarter of it and applying appropriate boundary conditions on the planes of symmetry (see Chapter 11). Figure 5.10(b) shows the idealisation of one quarter of the cube using two hexahedral elements.

There are three degrees of freedom at each node, namely, the linear displacements u, v and w in the x-, y- and z-directions. Since the base is fixed, all three degrees of freedom at nodes 1, 2, 3 and 4 are constrained to be zero.

The swaying nodes in the y-direction are calculated by applying symmetric boundary conditions at nodes 5, 7, 9 and 11 and antisymmetric conditions at nodes 5, 6, 9 and 10. These conditions are given in Table 5.3. The torsion modes about the z-axis are calculated by applying antisymmetric boundary conditions at nodes 5, 6, 7, 9, 10 and 11. This results in all three degrees of freedom being

Table 5.3. *Boundary conditions for symmetrical and antisymmetrical motion*

Plane	Boundary conditions	
	Symmetrical	Antisymmetrical
yz	$u = 0$	$v = 0, w = 0$
zx	$v = 0$	$w = 0, u = 0$

zero at nodes 5 and 9. Finally, the longitudinal modes in the z-direction are calculated by applying symmetric boundary conditions at nodes 5, 6, 7, 9, 10 and 11.

The predicted frequencies are compared with the exact ones in Table 5.4. It can be seen that the accuracy of the torsion and longitudinal modes is better than that of the swaying modes.

5.9 Isoparametric Hexahedron

The usefulness of the rectangular hexahedron element presented in the previous section can be increased by converting it into an isoparametric element. The general shape of the element is shown in Figure 5.11(a). Any point (ξ, η, ζ) within the square hexahedron, shown in Figure 5.11(b), having corners at (± 1, ± 1, ± 1), can be mapped onto a point (x, y, z) within the element in Figure 5.11(a) by means of the relationships

$$x = \sum_{j=1}^{8} N_j(\xi, \eta, \zeta) x_j,$$
$$y = \sum_{j=1}^{8} N_j(\xi, \eta, \zeta) y_j, \tag{5.104}$$
$$z = \sum_{j=1}^{8} N_j(\xi, \eta, \zeta) z_j$$

where (x_j, y_j, z_j) are the coordinates of node point j. The functions $N_j(\xi, \eta, \zeta)$ are defined by (5.78). The methods used in Section 4.3 can be extended to show that the faces in Figure 5.11(b) map onto the faces in Figure 5.11(a).

Table 5.4. *Comparison of predicted frequencies (kHz) of a cube with exact frequencies*

Mode	FEM [5.9]	Exact [5.9]	% Difference
S	2.399	2.212	8.5
T	3.250	3.020	7.6
L	5.511	5.239	5.2
S	6.830	5.915	15.5

S, Swaying mode; T, Torsion mode; L, Longitudinal mode.

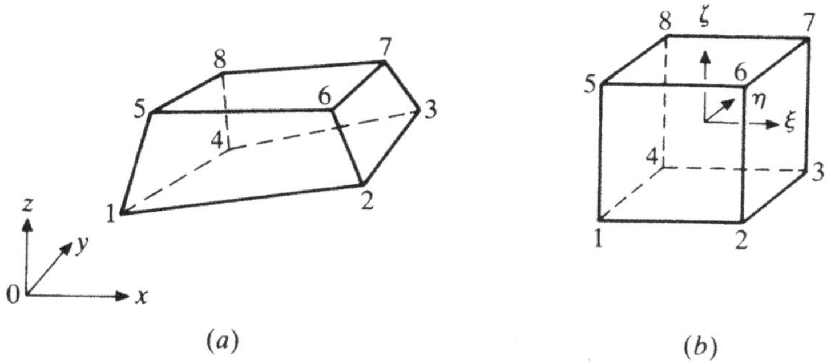

Figure 5.11. Geometry of an isoparametric hexahedron. (*a*) Physical coordinates, (*b*) isoparametric coordinates.

The variations in displacement within the element can be expressed in (ξ, η, ζ) coordinates using (5.77). Both the geometry and deformation of the element are, therefore, represented by the same functions in accordance with the definition of an isoparametric element.

The position and displacement of any point (η, ζ) on the face 2376 in Figure 5.11(*a*) is uniquely determined from the coordinates and displacements of nodes, 2, 3, 7 and 6 using (5.104) and (5.77). Two adjacent elements having the face 2376 as a common face, will have the same nodes and the same variation with η and ζ over the face. The displacements will, therefore, be continuous between elements.

The inertia matrix is given by equation (5.83), that is

$$[\mathbf{m}]_e = \int_{V_e} \rho[\mathbf{N}]^{\mathrm{T}}[\mathbf{N}]\,\mathrm{d}V \tag{5.105}$$

where $[\mathbf{N}]$ is defined by (5.81).

Because the N_i in (5.81) are expressed in (ξ, η, ζ) coordinates, it is necessary to transform the integral from (x, y, z) coordinates to (ξ, η, ζ) coordinates.

The element of volume in the (ξ, η, ζ) coordinates is given by the triple scalar product

$$\mathrm{d}V = \mathrm{d}\vec{\boldsymbol{\xi}} \cdot (\mathrm{d}\vec{\boldsymbol{\eta}} \wedge \mathrm{d}\vec{\boldsymbol{\zeta}}) \tag{5.106}$$

where the vectors on the right-hand side are defined by

$$\begin{aligned}
\mathrm{d}\vec{\boldsymbol{\xi}} &= \left(\frac{\partial x}{\partial \xi}, \frac{\partial y}{\partial \xi}, \frac{\partial z}{\partial \xi}\right)\mathrm{d}\xi \\
\mathrm{d}\vec{\boldsymbol{\eta}} &= \left(\frac{\partial x}{\partial \eta}, \frac{\partial y}{\partial \eta}, \frac{\partial z}{\partial \eta}\right)\mathrm{d}\eta \\
\mathrm{d}\vec{\boldsymbol{\zeta}} &= \left(\frac{\partial x}{\partial \zeta}, \frac{\partial y}{\partial \zeta}, \frac{\partial z}{\partial \zeta}\right)\mathrm{d}\zeta
\end{aligned} \tag{5.107}$$

Substituting (5.107) into (5.106) gives

$$\mathrm{d}V = \det[\mathbf{J}]\,\mathrm{d}\xi\,\mathrm{d}\eta\,\mathrm{d}\zeta \tag{5.108}$$

where

$$[\mathbf{J}] = \begin{bmatrix} \dfrac{\partial x}{\partial \xi} & \dfrac{\partial y}{\partial \xi} & \dfrac{\partial z}{\partial \xi} \\ \dfrac{\partial x}{\partial \eta} & \dfrac{\partial y}{\partial \eta} & \dfrac{\partial z}{\partial \eta} \\ \dfrac{\partial x}{\partial \zeta} & \dfrac{\partial y}{\partial \zeta} & \dfrac{\partial z}{\partial \zeta} \end{bmatrix} \tag{5.109}$$

is the Jacobian of the transformation.

Substituting (5.104) into (5.109) gives

$$[\mathbf{J}] = \begin{bmatrix} \dfrac{\partial N_1}{\partial \xi} & \cdots & \dfrac{\partial N_8}{\partial \xi} \\ \dfrac{\partial N_1}{\partial \eta} & \cdots & \dfrac{\partial N_8}{\partial \eta} \\ \dfrac{\partial N_1}{\partial \zeta} & \cdots & \dfrac{\partial N_8}{\partial \zeta} \end{bmatrix} \begin{bmatrix} x_1 & y_1 & z_1 \\ \vdots & \vdots & \vdots \\ x_8 & y_8 & z_8 \end{bmatrix} \tag{5.110}$$

The expression for the inertia matrix now becomes

$$[\mathbf{m}]_e = \int_{-1}^{+1}\int_{-1}^{+1}\int_{-1}^{+1} \rho[\mathbf{N}]^{\mathrm{T}}[\mathbf{N}]\det[\mathbf{J}]\,\mathrm{d}\xi\,\mathrm{d}\eta\,\mathrm{d}\zeta \tag{5.111}$$

The integral in (5.111) is evaluated using numerical integration, as described in the previous section. This process requires a knowledge of the order of the function det[**J**]. Substituting (5.78) into (5.110) gives

$$[\mathbf{J}] = \frac{1}{8}\begin{bmatrix} (e_1+e_2\eta+e_3\zeta+e_4\eta\zeta) & (f_1+f_2\eta+f_3\zeta+f_4\eta\zeta) & (g_1+g_2\eta+g_3\zeta+g_4\eta\zeta) \\ (e_5+e_2\xi+e_6\zeta+e_4\xi\zeta) & (f_5+f_2\xi+f_6\zeta+f_4\xi\zeta) & (g_5+g_2\xi+g_6\zeta+g_4\xi\zeta) \\ (e_7+e_3\xi+e_6\eta+e_4\xi\eta) & (f_7+f_3\xi+f_6\eta+f_4\xi\eta) & (g_7+g_3\xi+g_6\eta+g_4\xi\eta) \end{bmatrix} \tag{5.112}$$

where

$$\begin{aligned} e_1 &= (-x_1+x_2+x_3-x_4-x_5+x_6+x_7-x_8) \\ e_2 &= (x_1-x_2+x_3-x_4+x_5-x_6+x_7-x_8) \\ e_3 &= (x_1-x_2-x_3+x_4-x_5+x_6+x_7-x_8) \\ e_4 &= (-x_1+x_2-x_3+x_4+x_5-x_6+x_7-x_8) \\ e_5 &= (-x_1-x_2+x_3+x_4-x_5-x_6+x_7+x_8) \\ e_6 &= (x_1+x_2-x_3-x_4-x_5-x_6+x_7+x_8) \\ e_7 &= (-x_1-x_2-x_3-x_4+x_5+x_6+x_7+x_8) \end{aligned} \tag{5.113}$$

The f and g coefficients can be obtained from the e coefficients by replacing the x-coordinates by the y- and z-coordinates respectively.

Evaluating the determinant of [**J**] using (5.112) shows that, in general, it is tri-quadratic in ξ, η and ζ. [**N**] is a tri-linear function and so $[\mathbf{N}]^{\mathrm{T}}[\mathbf{N}]\det[\mathbf{J}]$ is a tri-quartic function. This means that (5.111) can be evaluated using a $(3 \times 3 \times 3)$ array of integration points.

The stiffness matrix is given by (5.89), that is

$$[\mathbf{k}]_e = \int_{V_e} [\mathbf{B}]^{\mathrm{T}}[\mathbf{D}][\mathbf{B}]\,\mathrm{d}V \tag{5.114}$$

where $[\mathbf{B}]$ is defined by (5.90) and (5.91). Transforming to (ξ, η, ζ) coordinates using (5.108) gives

$$[\mathbf{k}]_e = \int_{-1}^{+1}\int_{-1}^{+1}\int_{-1}^{+1} [\mathbf{B}]^{\mathrm{T}}[\mathbf{D}][\mathbf{B}]\det[\mathbf{J}]\,\mathrm{d}\xi\,\mathrm{d}\eta\,\mathrm{d}\zeta \tag{5.115}$$

Now

$$\begin{bmatrix} \partial/\partial x \\ \partial/\partial y \\ \partial/\partial z \end{bmatrix} = [\mathbf{J}]^{-1} \begin{bmatrix} \partial/\partial \xi \\ \partial/\partial \eta \\ \partial/\partial \zeta \end{bmatrix} \tag{5.116}$$

and so the elements of $[\mathbf{B}]$ are given by

$$\begin{bmatrix} \partial N_1/\partial x & \cdots & \partial N_8/\partial x \\ \partial N_1/\partial y & \cdots & \partial N_8/\partial y \\ \partial N_1/\partial z & \cdots & \partial N_8/\partial z \end{bmatrix} = [\mathbf{J}]^{-1} \begin{bmatrix} \partial N_1/\partial \xi & \cdots & \partial N_8/\partial \xi \\ \partial N_1/d\partial \eta & \cdots & \partial N_8/\partial \eta \\ \partial N_1/\partial \zeta & \cdots & \partial N_8/\partial \zeta \end{bmatrix} \tag{5.117}$$

Expressions (5.112) and (5.117) show that the elements of $[\mathbf{B}]$ are obtained by dividing a tri-quadratic function of ξ, η and ζ by another tri-quadratic function. Therefore, the elements of $[\mathbf{B}]^{\mathrm{T}}[\mathbf{D}][\mathbf{B}]\det[\mathbf{J}]$ are tri-quartic functions divided by tri-quadratic functions. This means that $[\mathbf{k}]_{\mathrm{e}}$ cannot be evaluated exactly using numerical integration. The minimum number of integration points that should be used is the number required to evaluate exactly the volume of the element (see Section 4.3). The volume is given by

$$V_e = \int_{-1}^{+1}\int_{-1}^{+1}\int_{-1}^{+1} \det[\mathbf{J}]\,\mathrm{d}\xi\,\mathrm{d}\eta\,\mathrm{d}\zeta \tag{5.118}$$

Since $\det[\mathbf{J}]$ is tri-quadratic the volume can be determined exactly using a $(2 \times 2 \times 2)$ array of integration points. Experience has shown that this number can be used in practice. However, as in the case of rectangular hexahedron, Section 5.8, the shear strains should be evaluated at $\xi = \eta = \zeta = 0$.

The equivalent nodal forces due to a distributed load over the face $\xi = 1$ are again obtained by substituting (5.79) into (5.75). This gives

$$\{\mathbf{f}\}_e = \int_{S_e} [\mathbf{N}]^{\mathrm{T}}_{\xi=1} \begin{bmatrix} p_x \\ p_y \\ p_z \end{bmatrix} \mathrm{d}S \tag{5.119}$$

When $\xi = 1$ the elements of $[\mathbf{N}]$ are given by (5.98).

The element of area, $\mathrm{d}S$, on $\xi = 1$ is given by

$$\mathrm{d}S = |\mathrm{d}\vec{\eta} \wedge \mathrm{d}\vec{\zeta}\,|_{\xi=1} = G\,\mathrm{d}\eta\,\mathrm{d}\zeta \tag{5.120}$$

where $\mathrm{d}\vec{\boldsymbol{\eta}}$ and $\mathrm{d}\vec{\boldsymbol{\zeta}}$ are defined in (5.107). The components of $\mathrm{d}\vec{\boldsymbol{\eta}}$ and $\mathrm{d}\vec{\boldsymbol{\zeta}}$ are, therefore, given by the second and third rows of the Jacobian given in (5.112). Substituting

Table 5.5. *Summary of the number of integration points required for a linear isoparametric hexahedron*

Matrix	Gauss-point array
$[\mathbf{m}]_e$	$3 \times 3 \times 3$
$[\mathbf{k}]_e$	$2 \times 2 \times 2$
$[\mathbf{f}]_e$	2×2

(5.120) into (5.119) and assuming that p_x, p_y, p_z are constant gives

$$\{\mathbf{f}\}_e = \int_{-1}^{+1}\int_{-1}^{+1} [\mathbf{N}]^{\mathrm{T}}_{\xi=1} G \,\mathrm{d}\eta\,\mathrm{d}\zeta \begin{bmatrix} p_x \\ p_y \\ p_z \end{bmatrix} \tag{5.121}$$

Expressions (5.98), (5.112) and (5.120) indicate that the integrand in (5.121) is bi-cubic in η and ζ. It can, therefore, be evaluated using a (2×2) array of integration points.

The stresses are calculated using (5.103) where the matrix $[\mathbf{B}]$ is defined by (5.90), (5.91) and (5.117). The best position to evaluate the stresses is at $\xi = 0$, $\eta = 0$, $\zeta = 0$.

Table 5.5 summarises the number of integration points required to evaluate the inertia, stiffness and equivalent nodal force matrices. These require 27, 8 and 4 respectively. Using 27 points to evaluate the inertia matrix can be quite time consuming. This can be overcome by using the following fourteen point integration scheme [5.10], which gives similar accuracy to that obtained with 27 Gauss points. There are six points placed at $(\mp b, 0, 0)$, $(0, \mp b, 0)$ and $(0, 0, \mp b)$ where $b = 0.795822426$, all with weight 0.886426593. The other eight points are placed at $(\mp c, \mp c, \mp c)$ where $c = 0.758786911$, all with weight 0.335180055.

5.10 Right Pentahedron

In some applications it is necessary to supplement the hexahedral element with a compatible triangular wedge-shaped element, a pentahedron, as shown in Figure 5.12. For this element it is convenient to use Cartesian coordinates in the

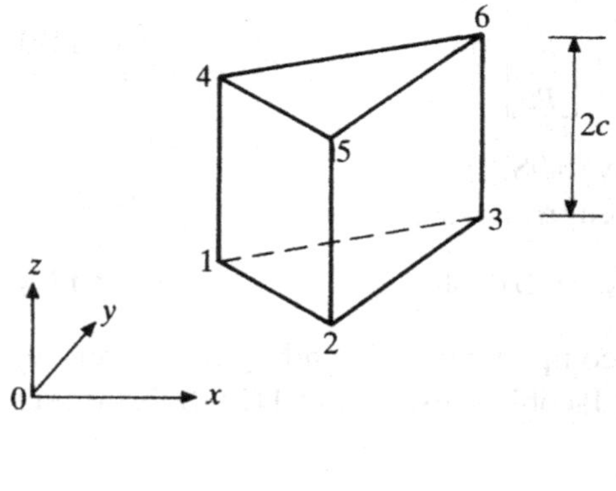

Figure 5.12. Geometry of a pentahedron.

z-direction and area coordinates in planes which are parallel to the xy-plane. The displacement functions are required in the form

$$u = \sum_{j=1}^{6} N_j u_j, \qquad v = \sum_{j=1}^{6} N_j v_j, \qquad w = \sum_{j=1}^{6} N_j w_j \tag{5.122}$$

where the functions N_j are required to have a unit value at node j and zero values at the other five nodes. Combining the results for two-dimensional triangles and three-dimensional rectangles, it can easily be seen that the functions N_j take the form

$$N_j = \tfrac{1}{2} L_j (1 + \zeta_j \zeta) \qquad j = 1, 2, \ldots, 6 \tag{5.123}$$

with

$$L_j = L_{j-3} \qquad j = 4, 5, 6$$

where L_j are area coordinates, as defined in Section 4.4, $\zeta = z/c$ and ζ_j is the value of ζ at node j.

On each face the variation of displacement is bilinear and uniquely determined by the values at the node points on it. There will, therefore, be continuity of displacement between adjacent elements.

The expressions (5.122) can be written in the combined form

$$\begin{bmatrix} u \\ v \\ w \end{bmatrix} = [\mathbf{N}]\{\mathbf{u}\}_e \tag{5.124}$$

where

$$\{\mathbf{u}\}_e^{\mathrm{T}} = \lfloor u_1 \quad v_1 \quad w_1 \quad \cdots \quad u_6 \quad v_6 \quad w_6 \rfloor \tag{5.125}$$

and

$$[\mathbf{N}] = \begin{bmatrix} N_1 & 0 & 0 & \cdots & N_6 & 0 & 0 \\ 0 & N_1 & 0 & \cdots & 0 & N_6 & 0 \\ 0 & 0 & N_1 & \cdots & 0 & 0 & N_6 \end{bmatrix} \tag{5.126}$$

The element inertia matrix will be given by (5.83), that is

$$[\mathbf{m}]_e = \int_{V_e} \rho [\mathbf{N}]^{\mathrm{T}} [\mathbf{N}] \, \mathrm{d}V \tag{5.127}$$

A typical element of this matrix is

$$\begin{aligned} &\rho c \int_A \int_{-1}^{+1} N_i N_j \, \mathrm{d}\zeta \, \mathrm{d}A \\ &= \frac{\rho c}{4} \int_{-1}^{+1} (1 + \zeta_i \zeta)(1 + \zeta_j \zeta) \, \mathrm{d}\zeta \int_A L_i L_j \, \mathrm{d}A \\ &= \frac{\rho c}{4} 2 (1 + \tfrac{1}{3} \zeta_i \zeta_j) \times \begin{cases} \dfrac{A}{6} & L_j = L_i \\[2ex] \dfrac{A}{12} & L_j \neq L_i \end{cases} \end{aligned} \tag{5.128}$$

where A is the area of the triangular cross-section.

Using this result gives the following inertia matrix

$$[\mathbf{m}]_e = \begin{bmatrix} \mathbf{m}_1 & \mathbf{m}_2 \\ \mathbf{m}_2 & \mathbf{m}_1 \end{bmatrix} \tag{5.129}$$

with

$$\mathbf{m}_1 = 2\mathbf{m}_2 \tag{5.130}$$

and

$$\mathbf{m}_2 = \frac{\rho c A}{72} \begin{bmatrix} 4 & 0 & 0 & 2 & 0 & 0 & 2 & 0 & 0 \\ 0 & 4 & 0 & 0 & 2 & 0 & 0 & 2 & 0 \\ 0 & 0 & 4 & 0 & 0 & 2 & 0 & 0 & 2 \\ 2 & 0 & 0 & 4 & 0 & 0 & 2 & 0 & 0 \\ 0 & 2 & 0 & 0 & 4 & 0 & 0 & 2 & 0 \\ 0 & 0 & 2 & 0 & 0 & 4 & 0 & 0 & 2 \\ 2 & 0 & 0 & 2 & 0 & 0 & 4 & 0 & 0 \\ 0 & 2 & 0 & 0 & 2 & 0 & 0 & 4 & 0 \\ 0 & 0 & 2 & 0 & 0 & 2 & 0 & 0 & 4 \end{bmatrix} \tag{5.131}$$

The element stiffness matrix will be given by (5.89), that is

$$[\mathbf{k}]_e = \int_{V_e} [\mathbf{B}]^{\mathrm{T}}[\mathbf{D}][\mathbf{B}]\,\mathrm{d}V \tag{5.132}$$

The strain matrix $[\mathbf{B}]$ is of the form

$$[\mathbf{B}] = \begin{bmatrix} \mathbf{B}_1 & \cdots & \mathbf{B}_6 \end{bmatrix} \tag{5.133}$$

where $\mathbf{B}_i$ is defined in (5.91). This matrix involves the derivatives $\partial N_i/\partial x$, $\partial N_i/\partial y$, $\partial N_i/\partial z$. Using (5.123) gives

$$\begin{aligned} \frac{\partial N_i}{\partial x} &= \frac{a_i}{4A}(1 + \zeta_i \zeta) \\ \frac{\partial N_i}{\partial y} &= \frac{b_i}{4A}(1 + \zeta_i \zeta) \\ \frac{\partial N_i}{\partial z} &= \frac{\zeta_i}{2c} L_i \end{aligned} \tag{5.134}$$

where

$$a_i = a_{i-3},\, b_i = b_{i-3}, \qquad i = 4, 5, 6 \tag{5.135}$$

a_i and b_i are defined in (4.10) and A, the area of the triangular cross-section, in (4.11). The derivation of the first two expressions in (5.134) is given in Section 4.5.

Substituting (5.133) and (5.134) into (5.132) and integrating will give the element stiffness matrix. The integration can be carried out analytically, but as this is a tedious process it is simpler to use numerical integration. The integrand contains terms which are quadratic in either L_i or ζ. Therefore, the exact value of the integral can be obtained using a (3×2) array of integration points. The positions of these points and the weight coefficients are given in Tables 5.1 and 3.7.

The equivalent nodal forces due to a distributed load over the face $\zeta = 1$ are given by an expression similar to (5.97), namely

$$\{\mathbf{f}\}_e = \int_A [\mathbf{N}]^{\mathrm{T}}_{\zeta=1} \begin{bmatrix} p_x \\ p_y \\ p_z \end{bmatrix} \mathrm{d}A \tag{5.136}$$

When $\zeta = 1$

$$N_i = \begin{cases} 0 & \text{for } i = 1, 2, 3 \\ L_{i-3} & \text{for } i = 4, 5, 6 \end{cases} \tag{5.137}$$

Assuming the p_x, p_y and p_z are constant, the equivalent nodal forces are

$$\{\mathbf{f}\}_e = \begin{bmatrix} \mathbf{f}_1 \\ \vdots \\ \mathbf{f}_6 \end{bmatrix} \tag{5.138}$$

where

$$\{\mathbf{f}_i\} = 0 \qquad \text{for } i = 1, 2, 3$$

and

$$\{\mathbf{f}_i\} = \frac{A}{3} \begin{bmatrix} p_x \\ p_y \\ p_z \end{bmatrix} \qquad \text{for } i = 4, 5, 6 \tag{5.139}$$

where A is the area of the triangular cross-section. The integration has been carried out using (4.90). Therefore, one third of the total force has been concentrated at each node of the face.

The equivalent nodal forces due to a distributed load over the face $L_1 = 0$ are given by

$$\{\mathbf{f}\}_e = \int_A [\mathbf{N}]^{\mathrm{T}}_{L_1=0} \begin{bmatrix} p_x \\ p_y \\ p_z \end{bmatrix} \mathrm{d}A \tag{5.140}$$

When $L_1 = 0$

$$N_i = \begin{cases} 0 & \text{for } i = 1, 4 \\ \frac{1}{2} L_i (1 + \zeta_i \zeta) & \text{for } i = 2, 3 \\ \frac{1}{2} L_{i-3} (1 + \zeta_i \zeta) & \text{for } i = 5, 6 \end{cases} \tag{5.141}$$

Assuming the p_x, p_y and p_z are constant, the equivalent nodal forces are given by (5.137) where

$$\begin{aligned} \{\mathbf{f}_i\} &= 0 \qquad \text{for } i = 1, 4 \\ \{\mathbf{f}_i\} &= \frac{c l_{2-3}}{2} \begin{bmatrix} p_x \\ p_y \\ p_z \end{bmatrix} \end{aligned} \tag{5.142}$$

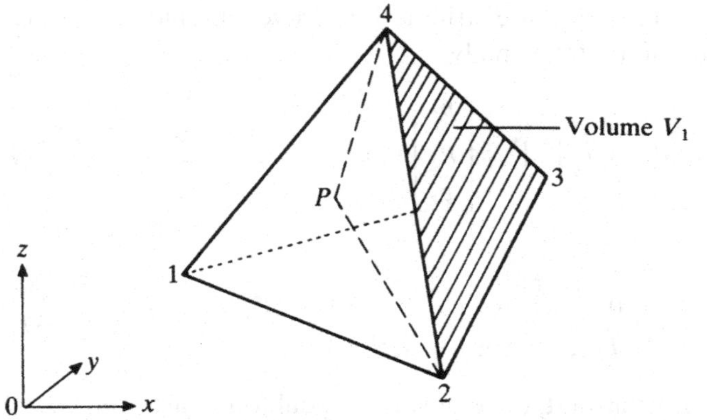

Figure 5.13. Definition of volume coordinates for a tetrahedron.

where $l_{2\text{–}3}$ is the length of the side 2–3. This integration has been carried out using (4.101). In this case one quarter of the total force has been concentrated at each node of the face.

The stresses within the element are given by (5.103). The best position to evaluate them is at the centroid of the element.

5.11 Volume Coordinates for Tetrahedra

When dealing with tetrahedra elements it is advantageous to use volume coordinates which can be defined in an analogous manner to area coordinates for a triangle (Section 4.4).

The first step in defining the volume coordinates of a point P of a tetrahedron is to join it to the four vertices of the tetrahedron. This will divide the tetrahedron into four sub-tetrahedra. The volume of the sub-tetrahedron defined by $P234$ is denoted by V_1, as shown in Figure 5.13. The volumes V_2, V_3 and V_4 are defined in a similar manner. The volume coordinates (L_1, L_2, L_3, L_4) of the point P are defined as

$$L_1 = \frac{V_1}{V}, \qquad L_2 = \frac{V_2}{V}, \qquad L_3 = \frac{V_3}{V}, \qquad L_4 = \frac{V_4}{V} \tag{5.143}$$

where V is the volume of the tetrahedron 1234. Since

$$V_1 + V_2 + V_3 + L_4 = V \tag{5.144}$$

the four volume coordinates are related by the expression

$$L_1 + L_2 + L_3 + L_4 = 1 \tag{5.145}$$

The coordinates of the four vertices are (1, 0, 0, 0), (0, 1, 0, 0), (0, 0, 1, 0) and (0, 0, 0, 1) respectively.

Volume coordinates can also be interpreted as the ratio of lengths. For example

$$L_1 = \frac{\text{distance from } P \text{ to plane 234}}{\text{distance from 1 to plane 234}} \tag{5.146}$$

This definition indicates that the plane $L_1 =$ constant is parallel to the plane 234 whose equation is $L_1 = 0$.

Cartesian and volume coordinates are related by

$$
\begin{aligned}
x &= x_1 L_1 + x_2 L_2 + x_3 L_3 + x_4 L_4 \\
y &= y_1 L_1 + y_2 L_2 + y_3 L_3 + y_4 L_4 \\
z &= z_1 L_1 + z_2 L_2 + z_3 L_3 + z_4 L_4
\end{aligned}
\tag{5.147}
$$

where (x_i, y_i) are the Cartesian coordinates of the vertex i. Combining (5.145) and (5.147) gives

$$
\begin{bmatrix} 1 \\ x \\ y \\ z \end{bmatrix} = \begin{bmatrix} 1 & 1 & 1 & 1 \\ x_1 & x_2 & x_3 & x_4 \\ y_1 & y_2 & y_3 & y_4 \\ z_1 & z_2 & z_3 & z_4 \end{bmatrix} \begin{bmatrix} L_1 \\ L_2 \\ L_3 \\ L_4 \end{bmatrix} \tag{5.148}
$$

Inverting this relationship gives

$$
\begin{bmatrix} L_1 \\ L_2 \\ L_3 \\ L_4 \end{bmatrix} = \frac{1}{6V} \begin{bmatrix} a_1 & b_1 & c_1 & d_1 \\ a_2 & b_2 & c_2 & d_2 \\ a_3 & b_3 & c_3 & d_3 \\ a_4 & b_4 & c_4 & d_4 \end{bmatrix} \begin{bmatrix} 1 \\ x \\ y \\ z \end{bmatrix} \tag{5.149}
$$

where

$$
V = \frac{1}{6} \begin{bmatrix} 1 & x_1 & y_1 & z_1 \\ 1 & x_2 & y_2 & z_2 \\ 1 & x_3 & y_3 & z_3 \\ 1 & x_4 & y_4 & z_4 \end{bmatrix} \tag{5.150}
$$

is the volume of tetrahedron 1234, and

$$
a_1 = \begin{vmatrix} x_2 & y_2 & z_2 \\ x_3 & y_3 & z_3 \\ x_4 & y_4 & z_4 \end{vmatrix} \tag{5.151}
$$

$$
b_1 = -\begin{vmatrix} 1 & y_2 & z_1 \\ 1 & y_3 & z_3 \\ 1 & y_4 & z_4 \end{vmatrix} \tag{5.152}
$$

$$
c_1 = \begin{vmatrix} 1 & x_2 & z_2 \\ 1 & x_3 & z_3 \\ 1 & x_4 & z_4 \end{vmatrix} \tag{5.153}
$$

$$
d_1 = -\begin{vmatrix} 1 & x_2 & y_2 \\ 1 & x_3 & y_3 \\ 1 & x_4 & y_4 \end{vmatrix} \tag{5.154}
$$

The other constants in (5.149) are obtained through a cyclic permutation of the subscripts 1, 2, 3 and 4. As these terms are the elements of the adjoint matrix of the coefficient matrix in (5.148) it is necessary to give the proper signs to them. Thus a_3

will have the same sign as a_1 and both a_2 and a_4 will have the opposite sign to a_1. A similar rule applies to the coefficients b, c and d.

5.12 Tetrahedron Element

The displacement functions for a tetrahedron element with nodes only at the four vertices take the form

$$u = \sum_{j=1}^{4} N_j u_j, \qquad v = \sum_{j=1}^{4} N_j v_j, \qquad w = \sum_{j=1}^{4} N_j w_j \tag{5.155}$$

where the functions N_j are required to have a unit value at node j and zero values at the other three nodes. These conditions are satisfied by the volume coordinates L_j. Therefore

$$N_j = L_j \tag{5.156}$$

On each face of the tetrahedron the variation of displacement is bilinear and uniquely determined by the values at the node points on it. There will, therefore, be continuity of displacement between adjacent elements.

The expressions (5.155) can be written in the combined form

$$\begin{bmatrix} u \\ v \\ w \end{bmatrix} = [\mathbf{N}]\,\{\mathbf{u}\}_e \tag{5.157}$$

where

$$\{\mathbf{u}\}_e^{\mathrm{T}} = \lfloor u_1 \quad v_1 \quad w_1 \quad \cdots \quad u_4 \quad v_4 \quad w_4 \rfloor \tag{5.158}$$

and

$$[\mathbf{N}] = \begin{bmatrix} N_1 & 0 & 0 & \cdots & N_4 & 0 & 0 \\ 0 & N_1 & 0 & \cdots & 0 & N_4 & 0 \\ 0 & 0 & N_1 & \cdots & 0 & 0 & N_4 \end{bmatrix} \tag{5.159}$$

The element inertia matrix will be given by (5.83), that is

$$[\mathbf{m}]_e = \int_{V_e} [\mathbf{N}]^{\mathrm{T}}[\mathbf{N}]\,\mathrm{d}V \tag{5.160}$$

A typical element of this matrix is of the form $\rho \int_{V_e} L_i L_j \,\mathrm{d}V$. Integrals of this form can be evaluated using the following formula [5.11]

$$\int_V L_1^m L_2^n L_3^p L_4^q \,\mathrm{d}V = \frac{m!n!p!q!}{(m+n+p+q+3)!} 6V \tag{5.161}$$

Therefore

$$\int_{V_e} L_i L_j \,\mathrm{d}V = \begin{cases} V/10 & j = i \\ V/20 & j \neq i \end{cases} \tag{5.162}$$

Using this result gives the following inertia matrix

$$[\mathbf{m}]_e = \frac{\rho V}{20}\begin{bmatrix} 2 & 0 & 0 & 1 & 0 & 0 & 1 & 0 & 0 & 1 & 0 & 0 \\ 0 & 2 & 0 & 0 & 1 & 0 & 0 & 1 & 0 & 0 & 1 & 0 \\ 0 & 0 & 2 & 0 & 0 & 1 & 0 & 0 & 1 & 0 & 0 & 1 \\ 1 & 0 & 0 & 2 & 0 & 0 & 1 & 0 & 0 & 1 & 0 & 0 \\ 0 & 1 & 0 & 0 & 2 & 0 & 0 & 1 & 0 & 0 & 1 & 0 \\ 0 & 0 & 1 & 0 & 0 & 2 & 0 & 0 & 1 & 0 & 0 & 1 \\ 1 & 0 & 0 & 1 & 0 & 0 & 2 & 0 & 0 & 1 & 0 & 0 \\ 0 & 1 & 0 & 0 & 1 & 0 & 0 & 2 & 0 & 0 & 1 & 0 \\ 0 & 0 & 1 & 0 & 0 & 1 & 0 & 0 & 2 & 0 & 0 & 1 \\ 1 & 0 & 0 & 1 & 0 & 0 & 1 & 0 & 0 & 2 & 0 & 0 \\ 0 & 1 & 0 & 0 & 1 & 0 & 0 & 1 & 0 & 0 & 2 & 0 \\ 0 & 0 & 1 & 0 & 0 & 1 & 0 & 0 & 1 & 0 & 0 & 2 \end{bmatrix} \tag{5.163}$$

The element stiffness matrix is given by (5.89), that is

$$[\mathbf{k}]_e = \int_{V_e} [\mathbf{B}]^{\mathrm{T}}[\mathbf{D}][\mathbf{B}]\, \mathrm{d}V \tag{5.164}$$

The strain matrix $[\mathbf{B}]$ is of the form

$$[\mathbf{B}] = [\mathbf{B}_1 \quad \cdots \quad \mathbf{B}_4] \tag{5.165}$$

where $\mathbf{B}_i$ is defined in (5.91). This matrix involves the derivatives $\partial N_i/\partial x$, $\partial N_i/\partial y$, $\partial N_i/\partial z$. From (5.149)

$$\frac{\partial N_i}{\partial x} = \frac{\partial L_i}{\partial x} = \frac{b_i}{6V} \tag{5.166}$$

Similarly

$$\frac{\partial N_i}{\partial y} = \frac{\partial L_i}{\partial y} = \frac{c_i}{6V} \tag{5.167}$$

and

$$\frac{\partial N_i}{\partial z} = \frac{\partial L_i}{\partial z} = \frac{d_i}{6V} \tag{5.168}$$

Combining (5.164) to (5.167) gives

$$\mathbf{B}_i = \frac{1}{6V}\begin{bmatrix} b_i & 0 & 0 \\ 0 & c_i & 0 \\ 0 & 0 & d_i \\ c_i & b_i & 0 \\ d_i & 0 & b_i \\ 0 & d_i & c_i \end{bmatrix} \tag{5.169}$$

Since $[\mathbf{B}]$ is a constant matrix (5.164) reduces to

$$[\mathbf{k}]_e = V[\mathbf{B}]^{\mathrm{T}}[\mathbf{D}][\mathbf{B}] \tag{5.170}$$

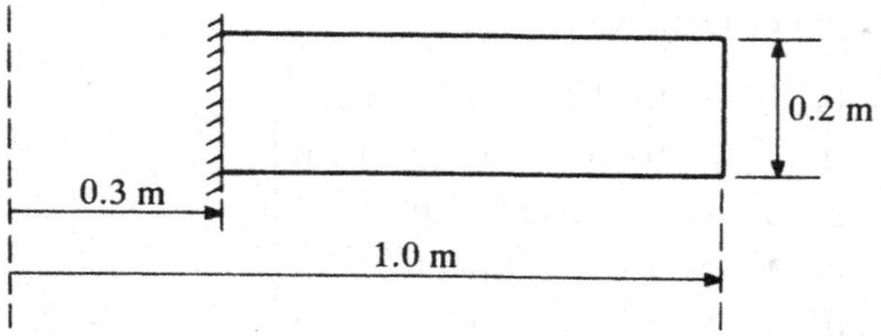

Figure 5.14. Geometry of a circular disc.

The equivalent nodal forces due to a distributed load over the face $L_1 = 0$ are given by

$$\{\mathbf{f}\}_e = \int_A [\mathbf{N}]^{\mathrm{T}}_{L_1=0} \begin{bmatrix} p_x \\ p_y \\ p_z \end{bmatrix} \mathrm{d}A \tag{5.171}$$

When integrating the volume coordinates L_2, L_3 and L_4 over the face 234 they reduce to area coordinates. The integrals in (5.171) can, therefore, be evaluated using (4.90).

Assuming the p_x, p_y and p_z are constant, the equivalent nodal forces are

$$\{\mathbf{f}\}_e = \begin{bmatrix} 0 \\ \mathbf{f}_2 \\ \mathbf{f}_3 \\ \mathbf{f}_4 \end{bmatrix} \tag{5.172}$$

where

$$\{\mathbf{f}_i\} = \frac{A}{3} \begin{bmatrix} p_x \\ p_y \\ p_z \end{bmatrix} \qquad \text{for } i = 2, 3, 4 \tag{5.173}$$

where A is the area of face 234. Therefore, one third of the total force has been concentrated at each node of the face.

The stresses within the element are given by (5.103). Since $[\mathbf{B}]$ is constant the stresses are constant within the element. It is usual to assign these constant values to the centroid of the element $(\frac{1}{4}, \frac{1}{4}, \frac{1}{4}, \frac{1}{4})$.

5.13 Increasing the Accuracy of Elements

Section 5.4 refers to the fact that any of the element shapes and associated displacement functions used for membrane analysis in Chapter 4 can be used to analyse axisymmetric solids. Triangular elements are illustrated in Figures 4.18 and 4.19(*b*), and quadrilateral elements in Figures 4.8, 4.12, 4.15, 4.17 and 4.19(*a*).

Reference [5.12] illustrates how increased accuracy can be obtained when using eight node rectangles rather than four node ones by considering the fixed-free circular disc shown in Figure 5.14. Two idealisations were used as shown in Figure 5.15. The four node element was evaluated using a (2×2) array of integration points and the eight node element with a (3×3) array. Taking $E = 196 \times 10^9$ N/m^2, $\nu = 0.3$ and $\rho = 7800$ kg/m^3, the results obtained are given in Table 5.6.

Table 5.6. *Comparison of predicted frequencies (Hz) of a fixed-free circular disc*

Nodal diameters	Idealisation I	Idealisation II	Exact [5.13]
0	349	305	312
1	342	290	276
2	360	341	323

The accuracy of three-dimensional elements can be increased using methods similar to the ones presented in Section 4.6 for membrane elements. The eight node hexahedron can be improved by including extra displacement functions when evaluating the stiffness matrix [5.14] and so the components of displacement are assumed to be given by

$$
\begin{aligned}
u &= \sum_{j=1}^{8} N_j u_j + \alpha_1(1-\xi^2) + \alpha_2(1-\eta^2) + \alpha_3(1-\zeta^2) \\
v &= \sum_{j=1}^{8} N_j v_j + \alpha_4(1-\xi^2) + \alpha_5(1-\eta^2) + \alpha_6(1-\zeta^2) \\
w &= \sum_{j=1}^{8} N_j w_j + \alpha_7(1-\xi^2) + \alpha_8(1-\eta^2) + \alpha_9(1-\zeta^2)
\end{aligned}
\tag{5.174}
$$

where the functions $N_j(\xi, \eta, \zeta)$ are defined by (5.78). After evaluating the stiffness matrix the parameters α_1 to α_9 are eliminated using static condensation, as described in Section 4.6.

If the cube shown in Figure 5.10 is analysed using this modified stiffness matrix, then the results obtained are as shown in Table 5.7. Comparing these with those given in Table 5.4 shows that the accuracy of the frequencies has increased, particularly those of the swaying modes.

Alternatively, the accuracy of the rectangular hexahedron can be increased by introducing one additional node point at the centre of each edge. This results in

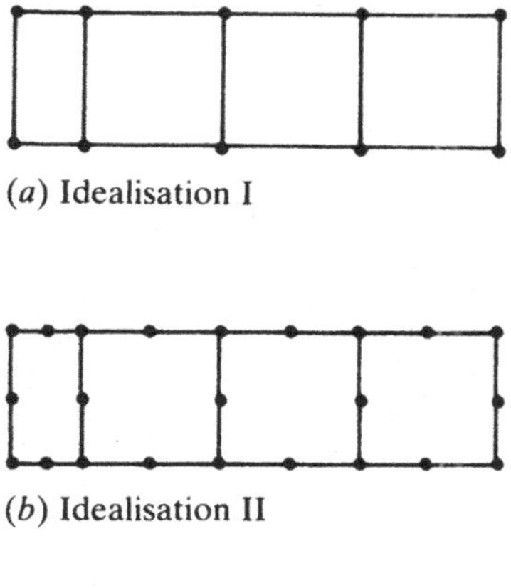

(*a*) Idealisation I

(*b*) Idealisation II

Figure 5.15. Idealisation of a circular disc.

Table 5.7. *Comparison of predicted frequencies (kHz) of a cube using extra displacement functions*

Mode	FEM	Exact [5.9]	% Difference
S	2.316	2.212	4.7
T	3.250	3.020	7.6
L	5.468	5.239	4.4
S	6.630	5.915	12.1

S, Swaying mode; T, torsion mode; L, Longitudinal mode.

twenty node points as shown in Figure 5.16. In this case the displacements are given by [5.15]

$$u = \sum_{j=1}^{20} N_j u_j, \qquad v = \sum_{j=1}^{20} N_j v_j, \qquad w = \sum_{j=1}^{20} N_j w_j \tag{5.175}$$

where

$$N_j(\xi, \eta, \zeta) = \tfrac{1}{8}(1 + \xi_j \xi)(1 + \eta_j \eta)(1 + \zeta_j \zeta)(\xi_j \xi + \eta_j \eta + \zeta_j \zeta - 2) \tag{5.176}$$

for nodes 1 to 8

$$N_j(\xi, \eta, \zeta) = \tfrac{1}{4}(1 - \xi^2)(1 + \eta_j \eta)(1 + \zeta_j \zeta) \tag{5.177}$$

for nodes 9, 11, 13 and 15

$$N_j(\xi, \eta, \zeta) = \tfrac{1}{4}(1 + \xi_j \xi)(1 - \eta^2)(1 + \zeta_j \zeta) \tag{5.178}$$

for nodes 10, 12, 14 and 16 and

$$N_j(\xi, \eta, \zeta) = \tfrac{1}{4}(1 + \xi_j \xi)(1 + \eta_j \eta)(1 - \zeta^2) \tag{5.179}$$

for nodes 17, 18, 19 and 20, where ξ_j, η_j and ζ_j are the coordinates of node j.

The stiffness matrix should be evaluated using a $(3 \times 3 \times 3)$ array of Gauss integration points or alternatively the fourteen point integration scheme referred to in Section 5.9. The inertia matrix can be evaluated exactly using either of these two schemes.

Reference [5.16] contains an analysis of the cantilever beam shown in Figure 5.17(*a*). Two idealisations are used, as shown in Figures 5.17(*b*) and (*c*). The first consists of 216 eight-node elements and the second 36 twenty-node elements. The following constraints are applied at the plane $z = 0$.

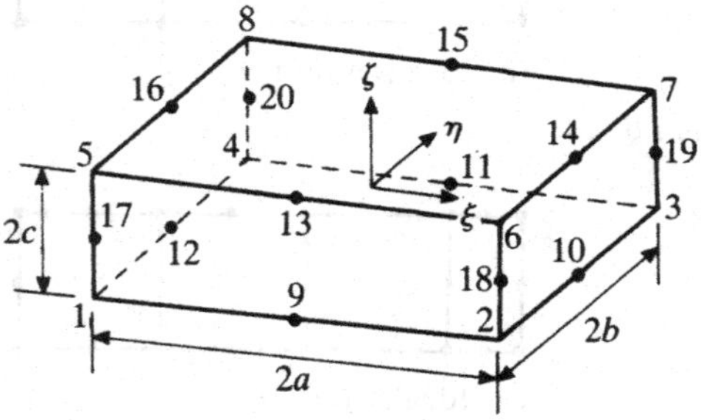

Figure 5.16. Geometry of twenty node rectangular hexahedron.

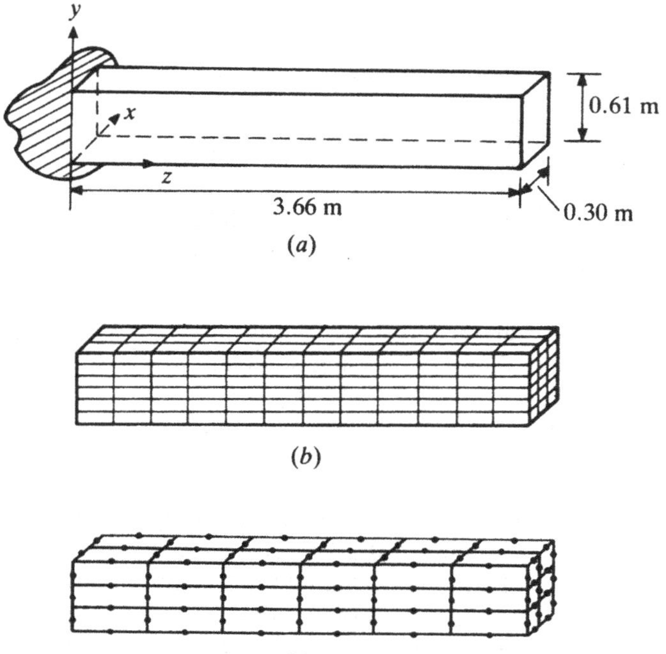

Figure 5.17. Geometry of a cantilever beam.

For model 5.17(b)

(1) $w = 0$ at every node
(2) $v = 0$ at every node along the line $y = 0.305$
(3) $u = 0$ at every node along the line $x = 0$

For model 5.17(c)

(4) as (1) above
(5) as (2) above
(6) $u = 0$ at every node along the line $x = 0.15$

The material constants used are $E = 2.068 \times 10^{11}$ N/m^2, $\nu = 0.3$ and $\rho = 8058$ kg/m^3. The frequencies obtained are compared with the exact values obtained using slender beam theory in Table 5.8. Model 5.17(b) has 1053 degrees of freedom whilst 5.17(c) has only 786. The increased accuracy has, therefore, been obtained with fewer elements and degrees of freedom.

Reference [5.17] presents an analysis of the drop hammer anvil shown in Figure 5.18. The idealisation consists of a $(2 \times 2 \times 1)$ array of twenty node elements. The anvil was considered to be completely free. The material constants are $E = 2.07 \times 10^{11}$ N/m^2, $\nu = 0.3$ and $\rho = 7860$ kg/m^3.

The frequencies of the first four modes are compared with measured values in Table 5.9.

Table 5.8. *Comparison of predicted frequencies (Hz) of a cantilever beam [5.16]*

Mode	Description	Idealisation 5.17(*b*)	5.17(*c*)	Exact
1	First bending in x-direction	22.0	18.6	18.6
2	First bending in y-direction	38.3	36.5	37.3
3	Second bending in x-direction	135.3	114.3	116.8

The twenty node hexahedron (Figure 5.16) can be transformed into a hexahedron with curved surfaces, as shown in Figure 5.19, using the relationships

$$x = \sum_{j=1}^{20} N_j x_j, \qquad y = \sum_{j=1}^{20} N_j y_j, \qquad z = \sum_{j=1}^{20} N_j z_j \tag{5.180}$$

where the functions $N_j(\xi, \eta, \zeta)$ are defined by (5.176) to (5.179). The determinant of the Jacobian, det $[\mathbf{J}]$, is an incomplete quintic [5.18]. Therefore, a $(5 \times 5 \times 5)$ array of integration points is required to evaluate the inertia matrix exactly. In practice, distortions are unlikely to be very great, especially with mesh refinement. Therefore, a $(3 \times 3 \times 3)$ (or equivalent 14 point) array may suffice. The stiffness matrix should be integrated with a $(3 \times 3 \times 3)$ or 14 point array [5.19].

The steam turbine blade shown in Figure 5.20 has been analysed in reference [5.20]. The idealisation consisted of a $(3 \times 1 \times 4)$ array of twenty node elements as shown in Figure 5.21. A blade was machined from mild steel and tested to provide frequencies for comparison.

Four different analyses were performed. In every case the inertia matrix was evaluated using a $(3 \times 3 \times 3)$ array of Gauss integration points. The integration schemes used for the stiffness matrix are as follows:

I 27*a*
II $(3 \times 3 \times 3)$
III 14
IV $(2 \times 2 \times 2)$

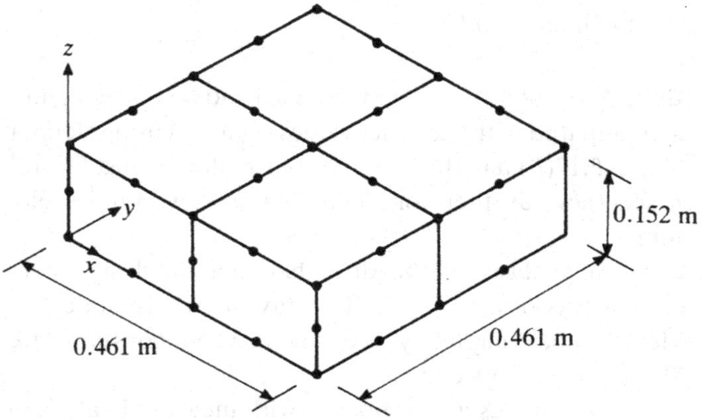

Figure 5.18. Geometry of an anvil.

Table 5.9. *Comparison of frequencies (kHz) of an anvil [5.17]*

Mode	Description	FEM	Experimental
1	Twist	1.90	1.82
2	Saddle	2.82	2.67
3	Umbrella	3.51	3.18
4	In-plane shear	4.37	4.0

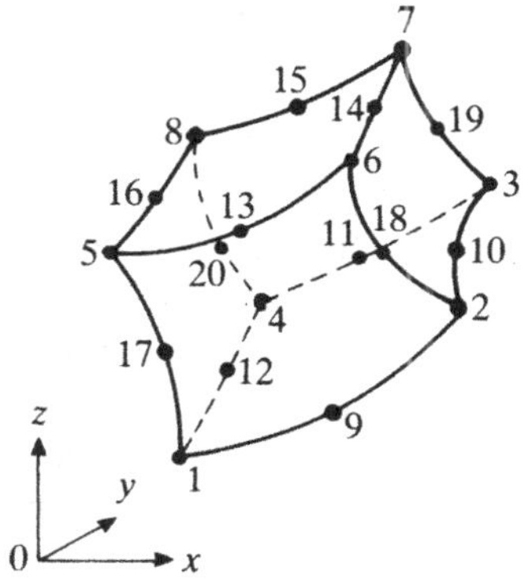

Figure 5.19. Geometry of a hexahedron with curved surfaces.

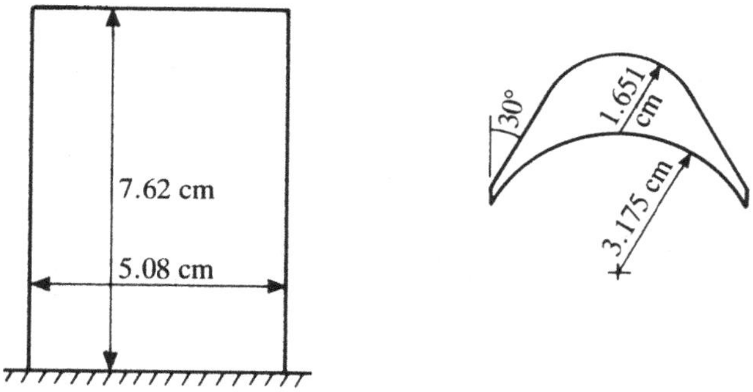

Figure 5.20. Geometry of a steam turbine blade.

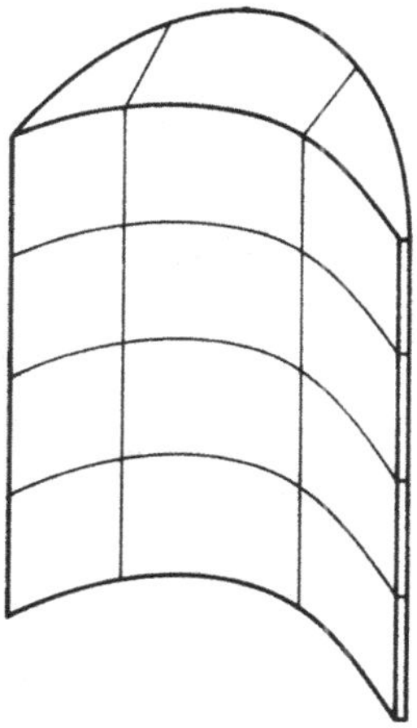

Figure 5.21. Idealisation of a steam turbine blade.

Table 5.10. *Comparison of the predicted natural frequencies (kHz) of a steam turbine blade [5.20]*

		Stiffness matrix integration scheme			
Mode	Experiment	I	II	III	IV
1	2.815	2.917	2.932	2.919	2.745
2	3.876	4.426	4.432	4.411	4.425
3	6.250	6.633	6.679	6.611	6.612
4		13.607	13.679	13.571	12.835
5		15.513	15.621	15.407	14.869

Scheme I, which consists of 27 integration points, is equivalent to a (4 × 4 × 4) array of Gauss points. Details are given in reference [5.10].

The frequencies obtained with the four analyses are compared with the measured frequencies in Table 5.10. These results suggest that the best scheme to adopt in this application is III, which consists of 14 integration points giving similar accuracy to a (3 × 3 × 3) array of Gauss points. In this case the percentage differences between predicted and measured frequencies are 3.4, 13.8 and 5.8.

The first five mode shapes can be described as follows:

(1) First flapwise bending
(2) First edgewise bending
(3) First torsion
(4) Second flapwise bending
(5) Coupled bending and torsion

The hexahedra shown in Figures 5.16 and 5.19 can be supplemented by compatible pentahedra as shown in Figure 5.22. The displacment functions for the right pentahedra in Figure 5.22(a) are of the form

$$u = \sum_{j=1}^{15} N_j u_j, \qquad v = \sum_{j=1}^{15} N_j v_j, \qquad w = \sum_{j=1}^{15} N_j w_j \tag{5.181}$$

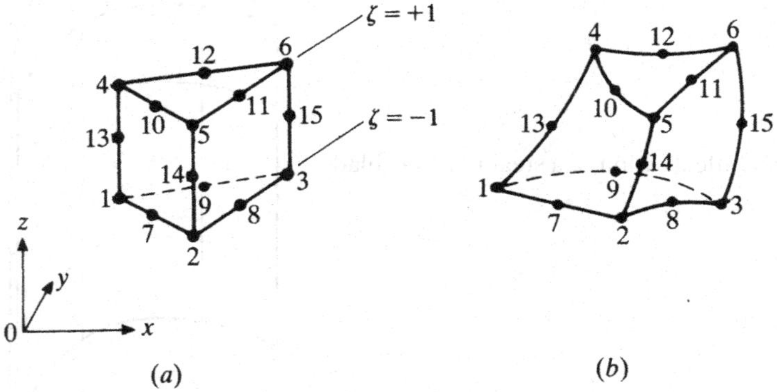

Figure 5.22. Geometry of pentahedra elements: (a) straight sides, (b) curved sides.

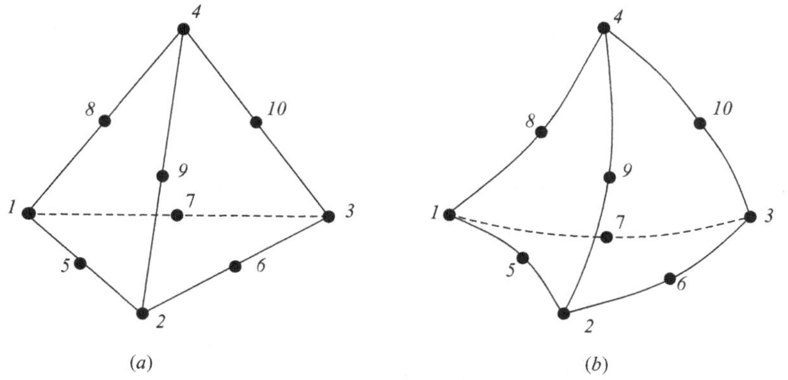

Figure 5.23. Geometry of tetrahedral elements (*a*) straight sides and (*b*) curved sides.

where the functions N_j are given by

$$N_j = \tfrac{1}{2}L_j(2L_j - 1)(1 + \zeta_j\zeta) - \tfrac{1}{2}L_j(1 - \zeta^2) \tag{5.182}$$

for nodes 1 to 6 with $L_j = L_{j-3}$ for $j = 4, 5, 6$

$$\begin{aligned} N_7 &= 2L_1L_2(1-\zeta), & N_{10} &= 2L_1L_2(1+\zeta) \\ N_8 &= 2L_2L_3(1-\zeta), & N_{11} &= 2L_2L_3(1+\zeta) \\ N_9 &= 2L_3L_1(1-\zeta), & N_{12} &= 2L_3L_1(1+\zeta) \end{aligned} \tag{5.183}$$

and

$$N_j = L_{j-12}(1 - \zeta^2) \tag{5.184}$$

for nodes 13 to 15.

Both the stiffness and inertia matrices can be evaluated using a (6 × 3) array of integration points, the positions of which are given in Tables 5.1 and 3.7 together with the corresponding weights.

This element can be transformed into one with curved sides, as shown in Figure 5.22(*b*), using the relationship

$$x = \sum_{j=1}^{15} N_j x_j, \qquad y = \sum_{j=1}^{15} N_j y_j, \qquad z = \sum_{j=1}^{15} N_j z_j \tag{5.185}$$

where the functions N_j are as defined in (5.182) to (5.184). The determinant of the Jacobian of the transformation is quartic in the area coordinates and quintic in ζ. This can be integrated exactly using a (6 × 3) array of integration points. This represents the minimum requirement for both the stiffness and inertia matrices.

The accuracy of the four-node tetrahedron shown in Figure 5.13 can also be improved by introducing one additional node point at the centre of each edge as shown in Figure 5.23(*a*). The displacement functions are of the form

$$u = \sum_{j=1}^{10} N_j u_j, \qquad v = \sum_{j=1}^{10} N_j v_j, \qquad w = \sum_{j=1}^{10} N_j w_j \tag{5.186}$$

where the functions N_j are given by

$$N_j = L_j(2L_j - 1) \qquad j = 1, 2, 3, 4 \tag{5.187}$$

$$N_5 = 4L_1L_2, \quad N_6 = 4L_2L_3, \quad N_7 = 4L_3L_1 \tag{5.188}$$

$$N_8 = 4L_1L_4, \quad N_9 = 4L_2L_4, \quad N_{10} = 4L_3L_4 \tag{5.189}$$

Both the inertia and stiffness matrices can be evaluated using four integration points. These are positioned at the points $(\alpha, \beta, \beta, \beta)$, $(\beta, \alpha, \beta, \beta)$, $(\beta, \beta, \alpha, \beta)$ and $(\beta, \beta, \beta, \alpha)$ where $\alpha = 0.58541020$ and $\beta = 0.13819660$ all with weight $\frac{1}{4}$.

The 10-node tetrahedron can be transformed into a tetrahedron with curved surfaces as shown in Figure 5.23(b) using the relationships

$$x = \sum_{j=1}^{10} N_j x_j, \qquad y = \sum_{j=1}^{10} N_j y_j, \qquad z = \sum_{j=1}^{10} N_j z_j \tag{5.190}$$

where the functions N_j (L_1, L_2, L_3, L_4) are defined by (5.187) to (5.189). Again, both the inertia and stiffness matrices can be evaluated using four integration points.

Problems

Note: Problems 5.7 and 5.8 require the use of a digital computer.

5.1 Find the Fourier series representation of the loading

$$p_r = \begin{cases} p & -\phi < \theta < +\phi \\ 0 & -\pi < \theta < -\phi, \phi < \theta < \pi \end{cases}$$

5.2 Find the Fourier series representation for a line load of magnitude P at $\theta = 0$. (Hint: Put $p = P/2a\phi$ in Problem 5.1 and then let $\phi \rightarrow 0$.)

5.3 Express the stiffness matrix of an axisymmetric triangular element in the form (5.43). (Hint: Use the fact that $[\mathbf{B}_{1i}]$ and $[\mathbf{B}_{2i}]$ can be expressed in the form $[\mathbf{B}_{1i}] = [\mathbf{B}^0_{1i}] + n[\mathbf{B}^n_{1i}]$ and $[\mathbf{B}_{2i}] = [\mathbf{B}^0_{2i}] + n[\mathbf{B}^n_{2i}]$).

5.4 Derive the element matrices for an axisymmetric, triangular core element when both nodes 1 and 3 are on the z-axis, as shown in Figure 5.6(b).

5.5 Derive the element matrices for an axisymmetric, linear rectangular element. Discuss the need for assuming the shear strains to be constant, in order to improve the accuracy of the element.

5.6 Derive the element matrices for an axisymmetric, linear quadrilateral element.

5.7 Figure 5.14 shows the cross-section of a circular disc which is fixed at the inner radius and free at the outer radius. Use the idealisation shown in Figure P5.7 to calculate the frequencies of the modes having 0, 1 and 2 nodal diameters and no nodal circles, other than the inner radius. Take $E = 196 \times 10^9$ N/m^2, $\nu = 0.3$ and $\rho = 7800$ kg/m^3. Compare these frequencies with the analytical values [5.13] 312.2, 276.3 and 322.5 Hz.

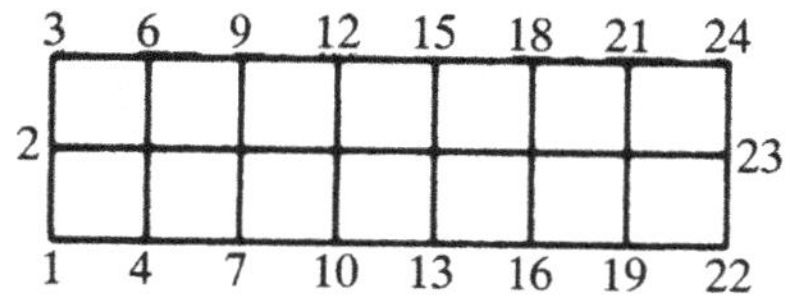

Figure P5.7. Idealisation of a circular disc.

5.8 Figure P5.8(a) shows the cross-section of a thick cylinder with shear diaphragm end conditions, that is, $u = 0$ and $v = 0$. Use the idealisation shown in Figure P5.8(b) to calculate the frequencies of the first four axisymmetric modes. Take $E = 207 \times 10^9$ N/m^2, $\nu = 0.3$ and $\rho = 7850$ kg/m^3. Compare these frequencies with the analytical values [5.21] 4985, 8095, 9538 and 9609 Hz which correspond to modes having 1, 2, 1 and 3 axial half-wavelengths respectively.

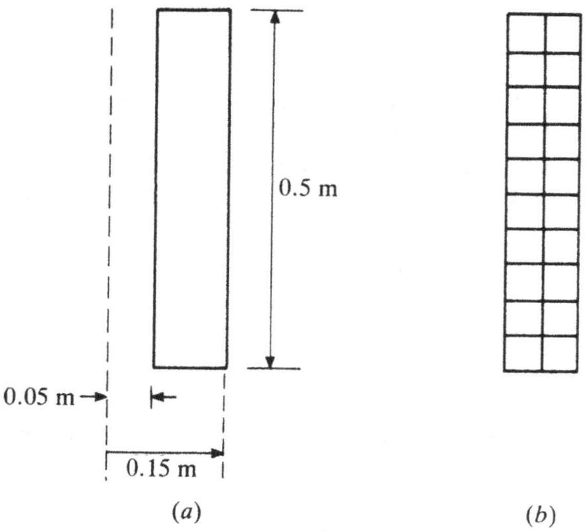

Figure P5.8. Geometry of a thick cylinder.

5.9 Show that if the six node pentahedron, presented in Section 5.10, is given a rigid body translation in the z-direction, then the inertia matrix, (5.129), gives the mass of the element.

5.10 Derive expressions for the inertia and stiffness matrices of a six-node isoparametric pentahedron element. How many integration points will be needed to evaluate these expressions?

5.11 Derive the equivalent nodal forces corresponding to a distributed load p_x per unit area in the x-direction over the face 2, 3, 7, 6 in Figure 5.16.

6 Flexural Vibration of Plates

Flat plate structures, such as the floors of aircraft, buildings and ships, bridge decks and enclosures surrounding machinery, are subject to dynamic loads normal to their plane. This results in flexural vibration. Such structures can be analysed by dividing the plate up into an assemblage of two-dimensional finite elements called plate bending elements. These elements may be either triangular, rectangular or quadrilateral in shape.

The energy expressions for a thin plate bending element are, from Section 2.6,

$$T_e = \frac{1}{2}\int_A \rho h \dot{w}^2 \,\mathrm{d}A \tag{6.1}$$

$$U_e = \frac{1}{2}\int_A \frac{h^3}{12} \{\boldsymbol{\chi}\}^{\mathrm{T}} [\mathbf{D}] \{\boldsymbol{\chi}\} \,\mathrm{d}A \tag{6.2}$$

with

$$\{\boldsymbol{\chi}\} = \begin{bmatrix} \partial^2 w/\partial x^2 \\ \partial^2 w/\partial y^2 \\ 2\partial^2 w/\partial x \partial y \end{bmatrix} \tag{6.3}$$

where $[\mathbf{D}]$ is defined by (2.45), (2.49) or (2.51) depending upon whether the material is anisotropic, orthotropic or isotropic. Also

$$\delta W_e = \int_A p_z \delta w \,\mathrm{d}A \tag{6.4}$$

The highest derivative appearing in these expressions is the second. Hence, for convergence, it will be necessary to ensure that w and its first derivatives $\partial w/\partial x$ and $\partial w/\partial y$ are continuous between elements. These three quantities are, therefore, taken as degrees of freedom at each node. Also, complete polynomials of at least degree two should be used (see Section 3.2). The assumed form of the displacement function, whatever the element shape, is

$$\begin{aligned} w = {} & \alpha_1 + \alpha_2 x + \alpha_3 y + \alpha_4 x^2 + \alpha_5 xy + \alpha_6 y^2 \\ & + \text{higher degree terms} \end{aligned} \tag{6.5}$$

Figure 6.1. Geometry of a rectangular element. $\xi = x/a$, $\eta = y/b$.

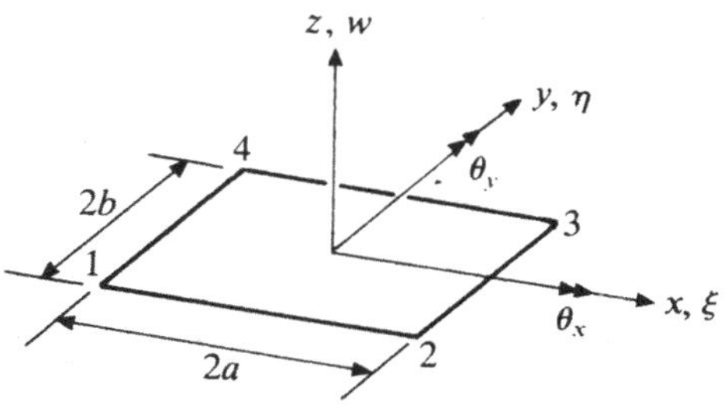

6.1 Thin Rectangular Element (non-conforming)

Figure 6.1 shows a rectangular element with four node points, one at each corner. There are three degrees of freedom at each node, namely, the component of displacement normal to the plane of the plate, w, and the two rotations $\theta_x = \partial w/\partial y$ and $\theta_y = -\partial w/\partial x$. In terms of the (ξ, η) coordinates, these become

$$\theta_x = \frac{1}{b}\frac{\partial w}{\partial \eta}, \qquad \theta_y = -\frac{1}{a}\frac{\partial w}{\partial \xi} \tag{6.6}$$

Since the element has tweleve degrees of freedom, the displacement function can be represented by a polynomial having twelve terms, that is

$$\begin{aligned} w = {} & \alpha_1 + \alpha_2\xi + \alpha_3\eta + \alpha_4\xi^2 + \alpha_5\xi\eta + \alpha_6\eta^2 \\ & + \alpha_7\xi^3 + \alpha_8\xi^2\eta + \alpha_9\xi\eta^2 + \alpha_{10}\eta^3 + \alpha_{11}\xi^3\eta + \alpha_{12}\xi\eta^3 \end{aligned} \tag{6.7}$$

Note that this function is a complete cubic to which has been added two quartic terms $\xi^3\eta$ and $\xi\eta^3$ which are symmetrically placed in Pascal's triangle (Figure 4.1). This will ensure that the element is geometrically invariant (see Chapter 4).

The expression (6.7) can be written in the following matrix form

$$\begin{aligned} w &= \lfloor 1 \quad \xi \quad \eta \quad \xi^2 \quad \xi\eta \quad \eta^2 \quad \xi^3 \quad \xi^2\eta \quad \xi\eta^2 \quad \eta^3 \quad \xi^3 n \quad \xi\eta^3 \rfloor \{\boldsymbol{\alpha}\} \\ &= \lfloor \mathbf{P}(\xi, \eta) \rfloor \{\boldsymbol{\alpha}\} \end{aligned} \tag{6.8}$$

where

$$\{\boldsymbol{\alpha}\}^{\mathrm{T}} = \lfloor \alpha_1 \quad \alpha_2 \quad \cdots \quad \alpha_{12} \rfloor \tag{6.9}$$

Differentiating (6.8) gives

$$\frac{\partial w}{\partial \xi} = \lfloor 0 \quad 1 \quad 0 \quad 2\xi \quad \eta \quad 0 \quad 3\xi^2 \quad 2\xi\eta \quad \eta^2 \quad 0 \quad 3\xi^2\eta \quad \eta^3 \rfloor \{\boldsymbol{\alpha}\} \tag{6.10}$$

and

$$\frac{\partial w}{\partial \eta} = \lfloor 0 \quad 0 \quad 1 \quad 0 \quad \xi \quad 2\eta \quad 0 \quad \xi^2 \quad 2\xi\eta \quad 3\eta^2 \quad \xi^3 \quad 3\xi\eta^2 \rfloor \{\boldsymbol{\alpha}\} \tag{6.11}$$

Evaluating (6.8), (6.10) and (6.11) at $\xi = \mp 1$, $\eta = \mp 1$ gives

$$\{\bar{\mathbf{w}}\}_e = [\mathbf{A}]_e \{\boldsymbol{\alpha}\} \tag{6.12}$$

where

$$\{\bar{\mathbf{w}}\}_e^{\mathrm{T}} = \lfloor w_1 \quad b\theta_{x1} \quad a\theta_{y1} \quad \cdots \quad w_4 \quad b\theta_{x4} \quad a\theta_{y4} \rfloor \tag{6.13}$$

and

$$[\mathbf{A}]_e = \begin{bmatrix} 1 & -1 & -1 & 1 & 1 & 1 & -1 & -1 & -1 & -1 & 1 & 1 \\ 0 & 0 & 1 & 0 & -1 & -2 & 0 & 1 & 2 & 3 & -1 & -3 \\ 0 & -1 & 0 & 2 & 1 & 0 & -3 & -2 & -1 & 0 & 3 & 1 \\ 1 & 1 & -1 & 1 & -1 & 1 & 1 & -1 & 1 & -1 & -1 & -1 \\ 0 & 0 & 1 & 0 & 1 & -2 & 0 & 1 & -2 & 3 & 1 & 3 \\ 0 & -1 & 0 & -2 & 1 & 0 & -3 & 2 & -1 & 0 & 3 & 1 \\ 1 & 1 & 1 & 1 & 1 & 1 & 1 & 1 & 1 & 1 & 1 & 1 \\ 0 & 0 & 1 & 0 & 1 & 2 & 0 & 1 & 2 & 3 & 1 & 3 \\ 0 & -1 & 0 & -2 & -1 & 0 & -3 & -2 & -1 & 0 & -3 & -1 \\ 1 & -1 & 1 & 1 & -1 & 1 & -1 & 1 & -1 & 1 & -1 & -1 \\ 0 & 0 & 1 & 0 & -1 & 2 & 0 & 1 & -2 & 3 & -1 & -3 \\ 0 & -1 & 0 & 2 & -1 & 0 & -3 & 2 & -1 & 0 & -3 & -1 \end{bmatrix} \tag{6.14}$$

Solving (6.12) for $\{\boldsymbol{\alpha}\}$ gives

$$\{\boldsymbol{\alpha}\} = [\mathbf{A}]_e^{-1}\{\bar{\mathbf{w}}\}_e \tag{6.15}$$

where

$$[\mathbf{A}]_e^{-1} = \frac{1}{8}\begin{bmatrix} 2 & 1 & -1 & 2 & 1 & 1 & 2 & -1 & 1 & 2 & -1 & -1 \\ -3 & -1 & 1 & 3 & 1 & 1 & 3 & -1 & 1 & -3 & 1 & 1 \\ -3 & -1 & 1 & -3 & -1 & -1 & 3 & -1 & 1 & 3 & -1 & -1 \\ 0 & 0 & 1 & 0 & 0 & -1 & 0 & 0 & -1 & 0 & 0 & 1 \\ 4 & 1 & -1 & -4 & -1 & -1 & 4 & -1 & 1 & -4 & 1 & 1 \\ 0 & -1 & 0 & 0 & -1 & 0 & 0 & 1 & 0 & 0 & 1 & 0 \\ 1 & 0 & -1 & -1 & 0 & -1 & -1 & 0 & -1 & 1 & 0 & -1 \\ 0 & 0 & -1 & 0 & 0 & 1 & 0 & 0 & -1 & 0 & 0 & 1 \\ 0 & 1 & 0 & 0 & -1 & 0 & 0 & 1 & 0 & 0 & -1 & 0 \\ 1 & 1 & 0 & 1 & 1 & 0 & -1 & 1 & 0 & -1 & 1 & 0 \\ -1 & 0 & 1 & 1 & 0 & 1 & -1 & 0 & -1 & 1 & 0 & -1 \\ -1 & -1 & 0 & 1 & 1 & 0 & -1 & 1 & 0 & 1 & -1 & 0 \end{bmatrix} \tag{6.16}$$

Substituting (6.15) into (6.8) gives

$$\begin{aligned} w &= \lfloor \mathbf{N}_1(\xi,\eta) \quad \mathbf{N}_2(\xi,\eta) \quad \mathbf{N}_3(\xi,\eta) \quad \mathbf{N}_4(\xi,\eta) \rfloor \{\mathbf{w}\}_e \\ &= \lfloor \mathbf{N}(\xi,\eta) \rfloor \{\mathbf{w}\}_e \end{aligned} \tag{6.17}$$

where

$$\{\mathbf{w}\}_e^{\mathrm{T}} = \lfloor w_1 \quad \theta_{x1} \quad \theta_{y1} \quad \cdots \quad w_4 \quad \theta_{x4} \quad \theta_{y4} \rfloor \tag{6.18}$$

and

$$\mathbf{N}_j^{\mathrm{T}}(\xi,\eta) = \begin{bmatrix} \frac{1}{8}(1+\xi_j\xi)(1+\eta_j\eta)(2+\xi_j\xi+\eta_j\eta-\xi^2-\eta^2) \\ (b/8)(1+\xi_j\xi)(\eta_j+\eta)(\eta^2-1) \\ -(a/8)(\xi_j+\xi)(\xi^2-1)(1+\eta_j\eta) \end{bmatrix} \tag{6.19}$$

(ξ_j, η_j) are the coordinates of node j. This element is commonly referred to as the ACM element [6.1, 6.2].

Evaluating (6.19) on the side 2–3 (i.e., $\xi = +1$) gives

$$\mathbf{N}_1^{\mathrm{T}} = \begin{bmatrix} 0 \\ 0 \\ 0 \end{bmatrix}, \qquad \mathbf{N}_2^{\mathrm{T}} \begin{bmatrix} \frac{1}{4}(1-\eta)(2-\eta-\eta^2) \\ (b/4)(-1+\eta)(\eta^2-1) \\ 0 \end{bmatrix}$$

$$\mathbf{N}_3^{\mathrm{T}} = \begin{bmatrix} \frac{1}{4}(1+\eta)(2+\eta-\eta^2) \\ (b/4)(1+\eta)(\eta^2-1) \\ 0 \end{bmatrix}, \qquad \mathbf{N}_4^{\mathrm{T}} = \begin{bmatrix} 0 \\ 0 \\ 0 \end{bmatrix} \tag{6.20}$$

This indicates that the displacement, and hence the rotation θ_x, is uniquely determined by the values of w and θ_x at nodes 2 and 3. Therefore, if the element is attached to another rectangular element at nodes 2 and 3, then w and θ_x will be continuous along the common side. The rotation θ_y is given by

$$\begin{aligned}\theta_y &= -\frac{1}{a}\frac{\partial w}{\partial \xi} \\ &= -\frac{1}{a}\left[\frac{\partial \mathbf{N}_1}{\partial \xi} \quad \cdots \quad \frac{\partial \mathbf{N}_4}{\partial \xi}\right]\{\mathbf{w}\}_e\end{aligned} \tag{6.21}$$

(see equations (6.6) and (6.17)). Substituting (6.19) into (6.21) and evaluating along $\xi = +1$ gives

$$\frac{\partial \mathbf{N}_j^{\mathrm{T}}}{\partial \xi} = \begin{bmatrix} \frac{1}{8}(1+\eta_j\eta)(-2+2\xi_j^2+\xi_j\eta_j\eta-\xi_j\eta^2) \\ (b/8)\xi_j(\eta_j+\eta)(\eta^2-1) \\ -(a/8)(2+2\xi_j)(1+\eta_j\eta) \end{bmatrix} \tag{6.22}$$

Evaluating (6.22) for $j = 1$ to 4 gives

$$\frac{\partial \mathbf{N}_1^{\mathrm{T}}}{\partial \xi} = \begin{bmatrix} \frac{1}{8}\eta(1-\eta^2) \\ -(b/8)(-1+\eta)(\eta^2-1) \\ 0 \end{bmatrix},$$

$$\frac{\partial \mathbf{N}_2^{\mathrm{T}}}{\partial \xi} = \begin{bmatrix} -\frac{1}{8}\eta(1-\eta^2) \\ (b/8)(-1+\eta)(\eta^2-1) \\ -(a/2)(1-\eta) \end{bmatrix}$$

$$\frac{\partial \mathbf{N}_3^{\mathrm{T}}}{\partial \xi} = \begin{bmatrix} \frac{1}{8}\eta(1-\eta^2) \\ (b/8)(1+\eta)(\eta^2-1) \\ -(a/2)(1+\eta) \end{bmatrix}, \tag{6.23}$$

$$\frac{\partial \mathbf{N}_4^{\mathrm{T}}}{\partial \xi} = \begin{bmatrix} -\frac{1}{8}\eta(1-\eta^2) \\ -(b/8)(1+\eta)(\eta^2-1) \\ 0 \end{bmatrix}$$

For θ_y to be continuous between elements it should be uniquely determined by its value at nodes 2 and 3. Expressions (6.21) and (6.23) indicate that in this case θ_y is determined by the values of w and θ_x at nodes 1, 2, 3 and 4 as well as θ_y at nodes 2 and 3. The element is, therefore, a non-conforming one. In spite of this, the element is used and will, therefore, be considered further and the effect of this lack of continuity indicated.

Substituting (6.17) into (6.1) gives

$$T_e = \tfrac{1}{2}\{\dot{\mathbf{w}}\}_e^{\mathrm{T}}[\mathbf{m}]_e[\dot{\mathbf{w}}]_e \tag{6.24}$$

where

$$\begin{aligned}[\mathbf{m}]_e &= \int_{A_e} \rho h \lfloor \mathbf{N} \rfloor^{\mathrm{T}} \lfloor \mathbf{N} \rfloor \,\mathrm{d}A \\ &= \rho hab \int_{-1}^{+1}\int_{-1}^{+1} \lfloor \mathbf{N}(\xi,\eta) \rfloor^{\mathrm{T}} \lfloor \mathbf{N}(\xi,\eta) \rfloor \,\mathrm{d}\xi\,\mathrm{d}\eta \end{aligned} \tag{6.25}$$

is the element inertia matrix. Substituting the functions $\mathbf{N}_j(\xi,\eta)$ from (6.19) and integrating gives

$$[\mathbf{m}]_e = \frac{\rho hab}{6300}\begin{bmatrix} \mathbf{m}_{11} & \mathbf{m}_{12}^{\mathrm{T}} \\ \mathbf{m}_{21} & \mathbf{m}_{22} \end{bmatrix} \tag{6.26}$$

where

$$\mathbf{m}_{11} = \begin{bmatrix} 3454 & & & & & \\ 922b & 320b^2 & & & \text{Sym} & \\ -922a & -252ab & 320a^2 & & & \\ 1226 & 398b & -548a & 3454 & & \\ 398b & 160b^2 & -168ab & 922b & 320b^2 & \\ 548a & 168ab & -240a^2 & 922a & 252ab & 320a^2 \end{bmatrix} \tag{6.27}$$

$$\mathbf{m}_{21} = \begin{bmatrix} 394 & 232b & -232a & 1226 & 548b & 398a \\ -232b & -120b^2 & 112ab & -548b & -240b^2 & -168ab \\ 232a & 112ab & -120a^2 & 398a & 168ab & 160a^2 \\ 1226 & 548b & -398a & 394 & 232b & 232a \\ -548b & -240b^2 & 168ab & -232b & -120b^2 & -112ab \\ -398a & -168ab & 160a^2 & -232a & -112ab & -120a^2 \end{bmatrix} \tag{6.28}$$

and

$$\mathbf{m}_{22} = \begin{bmatrix} 3454 & & & & & \\ -922b & 320b^2 & & & \text{Sym} & \\ 922a & -252ab & 320a^2 & & & \\ 1226 & -398b & 548a & 3454 & & \\ -398b & 160b^2 & -168ab & -922b & 320b^2 & \\ -548a & 168ab & -240a^2 & -922a & 252ab & 320a^2 \end{bmatrix} \tag{6.29}$$

In deriving this result, it is simpler to use the expression (6.8) for w and substitute for $\{\boldsymbol{\alpha}\}$ after performing the integration. A typical integral is then of the form

$$\int_{-1}^{+1}\int_{-1}^{+1} \xi^m \eta^n \, \mathrm{d}\xi \, \mathrm{d}\eta = \begin{cases} 0 & m \text{ or } n \text{ odd} \\ \dfrac{4}{(m+1)(n+1)} & m \text{ and } n \text{ even} \end{cases} \tag{6.30}$$

Substituting (6.17) into (6.3) and (6.2) gives

$$U_e = \tfrac{1}{2}\{\mathbf{w}\}_e^{\mathrm{T}}[\mathbf{k}]_e\{\mathbf{w}\}_e \tag{6.31}$$

where

$$[\mathbf{k}]_e = \int_{A_e} \frac{h^3}{12}[\mathbf{B}]^{\mathrm{T}}[\mathbf{D}][\mathbf{B}]\,\mathrm{d}A \tag{6.32}$$

is the element stiffness matrix, and

$$[\mathbf{B}] = \begin{bmatrix} \dfrac{\partial^2}{\partial x^2} \\ \dfrac{\partial^2}{\partial y^2} \\ 2\dfrac{\partial^2}{\partial x\,\partial y} \end{bmatrix} \lfloor \mathbf{N} \rfloor = \begin{bmatrix} \dfrac{1}{a^2} & \dfrac{\partial^2}{\partial \xi^2} \\ \dfrac{1}{b^2} & \dfrac{\partial^2}{\partial \eta^2} \\ \dfrac{2}{ab} & \dfrac{\partial^2}{\partial \xi\,\partial \eta} \end{bmatrix} \lfloor \mathbf{N}(\xi,\eta) \rfloor \tag{6.33}$$

Substituting the functions $\mathbf{N}_j(\xi, \eta)$ from (6.19) and integrating gives, for the isotropic case

$$[\mathbf{k}]_e = \frac{Eh^3}{48(1-\nu^2)ab} \begin{bmatrix} \mathbf{k}_{11} & & \text{Sym} & \\ \mathbf{k}_{21} & \mathbf{k}_{22} & & \\ \mathbf{k}_{31} & \mathbf{k}_{32} & \mathbf{k}_{33} & \\ \mathbf{k}_{41} & \mathbf{k}_{42} & \mathbf{k}_{43} & \mathbf{k}_{44} \end{bmatrix} \tag{6.34}$$

where

$$\mathbf{k}_{11} = \begin{bmatrix} \{4(\beta^2+\alpha^2)+\frac{2}{5}(7-2\nu)\} & & \text{Sym} \\ 2\{2\alpha^2+\frac{1}{5}(1+4\nu)\}b & 4\{\frac{4}{3}\alpha^2+\frac{4}{15}(1-\nu)\}b^2 & \\ 2\{-2\beta^2-\frac{1}{5}(1-4\nu)\}a & -4\nu ab & 4\{\frac{4}{3}\beta^2+\frac{4}{15}(1-\nu)\}a^2 \end{bmatrix} \tag{6.35}$$

$$\mathbf{k}_{21} = \begin{bmatrix} -\{2(2\beta^2-\alpha^2)+\frac{2}{5}(7-2\nu)\} & 2\{\alpha^2-\frac{1}{5}(1+4\nu)\}b & 2\{2\beta^2+\frac{1}{5}(1-\nu)\}a \\ 2\{\alpha^2-\frac{1}{5}(1+4\nu)\}b & 4\{\frac{2}{3}\alpha^2-\frac{4}{15}(1-\nu)\}b^2 & 0 \\ -2\{2\beta^2+\frac{1}{5}(1-\nu)\}a & 0 & 4\{\frac{2}{3}\beta^2-\frac{1}{15}(1-\nu)\}a^2 \end{bmatrix} \tag{6.36}$$

$$\mathbf{k}_{31} = \begin{bmatrix} -\left\{2(\beta^2+\alpha^2) - \frac{2}{5}(7-2\nu)\right\} & 2\left\{-\alpha^2 + \frac{1}{5}(1-\nu)\right\}b & 2\left\{\beta^2 - \frac{1}{5}(1-\nu)\right\}a \\ 2\left\{\alpha^2 - \frac{1}{5}(1-\nu)\right\}b & 4\left\{\frac{1}{3}\alpha^2 + \frac{1}{15}(1-\nu)\right\}b^2 & 0 \\ 2\left\{-\beta^2 + \frac{1}{5}(1-\nu)\right\}a & 0 & 4\left\{\frac{1}{3}\beta^2 + \frac{1}{15}(1-\nu)\right\}a^2 \end{bmatrix} \tag{6.37}$$

$$\mathbf{k}_{41} = \begin{bmatrix} \left\{2(\beta^2-2\alpha^2) - \frac{2}{5}(7-2\nu)\right\} & 2\left\{-2\alpha^2 - \frac{1}{5}(1-\nu)\right\}b & 2\left\{-\beta^2 + \frac{1}{5}(1+4\nu)\right\}a \\ 2\left\{2\alpha^2 + \frac{1}{5}(1-\nu)\right\}b & 4\left\{\frac{2}{3}\alpha^2 - \frac{1}{15}(1-\nu)\right\}b^2 & 0 \\ 2\left\{-\beta^2 + \frac{1}{5}(1+4\nu)\right\}a & 0 & 4\left\{\frac{2}{3}\beta^2 - \frac{4}{15}(1-\nu)\right\}a^2 \end{bmatrix} \tag{6.38}$$

and

$$\alpha = \frac{a}{b}, \qquad \beta = \frac{b}{a}. \tag{6.39}$$

Defining the following matrices

$$\mathbf{I}_1 = \begin{bmatrix} -1 & 0 & 0 \\ 0 & 1 & 0 \\ 0 & 0 & 1 \end{bmatrix}, \quad \mathbf{I}_2 = \begin{bmatrix} 1 & 0 & 0 \\ 0 & -1 & 0 \\ 0 & 0 & 1 \end{bmatrix}, \quad \mathbf{I}_3 = \begin{bmatrix} 1 & 0 & 0 \\ 0 & 1 & 0 \\ 0 & 0 & -1 \end{bmatrix} \tag{6.40}$$

the remaining sub-matrices of (6.34) are given by

$$\begin{aligned} &\mathbf{k}_{22} = \mathbf{I}_3^{\mathrm{T}}\mathbf{k}_{11}\mathbf{I}_3 \\ &\mathbf{k}_{32} = \mathbf{I}_3^{\mathrm{T}}\mathbf{k}_{41}\mathbf{I}_3, \quad \mathbf{k}_{33} = \mathbf{I}_1^{\mathrm{T}}\mathbf{k}_{11}\mathbf{I}_1 \\ &\mathbf{k}_{42} = \mathbf{I}_3^{\mathrm{T}}\mathbf{k}_{31}\mathbf{I}_3, \quad \mathbf{k}_{43} = \mathbf{I}_1^{\mathrm{T}}\mathbf{k}_{21}\mathbf{I}_1, \quad \mathbf{k}_{44} = \mathbf{I}_2^{\mathrm{T}}\mathbf{k}_{11}\mathbf{I}_2 \end{aligned} \tag{6.41}$$

These relationships are derived in reference [6.3].

As in the case of the inertia matrix, it is simpler to use the expression (6.8) for w and substitute for $\{\boldsymbol{\alpha}\}$ after performing the integration using (6.30). This procedure has been generalised for a number of plate elements with anisotropic material properties in reference [6.4].

Substituting (6.17) into (6.4) gives

$$\delta W_e = \{\delta \mathbf{w}\}_e^{\mathrm{T}} \{\mathbf{f}\}_e \tag{6.42}$$

where

$$\{\mathbf{f}\}_e = \int_A \lfloor \mathbf{N} \rfloor^{\mathrm{T}} p_z \, \mathrm{d}A \tag{6.43}$$

is the element equivalent nodal force matrix. Assuming p_z to be constant, substituting for $\lfloor \mathbf{N} \rfloor$ from (6.19) and integrating gives

$$\{\mathbf{f}\}_e = p_z \frac{ab}{3} \begin{bmatrix} 3 \\ b \\ -a \\ 3 \\ b \\ a \\ 3 \\ -b \\ a \\ 3 \\ -b \\ -a \end{bmatrix} \tag{6.44}$$

The stresses at any point in the plate are given by (2.63)

$$\begin{bmatrix} \sigma_x \\ \sigma_y \\ \tau_{xy} \end{bmatrix} = \{\boldsymbol{\sigma}\} = [\mathbf{D}]\{\boldsymbol{\varepsilon}\} \tag{6.45}$$

Substituting for the strains $\{\boldsymbol{\varepsilon}\}$ from (2.65) gives

$$\{\boldsymbol{\sigma}\} = -z[\mathbf{D}]\{\boldsymbol{\chi}\} \tag{6.46}$$

where $\{\boldsymbol{\chi}\}$ is defined in (6.3). Substituting for w in $\{\boldsymbol{\chi}\}$ using (6.17) gives

$$\{\boldsymbol{\sigma}\} = -z[\mathbf{D}][\mathbf{B}]\{\mathbf{w}\}_e \tag{6.47}$$

where $[\mathbf{B}]$ is defined in (6.33) and $\{\mathbf{w}\}_e$ in (6.18). Since $[\mathbf{B}]$ is a function of x and y (or ξ and η), then (6.47) gives the stresses at the point (x, y, z) in terms of the nodal displacements.

The bending moments M_x and M_y and twisting moments M_{xy} and M_{yx} per unit length are defined by

$$\begin{aligned} M_x &= \int_{-h/2}^{+h/2} \sigma_x z\,\mathrm{d}z, & M_y &= \int_{-h/2}^{+h/2} \sigma_y z\,\mathrm{d}z \\ M_{xy} &= -\int_{-h/2}^{+h/2} \tau_{xy} z\,\mathrm{d}z, & M_{yx} &= \int_{-h/2}^{+h/2} \tau_{yx} z\,\mathrm{d}z \end{aligned} \tag{6.48}$$

Since $\tau_{yx} = \tau_{xy}$ then $M_{yx} = -M_{xy}$. The directions of these moments are indicated in Figure 6.2.

Substituting (6.47) into (6.48) and integrating gives

$$\begin{bmatrix} M_x \\ M_y \\ M_{xy} \end{bmatrix} = -\frac{h^3}{12}[\mathbf{I}]_3[\mathbf{D}][\mathbf{B}]\{\mathbf{w}\}_e \tag{6.49}$$

where $[\mathbf{I}]_3$ is defined in (6.40).

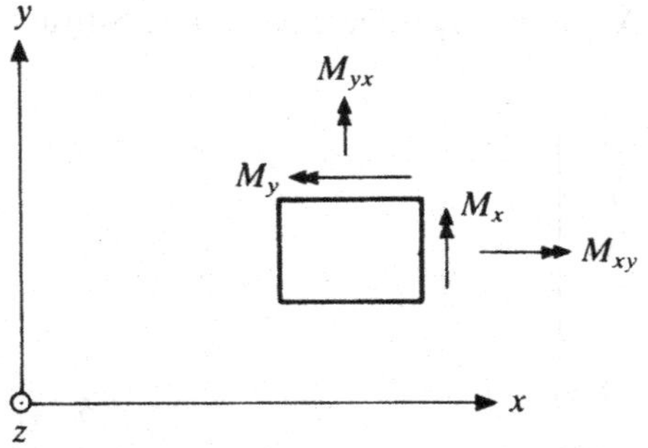

Figure 6.2. Sign convention for bending and twisting moments.

Both bending moments and stresses will be more accurate at a (2 × 2) array of integration points.

EXAMPLE 6.1 Use the ACM element to estimate the five lowest frequencies of a square plate which is simply supported on all four edges. Compare the results with the analytical solution $\pi^2(m^2+n^2)(D/\rho h L^4)^{1/2}$ rad/s, where L is the length of each side and (m, n) are the number of half-waves in the x- and y-directions.

Since the plate has two axes of symmetry, the modes which are symmetric or antisymmetric about each of these can be calculated separately by idealising one-quarter of the plate and applying appropriate boundary conditions on the axes of symmetry (Chapter 8).

Figure 6.3 shows one-quarter of the plate represented by four rectangular elements. Since side 1–3 is simply supported w, θ_y are zero at nodes 1, 2 and 3. Similarly, since side 1–7 is simply supported w, θ_x are zero at nodes 1, 4 and 7. The modes which are symmetric with respect to the side 3–9 are obtained by setting θ_y zero at 3, 6 and 9, and the antisymmetric modes by setting w, θ_x to be zero at 3, 6 and 9. Similarly, the modes which are symmetric with respect to the side 7–9 are obtained by setting θ_x to be zero at 7, 8 and 9, and the antisymmetric modes by setting w, θ_y to be zero at 7, 8 and 9. Therefore, the modes which are symmetric with respect to both axes of symmetry are obtained by considering a twelve degree of freedom model, the degrees of freedom being w at 9, θ_x at 2 and 3, θ_y at 4 and 7, w and θ_x at 6, w and θ_y at 8, and w, θ_x, θ_y at 5.

Analyses have been performed using (2 × 2), (3 × 3), (4 × 4) and (5 × 5) meshes of elements for the quarter plate. The results are compared with the analytical frequencies in Figure 6.4. Unlike the examples presented in the

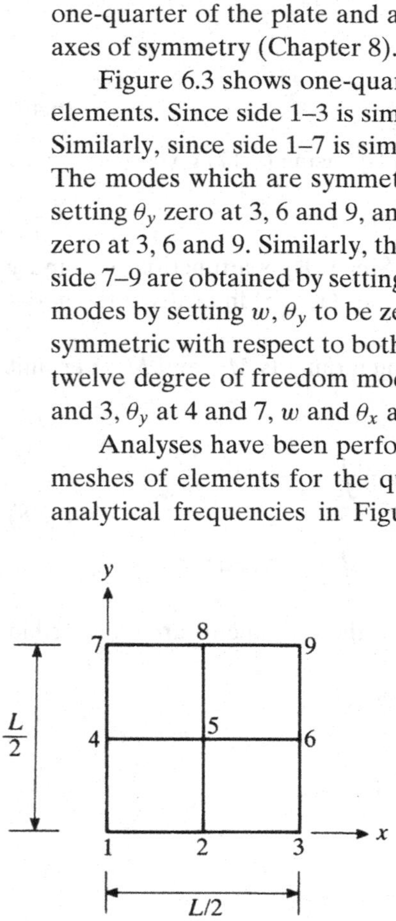

Figure 6.3. Idealisation of one-quarter of a square plate.

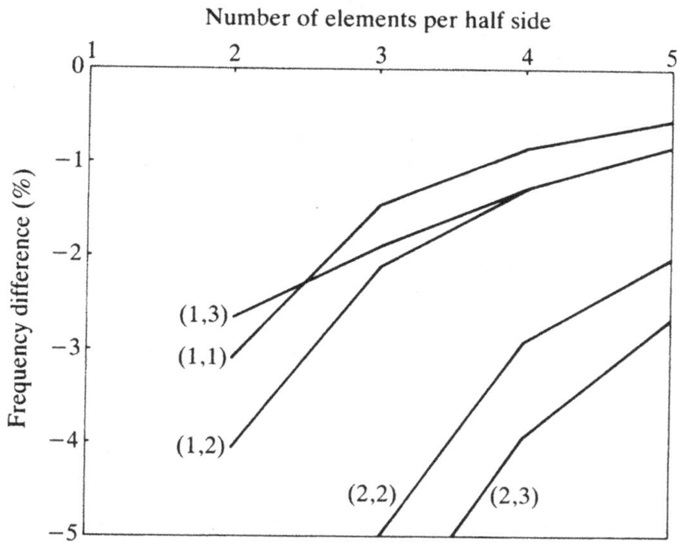

Figure 6.4. Flexural vibrations of a simply supported square plate. ACM element.

previous chapters, the frequencies predicted using the finite element method are less than the analytical frequencies. This is a consequence of the ACM element being a non-conforming one. However, as can be seen from the figure, this does not preclude the frequencies from converging to the analytical frequencies as the number of elements is increased. Results for a variety of other boundary conditions are presented in references [6.5, 6.6].

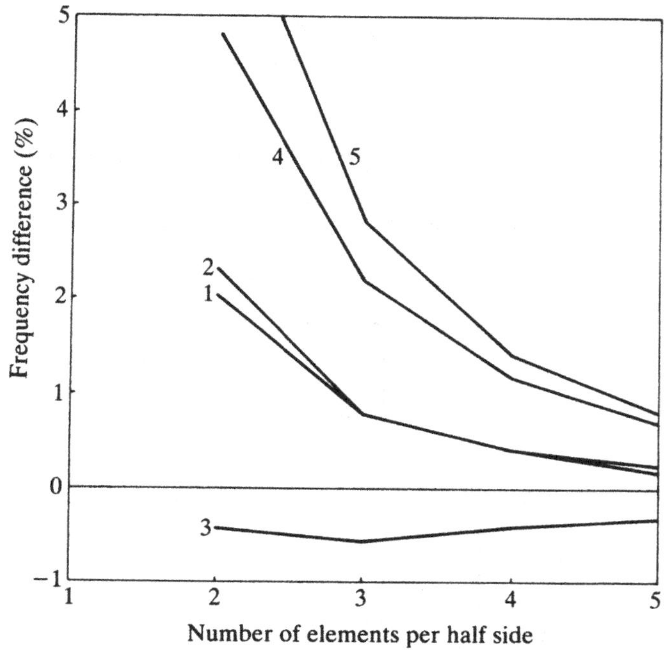

Figure 6.5. Flexural vibrations of a simply supported/free square plate. ACM element.

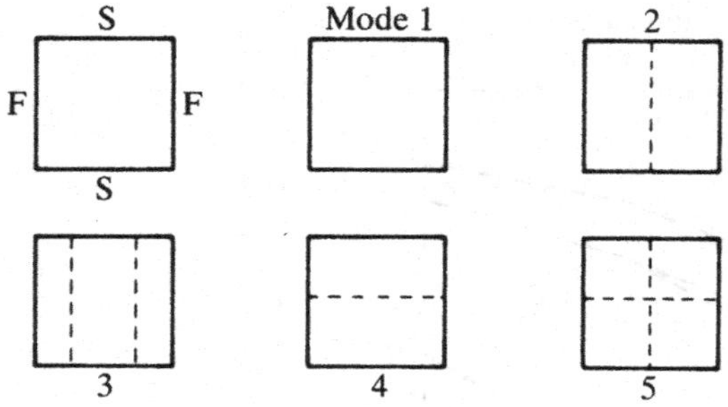

Figure 6.6. Modal patterns for a simply supported (S)/free (F) square plate.

It cannot be asserted that frequencies predicted by the ACM element will always be less than the correct ones. In fact, predictions can be either above or below the true ones. This is illustrated by the results for a square plate having one pair of opposite sides simply supported and the other pair free, as shown in Figure 6.5. The modal patterns are illustrated in Figure 6.6. Four of the five modes shown converge from above whilst the other converges from below.

EXAMPLE 6.2 Calculate the first six natural frequencies and modes of a square plate of side 0.3048 m and thickness 3.2766 mm which is point supported at its four corners. Compare the results with the analytical solutions given in references [6.8, 6.9] and the experimental results of [6.7]. Take $E = 73.084 \times 10^9$ N/m^2, $\nu = 0.3$, $\rho = 2821$ kg/m^3.

The plate has two axes of symmetry. A quarter plate was therefore represented by (2×2) and (4×4) meshes of elements. Either symmetric or antisymmetric boundary conditions were applied along the axes of symmetry. In addition, the displacement w at the corner node point was set to zero.

The predicted frequencies are compared with the analytical and experimental ones in Table 6.1. The modal patterns are given in Figure 6.7. Notice that there are two modes with different modal patterns 2(a) and 2(b) having identical frequencies.

EXAMPLE 6.3 Figure 6.8(a) shows a rectangular plate which is stiffened in one direction. The details of the stiffener are given in Figure 6.8(b). Calculate the frequencies of the first four modes by considering an equivalent orthotropic

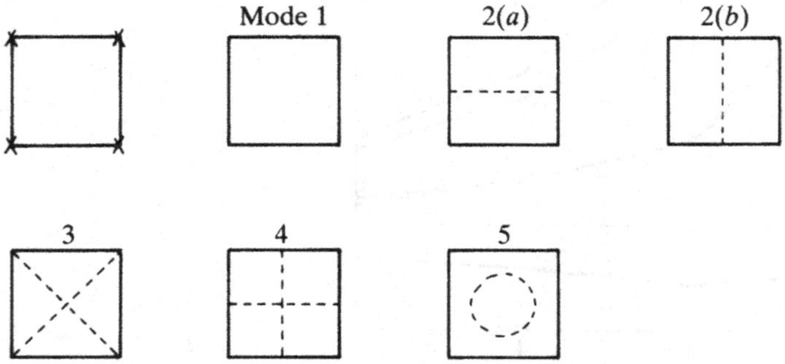

Figure 6.7. Modal patterns for a corner supported square plate.

Table 6.1. *Comparison of predicted frequencies of a corner supported square plate. ACM element*

Mode	FEM [6.9]		Analytical		Experimental
	(2 × 2)	(4 × 4)	[6.7]	[6.8]	[6.7]
1	62.15	62.09	61.4	61.11	62
2(*a*), (*b*)	141.0	138.5	136	134.6	134
3	169.7	169.7	170	166.3	169
4	343.7	340.0	333	331.9	330
5	397.4	396.0	385	383.1	383

plate and assuming all four edges to be simply supported. Compare the results with the analytical solutions given in reference [6.10]. Take $E = 206.84 \times 10^9$ N/m^2, $\nu = 0.3$ and $\rho = 7833$ kg/m^3.

The material axes, $\bar{x}$, $\bar{y}$, of the equivalent orthotropic plate are coincident with the geometric axes, x, y, shown in Figure 6.8(a). Using the method given in reference [6.11] the elastic constants of this equivalent plate are

$$D_x = 3.396D, \qquad D_y = D, \qquad H = 1.08D$$

where

$$D_x = \frac{E'_x h_e^3}{12}, \qquad D_y = \frac{E'_y h_e^3}{12}$$

$$H = \frac{E'_x \nu_{xy} h_e^3}{12} + \frac{G_{xy} h_e^3}{6}$$

$$D = \frac{Eh^3}{12(1-\nu^2)}$$

h = thickness of unstiffened plate

h_e = thickness of equivalent orthotropic plate

$= 1.125h$

E'_x, E'_y ν_{xy} are defined in Chapter 2

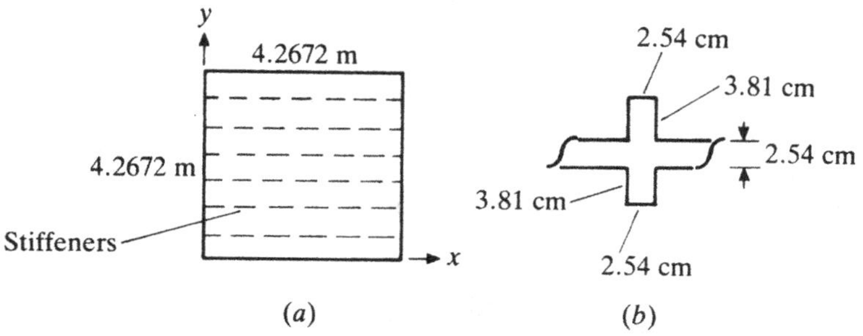

Figure 6.8. Geometry of a stiffened square plate.

Table 6.2. *Comparison of predicted frequencies of a stiffened, simply supported plate*

	Analytical (Hz) [6.10]		
Mode	Discrete	Smeared	FEM (ACM)
(1, 1)	8.089	8.258	8.142
(1, 2)	16.720	17.092	16.721
(2, 1)	25.249	25.824	25.373
(2, 2)	32.357	33.067	31.631

Using the values of E, ν and ρ given and the above relationships, it can be shown that the material properties of the equivalent orthotropic plate are:

$$
\begin{aligned}
E_x &= 493.313 \times 10^9 \text{ N/m}^2 \\
E_y &= 145.266 \times 10^9 \text{ N/m}^2 \\
\nu_{xy} &= 0.1628 \\
G_{xy} &= 42.072 \times 10^9 \text{ N/m}^2 \\
\rho &= 7833 \text{ kg/m}^3 \\
h_e &= 2.8575 \text{ cm}
\end{aligned}
$$

The full plate was analysed using an (8×8) mesh of elements. This means that there were 81 node points with three degrees of freedom each. Of the 243 degrees of freedom, 68 are zero due to the simply supported boundary conditions. Seventy-seven master degrees of freedom (see Chapter 11) were then selected automatically from the remaining 175 degrees of freedom. The frequencies obtained are given in Table 6.2. These are compared with two sets of analytically predicted frequencies. The first set was obtained by treating the plate as a discretely stiffened plate. The second set was obtained by considering the equivalent orthotropic plate and using the Rayleigh method.

It has been indicated that the element presented in this section is a non-conforming one, since the normal slope is not continuous between elements. There are several ways of overcoming this problem, namely:

(1) Introduce additional nodal degrees of freedom.
(2) Ensure that the normal slope varies linearly along an edge.
(3) Introduce additional node points.
(4) Use thick plate theory and reduced integration (see Section 3.10 for a similar treatment of a beam).

These methods will be presented in the following sections.

6.2 Thin Rectangular Element (conforming)

A conforming rectangular element can be obtained by taking products of the functions (3.126) for a uniform, slender beam. In this case the displacement function for the plate is of the form (6.17) with

$$\mathbf{N}_j^{\mathrm{T}}(\xi,\eta) = \begin{bmatrix} f_j(\xi)f_j(\eta) \\ bf_j(\xi)g_j(\eta) \\ -ag_j(\xi)f_j(\eta) \end{bmatrix} \tag{6.50}$$

where

$$f_j(\xi) = \tfrac{1}{4}(2 + 3\xi_j\xi - \xi_j\xi^3)$$

and

$$g_j(\xi) = \tfrac{1}{4}(-\xi_j - \xi + \xi_j\xi^2 + \xi^3) \tag{6.51}$$

The functions of η are obtained by replacing ξ_j, ξ by η_j, η respectively. (ξ_j, η_j) are the coordinates of node j.

A close inspection of the displacement function defined by (6.17), (6.50) and (6.51) reveals that the twist $\partial^2 w/\partial x \partial y$ is zero at the four node points. This means that, in the limit, as an increasing number of elements is used, the plate will tend towards a zero twist condition. This can be overcome by introducing $\partial^2 w/\partial x \partial y$ as an additional degree of freedom at each node point. In this case the displacement function is of the form (6.17) with

$$\{\mathbf{w}\}_e^{\mathrm{T}} = \lfloor w_1 \quad \theta_{x1} \quad \theta_{y1} \quad w_{xy1} \quad \cdots \quad w_4 \quad \theta_{x4} \quad \theta_{y4} \quad w_{xy4} \rfloor \tag{6.52}$$

where $w_{xy} \equiv \partial^2 w/\partial x \, \partial y$, and

$$\mathbf{N}_j^{\mathrm{T}}(\xi,\eta) = \begin{bmatrix} f_j(\xi)f_j(\eta) \\ bf_j(\xi)g_j(\eta) \\ -ag_j(\xi)f_j(\eta) \\ abg_j(\xi)g_j(\eta) \end{bmatrix} \tag{6.53}$$

This element is commonly referred to as the CR element [6.12].

By specifying nodal degrees of freedom which are consistent with rigid body displacements, corresponding to vertical translation and rotation about the x- and y-axes, it can be shown that this element can perform rigid body movement without deformation. Similarly for pure bending in the x- and y-directions. (Note that this is also true for the functions (6.50).) The nodal displacements which are consistent with a state of constant twist are

$$\begin{aligned} w_1, w_3 &= 1, & w_2, w_4 &= -1 \\ \theta_{x2}, \theta_{x3} &= \frac{1}{b}, & \theta_{x1}, \theta_{x4} &= -\frac{1}{b} \\ \theta_{y1}, \theta_{y2} &= \frac{1}{a}, & \theta_{y3}, \theta_{y4} &= -\frac{1}{a} \\ w_{xyj} &= \frac{1}{ab}, & j &= 1, \ldots, 4. \end{aligned} \tag{6.54}$$

Substituting these into (6.17) and (6.53) gives

$$w = \xi\eta \tag{6.55}$$

as required. Thus, the first six terms in (6.5) are present in the functions (6.53).

The element inertia, stiffness and equivalent nodal force matrices are given by (6.25), (6.32) and (6.43) where the matrix $\lfloor \mathbf{N} \rfloor$ is defined by (6.17) and (6.53). These expressions may be evaluated by the combined analytical/numerical method given in reference [6.4]. They can also be evaluated analytically. The burden of the calculations is considerably eased if the functions (6.51) are expressed in terms of Legendre polynomials [6.13] as follows:

$$\begin{aligned} f_j(\xi) &= \tfrac{1}{2} P_0 + \tfrac{3}{5}\xi_j P_1 - \tfrac{1}{10}\xi_j P_3 \\ g_j(\xi) &= -\tfrac{1}{6}\xi_j P_0 - \tfrac{1}{10} P_1 + \tfrac{1}{6}\xi_j P_2 + \tfrac{1}{10} P_3 \end{aligned} \tag{6.56}$$

where

$$\begin{aligned} &P_0 = 1, \qquad P_1 = \xi, \qquad P_2 = \tfrac{1}{2}(3\xi^2 - 1), \\ &P_3 = \tfrac{1}{2}(5\xi^3 - 3\xi) \end{aligned} \tag{6.57}$$

(see Section 3.10 for further details).

The derivatives of these functions can also be expressed in terms of Legendre polynomials, viz:

$$\begin{aligned} f_j'(\xi) &= \tfrac{1}{2}\xi_j P_0 - \tfrac{1}{2}\xi_j P_2 \\ f_j''(\xi) &= -\tfrac{3}{2}\xi_j P_1 \\ g_j'(\xi) &= \tfrac{1}{2}\xi_j P_1 + P_2 \\ g_j''(\xi) &= \tfrac{1}{2}\xi_j P_0 + \tfrac{3}{2} P_1 \end{aligned} \tag{6.58}$$

Integrals of products of the functions (6.51) and their derivatives can now be evaluated using the following relationships:

$$\int_{-1}^{+1} P_n(\xi) P_m(\xi)\, \mathrm{d}\xi = \begin{cases} \dfrac{2}{(2n+1)} & \text{when } m = n \\ 0 & \text{when } m \neq n \end{cases} \tag{6.59}$$

All three matrices are presented in reference [6.12] and the stiffness matrix in [6.13], both for the isotropic case. The extension to the orthotropic case can be found in reference [6.14].

EXAMPLE 6.4 Repeat Example 6.1 using the CR element.

The boundary conditions along a simply supported edge are the same as in Example 6.1 since $w_{xy} \neq 0$ there. Along an axis of symmetry the boundary conditions are the same as in Example 6.1 for antisymmetric modes, but in the case of symmetric modes there is the additional constraint that w_{xy} is zero.

The percentage differences between the finite element and analytical frequencies are presented in Table 6.3. The accuracy of this element is considerably better than the accuracy obtained with the ACM element (see Figure 6.5). In fact the present results would be insignificant if drawn on the same scale as the figure. Also, note that the CR element produces frequencies which are greater than the exact analytical frequencies. This is because all the requirements of the Rayleigh-Ritz method have been satisfied (see Section 3.1). Results for a variety of boundary conditions are presented in references [6.5, 6.6, 6.15 and 6.16].

Table 6.3. *Comparison of predicted and analytical frequencies for a simply supported square plate. CR element (% difference)*

	FEM grids ($\frac{1}{4}$ plate)			
Mode	2 × 2	3 × 3	4 × 4	5 × 5
(1, 1)	0.02	0.01	0.0	0.0
(1, 2), (2, 1)	0.26	0.05	0.02	0.01
(2, 2)	0.22	0.04	0.01	0.01
(1, 3), (3, 1)	1.51	0.32	0.11	0.04
(2, 3), (3, 2)	0.99	0.21	0.07	0.03

EXAMPLE 6.5 Repeat Example 6.2 using the CR element.

The analysis is exactly the same as in Example 6.2 except there are now four degrees of freedom per node instead of three. Also, for modes which are symmetric about an axis of symmetry there is the additional constraint that w_{xy} is zero.

The predicted frequencies are compared with the analytical and experimental ones in Table 6.4. A comparison with Table 6.1 indicates that the frequencies predicted with the CR element are lower than the ones predicted with the ACM element.

Although the CR element is more accurate than the ACM element, it does suffer from the disadvantage that it is difficult to use in conjunction with other elements when analysing built-up structures (see Chapter 7) due to the presence of the degree of freedom w_{xy}. Because of this, reference [6.17] introduces the approximations

$$\begin{aligned} w_{xy1} &= \frac{1}{2b}(\theta_{y1} - \theta_{y4}), & w_{xy2} &= \frac{1}{2a}(\theta_{x2} - \theta_{x1}) \\ w_{xy3} &= \frac{1}{2b}(\theta_{y2} - \theta_{y3}), & w_{xy4} &= \frac{1}{2a}(\theta_{x3} - \theta_{x4}) \end{aligned} \tag{6.60}$$

Substituting (6.60) into (6.17), (6.52) and (6.53) and simplifying, shows that w is of the form (6.17) with $\{\mathbf{w}\}_e$ given by (6.18) and

$$\mathbf{N}_j^{\mathrm{T}} = (\xi, \eta) = \begin{bmatrix} f_j(\xi) f_j(\eta) \\ b F_j(\xi) g_j(\eta) \\ -a g_j(\xi) F_j(\eta) \end{bmatrix} \tag{6.61}$$

Table 6.4. *Comparison of predicted frequencies of a corner supported square plate*

	FEM [6.9] (CR)		Analytical		Experimental
Mode	(2 × 2)	(4 × 4)	[6.7]	[6.8]	[6.7]
1	62.03	61.79	61.4	62.11	62
2(*a*), (*b*)	138.9	134.9	136	134.6	134
3	169.7	169.6	170	166.3	169
4	338.9	335.1	333	331.9	330
5	391.5	387.5	385	383.1	383

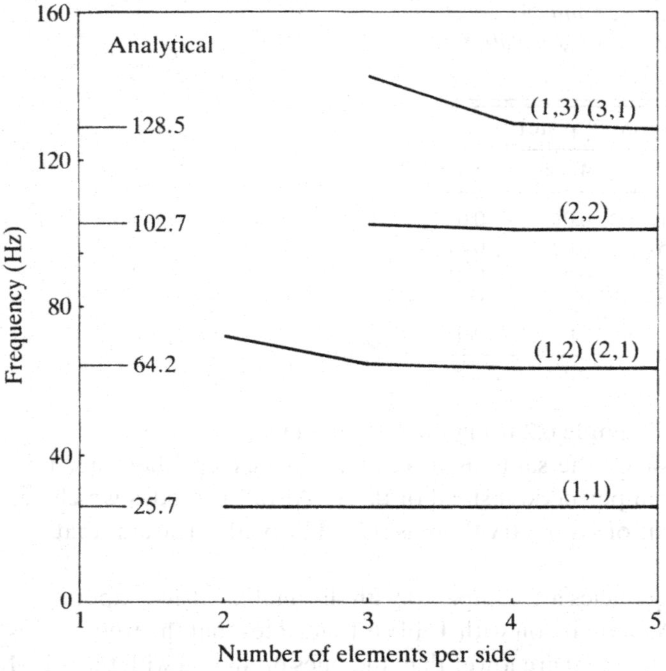

Figure 6.9. Flexural vibrations of a simply supported square plate. WB element [6.17].

where

$$F_j(\xi) = \begin{cases} \frac{1}{8}(5 + 5\xi_j\xi - \xi^2 - \xi_j\xi^3) & j = 1, 3 \\ \frac{1}{8}(3 + 5\xi_j\xi + \xi^2 - \xi_j\xi^3) & j = 2, 4 \end{cases} \tag{6.62}$$

and

$$F_j(\eta) = \begin{cases} \frac{1}{8}(3 + 5\eta_j\eta + \eta^2 - \eta_j\eta^3) & j = 1, 3 \\ \frac{1}{8}(5 + 5\eta_j\eta - \eta^2 - \eta_j\eta^3) & j = 2, 4 \end{cases} \tag{6.63}$$

(ξ_j, η_j) are the coordinates of node j. This element will be referred to as the WB element.

The effect of applying the constraints (6.60) to the CR element is to make it a non-conforming one. The displacement and tangential slope are continuous between elements but the normal slope is not.

EXAMPLE 6.6 Use the WB element to estimate the four lowest frequencies of a square plate of side 2.4 m and thickness 0.03 m, which is simply supported on all four edges. Take $E = 21 \times 10^{10}$ N/m^2, $\nu = 0.3$ and $\rho = 7800$ kg/m^3. Compare the results with the analytical solution $(\pi/2)(m^2 + n^2)(D/\rho h L^4)^{1/2}$ Hz, where L is the length of each side and (m, n) are the number of half-waves in the x- and y-directions.

The complete plate was represented by (2×2), (3×3), (4×4) and (5×5) meshes of elements. The frequencies obtained are presented in Figure 6.9. The frequencies obtained with the WB element rapidly converge to the analytical frequencies as the number of elements increases.

6.3 Thick Rectangular Element

The energy expressions for a thick plate element are, from Section 2.7

$$T_e = \frac{1}{2}\int_A \rho\left(h\dot{w}^2 + \frac{h^3}{12}\dot{\theta}_x^2 + \frac{h^3}{12}\dot{\theta}_y^2\right) \mathrm{d}A \tag{6.64}$$

$$U_e = \frac{1}{2}\int_A \frac{h^3}{12}\{\boldsymbol{\chi}\}^{\mathrm{T}}[\mathbf{D}]\{\boldsymbol{\chi}\}\,\mathrm{d}A + \frac{1}{2}\int_A \kappa h\{\boldsymbol{\gamma}\}^{\mathrm{T}}[\mathbf{D}^{\mathrm{s}}]\{\boldsymbol{\gamma}\}\,\mathrm{d}A \tag{6.65}$$

with

$$\{\boldsymbol{\chi}\} = \begin{bmatrix} -\partial\theta_y/\partial x \\ \partial\theta_x/\partial y \\ \partial\theta_x/\partial x - \partial\theta_y/\partial y \end{bmatrix}, \qquad \{\boldsymbol{\gamma}\} = \begin{bmatrix} \theta_y + \partial w/\partial x \\ -\theta_x + \partial w/\partial y \end{bmatrix} \tag{6.66}$$

where $[\mathbf{D}]$ is defined by (2.45), (2.49) or (2.51) and $[\mathbf{D}^{\mathrm{s}}]$ is defined by (2.77). Also

$$\delta W = \int_A p_z \delta w \,\mathrm{d}A \tag{6.67}$$

The highest derivative of w, θ_x and θ_y appearing in the energy expressions is the first. Therefore, w, θ_x and θ_y are the only degrees of freedom required at the node points. The displacement functions are of the form

$$w = \sum_{j=1}^{4} N_j w_j, \qquad \theta_x = \sum_{j=1}^{4} N_j \theta_{x_i}, \qquad \theta_y = \sum_{j=1}^{4} N_j \theta_{y_i} \tag{6.68}$$

where the functions N_j are defined by (4.36), that is

$$N_j = \tfrac{1}{4}(1 + \xi_j\xi)(1 + \eta_j\eta) \tag{6.69}$$

These functions ensure that w, θ_x and θ_y are continuous between elements. This element will be referred to as the HTK element [6.18]. Combining the expressions (6.68) gives

$$\begin{bmatrix} w \\ \theta_x \\ \theta_y \end{bmatrix} = [\mathbf{N}]\{\mathbf{w}\}_e \tag{6.70}$$

where

$$\{\mathbf{w}\}_e^{\mathrm{T}} = \lfloor w_1 \quad \theta_{x1} \quad \theta_{y1} \quad \cdots \quad w_4 \quad \theta_{x4} \quad \theta_{y4} \rfloor \tag{6.71}$$

and

$$[\mathbf{N}] = \begin{bmatrix} N_1 & 0 & 0 & \cdots & N_4 & 0 & 0 \\ 0 & N_1 & 0 & \cdots & 0 & N_4 & 0 \\ 0 & 0 & N_1 & \cdots & 0 & 0 & N_4 \end{bmatrix} \tag{6.72}$$

Substituting (6.70) into (6.64) gives

$$T_e = \tfrac{1}{2}\{\dot{\mathbf{w}}\}_e^{\mathrm{T}}[\mathbf{m}]_e\{\dot{\mathbf{w}}\}_e \tag{6.73}$$

where

$$[\mathbf{m}]_e = \int_A \rho\, [\mathbf{N}]^{\mathrm{T}} \begin{bmatrix} h & 0 & 0 \\ 0 & h^3/12 & 0 \\ 0 & 0 & h^3/12 \end{bmatrix} [\mathbf{N}]\, \mathrm{d}A \tag{6.74}$$

Substituting the functions (6.69) into (6.74) and integrating gives

$$[\mathbf{m}]_e = [\mathbf{m}^{\mathrm{t}}] + [\mathbf{m}^{\mathrm{r}}] \tag{6.75}$$

where

$$[\mathbf{m}^{\mathrm{t}}] = \frac{\rho hab}{108} \begin{bmatrix} 48 & & & & & & & & & & & \\ 0 & 0 & & & & & & & & \text{Sym} & & \\ 0 & 0 & 0 & & & & & & & & & \\ 24 & 0 & 0 & 48 & & & & & & & & \\ 0 & 0 & 0 & 0 & 0 & & & & & & & \\ 0 & 0 & 0 & 0 & 0 & 0 & & & & & & \\ 12 & 0 & 0 & 24 & 0 & 0 & 48 & & & & & \\ 0 & 0 & 0 & 0 & 0 & 0 & 0 & 0 & & & & \\ 0 & 0 & 0 & 0 & 0 & 0 & 0 & 0 & 0 & & & \\ 24 & 0 & 0 & 12 & 0 & 0 & 24 & 0 & 0 & 48 & & \\ 0 & 0 & 0 & 0 & 0 & 0 & 0 & 0 & 0 & 0 & 0 & \\ 0 & 0 & 0 & 0 & 0 & 0 & 0 & 0 & 0 & 0 & 0 & 0 \end{bmatrix} \tag{6.76}$$

$$[\mathbf{m}^{\mathrm{r}}] = \frac{\rho h^3 ab}{108} \begin{bmatrix} 0 & & & & & & & & & & & \\ 0 & 4 & & & & & & & & \text{Sym} & & \\ 0 & 0 & 4 & & & & & & & & & \\ 0 & 0 & 0 & 0 & & & & & & & & \\ 0 & 2 & 0 & 0 & 4 & & & & & & & \\ 0 & 0 & 2 & 0 & 0 & 4 & & & & & & \\ 0 & 0 & 0 & 0 & 0 & 0 & 0 & & & & & \\ 0 & 1 & 0 & 0 & 2 & 0 & 0 & 4 & & & & \\ 0 & 0 & 1 & 0 & 0 & 2 & 0 & 0 & 4 & & & \\ 0 & 0 & 0 & 0 & 0 & 0 & 0 & 0 & 0 & 0 & & \\ 0 & 2 & 0 & 0 & 1 & 0 & 0 & 2 & 0 & 0 & 4 & \\ 0 & 0 & 2 & 0 & 0 & 1 & 0 & 0 & 2 & 0 & 0 & 4 \end{bmatrix} \tag{6.77}$$

Substituting (6.70) into (6.66) and (6.65) gives

$$U_e = \tfrac{1}{2}\{\mathbf{w}\}_e^{\mathrm{T}}[\mathbf{k}]_e\{\mathbf{w}\}_e \tag{6.78}$$

where the stiffness matrix $[\mathbf{k}]_e$ is of the form

$$[\mathbf{k}]_e = [\mathbf{k}^{\mathrm{f}}] + [\mathbf{k}^{\mathrm{s}}] \tag{6.79}$$

where

$$[\mathbf{k}^{\mathrm{f}}] = \int_A \frac{h^3}{12}[\mathbf{B}^{\mathrm{f}}]^{\mathrm{T}}\,[\mathbf{D}]\,[\mathbf{B}^{\mathrm{f}}]\,\mathrm{d}A \tag{6.80}$$

and

$$[\mathbf{k}^{\mathrm{s}}] = \int_A \kappa h[\mathbf{B}^{\mathrm{s}}]^{\mathrm{T}}\,[\mathbf{D}^{\mathrm{s}}]\,[\mathbf{B}^{\mathrm{s}}]\,\mathrm{d}A \tag{6.81}$$

The strain matrix $[\mathbf{B}^{\mathrm{f}}]$ is of the form

$$[\mathbf{B}^{\mathrm{f}}] = [\mathbf{B}_1^{\mathrm{f}} \quad \mathbf{B}_2^{\mathrm{f}} \quad \mathbf{B}_3^{\mathrm{f}} \quad \mathbf{B}_4^{\mathrm{f}}] \tag{6.82}$$

where

$$\mathbf{B}_j^{\mathrm{f}} = \begin{bmatrix} 0 & 0 & -\partial N_j/\partial x \\ 0 & \partial N_j/\partial y & 0 \\ 0 & \partial N_j/\partial x & -\partial N_j/\partial y \end{bmatrix} \tag{6.83}$$

The strain matrix $[\mathbf{B}^{\mathrm{s}}]$ is of the form

$$[\mathbf{B}^{\mathrm{s}}] = [\mathbf{B}_1^{\mathrm{s}} \quad \mathbf{B}_2^{\mathrm{s}} \quad \mathbf{B}_3^{\mathrm{s}} \quad \mathbf{B}_4^{\mathrm{s}}] \tag{6.84}$$

where

$$\mathbf{B}_j^{\mathrm{s}} = \begin{bmatrix} \partial N_j/\partial x & 0 & N_j \\ \partial N_j/\partial y & -N_j & 0 \end{bmatrix} \tag{6.85}$$

Substituting (6.69) into (6.83) and (6.85) gives

$$\mathbf{B}_j^{\mathrm{f}} = \begin{bmatrix} 0 & 0 & -\xi_j(1+\eta_j\eta)/4a \\ 0 & (1+\xi_j\xi)\eta_j/4b & 0 \\ 0 & \xi_j(1+\eta_j\eta)/4a & -(1+\xi_j\xi)\eta_j/4b \end{bmatrix} \tag{6.86}$$

and

$$\mathbf{B}_j^{\mathrm{s}} = \begin{bmatrix} \xi_j(1+\eta_j\eta)/4a & 0 & (1+\xi_j\xi)(1+\eta_j\eta)/4 \\ (1+\xi_j\xi)\eta_j/4b & -(1+\xi_j\xi)(1+\eta_j\eta)/4 & 0 \end{bmatrix} \tag{6.87}$$

Substituting (6.86) and (6.87) into (6.82) and (6.84) and the resulting matrices into (6.80) and (6.81) will give the element stiffness matrix as defined by (6.79). Both (6.80) and (6.81) may be evaluated exactly using a (2×2) array of Gauss integration points. For thick plates this gives acceptable results. However, as the thickness of the plate is reduced, the element becomes over-stiff in the same way as the corresponding deep beam formulation discussed in Sections 3.9 and 3.10. This can be overcome by evaluating the shear energy term (6.81) using a one point Gauss integration scheme [6.18, 6.19].

For the isotropic case, the stiffness matrix due to flexure is of the form

$$[\mathbf{k}^{\mathrm{f}}] = \frac{Eh^3}{48ab(1-\nu)^2}\begin{bmatrix} \mathbf{k}_{11}^{\mathrm{f}} & & & \text{Sym} \\ \mathbf{k}_{21}^{\mathrm{f}} & \mathbf{k}_{22}^{\mathrm{f}} & & \\ \mathbf{k}_{31}^{\mathrm{f}} & \mathbf{k}_{32}^{\mathrm{f}} & \mathbf{k}_{33}^{\mathrm{f}} & \\ \mathbf{k}_{41}^{\mathrm{f}} & \mathbf{k}_{42}^{\mathrm{f}} & \mathbf{k}_{43}^{\mathrm{f}} & \mathbf{k}_{44}^{\mathrm{f}} \end{bmatrix} \tag{6.88}$$

where

$$\mathbf{k}_{11}^{\text{f}} = \begin{bmatrix} 0 & 0 & 0 \\ 0 & \frac{4}{3}\{\alpha^2 + \frac{1}{2}(1-\nu)\}\,b^2 & -\frac{1}{2}(1+\nu)ab \\ 0 & -\frac{1}{2}(1+\nu)ab & \frac{4}{3}\{\beta^2 + \frac{1}{2}(1-\nu)\}\,a^2 \end{bmatrix} \tag{6.89}$$

$$\mathbf{k}_{21}^{\text{f}} = \begin{bmatrix} 0 & 0 & 0 \\ 0 & \frac{2}{3}\{\alpha^2 - (1-\nu)\}\,b^2 & -\frac{1}{2}(3\nu-1)ab \\ 0 & \frac{1}{2}(3\nu-1)ab & \frac{1}{3}\{-4\beta^2 + (1-\nu)\}\,a^2 \end{bmatrix} \tag{6.90}$$

$$\mathbf{k}_{31}^{\text{f}} = \begin{bmatrix} 0 & 0 & 0 \\ 0 & \frac{2}{3}\{-\alpha^2 - \frac{1}{2}(1-\nu)\}\,b^2 & \frac{1}{2}(1+\nu)ab \\ 0 & \frac{1}{2}(1+\nu)ab & \frac{2}{3}\{-\beta^2 - \frac{1}{2}(1-\nu)\}\,a^2 \end{bmatrix} \tag{6.91}$$

$$\mathbf{k}_{41}^{\text{f}} = \begin{bmatrix} 0 & 0 & 0 \\ 0 & \frac{1}{3}\{-4\alpha^2 + (1-\nu)\}\,b^2 & \frac{1}{2}(3\nu-1)ab \\ 0 & -\frac{1}{2}(3\nu-1)ab & \frac{2}{3}\{\beta^2 - (1-\nu)\}\,a^2 \end{bmatrix} \tag{6.92}$$

and

$$\alpha = \frac{a}{b}, \qquad \beta = \frac{b}{a} \tag{6.93}$$

The remaining sub-matrices of (6.88) are given by relationships corresponding to (6.41).

The stiffness matrix due to shear is of the form

$$[\mathbf{k}^{\text{s}}] = \frac{Eh^3}{48ab\beta_{\text{s}}} \begin{bmatrix} \mathbf{k}_{11}^{\text{s}} & & & \\ \mathbf{k}_{21}^{\text{s}} & \mathbf{k}_{22}^{\text{s}} & & \text{Sym} \\ \mathbf{k}_{31}^{\text{s}} & \mathbf{k}_{32}^{\text{s}} & \mathbf{k}_{33}^{\text{s}} & \\ \mathbf{k}_{41}^{\text{s}} & \mathbf{k}_{42}^{\text{s}} & \mathbf{k}_{43}^{\text{s}} & \mathbf{k}_{44}^{\text{s}} \end{bmatrix} \tag{6.94}$$

where $\beta_{\text{s}} = Eh^2/12\kappa Gb^2$ is a shear parameter which is similar to the one defined in Sections 3.9 and 3.10 for a deep beam, also

$$\mathbf{k}_{11}^{\text{s}} = \begin{bmatrix} (1+\alpha^2) & \alpha^2 b & -a \\ \alpha^2 b & \alpha^2 b^2 & 0 \\ -a & 0 & a^2 \end{bmatrix} \tag{6.95}$$

$$\mathbf{k}_{21}^{\text{s}} = \begin{bmatrix} (-1+\alpha^2) & \alpha^2 b & a \\ \alpha^2 b & \alpha^2 b^2 & 0 \\ -a & 0 & a^2 \end{bmatrix} \tag{6.96}$$

$$\mathbf{k}^{s}_{31} = \begin{bmatrix} (-1-\alpha^2) & -\alpha^2 b & a \\ \alpha^2 b & \alpha^2 b & 0 \\ -a & 0 & a^2 \end{bmatrix} \tag{6.97}$$

$$\mathbf{k}^{s}_{41} = \begin{bmatrix} (1-\alpha^2) & -\alpha^2 b & -a \\ \alpha^2 b & \alpha^2 b^2 & 0 \\ -a & 0 & a^2 \end{bmatrix} \tag{6.98}$$

The remaining sub-matrices of (6.94) are given by relationships corresponding to (6.41).

Substituting for w from (6.68) into (6.67) gives

$$\delta W_e = \{\delta \mathbf{w}\}_e^{\mathrm{T}} \{\mathbf{f}\}_e \tag{6.99}$$

where

$$\{\mathbf{f}\}_e = \int_A [\mathbf{N}]^{\mathrm{T}} \begin{bmatrix} p_z \\ 0 \\ 0 \end{bmatrix} \mathrm{d}A \tag{6.100}$$

is the element equivalent nodal force matrix. Assuming p_z to be constant, substituting for $[\mathbf{N}]$ from (6.72) and (6.69) and integrating gives

$$\{\mathbf{f}\}_e = p_z ab \begin{bmatrix} 1 \\ 0 \\ 0 \\ 1 \\ 0 \\ 0 \\ 1 \\ 0 \\ 0 \\ 1 \\ 0 \\ 0 \end{bmatrix} \tag{6.101}$$

In this case one quarter of the total force is concentrated at each node.

The bending and twisting moments within the element are, from (6.49),

$$\begin{bmatrix} M_x \\ M_y \\ M_{xy} \end{bmatrix} = -\frac{h^3}{12} [\mathbf{I}]_3 [\mathbf{D}] [\mathbf{B}^{\mathrm{f}}]\{\mathbf{w}\}_e \tag{6.102}$$

Table 6.5. *Comparison of predicted non-dimensional frequencies of a simply supported square plate. HTK element*

	$\lambda^{1/2}$		
Mode	FEM [6.20]	Analytical [6.21]	% Difference
(1, 1)	0.0945	0.0930	1.61
(2, 1)	0.2347	0.2218	5.82
(2, 2)	0.3597	0.3402	5.73
(3, 1)	0.4729	0.4144	14.1
(3, 2)	0.5746	0.5197	10.6
(3, 3)	0.7520	0.6821	10.2

$\lambda = \rho h^2 \omega^2 / G.$

where $[\mathbf{I}]_3$ is defined by (6.40). These will be more accurate at a (2×2) array of integration points. The shear forces per unit length are

$$\begin{bmatrix} Q_x \\ Q_y \end{bmatrix} = \kappa h\,[\mathbf{D}^s]\,[\mathbf{B}^s]\{\mathbf{w}\}_e \tag{6.103}$$

where Q_x, Q_y act on the faces whose normals are in the x-, y-directions respectively. These will be accurate at the centre of the element.

EXAMPLE 6.7 Use the HTK element to estimate the six lowest frequencies of a moderately thick, simply supported square plate with a span/thickness ratio of 10. Take $\nu = 0.3$. Compare the results with the analytical solution given in reference [6.21].

The plate is represented by an (8×8) mesh of elements. Since the boundaries are simply supported, then w is zero at all boundary nodes. The results obtained for the non-dimensional frequency $\omega(\rho h^2/G)^{1/2}$ are compared with the analytical values in Table 6.5.

The analysis was repeated using (2×2), (4×4), (6×6) and (10×10) meshes of elements. The convergence of the lowest non-dimensional frequency with the increase in number of elements is shown in Figure 6.10.

EXAMPLE 6.8 Investigate the effect of changing the span/thickness ratio on the accuracy of the lowest frequency of a simply supported square plate. Take $\nu = 0.3$.

The plate is represented by an (8×8) mesh of elements. The lowest non-dimensional frequency $\omega(\rho h b^4/D)^{1/2}$ has been calculated for various values of the span/thickness ratio b/h. The results are shown in Figure 6.11. The analytical frequency shown is for a thin plate. As the plate gets thinner b/h increases and the non-dimensional frequency increases reaching an asymptotic value which is greater than the analytical value for a thin plate.

References [6.18, 6.19] present a number of static solutions using the HTK element. These indicate that accurate solutions can be obtained if the boundary conditions are simply supported or clamped. Reference [6.22] indicates that in the case of a cantilever plate large errors can occur.

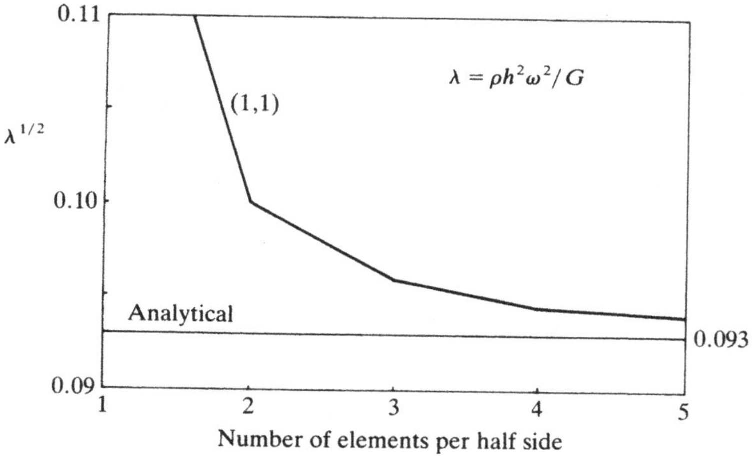

Figure 6.10. Convergence of the lowest non-dimensional frequency of a simply supported square plate. $b/h = 10$ [6.20].

6.4 Thin Triangular Element (non-conforming)

Figure 6.12 shows a triangular element with three node points, one at each vertex. There are three degrees of freedom at each node, namely, the component of displacement normal to the plane of the plate, w, and the two rotations $\theta_x = \partial w/\partial y$ and $\theta_y = -\partial w/\partial x$.

Since the element has nine degrees of freedom the displacement function can be represented by a polynomial having nine terms. A complete cubic has ten terms (see Figure 4.1). Equation (6.5) indicates that the constant, linear and quadratic terms should be retained. In order to maintain symmetry of the cubic terms the coefficients of x^2y and xy^2 are taken to be equal. Therefore

$$\begin{aligned} w = {} & \alpha_1 + \alpha_2 x + \alpha_3 y + \alpha_4 x^2 + \alpha_5 xy \\ & + \alpha_6 y^2 + \alpha_7 x^3 + \alpha_8 (x^2 y + xy^2) + \alpha_9 y^3 \end{aligned} \tag{6.104}$$

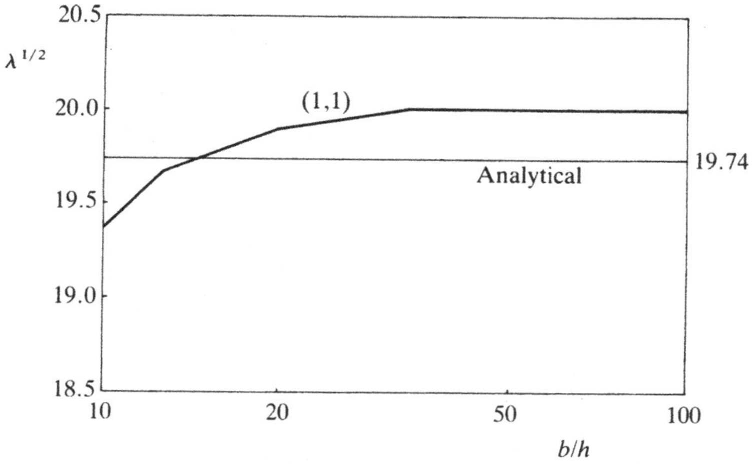

Figure 6.11. Effect of span/thickness ratio on lowest frequency of simply supported square plate [6.20].

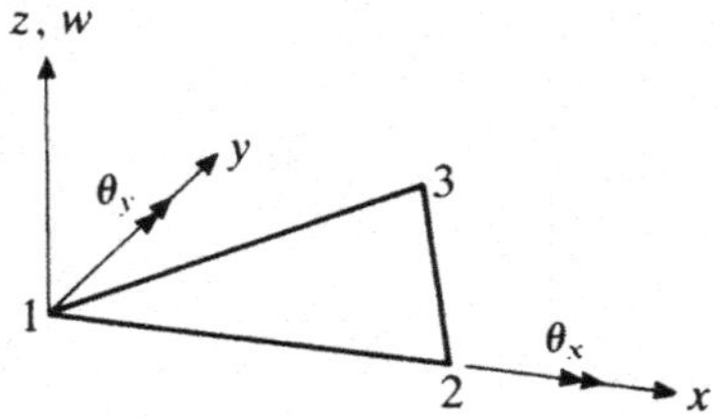

Figure 6.12. Geometry of a triangular element.

This expression can be written in the following matrix form

$$\begin{aligned} w &= \lfloor 1 \quad x \quad y \quad x^2 \quad xy \quad y^2 \quad x^3 \quad (x^2y+xy^2) \quad y^3 \rfloor \{\boldsymbol{\alpha}\} \\ &= \lfloor \mathbf{P}(x, y) \rfloor \{\boldsymbol{\alpha}\} \end{aligned} \tag{6.105}$$

where

$$\{\boldsymbol{\alpha}\}^{\mathrm{T}} = \lfloor \alpha_1 \quad \alpha_2 \quad \cdots \quad \alpha_9 \rfloor \tag{6.106}$$

Differentiating (6.105) with respect to x and y gives

$$\begin{bmatrix} w \\ \theta_x \\ \theta_y \end{bmatrix} = \begin{bmatrix} 1 & x & y & x^2 & xy & y^2 & x^3 & (x^2y+xy^2) & y^3 \\ 0 & 0 & 1 & 0 & x & 2y & 0 & (x^2+2xy) & 3y^2 \\ 0 & -1 & 0 & -2x & -y & 0 & -3x^2 & -(2xy+y^2) & 0 \end{bmatrix} \{\boldsymbol{\alpha}\} \tag{6.107}$$

Evaluating (6.107) at nodes 1, 2 and 3 with coordinates $(0, 0)$, $(x_2, 0)$ and (x_3, y_3) gives

$$\{\mathbf{w}\}_e = [\mathbf{A}]_e \{\boldsymbol{\alpha}\} \tag{6.108}$$

where

$$\{\mathbf{w}\}_e^{\mathrm{T}} = \lfloor w_1 \quad \theta_{x1} \quad \theta_{y1} \quad w_2 \quad \theta_{x2} \quad \theta_{y2} \quad w_3 \quad \theta_{x3} \quad \theta_{y3} \rfloor \tag{6.109}$$

and

$$[\mathbf{A}]_e = \begin{bmatrix} 1 & 0 & 0 & 0 & 0 & 0 & 0 & 0 & 0 \\ 0 & 0 & 1 & 0 & 0 & 0 & 0 & 0 & 0 \\ 0 & -1 & 0 & 0 & 0 & 0 & 0 & 0 & 0 \\ 1 & x_2 & 0 & x_2^2 & 0 & 0 & x_2^3 & 0 & 0 \\ 0 & 0 & 1 & 0 & x_2 & 0 & 0 & x_2^2 & 0 \\ 0 & -1 & 0 & -2x_2 & 0 & 0 & -3x_2^2 & 0 & 0 \\ 1 & x_3 & y_3 & x_3^2 & x_3y_3 & y_3^2 & x_3^3 & (x_3^2y_3+x_3y_3^2) & y_3^3 \\ 0 & 0 & 1 & 0 & x_3 & 2y_3 & 0 & (x_3^2+2x_3y_3) & 3y_3^2 \\ 0 & -1 & 0 & -2x_3 & -y_3 & 0 & -3x_3^2 & -(2x_3y_3+y_3^2) & 0 \end{bmatrix} \tag{6.110}$$

Solving (6.108) for $\{\boldsymbol{\alpha}\}$ gives

$$\{\boldsymbol{\alpha}\} = [\mathbf{A}]_e^{-1} \{\mathbf{w}\}_e \tag{6.111}$$

Substituting (6.111) into (6.105) gives

$$w = \lfloor \mathbf{P}(x, y) \rfloor [\mathbf{A}]_e^{-1} \{\mathbf{w}\}_e \tag{6.112}$$

Unfortunately the matrix $[\mathbf{A}]_e$ is singular whenever

$$x_2 - 2x_3 - y_3 = 0 \tag{6.113}$$

and, therefore, cannot be inverted. If this occurs the positions of the nodes should be altered to avoid this condition. This element is commonly referred to as element T [6.23].

Evaluating (6.107) along $y = 0$ gives

$$\begin{bmatrix} w \\ \theta_x \\ \theta_y \end{bmatrix} = \begin{bmatrix} 1 & x & 0 & x^2 & 0 & 0 & x^3 & 0 & 0 \\ 0 & 0 & 1 & 0 & x & 0 & 0 & x^2 & 0 \\ 0 & -1 & 0 & -2x & 0 & 0 & -3x^2 & 0 & 0 \end{bmatrix} \{\boldsymbol{\alpha}\} \tag{6.114}$$

From this it can be seen that w varies cubically and θ_y quadratically. The coefficients $\alpha_1, \alpha_2, \alpha_4$ and α_7 can be expressed in terms of w_1, θ_{y1}, w_2 and θ_{y2}, by evaluating these expressions at nodes 1 and 2. This means that the displacement and tangential slope will be continuous between elements.

On the other hand, θ_x is a quadratic function having coefficients α_3, α_5 and α_8. These cannot be determined using only the values of θ_x at nodes 1 and 2 only. Therefore, the normal slope will not be continuous between elements, and the element is a non-conforming one. The other disadvantage of this element is that the assumed function (6.104) is not invariant with respect to the choice of coordinate axes due to combining the x^2y and xy^2 terms.

Substituting (6.112) into (6.1) gives

$$T_e = \tfrac{1}{2} \{\dot{\mathbf{w}}\}_e^{\mathrm{T}} [\bar{\mathbf{m}}]_e \{\dot{\mathbf{w}}\}_e \tag{6.115}$$

where

$$[\bar{\mathbf{m}}]_e = [\mathbf{A}]_e^{-\mathrm{T}} \int_A \rho h \lfloor \mathbf{P} \rfloor^{\mathrm{T}} \lfloor \mathbf{P} \rfloor \, \mathrm{d}A \, [\mathbf{A}]_e^{-1} \tag{6.116}$$

is the element inertia matrix. A typical element in the integrand is of the form $\rho h \int_A x^m y^n \, \mathrm{d}A$. Integrals of this form can be evaluated using one of the following [6.24]

For $x_3 \neq 0, x_3 \neq x_2$:

$$\int_A x^m y^n \, \mathrm{d}A = \sum_{r=0}^{m+1} \sum_{s=0}^{r} \frac{(-1)^{r+s} m!}{(m+1-r)!(r-s)!s!(n+r+1)} x_2^{m+1-s} x_3^s y_3^{n+1} - \frac{x_3^{m+1} y_3^{n+1}}{(m+1)(m+n+2)} \tag{6.117}$$

For $x_3 = 0$:

$$\int_A x^m y^n \, \mathrm{d}A = \sum_{r=0}^{m+1} \frac{(-1)^r m!}{(m+1-r)!r!(n+r+1)} x_2^{m+1} y_3^{n+1} \tag{6.118}$$

For $x_3 = x_2$:

$$\int_A x^m y^n \, \mathrm{d}A = \frac{1}{(n+1)(m+n+2)} x_2^{m+1} y_3^{n+1} \tag{6.119}$$

Substituting (6.112) into (6.3) and (6.2) gives

$$U_e = \tfrac{1}{2}\{\mathbf{w}\}_e^{\mathrm{T}}[\bar{\mathbf{k}}]_e\{\mathbf{w}\}_e \tag{6.120}$$

where

$$[\bar{\mathbf{k}}]_e = [\mathbf{A}]_e^{-\mathrm{T}} \int_A \frac{h^3}{12}[\bar{\mathbf{B}}]^{\mathrm{T}}[\mathbf{D}][\bar{\mathbf{B}}] \, \mathrm{d}A \, [\mathbf{A}]_e^{-1} \tag{6.121}$$

where

$$[\bar{\mathbf{B}}] = \begin{bmatrix} 0 & 0 & 0 & 2 & 0 & 0 & 6x & 2y & 0 \\ 0 & 0 & 0 & 0 & 0 & 2 & 0 & 2x & 6y \\ 0 & 0 & 0 & 0 & 2 & 0 & 0 & 4(x+y) & 0 \end{bmatrix} \tag{6.122}$$

The integrand in (6.121) can also be evaluated using (6.117)–(6.119).

Substituting (6.112) into (6.4) gives

$$\delta W_e = \{\delta \mathbf{w}\}_e^{\mathrm{T}} \{\bar{\mathbf{f}}\}_e \tag{6.123}$$

where

$$\{\bar{\mathbf{f}}\}_e = [\mathbf{A}]_e^{-\mathrm{T}} \int_A \lfloor \mathbf{P} \rfloor^{\mathrm{T}} p_z \, \mathrm{d}A \tag{6.124}$$

Again (6.117)–(6.119) should be used to evaluate the integrand.

The next step is to transform the energy expressions (6.115), (6.120) and (6.123) into expressions involving nodal degrees of freedom relative to global axes (see Figure 6.13), w, θ_X and θ_Y. The relationship between displacement components in

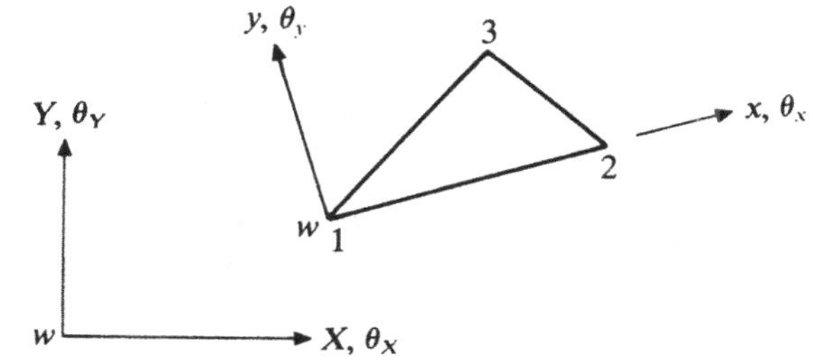

Figure 6.13. Orientation of a triangular element with respect to global axes.

local and global axes is (see Section (3.6))

$$\begin{bmatrix} w \\ \theta_x \\ \theta_y \end{bmatrix} = \begin{bmatrix} 1 & 0 & 0 \\ 0 & \cos(x, X) & \cos(x, Y) \\ 0 & \cos(y, X) & \cos(y, Y) \end{bmatrix} \begin{bmatrix} w \\ \theta_X \\ \theta_Y \end{bmatrix}$$

$$= [\mathbf{L}_2] \begin{bmatrix} w \\ \theta_X \\ \theta_Y \end{bmatrix} \tag{6.125}$$

Since the local x-axis lies along the side 1–2 and the local y-axis is perpendicular to it

$$\cos(x, X) = X_{21}/L_{12}, \qquad \cos(x, Y) = Y_{21}/L_{12}$$
$$\cos(y, X) = -Y_{21}/L_{12}, \qquad \cos(y, Y) = X_{21}/L_{12} \tag{6.126}$$

where

$$X_{21} = X_2 - X_1, \qquad Y_{21} = Y_2 - Y_1 \tag{6.127}$$

and

$$L_{12} = \left(X_{21}^2 + Y_{21}^2\right)^{1/2} \tag{6.128}$$

(X_1, Y_1) and (X_2, Y_2) are the global coordinates of nodes 1 and 2.

The degrees of freedom at all three nodes of the element can be transformed from local to global axes by means of the relation

$$\{\mathbf{w}\}_e = [\mathbf{R}]_e\{\mathbf{w}\}_e \tag{6.129}$$

where

$$\{\mathbf{w}\}_e^{\mathrm{T}} = \lfloor w_1 \quad \theta_{X1} \quad \theta_{Y1} \quad w_2 \quad \theta_{X2} \quad \theta_{Y2} \quad w_3 \quad \theta_{X3} \quad \theta_{Y3} \rfloor \tag{6.130}$$

and

$$[\mathbf{R}]_e = \begin{bmatrix} \mathbf{L}_2 & \mathbf{0} & \mathbf{0} \\ \mathbf{0} & \mathbf{L}_2 & \mathbf{0} \\ \mathbf{0} & \mathbf{0} & \mathbf{L}_2 \end{bmatrix} \tag{6.131}$$

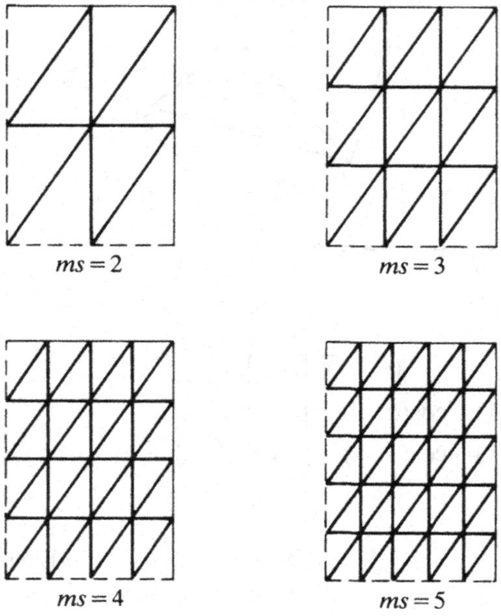

Figure 6.14. Idealisations of one quarter of a rectangular plate of aspect ratio 1.48:1. ms = mesh size.

Substituting (6.129) into (6.115), (6.120) and (6.123) gives

$$\begin{aligned} T_e &= \tfrac{1}{2}\{\dot{\mathbf{W}}\}_e^{\mathrm{T}}[\mathbf{m}]_e\{\dot{\mathbf{W}}\}_e \\ U_e &= \tfrac{1}{2}\{\mathbf{w}\}_e^{\mathrm{T}}[\mathbf{k}]_e\{\mathbf{w}\}_e \\ \delta W &= \{\delta W\}_e^{\mathrm{T}}\{\mathbf{f}\}_e \end{aligned} \tag{6.132}$$

where

$$\begin{aligned} [\mathbf{m}]_e &= [\mathbf{R}]_e^{\mathrm{T}}[\bar{\mathbf{m}}]_e[\mathbf{R}]_e \\ [\mathbf{k}]_e &= [\mathbf{R}]_e^{\mathrm{T}}[\bar{\mathbf{k}}]_e[\mathbf{R}]_e \\ \{\mathbf{f}\}_e &= [\mathbf{R}]_e^{\mathrm{T}}\{\bar{\mathbf{f}}\}_e \end{aligned} \tag{6.133}$$

When forming the element matrices referred to local axes the local coordinates of nodes 2 and 3 are required. These can be obtained from their global coordinates by means of the relation

$$\begin{bmatrix} x \\ y \end{bmatrix} = \begin{bmatrix} \cos(x, X) & \cos(x, Y) \\ \cos(y, X) & \cos(y, Y) \end{bmatrix} \begin{bmatrix} X - X_1 \\ Y - Y_1 \end{bmatrix} \tag{6.134}$$

EXAMPLE 6.9 Use the triangular element T to estimate the five lowest frequencies of a thin rectangular plate of aspect ratio 1.48:1 which is simply supported on all four edges. Compare the results with the analytical solution $\pi^2\{m^2 + (na/b)^2\}(D/\rho h a^4)^{1/2}$ rad/s, where a, b are the lengths of the sides.

One quarter of the plate was idealised in the ways indicated in Figure 6.14. The local axes were taken as indicated in Figure 6.15(a). Simply supported boundary conditions are applied along the outer boundaries and either symmetric or antisymmetric boundary conditions applied along the two axes of

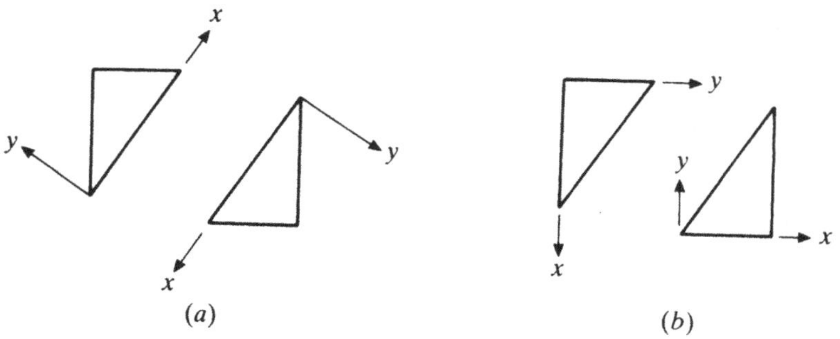

Figure 6.15. Choice of local axes for triangular elements.

symmetry. The results obtained are compared with the analytical frequencies in Figure 6.16. The frequencies obtained using the finite element method are all less than the analytical frequencies. All five modes converge monotonically for mesh sizes greater than three.

To illustrate the fact that the assumed function is not invariant with respect to the choice of local axes, the plate was analysed using a mesh size of five and the local axes as indicated in Figure 6.15(*b*). The percentage difference in frequencies when compared with the analytical solution is given for the two choices of axes shown in Figure 6.15 in Table 6.6. This indicates that quite different results are obtained depending upon the choice of local axes.

EXAMPLE 6.10 Figure 6.17 shows a triangular cantilever plate having a thickness of 1.55 mm. Use the triangular element T to calculate the six lowest frequencies and modes. Take $E = 200 \times 10^9$ N/m^2, $\nu = 0.3$ and $\rho = 7870$ kg/m^3. Compare these frequencies and mode shapes with the experimental measurements given in reference [6.26].

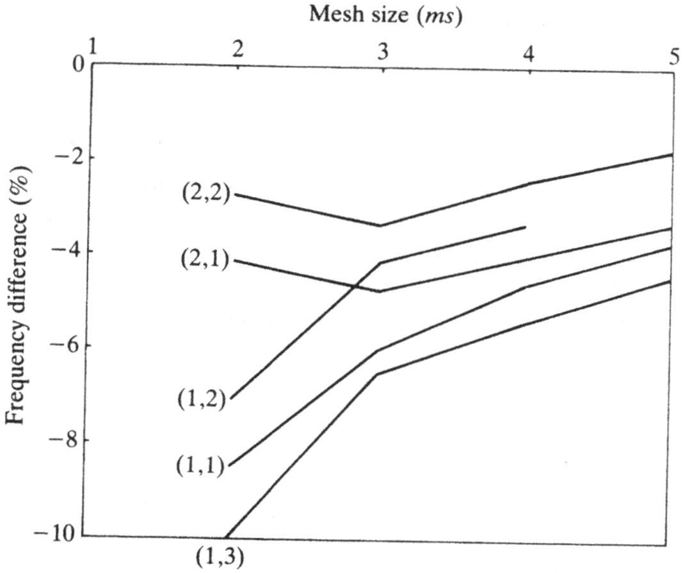

Figure 6.16. Flexural vibrations of a simply supported rectangular plate: aspect ratio 1.48:1. Element T [6.24, 6.25].

Table 6.6. *Comparison of predicted frequencies for different local axis systems of a simply supported rectangular plate: aspect ratio 1.48:1. Mesh size 5 [6.24]. Element T*

	Local axes	
Mode	6.15(*a*)	6.15(*b*)
(1, 1)	−3.77	−15.22
(2, 1)	−3.31	−12.35
(1, 3)	−4.42	−9.15
(2, 2)	−1.74	+6.05

Two idealisations were used in analysing the plate as shown in Figures 6.18(*a*) and (*b*). The mesh sizes 5 and 10 consisted of 25 and 100 elements respectively. The local axes used are the ones illustrated in Figure 6.15(*a*).

The percentage differences between the predicted and measured frequencies are given in Table 6.7. For a mesh size of 5 these values vary from −2.43 to +18.98. Increasing the mesh size to 10 reduces this range to (−3.26, +0.79). A comparison of the predicted and measured modal patterns is given in Figure 6.19. The first mode is the fundamental bending mode.

6.5 Thin Triangular Element (conforming)

One way of achieving continuity of the lateral displacement and both its first derivatives between elements, is to ensure that the normal slope varies linearly along an edge, as indicated in Section 6.1. This technique is used in reference [6.27] to derive a conforming thin triangular element (HCT) in Cartesian coordinates. The element has also been derived using area coordinates in reference [6.28] (where it is called LCCT-9). Both derivations are presented for comparison.

6.5.1 Cartesian Coordinates

Figure 6.20(*a*) shows a triangular element divided into three sub-triangles. The interior point 0 may be located arbitrarily, but it is convenient to position it at the centroid of the triangle. X, Y are the global axes of the system. It is convenient to use local axes x, y for each sub-triangle, where x is parallel to the exterior edge of the sub-triangle and y is perpendicular to it. Both the local and global axes for sub-triangle 1 are indicated in Figure 6.20(*b*).

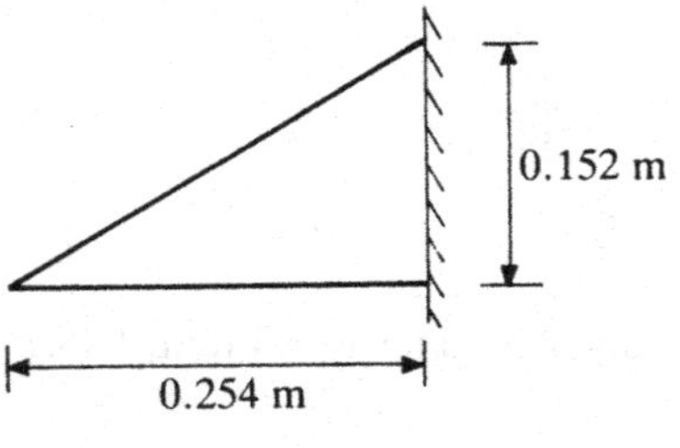

Figure 6.17. Geometry of a triangular cantilever plate.

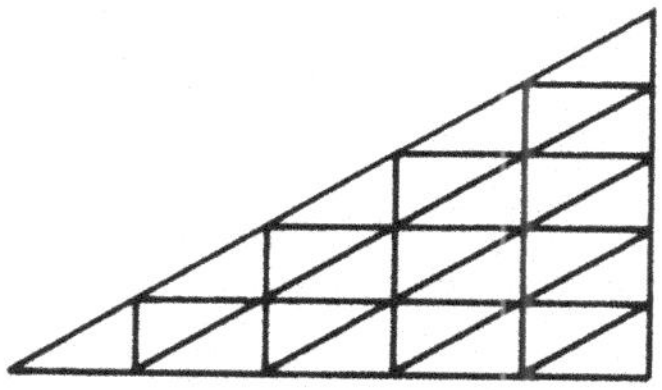

(*a*) Mesh size 5

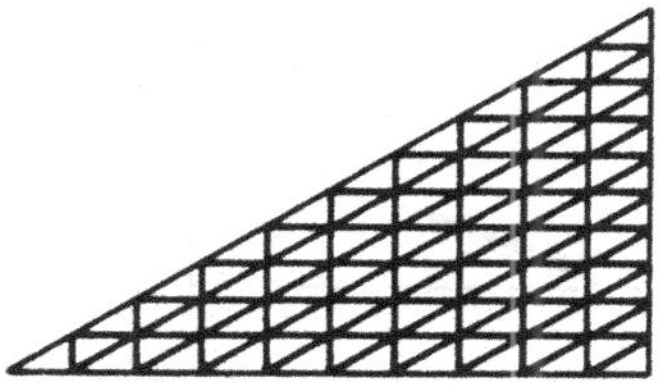

(*b*) Mesh size 10

Figure 6.18. Idealisation of a triangular plate using triangular elements.

Independent displacement functions are used for each sub-triangle. For example, for sub-triangle 1 it is assumed that

$$w^{(1)} = \lfloor \mathbf{P}^{(1)} \rfloor \{\boldsymbol{\alpha}^{(1)}\} \tag{6.135}$$

where

$$\lfloor \mathbf{P}^{(1)} \rfloor = \lfloor 1 \quad x \quad y \quad x^2 \quad xy \quad y^2 \quad x^3 \quad xy^2 \quad y^3 \rfloor \tag{6.136}$$

Note that the term x^2y has been omitted to ensure that $\partial w/\partial y$ varies linearly with x along the side 2–3.

Substituting (6.135) into (6.1) gives

$$T_e^{(1)} = \tfrac{1}{2}\{\dot{\boldsymbol{\alpha}}^{(1)}\}^{\mathrm{T}}[\bar{\mathbf{m}}^{(1)}]\{\dot{\boldsymbol{\alpha}}^{(1)}\} \tag{6.137}$$

where

$$[\bar{\mathbf{m}}^{(1)}] = \int_{A_1} \rho h \lfloor \mathbf{P}^{(1)} \rfloor^{\mathrm{T}} [\mathbf{P}^{(1)}] \,\mathrm{d}A \tag{6.138}$$

Table 6.7. *Comparison of predicted and measured frequencies of a triangular cantilever plate. Element T*

Mode number	Measured frequency (Hz) [6.26]	Predicted frequencies (% difference) [6.24]	
		$ms = 5$	$ms = 10$
1	37.5	−2.43	−2.88
2	161.0	−0.82	−3.26
3	243.0	+8.65	+0.75
4	392.0	+9.43	−1.48
5	592.0	+5.76	−0.76
6	744.0	+18.98	+0.79

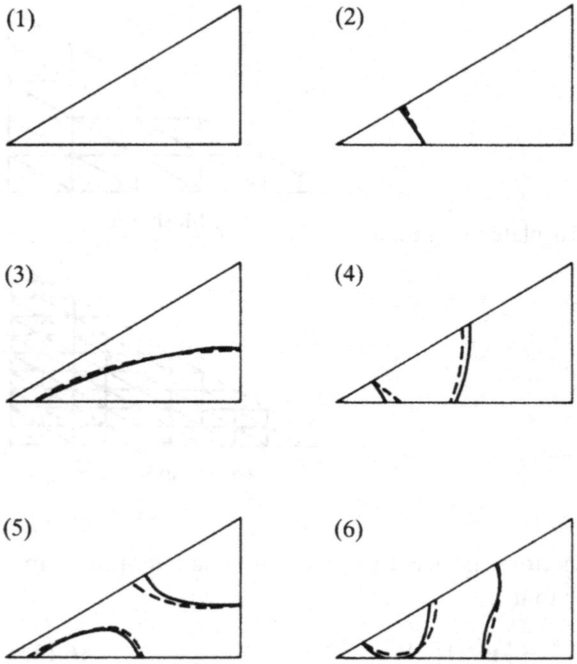

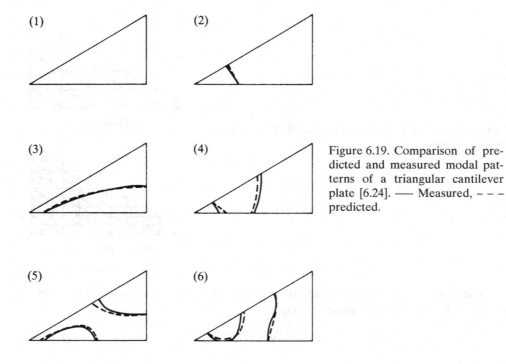

Figure 6.19. Comparison of predicted and measured modal patterns of a triangular cantilever plate [6.24]. —— Measured, – – – predicted.

and A_1 is the area of sub-triangle 1. This expression can be evaluated using the technique described in Section 6.4. This procedure is repeated for each of the three sub-triangles. Adding the kinetic energies together gives

$$T_e = \tfrac{1}{2}\{\dot{\boldsymbol{\alpha}}\}^{\mathrm{T}}\,[\bar{\mathbf{m}}]_e\,\{\dot{\boldsymbol{\alpha}}\} \tag{6.139}$$

where

$$[\bar{\mathbf{m}}]_e = \begin{bmatrix} [\bar{\mathbf{m}}^{(1)}] & & \\ & [\bar{\mathbf{m}}^{(2)}] & \\ & & [\bar{\mathbf{m}}^{(3)}] \end{bmatrix}, \qquad \{\boldsymbol{\alpha}\} = \begin{bmatrix} \{\boldsymbol{\alpha}^{(1)}\} \\ \{\boldsymbol{\alpha}^{(2)}\} \\ \{\boldsymbol{\alpha}^{(3)}\} \end{bmatrix} \tag{6.140}$$

The column matrix $\{\boldsymbol{\alpha}\}$ consists of 27 coefficients. Eighteen constraints are applied to ensure internal compatibility between the sub-triangles. This reduces the

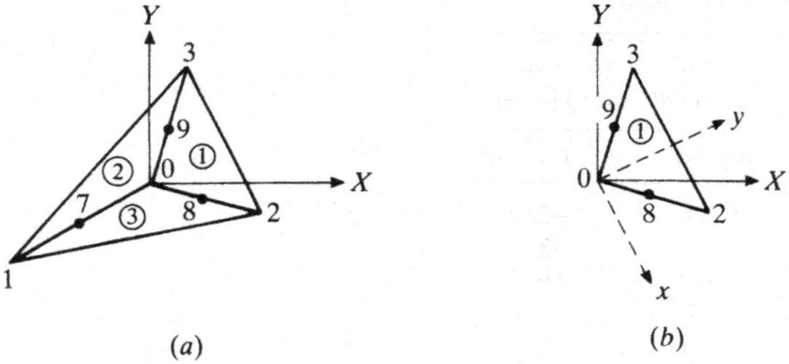

Figure 6.20. Geometry of a triangular element divided into three sub-triangles.

number of unknown coefficients to nine which are expressed in terms of the three degrees of freedom w, θ_X, θ_Y at the three nodes 1, 2 and 3.

The displacement and rotations with respect to local axes for sub-triangle 1 are

$$\begin{bmatrix} w \\ \theta_x \\ \theta_y \end{bmatrix} = \begin{bmatrix} 1 & x & y & x^2 & xy & y^2 & x^3 & xy^2 & y^3 \\ 0 & 0 & 1 & 0 & x & 2y & 0 & 2xy & 3y^2 \\ 0 & -1 & 0 & -2x & -y & 0 & -3x^2 & -y^2 & 0 \end{bmatrix} \{\boldsymbol{\alpha}^{(1)}\} \tag{6.141}$$

Evaluating (6.141) at node 2 gives

$$\begin{bmatrix} w \\ \theta_x \\ \theta_y \end{bmatrix}_2 = \{\bar{\mathbf{w}}^{(1)}\}_2 = [\bar{\mathbf{A}}^{(1)}]_2\{\boldsymbol{\alpha}^{(1)}\} \tag{6.142}$$

where

$$[\bar{\mathbf{A}}^{(1)}]_2 = \begin{bmatrix} 1 & x_2 & y_2 & x_2^2 & x_2y_2 & y_2^2 & x_2^3 & x_2y_2^2 & y_2^3 \\ 0 & 0 & 1 & 0 & x_2 & 2y_2 & 0 & 2x_2y_2 & 3y_2^2 \\ 0 & -1 & 0 & -2x_2 & -y_2 & 0 & -3x_2^2 & -y_2^2 & 0 \end{bmatrix} \tag{6.143}$$

and (x_2, y_2) are the local coordinates of node 2. These can be calculated from the global coordinates by the procedure described in the previous section.

Transforming (6.142) to global axes gives

$$\begin{bmatrix} w \\ \theta_X \\ \theta_Y \end{bmatrix}_2 = \{\mathbf{w}^{(1)}\}_2 = [\mathbf{L}_2^{(1)}]^{\mathrm{T}}\{\bar{\mathbf{w}}^{(1)}\}_2$$

$$= [\mathbf{L}_2^{(1)}]^{\mathrm{T}}[\bar{\mathbf{A}}^{(1)}]_2\{\boldsymbol{\alpha}^{(1)}\} = [\mathbf{A}^{(1)}]_2\{\boldsymbol{\alpha}^{(1)}\} \tag{6.144}$$

where $[\mathbf{L}_2^{(1)}]$ is of the form (6.125). Similar expressions may be written for the other nodes in this and the other two sub-triangles.

Compatibility of displacement and tangential slope is achieved by equating the degrees of freedom at common nodes of the three sub-triangles. This gives

$$\begin{aligned} \{\mathbf{w}^{(3)}\}_1 &= \{\mathbf{w}^{(2)}\}_1 \\ \{\mathbf{w}^{(1)}\}_2 &= \{\mathbf{w}^{(3)}\}_2 \\ \{\mathbf{w}^{(2)}\}_3 &= \{\mathbf{w}^{(1)}\}_3 \\ \{\mathbf{w}^{(2)}\}_0 &= \{\mathbf{w}^{(3)}\}_0 \\ \{\mathbf{w}^{(1)}\}_0 &= \{\mathbf{w}^{(2)}\}_0 \end{aligned} \tag{6.145}$$

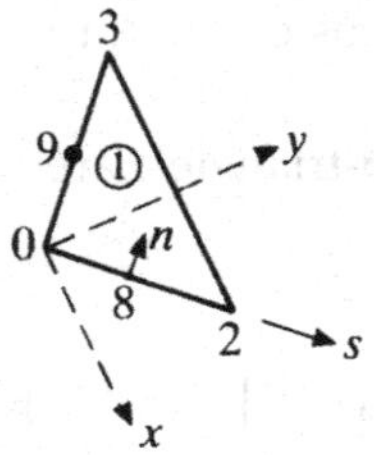

Figure 6.21. Geometry of a sub-triangle.

Compatibility of normal slope along the interior edges of the sub-triangles is ensured by equating the normal slopes at their mid-points 7, 8 and 9. The method of calculating the normal slope will be illustrated by considering the point 8 of sub-triangle 1 (see Figure 6.21). If s is the direction from 0 to the exterior node 2 and n is normal to this, then

$$\begin{aligned}\left(\frac{\partial w^{(1)}}{\partial n}\right)_8 &= \lfloor 0 \quad \cos(s,x) \quad \cos(s,y)\rfloor_8\{\bar{\mathbf{w}}^{(1)}\}_8 \\ &= \lfloor 0 \quad \cos(s,x) \quad \cos(s,y)\rfloor_5[\bar{\mathbf{A}}^{(1)}]_8\{\boldsymbol{\alpha}^{(1)}\} \\ &= \lfloor \mathbf{A}^{(1)}\rfloor_8\{\boldsymbol{\alpha}^{(1)}\}\end{aligned} \tag{6.146}$$

Similarly for the other interior node of sub-triangle 1 and the other sub-triangles. The three conditions imposing compatibility of normal slopes between the sub-triangles are

$$\begin{aligned}\left(\frac{\partial w^{(3)}}{\partial n}\right)_7 &= \left(\frac{\partial w^{(2)}}{\partial n}\right)_7 \\ \left(\frac{\partial w^{(1)}}{\partial n}\right)_8 &= \left(\frac{\partial w^{(3)}}{\partial n}\right)_8 \\ \left(\frac{\partial w^{(2)}}{\partial n}\right)_9 &= \left(\frac{\partial w^{(1)}}{\partial n}\right)_9\end{aligned} \tag{6.147}$$

The relationships (6.145) and (6.147) are the eighteen constraints required to ensure internal compatibility. The remaining nine coefficients in (6.140) are obtained by relating the sub-triangle degrees of freedom at the external nodes to the degrees of freedom of the complete triangle, that is

$$\begin{aligned}\{\mathbf{w}\}_1 &= \{\mathbf{w}^{(3)}\}_1 \\ \{\mathbf{w}\}_2 &= \{\mathbf{w}^{(1)}\}_2 \\ \{\mathbf{w}\}_3 &= \{\mathbf{w}^{(2)}\}_3\end{aligned} \tag{6.148}$$

Substituting (6.144) into (6.145), (6.146) into (6.147) and combining with (6.148) gives

$$\begin{bmatrix} \{\mathbf{w}\}_1 \\ \{\mathbf{w}\}_2 \\ \{\mathbf{w}\}_3 \\ 0 \\ 0 \\ 0 \\ 0 \\ 0 \\ 0 \\ 0 \\ 0 \end{bmatrix} = \begin{bmatrix} 0 & 0 & [\mathbf{A}^{(3)}]_1 \\ [\mathbf{A}^{(1)}]_2 & 0 & 0 \\ 0 & [\mathbf{A}^{(2)}]_3 & 0 \\ 0 & -[\mathbf{A}^{(2)}]_1 & [\mathbf{A}^{(3)}]_1 \\ [\mathbf{A}^{(1)}]_2 & 0 & -[\mathbf{A}^{(3)}]_2 \\ -[\mathbf{A}^{(1)}]_3 & [\mathbf{A}^{(2)}]_3 & 0 \\ 0 & [\mathbf{A}^{(2)}]_0 & -[\mathbf{A}^{(3)}]_0 \\ [\mathbf{A}^{(1)}]_0 & -[\mathbf{A}^{(2)}]_0 & 0 \\ 0 & -[\mathbf{A}^{(2)}]_7 & [\mathbf{A}^{(3)}]_7 \\ [\mathbf{A}^{(1)}]_8 & 0 & -[\mathbf{A}^{(3)}]_8 \\ -[\mathbf{A}^{(1)}]_9 & [\mathbf{A}^{(2)}]_9 & 0 \end{bmatrix} \begin{bmatrix} \{\boldsymbol{\alpha}^{(1)}\} \\ \{\boldsymbol{\alpha}^{(2)}\} \\ \{\boldsymbol{\alpha}^{(3)}\} \end{bmatrix} \tag{6.149}$$

This equation may be written symbolically as

$$\begin{bmatrix} \{\mathbf{w}\}_e \\ 0 \end{bmatrix} = \begin{bmatrix} [\mathbf{A}_{11}] & [\mathbf{A}_{10}] \\ [\mathbf{A}_{01}] & [\mathbf{A}_{00}] \end{bmatrix} \begin{bmatrix} \{\boldsymbol{\alpha}^{(1)}\} \\ \{\boldsymbol{\alpha}^{(0)}\} \end{bmatrix} \tag{6.150}$$

Solving for $\{\boldsymbol{\alpha}^{(0)}\}$ from the second of these two matrix equations gives

$$\{\boldsymbol{\alpha}^{(0)}\} = -[\mathbf{A}_{00}]^{-1}[\mathbf{A}_{01}]\{\boldsymbol{\alpha}^{(1)}\} \tag{6.151}$$

Substituting into the first equation in (6.150) gives

$$\begin{aligned} \{\mathbf{w}\}_e &= [[\mathbf{A}_{11}] - [\mathbf{A}_{10}][\mathbf{A}_{00}]^{-1}[\mathbf{A}_{01}]]\{\boldsymbol{\alpha}^{(1)}\} \\ &= [\bar{\mathbf{A}}]\{\boldsymbol{\alpha}^{(1)}\} \end{aligned} \tag{6.152}$$

Solving for $\{\boldsymbol{\alpha}^{(1)}\}$ gives

$$\{\boldsymbol{\alpha}^{(1)}\} = [\bar{\mathbf{A}}]^{-1}\{\mathbf{w}\}_e \tag{6.153}$$

Combining (6.151) and (6.153) gives

$$\begin{aligned} \begin{bmatrix} \{\boldsymbol{\alpha}^{(1)}\} \\ \{\boldsymbol{\alpha}^{(0)}\} \end{bmatrix} &= \begin{bmatrix} [\bar{\mathbf{A}}]^{-1} \\ -[\mathbf{A}_{00}]^{-1}[\mathbf{A}_{01}][\bar{\mathbf{A}}]^{-1} \end{bmatrix} \{\mathbf{w}\}_e \\ &= [\bar{\bar{\mathbf{A}}}]\{\mathbf{w}\}_e \end{aligned} \tag{6.154}$$

Substituting (6.154) into (6.139) gives the element inertia matrix in terms of the nine nodal degrees of freedom

$$[\mathbf{m}]_e = [\bar{\bar{\mathbf{A}}}]^{\mathrm{T}}[\bar{\mathbf{m}}]_e[\bar{\bar{\mathbf{A}}}] \tag{6.155}$$

The stiffness and equivalent nodal force matrices can be derived in a similar manner. The stiffness matrix is given by

$$[\mathbf{k}]_e = [\bar{\bar{\mathbf{A}}}]^{\mathrm{T}}[\bar{\mathbf{k}}]_e[\bar{\bar{\mathbf{A}}}] \tag{6.156}$$

where

$$[\bar{\mathbf{k}}]_e = \begin{bmatrix} [\bar{\mathbf{k}}^{(1)}] & & \\ & [\bar{\mathbf{k}}^{(2)}] & \\ & & [\bar{\mathbf{k}}^{(3)}] \end{bmatrix} \tag{6.157}$$

A typical element of this matrix is

$$[\bar{\mathbf{k}}^{(1)}] = \int_{A_1} \frac{h^3}{12}[\bar{\mathbf{B}}^{(1)}]^{\mathrm{T}}[\mathbf{D}][\bar{\mathbf{B}}^{(1)}]\,\mathrm{d}A \tag{6.158}$$

where

$$[\bar{\mathbf{B}}^{(1)}] = \begin{bmatrix} 0 & 0 & 0 & 2 & 0 & 0 & 6x & 0 & 0 \\ 0 & 0 & 0 & 0 & 0 & 2 & 0 & 2x & 6y \\ 0 & 0 & 0 & 0 & 2 & 0 & 0 & 4y & 0 \end{bmatrix} \tag{6.159}$$

The equivalent nodal force matrix is

$$\{\mathbf{f}\}_e = [\bar{\bar{\mathbf{A}}}]^{\mathrm{T}}\{\bar{\mathbf{f}}\}_e \tag{6.160}$$

where

$$\{\bar{\mathbf{f}}\}_e = \begin{bmatrix} \{\bar{\mathbf{f}}^{(1)}\} \\ \{\bar{\mathbf{f}}^{(2)}\} \\ \{\bar{\mathbf{f}}^{(3)}\} \end{bmatrix} \tag{6.161}$$

A typical element of this matrix is

$$\{\bar{\mathbf{f}}^{(1)}\} = \int_{A_1} \lfloor \mathbf{P}^{(1)} \rfloor^{\mathrm{T}} p_z\,\mathrm{d}A \tag{6.162}$$

6.5.2 Area Coordinates

When using area coordinates the displacement function for a sub-triangle is initially assumed to be a complete cubic which has ten terms. Therefore, the matrix $\lfloor \mathbf{P}^{(1)} \rfloor$ in (6.135) is

$$[\mathbf{P}^{(1)}] = \lfloor L_1^3 \quad L_2^3 \quad L_3^3 \quad L_1^2L_2 \quad L_1^2L_3 \quad L_2^2L_3 \quad L_1L_2^2 \quad L_1L_3^2 \quad L_2L_3^2 \quad L_1L_2L_3 \rfloor \tag{6.163}$$

The inertia matrix $[\bar{\mathbf{m}}^{(1)}]$ expressed in terms of the coefficients $\{\boldsymbol{\alpha}^{(1)}\}$ (see expressions (6.137) and (6.138)) can be evaluated using (4.90). This results in the

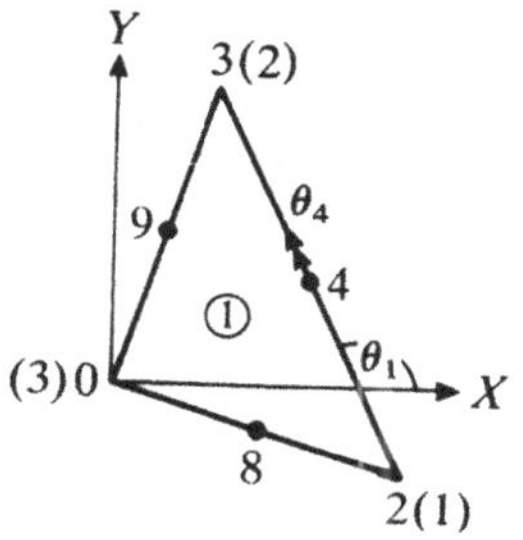

Figure 6.22. Geometry of sub-triangle 1.

following matrix.

$$
[\bar{\mathbf{m}}^{(1)}] = \frac{\rho h A^{(1)}}{5040}
\begin{bmatrix}
180 & & & & & & & & & \\
9 & 180 & & & & & & \text{Sym} & & \\
9 & 9 & 180 & & & & & & & \\
30 & 12 & 3 & 12 & & & & & & \\
30 & 3 & 12 & 6 & 12 & & & & & \\
3 & 30 & 12 & 3 & 2 & 12 & & & & \\
12 & 30 & 3 & 9 & 3 & 6 & 12 & & & \\
12 & 3 & 30 & 3 & 9 & 3 & 2 & 12 & & \\
3 & 12 & 30 & 2 & 3 & 9 & 3 & 6 & 12 & \\
6 & 6 & 6 & 3 & 3 & 3 & 3 & 3 & 3 & 2
\end{bmatrix}
\tag{6.164}
$$

The ten coefficients in $\{\boldsymbol{\alpha}^{(1)}\}$ are expressed in terms of w, θ_X, θ_Y at nodes 0, 2 and 3 (see Figure 6.22) and the normal slope, θ_4, at node 4 which is located at the mid-point of side 2–3. The node numbers in brackets are the numbers used for the area coordinates. This calculation is carried out using the relations

$$
\begin{aligned}
\theta_X &= \frac{\partial w^{(1)}}{\partial Y} = \frac{1}{2A_1}\sum_{j=1}^{3} b_j \frac{\partial w^{(1)}}{\partial L_j} \\
\theta_Y &= -\frac{\partial w^{(1)}}{\partial X} = \frac{1}{2A_1}\sum_{j=1}^{3} a_j \frac{\partial w^{(1)}}{\partial L_j}
\end{aligned}
\tag{6.165}
$$

from (4.96), where a_j and b_j are defined in (4.10) and

$$
\begin{aligned}
\theta_4 &= (\cos\theta_1 \theta_X + \sin\theta_1 \theta_Y)_4 \\
&= (\frac{b_3}{l_3}\theta_X - \frac{a_3}{l_3}\theta_Y)_4
\end{aligned}
\tag{6.166}
$$

where

$$
l_3 = \left(a_3^2 + b_3^2\right)^{1/2} \tag{6.167}
$$

The result is

$$
\{\boldsymbol{\alpha}^{(1)}\} = [\bar{\mathbf{c}}]\{\mathbf{w}^{(1)}\} \tag{6.168}
$$

where

$$\{\mathbf{w}^{(1)}\}^{\mathrm{T}} = \lfloor w_2 \quad \theta_{X2} \quad \theta_{Y2} \quad w_3 \quad \theta_{X3} \quad \theta_{Y3} \quad \theta_4 \quad w_0 \quad \theta_{X0} \quad \theta_{Y0} \rfloor \tag{6.169}$$

The non-zero elements of $[\bar{\mathbf{c}}]$ are

$$\begin{aligned}
\bar{c}_{11} &= \bar{c}_{24} = \bar{c}_{36} = 1 \\
\bar{c}_{41} &= \bar{c}_{51} = \bar{c}_{64} = \bar{c}_{74} = \bar{c}_{88} = \bar{c}_{98} = 3 \\
-\bar{c}_{65} &= \bar{c}_{99} = a_1 \\
\bar{c}_{52} &= -\bar{c}_{89} = a_2 \\
-\bar{c}_{42} &= \bar{c}_{75} = a_3 \\
-\bar{c}_{66} &= \bar{c}_{9,10} = b_1 \\
\bar{c}_{53} &= -\bar{c}_{8,10} = b_2 \\
-\bar{c}_{43} &= \bar{c}_{76} = b_3 \\
\bar{c}_{10,1} &= 6\mu_3 \\
\bar{c}_{10,2} &= (a_1 - a_3\mu_3) \\
\bar{c}_{10,3} &= (b_1 - b_3\mu_3) \\
\bar{c}_{10,4} &= 6\lambda_3 \\
\bar{c}_{10,5} &= (a_3\lambda_3 - a_2) \\
\bar{c}_{10,6} &= (b_3\lambda_3 - b_2) \\
\bar{c}_{10,7} &= 4h_3
\end{aligned} \tag{6.170}$$

where

$$\begin{aligned}
\lambda_3 &= -(a_2a_3 + b_2b_3)/l_3^2 \\
\mu_3 &= 1 - \lambda_3 \\
h_3 &= 2A_1/l_3 \\
A_1 &= \tfrac{1}{2}(a_1b_2 - a_2b_1)
\end{aligned} \tag{6.171}$$

Substituting (6.168) into (6.137) gives

$$T_e^{(1)} = \tfrac{1}{2}\{\dot{\mathbf{w}}^{(1)}\}^{\mathrm{T}}[\mathbf{m}^{(1)}]\{\dot{\mathbf{w}}^{(1)}\} \tag{6.172}$$

where

$$[\mathbf{m}^{(1)}] = [\bar{\mathbf{c}}]^{\mathrm{T}}[\bar{\mathbf{m}}^{(1)}][\bar{\mathbf{c}}] \tag{6.173}$$

The inertia matrices for sub-triangles 2 and 3 are obtained in a similar manner. These matrices are then assembled together to give the inertia matrix, $[\bar{\mathbf{m}}]_e$, of the

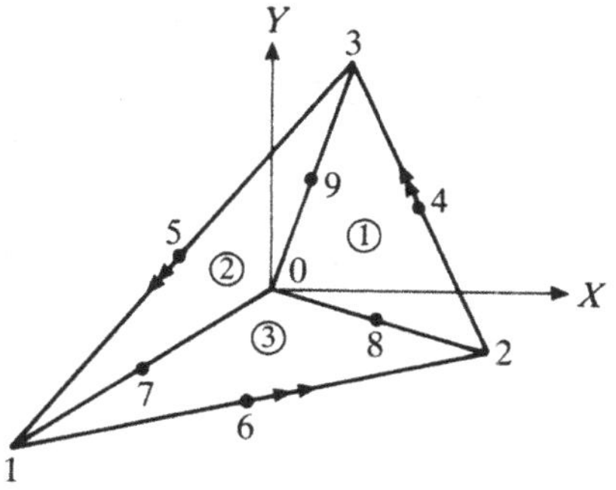

Figure 6.23. Geometry of triangular element divided into three sub-triangles.

complete triangle shown in Figure 6.23 in terms of the degrees of freedom.

$$\lfloor w_1\ \ \theta_{X1}\ \ \theta_{Y1}\ \ w_2\ \ \theta_{X2}\ \ \theta_{Y2}\ \ w_3\ \ \theta_{X3}\ \ \theta_{Y3}\ \ \theta_4\ \ \theta_5\ \ \theta_6\ \ w_0\ \ \theta_{X0}\ \ \theta_{Y0}\rfloor = \begin{bmatrix} \{\bar{\mathbf{w}}\} \\ \{\mathbf{w}^{(0)}\} \end{bmatrix}^{\mathrm{T}} \tag{6.174}$$

In assembling these matrices it is assumed that displacement and rotations at common nodes of the three sub-triangles are equal. This will ensure compatibility of displacement and tangential slope along their common edges. Compatibility of normal slope along these same edges (0–1, 0–2 and 0–3 in Figure 6.23) is ensured by equating the normal slopes at their mid-points 7, 8 and 9. These can be obtained using relationships similar to (6.166). This will give three equations of the form

$$[\mathbf{B}_n\quad \mathbf{B}_0]\begin{bmatrix} \{\bar{\mathbf{w}}\} \\ \{\mathbf{w}^{(0)}\} \end{bmatrix} = 0 \tag{6.175}$$

which can be used to eliminate $\{\mathbf{w}^{(0)}\}$. Rearranging (6.175) gives

$$\{\mathbf{w}^{(0)}\} = -[\mathbf{B}_0]^{-1}[\mathbf{B}_n]\{\bar{\mathbf{w}}\} \tag{6.176}$$

Therefore

$$\begin{bmatrix} \{\bar{\mathbf{w}}\} \\ \{\mathbf{w}^{(0)}\} \end{bmatrix} = \begin{bmatrix} [\mathbf{I}] \\ -[\mathbf{B}_0]^{-1}[\mathbf{B}_n] \end{bmatrix}\{\bar{\mathbf{w}}\} = [\bar{\mathbf{A}}]\{\bar{\mathbf{w}}\} \tag{6.177}$$

The inertia matrix in terms of degrees of freedom at nodes 1 to 6 only is, therefore

$$[\mathbf{m}^{12}]_e = [\bar{\mathbf{A}}]^{\mathrm{T}}[\bar{\mathbf{m}}]_e[\bar{\mathbf{A}}] \tag{6.178}$$

This matrix is of order (12×12) and is referred to as the LCCT-12 element in reference [6.28]. The LCCT-9 element is obtained by constraining the normal slopes to vary linearly along each side of the complete triangle. At node 4 (see equation (6.166))

$$\begin{aligned} \theta_4 &= \frac{b_3}{l_3}\theta_{X4} - \frac{a_3}{l_3}\theta_{Y4} \\ &= \frac{b_3}{2l_3}(\theta_{X2} + \theta_{X3}) - \frac{a_3}{2l_3}(\theta_{Y2} + \theta_{Y3}) \end{aligned} \tag{6.179}$$

Similar relationships can be derived for θ_5 and θ_6. Defining

$$\{\mathbf{w}\}^{\mathrm{T}} = \lfloor w_1 \quad \theta_{X1} \quad \theta_{Y1} \quad w_2 \quad \theta_{X2} \quad \theta_{Y2} \quad w_3 \quad \theta_{X3} \quad \theta_{Y3} \rfloor \tag{6.180}$$

and

$$\{\theta\}^{\mathrm{T}} = \lfloor \theta_4 \quad \theta_5 \quad \theta_6 \rfloor \tag{6.181}$$

then these relationships can be written in the form

$$\{\boldsymbol{\theta}\} = [\mathbf{A}_c]\{\mathbf{w}\} \tag{6.182}$$

Therefore

$$\{\bar{\mathbf{w}}\} = \begin{bmatrix} \{\mathbf{w}\} \\ \{\boldsymbol{\theta}\} \end{bmatrix} = \begin{bmatrix} [\mathbf{I}] \\ [\mathbf{A}_c] \end{bmatrix} \{\mathbf{w}\} = [\bar{\bar{\mathbf{A}}}]\,\{\mathbf{w}\} \tag{6.183}$$

The inertia matrix in terms of degrees of freedom at nodes 1 to 3 is, therefore

$$[\mathbf{m}^9]_e = [\bar{\bar{\mathbf{A}}}]^{\mathrm{T}}[\mathbf{m}^{12}]_e[\bar{\bar{\mathbf{A}}}] \tag{6.184}$$

The stiffness and equivalent nodal force matrices can be derived in a similar manner. The stiffness matrix of sub-triangle 1 in terms of the coefficients $\{\boldsymbol{\alpha}^{(1)}\}$ is

$$[\bar{\mathbf{k}}^{(1)}] = \int_{A_1} \frac{h^3}{12}[\mathbf{B}^{(1)}]^{\mathrm{T}}[\mathbf{D}][\mathbf{B}^{(1)}]\,\mathrm{d}A \tag{6.185}$$

The components of strain are defined as

$$\{\boldsymbol{\chi}^{(1)}\} = \begin{bmatrix} \partial^2 w^{(1)}/\partial x^2 \\ \partial^2 w^{(1)}/\partial y^2 \\ 2\partial^2 w^{(1)}/\partial x \partial y \end{bmatrix} \tag{6.186}$$

Using (4.96) to convert to derivatives with respect to area coordinates gives

$$\begin{aligned} \frac{\partial^2}{\partial x^2} &= \frac{1}{4A_1^2}\sum_{j=1}^{3}\sum_{k=1}^{3} a_j a_k \frac{\partial^2}{\partial L_j \partial L_k} \\ \frac{\partial^2}{\partial y^2} &= \frac{1}{4A_1^2}\sum_{j=1}^{3}\sum_{k=1}^{3} b_j b_k \frac{\partial^2}{\partial L_j \partial L_k} \\ \frac{\partial^2}{\partial x \partial y} &= \frac{1}{4A_1^2}\sum_{j=1}^{3}\sum_{k=1}^{3} a_j b_k \frac{\partial^2}{\partial L_j \partial L_k} \end{aligned} \tag{6.187}$$

Since $\lfloor \mathbf{P}^{(1)} \rfloor$ is cubic, the strain matrix will be linear. It can, therefore, be written in the form

$$[\mathbf{B}^{(1)}] = \frac{1}{4A_1^2}[L_1[\mathbf{W}_1] + L_2[\mathbf{W}_2] + L_3[\mathbf{W}_3]] \tag{6.188}$$

where $[\mathbf{W}_1]$, $[\mathbf{W}_2]$ and $[\mathbf{W}_3]$ are each (3×10) matrices of constants. This form will facilitate the evaluation of (6.185).

Table 6.8. *Percentage difference between finite element and analytical frequencies of a simply supported plate of aspect ratio 1.48:1*

Mesh size	2		3		4	
Mode	(1, 1)	(1, 3)	(1, 1)	(1, 3)	(1, 1)	(1, 3)
T [6.24]	−8.5	−10.0	−6.1	−6.6	−4.7	−5.4
HCT [6.29]	4.7	16.1	1.9	6.6	1.1	3.5

The equivalent force matrix in terms of the parameters $\{\boldsymbol{\alpha}^{(1)}\}$ is

$$\{\bar{\mathbf{f}}^{(1)}\} = \int_{A_1} \lfloor \mathbf{P}^{(1)} \rfloor^{\mathrm{T}} p_z \, \mathrm{d}A \tag{6.189}$$

EXAMPLE 6.11 Use the triangular elements T and HCT to estimate the two lowest frequencies of doubly symmetric modes of a thin rectangular plate of aspect ratio 1.48:1 which is simply supported on all four edges. Compare the results with the analytical solution $\pi^2[m^2 + (na/b)^2](D/\rho h a^4)^{1/2}$ rad/s where a, b are the lengths of the sides.

The plate was analysed using mesh sizes 2, 3 and 4 indicated in Figure 6.14. In the case of element T the local axes used are the ones shown in Figure 6.15(a). The frequencies predicted using the finite element method are compared with the analytical frequencies in Table 6.8. Element T underestimates the frequencies of these two modes whilst element HCT overestimates them. The HCT element results are seen to converge faster than those obtained with the element T.

EXAMPLE 6.12 Use the triangular elements LCCT-9 and LCCT-12 to estimate the two lowest frequencies of doubly symmetric modes of a thin square plate which is simply supported on all four edges. Compare the results with the analytical solution given in Example 6.1.

The plate was analysed using mesh sizes 2, 3 and 4 indicated in Figure 6.14. The frequencies are compared in Table 6.9. Both elements give results which converge from above, but the results for the LCCT-12 element are considerably more accurate than the LCCT-9 results. Reference [6.33] quotes results obtained using the HCT element which agree with those quoted in Table 6.9 for the LCCT-9 element confirming that the two elements are equivalent.

Table 6.9. *Percentage difference between finite element and analytical frequencies of a simply supported square plate*

Mesh size	2		3		4	
Mode	(1, 1)	(1, 3)	(1, 1)	(1, 3)	(1, 1)	(1, 3)
LCCT-9	2.84	15.5	1.24	5.95	0.69	3.12
LCCT-12	0.20	1.70	0.06	0.55	0.02	0.21

6.6 Thick Triangular Element

The energy expressions for a thick plate element are given by expressions (6.64) to (6.67) of Section 6.3. The displacement functions

$$w = \sum_{j=1}^{3} L_j w_j, \qquad \theta_X = \sum_{j=1}^{3} L_j \theta_{X_j}, \qquad \theta_Y = \sum_{j=1}^{3} L_j \theta_{Y_j} \tag{6.190}$$

where $(w_j, \theta_{X_j}, \theta_{Y_j})$ are the degrees of freedom at node j and L_1, L_2, L_3 are area coordinates for the triangle, ensure that w, θ_X, θ_Y are continuous between elements. This element will be referred to as element THT.

Combining expressions (6.190) gives

$$\begin{bmatrix} w \\ \theta_X \\ \theta_Y \end{bmatrix} = [\mathbf{N}]\,\{\mathbf{w}\}_e \tag{6.191}$$

where

$$\{\mathbf{w}\}_e^{\mathrm{T}} = \lfloor w_1 \quad \theta_{X1} \quad \theta_{Y1} \quad w_2 \quad \theta_{X2} \quad \theta_{Y2} \quad w_3 \quad \theta_{X3} \quad \theta_{Y3} \rfloor \tag{6.192}$$

and

$$[\mathbf{N}] = \begin{bmatrix} L_1 & 0 & 0 & L_2 & 0 & 0 & L_3 & 0 & 0 \\ 0 & L_1 & 0 & 0 & L_2 & 0 & 0 & L_3 & 0 \\ 0 & 0 & L_1 & 0 & 0 & L_2 & 0 & 0 & L_3 \end{bmatrix} \tag{6.193}$$

Substituting (6.193) into the expression for the inertia matrix (6.74) and integrating using (4.90) gives

$$[\mathbf{m}]_e = [\mathbf{m}^{\mathrm{t}}] + [\mathbf{m}^{\mathrm{r}}] \tag{6.194}$$

where

$$[\mathbf{m}^{\mathrm{t}}] = \frac{\rho h A}{144} \begin{bmatrix} 24 & & & & & & & & \\ 0 & 0 & & & & & & & \\ 0 & 0 & 0 & & & & \mathrm{Sym} & & \\ 12 & 0 & 0 & 24 & & & & & \\ 0 & 0 & 0 & 0 & 0 & & & & \\ 0 & 0 & 0 & 0 & 0 & 0 & & & \\ 12 & 0 & 0 & 12 & 0 & 0 & 24 & & \\ 0 & 0 & 0 & 0 & 0 & 0 & 0 & 0 & \\ 0 & 0 & 0 & 0 & 0 & 0 & 0 & 0 & 0 \end{bmatrix} \tag{6.195}$$

$$[\mathbf{m}^{\mathrm{r}}] = \frac{\rho h^3 A}{144}\begin{bmatrix} 0 & & & & & & & & \\ 0 & 2 & & & & & \text{Sym} & & \\ 0 & 0 & 2 & & & & & & \\ 0 & 0 & 0 & 0 & & & & & \\ 0 & 1 & 0 & 0 & 2 & & & & \\ 0 & 0 & 1 & 0 & 0 & 2 & & & \\ 0 & 0 & 0 & 0 & 0 & 0 & 0 & & \\ 0 & 1 & 0 & 0 & 1 & 0 & 0 & 2 & \\ 0 & 0 & 1 & 0 & 0 & 1 & 0 & 0 & 2 \end{bmatrix} \tag{6.196}$$

Substituting (6.191) into (6.66) and using (4.96) shows that the strain matrix due to flexure is

$$[\mathbf{B}^{\mathrm{f}}] = \frac{1}{2A}\begin{bmatrix} 0 & 0 & -a_1 & 0 & 0 & -a_2 & 0 & 0 & -a_3 \\ 0 & b_1 & 0 & 0 & b_2 & 0 & 0 & b_3 & 0 \\ 0 & a_1 & -b_1 & 0 & a_2 & -b_2 & 0 & a_3 & -b_3 \end{bmatrix} \tag{6.197}$$

As this is constant the integration of (6.80) is trivial. The stiffness matrix due to the flexure is therefore

$$[\mathbf{k}^{\mathrm{f}}] = \frac{h^3}{12}A[\mathbf{B}^{\mathrm{f}}]^{\mathrm{T}}[\mathbf{D}][\mathbf{B}^{\mathrm{f}}] \tag{6.198}$$

Substituting (6.191) into (6.66) and using (4.96) shows that the strain matrix due to shear is

$$[\mathbf{B}^{\mathrm{s}}] = \begin{bmatrix} \dfrac{a_1}{2A} & 0 & L_1 & \dfrac{a_2}{2A} & 0 & L_2 & \dfrac{a_3}{2A} & 0 & L_3 \\ \dfrac{b_1}{2A} & -L_1 & 0 & \dfrac{b_2}{2A} & -L_2 & 0 & \dfrac{b_3}{2A} & -L_3 & 0 \end{bmatrix} \tag{6.199}$$

Substituting (6.199) into (6.81) and integrating using (4.90) gives the following result for an isotropic material.

$$[\mathbf{k}^{\mathrm{s}}] = \kappa G h \begin{bmatrix} \mathbf{k}^{\mathrm{s}}_{11} & & \text{Sym} \\ \mathbf{k}^{\mathrm{s}}_{21} & \mathbf{k}^{\mathrm{s}}_{22} & \\ \mathbf{k}^{\mathrm{s}}_{31} & \mathbf{k}^{\mathrm{s}}_{32} & \mathbf{k}^{\mathrm{s}}_{33} \end{bmatrix} \tag{6.200}$$

where

$$\mathbf{k}^{\mathrm{s}}_{11} = \begin{bmatrix} \dfrac{(a_1^2 + b_1^2)}{4A} & -\dfrac{b_1}{6} & \dfrac{a_1}{6} \\ -\dfrac{b_1}{6} & \dfrac{A}{6} & 0 \\ \dfrac{a_1}{6} & 0 & \dfrac{A}{6} \end{bmatrix} \tag{6.201}$$

$$\mathbf{k}^{\mathrm{s}}_{21} = \begin{bmatrix} \dfrac{(a_2 a_1 + b_2 b_1)}{4A} & -\dfrac{b_2}{6} & \dfrac{a_2}{6} \\ -\dfrac{b_1}{6} & \dfrac{A}{12} & 0 \\ \dfrac{a_1}{6} & 0 & \dfrac{A}{12} \end{bmatrix} \tag{6.202}$$

$$\mathbf{k}_{31}^{s} = \begin{bmatrix} \dfrac{(a_3a_1 + b_3b_1)}{4A} & -\dfrac{b_3}{6} & \dfrac{a_3}{6} \\ -\dfrac{b_1}{6} & \dfrac{A}{12} & 0 \\ \dfrac{a_1}{6} & 0 & \dfrac{A}{12} \end{bmatrix} \tag{6.203}$$

$$\mathbf{k}_{22}^{s} = \begin{bmatrix} \dfrac{(a_2^2 + b_2^2)}{4A} & -\dfrac{b_2}{6} & \dfrac{a_2}{6} \\ -\dfrac{b_2}{6} & \dfrac{A}{6} & 0 \\ \dfrac{a_2}{6} & 0 & \dfrac{A}{6} \end{bmatrix} \tag{6.204}$$

$$\mathbf{k}_{32}^{s} = \begin{bmatrix} \dfrac{(a_3a_2 + b_3b_2)}{4A} & -\dfrac{b_3}{6} & \dfrac{a_3}{6} \\ -\dfrac{b_2}{6} & \dfrac{A}{12} & 0 \\ \dfrac{a_2}{6} & 0 & \dfrac{A}{12} \end{bmatrix} \tag{6.205}$$

$$\mathbf{k}_{33}^{s} = \begin{bmatrix} \dfrac{(a_3^2 + b_3^2)}{4A} & -\dfrac{b_3}{6} & \dfrac{a_3}{6} \\ -\dfrac{b_3}{6} & \dfrac{A}{6} & 0 \\ \dfrac{a_3}{6} & 0 & \dfrac{A}{6} \end{bmatrix} \tag{6.206}$$

The complete stiffness matrix for the element is

$$[\mathbf{k}]_e = [\mathbf{k}^{\mathrm{f}}] + [\mathbf{k}^{\mathrm{s}}] \tag{6.207}$$

The use of reduced integration for the shear stiffness matrix, as in the case of the thick rectangular element (Section 6.3), is not recommended [6.30].

Substituting (6.193) into (6.100) and assuming p_z to be constant gives the equivalent nodal force matrix

$$\{\mathbf{f}\}_e = \frac{p_z A}{3} \begin{bmatrix} 1 \\ 0 \\ 0 \\ 1 \\ 0 \\ 0 \\ 1 \\ 0 \\ 0 \end{bmatrix} \tag{6.208}$$

Therefore, one third of the total force is concentrated at each node.

Reference [6.31] presents a static solution for a simply supported square plate, of thickness to span ratio 0.1, which is subjected to a uniform load. It is demonstrated that the convergence is slow as the number of elements is increased. This

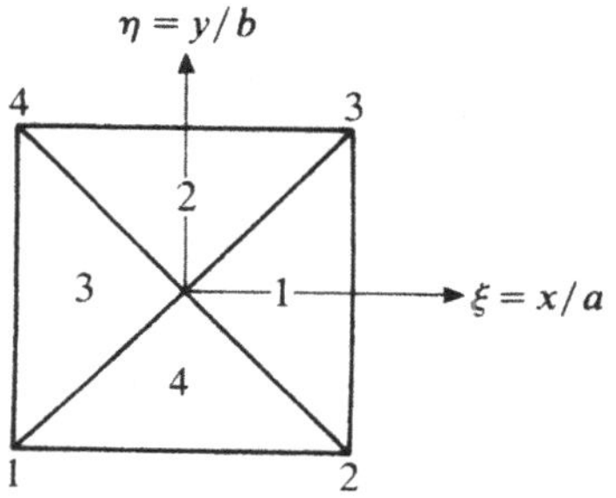

Figure 6.24. Geometry of a rectangular plate bending element.

reference also demonstrates that an improved rate of convergence can be obtained by representing the lateral displacement, w, by a quadratic function whilst still representing θ_X, θ_Y by linear functions. This rapid rate of convergence is also obtained when calculating the first two doubly symmetric frequencies of this same plate.

6.7 Other Plate Bending Elements

A conforming rectangular element has been developed in reference [6.32] using the smooth surface interpolation functions of reference [6.33]. Using the non-dimensional coordinates (ξ, η) defined in Figure 6.24, the displacement function is expressed in the form

$$w = \sum_{j=1}^{12} F_j(\xi, \eta)\alpha_j \tag{6.209}$$

The functions $F_j(\xi, \eta)$ are defined as follows

$$\begin{aligned}
&F_1 = 1, \quad && F_2 = \xi^2, \quad && F_3 = \eta^2, \\
&F_4 = \xi, && F_5 = \xi^3, && \\
&F_7 = \eta, && F_8 = \eta^3, && \\
&F_{10} = \xi\eta, && F_{11} = 3\xi^3\eta + 3\xi\eta^3 - \xi^3\eta^3 - 5\xi\eta &&
\end{aligned} \tag{6.210}$$

The remaining functions are defined by dividing the rectangle into four triangles by inserting the diagonals as shown in Figure 6.24.

$$F_6 = \begin{cases} \xi^2 - 2\xi + \eta^2 & \text{in region 1} \\ 2\xi\eta - 2\xi & \text{in region 2} \\ -\xi^2 - 2\xi - \eta^2 & \text{in region 3} \\ -2\xi\eta - 2\xi & \text{in region 4} \end{cases} \tag{6.211}$$

$$F_9 = \begin{cases} 2\xi\eta - 2\eta & \text{in region 1} \\ \eta^2 - 2\eta + \xi^2 & \text{in region 2} \\ -2\xi\eta - 2\eta & \text{in region 3} \\ -\eta^2 - 2\eta - \xi^2 & \text{in region 4} \end{cases} \tag{6.212}$$

$$F_{12} = \begin{cases} \frac{1}{4}(\xi^3\eta^3 - \xi^5\eta - 3\xi\eta^3 + 3\xi^3\eta) & \text{in regions 1, 3} \\ \frac{1}{4}(\xi\eta^5 - \xi^3\eta^3 - 3\xi\eta^3 + 3\xi^3\eta) & \text{in regions 2, 4} \end{cases} \tag{6.213}$$

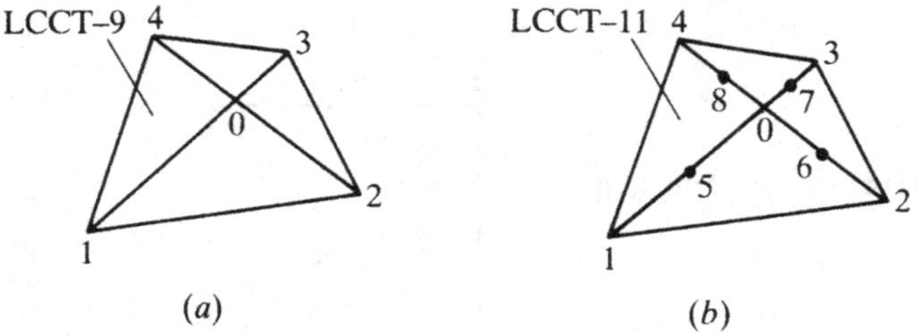

Figure 6.25. Geometry of element Q19: (*a*) inertia, (*b*) stiffness.

Functions F_1, F_4 and F_7 correspond to rigid body translation and rotation whilst F_2, F_3 and F_{10} correspond to constant curvature and twist. Also, functions F_7, F_8 and F_9 are the same as functions F_4, F_5 and F_6 rotated through 90° about the z-axis.

The coefficients α_j in (6.209) are expressed in terms of w, θ_x and θ_y at the four node points by the usual technique. The functions (6.210) to (6.213) define a cubic variation of w and a linear variation of the normal slope along the edges of the element. Therefore, the element is a conforming one. It will be referred to as the DP element.

Reference [6.28] presents a conforming quadrilateral (Q19) which has a linear variation of normal slope along each edge. The inertia matrix is obtained by assembling four LCCT-9 elements (Section 6.5.2). The element has, therefore, fifteen degrees of freedom (three at nodes 1, 2, 3, 4, 0 in Figure 6.25(*a*)). The stiffness matrix is obtained by assembling four LCCT-11 elements. An LCCT-11 element is derived from an LCCT-12 element (Section 6.5.2) by constraining the normal slope to vary linearly along one side. Therefore, the stiffness matrix has nineteen degrees of freedom, three at nodes 1, 2, 3, 4, 0 and one at nodes 5, 6, 7 and 8 in Figure 6.25(*b*). The degrees of freedom at nodes 5 to 8 are removed by static condensation (Section 4.6) leaving the matrix with fifteen degrees of freedom which is the same as the inertia matrix.

Reference [6.34] presents a conforming quadrilateral (CQ) with a quadratic variation of normal slope along each edge. The degrees of freedom are w, θ_x, θ_y at nodes 1, 2, 3, 4 and a normal slope at nodes 5, 6, 7, 8 (Figure 6.26). The element has, therefore, sixteen degrees of freedom. It is derived by dividing the quadrilateral into four sub-triangles and a cubic displacement function defined within each triangle. Reference [6.34] uses oblique axes for each triangle. The element has been rederived in reference [6.35] using area coordinates which are more convenient.

The displacement functions for each sub-triangle are expressed in terms of the seven degrees of freedom at the three external nodes and w, θ_x, θ_y at the internal

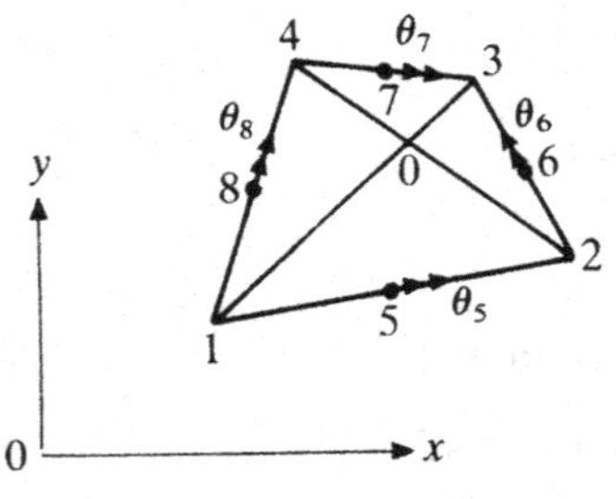

Figure 6.26. Geometry of element CQ.

Table 6.10. *Percentage difference between finite element and analytical frequencies of a simply supported square plate*

FEM grids ($\frac{1}{4}$ plate)	2 × 2		4 × 4	
Mode	(1, 1)	(1, 3)	(1, 1)	(1, 3)
DP	2.6	4.2	0.61	0.72
Q19	1.5	3.9	0.35	0.78
CQ	0.03	1.3	0.00	0.10

node 0. After assembling the four sub-triangles the three degrees of freedom at 0 are eliminated using continuity of normal slope between three pairs of sub-triangles.

The frequencies of the first two doubly symmetric modes of a simply supported square plate have been calculated using the above three elements. These frequencies are compared with the analytical frequencies in Table 6.10. Element Q19, which has three degrees of freedom more than element DP, produces only slightly better results. However, element CQ which has four degrees of freedom more than DP, gives much better accuracy.

Reference [6.36] presents a thick eight-node isoparametric element (RH) as shown in Figure 6.27. The displacements and geometry are represented by

$$w = \sum_{j=1}^{8} N_j w_j, \qquad \theta_x = \sum_{j=1}^{8} N_j \theta_{xj}, \qquad \theta_y = \sum_{j=1}^{8} N_j \theta_{y_j} \tag{6.214}$$

and

$$x = \sum_{j=1}^{8} N_j x_j, \qquad y = \sum_{j=1}^{8} N_j y_j \tag{6.215}$$

where the functions are identical to the ones defined for an eight-node membrane, that is (4.108) to (4.110). The inertia and stiffness matrices are evaluated using (3 × 3) and (2 × 2) arrays of integration points.

A simply supported square plate with a span/thickness ratio of 10 and $\nu = 0.3$ has been analysed using a (4 × 4) mesh of elements. The results obtained are compared with analytical values in Table 6.11. These show that greater accuracy is obtained than with an (8 × 8) mesh of four node elements (Table 6.5).

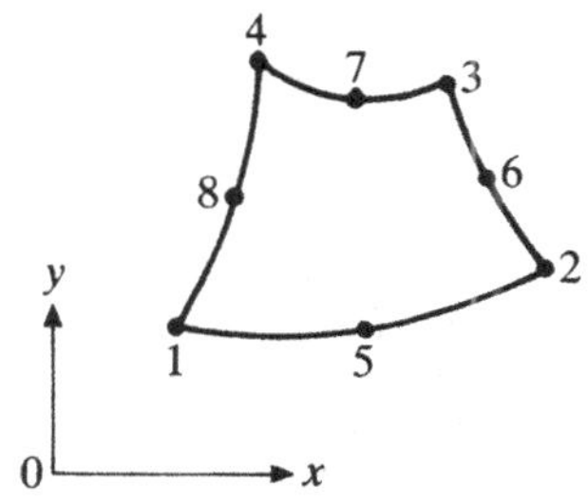

Figure 6.27. Geometry of an isoparametric element.

Table 6.11. *Comparison of predicted non-dimensional frequencies of a simply supported square plate. Element RH*

Mode	$\lambda^{1/2}$ FEM [6.36]	Analytical [6.21]	% Difference
(1, 1)	0.0931	0.0930	0.11
(2, 1)	0.2237	0.2218	0.86
(2, 2)	0.3384	0.3402	−0.53
(3, 1)	0.4312	0.4144	4.1
(3, 2)	0.5379	0.5197	3.5
(3, 3)	0.7661	0.6821	12.2

$\lambda = \rho h^2 \omega^2 / G$.

The main difference between thick and thin plates is that with thin plates the transverse shear strains γ_{xz}, γ_{yz} are negligible whilst for thick plates they are not. In developing a finite element model for thick plates it has been shown that continuity of w, θ_x, θ_y can easily be obtained by assuming independent functions for each. In the case of thin plates the vanishing of the transverse shear strains means that $\theta_x = \partial w/\partial y$ and $\theta_y = -\partial w/\partial x$. Thus w, θ_x, θ_y cannot be treated as independent and only a single function assumed for w. This leads to difficulties in ensuring that the normal slope is continuous between elements.

Another approach to developing a finite element for thin plates is the discrete Kirchhoff shear approach. This technique starts by assuming independent functions for w, θ_x, θ_y and then applies constraints to ensure that the transverse shear strains are zero at a discrete set of points.

Reference [6.37] presents several quadrilateral elements based upon this approach. For one of them (DKQ2), it is assumed that initially it has eight nodes, as shown in Figure 6.28. The displacement functions for w, θ_x, θ_y are taken to be (6.214). The degrees of freedom w, θ_n at nodes 5, 6, 7, 8 are eliminated by applying the constraints $\gamma_{xz} = 0$, $\gamma_{yz} = 0$ at a (2×2) array of Gauss points within the element. The degree of freedom θ_s at 5, 6, 7 and 8 is then eliminated by applying the constraint that θ_s varies linearly along each edge. In deriving the stiffness matrix the shear strain energy is ignored.

The accuracy of the frequency of the first two doubly-symmetric modes of a simply supported square plate obtained with the DKQ2 element and a consistent inertia matrix is indicated in Table 6.12. The meshes used for a quarter plate are (2×2) and (4×4). Reference [6.37] comments that the element is good for rectangular and parallelogram shapes but is not recommended for use as a general quadrilateral.

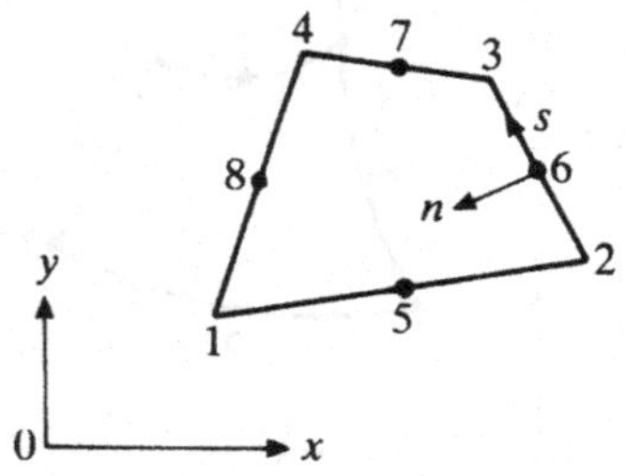

Figure 6.28. Geometry of the DKQ2 element.

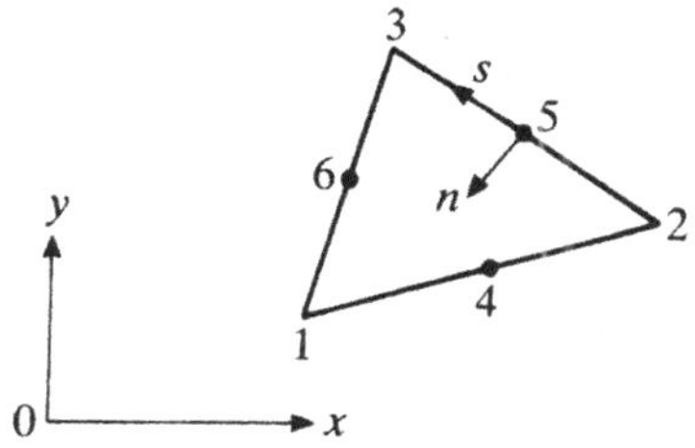

Figure 6.29. Geometry of the DKT element.

Reference [6.30] presents a triangular discrete Kirchhoff shear element (DKT). Initially, it is assumed that it has six nodes, as shown in Figure 6.29. The rotations θ_x, θ_y are assumed to vary quadratically over the element; therefore

$$\theta_x = \sum_{j=1}^{6} N_j \theta_{x_j}, \qquad \theta_y = \sum_{j=1}^{6} N_j \theta_{y_j} \tag{6.216}$$

where the functions N_j are given by (4.112). The degree of freedom θ_n at nodes 4, 5 and 6 is eliminated by requiring $\gamma_{sz} = 0$ at these points. In addition, the lateral displacement, w, is assumed to vary cubically along each edge. Thus for edge 2–3 (using (3.124))

$$\left(\frac{\partial w}{\partial s}\right)_6 = -\frac{3}{2l_{2-3}} w_2 - \frac{1}{4}\left(\frac{\partial w}{\partial s}\right)_2 + \frac{3}{2l_{2-3}} w_3 - \frac{1}{4}\left(\frac{\partial w}{\partial s}\right)_3 \tag{6.217}$$

The degree of freedom θ_s at nodes 4, 5 and 6 is eliminated by requiring this quantity to vary linearly along each side. Finally, the constraints $\gamma_{xz} = 0$ and $\gamma_{yz} = 0$ are applied at nodes 1, 2 and 3. The element is left with the degrees of freedom w, θ_x, θ_y at nodes 1, 2 and 3. The stiffness matrix is evaluated exactly using 3 integration points.

The results for a simply supported square plate using mesh sizes 2 and 4 (Figure 6.14) are given in Table 6.12. These results were obtained using a lumped diagonal mass matrix (more details of such representations are given in Chapter 12).

Reference [6.38] describes an improved discrete Kirchhoff triangular element which is referred to as the IMDKT element. Once again the rotations θ_x, θ_y are assumed to vary quadratically (6.212). θ_x, θ_y at nodes 4, 5, 6 (see Figure 6.29) are eliminated by assuming w to vary cubically and θ_s to vary linearly along each side. This gives

$$\begin{bmatrix} \bar{\theta}_x \\ \bar{\theta}_y \end{bmatrix} = \bar{\mathbf{N}}\mathbf{q} \tag{6.218}$$

Table 6.12. *Percentage difference between finite element and analytical frequencies of a simply supported square plate using discrete Kirchhoff shear elements*

Mesh size	2		3		4		5	
Mode	(1, 1)	(1, 3)	(1, 1)	(1, 3)	(1, 1)	(1, 3)	(1, 1)	(1, 3)
DKQ2	−1.1	+5.6			−0.3	−0.05		
DKT	−6.0	−16.0			−1.6	−3.2		
IMDKT	−0.45	+3.38	−0.3	+1.53			−0.15	+0.48

where

$$\mathbf{q}^{\mathrm{T}} = \lfloor w_1\ \theta_{x1}\ \theta_{y1} \quad w_2\ \theta_{x2}\ \theta_{y2} \quad w_3\ \theta_{x3}\ \theta_{y3} \rfloor. \tag{6.219}$$

Assuming both θ_x, θ_y to vary linearly gives

$$\begin{bmatrix} \theta_x^* \\ \theta_y^* \end{bmatrix} = \mathbf{N}^*\mathbf{q} \tag{6.220}$$

It has been found that the best results are obtained using a combination of the two element displacement expressions (6.218) and (6.220), namely

$$\begin{bmatrix} \theta_x \\ \theta_y \end{bmatrix} = \hat{\mathbf{N}}\mathbf{q} \tag{6.221}$$

where

$$\hat{\mathbf{N}} = \bar{\mathbf{N}} + \lambda(\bar{\mathbf{N}} - \mathbf{N}^*) \tag{6.222}$$

and λ is a correction factor. The element stiffness matrix is now evaluated using (6.221) ignoring the shear strain energy. The inertia matrix is calculated using a combined displacement function

$$w = \bar{w} + \alpha(\bar{w} - w^*) \tag{6.223}$$

where $\bar{w}$ is assumed to vary cubically (see [6.39]) and w^* is a linear function. The results for a simply supported square plate using mesh sizes 2, 3, 4, 5 (Figure 6.14) and $\lambda = 0.08$, $\alpha = -0.1$ are given in Table 6.12.

Another technique of avoiding shear locking is to use mixed interpolation of the transverse displacements, rotations and transverse shear strains. A family of elements which use this approach is presented in reference [6.39]. These are referred to as MITC*n* elements, where *n* indicates the number of nodes.

The displacements of the rectangular MITC4 element, shown in Figure 6.30, are assumed to be of the form (6.68), namely

$$w = \sum_{j=1}^{4} N_j w_j, \qquad \theta_x = \sum_{j=1}^{4} N_j \theta_{xj}, \qquad \theta_y = \sum_{j=1}^{4} N_j \theta_{yj} \tag{6.224}$$

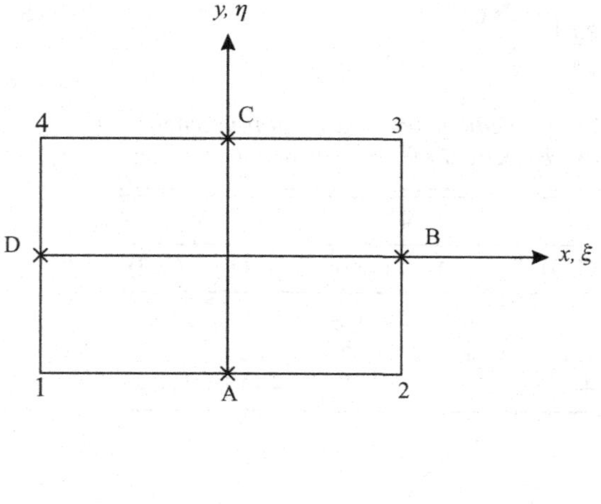

Figure 6.30. Geometry of the MITC4 element.

Table 6.13. *Percentage differences between MITC4 finite element and the results of reference [6.41] for a square plate of side L with mixed boundary conditions h/L = 0.1*

Mesh size whole plate	16	32	64
Mode			
1, 1	0.40	0.05	−0.005
2, 1	0.84	0.20	0.03
1, 2	1.76	0.44	0.10
2, 3	1.42	0.33	0.05

where the functions N_j are of bilinear form as given in equation (6.69). The shear strains are first calculated at the points A, B, C and D using equations (6.66) and (6.224). The shear strains at any other point within the element are then obtained from these by interpolation using

$$\begin{aligned}\gamma_{xz} &= \tfrac{1}{2}(1-\eta)\gamma_{xz}^{A} + \tfrac{1}{2}(1+\eta)\gamma_{xz}^{C}\\ \gamma_{yz} &= \tfrac{1}{2}(1-\xi)\gamma_{yz}^{D} + \tfrac{1}{2}(1+\xi)\gamma_{yz}^{B}\end{aligned} \tag{6.225}$$

The inertia and stiffness matrices are then formulated by substituting these expressions into (6.64) and (6.65) and using full Gaussian integration.

Reference [6.40] presents results for both a moderately thick plate ($h/L = 0.1$) and a thin plate ($h/L = 0.0001$) having simply supported, clamped, simply-supported, free boundaries. These are given in Tables 6.13 and 6.14. It can be seen that this element can be used for both moderately thick and thin plates.

In reference [6.43] the displacement function for a triangle is taken to be a complete quintic in x and y, a total of 21 terms. Three constraints are applied to ensure that the normal slope varies cubically over each edge of the triangle. The remaining 18 coefficients are expressed in terms of w, $\partial w/\partial x$, $\partial w/\partial y$, $\partial^2 w/\partial x^2$, $\partial^2 w/\partial x\partial y$ and $\partial^2 w/\partial y^2$ at the three vertices. The resulting element is a conforming one (NRCC).

The results for a simply supported square plate are compared with the analytical solution in Table 6.15. Although the results are extremely accurate, the predicted frequencies of the (1, 3) and (3, 1) modes are not equal as expected. Reference [6.44] presents an alternative formulation of this element and applies it to a number of polygonal plates.

Table 6.14. *Percentage differences between MITC4 finite element and the results of reference [6.42] for a square plate of side L with mixed boundary conditions h/L = 0.0001*

Mesh size whole plate	16	32	64
Mode			
1, 1	0.37	0.10	0.03
2, 1	0.79	0.23	0.09
1, 2	1.87	0.49	0.15
2, 2	1.51	0.40	0.13

Table 6.15. *Percentage difference between predicted and analytical frequencies for a simply supported square plate. NRCC element*

	Mesh size (Figure 6.14)		
Mode	1	2	3
(1, 1)	0.025	0.0005	0.0
(1, 2), (2, 1)	1.76	0.024	0.002
(2, 2)	2.44	0.036	0.003
(1, 3)	1.74	0.16	0.014
(3, 1)	2.09	0.26	0.023
(2, 3)	11.3	0.28	0.024

Table 6.16. *Percentage differences between predicted and analytical frequencies for a simply supported plate of aspect ratio 1.48:1. UM6 element*

	FEM grids ($\frac{1}{4}$ plate)		
Mode	1×1	2×2	3×3
1	0.014	0.0001	0.0
2	0.48	0.004	0.0005
3	0.61	0.002	0.0002
4	1.76	0.024	0.003
5	6.94	0.014	0.001

Table 6.17. *Natural frequencies (Hz) of a cantilever triangular plate of aspect ratio 1:1. Idealisation Figure 6.31(a)*

	Element			
Mode	DKT	LCCT-12	NRCC	Experimental [6.26]
1	34.5	36.6	36.6	34.5
2	117.6	140.7	139.3	136
3	155.6	196.0	194.0	190
4	271.1	344.0	333.4	325
5	331.2	475.8	454.2	441
6	403.7	629.7	590.5	578

$E = 206.7 \times 10^9$ N/m^2; $\nu = 0.3$; $\rho = 7890$ kg/m^3.

Table 6.18. *Natural frequencies (Hz) of a cantilever triangular plate of aspect ratio 1:1. Idealisation Figure 6.31(b)*

	Element		
Mode	LCCT-12/CQ	NRCC/UM6	Experimental [6.26]
1	36.6	36.5	34.5
2	139.7	139.0	136
3	194.7	193.6	190
4	339.7	332.6	325
5	463.3	452.9	441
6	607.6	588.7	578

E, ν, ρ as in Table 6.17.

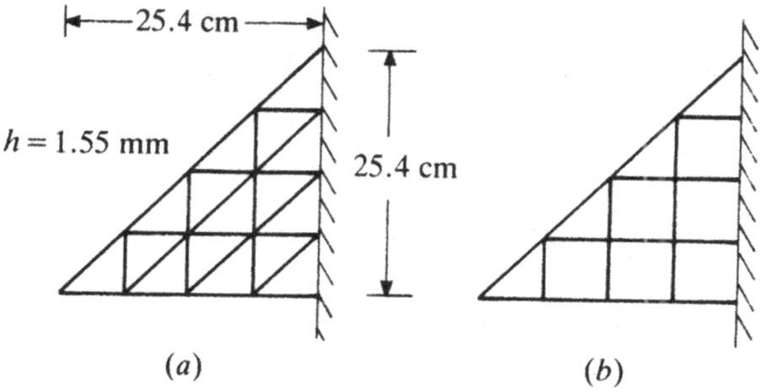

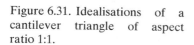

Figure 6.31. Idealisations of a cantilever triangle of aspect ratio 1:1.

Reference [6.45] presents a rectangular element, UM6, which is compatible with the NRCC element. It has the same degrees of freedom at the four corners and the same variation of displacement and normal slope over each edge. The results in Table 6.16 for a simply supported rectangular plate indicate its accuracy.

The natural frequencies of a cantilever triangle of aspect ratio 1:1 have been analysed using various triangular elements and the idealisation shown in Figure 6.31(*a*). The results are compared with measured frequencies in Table 6.17. The element DKT underestimates the frequencies whilst the other two overestimate them.

Figure 6.31(*b*) shows an idealisation which consists of a mixture of square and triangular elements. The frequencies predicted using two different pairs of compatible elements are compared with the measured ones in Table 6.18. These results show that they are more accurate than the corresponding ones in Table 6.17.

Problems

Note: Problems 6.2, 6.3, 6.5, 6.6 require the use of a digital computer.

6.1 Show that the stiffness matrix of the ACM element can also be expressed in the form $[\mathbf{C}]^T[\mathbf{H}][\mathbf{C}]$ where $[\mathbf{C}] = [\mathbf{A}]_e^{-1}[\mathbf{d}]$. The matrix $[\mathbf{d}]$ is a diagonal matrix whose elements are 1, b, a repeated four times. Find $[\mathbf{H}]$ for the anisotropic case.

6.2 Use a (4 × 4) mesh of ACM elements to predict the six lowest frequencies of a square plate of side 1 m and thickness 2 mm which has all four edges fully clamped. Take $E = 207 \times 10^9$ N/m^2, $\nu = 0.3$, $\rho = 7850$ kg/m^3. Compare these frequencies with the analytical frequencies [6.46] 17.800, 36.304, 36.304, 53.528, 65.085, 65.391 Hz.

6.3 Figure P6.3(*a*) shows a square cantilever plate which has a stepped thickness as indicated. Use the idealisation of ACM elements shown in Figure P6.3(*b*) to calculate the nine lowest frequencies. Take $E = 206.84 \times 10^9$ N/m^2, $\nu = 0.3$ and

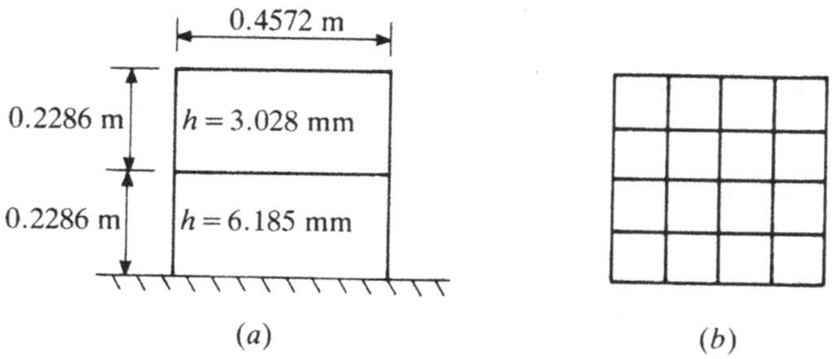

Figure P6.3. Geometry (*a*) and idealisation (*b*) of a stepped cantilever plate.

$\rho = 7853$ kg/m^3. Compare these frequencies with the experimental frequencies [6.47] 29.5, 56.6, 102.7, 129.8, 149.8, 264.4, 269.9, 308.5, 344.5 Hz.

6.4 If the thickness of the ACM element varies linearly, how many Gauss integration points are required to evaluate the inertia and stiffness matrices?

6.5 Figure P6.5(a) shows a rectangular cantilever with a wedge-shape cross-section. Use the idealisation of tapered ACM elements shown in Figure P6.5(b) to calculate the five lowest frequencies. Take $E = 206.84 \times 10^9$ N/m^2, $\nu = 0.3$, $\rho = 7861$ kg/m^3. Compare these frequencies with the experimental frequencies [6.48] 155.8, 668.4, 914.3, 1809.7, 2169.2 Hz.

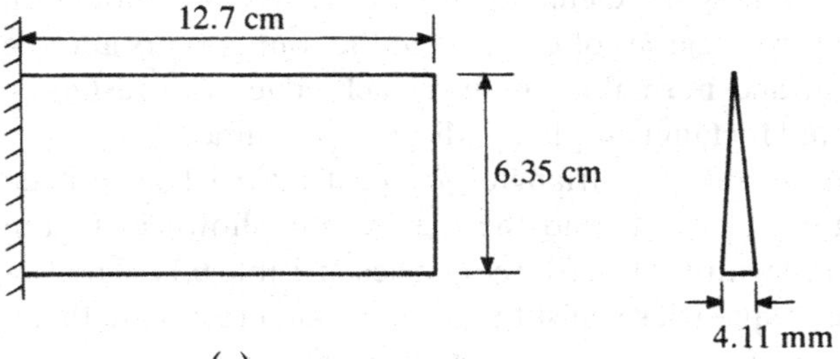

(a)

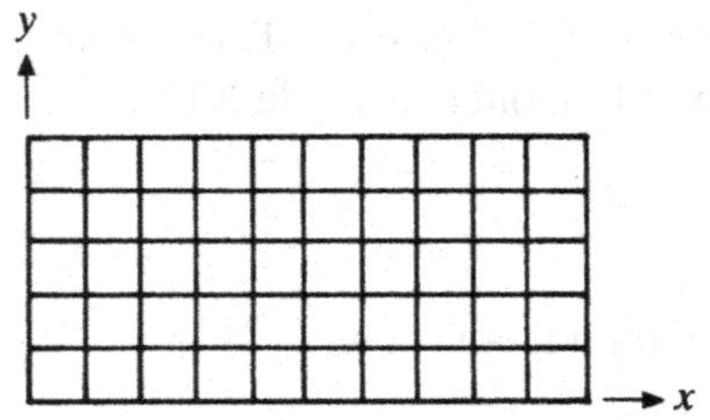

(b)

Figure P6.5. Geometry of a cantilever with a wedge shaped cross-section.

6.6 repeat Problem 6.5 using constant thickness ACM elements. Define the thickness of an element to be its average thickness.

6.7 Show that the displacement function defined by (6.17), (6.50) and (6.51) is a conforming one by evaluating w and θ_y on the side $\xi = 1$.

6.8 Show that the displacement function defined by (6.17), (6.52) and (6.53) gives continuity of w, θ_x, θ_y and w_{xy}.

6.9 Show that the displacement function defined by (6.17), (6.51), (6.61), (6.62) and (6.63) is a non-conforming one by evaluating w and θ_y on the side $\xi = 1$.

6.10 Show that the first six terms in (6.5) are present in the displacement function defined in Problem 6.9.

6.11 Show that the inertia matrix (6.75) gives the correct mass and moments of inertia.

6.12 Derive the stiffness matrix due to transverse shear strain energy for a rectangular thick plate bending element using equations (6.66), (6.224) and (6.225).

6.13 Derive expressions for the inertia, stiffness and equivalent nodal force matrices for a quadrilateral version of the HTK thick plate element, presented in Section 6.3.

6.14 Show that the inertia matrix (6.194) gives the correct mass and moment of inertia about an axis through nodes 1 and 2.

6.15 The displacement functions for a thick triangular element are assumed to be

$$w = \alpha_1 + \alpha_2 x + \alpha_3 y + \alpha_4 x^2 + \alpha_5 xy + \alpha_6 y^2$$

$$\theta_x = \beta_1 + \beta_2 x + \beta_3 y$$

$$\theta_y = \gamma_1 + \gamma_2 x + \gamma_3 y$$

Show that the assumption that Q_s is constant along each side, where s is along a side, leads to the relationships

$$\alpha_4 = -\tfrac{1}{2}\gamma_5 \qquad \alpha_5 = \tfrac{1}{2}(\beta_2 - \gamma_3), \qquad \alpha_6 = \tfrac{1}{2}\beta_3$$

7 Vibration of Stiffened Plates and Folded Plate Structures

Plates stiffened by beams can be found in many light-weight structures such as bridge-decks, building floors, ships' hulls and decks, and aircraft. The stiffeners may be either of solid cross-section or thin-walled, both open and closed and attached in either one or two directions, eccentric to the plate middle-surface. This means that the membrane and flexural motion of the plate become coupled. The solution of such structures involves combining the framework, membrane and plate bending elements described in previous chapters.

Many light-weight structures consist of plates which meet at angles to one another. This also has the effect of coupling the membrane and flexural motion of the plates.

7.1 Stiffened Plates I

The membrane displacements of a plate are usually much smaller than the bending displacements. Therefore, as a first approximation, they may be neglected. In modelling stiffened plates using finite elements, it is usual to assume that a stiffener is attached to the plate along a single line. For simplicity, the stiffener cross-section will be assumed to be symmetric about the z-axis (see Figure 7.1). The centroid c of the cross-section is a distance e from the plate middle surface.

Bending of the plate will induce bending of the beam stiffener in the xz-plane, which in turn causes extension of the centroidal axis, and also torsion of the beam. The energy expressions in terms of centroidal displacements are, therefore (Section 3.7),

$$
\begin{aligned}
T_e &= \frac{1}{2}\int_{-a}^{+a} \rho A\left(\dot{u}_c^2 + \dot{v}_c^2 + \dot{w}_c^2\right)\,\mathrm{d}x + \frac{1}{2}\int_{-a}^{+a} \rho I_x \dot{\theta}_x^2\,\mathrm{d}x \\
U_e &= \frac{1}{2}\int_{-a}^{+a} EA\left(\frac{\partial u_c}{\partial x}\right)^2 \mathrm{d}x + \frac{1}{2}\int_{-a}^{+a} EI_y\left(\frac{\partial^2 w_c}{\partial x^2}\right)^2 \mathrm{d}x \\
&\quad + \frac{1}{2}\int_{-a}^{+a} GJ\left(\frac{\partial \theta_x}{\partial x}\right)^2 \mathrm{d}x
\end{aligned}
\tag{7.1}
$$

It has been assumed that resistance to lateral motion is negligible.

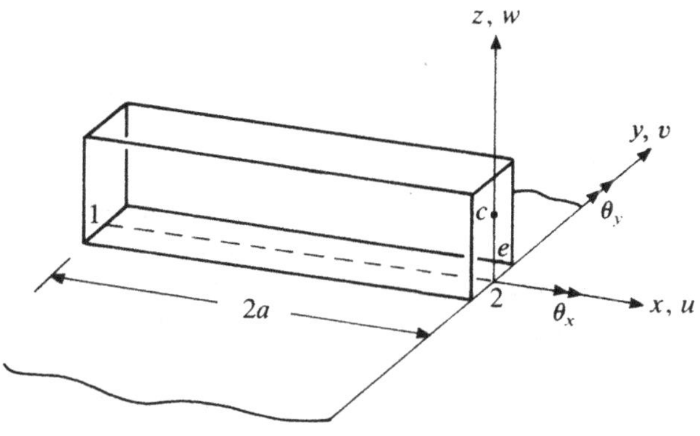

Figure 7.1. Geometry of a stiffened plate.

The centroidal displacements can be expressed in terms of the displacements at the attachment line as follows

$$\begin{aligned} u_c &= e\theta_y = -e\partial w/\partial x \\ v_c &= -e\theta_x \\ w_c &= w \end{aligned} \tag{7.2}$$

Substituting (7.2) into (7.1) gives

$$\begin{aligned} T_e = \frac{1}{2}\int_{-a}^{+a} \rho A e^2 \left(\frac{\partial \dot{w}}{\partial x}\right)^2 \mathrm{d}x + \frac{1}{2}\int_{-a}^{+a} \rho A \dot{w}^2 \,\mathrm{d}x \\ + \frac{1}{2}\int_{-a}^{+a} \rho (I_x + Ae^2)\dot{\theta}_x^2 \,\mathrm{d}x \end{aligned} \tag{7.3}$$

$$U_e = \frac{1}{2}\int_{-a}^{+a} E(I_y + Ae^2)\left(\frac{\partial^2 w}{\partial x^2}\right)^2 \mathrm{d}x + \frac{1}{2}\int_{-a}^{+a} GJ\left(\frac{\partial \theta_x}{\partial x}\right)^2 \mathrm{d}x \tag{7.4}$$

The displacements of the beam should be compatible with the plate displacements along the attachment line. Many of the thin plate bending elements presented in Chapter 6 have a cubic variation of normal displacement along an edge. Therefore, the variation of displacement in the z-direction along the attachment line should also be cubic. The displacement function in bending is

$$w = \lfloor \mathbf{N}_w(\xi) \rfloor \{\mathbf{w}\}_e \tag{7.5}$$

where

$$\{\mathbf{w}\}_e^{\mathrm{T}} = \lfloor w_1 \quad \theta_{y1} \quad w_2 \quad \theta_{y2} \rfloor \tag{7.6}$$

and

$$\lfloor N_w(\xi) \rfloor = \lfloor N_1(\xi) \quad -aN_2(\xi) \quad N_3(\xi) \quad -aN_4(\xi) \rfloor \tag{7.7}$$

The functions $N_1(\xi)$ to $N_4(\xi)$ are defined by equations (3.126).

Elements HCT, LCCT-9, DP and Q19 presented in Chapter 6 have a linear variation of normal slope along an edge. In these cases the rotation of the stiffener about the attachment line should be taken to be

$$\theta_x = \lfloor \mathbf{N}_x(\xi) \rfloor \{\boldsymbol{\theta}_x\}_e \tag{7.8}$$

where

$$\{\boldsymbol{\theta}_x\}_e^{\mathrm{T}} = \lfloor \theta_{x1} \quad \theta_{x2} \rfloor \tag{7.9}$$

and

$$\lfloor \mathbf{N}_x(\xi) \rfloor = \lfloor N_1(\xi) \quad N_2(\xi) \rfloor \tag{7.10}$$

The functions $N_1(\xi)$ and $N_2(\xi)$ are defined by equation (3.51). Equation (7.8) could also be used in conjunction with the various non-conforming elements presented in Chapter 6.

The second and third integrals in (7.3) are similar to the integrals in expressions (3.127) and (3.92). Therefore, when (7.5) and (7.8) are substituted into (7.3) the second and third integrals can be deduced from expressions (3.132) and (3.101), since the same displacement functions have been used. This gives

$$\frac{1}{2}\int_{-a}^{+a} \rho A \dot{w}^2 \,\mathrm{d}x = \frac{1}{2}\{\dot{\mathbf{w}}\}_e^{\mathrm{T}} \frac{\rho A a}{105} \begin{bmatrix} 78 & & & \\ -22a & 8a^2 & & \text{Sym} \\ 27 & -13a & 78 & \\ 13a & -6a & 22a & 8a^2 \end{bmatrix} \{\dot{\mathbf{w}}\}_e \tag{7.11}$$

and

$$\frac{1}{2}\int_{-a}^{+a} \rho (I_x + Ae^2)\, \dot{\theta}_x^2 \,\mathrm{d}x = \frac{1}{2}\{\dot{\boldsymbol{\theta}}\}_e^{\mathrm{T}} \rho A \left(r_x^2 + e^2\right) \frac{a}{3} \begin{bmatrix} 2 & 1 \\ 1 & 2 \end{bmatrix} \{\dot{\boldsymbol{\theta}}\}_e \tag{7.12}$$

where $r_x^2 = I_x / A$.

Substituting (7.5) into the first integral of (7.3) and evaluating gives

$$\frac{1}{2}\int_{-a}^{+a} \rho A e^2 \left(\frac{\partial \dot{w}}{\partial x}\right)^2 \mathrm{d}x$$

$$= \frac{1}{2}\{\dot{\mathbf{w}}\}_e^{\mathrm{T}} \frac{\rho A e^2 a}{30} \begin{bmatrix} 18 & & & \\ -3a & 8a^2 & & \text{Sym} \\ -18 & 3a & 18 & \\ -3a & -2a^2 & 3a & 8a^2 \end{bmatrix} \{\dot{\mathbf{w}}\}_e \tag{7.13}$$

Combining (7.11) to (7.13) gives

$$T_e = \tfrac{1}{2}\{\dot{\mathbf{u}}\}_e^{\mathrm{T}} [\mathbf{m}]_e \{\dot{\mathbf{u}}\}_e \tag{7.14}$$

where

$$\{\mathbf{u}\}_e^{\mathrm{T}} = \lfloor w_1 \quad \theta_{x1} \quad \theta_{y1} \quad w_2 \quad \theta_{x2} \quad \theta_{y2} \rfloor \tag{7.15}$$

and

$$[\mathbf{m}]_e = \begin{bmatrix} \mathbf{m}_{11} & \mathbf{m}_{12} \\ \mathbf{m}_{12}^{\mathrm{T}} & \mathbf{m}_{22} \end{bmatrix} \tag{7.16}$$

where

$$\mathbf{m}_{11} = \frac{\rho A a}{210}\begin{bmatrix} 282 & 0 & -63a \\ 0 & 140e_x^2 & 0 \\ -63a & 0 & 72a^2 \end{bmatrix} \tag{7.17}$$

$$\mathbf{m}_{12} = \frac{\rho A a}{210}\begin{bmatrix} -72 & 0 & 5a \\ 0 & 70e_x^2 & 0 \\ -5a & 0 & -26a^2 \end{bmatrix} \tag{7.18}$$

$$\mathbf{m}_{22} = \frac{\rho A a}{210}\begin{bmatrix} 282 & 0 & 63a \\ 0 & 140e_x^2 & 0 \\ 63a & 0 & 72a^2 \end{bmatrix} \tag{7.19}$$

where $e_x^2 = r_x^2 + e^2$.

The two integrals in (7.4) are similar to the integrals in expressions (3.128) and (3.93). The evaluation of these integrals, after substituting for w and θ_x from (7.5) and (7.8), can be deduced from expressions (3.135) and (3.102) since the same displacement functions have been used. This gives

$$\frac{1}{2}\int_{+a}^{+a} E(I_y + Ae^2)\left(\frac{\partial^2 w}{\partial x^2}\right)^2 \mathrm{d}x$$

$$= \frac{1}{2}\{\mathbf{w}\}_e^{\mathrm{T}} EA(r_y^2 + e^2)\frac{1}{2a^3}\begin{bmatrix} 3 & & & \text{Sym} \\ -3a & 4a^2 & & \\ -3 & 3a & 3 & \\ -3a & 2a^2 & 3a & 4a^2 \end{bmatrix}\{\mathbf{w}\}_e \tag{7.20}$$

where $r_y^2 = I_y/A$ and

$$\frac{1}{2}\int_{-a}^{+a} GJ\left(\frac{\partial \theta_x}{\partial x}\right)^2 \mathrm{d}x = \frac{1}{2}\{\boldsymbol{\theta}_x\}_e^{\mathrm{T}}\frac{GJ}{2a}\begin{bmatrix} 1 & -1 \\ -1 & 1 \end{bmatrix}\{\boldsymbol{\theta}_x\}_e \tag{7.21}$$

Combining (7.20) and (7.21) gives

$$U_e = \tfrac{1}{2}\{\mathbf{u}\}_e^{\mathrm{T}}[\mathbf{k}]_e\{\mathbf{u}\}_e \tag{7.22}$$

where

$$[\mathbf{k}]_e = \begin{bmatrix} \mathbf{k}_{11} & \mathbf{k}_{12} \\ \mathbf{k}_{12}^{\mathrm{T}} & \mathbf{k}_{22} \end{bmatrix} \tag{7.23}$$

with

$$\mathbf{k}_{11} = \frac{EA}{4a^3}\begin{bmatrix} 6e_y^2 & 0 & -6ae_y^2 \\ 0 & a^2r_J^2(1+v) & 0 \\ -6ae_y^2 & 0 & 8a^2e_y^2 \end{bmatrix} \tag{7.24}$$

$$\mathbf{k}_{12} = \frac{EA}{4a^3}\begin{bmatrix} -6e_y^2 & 0 & -6ae_y^2 \\ 0 & a^2r_J^2/(1+v) & 0 \\ 6ae_y^2 & 0 & 4a^2e_y^2 \end{bmatrix} \tag{7.25}$$

and

$$\mathbf{k}_{22} = \frac{EA}{4a^3}\begin{bmatrix} 6e_y^2 & 0 & 6ae_y^2 \\ 0 & a^2r_J^2/(1+v) & 0 \\ 6ae_y^2 & 0 & 8a^2e_y^2 \end{bmatrix} \tag{7.26}$$

where $r_J^2 = J/A$ and $e_y^2 = r_y^2 + e^2$.

The nodal degrees of freedom of this stiffener element w, θ_x, θ_y are the same as the nodal degrees of freedom for the plate bending elements ACM, WB, T, HCT, LCCT-9, DP, Q19 and BCIZ1 presented in Chapter 6. The assembly of these plate and stiffener elements is, therefore, straightforward.

The plate elements LCCT-12 and CQ presented in Chapter 6 have a cubic variation for w and a quadratic variation of normal slope along an edge. There are three degrees of freedom w, θ_x, θ_y at each end of the side and one degree of freedom, the normal rotation, at the mid-point. A compatible beam stiffener element can, therefore, be derived using (7.5) for w and

$$\theta_x = \lfloor N_1(\xi) \quad N_2(\xi) \quad N_3(\xi) \rfloor \begin{bmatrix} \theta_{x1} \\ \theta_{x2} \\ \theta_{x3} \end{bmatrix} \tag{7.27}$$

for θ_x, with $N_1(\xi)$ to $N_3(\xi)$ being defined by (3.215). Node 3 is half way between nodes 1 and 2 in Figure 7.1.

The plate element CR, Section 6.2, has a cubic variation for w and also a cubic variation of normal slope. There are four degrees of freedom at the two ends of a side, namely w, θ_x, θ_y and w_{xy}. The normal displacement w should be represented by a cubic function expressed in terms of w and θ_y at nodes 1 and 2 and the rotation θ_x by a cubic function expressed in terms of θ_x and w_{xy} at the two nodes.

The plate elements NRCC and UM6 presented in Chapter 6 have a quintic variation for w and a cubic variation of normal slope over each edge. There are six degrees of freedom at the two ends of a side, namely w, $\partial w/\partial x, \partial w/\partial y$, $\partial^2 w/\partial x^2, \partial^2 w/\partial x \partial y$ and $\partial^2 w/\partial y^2$. A compatible beam stiffener element can be obtained by taking a quintic variation for w expressed in terms of w, $\partial w/\partial x$ and $\partial^2 w/\partial x^2$ and a cubic variation for θ_x expressed in terms of $\partial w/\partial y$ and $\partial^2 w/\partial x \partial y$ [7.1].

Reference [7.2] extends this type of analysis to plates stiffened by thin-walled open section beams. In this case the strain energy due to warping of the cross-sections of the stiffeners has to be taken into account. Both w and θ_x are represented by cubic functions. The coefficients of the cubic function for θ_x are expressed in terms of the values of θ_x and $\partial^2 \theta_x/\partial x^2$ at the two ends of the element.

7.2 Stiffened Plates II

In order to include the membrane displacements of the plate in the analysis, it is first necessary to derive a plate element which includes both membrane and bending deformations. Such elements are referred to as facet shell elements. The procedure will be illustrated for the rectangular element shown in Figure 7.2.

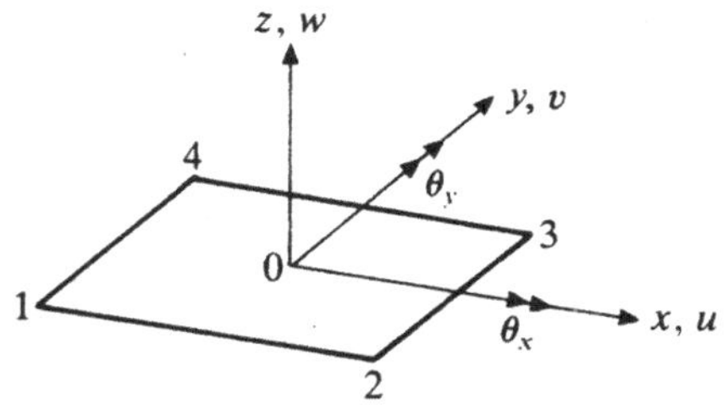

Figure 7.2. Geometry of a rectangular facet shell element.

The kinetic energy of a rectangular membrane element with four nodes is of the form (see Chapter 4)

$$T_{\mathrm{m}} = \tfrac{1}{2}\{\dot{\mathbf{u}}\}^{\mathrm{T}}[\mathbf{m}]^{\mathrm{m}}\{\dot{\mathbf{u}}\} \tag{7.28}$$

where the subscript e has been omitted for convenience and a superscript m has been introduced to denote membrane motion. Also

$$\{\mathbf{u}\}^{\mathrm{T}} = \lfloor u_1 \quad v_1 \quad u_2 \quad v_2 \quad u_3 \quad v_3 \quad u_4 \quad v_4 \rfloor \tag{7.29}$$

and

$$[\mathbf{m}]^{\mathrm{m}} = \begin{bmatrix} \mathbf{m}^{\mathrm{m}}_{11} & & & \mathrm{Sym} \\ \mathbf{m}^{\mathrm{m}}_{21} & \mathbf{m}^{\mathrm{m}}_{22} & & \\ \mathbf{m}^{\mathrm{m}}_{31} & \mathbf{m}^{\mathrm{m}}_{32} & \mathbf{m}^{\mathrm{m}}_{33} & \\ \mathbf{m}^{\mathrm{m}}_{41} & \mathbf{m}^{\mathrm{m}}_{42} & \mathbf{m}^{\mathrm{m}}_{43} & \mathbf{m}^{\mathrm{m}}_{44} \end{bmatrix} \tag{7.30}$$

Each submatrix $\mathbf{m}^{\mathrm{m}}_{ij}$ is of order (2×2).

The kinetic energy of a rectangular plate-bending element with four nodes is of the form (see Chapter 6)

$$T_{\mathrm{b}} = \tfrac{1}{2}\{\dot{\mathbf{w}}\}^{\mathrm{T}}[\mathbf{m}]^{\mathrm{b}}\{\dot{\mathbf{w}}\} \tag{7.31}$$

where

$$\{\mathbf{w}\}^{\mathrm{T}} = \lfloor w_1 \quad \theta_{x1} \quad \theta_{y1} \quad w_2 \quad \theta_{x2} \quad \theta_{y2} \quad w_3 \quad \theta_{x3} \quad \theta_{y3} \quad w_4 \quad \theta_{x4} \quad \theta_{y4} \rfloor \tag{7.32}$$

and

$$[\mathbf{m}]^{\mathrm{b}} = \begin{bmatrix} \mathbf{m}^{\mathrm{b}}_{11} & & & \mathrm{Sym} \\ \mathbf{m}^{\mathrm{b}}_{21} & \mathbf{m}^{\mathrm{b}}_{22} & & \\ \mathbf{m}^{\mathrm{b}}_{31} & \mathbf{m}^{\mathrm{b}}_{32} & \mathbf{m}^{\mathrm{b}}_{33} & \\ \mathbf{m}^{\mathrm{b}}_{41} & \mathbf{m}^{\mathrm{b}}_{42} & \mathbf{m}^{\mathrm{b}}_{43} & \mathbf{m}^{\mathrm{b}}_{44} \end{bmatrix} \tag{7.33}$$

Each sub-matrix is of order (3×3).

Combining (7.28) and (7.31) gives

$$T_{\mathrm{s}} = \tfrac{1}{2}\{\dot{\mathbf{u}}\}^{\mathrm{T}}_{\mathrm{s}}[\mathbf{m}]^{\mathrm{s}}\{\dot{\mathbf{u}}\}_{\mathrm{s}} \tag{7.34}$$

where

$$\{\mathbf{u}\}^{\mathrm{T}}_{\mathrm{s}} = \lfloor u_1 \quad v_1 \quad w_1 \quad \theta_{x1} \quad \theta_{y1} \quad \cdots \quad u_4 \quad v_4 \quad w_4 \quad \theta_{x4} \quad \theta_{y4} \rfloor \tag{7.35}$$

and

$$[\mathbf{m}]^{\mathrm{s}} = \begin{bmatrix} \mathbf{m}_{11}^{\mathrm{s}} & & & \text{Sym} \\ \mathbf{m}_{21}^{\mathrm{s}} & \mathbf{m}_{22}^{\mathrm{s}} & & \\ \mathbf{m}_{31}^{\mathrm{s}} & \mathbf{m}_{32}^{\mathrm{s}} & \mathbf{m}_{33}^{\mathrm{s}} & \\ \mathbf{m}_{41}^{\mathrm{s}} & \mathbf{m}_{42}^{\mathrm{s}} & \mathbf{m}_{43}^{\mathrm{s}} & \mathbf{m}_{44}^{\mathrm{s}} \end{bmatrix} \tag{7.36}$$

In this case each sub-matrix is of order (5 × 5) and is of the form

$$\mathbf{m}_{ij}^{\mathrm{s}} = \begin{bmatrix} \mathbf{m}_{ij}^{\mathrm{m}} & 0 \\ 0 & \mathbf{m}_{ij}^{\mathrm{b}} \end{bmatrix} \tag{7.37}$$

Similarly, the strain energy of a facet shell element is of the form

$$U_{\mathrm{s}} = \tfrac{1}{2}\{\mathbf{u}\}_{\mathrm{s}}^{\mathrm{T}}[\mathbf{k}]^{\mathrm{s}}\{\mathbf{u}\}_{\mathrm{s}} \tag{7.38}$$

where

$$[\mathbf{k}]^{\mathrm{s}} = \begin{bmatrix} \mathbf{k}_{11}^{\mathrm{s}} & & & \text{Sym} \\ \mathbf{k}_{21}^{\mathrm{s}} & \mathbf{k}_{22}^{\mathrm{s}} & & \\ \mathbf{k}_{31}^{\mathrm{s}} & \mathbf{k}_{32}^{\mathrm{s}} & \mathbf{k}_{33}^{\mathrm{s}} & \\ \mathbf{k}_{41}^{\mathrm{s}} & \mathbf{k}_{42}^{\mathrm{s}} & \mathbf{k}_{43}^{\mathrm{s}} & \mathbf{k}_{44}^{\mathrm{s}} \end{bmatrix} \tag{7.39}$$

and

$$\mathbf{k}_{ij}^{\mathrm{s}} = \begin{bmatrix} \mathbf{k}_{ij}^{\mathrm{m}} & 0 \\ 0 & \mathbf{k}_{ij}^{\mathrm{b}} \end{bmatrix} \tag{7.40}$$

The sub-matrices $\mathbf{k}_{ij}^{\mathrm{m}}$ and $\mathbf{k}_{ij}^{\mathrm{b}}$, of order (2 × 2) and (3 × 3) respectively, are the appropriate sub-matrices of the membrane and bending stiffness matrices.

If it is assumed that the resistance of the beam stiffener to lateral motion is negligibly small compared to that of the plate, then the energy expressions of the stiffener in terms of centroidal displacements (7.1) can again be used. In this case, the relationships between the centroidal displacements and displacements at the attachment line are

$$\begin{aligned} u_c &= u + e\theta_y = u - e\partial w/\partial x \\ v_c &= v - e\theta_x \\ w_c &= w \end{aligned} \tag{7.41}$$

Substituting (7.41) into (7.1) gives

$$\begin{aligned} T_e = {}& \frac{1}{2}\int_{-a}^{+a} \rho A(\dot{u}^2 + \dot{v}^2 + \dot{w}^2)\,\mathrm{d}x + \frac{1}{2}\int_{-a}^{+a} \rho A e^2 \left(\frac{\partial \dot{w}}{\partial x}\right)^2 \mathrm{d}x \\ & + \frac{1}{2}\int_{-a}^{+a} \rho(I_x + Ae^2)\,\dot{\theta}_x^2\,\mathrm{d}x - \int_{-a}^{+a} \rho A e \dot{u}\frac{\partial \dot{w}}{\partial x}\mathrm{d}x - \int_{-a}^{+a} \rho A e \dot{v}\dot{\theta}_x\,\mathrm{d}x \end{aligned} \tag{7.42}$$

$$\begin{aligned} U_e = {}& \frac{1}{2}\int_{-a}^{+a} EA\left(\frac{\partial u}{\partial x}\right)^2 \mathrm{d}x + \frac{1}{2}\int_{-a}^{+a} E(I_y + Ae^2)\left(\frac{\partial^2 w}{\partial x^2}\right)^2 \mathrm{d}x \\ & + \frac{1}{2}\int_{-a}^{+a} GJ\left(\frac{\partial \theta_x}{\partial x}\right)^2 \mathrm{d}x - \int_{-a}^{+a} EAe\frac{\partial u}{\partial x}\frac{\partial^2 w}{\partial x^2}\,\mathrm{d}x \end{aligned} \tag{7.43}$$

The displacements w and θ_x will again be assumed to be given by (7.5) and (7.8) respectively. Many of the membrane elements presented in Chapter 4 have a linear variation of u and v along each side. In these cases the following expressions should be used for the stiffener

$$u = \lfloor \mathbf{N}(\xi) \rfloor \{\mathbf{u}\}_e \tag{7.44}$$

and

$$v = \lfloor \mathbf{N}(\xi) \rfloor \{\mathbf{v}\}_e \tag{7.45}$$

where

$$\{\mathbf{u}\}_e^{\mathrm{T}} = \lfloor u_1 \quad u_2 \rfloor \tag{7.46}$$

$$\{\mathbf{v}\}_e^{\mathrm{T}} = \lfloor v_1 \quad v_2 \rfloor \tag{7.47}$$

and

$$\lfloor \mathbf{N}(\xi) \rfloor = \lfloor N_1(\xi) \quad N_2(\xi) \rfloor \tag{7.48}$$

The functions $\mathrm{N}_1(\xi)$ and $N_2(\xi)$ are defined by equation (3.51).

Substituting (7.5), (7.8), (7.44) and (7.45) into (7.42) gives

$$T_e = \tfrac{1}{2}\{\dot{\mathbf{u}}\}_{\mathrm{b}}^{\mathrm{T}}[\mathbf{m}]^{\mathrm{b}}\{\dot{\mathbf{u}}\}_{\mathrm{b}} \tag{7.49}$$

where

$$\{\mathbf{u}\}_{\mathrm{b}}^{\mathrm{T}} = \lfloor u_1 \quad v_1 \quad w_1 \quad \theta_{x1} \quad \theta_{y1}, \quad u_2 \quad v_2 \quad w_2 \quad \theta_{x2} \quad \theta_{y2} \rfloor \tag{7.50}$$

and

$$[\mathbf{m}]^{\mathrm{b}} = \begin{bmatrix} \mathbf{m}_{11} & \mathbf{m}_{12} \\ \mathbf{m}_{12}^{\mathrm{T}} & \mathbf{m}_{22} \end{bmatrix} \tag{7.51}$$

where

$$\mathbf{m}_{11} = \frac{\rho A a}{210}\begin{bmatrix} 140 & 0 & -105e/a & 0 & -35e \\ 0 & 140 & 0 & 140e & 0 \\ -105e/a & 0 & 282 & 0 & -63a \\ 0 & 140e & 0 & 140\left(r_x^2 + e^2\right) & 0 \\ -35e & 0 & -63a & 0 & 72a^2 \end{bmatrix} \tag{7.52}$$

$$\mathbf{m}_{12} = \frac{\rho A a}{210}\begin{bmatrix} 70 & 0 & 105e/a & 0 & 35e \\ 0 & 70 & 0 & 70e & 0 \\ -105e/a & 0 & -72 & 0 & 5a \\ 0 & 70e & 0 & 70\left(r_x^2 + e^2\right) & 0 \\ 35e & 0 & -5a & 0 & -26a^2 \end{bmatrix} \tag{7.53}$$

$$\mathbf{m}_{22} = \frac{\rho A a}{210}\begin{bmatrix} 140 & 0 & 105e/a & 0 & -35e \\ 0 & 140 & 0 & 140e & 0 \\ 105e/a & 0 & 282 & 0 & 63a \\ 0 & 140e & 0 & 140\left(r_x^2 + e^2\right) & 0 \\ -35e & 0 & 63a & 0 & 72a^2 \end{bmatrix} \tag{7.54}$$

Similarly, substituting (7.5), (7.8), (7.44) and (7.45) into (7.43) gives

$$U_e = \tfrac{1}{2}\{\mathbf{u}\}_{\mathrm{b}}^{\mathrm{T}}[\mathbf{k}]^{\mathrm{b}}\{\mathbf{u}\}_{\mathrm{b}} \tag{7.55}$$

where

$$[\mathbf{k}]^{\mathrm{b}} = \begin{bmatrix} \mathbf{k}_{11} & \mathbf{k}_{12} \\ \mathbf{k}_{12}^{\mathrm{T}} & \mathbf{k}_{22} \end{bmatrix} \tag{7.56}$$

and

$$\mathbf{k}_{11} = \frac{EA}{4a^3}\begin{bmatrix} 2a^2 & 0 & 0 & 0 & -2ea^2 \\ 0 & 0 & 0 & 0 & 0 \\ 0 & 0 & 6e_y^2 & 0 & -6ae_y^2 \\ 0 & 0 & 0 & a^2r_J^2/(1+v) & 0 \\ -2ea^2 & 0 & -6ae_y^2 & 0 & 8a^2e_y^2 \end{bmatrix} \tag{7.57}$$

$$\mathbf{k}_{12} = \frac{EA}{4a^3}\begin{bmatrix} -2a^2 & 0 & 0 & 0 & 2ea^2 \\ 0 & 0 & 0 & 0 & 0 \\ 0 & 0 & -6e_y^2 & 0 & -6ae_y^2 \\ 0 & 0 & 0 & -a^2r_J^2/(1+v) & 0 \\ 2ea^2 & 0 & 6ae_y^2 & 0 & 4a^2e_y^2 \end{bmatrix} \tag{7.58}$$

$$\mathbf{k}_{22} = \frac{EA}{4a^3}\begin{bmatrix} 2a^2 & 0 & 0 & 0 & -2ea^2 \\ 0 & 0 & 0 & 0 & 0 \\ 0 & 0 & 6e_y^2 & 0 & 6ae_y^2 \\ 0 & 0 & 0 & a^2r_J^2/(1+v) & 0 \\ -2ea^2 & 0 & 6ae_y^2 & 0 & 8a^2e_y^2 \end{bmatrix} \tag{7.59}$$

Reference [7.3] uses the plate bending element CQ, which is a quadrilateral having three degrees of freedom at the four vertices and a single rotation at the mid-points of the sides (see Section 6.7). It is combined with a quadrilateral membrane element which is obtained by assembling together four six-node triangles (see Section 4.6) with two degrees of freedom at each node. The degrees of freedom at five internal node points are removed using static condensation leaving the same eight node points as the bending element CQ. A compatible beam element is constructed by assuming a cubic variation for w and a quadratic variation for u and θ_x. It has three nodes with u, w, θ_x, θ_y as degrees of freedom at the two end points and u, θ_x at its mid-point. Only static analysis is performed.

Reference [7.4] uses the plate bending element NRCC, which is a triangle having six degrees of freedom at the three vertices (see Section 6.7). It is combined with a triangular membrane element also having six degrees of freedom at the three vertices, namely, u, $\partial u/\partial x$, $\partial u/\partial y$, v, $\partial v/\partial x$, $\partial v/\partial y$. A compatible beam element is

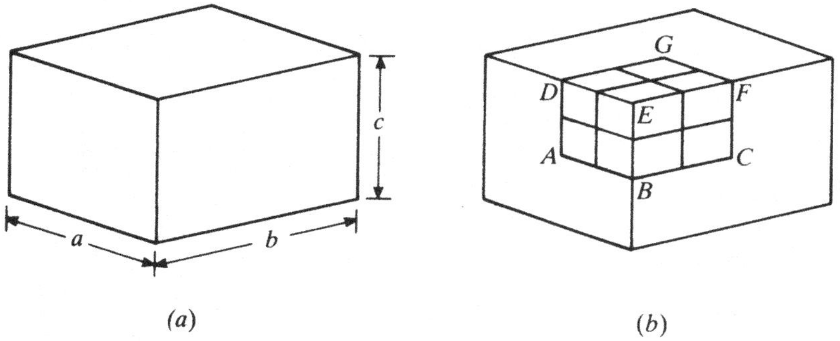

Figure 7.3. Geometry of a box structure.

constructed by assuming a quintic variation for w and cubic variations for u, v, θ_x. The element has two nodes with u, $\partial u/\partial x$, v, $\partial v/\partial x$, w, $\partial w/\partial x$, $\partial^2 w/\partial x^2$, θ_x, $\partial\theta_x/\partial x$ as degrees of freedom. Since $\theta_x = \partial w/\partial y$, then $\partial\theta_x/\partial x = \partial^2 w/\partial x\partial y$. The strain energy of the beam due to lateral motion is included. However, the kinetic energy corresponding to in-plane motion of both the plate and stiffener is neglected. This formulation has the added complexity that $\partial u/\partial y$, $\partial v/\partial y$, $\partial^2 w/\partial y^2$ are not continuous across a stiffener which is parallel to the x-axis and, therefore, cannot be equated at stiffener nodes. Results are given and compared with experimental measurements for two square plates with two stiffeners each. In all cases, all four boundaries are clamped. A large number of frequencies and modes are predicted accurately.

Reference [7.5] treats plates which are stiffened with thin-walled open-section beams. Both bending and membrane motion of the plate is included. Extension, bending in two directions and torsion, including the effect of cross-sectional warping, of the beams is included in the analysis. Very close agreement is obtained with frequencies produced by finite difference and transfer matrix analyses.

Further applications of stiffened plate analysis can be found in references [7.6–7.17].

7.3 Folded Plates I

The membrane displacements of a box structure whose side ratios are close to unity, such as the one shown in Figure 7.3(*a*), are usually much smaller than the bending displacements. Therefore, they may be neglected. This means that lines of intersection of two faces cannot deform and only rotation about such a line is possible. In addition, neither displacements nor rotations are possible at corners.

If the box has three planes of symmetry, then only one-eighth of the box need be idealised, as shown in Figure 7.3(*b*). All the natural frequencies and modes can be calculated using eight combinations of symmetric and antisymmetric boundary conditions about the three planes of symmetry.

As a consequence of the above assumptions, the box may be treated as a flat plate, as shown in Figure 7.4, provided that additional box constraints are applied.

To illustrate the procedure, consider the modes which are symmetric with respect to all three planes shown in Figure 7.3. This means that the X- and Y-axes in Figure 7.4 are lines of symmetry, as are the lines GF and CF. Using a four-node

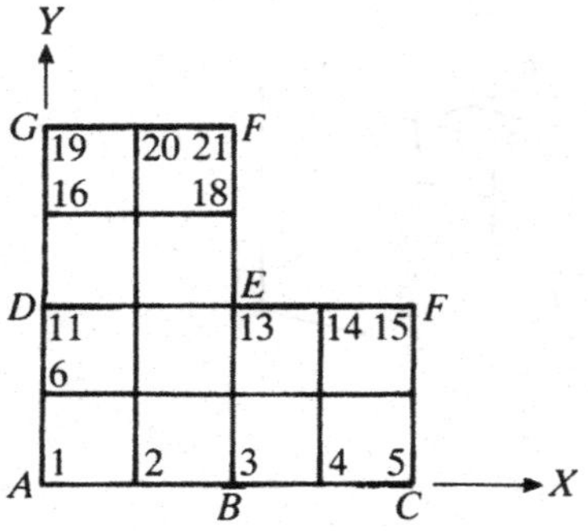

Figure 7.4. Flat plate idealisation of one-eighth of a box (see Figure 7.3(*b*)).

plate-bending element with w, θ_X and θ_Y as degrees of freedom at each node, the symmetrical boundary conditions are:

(1) $\theta_X = 0$ at nodes 1 to 5
(2) $\theta_Y = 0$ at nodes 1, 6, 11, 16, 19
(3) $\theta_X = 0$ at nodes 19 to 21
(4) $\theta_Y = 0$ at nodes 5, 10, 15

Because the edges of the box cannot deform, then:

(5) $w = 0, \theta_Y = 0$ at nodes 11 to 15
(6) $w = 0, \theta_X = 0$ at nodes 3, 8, 13, 18, 21.

This flat plate will represent the behaviour of a box if the displacements at nodes 18 and 21 are constrained to have the same displacements as nodes 14 and 15 respectively, that is

(7) $(\theta_Y)_{18} = (\theta_X)_{14}, (\theta_Y)_{21} = (\theta_X)_{15}$

Reference [7.14] uses this technique to analyse a box having dimensions 24.38 cm ×30.48 cm ×36.58 cm. The thickness of the walls is 0.3175 cm and the material properties are

$$E = 207 \times 10^9 \text{ N/m}^2, \qquad \nu = 0.3, \qquad \rho = 7861 \text{ kg/m}^3$$

The box was analysed using the CR plate bending element (Section 6.2) which has the additional degree of freedom $w_{xy} (\equiv \partial^2 w/\partial x \partial y)$ at each node. This requires the additional constraints

(8) $(w_{xy})_{13} = 0, (w_{xy})_{18} = (w_{xy})_{14}$

Also the condition $w_{xy} = 0$ should be inserted into (1) to (4).

The frequencies are compared with the analytical frequencies given in reference [7.19]. The results for the first three modes which are symmetrical about all three planes of symmetry are given in Table 7.1. Using all eight combinations of symmetric and antisymmetric boundary conditions, reference [7.18] shows that 16 frequencies differ from the analytical frequencies by less than 1%.

Table 7.1. *Natural frequencies of the symmetric modes of a closed box*

Mode no.	FEM [7.18]	Analytical [7.19]	% Difference
1	179.7	179	0.4
2	272.9	272	0.3
3	334.0	333	0.3

Reference [7.20] uses this technique to analyse the vibration characteristics of a rectangular box structure having two sloping roofs. Angles of 5°, 15° and 22.5° are considered. The structure was modelled using the UM6 rectangular element and NRCC triangular element (Section 6.7). The first six calculated frequencies are compared with experimentally measured frequencies. The two sets agree to within 9%.

7.4 Folded Plates II

In the case of slender structures, such as the box-beam shown in Figure 7.5, the membrane displacements are of the same order of magnitude as the bending displacements and so cannot be ignored. When idealising such a structure facet shell elements, which include both membrane and bending action, should be used (see Section 7.2).

In order to illustrate the procedure, consider the problem of determining the flexural modes of the structure shown in Figure 7.5 when one end is fixed and the other free. Take $a = 20$ m, $b = c = 1$ m.

The structure has two planes of symmetry and so it is only necessary to idealise one-quarter of the structure, as shown in Figure 7.6. There are 40 elements in the X-direction, making a total of 80 elements. The flexural modes can be calculated by taking the XY-plane through AB as a plane of symmetry and the ZX-plane through EF as a plane of antisymmetry.

Using the facet shell element as described in Section 7.2, each element has four nodes with five degrees of freedom at each node, namely $u, v, w, \theta_x, \theta_y$. As there are two sets of elements to be joined at right angles, it is convenient to increase the

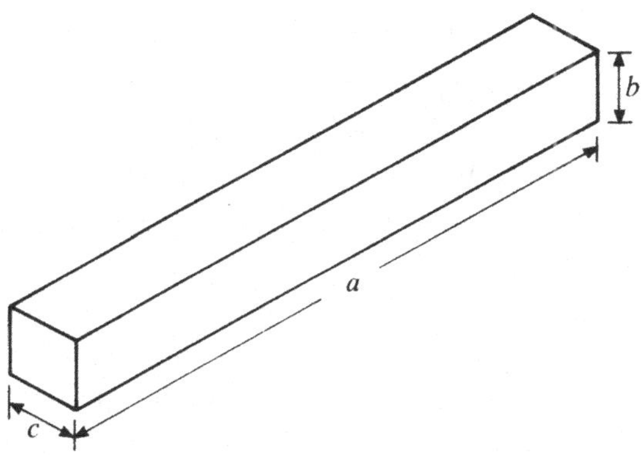

Figure 7.5. Geometry of a hollow box-beam ($a \gg b, c$).

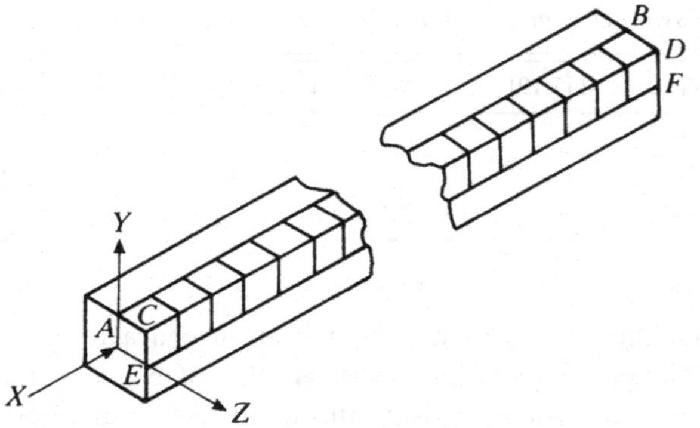

Figure 7.6. Idealisation of one-quarter of a hollow box-beam.

number of nodal degrees of freedom to six by including the rotation θ_z. In this case equations (7.37) and (7.40) become

$$\mathbf{m}^{\mathrm{s}}_{ij} = \begin{bmatrix} \mathbf{m}^{\mathrm{m}}_{ij} & \mathbf{0} & \mathbf{0} \\ \mathbf{0} & \mathbf{m}^{\mathrm{b}}_{ij} & \mathbf{0} \\ \mathbf{0} & \mathbf{0} & \mathbf{0} \end{bmatrix} \tag{7.60}$$

$$\mathbf{k}^{\mathrm{s}}_{ij} = \begin{bmatrix} \mathbf{k}^{\mathrm{m}}_{ij} & \mathbf{0} & \mathbf{0} \\ \mathbf{0} & \mathbf{k}^{\mathrm{b}}_{ij} & \mathbf{0} \\ \mathbf{0} & \mathbf{0} & \mathbf{0} \end{bmatrix} \tag{7.61}$$

and are of order (6×6).

The local axes of the elements in the plane $CDEF$ are parallel to the global axes. However, the local axes of the elements in the plane $ABCD$ are rotated through $-90°$ about the X-axis from the global axes. The inertia and stiffness matrices for these elements are given by $\mathbf{R}^{\mathrm{T}}\mathbf{m}^{\mathrm{s}}_{ij}\mathbf{R}$ and $\mathbf{R}^{\mathrm{T}}\mathbf{k}^{\mathrm{s}}_{ij}\mathbf{R}$ where

$$\mathbf{R} = \left[\begin{array}{ccc|ccc} 1 & 0 & 0 & & & \\ 0 & 0 & -1 & & & \\ 0 & 1 & 0 & & & \\ & & & 1 & 0 & 0 \\ & \mathbf{0} & & 0 & 0 & -1 \\ & & & 0 & 1 & 0 \end{array}\right] \tag{7.62}$$

The constraints which are to be applied are as follows:

(1) $U, V, W, \theta_X, \theta_Y, \theta_Z = 0$ at A, C and E
(2) $W = 0, \theta_X = 0, \theta_Y = 0$ at nodes along AB
(3) $U = 0, W = 0, \theta_Y = 0$ at nodes along EF

The wall thickness is 0.05 m and the material properties are

$$E = 207 \times 10^9\ \mathrm{N/m^2}, \qquad \nu = 0.3, \qquad \rho = 7861\ \mathrm{kg/m^3}$$

Using the linear rectangle (Section 4.2) for the membrane part and the ACM element (Section 6.1) for the bending part to form the facet shell element, produces the frequencies given in Table 7.2.

Table 7.2. *Natural frequencies (Hz) of a cantilever box-beam*

Mode no.	FEM	Analytical [7.21]	% Difference
1	2.961	2.912	+1.7
2	17.868	17.565	+1.7
3	47.077	46.558	+1.1
4	84.106	85.081	−1.1

The frequencies obtained are compared with analytical frequencies for a deep beam (see Example 3.10). The comparison is good because the beam-type modes of the structure occur at much lower frequencies than the plate-type modes. This is a consequence of the side ratios a/b and a/c being large.

Although this example indicates an accurate prediction of the natural frequencies, the facet shell element employed should be used with care. This is because continuity of displacement along the common edge between two elements meeting at right angles is lost. This is illustrated in Figure 7.7(*a*). The normal component of membrane displacement in the vertical element should be equal to the bending displacement of the horizontal element. This is clearly not true between nodes since membrane displacements vary linearly whilst bending displacements have a cubic variation. Reference [7.22] develops a special membrane element to overcome this problem. The 24 degree of freedom membrane element of reference [7.23] is taken as the starting point. The nodal degrees of freedom are $u, \partial u/\partial x, \partial u/\partial y, v, \partial v/\partial x, \partial v/\partial y$. Both u and v vary cubically. Constraints are then applied at the four-node points to ensure that all in-plane displacements vary linearly, except those which are required to be cubic to ensure compatibility with the transverse displacements in adjacent elements. These constraints reduce the number of degrees of freedom to 12, that is three at each node. This element is then combined with the WB element described in Section 6.2. However, a simpler way is to combine the eight node membrane and thick plate bending elements referred to in Sections 4.6 and 6.7 respectively. In this case both membrane and bending displacements vary quadratically (see Figure 7.7(*b*)). Such a facet

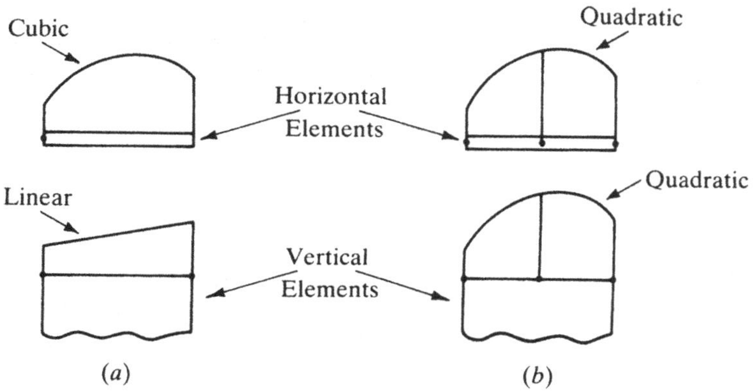

Figure 7.7. Joining facet shell elements at right angles.

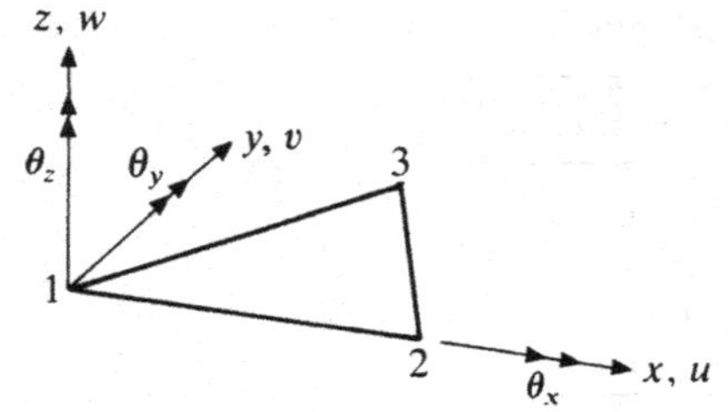

Figure 7.8. Geometry of a triangular facet-shell element.

shell element has been used successfully in analysing models of diesel engines in reference [7.25].

7.5 Folded Plates III

General folded plate structures can be analysed using a triangular facet-shell element. Figure 7.8 shows one such element having three nodes with six degrees of freedom at each node, namely, u, v, w, θ_x, θ_y, θ_z with respect to the local axes shown.

The inertia and stiffness matrices of such an element will be of the form

$$[\mathbf{m}]^s = \begin{bmatrix} \mathbf{m}^s_{11} & & \text{Sym} \\ \mathbf{m}^s_{21} & \mathbf{m}^s_{22} & \\ \mathbf{m}^s_{31} & \mathbf{m}^s_{32} & \mathbf{m}^s_{33} \end{bmatrix} \tag{7.63}$$

and

$$[\mathbf{k}]^s = \begin{bmatrix} \mathbf{k}^s_{11} & & \text{Sym} \\ \mathbf{k}^s_{21} & \mathbf{k}^s_{22} & \\ \mathbf{k}^s_{31} & \mathbf{k}^s_{32} & \mathbf{k}^s_{33} \end{bmatrix} \tag{7.64}$$

where $\mathbf{m}^s_{ij}$ and $\mathbf{k}^s_{ij}$ are of the form given by (7.60) and (7.61).

The next step is to transform the energy expressions into expressions involving nodal degrees of freedom relative to global axes (see Figure 7.9). In this case the

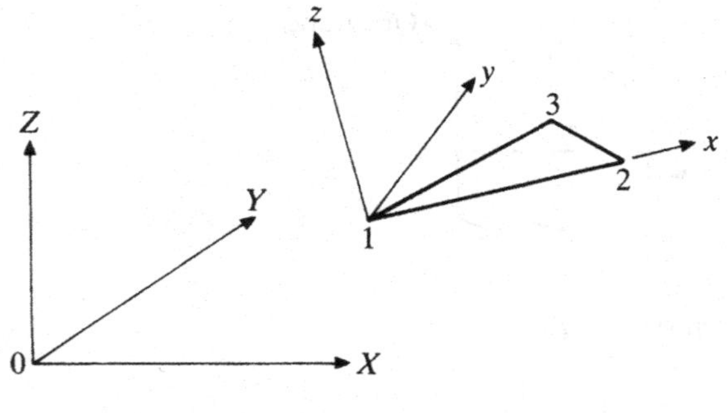

Figure 7.9. Orientation of a triangular element with respect to global axes.

inertia and stiffness matrices are $[\mathbf{R}]^{\mathrm{T}}[\mathbf{m}]^{\mathrm{s}}[\mathbf{R}]$ and $[\mathbf{R}]^{\mathrm{T}}[\mathbf{k}]^{\mathrm{s}}[\mathbf{R}]$ respectively, where

$$[\mathbf{R}] = \begin{bmatrix} \mathbf{L}_3 & & & & & \\ & \mathbf{L}_3 & & & & \\ & & \mathbf{L}_3 & & & \\ & & & \mathbf{L}_3 & & \\ & & & & \mathbf{L}_3 & \\ & & & & & \mathbf{L}_3 \end{bmatrix} \tag{7.65}$$

The matrix $\mathbf{L}_3$ is a matrix of direction cosines as defined by equation (3.195). If the local x-axis is defined by nodes 1 and 2 and the local y-axis lies in the plane of the element, then $\mathbf{L}_3$ can be calculated by the technique described in Section 3.7.

The fact that there are zero contributions to the inertia and stiffness matrices in the θ_z degrees of freedom with respect to local axes, can produce problems which require attention. If all the elements which meet at a node are coplanar, then the inertia and stiffness matrices will be singular. This can be overcome by removing the rotation about the normal to the plane. If the normal is in the direction of one of the global axes, then this can be achieved by applying a zero constraint. Otherwise, a linear constraint equation will have to be used (see Section 1.5). If elements at a node are almost coplanar, the inertia and stiffness matrices are nearly singular. This may lead to inaccuracies in the solution. In such a situation a 'normal' cannot easily be specified in order to define a zero rotation. One way of avoiding such a difficulty is to insert an arbitrary stiffness corresponding to the local θ_z degree of freedom [7.26]. In this case (7.61) becomes

$$\mathbf{k}_{ij}^{\mathrm{s}} = \begin{bmatrix} \mathbf{k}_{ij}^{\mathrm{m}} & \mathbf{0} & \mathbf{0} \\ \mathbf{0} & \mathbf{k}_{ij}^{\mathrm{b}} & \mathbf{0} \\ \mathbf{0} & \mathbf{0} & \mathbf{k}_{ij}^{\mathrm{r}} \end{bmatrix} \tag{7.66}$$

where

$$k_{ij}^{\mathrm{r}} = \begin{cases} EAh\alpha & j = i \\ EAh\alpha/2 & j \neq i \end{cases} \tag{7.67}$$

where α is a small parameter (say 10^{-5}). The offending degrees of freedom are then removed by means of the reduction technique (see Chapter 11).

Several attempts have been made to determine the real stiffness and inertia coefficients for the θ_z degree of freedom. For example, reference [7.27] converts the six-node triangle with two degrees of freedom per node, shown in Figure 4.18, to a three-node triangle with three degrees of freedom per node, namely u, v, θ_z. This is obtained by assuming that the tangential displacement along each side varies linearly and the normal displacement varies quadratically. By this means the components of displacement at the mid side nodes can then be expressed in terms of the corner node displacements. This procedure can also be applied to the eight-node quadrilateral shown in Figure 4.17. The performance of this latter element has been improved in references [7.28, 7.29, 7.30].

Table 7.3. *Natural frequencies (Hz) of a singly curved rectangular plate with clamped boundaries*

Mode no.	*m, n*	FEM [7.31]	Analytical [7.32, 7.33]	% Difference
1	1, 2	830	870	−4.6
2	1, 3	944	958	−1.5
3	1, 3	1288	1288	0.0
4	2, 1	1343	1364	−1.5

If two facet elements display continuity of displacements and slopes when joined together in the same plane, they may lose this feature when joined at an angle. This problem is particularly severe when two elements meet at right angles, as described in Section 7.4.

Facet shell elements can also be used to analyse curved shells. However, it should be remembered that membrane and bending actions within a single element are uncoupled, simply because the element is flat. The necessary coupling for the entire shell comes about when elements are joined at angles to one another. It is, therefore, necessary to use many elements to obtain good accuracy.

Reference [7.31] uses a combination of the linear triangular membrane (Section 4.1) and the plate bending element T (Section 6.4) to analyse a singly curved rectangular plate which is clamped on all four edges. The dimensions of the plate are 7.62 cm along the straight edges and 10.16 cm along the curved. The radius of curvature is 76.2 cm and the thickness 0.33 mm. The material properties are

$$E = 68.95 \times 10^9 \text{ N/m}^2, \qquad \nu = 0.33, \qquad \rho = 2657 \text{ kg/m}^3$$

The idealisation used is shown in Figure 7.10.

The four lowest natural frequencies are compared with the frequencies obtained using the extended Rayleigh–Ritz method in Table 7.3. In this table m, n denote the number of half-waves in the straight and curved directions respectively.

Further examples are given in reference [7.34]. A free cylindrical shell and a clamped hemispherical one are analysed using rectangular and quadrilateral MITC4 shell elements respectively (see Chapter 6).

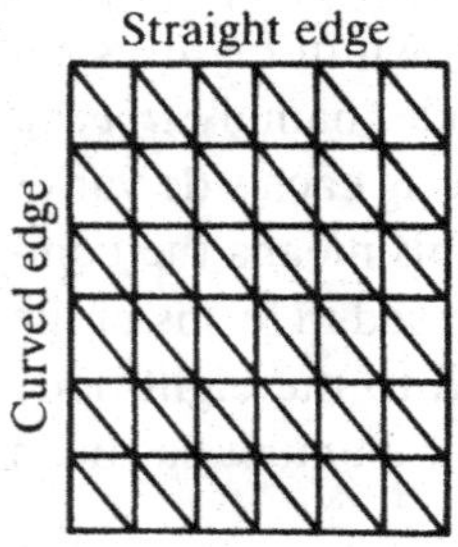

Figure 7.10. Idealisation of a singly curved rectangular plate.

Problems

Note: Problems 7.4 to 7.6 require the use of a digital computer.

7.1. Derive the inertia and stiffness matrices for the beam stiffener element in Figure 7.1 using the technique described in Section 3.11. Compare the results with (7.16) and (7.23).

7.2. Derive the inertia and stiffness matrices for the beam stiffener element in Figure 7.1 using the displacement functions (7.5) and (7.27).

7.3. Derive the displacement functions for the beam stiffener element in Figure 7.1 which are compatible with the plate element CR.

7.4. Figure P7.4 shows a square plate which is clamped on all four boundaries and stiffened by two rectangular cross-section beams. The material properties are $E = 69 \times 10^9$ N/m^2, $\nu = 0.33$ and $\rho = 2600$ kg/m^3. Calculate the four lowest natural frequencies and compare them with the experimental values 859, 1044, 1292 and 1223 Hz [7.4].

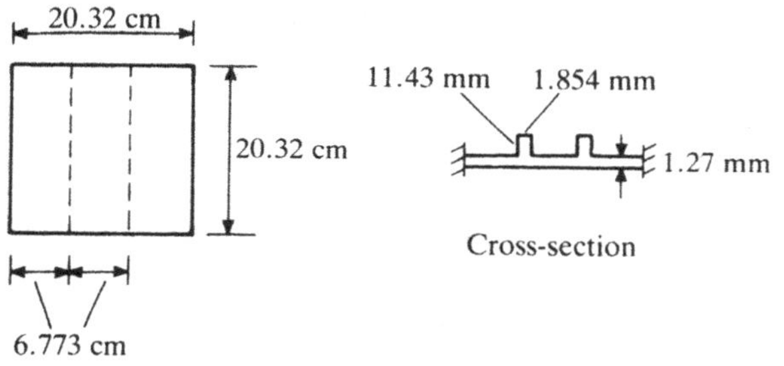

Figure P7.4

7.5. Figure P7.5 shows a cantilever plate which is stiffened by a square cross-section beam. The material properties are $E = 207 \times 10^9$ N/m^2, $\nu = 0.3$ and $\rho = 7861$ kg/m^3. Calculate the seven lowest natural frequencies and compare them with the experimental values 160, 355, 831, 893, 1257, 1630 and 2000 Hz [7.6].

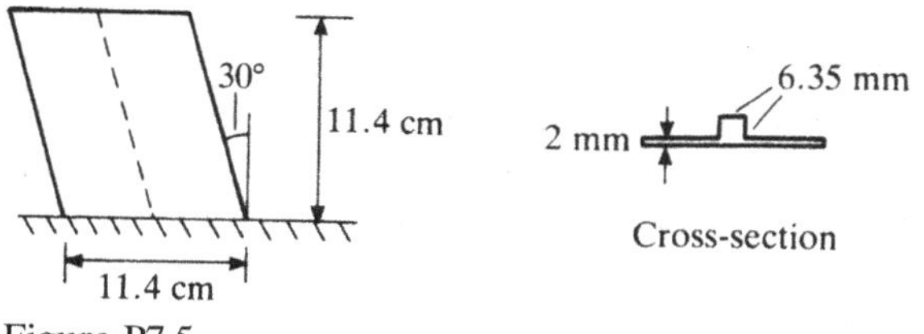

Figure P7.5

7.6. Figure P7.6 shows a cantilever folded plate of thickness 4 mm. The material properties are $E = 207 \times 10^9$ N/m^2, $\nu = 0.3$ and $\rho = 7861$ kg/m^3. Calculate the eight lowest natural frequencies and compare them with the analytical values 211(A), 418(S), 768(A), 899(S), 1529(S), 1553(A), 1757(A), 2306(S) Hz [7.35]

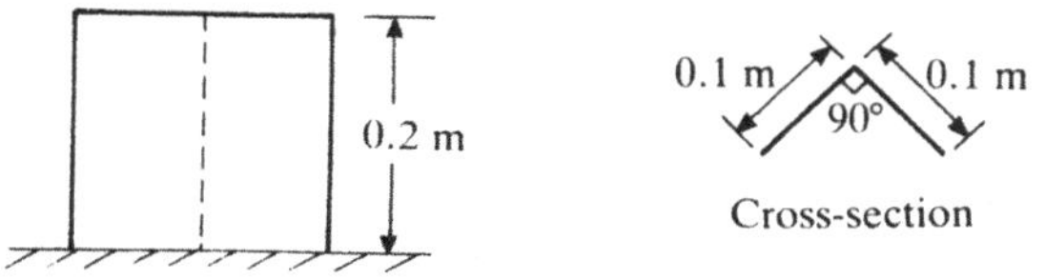

Figure P7.6

8 Vibration of Shells

Curved plates and shells are widely used in naval, aeronautical, mechanical and civil engineering. They appear extensively in the construction of ships, submarines, aircraft, aerospace structures, fan blades, modern buildings and cooling towers. A considerable amount of literature is devoted to the analysis of shells, see for example references [8.1–8.10].

There are three methods of formulating shell elements:

(1) Combining a membrane element with a plate bending element to form a flat (facet) shell element (see Chapter 7).
(2) Deriving a curved element using classical thin shell theory.
(3) Deriving a curved element which is a degenerate solid element to form a thick shell element.

8.1 Thin Shell Elements

One of the problems associated with developing thin shell elements is to decide which of the many available shell theories to use. Shallow shell theory is simpler to use than general shell theory because calculations are performed in a reference plane onto which the shell is projected. But shell curvature is not discarded, so membrane and bending actions interact, as they should. Reviews of many of these can be found in references [8.11, 8.12]. For convenience, it is better to choose one of the simpler theories. However, some theories do not yield zero strain when rigid body motions are prescribed [8.13]. Such a theory should not be used. One theory, which is commonly used for shell elements, by Novozhilov [8.14], yields the following energy expressions

$$U_e = \frac{1}{2} \iint_{A_e} \int_{-h/2}^{+h.2} \{\boldsymbol{\varepsilon}\}^{\mathrm{T}} [\mathbf{D}] \{\boldsymbol{\varepsilon}\} \, d\zeta \, \mathrm{d}x \, \mathrm{d}y \tag{8.1}$$

where h is the thickness of the shell, $[\mathbf{D}]$ is a matrix of material constants and

$$\{\boldsymbol{\varepsilon}\} = \begin{bmatrix} \varepsilon_x \\ \varepsilon_y \\ \gamma_{xy} \end{bmatrix} \tag{8.2}$$

A_e is the area of projection of the element onto the base plane; x, y are Cartesian coordinates in the base plane and ζ the thickness coordinate normal to the middle surface.

The strain–displacement relationships for a shallow shell are

$$\begin{aligned}\varepsilon_x &= \frac{\partial u}{\partial x} - k_{xx}w - \zeta\frac{\partial^2 w}{\partial x^2}\\ \varepsilon_y &= \frac{\partial v}{\partial y} - k_{yy}w - \zeta\frac{\partial^2 w}{\partial y^2}\\ \gamma_{xy} &= \frac{\partial u}{\partial y} + \frac{\partial v}{\partial x} - 2k_{xy}w - 2\zeta\frac{\partial^2 w}{\partial x\partial y}\end{aligned} \tag{8.3}$$

where k_{xx}, k_{yy}, k_{xy} are the curvatures and twist of the middle surface, w is the normal displacement and u, v the tangential displacements of a point in the middle surface of the shell.

The kinetic energy of the shell is

$$T_e = \frac{1}{2}\iint_{A_e} \rho h[\dot{u}^2 + \dot{v}^2 + \dot{w}^2]\,\mathrm{d}x\,\mathrm{d}y \tag{8.4}$$

where ρ is the density of the material.

The highest derivative of w in equations (8.1) to (8.4) is the second, as is the case for flat plate bending elements. Therefore, any of the displacement functions described in Chapter 6 can be used directly to represent the normal displacement. All that remains is to develop suitable displacement functions for u and v. The highest derivative of these components in equations (8.1) to (8.4) is the first. This suggests that the displacement functions developed for flat membrane elements in Chapter 4 are suitable. However, these lead to a poor approximation for rigid body displacements. Experience has shown that a better approximation is obtained if all three displacement functions are represented by polynomials of equal degree. This involves having derivatives of u and v as additional nodal degrees of freedom. This creates problems when trying to model such as step changes in thickness.

Reference [8.15] develops a quadrilateral shell element using the functions of the plate bending element CQ defined in Section 6.7. It is a fully conforming element having 48 degrees of freedom, namely, the three displacements, u, v, w, and their first derivatives, u_x, u_y, v_x, v_y, w_x, w_y, at the four corners together with the normal derivatives, u_n, v_n, w_n, at the midpoints of the sides. This reference also analyses a singly curved rectangular plate which is clamped on all four edges. The dimensions of the plate are 7.62 cm along the straight edges and 10.16 cm along the curved edges. The radius of curvature is 76.2 cm and the thickness 0.33 mm. The material properties are

$$E = 68.95 \times 10^9\ \mathrm{N/m^2}, \qquad v = 0.33, \qquad \rho = 2657\ \mathrm{kg/m^3}$$

The first four lowest natural frequencies obtained using an 8×8 mesh of elements are compared with the frequencies obtained using the extended Rayleigh–Ritz method in Table 8.1. In this table, m and n denote the number of half waves in the straight and curved directions respectively. It can be seen that the two sets of frequencies are in very close agreement. Comparison with Table 7.3 indicates the increase in accuracy obtained using curved shell elements rather than facet ones.

Table 8.1. *Natural frequencies (Hz) of a singly curved rectangular plate with clamped boundaries*

Mode no.	m, n	FEM [8.15]	Analytical [8.16]
1	1, 2	869	870
2	1, 3	958	958
3	1, 3	1289	1288
4	2, 1	1364	1364

Reference [8.17] derives a doubly curved triangular element and applies it to a number of shell vibration problems.

The analysis of axisymmetric shells can be carried out by means of a sequence of one-dimensional problems in a similar way to axisymmetric solids presented in Chapter 5. Elements having straight or curved meridians are available. Details of such elements and applications can be found in references [8.18–8.22].

8.2 Thick Shell Elements

Thick shells can be analysed using either middle surface or solid-type shell elements. Middle surface elements have, as the name implies, node points lying only in the middle surface of the element (see Figure 8.1(*a*)). The nodal degrees of freedom consist of three components of displacement and two components of rotation. These elements are constructed by methods which are similar to the ones used for thick plate elements. Variable thickness is usually allowed for in the formulation. The elements are, therefore, of the superparametric-type since more nodes are required to define the geometry than the deformation. They can be used to model thin shells provided reduced integration is used.

Solid-type shell elements are constructed by modifying three-dimensional solid elements. The node points are situated in both the top and the bottom surfaces (see Figure 8.1(*b*)) and have three degrees of freedom each, namely, three components of displacement. With such an element, numerical difficulties arise when the shell is thin. This can be overcome by constraining the relative displacement between the nodes on the same shell normal. Although solid-type shell elements have twice as many nodes as middle surface shell elements of the same order, the total number of

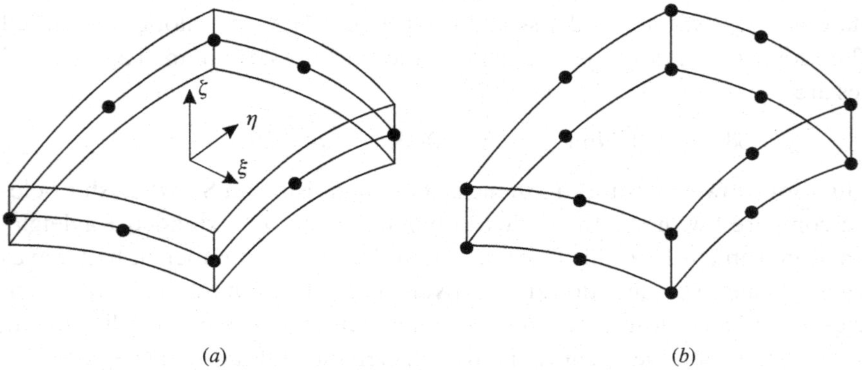

Figure 8.1. Shell elements: (*a*) middle surface and (*b*) solid type.

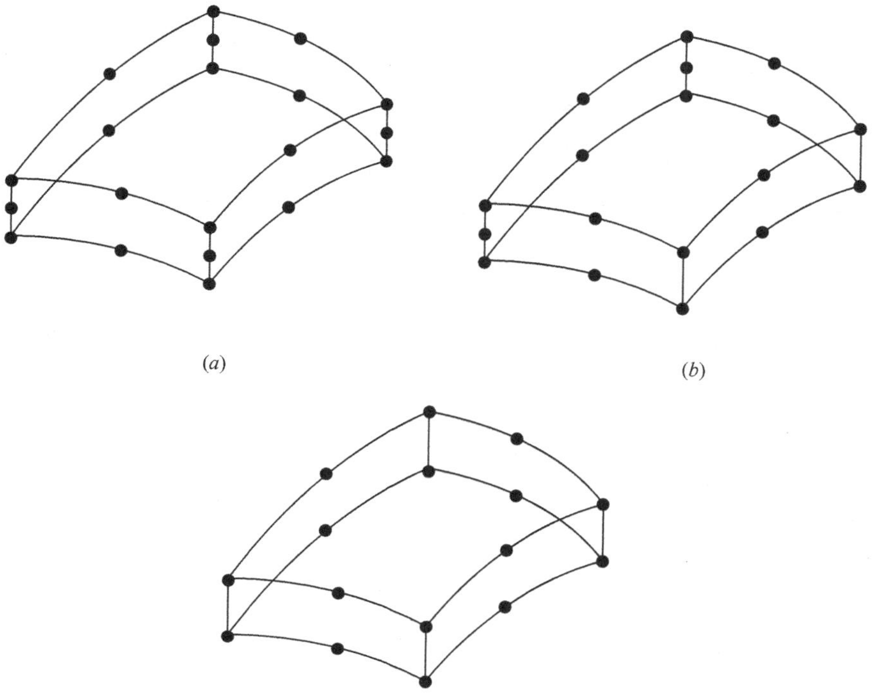

Figure 8.2. (*a*) A 20-node solid element, (*b*) an 18-node transition element and (*c*) a 16-node solid-type shell element.

degrees of freedom per element remains unchanged, if a drilling degree of freedom is included at each node in the latter case.

In some problems, shells may be connected to solids. Both types of shell element may be connected to solid elements via transition elements. For example, a 16-node, solid-type shell element can be connected to a 20-node solid element by means of an 18-node transition element (see Figure 8.2). Again, an 8-node middle surface shell element can be connected to a 20-node solid element by means of a 13-node transition element (see Figure 8.3). Also, an 8-node middle surface element can be connected to a 16-node solid shell element by means of an 11-node transition element (see Figure 8.4). Details of these elements are given in reference [8.23].

8.2.1 Middle Surface Shell Element

The development of a middle surface shell element is similar to the technique used for a thick plate bending element (see Chapters 2 and 6). That is, it is assumed that the direct stress in the normal direction is zero, and that normals to the middle surface before deformation remain straight during deformation.

With reference to Figure 8.5, at a typical node i, a thickness-direction vector $\vec{V}_{3i}$ is given by

$$\vec{V}_{3i} = h_i \begin{bmatrix} l_{3i} \\ m_{3i} \\ n_{3i} \end{bmatrix} \tag{8.5}$$

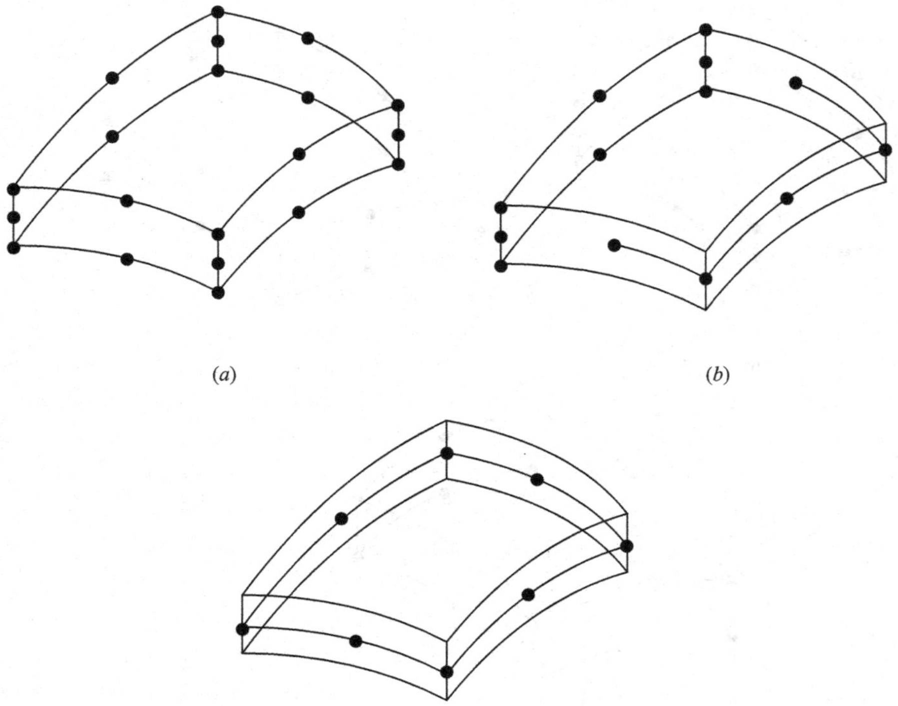

(a) (b) (c)

Figure 8.3. (a) A 20-node solid element, (b) a 13-node transition element and (c) an 8-node middle surface shell element.

where h_i is the thickness at node i and (l_{3i}, m_{3i}, n_{3i}) are the direction cosines of $\vec{V}_{3i}$ with respect to global axes (x, y, z), given by

$$\begin{bmatrix} l_{3i} \\ m_{3i} \\ n_{3i} \end{bmatrix} = \frac{1}{h_i}\left[\begin{bmatrix} x_i \\ y_i \\ z_i \end{bmatrix}_{\text{top}} - \begin{bmatrix} x_i \\ y_i \\ z_i \end{bmatrix}_{\text{bot}}\right] \tag{8.6}$$

Assuming that the element has eight nodes as shown in Figure 8.1(a), the global coordinates of an arbitrary point in the element are

$$\begin{bmatrix} x \\ y \\ z \end{bmatrix} = \sum_{i=1}^{8} N_i(\xi, \eta) \begin{bmatrix} x_i \\ y_i \\ z_i \end{bmatrix}_{\text{mid}} + \sum_{i=1}^{8} N_i(\xi, \eta)\zeta \frac{h_i}{2} \begin{bmatrix} l_{3i} \\ m_{3i} \\ n_{3i} \end{bmatrix} \tag{8.7}$$

where the functions $N_i(\xi, \eta)$ are given by equations (4.108) to (4.110) and $-1 \leq \zeta \leq +1$. $(x_i\ y_i\ z_i)_{\text{mid}}$ are the coordinates of a node on the middle surface and are given by

$$\begin{bmatrix} x_i \\ y_i \\ z_i \end{bmatrix}_{\text{mid}} = \frac{1}{2}\left[\begin{bmatrix} x_i \\ y_i \\ z_i \end{bmatrix}_{\text{top}} + \begin{bmatrix} x_i \\ y_i \\ z_i \end{bmatrix}_{\text{bot}}\right] \tag{8.8}$$

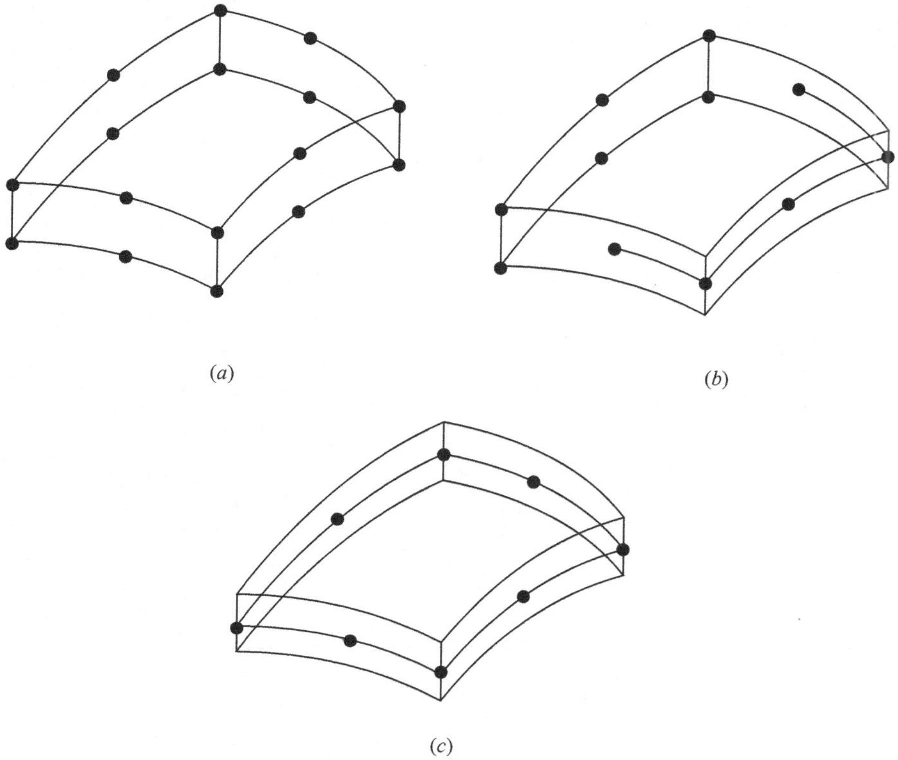

Figure 8.4. (*a*) A 16-node solid shell element, (*b*) an 11-node transition element and (*c*) an 8-node middle surface shell element.

The degrees of freedom at each middle surface node are three displacements (u, v, w) in the directions of the global axes (x, y, z) and two rotations α and β about $\vec{V}_{1i}$ and $\vec{V}_{2i}$ respectively. These two vectors are perpendicular to $\vec{V}_{3i}$ and to each other. Note that they are tangential to the element mid-surface at node i. In order to specify a specific direction, $\vec{V}_{1i}$ is taken to be normal to both $\vec{V}_{3i}$ and the global y-direction. That is

$$\vec{V}_{1i} = \vec{j} \wedge \vec{V}_{3i} \tag{8.9}$$

Figure 8.5. Local and global coordinates of a middle surface shell element. ○ Geometric nodes and • displacement nodes.

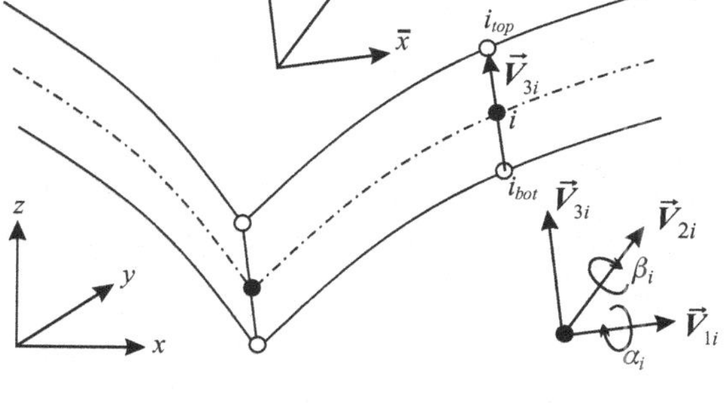

where $\vec{j}$ is a unit vector in the y-direction. Then

$$\vec{V}_{2i} = \vec{V}_{3i} \wedge \vec{V}_{1i} \tag{8.10}$$

If $\vec{j}$ and $\vec{V}_{3i}$ are parallel, then take

$$\vec{V}_{2i} = \vec{V}_{3i} \wedge \vec{i} \tag{8.11}$$

and

$$\vec{V}_{1i} = \vec{V}_{2i} \wedge \vec{V}_{3i} \tag{8.12}$$

where $\vec{i}$ is a unit vector in the x-direction.

Assuming that normals to the middle surface before deformation remain straight during deformation, the displacements of an arbitrary point in the element are

$$\begin{bmatrix} u \\ v \\ w \end{bmatrix} = \sum_{i=1}^{8} N_i(\xi, \eta) \left[\begin{bmatrix} u_i \\ v_i \\ w_i \end{bmatrix} + \zeta \frac{h_i}{2} [\boldsymbol{\mu}_i] \begin{bmatrix} \alpha_i \\ \beta_i \end{bmatrix} \right] \tag{8.13}$$

where

$$[\boldsymbol{\mu}_i] = \begin{bmatrix} -l_{2i} & l_{1i} \\ -m_{2i} & m_{1i} \\ -n_{2i} & n_{1i} \end{bmatrix} \tag{8.14}$$

the columns of which are the direction cosines of the vectors $-\vec{V}_{2i}$ and $\vec{V}$ respectively.

Equation (8.13) can be written in the alternative form

$$\begin{bmatrix} u \\ v \\ w \end{bmatrix} = \sum_{i=1}^{8} [\mathbf{N}_{Ai} + \zeta \mathbf{N}_{Bi}]\{\mathbf{u}_i\} \tag{8.15}$$

where

$$\{\mathbf{u}_i\}^T = \lfloor u \; v \; w \; \alpha \; \beta \rfloor_i \tag{8.16}$$

$$[\mathbf{N}_{Ai}] = \begin{bmatrix} 1 & 0 & 0 & 0 & 0 \\ 0 & 1 & 0 & 0 & 0 \\ 0 & 0 & 1 & 0 & 0 \end{bmatrix} N_i(\xi, \eta) \tag{8.17}$$

$$[\mathbf{N}_{Bi}] = \begin{bmatrix} 0 & 0 & 0 & -l_{2i} & l_{1i} \\ 0 & 0 & 0 & -m_{2i} & m_{1i} \\ 0 & 0 & 0 & -n_{2i} & n_{1i} \end{bmatrix} \frac{h_i}{2} N_i(\xi, \eta) \tag{8.18}$$

In order to introduce the assumption that the direct stress in the normal direction is zero, the strain components $\{\bar{\boldsymbol{\varepsilon}}\}$ with respect to the local axes $\bar{x}, \bar{y}, \bar{z}$ in Figure 8.5 are used. Here

$$\{\bar{\boldsymbol{\varepsilon}}\} = \lfloor \bar{\varepsilon}_x \; \bar{\varepsilon}_y \; \bar{\varepsilon}_z \; \bar{\gamma}_{xy} \; \bar{\gamma}_{xz} \; \bar{\gamma}_{yz} \rfloor \tag{8.19}$$

The strain energy is given by (see equation (2.86))

$$U_e = \frac{1}{2}\int_{V_e} \{\bar{\boldsymbol{\varepsilon}}\}^{\mathrm{T}} [\bar{\mathbf{D}}] \{\bar{\boldsymbol{\varepsilon}}\}\, \mathrm{d}V \tag{8.20}$$

Assuming that the direct stress in the normal direction is zero, then

$$[\bar{\mathbf{D}}] = \frac{E}{(1-\nu^2)} \begin{bmatrix} 1 & \nu & 0 & 0 & 0 & 0 \\ \nu & 1 & 0 & 0 & 0 & 0 \\ 0 & 0 & 0 & 0 & 0 & 0 \\ 0 & 0 & 0 & \frac{1}{2}(1-\nu) & 0 & 0 \\ 0 & 0 & 0 & 0 & \frac{\kappa}{2}(1-\nu) & 0 \\ 0 & 0 & 0 & 0 & 0 & \frac{\kappa}{2}(1-\nu) \end{bmatrix} \tag{8.21}$$

for an isotropic material, where $\kappa = 5/6$ or $\pi^2/12$. The strain components $\{\bar{\boldsymbol{\varepsilon}}\}$ can be expressed in terms of strain components $\{\boldsymbol{\varepsilon}\}$ with respect to global axes x, y, z by means of the transformation (see Appendix 2)

$$\{\bar{\boldsymbol{\varepsilon}}\} = [\mathbf{R}_\varepsilon]\{\boldsymbol{\varepsilon}\} \tag{8.22}$$

Substituting equation (8.22) into equation (8.20) gives

$$\begin{aligned} U_e &= \frac{1}{2}\int_{V_e} \{\boldsymbol{\varepsilon}\}^{\mathrm{T}} [\mathbf{R}_\varepsilon]^{\mathrm{T}} [\bar{\mathbf{D}}] [\mathbf{R}_\varepsilon] \{\boldsymbol{\varepsilon}\}\, \mathrm{d}V \\ &= \frac{1}{2}\int_{V_e} \{\boldsymbol{\varepsilon}\}^{\mathrm{T}} [\mathbf{D}] \{\boldsymbol{\varepsilon}\}\, \mathrm{d}V \end{aligned} \tag{8.23}$$

Thus, the matrix of material constants $[\mathbf{D}]$ with respect to global axes is

$$[\mathbf{D}] = [\mathbf{R}_\varepsilon]^{\mathrm{T}}[\bar{\mathbf{D}}][\mathbf{R}_\varepsilon] \tag{8.24}$$

The next step is to express equation (8.23) in terms of (ξ, η, ζ) coordinates along the lines presented in Section 5.9.

From (5.108),

$$\mathrm{d}V = \det[\mathbf{J}]\, \mathrm{d}\xi\, \mathrm{d}\eta\, \mathrm{d}\zeta \tag{8.25}$$

where

$$[\mathbf{J}] = \begin{bmatrix} \dfrac{\partial x}{\partial \xi} & \dfrac{\partial y}{\partial \xi} & \dfrac{\partial z}{\partial \xi} \\ \dfrac{\partial x}{\partial \eta} & \dfrac{\partial y}{\partial \eta} & \dfrac{\partial z}{\partial \eta} \\ \dfrac{\partial x}{\partial \zeta} & \dfrac{\partial y}{\partial \zeta} & \dfrac{\partial z}{\partial \zeta} \end{bmatrix} \tag{8.26}$$

Substituting equation (8.7) into equation (8.26) gives the first column of [**J**] as

$$\begin{aligned}\frac{\partial x}{\partial \xi} &= \sum_{i=1}^{8} \frac{\partial N_i}{\partial \xi}(x_{im} + \zeta h_i l_{3i}/2)\\ \frac{\partial x}{\partial \eta} &= \sum_{i=1}^{8} \frac{\partial N_i}{\partial \eta}(x_{im} + \zeta h_i l_{3i}/2)\\ \frac{\partial x}{\partial \zeta} &= \sum_{i=1}^{8} N_i(h_i l_{3i}/2)\end{aligned} \tag{8.27}$$

Expressions for the second and third columns of [**J**] are similar.

From equation (5.74)

$$\{\boldsymbol{\varepsilon}\} = [\mathbf{H}]\left\lfloor \frac{\partial u}{\partial x}\ \frac{\partial u}{\partial y}\ \frac{\partial u}{\partial z}\ \frac{\partial v}{\partial x} \cdots \frac{\partial w}{\partial z}\right\rfloor^{\mathrm{T}} \tag{8.28}$$

where

$$[\mathbf{H}] = \begin{bmatrix} 1 & 0 & 0 & 0 & 0 & 0 & 0 & 0 & 0\\ 0 & 0 & 0 & 0 & 1 & 0 & 0 & 0 & 0\\ 0 & 0 & 0 & 0 & 0 & 0 & 0 & 0 & 1\\ 0 & 1 & 0 & 1 & 0 & 0 & 0 & 0 & 0\\ 0 & 0 & 1 & 0 & 0 & 0 & 1 & 0 & 0\\ 0 & 0 & 0 & 0 & 0 & 1 & 0 & 1 & 0 \end{bmatrix} \tag{8.29}$$

From equation (5.116)

$$\begin{bmatrix} \dfrac{\partial}{\partial x}\\ \dfrac{\partial}{\partial y}\\ \dfrac{\partial}{\partial z} \end{bmatrix} = [\mathbf{J}]^{-1} \begin{bmatrix} \dfrac{\partial}{\partial \xi}\\ \dfrac{\partial}{\partial \eta}\\ \dfrac{\partial}{\partial \zeta} \end{bmatrix} \tag{8.30}$$

Therefore,

$$\begin{bmatrix} \dfrac{\partial u}{\partial x}\\ \dfrac{\partial u}{\partial y}\\ \dfrac{\partial u}{\partial z}\\ \dfrac{\partial v}{\partial x}\\ \vdots\\ \dfrac{\partial w}{\partial z} \end{bmatrix} = \begin{bmatrix} \mathbf{J}^{-1} & \mathbf{0} & \mathbf{0}\\ \mathbf{0} & \mathbf{J}^{-1} & \mathbf{0}\\ \mathbf{0} & \mathbf{0} & \mathbf{J}^{-1} \end{bmatrix} \begin{bmatrix} \dfrac{\partial u}{\partial \xi}\\ \dfrac{\partial u}{\partial \eta}\\ \dfrac{\partial u}{\partial \zeta}\\ \dfrac{\partial v}{\partial \xi}\\ \vdots\\ \dfrac{\partial w}{\partial \zeta} \end{bmatrix} \tag{8.31}$$

Substituting equation (8.13) into the right-hand side gives

$$\begin{bmatrix} \dfrac{\partial u}{\partial \xi} \\ \dfrac{\partial u}{\partial \eta} \\ \dfrac{\partial u}{\partial \zeta} \\ \dfrac{\partial v}{\partial \xi} \\ \vdots \\ \dfrac{\partial w}{\partial \zeta} \end{bmatrix} = \sum_{i=1}^{8} \begin{bmatrix} \dfrac{\partial N_i}{\partial \xi} & 0 & 0 \\ \dfrac{\partial N_i}{\partial \eta} & 0 & 0 \\ 0 & 0 & 0 \\ 0 & \dfrac{\partial N_i}{\partial \xi} & 0 \\ \vdots & \vdots & \vdots \\ 0 & 0 & 0 \end{bmatrix} \begin{bmatrix} u_i \\ v_i \\ w_i \end{bmatrix}$$

$$+ \sum_{i=1}^{8} \begin{bmatrix} -\zeta h_i \dfrac{\partial N_i}{\partial \xi} l_{2i}/2 & \zeta h_i \dfrac{\partial N_i}{\partial \xi} l_{1i}/2 \\ -\zeta h_i \dfrac{\partial N_i}{\partial \eta} l_{2i}/2 & \zeta h_i \dfrac{\partial N_i}{\partial \eta} l_{1i}/2 \\ -h_i N_i l_{2i}/2 & -h_i N_i l_{1i}/2 \\ -\zeta h_i \dfrac{\partial N_i}{\partial \xi} m_{2i}/2 & \zeta h_i \dfrac{\partial N_i}{\partial \xi} m_{1i}/2 \\ \vdots & \vdots \\ -h_i N_i n_{2i}/2 & h_i N_i n_{1i}/2 \end{bmatrix} \tag{8.32}$$

Combining equations (8.28), (8.31) and (8.32) yields an expression of the form

$$\begin{aligned} \{\boldsymbol{\varepsilon}\} &= \sum_{i=1}^{8} [\mathbf{B}_i]\{\mathbf{u}_i\} \\ &= [\mathbf{B}]\{\mathbf{u}\}_e \end{aligned} \tag{8.33}$$

where

$$[\mathbf{B}] \; [\mathbf{B}_1 \, \mathbf{B}_2 \cdots \mathbf{B}_8] \tag{8.34}$$

$$\{\mathbf{u}\}_e^{\mathrm{T}} = \lfloor \mathbf{u}_1^{\mathrm{T}} \mathbf{u}_2^{\mathrm{T}} \cdots \mathbf{u}_8^{\mathrm{T}} \rfloor \tag{8.35}$$

and $\mathbf{u}_i^T$ is defined in equation (8.16).

Substituting equations (8.25) and (8.33) into equation (8.23) gives

$$U_e = \frac{1}{2}\{\mathbf{u}\}_e^T \int_{-1}^{+1}\int_{-1}^{+1}\int_{-1}^{+1} [\mathbf{B}]^{\mathrm{T}}[\mathbf{D}][\mathbf{B}] \det[\mathbf{J}] \, d\xi \, d\eta \, d\zeta \{\mathbf{u}\}_e \tag{8.36}$$

which is of the form

$$U_e = \tfrac{1}{2}\{\mathbf{u}\}_e^{\mathrm{T}}[\mathbf{k}]_e\{\mathbf{u}\}_e \tag{8.37}$$

where the element stiffness matrix is

$$[\mathbf{k}]_e = \int_{-1}^{+1}\int_{-1}^{+1}\int_{-1}^{+1} [\mathbf{B}]^{\mathrm{T}}[\mathbf{D}][\mathbf{B}] \det[\mathbf{J}] \, d\xi \, d\eta \, d\zeta \tag{8.38}$$

The expression (8.38) is evaluated using Gauss–Legendre numerical integration. Reference [8.24] recommends 2 integration points in the ζ-direction and a 3×3 array in the (ξ, η) coordinates. It is demonstrated that this gives accurate results for thick shells but not for thin ones. A further reference [8.25] shows that reducing the integration to a 2×2 array in the (ξ, η) coordinates gives accurate results for thin shells also.

The kinetic energy is given by (see equation (2.88))

$$T_e = \frac{1}{2} \int_{V_e} \rho (\dot{u}^2 + \dot{v}^2 + \dot{w}^2)\, \mathrm{d}V \tag{8.39}$$

where ρ is the mass per unit volume.

Now, equation (8.15) can be expressed in the alternative form

$$\begin{bmatrix} u \\ v \\ w \end{bmatrix} = [\mathbf{N}_A + \zeta \mathbf{N}_B]\{\mathbf{u}\}_e \tag{8.40}$$

where $\mathbf{u}_e$ is defined in equation (8.35) and

$$\begin{aligned} [\mathrm{N}_A] &= [\mathrm{N}_{A1} \ \ldots \ \mathrm{N}_{A8}] \\ [\mathrm{N}_B] &= [\mathrm{N}_{B1} \ \ldots \mathrm{N}_{B8}] \end{aligned} \tag{8.41}$$

Substituting equations (8.25) and (8.40) into (8.39) gives

$$T_e = \frac{1}{2}\{\dot{\mathbf{u}}\}_e^{\mathrm{T}} \int_{-1}^{+1}\int_{-1}^{+1}\int_{-1}^{+1} \rho\, [\mathbf{N}_A + \zeta \mathbf{N}_B]^{\mathrm{T}} [\mathbf{N}_A + \zeta \mathbf{N}_B] \det[\mathbf{J}]\, \mathrm{d}\xi\, \mathrm{d}\eta\, \mathrm{d}\zeta\ \{\dot{\mathbf{u}}\}_e \tag{8.42}$$

which is of the form

$$T_e = \tfrac{1}{2}\{\dot{\mathbf{u}}\}_e^{\mathrm{T}}[\mathbf{m}]_e\{\dot{\mathbf{u}}\}_e \tag{8.43}$$

where the element inertia matrix is

$$[\mathbf{m}]_e = \int_{-1}^{+1}\int_{-1}^{+1}\int_{-1}^{+1} \rho\, [\mathbf{N}_A + \zeta \mathbf{N}_B]^{\mathrm{T}} [\mathbf{N}_A + \zeta \mathbf{N}_B] \det[\mathbf{J}]\, \mathrm{d}\xi\, \mathrm{d}\eta\, \mathrm{d}\zeta \tag{8.44}$$

The expression in equation (8.44) is evaluated using 2 integration points in the ζ-direction and a 3×3 array in the (ξ, η) coordinates.

Note that in the case of thin shells the dependence of the matrix $[\mathbf{J}]$ on ζ can be neglected leading to simplified expressions for both the stiffness and inertia matrices.

If p_x, p_y, p_z are the components of the applied surface forces per unit area over the surface $\zeta = +1$, then the virtual work is

$$\begin{aligned} \delta W_e &= \int_{S_e} (p_x \delta u + p_y \delta v + p_z \delta w)\, \mathrm{d}s \\ &= \int_{S_e} \lfloor \delta u \ \ \delta v \ \ \delta w \rfloor \begin{bmatrix} p_x \\ p_y \\ p_z \end{bmatrix} \mathrm{d}s \end{aligned} \tag{8.45}$$

Table 8.2. *Natural frequencies (Hz) of a singly curved square cantilever plate*

Mode	FEM [8.29]	Analytical [8.30]
First torsion	87	87
First bending	143	136
Second bending	252	259
Second torsion	367	351
Third bending	412	395

Substituting equation (8.40), evaluated on $\zeta = +1$, into equation (8.45) gives

$$\delta W_e = \{\mathbf{u}\}_e^{\mathrm{T}} \int_{-1}^{+1}\int_{-1}^{+1} [\mathbf{N}_A + \mathbf{N}_B]^{\mathrm{T}} \begin{bmatrix} p_x \\ p_y \\ p_z \end{bmatrix} \mathrm{d}s \tag{8.46}$$

Now, from Section 5.9

$$\mathrm{d}s = |\mathbf{d\xi} \wedge \mathbf{d\eta}|_{\zeta=1} = G\,\mathrm{d}\xi\,\mathrm{d}\eta \tag{8.47}$$

where $\mathbf{d\xi}$ and $\mathbf{d\eta}$ are given by the first two rows of the Jacobian $[\mathbf{J}]$ in equation (8.26).

Therefore, the equivalent nodal forces are

$$\{\mathbf{f}\}_e = \int_{-1}^{+1}\int_{-1}^{+1} [\mathbf{N}_A + \mathbf{N}_B]^{\mathrm{T}} \begin{bmatrix} p_x \\ p_y \\ p_z \end{bmatrix} G\,\mathrm{d}\xi\,\mathrm{d}\eta \tag{8.48}$$

Additional details may be found in references [8.26–8.28].

Reference [8.29] analyses a singly curved (radius 60.96 cm) square (30.48 × 30.48 cm) cantilever plate made of steel 0.3048 cm thick using a 3 × 2 mesh of elements. The frequencies obtained are compared with measured values in Table 8.2. These results indicate that accurate frequencies for a thin shell can be obtained using reduced integration. Further examples of the use of this element can be found in references [8.29, 8.31].

8.3 Thick Axisymmetric Shell Elements

Thick axisymmetric shells can be analysed using either solid type (Figure 8.6(a)) or middle surface (Figure 8.6(c)) axisymmetric ring elements. These can be connected to one another via transition elements (Figure 8.6(b)). Details of these latter elements are given in reference [8.32].

8.3.1 Middle Surface Axisymmetric Shell Element

A middle surface axisymmetric shell element is derived by combining the techniques used in Sections 5.1 to 5.4, for axisymmetric solids, and Section 8.2.1 for a doubly curved, middle surface shell element.

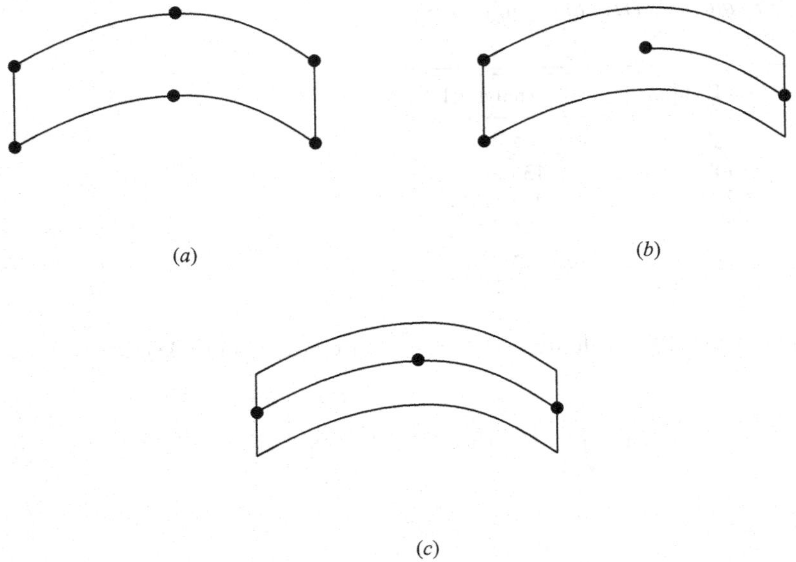

Figure 8.6. Axisymmetric shell elements: (*a*) a 6-node solid, (*b*) a 4-node transition and (*c*) a 3-node middle surface shell element.

With reference to Figure 8.7(*b*), at a typical node *i*, a thickness-direction vector $\vec{V}_{3i}$ is given by

$$\vec{V}_{3i} = h_i \begin{bmatrix} l_{3i} \\ m_{3i} \end{bmatrix} \tag{8.49}$$

where h_i is the thickness at node *i* and $(l_{3i}\ m_{3i})$ are the direction cosines of $\vec{V}_{3i}$ with respect to the global axes, (r, θ, z) given by

$$\begin{bmatrix} l_{3i} \\ m_{3i} \end{bmatrix} = \frac{1}{h_i} \left[\begin{bmatrix} r_i \\ z_i \end{bmatrix}_{\text{top}} - \begin{bmatrix} r_i \\ z_i \end{bmatrix}_{\text{bot}} \right] \tag{8.50}$$

Assuming that the element has three nodes as shown in Figure 8.7(*a*), the global coordinates of an arbitrary point in the element are

$$\begin{bmatrix} r \\ z \end{bmatrix} = \sum_{i=1}^{3} N_i(\xi) \begin{bmatrix} r_i \\ z_i \end{bmatrix}_{\text{mid}} + \sum_{i=1}^{3} N_i(\xi)\eta \frac{1}{h_i} \begin{bmatrix} l_{3i} \\ m_{3i} \end{bmatrix} \tag{8.51}$$

where the functions $N_i(\xi)$ are given by equations (3.215) and $-1 \leq \xi \leq +1$. $(r_i, z_i)_{\text{mid}}$ are the coordinates of a node on the middle surface and are given by

$$\begin{bmatrix} r_i \\ z_i \end{bmatrix}_{\text{mid}} = \frac{1}{2} \left[\begin{bmatrix} r_i \\ z_i \end{bmatrix}_{\text{top}} + \begin{bmatrix} r_i \\ z_i \end{bmatrix}_{\text{bot}} \right] \tag{8.52}$$

The degrees of freedom at each middle surface node are, for both symmetric and antisymmetric motion, about $\theta = 0$, (u, v, w) in the directions of the global axes (r, θ, z) and two rotations α and β about $\vec{V}_{2i}$ and $\vec{V}_{1i}$ respectively. Note that $\vec{V}_{1i}$ is

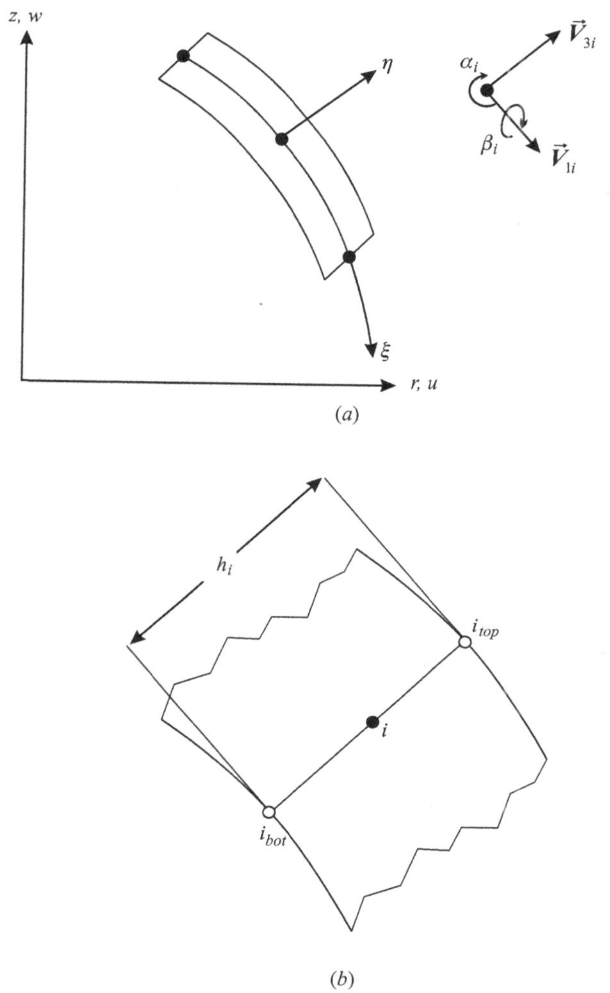

Figure 8.7. Local and global coordinates of an axisymmetric middle surface shell element. ○ Geometric nodes and • displacement nodes.

normal to $\vec{V}_{3i}$ and both lie in the (r, z) plane (see Figure 8.7(a)) and $\vec{V}_{2i}$ is normal to both $\vec{V}_{1i}$ and $\vec{V}_{3i}$. θ is in the circumferential direction (see Figure 5.1(b)).

Assuming that normals to the middle surface before deformation remain straight during deformation, the displacements of an arbitrary point in the element are

$$\begin{bmatrix} u \\ v \\ w \end{bmatrix} = \sum_{i=1}^{3} N_i(\xi) \left[\begin{bmatrix} u_i \\ v_i \\ w_i \end{bmatrix} + \eta \frac{h_i}{2} [\boldsymbol{\mu}_i] \begin{bmatrix} \alpha_i \\ \beta_i \end{bmatrix} \right] \tag{8.53}$$

where

$$[\boldsymbol{\mu}_i] = \begin{bmatrix} l_{1i} & 0 \\ 0 & -1 \\ m_{1i} & 0 \end{bmatrix} \tag{8.54}$$

and $(l_{1i}\ m_{1i})$ are the direction cosines of the vector $\vec{V}_{1i}$ which are given by

$$l_{1i} = -m_{3i}, m_{1i} = l_{3i} \tag{8.55}$$

Equation (8.53) can be written in the alternative form

$$\begin{bmatrix} u \\ v \\ w \end{bmatrix} = \sum_{i=1}^{3} [\mathbf{N}_{Ai} + \eta \mathbf{N}_{Bi}] \{\mathbf{u}\}_i \tag{8.56}$$

where

$$\mathbf{u}_i^{\mathrm{T}} = \lfloor u\ v\ w\ \alpha\ \beta \rfloor_i \tag{8.57}$$

$$\mathbf{N}_{Ai} = \begin{bmatrix} 1 & 0 & 0 & 0 & 0 \\ 0 & 1 & 0 & 0 & 0 \\ 0 & 0 & 1 & 0 & 0 \end{bmatrix} N_i(\xi) \tag{8.58}$$

$$\mathbf{N}_{Bi} = \begin{bmatrix} 0 & 0 & 0 & l_{1i} & 0 \\ 0 & 0 & 0 & 0 & -1 \\ 0 & 0 & 0 & m_{1i} & 0 \end{bmatrix} \frac{h_i}{2} N_i(\xi) \tag{8.59}$$

For $n \neq 0$, the energy expressions for a single harmonic component are (see equations (5.11)

$$T_e = \frac{1}{2}\pi \int_{A_e} \rho(\dot{u}^2 + \dot{v}^2 + \dot{w}^2) r \,\mathrm{d}A \tag{8.60}$$

$$U_e = \frac{1}{2}\pi \int_{A_e} \{\bar{\boldsymbol{\varepsilon}}\}^{\mathrm{T}} [\bar{\mathbf{D}}] \{\bar{\boldsymbol{\varepsilon}}\} r \,\mathrm{d}A \tag{8.61}$$

$$\delta W = \frac{1}{2}\pi \int_{S_e} (p_r \delta u + p_\theta \delta v + p_z \delta w) r_s \,\mathrm{d}_s \tag{8.62}$$

where the subscript n has been omitted for convenience. The components of strain with respect to local axes $(\bar{r}, \theta, \bar{z})$ in equation (8.61) are

$$\{\bar{\boldsymbol{\varepsilon}}\}^{\mathrm{T}} = \lfloor \bar{\varepsilon}_r \bar{\varepsilon}_\theta\ \bar{\varepsilon}_z \bar{\gamma}_{zr}\ \bar{\gamma}_{r\theta}\ \bar{\gamma}_{\theta z} \rfloor \tag{8.63}$$

Assuming that the direct stress in the normal direction is zero, then

$$[\bar{\mathbf{D}}] = \frac{E}{(1-\nu^2)} \begin{bmatrix} 1 & \nu & 0 & 0 & 0 & 0 \\ \nu & 1 & 0 & 0 & 0 & 0 \\ 0 & 0 & 0 & 0 & 0 & 0 \\ 0 & 0 & 0 & \frac{\kappa}{2}(1-\nu) & 0 & 0 \\ 0 & 0 & 0 & 0 & \frac{1}{2}(1-\nu) & 0 \\ 0 & 0 & 0 & 0 & 0 & \frac{\kappa}{2}(1-\nu) \end{bmatrix} \tag{8.64}$$

for an isotropic material, and $\kappa = 5/6$ or $\pi^2/12$.

The strain components $\{\bar{\boldsymbol{\varepsilon}}\}$ can be expressed in terms of strain components $\{\boldsymbol{\varepsilon}\}$ with respect to global axes (r, θ, z) by means of the transformation

$$\{\bar{\boldsymbol{\varepsilon}}\} = [\mathbf{R}_\varepsilon^A]\{\boldsymbol{\varepsilon}\} \tag{8.65}$$

Here, $[\mathbf{R}_\varepsilon^A]$ is given by equation (A2.6) with rows and columns four and five interchanged [8.18, 8.26]. Also, $m_1 = l_2 = n_2 = m_3 = 0$ and $m_2 = 1$ since $\vec{V}_2$ does not change direction. Substituting equation (8.65) into equation (8.61) gives

$$U_e = \frac{\pi}{2}\int_{A_e}\{\boldsymbol{\varepsilon}\}^{\mathrm{T}}[\mathbf{D}]\{\boldsymbol{\varepsilon}\}r\,\mathrm{d}A \tag{8.66}$$

where

$$[\mathbf{D}] = [\mathbf{R}_\varepsilon^A]^{\mathrm{T}}[\bar{\mathbf{D}}][\mathbf{R}_\varepsilon^A] \tag{8.67}$$

From Section 4.3

$$\mathrm{d}A = \det[\mathbf{J}]\,\mathrm{d}\xi\,\mathrm{d}\eta \tag{8.68}$$

where

$$[\mathbf{J}] = \begin{bmatrix} \dfrac{\partial r}{\partial \xi} & \dfrac{\partial z}{\partial \xi} \\ \dfrac{\partial r}{\partial \eta} & \dfrac{\partial z}{\partial \eta} \end{bmatrix} \tag{8.69}$$

Substituting equation (8.51) into equation (8.69) gives

$$\begin{aligned}
\frac{\partial r}{\partial \xi} &= \sum_{i=1}^{3}\frac{\partial N_i}{\partial \xi}(r_{im} + \eta h_i l_{3i}/2) \\
\frac{\partial r}{\partial \eta} &= \sum_{i=1}^{3} N_i(h_i l_{3i}/2) \\
\frac{\partial z}{\partial \xi} &= \sum_{i=1}^{3}\frac{\partial N_i}{\partial \xi}(z_{im} + \eta h_i m_{3i}/2) \\
\frac{\partial z}{\partial \eta} &= \sum_{i=1}^{3} N_i(h_i m_{3i}/2)
\end{aligned} \tag{8.70}$$

Now, from equation (2.92)

$$[\boldsymbol{\varepsilon}] = \begin{bmatrix} \dfrac{\partial u}{\partial r} \\ \dfrac{u}{r} + \dfrac{1}{r}\dfrac{\partial v}{\partial \theta} \\ \dfrac{\partial w}{\partial z} \\ \dfrac{\partial u}{\partial z} + \dfrac{\partial w}{\partial r} \\ \dfrac{1}{r}\dfrac{\partial u}{\partial \theta} + \dfrac{\partial v}{\partial r} - \dfrac{v}{r} \\ \dfrac{\partial v}{\partial z} + \dfrac{1}{r}\dfrac{\partial w}{\partial \theta} \end{bmatrix} \tag{8.71}$$

The derivatives with respect to r, z can be converted to derivatives with respect to ξ, η by means of a relationship similar to equation (4.74), namely

$$\begin{bmatrix} \frac{\partial}{\partial r} \\ \frac{\partial}{\partial z} \end{bmatrix} = [\mathbf{J}]^{-1} \begin{bmatrix} \frac{\partial}{\partial \xi} \\ \frac{\partial}{\partial \eta} \end{bmatrix} \tag{8.72}$$

Combining equations (8.69), (8.70) and (8.72) and incorporating the result in equation (8.71) yields an expression of the form

$$\begin{aligned} \{\boldsymbol{\varepsilon}\} &= \sum_{i=1}^{3} [\mathbf{B}_i]\{\mathbf{u}_i\} \\ &= [\mathbf{B}]\{\mathbf{u}\}_e \end{aligned} \tag{8.73}$$

where

$$[\mathbf{B}] = [\mathbf{B}_1 \ \mathbf{B}_2 \ \mathbf{B}_3] \tag{8.74}$$

$$\{\mathbf{u}\}_e^{\mathrm{T}} = \lfloor \mathbf{u}_1^{\mathrm{T}} \ \mathbf{u}_2^{\mathrm{T}} \mathbf{u}_3^{\mathrm{T}} \rfloor \tag{8.75}$$

and $\mathbf{u}_i^T$ is defined in equation (8.57).

Substituting equations (8.68) and (8.73) into equation (8.66) gives

$$U_e = \tfrac{1}{2}\{\mathbf{u}\}_e^{\mathrm{T}}[\mathbf{k}]_e\{\mathbf{u}\}_e \tag{8.76}$$

where the element stiffness matrix is

$$[\mathbf{k}]_e = \pi \int_{-1}^{+1}\int_{-1}^{+1} [\mathbf{B}]^{\mathrm{T}}[\mathbf{D}][\mathbf{B}] r \det[\mathbf{J}]\, \mathrm{d}\xi\, \mathrm{d}\eta \tag{8.77}$$

The expression (8.77) is evaluated using Gauss–Legendre numerical integration with 2 integration points in each direction.

Expressing equation (8.56) in the alternative form

$$\begin{bmatrix} u \\ v \\ w \end{bmatrix} = [\mathbf{N}_A + \eta \mathbf{N}_B]\{\mathbf{u}\}_e \tag{8.78}$$

and substituting into equation (8.60) gives

$$T_e = \tfrac{1}{2}\{\dot{\mathbf{u}}\}_e^{\mathrm{T}}[\mathbf{m}]_e\{\dot{\mathbf{u}}\}_e \tag{8.79}$$

where the element inertia matrix is

$$[\mathbf{m}]_e = \pi \int_{-1}^{+1}\int_{-1}^{+1} \rho\, [\mathbf{N}_A + \eta \mathbf{N}_B]^{\mathrm{T}} [\mathbf{N}_A + \eta \mathbf{N}_B]\, r \det[\mathbf{J}]\, \mathrm{d}\xi\, \mathrm{d}\eta \tag{8.80}$$

The expression in equation (8.80) is evaluated using 3 integration points in the ξ-direction and 2 in the η-direction.

Table 8.3. *Non-dimensional frequencies* $\lambda_{mn} = \omega_{mn}(\rho/E)^{1/2}$ *for a clamped hemispherical dome*

Mode		λ_{mn}	
m	n	FEM [8.33]	Analytical [8.34]
	1	0.761	0.760
	2	0.938	0.938
0	3	0.983	0.984
	4	1.020	1.020
	5	1.070	1.071

Substituting equation (8.78) into equation (8.62) yields the equivalent nodal forces

$$\{\mathbf{f}\}_e = \pi \int_{-1}^{+1} [\mathbf{N}_A + \eta \mathbf{N}_B]^{\mathrm{T}} \begin{bmatrix} p_r \\ p_\theta \\ p_z \end{bmatrix} r_s \, \mathrm{d}s \tag{8.81}$$

From equation (8.51)

$$r_s = \sum_{i=1}^{3} N_i(\xi)(r_i + h_i l_{3i}/2) \tag{8.82}$$

and

$$\mathrm{d}s = \left[\left(\frac{\partial r}{\partial \xi} \right)^2 + \left(\frac{\partial z}{\partial \xi} \right)^2 \right]^{\frac{1}{2}} \mathrm{d}\xi \tag{8.83}$$

Additional details may be found in references [8.18, 8.26, 8.33].

Reference [8.33] analyses a clamped hemispherical dome having a thickness-to-radius ration (h/a) of 0.01 using 27 elements. The non-dimensional frequencies obtained are compared with an analytical solution in Table 8.3. The two sets of results are in close agreement. Further examples of the use of this element can be found in reference [8.33].

Problems

8.1 Derive an expression for the matrix $[\mathbf{B}_i]$ using equations (8.28), (8.31) and (8.32).

8.2 Neglecting the dependence of the matrix $[\mathbf{J}]$ on ζ, derive a simplified expression for the stiffness matrix $[\mathbf{k}]_e$ as given by equation (8.38).

8.3 Neglecting the dependence of the matrix $[\mathbf{J}]$ on ζ, derive a simplified expression for the inertia matrix $[\mathbf{m}]_e$ as given by equation (8.44)

8.4 Derive an expression for G in equation (8.47).

8.5 Derive the inertia, stiffness and equivalent nodal force matrices for the element in Figure 8.7(a) when the motion is axisymmetric.

9 Vibration of Laminated Plates and Shells

Laminated plates and shells, especially laminated composites, are being used increasingly in various engineering applications including aerospace, mechanical, marine and automotive engineering. A laminated composite consists of several layers, laminae or plys of composite material bonded together. By making appropriate choice of composite material and stacking sequence or layup the laminate can be tailored to meet various design requirements such as strength and stiffness. A composite lamina, which consists of two or more materials, can take various forms. Only unidirectional fibres embedded in a matrix material will be considered here. The stacking sequence of a laminate consisting of layers of the same material with equal thickness can be defined by listing the orientation of the fibres of each layer. For example,

$$[0°/\theta°/0°]$$

is a three-layer laminate with laminae having fibre directions of 0°, $\theta°$ and 0° respectively. Repeated layers are indicated as follows:

$$[0°/\theta_3°/0°] = [0°/\theta°/\theta°/\theta°/0°]$$

If the stacking sequence below the midplane is a mirror image of that above it, then the laminate is symmetric and is indicated by

$$[0°/\theta°]_s = [0°/\theta°/\theta°/0°]$$

An antisymmetric laminate is one whose stacking sequence is antisymmetric, that is

$$[0°/\theta°/-\theta°/0°]$$

The analysis of laminated composite plates can be classified as follows [9.1]:

1. Equivalent single-layer theories
   (a) Classical laminated plate theory
   (b) Shear deformation laminated plate theories
2. Three-dimensional elasticity theory
   (a) Traditional three-dimensional elasticity formulations
   (b) Layer-wise theories
3. Multiple model methods

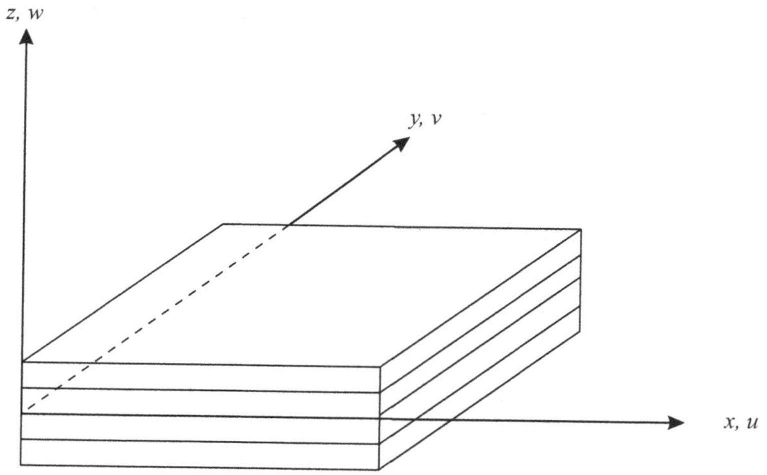

Figure 9.1. Laminated plate.

Only 1(a) and 1(b) will be treated here. Information regarding the other methods can be found in reference [9.1]. A number of reviews have appeared in the past [9.2–9.9]. A more thorough treatment of the topic can be found in references [9.1, 9.10, 9.11].

9.1 Laminated Plate Elements

9.1.1 Classical Laminated Plate Theory

Classical laminated plate theory (CLPT) makes the same assumptions as those made for a thin isotropic plate (see Section 2.6). That is, referring to Figure 9.1:

1. The direct stress in the z-direction, σ_z, is zero.
2. Normals to the middle surface of the undeformed plate remain straight and normal to the middle surface during deformation and are inextensible.

Thus, the components of displacement at a point (x, y, z) are given by

$$
\begin{aligned}
u(x, y, z) &= u_0(x, y) - z\frac{\partial w_0}{\partial x} \\
v(x, y, z) &= v_0(x, y) - z\frac{\partial w_0}{\partial y} \qquad (9.1) \\
w(x, y, z) &= w_0(x, y)
\end{aligned}
$$

where u_0, v_0 and w_0 are the components of displacement at a point (x, y) in the middle surface. It is necessary to include both the in-plane and bending displacements as they are coupled for some stacking sequences. The components of strain are as follows:

$$\begin{aligned}
\varepsilon_x &= \frac{\partial u}{\partial x} = \frac{\partial u_0}{\partial x} - z\frac{\partial^2 w_0}{\partial x^2} \\
\varepsilon_y &= \frac{\partial v}{\partial y} = \frac{\partial v_0}{\partial y} - z\frac{\partial^2 w_0}{\partial y^2} \\
\varepsilon_z &= \frac{\partial w}{\partial z} = \frac{\partial w_0}{\partial z} = 0 \\
\gamma_{xy} &= \frac{\partial u}{\partial y} + \frac{\partial v}{\partial x} = \frac{\partial u_0}{\partial y} + \frac{\partial v_0}{\partial x} - 2z\frac{\partial^2 w_0}{\partial x \partial y} \\
\gamma_{xz} &= \frac{\partial u}{\partial z} + \frac{\partial w}{\partial x} = 0 \\
\gamma_{yz} &= \frac{\partial v}{\partial z} + \frac{\partial w}{\partial y}
\end{aligned} \tag{9.2}$$

In matrix form

$$\begin{bmatrix} \varepsilon_x \\ \varepsilon_y \\ \gamma_{xy} \end{bmatrix} = \begin{bmatrix} \dfrac{\partial u_0}{\partial x} \\ \dfrac{\partial v_0}{\partial y} \\ \dfrac{\partial u_0}{\partial y} + \dfrac{\partial v_0}{\partial x} \end{bmatrix} - z \begin{bmatrix} \dfrac{\partial^2 w_0}{\partial x^2} \\ \dfrac{\partial^2 w_0}{\partial y^2} \\ 2\dfrac{\partial^2 w_0}{\partial x \partial y} \end{bmatrix} \tag{9.3}$$

That is

$$\{\boldsymbol{\varepsilon}\} = \{\boldsymbol{\varepsilon}_0\} - z\{\boldsymbol{\chi}_0\} \tag{9.4}$$

Extending equation (2.6.4) to a laminated plate gives

$$U = \frac{1}{2}\int_A \sum_{k=1}^{NL} \int_{z_{k-1}}^{z_k} \{\boldsymbol{\varepsilon}\}^{\mathrm{T}}[\mathbf{D}^{\mathrm{k}}]\{\boldsymbol{\varepsilon}\}\,\mathrm{dz}\,\mathrm{dA} \tag{9.5}$$

for the strain energy, where NL is the number of laminae and z_k, z_{k-1} are the top and bottom coordinates of the kth lamina. $[\mathbf{D}^k]$ is given by equations (2.46)–(2.50). Here, $\bar{x}$ lies in the direction of the unidirectional fibres and β is the angle between the axes $\bar{x}$ and x. Substituting equation (9.4) into equation (9.5) and integrating with respect to z gives

$$\begin{aligned}
U = \frac{1}{2}\int_A (&\{\boldsymbol{\varepsilon}_0\}^{\mathrm{T}}[\mathbf{A}]\{\boldsymbol{\varepsilon}_0\} - \{\boldsymbol{\varepsilon}_0\}^{\mathrm{T}}[\mathbf{B}]\{\boldsymbol{\chi}_0\} \\
&- \{\boldsymbol{\chi}_0\}^{\mathrm{T}}[\mathbf{B}]\{\boldsymbol{\varepsilon}_0\} + \{\boldsymbol{\chi}_0\}^{\mathrm{T}}[\mathbf{C}]\{\boldsymbol{\chi}_0\})\,\mathrm{dA}
\end{aligned} \tag{9.6}$$

where

$$\begin{aligned}
[\mathbf{A}] &= \sum_{k=1}^{NL} [\mathbf{D}^k](z_k - z_{k-1}) \\
[\mathbf{B}] &= \frac{1}{2}\sum_{k=1}^{NL} [\mathbf{D}^k](z_k^2 - z_{k-1}^2) \\
[\mathbf{C}] &= \frac{1}{3}\sum_{k=1}^{NL} [\mathbf{D}^k](z_k^3 - z_{k-1}^3)
\end{aligned} \tag{9.7}$$

Since $\bar{x}$ lies in the direction of the fibres and $\bar{y}$ is perpendicular to them, the material constants in equation (2.46) are given by

$$\begin{aligned} E_{\bar{x}} &= E_f v_f + E_m v_m, & \nu_{\overline{xy}} &= \nu_f v_f + \nu_m v_m \\ E_{\bar{y}} &= \frac{E_f E_m}{(E_f v_m + E_m v_f)}, & G_{\overline{xy}} &= \frac{G_f G_m}{(G_f v_m + G_m v_f)} \end{aligned} \tag{9.8}$$

and

$$G_f = \frac{E_f}{2(1+\nu_f)}, \qquad G_m = \frac{E_m}{2(1+\nu_m)}$$

where E_f is the modulus of the fibre, E_m the modulus of the matrix, ν_f the Poisson ratio of the fibre, ν_m the Poisson ratio of the matrix, v_f the fibre volume fraction and v_m the matrix volume fraction.

Note that the notation used here is compatible with that in Chapter 2. In the literature on laminated composites it is usual to use $[\mathbf{Q}^k]$ for $[\mathbf{D}^k]$ and $[\mathbf{D}]$ for $[\mathbf{C}]$.

Extending equation (2.68) to a laminated plate gives

$$T = \frac{1}{2}\int_A \sum_{k=1}^{NL} \int_{z_{k-1}}^{z_k} \rho^k \dot{w}_0^2 \, \mathrm{d}z \, \mathrm{d}A \tag{9.9}$$

for the kinetic energy, where ρ^k is the density of the kth layer. In using this expression it has been assumed that in-plane and rotary inertias are negligible. Now

$$\begin{aligned} \sum_{k=1}^{NL} \int_{z_{k-1}}^{z_k} \rho^k \mathrm{d}z &= \sum_{k=1}^{NL} \rho^k (z_k - z_{k-1}) \\ &= \rho h \end{aligned} \tag{9.10}$$

where h is the thickness of the laminate and ρ the average density. Therefore, equation (9.9) becomes

$$T = \frac{1}{2}\int_A \rho h \, \dot{w}_0^2 \, \mathrm{d}A \tag{9.11}$$

The virtual work of the transverse loading is (see equation (2.69))

$$\delta w = \int_A p_z \, \delta w_0 \, \mathrm{d}A \tag{9.12}$$

The highest derivative of u_0, v_0 is the first and of w_0 the second in equation (9.6). Therefore, any of the displacement functions presented in Chapter 4 can be used to represent the in-plane displacements of the laminate and any of those presented in Chapter 6 for the flexural displacements.

It should be emphasised that CLPT should only be used to analyse thin plates. Reference [9.12] suggests an in-plane dimension to thickness ratio of $a/h > 50$.

Reference [9.13] analyses several square cantilevered composite panels of dimensions 0.2032 m. The material properties are

$$\begin{aligned} E_{\bar{x}} &= 160.648 \times 10^9 \,\mathrm{N/m^2} \\ E_{\bar{y}} &= 12.479 \times 10^9 \,\mathrm{N/m^2} \\ G_{\overline{xy}} &= 0.22 \,\mathrm{N/m^2} \\ \rho &= 1910 \,\mathrm{kg/m^2} \end{aligned}$$

The natural frequencies and modes of free vibration were calculated using an 8×8 mesh of rectangular elements. The in-plane displacement functions used are those

Table 9.1. *Natural frequencies of square cantilevered composite panels [9.13]*

Stacking sequence	$[45_4]_s$		$[0_4/67.5_4]$	
Thickness (mm)	1.123		1.194	
	Frequencies (Hz)			
	FEM	Experimental	FEM	Experimental
	17.1	17.4	21.8	21.8
	50.7	56.1	43.4	43.3
	103	107	130	132
	124	143	138	149
	193	209	177	184
	237	238	254	177

described in Section 4.2, whilst the out-of-plane displacements were taken to be the ones presented in Section 6.1. The frequencies obtained for two of the panels are compared with experimentally measured ones in Table 9.1. In most cases, the two sets are in close agreement.

9.1.2 First-Order Shear Deformation Plate Theory

First-order shear deformation plate theory (FSDT) makes the same assumptions as those made for a thick isotropic plate (see Section 2.7). That is, referring to Figure 9.1:

1. The direct stress in the z-direction, σ_z, is zero.
2. Normals to the middle surface of the undeformed plate remain straight and are inextensible. However, they do not remain normal to the middle surface during deformation.

Thus, the components of displacement at a point (x, y, z) are given by

$$\begin{aligned} u(x, y, z) &= u_0(x, y) + z\,\theta_y(x, y) \\ v(x, y, z) &= v_0(x, y) - z\,\theta_x(x, y) \\ w(x, y, z) &= w_0(x, y) \end{aligned} \tag{9.13}$$

where θ_x, θ_y are rotations about the x- and y-axes of lines originally normal to the middle plane before deformation. The components of strain are as follows:

$$\begin{aligned} \varepsilon_x &= \frac{\partial u}{\partial x} = \frac{\partial u_0}{\partial x} + z\frac{\partial \theta_y}{\partial x} \\ \varepsilon_y &= \frac{\partial v}{\partial y} = \frac{\partial v_0}{\partial y} - z\frac{\partial \theta_x}{\partial y} \\ \varepsilon_z &= \frac{\partial w}{\partial z} = \frac{\partial w_0}{\partial z} = 0 \\ \gamma_{xy} &= \frac{\partial u}{\partial y} + \frac{\partial v}{\partial x} = \frac{\partial u_0}{\partial y} + \frac{\partial v_0}{\partial x} - z\left(\frac{\partial \theta_x}{\partial x} - \frac{\partial \theta_y}{\partial y}\right) \\ \gamma_{xz} &= \frac{\partial u}{\partial z} + \frac{\partial w}{\partial x} = \theta_y + \frac{\partial w_0}{\partial x} \\ \gamma_{yz} &= \frac{\partial v}{\partial z} + \frac{\partial w}{\partial y} = -\theta_x + \frac{\partial w_0}{\partial y} \end{aligned} \tag{9.14}$$

Note that for a thin plate, γ_{xz} and γ_{yz} are negligible. Therefore,

$$\theta_y = -\frac{\partial w_0}{\partial x}, \qquad \theta_x = \frac{\partial w_0}{\partial y}$$

In this case, equations (9.14) reduce to equation (9.2).

In matrix form equation (9.14) becomes

$$\begin{bmatrix} \varepsilon_x \\ \varepsilon_y \\ \gamma_{xy} \end{bmatrix} = \begin{bmatrix} \partial u_0/\partial x \\ \partial v_0/\partial y \\ \partial u_0/\partial y + \partial v_0/\partial x \end{bmatrix} - z \begin{bmatrix} -\partial \theta_y/\partial x \\ \partial \theta_x/\partial y \\ \partial \theta_x/\partial x - \partial \theta_y/\partial y \end{bmatrix} \tag{9.15}$$

That is

$$\{\boldsymbol{\varepsilon}\} = \{\boldsymbol{\varepsilon}_0\} - z\{\boldsymbol{\chi}_0\} \tag{9.16}$$

and

$$\{\boldsymbol{\gamma}\} = \begin{bmatrix} \gamma_{xz} \\ \gamma_{yz} \end{bmatrix} = \begin{bmatrix} \theta_y + \partial w_0/\partial x \\ -\theta_x + \partial w_0/\partial y \end{bmatrix} = \{\boldsymbol{\gamma}_0\} \tag{9.17}$$

Extending equation (2.75) to a laminated plate gives

$$U = \frac{1}{2}\int_A \sum_{k=1}^{NL} \int_{z_{k-1}}^{z_k} \{\boldsymbol{\varepsilon}\}^T[\mathbf{D}^k]\{\boldsymbol{\varepsilon}\}\,\mathrm{d}z\,\mathrm{d}A + \frac{1}{2}\int_A \sum_{k=1}^{NL} \int_{z_{k-1}}^{z_k} \{\boldsymbol{\tau}\}^T\{\boldsymbol{\gamma}_0\}\,\mathrm{d}z\,\mathrm{d}A \tag{9.18}$$

$[\mathbf{D}^k]$ is given by equations (2.46)–(2.50).

The relationship between the average shear stresses, $\{\boldsymbol{\tau}\}$, and the shear strains, as given by equation (9.17) is

$$\{\boldsymbol{\tau}\} = \begin{bmatrix} \tau_{xz} \\ \tau_{yz} \end{bmatrix} = \boldsymbol{\kappa}[\mathbf{D}^s]\{\boldsymbol{\gamma}_0\} \tag{9.19}$$

where κ is a constant which is introduced to account for the variation of the shear stresses and strains through the thickness. It is difficult to determine for arbitrary laminated plates (see, for example, references [9.14–9.16]). $[\mathbf{D}^s]$ is given by equation (2.77).

Substituting equations (9.16) and (9.19) into equation (9.18) and integrating with respect to z gives

$$U = \frac{1}{2}\int_A \left(\{\boldsymbol{\varepsilon}_0\}^{\mathrm{T}}[\mathbf{A}]\{\boldsymbol{\varepsilon}_0\} - \{\boldsymbol{\varepsilon}_0\}^{\mathrm{T}}[\mathbf{B}]\{\boldsymbol{\chi}_0\} - \{\boldsymbol{\chi}_0\}^{\mathrm{T}}[\mathbf{B}]\{\boldsymbol{\varepsilon}_0\} + \{\boldsymbol{\chi}_0\}^{\mathrm{T}}[\mathbf{C}]\{\boldsymbol{\chi}_0\}\right)\mathrm{d}A + \frac{1}{2}\int_A \kappa\{\boldsymbol{\gamma}_0\}^{\mathrm{T}}[\mathbf{A}^s]\{\boldsymbol{\gamma}_0\}\,\mathrm{d}A \tag{9.20}$$

$[\mathbf{A}]$, $[\mathbf{B}]$ and $[\mathbf{C}]$ are given by equation (9.7) and

$$[\mathbf{A}^s] = \sum_{k=1}^{NL} [\mathbf{D}^s](z_k - z_{k-1}) \tag{9.21}$$

Extending equation (2.79) to a laminated plate gives

$$T = \frac{1}{2}\int_A \sum_{k=1}^{NL} \int_{z_{k-1}}^{z_k} \rho^k (\dot{u}^2 + \dot{v}^2 + \dot{w}^2)\,\mathrm{d}z\,\mathrm{d}A \tag{9.22}$$

Neglecting in-plane inertias equations (9.13) reduce to

$$\begin{aligned} u(x, y, z) &= z\theta_y(x, y) \\ v(x, y, z) &= -z\theta_x(x, y) \\ w(x, y, z) &= w_0(x, y, z) \end{aligned} \tag{9.23}$$

Substituting equation (9.23) into equation (9.22) and integrating with respect to z gives

$$T = \frac{1}{2}\int_A (\rho h \dot{w}_0^2 + I_2\dot{\theta}_x^2 + I_2\dot{\theta}_y^2)\,\mathrm{d}A \tag{9.24}$$

where ρh is given by equation (9.10) and

$$I_2 = \sum_{k=1}^{NL} \int_{z_{k-1}}^{z_k} \rho^k z^2\,\mathrm{d}z = \frac{1}{3}\sum_{k=1}^{NL} \rho^k (z_k^3 - z_{k-1}^3) \tag{9.25}$$

The virtual work of the transverse loading is (see equation (2.69))

$$\delta w = \int_A p_z \delta w_0\,\mathrm{d}A \tag{9.26}$$

The highest derivative of u_0, v_0 w_0 θ_x, θ_y in equation (9.20) is the first. Therefore, any of the displacement functions presented in Chapter 4 can be used to represent all five components of displacement.

Reference [9.17] analyses several square and rectangular, simply supported plates. One square plate has a stacking sequence of [45/−45/45/−45] and the following dimensionless material properties

$$\begin{aligned} E_{\bar{x}}/E_{\bar{y}} &= 40 \\ G_{\overline{xy}}/E_{\bar{y}} &= 0.6 \\ \nu_{\overline{xy}} &= 0.25 \\ G_{\overline{xz}}/E_{\bar{y}} &= G_{y\bar{z}}/E_{\bar{y}} = 0.5 \end{aligned}$$

The non-dimensional natural frequencies $\lambda = \omega a^2(\rho/E_{\bar{y}}h^2)^{1/2}$ were calculated using a 2×2 mesh for a half plate and then a 2×2 mesh for a quarter plate. Here, a is the side length, h the thickness and ρ the average density. The displacement functions used are those described for an eight-node rectangle in Section 4.6. The frequencies obtained are compared with an analytical solution [9.18] in Table 9.2. Here, m, n denote the number of half-waves in the x- and y-directions. It can be seen that increasing the number of elements improves the accuracy, especially the (3, 3) mode.

9.1.3 Third-Order Shear Deformation Plate Theory

The assumptions made in the FSDT lead to constant shear stresses through the thickness, whereas three-dimensional analysis indicates that the variation should be

Table 9.2. *Non-dimensional natural frequencies of a simply supported square laminated plate (a/h = 10) [45/−45/45/−45]*

Mode m, n	FEM [9.17]		Analytical [9.18]
	Half plate	Quarter plate	
1, 1	18.259	18.609	18.46
1, 2	35.585	–	34.87
1, 3	54.3675	54.360	54.27
2, 3	70.315	–	67.17
1, 4	79.315	–	75.58
3, 3	99.5975	83.975	82.84

at least quadratic. In an attempt to overcome this, a shear correction factor is introduced. The need for such a factor can be overcome by expanding the components of displacement in terms of the thickness coordinate to a higher degree. For example, for a third-order theory the displacements are expressed as follows:

$$\begin{aligned} u &= u_0 + z\theta_y + z^2\phi_x + z^3\lambda_x \\ v &= v_0 - z\theta_x + z^2\phi_y + z^3\lambda_y \\ w &= w_0 \end{aligned} \tag{9.27}$$

These expressions introduce additional unknowns (ϕ_x, ϕ_y) and (λ_x, λ_y) which are difficult to interpret in physical terms. References [9.19, 9.20] overcome this by requiring the shear stresses (τ_{xz}, τ_{yz}) to vanish on the top and bottom faces of the laminate. That is

$$\tau_{xz}(x, y, \pm h/2) = 0, \qquad \tau_{yz}(x, y, \pm h/2) = 0 \tag{9.28}$$

Expressing equation (9.28) in terms of strains gives

$$\begin{aligned} \tau_{xz}(x, y, \pm h/2) &= D^s_{11}\gamma_{xz}(x, y, \pm h/2) + D^s_{12}\gamma_{yz}(x, y, \pm h/2) \\ \tau_{yz}(x, y, \pm h/2) &= D^s_{21}\gamma_{xz}(x, y, \pm h/2) + D^s_{22}\gamma_{yz}(x, y, \pm h/2) \end{aligned} \tag{9.29}$$

For the stresses to vanish for arbitrary D^s_{ij}

$$\gamma_{xz}(x, y, \pm h/2) = 0, \qquad \gamma_{yz}(x, y, \pm h/2) = 0 \tag{9.30}$$

Substituting equation (9.27) into equation (9.30) gives

$$\begin{aligned} \theta_y + \frac{\partial w_0}{\partial x} + (2z\phi_x + 3z^2\lambda_x)_{z=\pm h/2} &= 0 \\ -\theta_x + \frac{\partial w_0}{\partial y} + (2z\phi_y + 3z^2\lambda_y)_{z=\pm h/2} &= 0 \end{aligned} \tag{9.31}$$

From equation (9.31) the following relationships can be deduced

$$\begin{aligned} \phi_x &= 0, \qquad \lambda_x = -\frac{4}{3h^2}\left(\theta_y + \frac{\partial w_0}{\partial x}\right) \\ \phi_y &= 0, \qquad \lambda_y = -\frac{4}{3h^2}\left(\theta_x + \frac{\partial w_0}{\partial y}\right) \end{aligned} \tag{9.32}$$

Substituting equation (9.32) into equation (9.27) gives

$$
\begin{aligned}
u &= u_0 + z\theta_y - \frac{4}{3h^2}z^3\left(\theta_y + \frac{\partial w_0}{\partial x}\right)\\
v &= v_0 + z\theta_x - \frac{4}{3h^2}z^3\left(-\theta_x + \frac{\partial w_0}{\partial y}\right)\\
w &= w_0
\end{aligned}
\tag{9.33}
$$

Expressions (9.33) now involve only the components of displacement (u_0, v_0, w_0) and the rotations (θ_x, θ_y), as is the case for the FSDT. Note that in the case of a thin plate

$$
\theta_y = -\frac{\partial w_0}{\partial x}, \qquad \theta_x = \frac{\partial w_0}{\partial y}
\tag{9.34}
$$

and so equations (9.33) reduce to equation (9.1) for the CLPT.

Using equations (9.33) the components of strain are

$$
\begin{aligned}
\varepsilon_x &= \frac{\partial u}{\partial x} = \frac{\partial u_0}{\partial x} + z\frac{\partial \theta_y}{\partial x} - c_1 z^3\left(\frac{\partial \theta_y}{\partial x} + \frac{\partial^2 w_0}{\partial x^2}\right)\\
\varepsilon_y &= \frac{\partial v}{\partial y} = \frac{\partial v_0}{\partial y} - z\frac{\partial \theta_x}{\partial y} - c_1 z^3\left(-\frac{\partial \theta_x}{\partial y} + \frac{\partial^2 w_0}{\partial y^2}\right)\\
\varepsilon_z &= \frac{\partial w}{\partial z} = \frac{\partial w_0}{\partial z} = 0\\
\gamma_{xy} &= \frac{\partial u}{\partial y} + \frac{\partial v}{\partial x} = \frac{\partial u_0}{\partial y} + \frac{\partial v_0}{\partial x} - z\left(\frac{\partial \theta_x}{\partial x} - \frac{\partial \theta_y}{\partial y}\right)\\
&\quad - c_1 z^3\left(\frac{\partial \theta_y}{\partial y} - \frac{\partial \theta_x}{\partial x} + 2\frac{\partial^2 w_0}{\partial x \partial y}\right)
\end{aligned}
\tag{9.35}
$$

where $c_1 = 4/3h^2$;

$$
\begin{aligned}
\gamma_{xz} &= \frac{\partial u}{\partial z} + \frac{\partial w}{\partial x} = \theta_y + \frac{\partial w_0}{\partial x} - c_2 z^2\left(\theta_y + \frac{\partial w_0}{\partial x}\right)\\
\gamma_{yz} &= \frac{\partial v}{\partial z} + \frac{\partial w}{\partial y} = -\theta_x + \frac{\partial w_0}{\partial y} - c_2 z^2\left(-\theta_x + \frac{\partial w_0}{\partial y}\right)
\end{aligned}
\tag{9.36}
$$

where $c_2 = 3c_1$.

In matrix form

$$
\{\boldsymbol{\varepsilon}\} = \{\boldsymbol{\varepsilon}_0\} - z\{\boldsymbol{\chi}_0\} - z^3\{\boldsymbol{\chi}_2\}
\tag{9.37}
$$

where $\{\boldsymbol{\varepsilon}\}$, $\{\boldsymbol{\varepsilon}_0\}$ and $\{\boldsymbol{\chi}_0\}$ are defined by equations (9.15) and (9.16) and

$$
\{\boldsymbol{\chi}_2\} = c_1 \begin{bmatrix} \dfrac{\partial \theta_y}{\partial x} + \dfrac{\partial^2 w_0}{\partial x^2} \\[2ex] -\dfrac{\partial \theta_x}{\partial y} + \dfrac{\partial^2 w_0}{\partial y^2} \\[2ex] \dfrac{\partial \theta_y}{\partial y} - \dfrac{\partial \theta_x}{\partial x} + 2\dfrac{\partial^2 w_0}{\partial x \partial y} \end{bmatrix}
\tag{9.38}
$$

Also,

$$\{\boldsymbol{\gamma}\} = \{\boldsymbol{\gamma}_0\} - z^2\{\boldsymbol{\gamma}_2\} \tag{9.39}$$

where $\{\boldsymbol{\gamma}\}$ and $\{\boldsymbol{\gamma}_0\}$ are defined by equation (9.17) and

$$\{\boldsymbol{\gamma}_2\} = c_2 \begin{bmatrix} \theta_y + \dfrac{\partial w_0}{\partial x} \\ -\theta_x + \dfrac{\partial w_0}{\partial y} \end{bmatrix}. \tag{9.40}$$

Note that equation (9.39) indicates that the shear strains, and hence the shear stresses, vary quadratically.

Introducing

$$\{\boldsymbol{\tau}\} = [\mathbf{D}^s]\{\boldsymbol{\gamma}\} \tag{9.41}$$

into equation (2.75) produces the following expression for the strain energy

$$\begin{aligned} U = {} & \frac{1}{2}\int_A \sum_{k=1}^{NL} \int_{z_{k-1}}^{z_k} \{\boldsymbol{\varepsilon}\}^{\mathrm{T}}[\mathbf{D}^{\mathrm{k}}]\{\boldsymbol{\varepsilon}\}\,\mathrm{d}z\,\mathrm{dA} \\ & + \frac{1}{2}\int_A \sum_{k=1}^{NL} \int_{z_{k-1}}^{z_k} \{\boldsymbol{\gamma}\}^{\mathrm{T}}[\mathbf{D}^{\mathrm{s}}]\{\boldsymbol{\gamma}\}\,\mathrm{d}z\,\mathrm{dA} \end{aligned} \tag{9.42}$$

Substituting equations (9.37) and (9.39) into equation (9.42) and integrating with respect to z gives

$$\begin{aligned} U = {} & \frac{1}{2}\int_A (\{\boldsymbol{\varepsilon}_0\}^{\mathrm{T}}[\mathbf{A}]\{\boldsymbol{\varepsilon}_0\} - \{\boldsymbol{\varepsilon}_0\}^{\mathrm{T}}[\mathbf{B}]\{\boldsymbol{\chi}_0\} - \{\boldsymbol{\varepsilon}_0\}^{\mathrm{T}}[\mathbf{E}]\{\boldsymbol{\chi}_2\} \\ & - \{\boldsymbol{\chi}_0\}^{\mathrm{T}}[\mathbf{B}]\{\boldsymbol{\varepsilon}_0\} + \{\boldsymbol{\chi}_0\}^{\mathrm{T}}[\mathbf{C}]\{\boldsymbol{\chi}_0\} - \{\boldsymbol{\chi}_0\}^{\mathrm{T}}[\mathbf{F}]\{\boldsymbol{\chi}_2\} \\ & - \{\boldsymbol{\chi}_2\}^{\mathrm{T}}[\mathbf{E}]\{\boldsymbol{\varepsilon}_0\} + \{\boldsymbol{\chi}_2\}^{\mathrm{T}}[\mathbf{F}]\{\boldsymbol{\chi}_0\} + \{\boldsymbol{\chi}_2\}^{\mathrm{T}}[\mathbf{H}]\{\boldsymbol{\chi}_2\})\,\mathrm{dA} \\ & + \frac{1}{2}\int_A (\{\boldsymbol{\gamma}_0\}^{\mathrm{T}}[\mathbf{A}^{\mathrm{s}}]\{\boldsymbol{\gamma}_0\} - \{\boldsymbol{\gamma}_0\}^{\mathrm{T}}[\mathbf{C}^{\mathrm{s}}]\{\boldsymbol{\gamma}_2\} \\ & - \{\boldsymbol{\gamma}_2\}^{\mathrm{T}}[\mathbf{C}^{\mathrm{s}}]\{\boldsymbol{\gamma}_0\} + \{\boldsymbol{\gamma}_2\}^{\mathrm{T}}[\mathbf{F}^{\mathrm{s}}]\{\boldsymbol{\gamma}_2\})\,\mathrm{dA} \end{aligned} \tag{9.43}$$

where $[\mathbf{A}]$, $[\mathbf{B}]$ and $[\mathbf{C}]$ are given by equation (9.7) and

$$\begin{aligned} [\mathbf{E}] &= \frac{1}{4}\sum_{k=1}^{NL}[\mathbf{D}^k](z_k^4 - z_{k-1}^4) \\ [\mathbf{F}] &= \frac{1}{5}\sum_{k=1}^{NL}[\mathbf{D}^k](z_k^5 - z_{k-1}^5) \\ [\mathbf{H}] &= \frac{1}{7}\sum_{k=1}^{NL}[\mathbf{D}^k](z_k^7 - z_{k-1}^7) \end{aligned} \tag{9.44}$$

Also, $[\mathbf{A}^s]$ is given by equation (9.21) and

$$[\mathbf{C}^s] = \frac{1}{3}\sum_{k=1}^{NL}[\mathbf{D}^s](z_k^3 - z_{k-1}^3)$$
$$[\mathbf{F}^s] = \frac{1}{5}\sum_{k=1}^{NL}[\mathbf{D}^s](z_k^5 - z_{k-1}^5) \tag{9.45}$$

Substituting equation (9.33) into equation (9.22), neglecting in-plane inertias and integrating with respect to z gives

$$T = \frac{1}{2}\int_A \left(K_2\dot{\theta}_x^2 + K_2\dot{\theta}_y^2 + I_0\dot{w}^2 + c_1^2 I_6 \left(\frac{\partial \dot{w}}{\partial x}\right)^2 + c_1^2 I_6 \left(\frac{\partial \dot{w}}{\partial y}\right)^2 + 2c_1 J_4 \dot{\theta}_x \frac{\partial \dot{w}_0}{\partial y} - 2c_1 J_4 \dot{\theta}_y \frac{\partial \dot{w}_0}{\partial x} \right) \mathrm{d}A \tag{9.46}$$

where

$$K_2 = I_2 - 2c_1 I_4 + c_1^2 I_6$$
$$J_4 = I_4 - c_1 I_6 \tag{9.47}$$
$$I_i = \sum_{k=1}^{NL}\int_{z_{k-1}}^{z_k} \rho^k (z)^i \, \mathrm{d}z \qquad (i = 0, 2, 4, 6)$$

Once again the virtual work of the transverse loading is given by equation (9.26).

The highest derivative of u_0, v_0, θ_x, θ_y is the first and of w_0 is the second in equation (9.43). Therefore, any of the displacement functions presented in Chapter 4 can be used to represent u, v, θ_x, θ_y and any of the ones presented in Chapter 6 for w.

Reference [9.1] analyses the fundamental frequency of several anti-symmetric cross-ply square plates having various boundary conditions. The dimensionless material properties of a lamina are

$$E_{\bar{x}}/E_{\bar{y}} = 40$$
$$G_{\overline{xy}}/E_{\bar{y}} = G_{\overline{xz}}/E_{\bar{y}} = 0.6$$
$$\nu_{\overline{xy}} = 0.25$$
$$G_{\overline{yz}}/E_{\bar{y}} = 0.5$$

The non-dimensional frequencies $\lambda = \omega a^2(\rho/E_{\bar{y}}h^2)^{1/2}$ were calculated using a 4×4 mesh of elements. Here, a is the side length, h the thickness and ρ the average density. The displacement functions used are those described for a four-node element in Section 4.2 for $u_0, v_0, \theta_x, \theta_y$ and the four-node conforming element in Section 6.2 for w_0. The frequencies obtained are compared with an analytical solution in Table 9.3 for some of the boundary conditions considered. All plates have one pair of opposite sides simply supported. The other pair are either simply supported (SS) or clamped (CC). Reference [9.22] has also produced some results using the four-node non-conforming element in Section 6.1 for w_0 which are also included in Table 9.3.

Table 9.3. *Non-dimensional fundamental frequencies of antisymmetric cross-ply square plates*

NL	a/h	Solution	SS	CC
2	5	Analytical	9.087	11.890
		FEM [9.1]	9.103	12.053
		FEM [9.22]	9.194	11.972
	10	Analytical	10.568	15.709
		FEM [9.1]	10.594	15.914
10	5	Analytical	11.673	13.569
		FEM [9.1]	11.664	13.710
	10	Analytical	15.771	20.831
		FEM [9.1]	15.787	20.493

Further results may be found in references [9.23, 9.24]. References [9.1, 9.25] compare the natural frequencies obtained using all three theories presented in this section.

Some laminates have piezo electric layers either on the surface or embedded in them. They act as sensors and actuators for use in active control of vibration and noise. Details of the analysis of such configurations can be found in references [9.26–9.28].

9.2 Laminated Shell Elements

Finite elements for laminated shells can be derived by combining the techniques presented in Sections 8.2 and 9.1. Details and reviews are contained in references [9.1, 9.7, 9.9, 9.11, 9.29].

9.3 Sandwich Plate and Shell Elements

A sandwich plate or shell consists of two skins made of metal or laminate composite and a core. The traditional core is usually made of metallic honeycomb, which is transversely stiff and very flexible in-plane. Modern cores are made of light materials, such as plastic foams or non-metallic honeycombs, which generally have low rigidity. They are regarded as being transversely flexible (soft) when the ratio of Young's moduli of the face sheets to the core is high, usually between 500 and 1000.

The analysis of sandwich plates can be grouped into two classes: classical and modern. In the former, it is assumed that

1. the core has negligible in-plane stiffness and supports only shear
2. the transverse displacement at any cross section is constant
3. the bonding between face sheets and the core is thin and perfect
4. the deformations and strains are small

It has been shown [9.30] that classical laminate and first-order shear deformation theories can be inadequate for predicting the dynamic characteristics of sandwich plates. Because of this the modern approaches are higher order displacement models. For transversely stiff cores it is still assumed that the transverse

Table 9.4. *Natural frequencies (Hz) of a three-layer simply supported rectangular sandwich plate*

Mode	Analytical [9.37]	Experimental [9.37]	FEM [9.36]
1	23	–	23.63
2	45	45	46.10
3	71	69	74.20
4	80	78	86.19
5	91	92	96.36

displacement is constant. However, for soft cores it is assumed that the core transverse displacement varies through the core thickness as a polynomial of at least second order. Further information regarding sandwich construction can be found in references [9.31–9.33] whilst references [9.34, 9.35] are more recent reviews of analysis methods.

Reference [9.36] analyses a three layer, simply supported, sandwich plate of dimensions 1.83 × 1.22 m. The thickness of the face plates is 0.406 mm and of the core 6.44 mm. The material properties for the face plates are

$$E_x = E_y = 68.9 \times 10^9 \text{ N/m}^2$$

$$\nu_{xy} = 0.30$$

$$\rho = 2770 \text{ kg/m}^3$$

and for the core are

$$G_{xz} = 0.134 \times 10^9 \text{ N/m}^2$$

$$G_{yz} = 0.052 \times 10^9 \text{ N/m}^2$$

$$\rho = 122 \text{ kg/m}^3$$

The natural frequencies were calculated using a 6 × 6 mesh of rectangular elements. The third-order shear deformation theory described in Section 9.1.3 was used. However, by putting

$$-\theta_x + \frac{\partial w_0}{\partial y} = \psi_x$$

$$\theta_y + \frac{\partial w_0}{\partial x} = \psi_y$$

in equation (9.33) and taking u_0, v_0, w_0, θ_x, θ_y, ψ_x, ψ_y as variables, it is possible to use any of the displacement functions presented in Chapter 4 for all of them. In this case, the ones described in Section 4.7 for a four-node rectangle were used. In addition, the transverse shear strains were treated in the same way as the MITC4 element in Section 6.7. The frequencies obtained are compared with analytical and experimental ones in Table 9.4.

10 Hierarchical Finite Element Method

In previous chapters, it is demonstrated that increased accuracy can be obtained by refining the mesh. As a result, the number of finite elements increases and their width, h, decreases. Consequently, this approach is often referred to as the h-version.

Another way of improving the accuracy is to keep the finite element mesh constant and to increase the number of displacement functions within each element. When polynomials are used, this approach implies an increase in their degree p, and so is referred to as the p-version. If the set of functions corresponding to an approximation of order p constitutes a subset of the set of functions corresponding to an approximation of order $(p+1)$, then the p-version is called the 'hierarchical finite element method' (HFEM).

The HFEM has several advantages over the h-version:

(1) It does not require a change in the mesh to improve the accuracy of the solution.
(2) The stiffness and inertia matrices possess the embedding property, that is, the associated matrices for p_1 displacement functions are always submatrices of those for $p_2 > p_1$ functions [10.1].
(3) The embedding property can be used to prove that the eigenvalues corresponding to (p_1+1) displacement functions bracket the eigenvalues corresponding to p_1 functions. This is known as the inclusion principle [10.1].
(4) Simple structures can be modelled using just one element which avoids the assembly of elements.
(5) Joining elements of different polynomial degree is not difficult. Therefore, it is possible to include additional degrees of freedom where needed [10.2].
(6) The HFEM tends to give accurate results with far fewer degrees of freedom than the h-version [10.1, 10.3].

10.1 Polynomial Functions

The most commonly used hierarchical functions are derived from Rodrigues form of Legendre polynomials [10.3–10.5], namely

$$P_m(\xi) = \frac{1}{m!(-2)^m} \frac{\mathrm{d}^m}{\mathrm{d}\xi^m}\{(1-\xi^2)^m\} \qquad -1 \le \xi \le +1 \tag{10.1}$$

Expanding equation (10.1) using the binomial theorem gives the following expression [10.3]

$$P_m(\xi) = \sum_{n=0}^{m/2} \frac{(-1)^n}{2^n n!} \frac{(2m-2n-1)!!}{(m-2n)!} \xi^{m-2n} \tag{10.2}$$

where

$$m!! = m(m-2)\ldots(2 \text{ or } 1)$$

$$0!! = 1, \; (-1)!! = 1$$

and $m/2$ denotes its own integer part.

Let $P_m^s(\xi)$ denote the following s-multiple integral of $P_{m-s}(\xi)$

$$P_m^s(\xi) = \int_{-1}^{\xi} \cdots \int_{-1}^{\xi} P_{m-s}(\xi)\,\mathrm{d}\xi \cdots \mathrm{d}\xi \tag{10.3}$$

Substituting equation (10.2) into equation (10.3) gives

$$P_m^s(\xi) = \sum_{n=0}^{m/2} \frac{(-1)^n}{2^n n!} \frac{(2m-2n-2s-1)!!}{(m-2n)!} \xi^{m-2n} \tag{10.4}$$

These functions have the property that all their derivatives of order lower than s vanish at $\xi = \pm 1$. Not only are they orthogonal, but it is demonstrated in the following sections that they are **k**-orthogonal.

10.1.1 Axial Vibration of Rods

The axial displacement, u, of a rod is expressed in the form

$$u = \sum_{r=1}^{p} g_r(\xi) q_r \tag{10.5}$$

where p is the number of assumed functions and $-1 \le \xi \le +1$. The first two functions are given by equation (3.36). That is

$$g_1(\xi) = \tfrac{1}{2}(1-\xi), \qquad g_2(\xi) = \tfrac{1}{2}(1+\xi) \tag{10.6}$$

The hierarchical functions $g_r(\xi)$ for $r > 2$ are obtained by putting $s = 1$ and $m = (r-1)$ into equation (10.4). This gives

$$g_r(\xi) = \sum_{n=0}^{(r-1)/2} \frac{(-1)^n}{2^n n!} \frac{(2r-2n-5)!!}{(r-2n-1)!} \xi^{r-2n-1} \tag{10.7}$$

The first four of these functions are

$$\begin{aligned}
g_3(\xi) &= -(1/2) + (1/2)\xi^2 \\
g_4(\xi) &= -(1/2)\xi + (1/2)\xi^3 \\
g_5(\xi) &= (1/8) - (3/4)\xi^2 + (5/8)\xi^4 \\
g_6(\xi) &= (3/8)\xi - (5/4)\xi^3 + (7/8)\xi^5
\end{aligned} \tag{10.8}$$

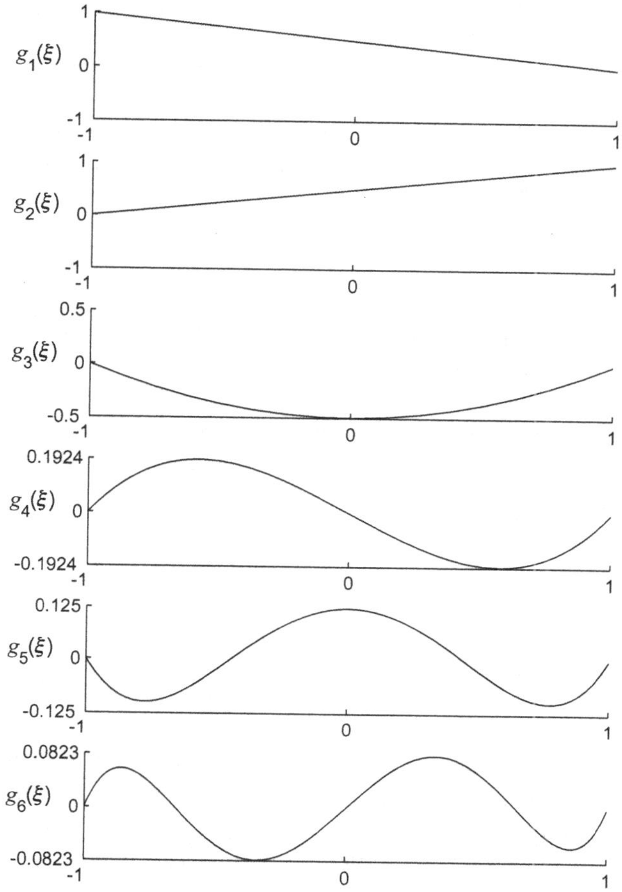

Figure 10.1. In-plane polynomial functions.

Note that all these functions are zero at $\xi = \pm 1$. Therefore, there is no increase in the number of nodal degrees of freedom and $q_1 = u_1$, and $q_2 = u_2$. The first six functions are shown in Figure 10.1.

Equation (10.5) can be written more concisely in the matrix form

$$u = \lfloor \mathbf{N}(\xi) \rfloor \{\mathbf{q}\} \tag{10.9}$$

where the row-matrix $\lfloor \mathbf{N}(\xi) \rfloor$ contains the functions $g_r(\xi)$ and the column matrix $\{\mathbf{q}\}$ the degrees of freedom q_r.

From Chapter 3, the element stiffness matrix is given by

$$[\mathbf{k}]_e = \frac{2EA}{a} \int_{-1}^{+1} \lfloor \mathbf{N}'(\xi) \rfloor^{\mathrm{T}} \lfloor \mathbf{N}'(\xi) \rfloor \, d\xi \tag{10.10}$$

where a now denotes the length of the element. Substituting equations (10.6) and (10.8) into equation (10.10) gives

$$[\mathbf{k}]_e = \frac{2EA}{a} \begin{bmatrix} 1/2 & -1/2 & 0 & 0 \\ -1/2 & 1/2 & 0 & 0 \\ 0 & 0 & 2/3 & 0 \\ 0 & 0 & 0 & 2/5 \end{bmatrix} \tag{10.11}$$

Table 10.1. *Non-dimensional natural frequencies of a fixed-free rod* $\omega(\rho L^2/E)^{1/2}$. *Polynomial functions*

Mode	FEM solutions: 1 Element	1 Element +1 Function	1 Element +2 Functions	Exact solution
1	1.732	1.577	1.571	1.571
2	–	5.673	4.837	4.712

when only two hierarchical functions are included. This illustrates the fact that the hierarchical functions used are **k**-orthogonal. The leading (2 × 2) submatrix is the stiffness matrix without the hierarchical functions as derived in Chapter 3 and the remaining diagonal terms are due to the hierarchical functions. Note that there is no coupling between themselves or between them and the nodal degrees of freedom. This is because

$$\int_{-1}^{+1} P_n(\xi)P_m(\xi)\,d\xi = 0 \qquad n \neq m \tag{10.12}$$

and

$$\int_{-1}^{+1} \xi^n P_m(\xi) = 0 \qquad 0 \leq n \leq (m-1) \tag{10.13}$$

The embedding property is also illustrated. The inclusion of one more hierarchical function gives rise to one additional row and column leaving the previous rows and columns unchanged.

From Chapter 3, the inertia matrix is given by

$$[\mathbf{m}]_e = \frac{\rho A a}{2}\int_{-1}^{+1} \lfloor \mathbf{N}(\xi)\rfloor^{\mathrm{T}} \lfloor \mathbf{N}(\xi)\rfloor\, d\xi \tag{10.14}$$

Substituting equations (10.6) and (10.8) into equation (10.14) gives

$$[\mathbf{m}]_e = \frac{\rho A a}{2}\begin{bmatrix} 2/3 & 1/3 & -1/3 & 1/15 \\ 1/3 & 2/3 & -1/3 & -1/15 \\ -1/3 & -1/3 & 4/15 & 0 \\ 1/15 & -1/15 & 0 & 4/105 \end{bmatrix} \tag{10.15}$$

Note that the functions $g_r(\xi)$ are not **m**-orthogonal.

The first two non-dimensional natural frequencies of the fixed-free rod shown in Figure 3.1 are given in Table 10.1.

10.1.2 Bending Vibration of Beams

The lateral displacement, v, of a beam is expressed in the form

$$v = \sum_{r=1}^{p} f_r(\xi)q_r \tag{10.16}$$

where p is the number of assumed functions and $-1 \leq \xi \leq +1$. The first four functions are given by equations (3.125) and (3.126) with a being replaced by $a/2$. That is

$$\begin{aligned}
f_1(\xi) &= (1/2) - (3/4)\xi + (1/4)\xi^3 \\
f_2(\xi) &= a((1/8) - (1/8)\xi - (1/8)\xi^2 + (1/8)\xi^3) \\
f_3(\xi) &= (1/2) + (3/4)\xi - (1/4)\xi^3 \\
f_4(\xi) &= a(-(1/8) - (1/8)\xi + (1/8)\xi^2 + (1/8)\xi^3)
\end{aligned} \quad (10.17)$$

The hierarchical functions $f_r(\xi)$ for $r > 4$ are obtained by putting $s = 2$ and $m = (r-1)$ into equation (10.4). This gives

$$f_r(\xi) = \sum_{n=0}^{(r-1)/2} \frac{(-1)^n}{2^n n!} \frac{(2r-2n-7)!!}{(r-2n-1)!} \xi^{(r-2n-1)} \quad (10.18)$$

The first four of these functions are

$$\begin{aligned}
f_5(\xi) &= (1/8) - (1/4)\xi^2 + (1/8)\xi^4 \\
f_6(\xi) &= (1/8)\xi - (1/4)\xi^3 + (1/8)\xi^5 \\
f_7(\xi) &= -(1/48) + (3/16)\xi^2 - (5/16)\xi^4 + (7/48)\xi^6 \\
f_8(\xi) &= -(1/16)\xi - (5/16)\xi^3 - (7/16)\xi^5 + (3/16)\xi^7
\end{aligned} \quad (10.19)$$

Note that the functions and their first derivatives are zero at $\xi = \pm 1$, and so the first four degrees of freedom are

$$q_1 = v_1, \qquad q_2 = \theta_{z1}, \qquad q_3 = v_2, \qquad q_4 = \theta_{z2} \quad (10.20)$$

The first eight functions are shown in Figure 10.2. Equation (10.16) can be written more concisely in the matrix form

$$v = \lfloor \mathbf{N}(\xi) \rfloor \{\mathbf{q}\} \quad (10.21)$$

where the row matrix $\lfloor \mathbf{N}(\xi) \rfloor$ contains the function $f_r(\xi)$ and the column matrix $\{\mathbf{q}\}$ the degrees of freedom q_r.

From Chapter 3, the element stiffness matrix is given by

$$[\mathbf{k}]_e = \frac{9EI_z}{a^3} \int_{-1}^{+1} \lfloor \mathbf{N}''(\xi) \rfloor^{\mathrm{T}} \lfloor \mathbf{N}''(\xi) \rfloor \, d\xi \quad (10.22)$$

where a denotes the length of the element.

Substituting equations (10.17) and (10.19) into equation (10.22) gives, using the first six functions

$$[\mathbf{k}]_e = \frac{EI_z}{a^3} \begin{bmatrix} 12 & 6a & -12 & 6a & 0 & 0 \\ 6a & 4a^2 & -6a & 2a^2 & 0 & 0 \\ -12 & -6a & 12 & -6a & 0 & 0 \\ 6a & 2a^2 & -6a & 4a^2 & 0 & 0 \\ 0 & 0 & 0 & 0 & 16/5 & 0 \\ 0 & 0 & 0 & 0 & 0 & 16/7 \end{bmatrix} \quad (10.23)$$

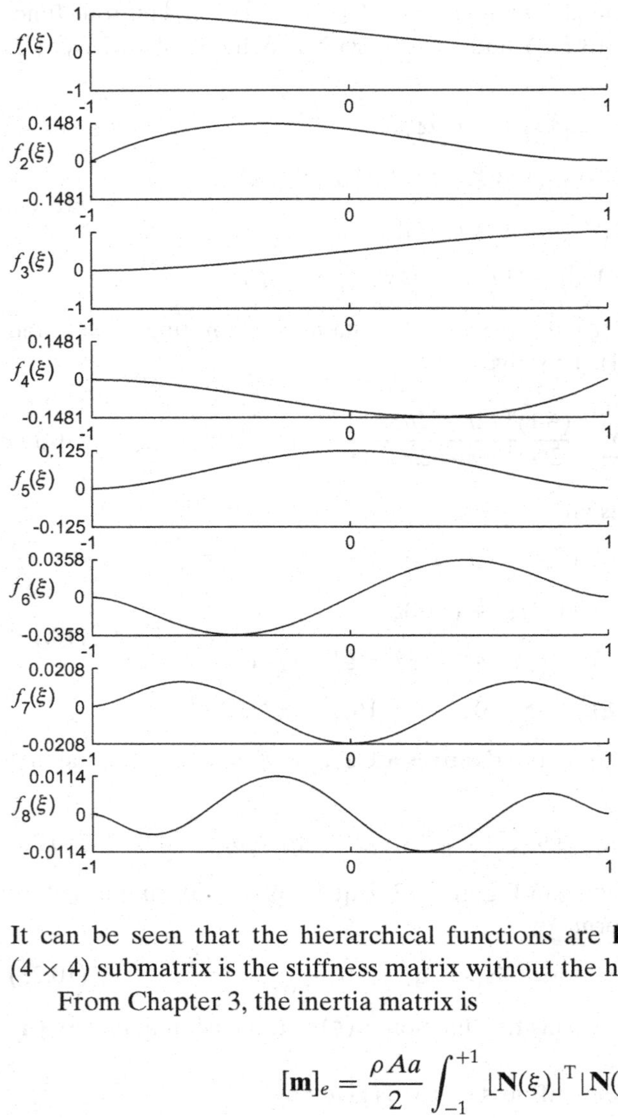

Figure 10.2. Out-of-plane polynomial functions.

It can be seen that the hierarchical functions are **k**-orthogonal and the leading (4×4) submatrix is the stiffness matrix without the hierarchical functions.

From Chapter 3, the inertia matrix is

$$[\mathbf{m}]_e = \frac{\rho A a}{2} \int_{-1}^{+1} \lfloor \mathbf{N}(\xi) \rfloor^{\mathrm{T}} \lfloor \mathbf{N}(\xi) \rfloor \, \mathrm{d}\xi \tag{10.24}$$

Substituting equations (10.17) and (10.19) into equation (10.24) gives, for six functions

$$[\mathbf{m}]_e = \frac{\rho A a}{210} \begin{bmatrix} 78 & 11a & 27 & -(13/2)a & 7 & -4/3 \\ 11a & 2a^2 & (13/2)a & -(3/2)a^2 & 15a & -a/6 \\ 27 & (13/2)a & 78 & -11a & 7 & 4/3 \\ -(13/2)a & -(3/2)a^2 & -11a & 2a^2 & -15a & -a/6 \\ 7 & 15a & 7 & -15a & 4/3 & 0 \\ -4/3 & -a/6 & 4/3 & -a/6 & 0 & 4/33 \end{bmatrix} \tag{10.25}$$

Table 10.2. *Non-dimensional natural frequencies of a cantilever beam* $\omega(\rho AL^4/EI)^{1/2}$ *[10.1]. Polynomial functions*

	FEM solutions				
	Number of elements/number of functions				
Mode	4/0	4/1	4/2	4/3	Exact solution
1	3.516	3.516	3.516	3.516	3.516
2	22.060	22.035	22.035	22.035	22.034
3	62.175	61.719	61.698	61.697	61.697
4	122.658	121.260	120.905	120.902	120.902

Reference [10.1] has studied the convergence of the natural frequencies of a cantilever beam with an increasing number of finite elements and hierarchical functions. Table 10.2 gives the results for four elements and an increasing number of hierarchical functions on all elements.

10.1.3 Flexural Vibration of Plates

The lateral displacement, w, of a thin plate is expressed in the form

$$w(\xi, \eta) = \sum_{r=1}^{p} \sum_{s=1}^{p} f_r(\xi) f_s(\eta) W_{rs} \tag{10.26}$$

where the functions $f_r(\xi)$, $f_s(\eta)$ are given by equations (10.17) and (10.19). The first 16 degrees of freedom (i.e. when both $r, s \leq 4$) are the same as the CR element in Section 6.2. The hierarchical functions also give rise to edge degrees of freedom (i.e. when one of $r, s > 4$) and internal degrees of freedom (i.e. when both $r, s > 4$). Equation (10.26) can be written more concisely in the matrix form

$$w = \lfloor \mathbf{N}(\xi, \eta) \rfloor \{\mathbf{W}\} \tag{10.27}$$

where the row matrix $\lfloor \mathbf{N}(\xi, \eta) \rfloor$ contains the functions $f_r(\xi)$ and $f_s(\eta)$ and the column matrix $\{\mathbf{W}\}$ the degrees of freedom W_{rs}.

Substituting equation (10.27) into equations (6.1) and (6.2) will produce the inertia and stiffness matrices. This will involve evaluating the integrals of products of the functions $f_r(\xi)$ and their derivatives. Similarly for the functions $f_s(\eta)$. This process has already been carried out using symbolic computing and the results tabulated in reference [10.6]. Interelement compatibility is achieved by matching the degrees of freedom at common nodes and edges.

Reference [10.7] has analysed rectangular plates, with sides a, b and thickness h, having 10 different boundary conditions. Each plate is modelled with just one finite element with various hierarchical mode order p. Table 10.3 shows the results for the first four modes of a simply supported square plate with increasing value of p which are compared with an exact solution [10.8]. Note that modes (2, 1) and (3, 1) are the same as modes (1, 2) and (1, 3), respectively.

The HFEM may also be used to calculate the vibration characteristics of non-rectangular plates. For example, reference [10.9] analyses a number of skew plates.

Table 10.3. *Non-dimensional natural frequencies of a square plate* ($a/b = 1$) *with simply supported boundaries* $\omega(\rho h a^4/D)^{1/2}$ *[10.7]. Polynomial functions*

	p			
Mode	6	8	10	Exact solution [10.8]
1, 1	19.7425	19.7392	19.7392	19.7392
1, 2	49.4916	49.3486	49.3480	49.3480
2, 2	79.1668	78.9577	78.9568	78.9568
1, 3	139.5994	100.1167	98.7162	98.6960

Figure 10.3 shows a skew plate with sides a, b, coordinates x', y' are rectangular coordinates whilst x, y are oblique coordinates. They are related as follows:

$$x = x' - y' \tan \phi, \qquad y = y' \sec \phi \tag{10.28}$$

ξ, η are now non-dimensional oblique coordinates. The lateral displacement, w, is still expressed in the form (10.26). Also, equations (6.1) and (6.2) are now expressed in terms of ξ, η. Because the displacement (10.26) is expressed in oblique coordinates, the hierarchical functions are non-orthogonal resulting in slower convergence. Table 10.4 shows the results for the first four modes of simply supported plates having skew angles of 15° and 45°. These have been calculated using a hierarchical mode order of 20. Also shown are the frequencies obtained in reference [10.10] using a different set of orthogonal functions. The two sets are in close agreement.

The HFEM can also be used to calculate the vibration characteristics of laminated plates. Reference [10.11] uses classical laminated plate theory (see Section 9.1.1) to analyse a number of square and rectangular, symmetric five-layer angle-ply plates with fully clamped edges. When formulating the HFEM the in-plane displacements of the middle-surface were assumed to be negligible. Therefore, only the transverse displacement, w, was considered which is expressed in the form (10.26).

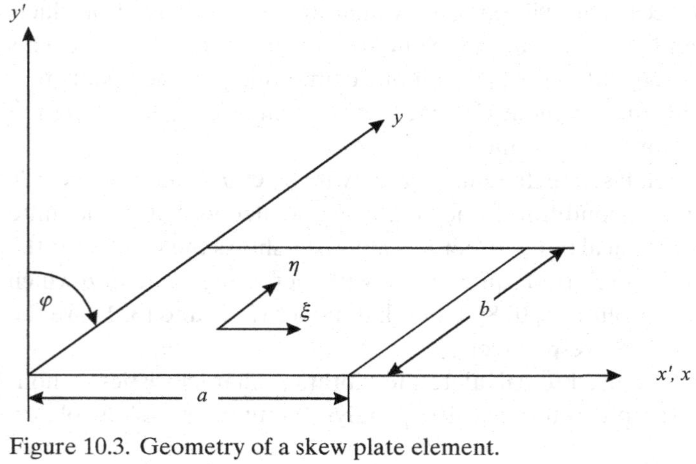

Figure 10.3. Geometry of a skew plate element.

Table 10.4. *Non-dimensional natural frequencies of a skew plate* ($a/b = 1$) *with simply supported boundaries* $\omega(\rho h a^4/D)^{1/2}$ *[10.9]. Polynomial functions*

		Mode number			
Skew angle ϕ	Method	1	2	3	4
	HFEM	20.869	48.205	56.109	79.044
15°	[10.10]	20.825	48.164	56.158	79.055
45°	HFEM	35.035	66.278	100.338	107.658
	[10.10]	34.938	66.422	100.867	107.776

Table 10.5 shows the convergence of the first four natural frequencies of a 45°, five-layer symmetrical angle-ply square plate with clamped boundaries having the following dimensionless material properties

$$E_{\bar{x}}/E_{\bar{y}} = 15.4$$
$$G_{\overline{xy}}/E_{\bar{y}} = 0.79$$
$$\nu_{\overline{xy}} = 0.30$$

The non-dimensional natural frequencies $\omega(\rho h a^4/D_0)^{1/2}$, where $D_0 = E_{\bar{x}}h^3/12(1 - \nu_{\overline{xy}}\nu_{\overline{yx}})$ were calculated using one finite element and an increasing number of hierarchical functions from 3 to 7. The frequencies rapidly converge and are in close agreement with results given in reference [10.12].

Reference [10.13] analyses thin rectangular laminated plate assemblies using classical laminated plate theory. In this case the in-plane displacements of the middle surface are retained. The three components of displacement were expressed in the form

$$u(\xi,\eta) = \sum_{r=1}^{p_{ux}}\sum_{s=1}^{p_{uy}} f_r(\xi) f_s(\eta) U_{rs}$$
$$v(\xi,\eta) = \sum_{r=1}^{p_{vx}}\sum_{s=1}^{p_{vy}} f_r(\xi) f_s(\eta) V_{rs} \tag{10.29}$$
$$w(\xi,\eta) = \sum_{r=1}^{p_{wx}}\sum_{s=1}^{p_{wy}} f_r(\xi) f_s(\eta) W_{rs}$$

Table 10.5. *Non-dimensional natural frequencies of a fully clamped, square laminated plate [45/ − 45/45/ − 45/45]* $\omega(\rho h a^4/D_0)^{1/2}$ *[10.11]. Polynomial functions*

	Mode			
p	1	2	3	4
3	22.447	41.972	49.428	66.330
5	22.383	41.660	48.341	65.135
7	22.381	41.645	48.316	65.086
[10.12]	22.40	41.64	48.32	65.09

Table 10.6. *Natural frequencies (Hz) of a cantilevered, rectangular laminated plate [45/45/0]$_s$ [10.13]. Polynomial functions*

Mode	p	h–p	Experiment [10.14]
1	4.875	4.867	4.8
2	30.02	29.99	29.8
3	48.57	49.27	51.3

where the first subscript on p denotes whether the displacement is in-plane or out-of-plane, whilst the second denotes whether it is in the x- or y-direction. The functions $f_r(\xi)$, $f_s(\eta)$ are once again defined by equations (10.17) and (10.19). It is advantageous to represent all three components of displacement by the same set of functions as it greatly reduces the computational effort and it simplifies the assembly process.

Table 10.6 gives the first three natural frequencies of a 305 mm long $\times$ 76.2 mm wide $\times$ 0.804 mm thick cantilever plate consisting of six layers with a lay-up of [45/45/0]$_s$. The material properties of each layer are

$$\begin{aligned} E_{\bar{x}} &= 98.0 \times 10^9 \text{ N/m}^2 \\ E_{\bar{y}} &= 7.9 \times 10^9 \text{ N/m}^2 \\ \nu_{\overline{xy}} &= 0.28 \\ G_{\overline{xy}} &= 5.6 \times 10^9 \text{ N/m}^2 \\ \rho &= 1520 \text{ kg/m}^3 \end{aligned}$$

The modes shown are the first and second bending and the first torsion. The results in the second column marked p were obtained using one element and 10 functions in each direction. The results in the third column marked $h - p$ were obtained using three elements along the length and with $p_{wx} = p_{wy} = 9$ and $p_{ux} = p_{uy} = p_{vx} = p_{vy} = 6$ in each element. The last column gives experimentally measured values [10.14].

10.1.4 Vibration of Shells

As indicated in Section 8.1, the components of displacement of the middle surface of a thin shell are the tangential components u, v and the normal component w. These are expressed in the form

$$\begin{aligned} u(\xi,\eta) &= \sum_{r=1}^{p}\sum_{s=1}^{p} g_r(\xi) g_s(\eta) U_{rs} \\ v(\xi,\eta) &= \sum_{r=1}^{p}\sum_{s=1}^{p} g_r(\xi) g_s(\eta) V_{rs} \\ w(\xi,\eta) &= \sum_{r=1}^{p}\sum_{s=1}^{p} f_r(\xi) f_s(\eta) W_{rs} \end{aligned} \tag{10.30}$$

Table 10.7. *Non-dimensional natural frequencies of a simply supported, singly curved, isotropic plate* $\omega(\rho R^2(1-\nu^2/E)^{1/2}$ *[10.15].* $(a/b = 1.0;\ R/h = 1000.0;\ R/a = 10.0;\ \nu = 0.3)$. *Polynomial functions*

Mode	p = 6	8	10	Analytical [10.16]
1, 1	0.7427	0.7426	0.7426	0.7426
1, 2	1.4429	1.4368	1.4367	1.4367
2, 1	1.6195	1.6158	1.6158	1.6158
2, 2	2.3350	2.3282	2.3281	2.3281

They are then substituted into a suitable thin shell theory such as the one by Novozhilov indicated in Section 8.1.

Reference [10.15] uses this approach to calculate the natural frequencies of cylindrically curved rectangular panels using Love and Timoshenko theory [10.16]. Table 10.7 shows some of the results for a simply supported, square, isotropic plate. The frequencies predicted using the HFEM rapidly converge with increasing number of hierarchical functions, p, to the analytical solution.

10.2 Trigonometric Functions

Reference [10.17] demonstrates that there is a limit to the number of polynomial hierarchical functions which can be used due to rounding errors. The principal source of such errors is the wide dynamic range in the coefficients that define a given function. It is therefore recommended that the number of functions be limited to (a) 24 or less in one-dimensional applications and (b) 14 or less in two-dimensional applications.

In order to overcome the problems associated with polynomial functions, reference [10.5] suggests using trigonometric functions which produce similar shapes to the polynomials. It is shown that they are numerically more stable than polynomials and so higher order functions can be used.

The most commonly used trigonometric functions are

$$g_r(\xi) = \sin\Big(\frac{\pi}{2}(r-2)(1+\xi)\Big) \quad \text{for } r > 2 \tag{10.31}$$

and

$$f_r(\xi) = \sin\Big(\frac{\pi}{2}(r-4)(1+\xi)\Big)\ \sin\Big(\frac{\pi}{2}(1+\xi)\Big) \quad \text{for } r > 4 \tag{10.32}$$

In both cases, $-1 \le \xi \le +1$. Note that $g_r(\xi) = 0$ for $\xi = \pm 1$ and $f_r(\xi) = 0,\ f'_r(\xi) = 0$ for $\xi = \pm 1$. The first four functions given by equation (10.32) along with the ones given by equation (10.17) are shown in Figure 10.4.

10.2.1 Axial Vibration of Rods

Following Section 10.1.1, the axial displacement, u, of a rod is expressed in the form

$$u = \lfloor \mathbf{N}(\xi) \rfloor \{\mathbf{q}\} \tag{10.33}$$

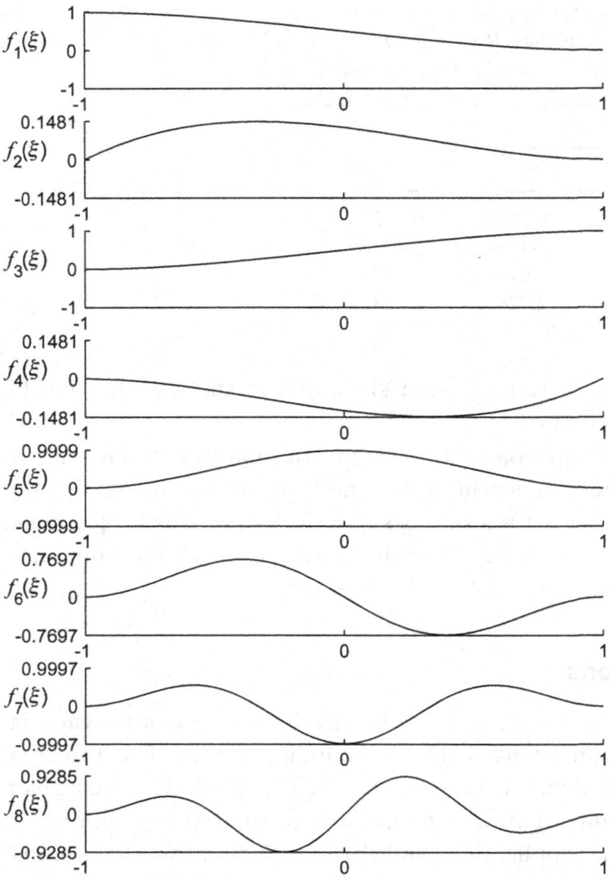

Figure 10.4. Out-of-plane trigonometric functions.

where the row matrix $\lfloor \mathbf{N}(\xi) \rfloor$ contains the functions $g_r(\xi)$ which are given by equations (10.6) and (10.31). Substituting equation (10.33) into the expression (10.10) for the element stiffness matrix gives

$$[\mathbf{k}]_e = \frac{EA}{A}\begin{bmatrix} 1 & -1 & 0 & 0 \\ -1 & 1 & 0 & 0 \\ 0 & 0 & \pi^2/2 & 0 \\ 0 & 0 & 0 & 2\pi^2 \end{bmatrix} \tag{10.34}$$

when only two hierarchical functions are included. Once again the hierarchical functions are **k**-orthogonal.

Substituting equation (10.33) into the expression (10.14) for the element inertia matrix gives

$$[\mathbf{m}]_e = \rho A a \begin{bmatrix} 1/3 & 1/6 & 1/\pi & 1/2\pi \\ 1/6 & 1/3 & 1/\pi & -1/2\pi \\ 1/\pi & 1/\pi & 1/2 & 0 \\ 1/2\pi & -1/2\pi & 0 & 1/2 \end{bmatrix} \tag{10.35}$$

Table 10.8. *Non-dimensional natural frequencies of a fixed-free rod* $\omega(\rho L^2/E)^{1/2}$. *Trigonometric functions*

	FEM solutions			
Mode	1 Element	1 Element + 1 function	1 Element + 2 functions	Exact solution
1	1.732	1.578	1.572	1.571
2	–	5.508	4.768	4.712

The first two non-dimensional natural frequencies of the fixed-free rod shown in Figure 3.1 are given in Table 10.8.

10.2.2 Bending Vibration of Beams

Following Section 10.1.2, the lateral displacement, v, of a beam is expressed in the form

$$v = \lfloor \mathbf{N}\{\xi\} \rfloor \{\mathbf{q}\} \tag{10.36}$$

where the row matrix $\lfloor \mathbf{N}\{\xi\} \rfloor$ contains the functions $f_r(\xi)$, the first four of which are given by equation (10.17). Instead of taking equation (10.32) as the hierarchical functions, reference [10.18] uses a mixture of polynomial and trigonometric functions, namely

$$f_r(\xi) = \frac{1}{4}(1-\xi^2)\,\sin\left(\frac{\pi}{2}(r-4)(1+\xi)\right) \tag{10.37}$$

for $-1 \leq \xi \leq +1$ and $r > 4$. These functions also give $f_r(\xi)$ and $f_r'(\xi) = 0$ for $\xi = \pm 1$. The first four functions given by equation (10.37) together with the ones given by equation (10.17) are shown in Figure 10.5. Table 10.9 shows the convergence of the non-dimensional natural frequencies of a cantilever beam using one finite element and an increasing number of hierarchical functions.

Table 10.9. *Non-dimensional natural frequencies of a cantilever beam.* $\omega(\rho A L^4/EI)^{1/2}$ *[10.18]. Mixed functions*

	FEM solutions				
	Number of elements/number of functions				
Mode	1/0	1/1	1/2	1/3	Exact solution
1	3.533	3.518	3.516	3.516	3.516
2	34.806	22.108	22.047	22.047	22.034
3	–	116.344	62.024	61.885	61.697
4	–	–	234.886	121.645	120.902

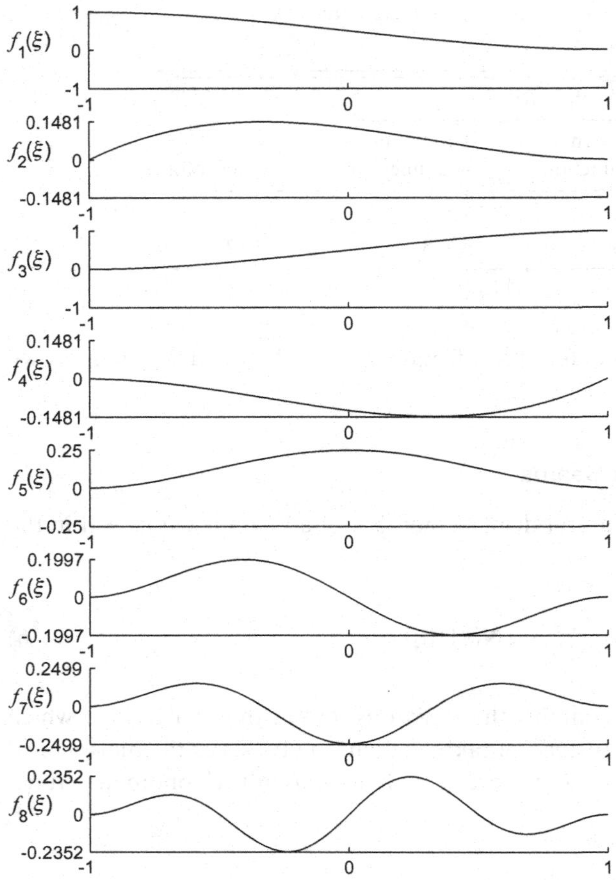

Figure 10.5. Out-of-plane mixed polynomial and trigonometric functions.

10.2.3 In-plane Vibration of Plates

The in-plane components of displacement are expressed in the form

$$u(\xi,\eta) = \sum_{r=1}^{p}\sum_{s=1}^{p} g_r(\xi)g_s(\eta)U_{rs}$$
$$v(\xi,\eta) = \sum_{r=1}^{p}\sum_{s=1}^{p} g_r(\xi)g_s(\eta)V_{rs} \tag{10.38}$$

where the functions $g_r(\xi)$, $g_s(\eta)$ are given by equations (10.6) and (10.31).

Reference [10.19] derives a trapezoidal element using the above functions. A number of numerical examples are presented including the analysis of the cantilever shear walls considered in Chapter 4. The material properties are $E = 34.474 \times 10^9$ N/m^2, $\nu = 0.11$, $\rho = 568.2$ kg/m^3 and the thickness is 0.2289 m. The wall was divided into two trapezoidal elements, as shown in Figure 10.6. Table 10.10 shows the convergence of the natural frequencies with the increase in the number p of hierarchical functions. The reason that some of the predicted frequencies are lower

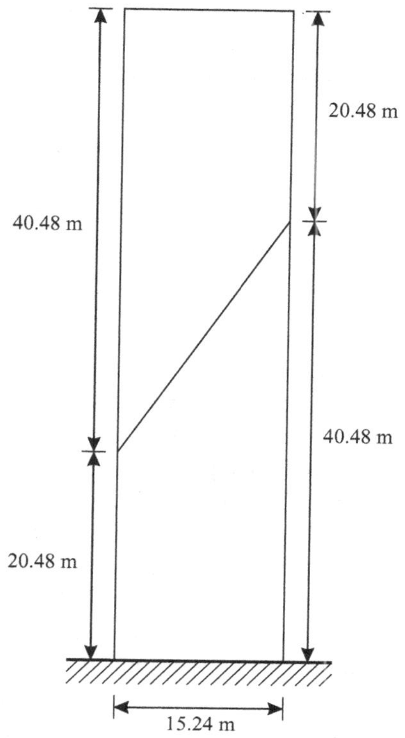

Figure 10.6. Geometry of a cantilever shear wall.

than the analytical solution is that it was obtained using deep beam theory and is therefore not the exact solution.

10.2.4 Flexural Vibration of Plates

As indicated in Section 10.1.3, the lateral displacement, w, of a thin plate is expressed in the form

$$w(\xi, \eta) = \sum_{r=1}^{p} \sum_{s=1}^{p} f_r(\xi) f_s(\eta) W_{rs} \tag{10.39}$$

Table 10.10. *Natural frequencies (Hz) of a cantilever shear wall. Trigonometric functions*

Method	Mode				
	1	2	3	4	5
Rectangular 8-node (Table 4.4)	4.986	26.327	31.964	62.648	95.955
HFEM [10.19]					
$p = 1$	5.358	31.967	32.175	82.412	99.414
3	4.991	26.193	31.950	60.640	95.942
5	4.963	25.753	31.945	59.745	95.810
Analytical	4.973	26.391	31.944	62.066	95.832

Table 10.11. *Non-dimensional natural frequencies of a thin square plate* ($a/b = 1$) *with clamped boundaries* $\omega(\rho h a^4/D)^{1/2}$ *[10.18]. Mixed functions*

Mode	p = 6	8	10	Exact solution [10.8]
1, 1	36.0240	36.0048	35.9964	35.9856
1, 2	73.5495	73.4758	73.4432	73.3935
2, 2	109.1022	108.6181	108.4327	108.2162
1, 3	132.4410	131.9948	131.8088	131.5815

Reference [10.18] uses equation (10.17) for the first four and equation (10.37) for the remainder. Results are given for square plates having various boundary conditions using one finite element and five hierarchical functions. The convergence of the natural frequencies of a clamped plate with increasing number of hierarchical functions is investigated. Table 10.11 shows a selection of those results.

Reference [10.20] derives a trapezoidal thick plate element. A trapezium can be mapped onto the square shown in Figure 4.12(b) by means of the transformation (4.55). Use can be made of the fact that it is a trapezium and not a quadrilateral to obtain an expression for the Jacobian of the transformation in analytic form thus enabling the element matrices to be analytically integrated in closed form.

The expressions for kinetic and strain energy are given by equations (6.64) to (6.66). The components of displacement are expressed in the form

$$\begin{aligned} w(\xi,\eta) &= \sum_{r=1}^{p}\sum_{s=1}^{p} g_r(\xi)g_s(\eta)W_{rs} \\ \theta_x(\xi,\eta) &= \sum_{r=1}^{p}\sum_{s=1}^{p} g_r(\xi)g_s(\eta)\theta_{xrs} \\ \theta_y(\xi,\eta) &= \sum_{r=1}^{p}\sum_{s=1}^{p} g_r(\xi)g_s(\eta)\theta_{yrs} \end{aligned} \tag{10.40}$$

where the functions $g_r(\xi)$, $g_s(\eta)$ are given by equations (10.6) and (10.31).

Reference [10.20] analyses various shaped plates. Table 10.12 shows the convergence of a square simply supported plate using one finite element and an increasing number of hierarchical functions.

Table 10.12. *Non-dimensional natural frequencies of a thick square plate* ($a/b = 1$, $a/h = 10$) *with simply supported boundaries* $\omega(\rho h a^4/D)^{1/2}$ *[10.20]. Trigonometric functions*

Mode	p = 6	8	10	Analytical solution [10.21]
1, 1	19.0691	19.0671	19.0661	19.0651
1, 2	45.5591	45.5127	45.4979	45.4831
2, 2	69.8817	69.8284	68.8107	69.7939
1, 3	85.3356	85.1569	85.0977	85.0385

Table 10.13. *Natural frequencies (Hz) of a rectangular simply supported sandwich panel [10.22]. Trigonometric functions*

Mode	p			Exact [10.23]	Experimental [10.23]
	2	8	12		
1	23.30	23.11	23.07	23	–
2	44.98	44.01	43.94	44	45
3	73.52	71.14	71.08	71	69
4	89.12	78.54	78.42	80	78
5	93.53	91.03	90.90	91	92
6	133.72	124.15	123.92	126	125

Reference [10.22] analyses sandwich panels using first-order shear deformation theory for the core. The components of displacement are expressed in the form

$$
\begin{aligned}
u_0(\xi,\eta) &= \sum_{r=1}^{p}\sum_{s=1}^{p} f_r(\xi)\,f_s(\eta)\,U_{rs}\\
v_0(\xi,\eta) &= \sum_{r=1}^{p}\sum_{s=1}^{p} f_r(\xi)\,f_s(\eta)\,V_{rs}\\
w_0(\xi,\eta) &= \sum_{r=1}^{p}\sum_{s=1}^{p} f_r(\xi)\,f_s(\eta)\,W_{rs}\\
\theta_x(\xi,\eta) &= \sum_{r=1}^{p}\sum_{s=1}^{p} f_r(\xi)\,f_s(\eta)\theta_{xrs}\\
\theta_y(\xi,\eta) &= \sum_{r=1}^{p}\sum_{s=1}^{p} f_r(\xi)\,f_s(\eta)\theta_{yrs}
\end{aligned}
\tag{10.41}
$$

where the functions $f_r(\xi)$, $f_s(\eta)$ are defined by equations (10.17) and (10.32).

Table 10.13 gives the first six natural frequencies of an aluminium honeycomb sandwich plate of dimensions 182.9 × 121.9 cm having all four edges simply supported. The material properties are as follows:

Face plates

$h = 0.41$ mm
$E = 68.95 \times 10^9$ N/m^2, $G = 25.92 \times 10^9$ N/m^2
$\rho = 2768$ kg/m^3

Core

$h = 6.35$ mm
$G_{xz} = 0.0517 \times 10^9$ N/m^2, $G_{yz} = 0.1345 \times 10^9$ N/m^2,
$\rho = 121.8$ kg/m^3.

One element and an increasing number of functions were used. The results are compared with exact frequencies and experimental measurements [10.23].

Table 10.14. *Natural frequencies (Hz) of a completely free conically curved panel [10.24]. Trigonometric functions*

	p		
Mode	8	16	Experiment
1	7.25	7.21	7.5
2	12.70	12.32	12.7
3	18.59	18.21	18.2
4	34.49	34.40	35.6
5	44.48	44.32	46.0

10.2.5 Vibration of Shells

Reference [10.24] analyses completely free, open, thin, conically curved panels. The components of displacement are expressed in the form

$$\begin{aligned} u(\xi,\eta) &= \sum_{r=1}^{p}\sum_{s=1}^{p} f_r(\xi) f_s(\eta) U_{rs} \\ v(\xi,\eta) &= \sum_{r=1}^{p}\sum_{s=1}^{p} f_r(\xi) f_s(\eta) V_{rs} \\ w(\xi,\eta) &= \sum_{r=1}^{p}\sum_{s=1}^{p} f_r(\xi) f_s(\eta) W_{rs} \end{aligned} \tag{10.42}$$

where the functions $f_r(\xi)$, $f_s(\eta)$ are defined by equations (10.17) and (10.32).

The geometry of one of the shells considered is defined by

Semi-vertex angle 3.8°
Radius of curvature at the small end 0.34 m
Total included angle 130°
Short length 1.14 m
Thickness 2.0 mm

Material properties

$E = 70 \times 10^9$ N/m^2,
$\nu = 0.3$
$\rho = 2700$ kg/m^3.

Table 10.14 shows the convergence of the natural frequencies with an increase in the number of hierarchical functions using only one element. A comparison is also made with experimentally measured frequencies.

Reference [10.25] analyses both open and closed conical sandwich shells using first-order shear deformation theory for the core. The components of displacement are expressed in the form (10.41) where the functions $f_r(\xi)$, $f_s(\eta)$ are defined by equations (10.17) and (10.32).

Table 10.15. *Natural frequencies (Hz) of a sandwich conical frustum with shear-diaphragms around the circumference [10.25]. Trigonometric functions*

Mode	FE Mesh		Analytical [10.26]
	8	16	
3, 1	84.4	84.1	86.8
4, 1	84.7	84.6	85.7
5, 1	114.3	113.6	113.3
2, 1	134.7	134.6	134.8
6, 1	157.2	155.8	–
5, 2	164.4	163.2	160.6

The geometry of one of the closed conical shells considered is, with reference to the mid-surface of the core, defined by

Semi-vertex angle 5.07°
Radius of curvature of the small end 570.2 mm
Radius of curvature at the large end 732.9 mm
Total included angle 360°
Slant length 1834.3 mm
Thickness of core 7.62 mm
Thickness of face plates 0.535 mm

The material properties are

$E = 25.59 \times 10^9$ N/m^2,
$\nu = 0.2$, $G = 7.03 \times 10^9$ N/m^2,
$\rho = 2800$ kg/m^3.

Core

$G_{xz} = 0.2249 \times 10^9$ N/m^2, $G_{\theta z} = 0.1286 \times 10^9$ N/m^2,
$\rho = 36.8$ kg/m^3.

Table 10.15 shows the frequencies obtained using eight elements circumferentially and both one and two elements axially. In both cases six hierarchical functions were used. A comparison is made with frequencies calculated analytically.

11 Analysis of Free Vibration

In order to determine the frequencies, ω_i, and modes, $\{\boldsymbol{\phi}\}_i$, of free vibration of a structure, it is necessary to solve the linear eigenproblem (see Chapter 3).

$$\left[\mathbf{K} - \omega_i^2\mathbf{M}\right]\{\boldsymbol{\phi}\}_i = 0 \qquad i = 1, 2, \ldots \tag{11.1}$$

where **K** and **M** are the stiffness and inertia matrices respectively. This chapter briefly describes some of the numerical methods used in finite element analysis for solving equation (11.1). Fuller details may be found in references [11.1–11.12].

11.1 Some Preliminaries

Dropping the suffix i in equation (11.1) for convenience and putting $\omega^2 = \lambda$ gives

$$[\mathbf{K} - \lambda\mathbf{M}]\{\boldsymbol{\phi}\} = 0 \tag{11.2}$$

If **K** and **M** are of order $(n \times n)$, then equation (11.2) represents a set of n linear homogeneous equations. The condition that these equations should have a non-zero solution is that the determinant of coefficients should vanish, that is

$$\det[\mathbf{K} - \lambda\mathbf{M}] = |\mathbf{K} - \lambda\mathbf{M}| = 0 \tag{11.3}$$

Equation (11.3) can be expanded to give a polynomial of degree n in λ. This polynomial equation will have n roots, $\lambda_1, \ \lambda_2, \ldots, \lambda_n$. Such roots are called eigenvalues. Since **M** is positive definite and **K** is either positive definite or positive semi-definite (see Chapter 1), the eigenvalues are all real and either positive or zero. However, they are not necessarily all different from one another. If they are not distinct, then the eigenproblem is said to have multiple eigenvalues. An eigenvalue which occurs exactly m times is said to be of multiplicity m.

If all the eigenvalues are distinct, then corresponding to each one, there exists a non-trivial solution to equation (11.2) for $\{\boldsymbol{\phi}\}$. These solutions are known as eigenvectors. An eigenvector is arbitrary to the extent that a scalar multiple of it is also a solution of (11.2). It is convenient to choose this multiplier in such a way that $\{\boldsymbol{\phi}\}$ has some desirable property. Such eigenvectors are called normalised eigenvectors. The most common procedures are to either scale $\{\boldsymbol{\phi}\}$ such that its largest component is unity or so that

$$\{\boldsymbol{\phi}\}^{\mathrm{T}}[\mathbf{M}]\{\boldsymbol{\phi}\} = 1 \tag{11.4}$$

If λ_s is an eigenvalue of multiplicity m, it can be shown that there are exactly m eigenvectors corresponding to it which satisfy equations (11.18) and (11.19) below [11.2 and 11.13]. Thus the statements in the previous paragraph are true, even when the eigenvalues are not distinct.

EXAMPLE 11.1 Calculate the eigenvalues and eigenvectors for the system shown in Figure P1.1 with $k_1 = k_4 = 3$, $k_2 = k_3 = 1$, $m_1 = m_3 = 2$ and $m_2 = 1$.

The stiffness and inertia matrices are

$$\mathbf{K} = \begin{bmatrix} 4 & -1 & 0 \\ -1 & 2 & -1 \\ 0 & -1 & 4 \end{bmatrix}, \qquad \mathbf{M} = \begin{bmatrix} 2 & 0 & 0 \\ 0 & 1 & 0 \\ 0 & 0 & 2 \end{bmatrix}$$

Equation (11.2) becomes

$$\begin{bmatrix} (4-2\lambda) & -1 & 0 \\ -1 & (2-\lambda) & -1 \\ 0 & -1 & (4-2\lambda) \end{bmatrix} \begin{bmatrix} \phi_1 \\ \phi_2 \\ \phi_3 \end{bmatrix} = 0 \tag{11.5}$$

Equating the determinant of coefficients to zero gives

$$(4-2\lambda)\{(2-\lambda)(4-2\lambda)-1\} - (4-2\lambda) = 0$$

that is,

$$(4-2\lambda)(2\lambda-2)(\lambda-3) = 0$$

and

$$\lambda = 1, 2, \quad \text{or} \quad 3.$$

When $\lambda = 1$, the first and third equations of (11.5) give

$$2\phi_1 - \phi_2 = 0$$
$$-\phi_2 + 2\phi_3 = 0$$

Therefore

$$\phi_2 = 2\phi_3$$

and

$$\phi_1 = \phi_3$$

The eigenvector is

$$\begin{bmatrix} 1 \\ 2 \\ 1 \end{bmatrix} \phi_3$$

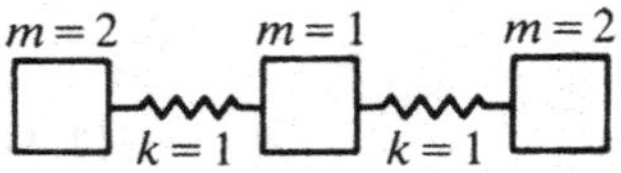

Figure 11.1. Three degree of freedom spring-mass system.

Taking

$$\phi_3 = \frac{1}{2} \text{ gives } \begin{bmatrix} \frac{1}{2} \\ 1 \\ \frac{1}{2} \end{bmatrix}$$

Alternatively, substitute the eigenvector into (11.4). This gives

$$8\phi_3^2 = 1$$

and so

$$\phi_3 = 1/2(2)^{1/2}$$

The eigenvector in this instance is

$$\begin{bmatrix} 1/2(2)^{1/2} \\ 1/2^{1/2} \\ 1/2(2)^{1/2} \end{bmatrix}$$

Similarly, the eigenvectors corresponding to the other two eigenvalues are

$$\lambda = 2, \quad \{\boldsymbol{\phi}\} = \begin{bmatrix} 1 \\ 0 \\ -1 \end{bmatrix} \quad \text{or} \quad \begin{bmatrix} \frac{1}{2} \\ 0 \\ -\frac{1}{2} \end{bmatrix}$$

$$\lambda = 3, \quad \{\boldsymbol{\phi}\} = \begin{bmatrix} \frac{1}{2} \\ -1 \\ \frac{1}{2} \end{bmatrix} \quad \text{or} \quad \begin{bmatrix} 1/2(2)^{1/2} \\ -1/2^{1/2} \\ 1/2(2)^{1/2} \end{bmatrix}$$

EXAMPLE 11.2 Calculate the eigenvalues and eigenvectors of the system shown in Figure 11.1.

The stiffness and inertia matrices are

$$\mathbf{K} = \begin{bmatrix} 1 & -1 & 0 \\ -1 & 2 & -1 \\ 0 & -1 & 1 \end{bmatrix}, \qquad \mathbf{M} = \begin{bmatrix} 2 & 0 & 0 \\ 0 & 1 & 0 \\ 0 & 0 & 2 \end{bmatrix}$$

Equation (11.2) becomes

$$\begin{bmatrix} (1-2\lambda) & -1 & 0 \\ -1 & (2-\lambda) & -1 \\ 0 & -1 & (1-2\lambda) \end{bmatrix} \begin{bmatrix} \phi_1 \\ \phi_2 \\ \phi_3 \end{bmatrix} = 0 \qquad (11.6)$$

Equating the determinant of coefficients to zero gives

$$(1-2\lambda)\{(2-\lambda)(1-2\lambda)-1\} - (1-2\lambda) = 0$$

that is

$$(1-2\lambda)\lambda(2\lambda-5)=0$$

and

$$\lambda = 0, 1/2 \quad \text{or} \quad 5/2.$$

When $\lambda = 0$, the first and third equations of (11.6) give

$$\phi_1 - \phi_2 = 0$$
$$-\phi_2 + \phi_3 = 0$$

The eigenvector is

$$\begin{bmatrix} 1 \\ 1 \\ 1 \end{bmatrix} \phi_3$$

Therefore, ϕ_3 can either be taken to be unity or a value to satisfy (11.4). This gives

$$5\phi_3^2 = 1 \qquad \text{or} \qquad \phi_3 = 1/5^{1/2}.$$

Similarly, the eigenvectors corresponding to the other two eigenvalues are

$$\lambda = 1/2, \qquad \{\boldsymbol{\phi}\} = \begin{bmatrix} -1 \\ 0 \\ 1 \end{bmatrix} \quad \text{or} \quad \begin{bmatrix} -\frac{1}{2} \\ 0 \\ \frac{1}{2} \end{bmatrix}$$

$$\lambda = 5/2, \qquad \{\boldsymbol{\phi}\} = \begin{bmatrix} -\frac{1}{4} \\ 1 \\ -\frac{1}{4} \end{bmatrix} \quad \text{or} \quad \begin{bmatrix} -1/2(5)^{1/2} \\ 2/5^{1/2} \\ -1/2(5)^{1/2} \end{bmatrix}$$

In this example one of the eigenvalues is zero and the corresponding eigenvector represents a rigid body displacement (note that the springs do not deform). The presence of zero eigenvalues is indicated by the fact that the determinant of the stiffness matrix is zero. The multiplicity of the eigenvalue $\lambda = 0$ is equal to the number of rigid body motions the system is capable of undergoing.

EXAMPLE 11.3 Calculate the eigenvalues and eigenvectors for the system shown in Figure P1.1 with $k_1 = 3, k_2 = 2, k_3 = 0, k_4 = 1, m_1 = m_2 = m_3 = 1$.

$$\mathbf{K} = \begin{bmatrix} 5 & -2 & 0 \\ -2 & 2 & 0 \\ 0 & 0 & 1 \end{bmatrix}, \qquad \mathbf{M} = \begin{bmatrix} 1 & 0 & 0 \\ 0 & 1 & 0 \\ 0 & 0 & 1 \end{bmatrix}$$

Equation (11.2) becomes

$$\begin{bmatrix} 5-\lambda & -2 & 0 \\ -2 & (2-\lambda) & 0 \\ 0 & 0 & (1-\lambda) \end{bmatrix} \begin{bmatrix} \phi_1 \\ \phi_2 \\ \phi_3 \end{bmatrix} = 0$$

The eigenvalues and corresponding eigenvectors are

$$\lambda = 1, \qquad \{\boldsymbol{\phi}\} = \begin{bmatrix} 1/5^{1/2} \\ 2/5^{1/2} \\ 0 \end{bmatrix}, \qquad \lambda = 1, \qquad \{\boldsymbol{\phi}\} = \begin{bmatrix} 0 \\ 0 \\ 1 \end{bmatrix}$$

and

$$\lambda = 6, \qquad \{\boldsymbol{\phi}\} = \begin{bmatrix} 2/5^{1/2} \\ -1/5^{1/2} \\ 0 \end{bmatrix}$$

Notice that $\lambda = 1$ is an eigenvalue of multiplicity 2 and that two eigenvectors have been found. However, these two eigenvectors are not unique. Any linear combination of them which also satisfies (11.18) and (11.19) is also an eigenvector. For example,

$$\{\boldsymbol{\phi}\} = \begin{bmatrix} 1/(10)^{1/2} \\ 2/10^{1/2} \\ 1/2^{1/2} \end{bmatrix} \quad \text{or} \quad \begin{bmatrix} 1/10^{1/2} \\ 2/10^{1/2} \\ -1/2^{1/2} \end{bmatrix}$$

If an eigenproblem solution procedure fails because **K** is singular, this difficulty can easily be overcome by a method called shifting. λ is replaced by $(\mu + \eta)$, where η is specified, in equation (11.2), to give

$$[(\mathbf{K} - \eta\mathbf{M}) - \mu\mathbf{M}]\,\{\boldsymbol{\phi}\} = 0 \tag{11.7}$$

η is chosen to ensure that $(\mathbf{K} - \eta\mathbf{M})$ is not singular. The eigenproblem (11.7) is then solved for the eigenvalues μ and associated eigenvectors. The eigenvalues of (11.2) will then be given by $(\mu + \eta)$.

EXAMPLE 11.4 Repeat Example 11.2 using the method of shifting.

Taking $\eta = -2$, then

$$(\mathbf{K} - \eta\mathbf{M}) = \begin{bmatrix} 5 & -1 & 0 \\ -1 & 4 & -1 \\ 0 & -1 & 5 \end{bmatrix}$$

Note that

$$|\mathbf{K} - \eta\mathbf{M}| = 5(20 - 1) + 1(-5) = 70 \neq 0$$

Equation (11.7) now becomes

$$\begin{bmatrix} (5 - 2\mu) & -1 & 0 \\ -1 & (4 - \mu) & -1 \\ 0 & -1 & (5 - 2\mu) \end{bmatrix} \begin{bmatrix} \phi_1 \\ \phi_2 \\ \phi_3 \end{bmatrix} = 0 \tag{11.8}$$

Equating the determinant of coefficients to zero gives

$$(5 - 2\mu)\{(4 - \mu)(5 - 2\mu) - 1\} - (5 - 2\mu) = 0$$

That is

$$(5 - 2\mu)(2\mu - 9)(\mu - 2) = 0$$

Therefore

$$\mu = 2, 5/2, \quad \text{or} \quad 9/2,$$

giving

$$\lambda = 0, 1/2 \quad \text{or} \quad 5/2$$

as before. The eigenvectors of (11.8) are also the same as those of equation (11.6).

Sometimes the inertia matrix **M** is also singular because of the presence of massless degrees of freedom. If the eigenproblem solution technique fails because of this, then this difficulty can be overcome by partitioning (11.2) in the following manner.

$$\begin{bmatrix} \mathbf{K}_{11} & \mathbf{K}_{12} \\ \mathbf{K}_{21} & \mathbf{K}_{22} \end{bmatrix} \begin{bmatrix} \boldsymbol{\phi}_1 \\ \boldsymbol{\phi}_2 \end{bmatrix} - \lambda \begin{bmatrix} \mathbf{M}_{11} & \mathbf{0} \\ \mathbf{0} & \mathbf{0} \end{bmatrix} \begin{bmatrix} \boldsymbol{\phi}_1 \\ \boldsymbol{\phi}_2 \end{bmatrix} = 0 \tag{11.9}$$

The second of the two equations in (11.9) gives

$$\mathbf{K}_{21}\boldsymbol{\phi}_1 + \mathbf{K}_{22}\boldsymbol{\phi}_2 = 0 \tag{11.10}$$

Solving for $\boldsymbol{\phi}_2$ gives

$$\boldsymbol{\phi}_2 = -\mathbf{K}_{22}^{-1}\mathbf{K}_{21}\boldsymbol{\phi}_1 \tag{11.11}$$

The relationship (11.11) is now substituted into the first of the two equations in (11.9). This results in

$$\left[(\mathbf{K}_{11} - \mathbf{K}_{12}\mathbf{K}_{22}^{-1}\mathbf{K}_{21}) - \lambda\mathbf{M}_{11}\right]\{\boldsymbol{\phi}_1\} = 0 \tag{11.12}$$

The inertia matrix in (11.12) is now non-singular. This process is known as static condensation (see also Section 4.6).

EXAMPLE 11.5 Calculate the eigenvalues of the system shown in Figure P1.1 with $k_1 = k_4 = 3$, $k_2 = k_3 = 1$, $m_1 = m_3 = 0$ and $m_2 = 1$.

The stiffness and inertia matrices are

$$\mathbf{K} = \begin{bmatrix} 4 & -1 & 0 \\ -1 & 2 & -1 \\ 0 & -1 & 4 \end{bmatrix}, \qquad \mathbf{M} = \begin{bmatrix} 0 & 0 & 0 \\ 0 & 1 & 0 \\ 0 & 0 & 0 \end{bmatrix}$$

Rearranging the degrees of freedom in the order u_2, u_1, u_3 gives

$$\mathbf{K} = \begin{bmatrix} 2 & -1 & -1 \\ -1 & 4 & 0 \\ -1 & 0 & 4 \end{bmatrix}, \qquad \mathbf{M} = \begin{bmatrix} 1 & 0 & 0 \\ 0 & 0 & 0 \\ 0 & 0 & 0 \end{bmatrix}$$

The partitioned matrices are

$$\mathbf{K}_{11} = [2] \qquad \mathbf{K}_{12} = \mathbf{K}_{21}^{\mathrm{T}} = \lfloor -1 \quad -1 \rfloor$$

$$\mathbf{K}_{22} = \begin{bmatrix} 4 & 0 \\ 0 & 4 \end{bmatrix} \qquad \mathbf{M}_{11} = [1]$$

Equation (11.12) now becomes

$$(3/2 - \lambda)\,\phi_2 = 0$$

Therefore $\lambda = 3/2$.

This result can be verified by solving (11.2) directly.

11.1.1 Orthogonality of Eigenvectors

If $\{\boldsymbol{\phi}\}_r$ and $\{\boldsymbol{\phi}\}_s$ are two eigenvectors corresponding to the eigenvalues λ_r and λ_s then

$$[\mathbf{K}]\{\boldsymbol{\phi}\}_r - \lambda_r[\mathbf{M}]\{\boldsymbol{\phi}\}_r = 0 \qquad (11.13)$$

and

$$[\mathbf{K}]\{\boldsymbol{\phi}\}_s - \lambda_s[\mathbf{M}]\{\boldsymbol{\phi}\}_s = 0 \qquad (11.14)$$

Premultiplying (11.14) by $\{\boldsymbol{\phi}\}_r^{\mathrm{T}}$ gives

$$\{\boldsymbol{\phi}\}_r^{\mathrm{T}}[\mathbf{K}]\{\boldsymbol{\phi}\}_s - \lambda_s\{\boldsymbol{\phi}\}_r^{\mathrm{T}}[\mathbf{M}]\{\boldsymbol{\phi}\}_s = 0 \qquad (11.15)$$

Similarly, premultiplying (11.13) by $\{\boldsymbol{\phi}\}_s^{\mathrm{T}}$ and transposing the result gives

$$\{\boldsymbol{\phi}\}_r^{\mathrm{T}}[\mathbf{K}]\{\boldsymbol{\phi}\}_s - \lambda_r\{\boldsymbol{\phi}\}_r^{\mathrm{T}}[\mathbf{M}]\{\boldsymbol{\phi}\}_s = 0 \qquad (11.16)$$

since both $[\mathbf{K}]$ and $[\mathbf{M}]$ are symmetric.

Subtracting (11.16) from (11.15) gives

$$(\lambda_r - \lambda_s)\,\{\boldsymbol{\phi}\}_r^{\mathrm{T}}[\mathbf{M}]\{\boldsymbol{\phi}\}_s = 0 \qquad (11.17)$$

For $s \neq r$ and $\lambda_s \neq \lambda_r$

$$\{\boldsymbol{\phi}\}_r^{\mathrm{T}}[\mathbf{M}]\{\boldsymbol{\phi}\}_s = 0 \qquad (11.18)$$

Substituting (11.18) into (11.16) gives

$$\{\boldsymbol{\phi}\}_r^{\mathrm{T}}[\mathbf{K}]\{\boldsymbol{\phi}\}_s = 0 \qquad (11.19)$$

Equations (11.18) and (11.19) are the orthogonality conditions for the eigenvectors. It can happen that $\lambda_s = \lambda_r$ with $s \neq r$. However, as already mentioned, eigenvectors can still be found which satisfy (11.18) and (11.19).

11.1.2 Transformation to Standard Form

Many texts consider the eigenproblem

$$[\mathbf{A} - \lambda\mathbf{I}]\{\boldsymbol{\psi}\} = 0 \qquad (11.20)$$

where $\mathbf{I}$ is a unit, diagonal matrix.

Equation (11.2) can be transformed into this form by first expressing the inertia matrix $\mathbf{M}$ as

$$\mathbf{M} = \mathbf{L}\mathbf{L}^{\mathrm{T}} \qquad (11.21)$$

where $\mathbf{L}$ is a lower triangular matrix. This is possible since $\mathbf{M}$ is symmetric and positive definite. Substituting (11.21) into (11.2) gives

$$[\mathbf{K} - \lambda \mathbf{L}\mathbf{L}^{\mathrm{T}}]\{\boldsymbol{\phi}\} = 0 \tag{11.22}$$

Premultiplying (11.22) by $\mathbf{L}^{-1}$ and substituting

$$\{\boldsymbol{\phi}\} = [\mathbf{L}]^{-\mathrm{T}}\{\boldsymbol{\psi}\} \tag{11.23}$$

gives

$$[\mathbf{L}^{-1}\mathbf{K}\mathbf{L}^{-\mathrm{T}} - \lambda \mathbf{I}]\{\boldsymbol{\psi}\} = 0 \tag{11.24}$$

This is of the form (11.20) with

$$\mathbf{A} = \mathbf{L}^{-1}\mathbf{K}\mathbf{L}^{-\mathrm{T}} \tag{11.25}$$

$\mathbf{A}$ is a symmetric matrix and the eigenvalues of (11.24) are the same as those of (11.2) since

$$\begin{aligned} |\mathbf{L}^{-1}\mathbf{K}\mathbf{L}^{-\mathrm{T}} - \lambda \mathbf{I}| &= |\mathbf{L}^{-1}(\mathbf{K} - \lambda \mathbf{M})\mathbf{L}^{-\mathrm{T}}| \\ &= |\mathbf{L}^{-1}||\mathbf{K} - \lambda \mathbf{M}||\mathbf{L}^{-\mathrm{T}}| \end{aligned} \tag{11.26}$$

and $|\mathbf{L}^{-1}| = |\mathbf{L}^{-\mathrm{T}}| \neq 0$. The eigenvectors of (11.2) are related to those of (11.24) by (11.23)

The elements l_{ij} of the matrix $\mathbf{L}$ can be determined using Cholesky's symmetric decomposition [11.1, 11.2], namely

$$\begin{aligned} l_{jj} &= \left(M_{jj} - \sum_{k=1}^{j-1} l_{jk}^2 \right)^{1/2} & j &= 1, 2, \ldots, n \\ l_{ij} &= \left(M_{ij} - \sum_{k=1}^{j-1} l_{ik} l_{jk} \right) \bigg/ l_{jj} & j &= 1, 2 \ldots, (n-1) \\ & & i &= (j+1), \ldots, n \end{aligned} \tag{11.27}$$

where n is the order of the matrix $\mathbf{M}$.

The matrix $\mathbf{A}$ as defined by (11.25) is obtained in two steps. First, the equation

$$\mathbf{L}\mathbf{B} = \mathbf{K} \tag{11.28}$$

is solved for $\mathbf{B}$ by forward substitution. This gives $\mathbf{L}^{-1}\mathbf{K}$. The second step consists of solving the equation

$$\mathbf{L}\mathbf{A} = \mathbf{B}^{\mathrm{T}} \tag{11.29}$$

by forward substitution for the matrix $\mathbf{A}$. This will give $\mathbf{L}^{-1}\mathbf{K}\mathbf{L}^{-\mathrm{T}}$.

The elements of $\mathbf{B}$ are given by

$$B_{ij} = \left(K_{ij} - \sum_{k=1}^{i-1} l_{ik} B_{kj} \right) \bigg/ l_{ii} \qquad i, j = 1, \ldots, n \tag{11.30}$$

The elements of $\mathbf{A}$ are obtained in a similar manner.

EXAMPLE 11.6 Decompose the inertia matrix of a fixed-free rod of length L, which is represented by four elements, using the Cholesky decomposition.

The inertia matrix is

$$\mathbf{M} = \frac{\rho AL}{24}\begin{bmatrix} 4 & 1 & 0 & 0 \\ 1 & 4 & 1 & 0 \\ 0 & 1 & 4 & 1 \\ 0 & 0 & 1 & 2 \end{bmatrix}$$

where ρ is density and A cross-sectional area.

Let

$$\mathbf{M} = \frac{\rho AL}{24}\mathbf{LL}^{\mathrm{T}}$$

This means that

$$\begin{bmatrix} l_{11} & 0 & 0 & 0 \\ l_{21} & l_{22} & 0 & 0 \\ l_{31} & l_{32} & l_{33} & 0 \\ l_{41} & l_{42} & l_{43} & l_{44} \end{bmatrix}\begin{bmatrix} l_{11} & l_{21} & l_{31} & l_{41} \\ 0 & l_{22} & l_{32} & l_{42} \\ 0 & 0 & l_{33} & l_{43} \\ 0 & 0 & 0 & l_{44} \end{bmatrix} = \begin{bmatrix} 4 & 1 & 0 & 0 \\ 1 & 4 & 1 & 0 \\ 0 & 1 & 4 & 1 \\ 0 & 0 & 1 & 2 \end{bmatrix}$$

Multiplying row one, r(1), of [**L**] by column one, c(1), of $[\mathbf{L}]^{\mathrm{T}}$ gives

$$\mathrm{r}(1) \times \mathrm{c}(1) = l_{11}^2 = 4 \qquad \text{therefore } l_{11} = 2$$

Similarly

$$\mathrm{r}(2) \times \mathrm{c}(1) = l_{21}l_{11} = 1 \qquad l_{21} = 0.5$$

$$\mathrm{r}(3) \times \mathrm{c}(1) = l_{31}l_{11} = 0 \qquad l_{31} = 0$$

$$\mathrm{r}(4) \times \mathrm{c}(1) = l_{41}l_{11} = 0 \qquad l_{41} = 0$$

The first column of **L** has now been determined.

$$\mathrm{r}(2) \times \mathrm{c}(2) = l_{21}^2 + l_{22}^2 = 4 \qquad l_{22} = 1.9365$$

$$\mathrm{r}(3) \times \mathrm{c}(2) = l_{31}l_{21} + l_{32}l_{22} = 1 \qquad l_{32} = 0.5164$$

$$\mathrm{r}(4) \times \mathrm{c}(2) = l_{41}l_{21} + l_{42}l_{22} = 0 \qquad l_{42} = 0$$

The second column of **L** has now been determined.

$$\mathrm{r}(3) \times \mathrm{c}(3) = l_{31}^2 + l_{32}^2 + l_{33}^2 = 4 \qquad l_{33} = 1.9322$$

$$\mathrm{r}(4) \times \mathrm{c}(3) = l_{41}l_{31} + l_{42}l_{32} + l_{42}l_{33} = 1 \qquad l_{43} = 0.5175$$

The third column of **L** has now been determined

$$\mathrm{r}(4) \times \mathrm{c}(4) = l_{41}^2 + l_{42}^2 + l_{43}^2 + l_{44}^2 = 2 \qquad l_{44} = 1.3161$$

and so

$$\mathbf{L} = \begin{bmatrix} 2 & 0 & 0 & 0 \\ 0.5 & 1.9365 & 0 & 0 \\ 0 & 0.5164 & 1.9322 & 0 \\ 0 & 0 & 0.5175 & 1.3161 \end{bmatrix}$$

EXAMPLE 11.7 Transform the equation of motion of a fixed-free rod of length L, which is represented by four elements, to standard form.

Defining

$$\lambda = \frac{\omega^2 \rho L^2}{96E}$$

then **K** and **M** in equation (11.2) are

$$\mathbf{K} = \begin{bmatrix} 2 & -1 & 0 & 0 \\ -1 & 2 & -1 & 0 \\ 0 & -1 & 2 & -1 \\ 0 & 0 & -1 & 1 \end{bmatrix}, \qquad \mathbf{M} = \begin{bmatrix} 4 & 1 & 0 & 0 \\ 1 & 4 & 1 & 0 \\ 0 & 1 & 4 & 1 \\ 0 & 0 & 1 & 2 \end{bmatrix}$$

The decomposition of **M** into the form $\mathbf{LL}^{\mathrm{T}}$ is given in Example 11.6. Solving (11.28) gives

$$\begin{bmatrix} 2 & 0 & 0 & 0 \\ 0.5 & 1.9365 & 0 & 0 \\ 0 & 0.5164 & 1.9322 & 0 \\ 0 & 0 & 0.5175 & 1.3161 \end{bmatrix} [B_{ij}] = \begin{bmatrix} 2 & -1 & 0 & 0 \\ -1 & 2 & -1 & 0 \\ 0 & -1 & 2 & -1 \\ 0 & 0 & -1 & 1 \end{bmatrix}$$

$$\mathrm{r}(1) \times \mathrm{c}(j) = 2B_{1j} = K_{1j}$$

Therefore

$$B_{11} = 1, \qquad B_{12} = -0.5, \qquad B_{13} = 0, \qquad B_{14} = 0$$

$$\mathrm{r}(2) \times \mathrm{c}(j) = 0.5B_{1j} + 1.9365B_{2j} = K_{2j}$$

and so

$$B_{2j} = (K_{2j} - 0.5B_{1j})/1.9365$$

This gives

$$B_{21} = -0.7746, \qquad B_{22} = 1.1619, \qquad B_{23} = -0.5164, \qquad B_{24} = 0$$

$$\mathrm{r}(3) \times \mathrm{c}(j) = 0.5164B_{2j} + 1.9322B_{3j} = K_{3j}$$

Therefore

$$B_{3j} = (K_{3j} - 0.5164B_{2j})/1.9322$$

$$B_{31} = 0.2070, \qquad B_{32} = -0.8281, \qquad B_{33} = 1.1731, \qquad B_{34} = -0.5175$$

$$\mathrm{r}(4) \times \mathrm{c}(j) = 0.5175B_{3j} + 1.3161B_{4j} = K_{4j}$$

That is

$$B_{4j} = (K_{4j} - 0.5175B_{3j})/1.3161$$

$$B_{41} = -0.0814, \qquad B_{42} = 0.3256, \qquad B_{43} = -1.2211, \qquad B_{44} = 0.9633$$

This gives

$$\mathbf{B} = \begin{bmatrix} 1 & -0.5 & 0 & 0 \\ -0.7746 & 1.1619 & -0.5164 & 0 \\ 0.2070 & -0.8281 & 1.1731 & -0.5175 \\ -0.0814 & 0.3256 & -1.2211 & 0.9633 \end{bmatrix}$$

Solving (11.29) gives

$$\begin{bmatrix} 2 & 0 & 0 & 0 \\ 0.5 & 1.9365 & 0 & 0 \\ 0 & 0.5164 & 1.9322 & 0 \\ 0 & 0 & 0.5175 & 1.3161 \end{bmatrix} [A_{ij}]$$

$$= \begin{bmatrix} 1 & -0.7746 & 0.2070 & -0.0814 \\ -0.5 & 1.1619 & -0.8281 & 0.3256 \\ 0 & -0.5164 & 1.1731 & -1.2211 \\ 0 & 0 & -0.5175 & 0.9633 \end{bmatrix}$$

From above

$$A_{1j} = B_{j1}/2$$

$$A_{11} = 0.5, \qquad A_{12} = -0.3873, \qquad A_{13} = 0.1035,$$

$$A_{14} = -0.0407$$

$$A_{2j} = (B_{j2} - 0.5A_{1j})/1.9365$$

$$A_{21} = -0.3873, \qquad A_{22} = 0.7, \qquad A_{23} = -0.4544,$$

$$A_{24} = 0.1786$$

$$A_{3j} = (B_{j3} - 0.5164A_{2j})/1.9322$$

$$A_{31} = 0.1035, \qquad A_{32} = -0.4543, \qquad A_{33} = 0.7286,$$

$$A_{34} = -0.6797$$

$$A_{4j} = (B_{j4} - 0.5175A_{3j})/1.3161$$

$$A_{41} = -0.0407, \qquad A_{42} = 0.1786, \qquad A_{43} = -0.6797,$$

$$A_{44} = 0.9992$$

and so

$$\mathbf{A} = \begin{bmatrix} 0.5 & -0.3873 & 0.1035 & -0.0407 \\ -0.3873 & 0.7 & -0.4543 & 0.1786 \\ 0.1035 & -0.4543 & 0.7286 & -0.6797 \\ -0.0407 & 0.1786 & -0.6797 & 0.9992 \end{bmatrix}$$

11.2 Sturm Sequences

Whichever method is used to calculate the eigenvalues and eigenvectors, it is useful to be able to determine the number of eigenvalues in a specified range. This is a useful piece of information when designing a structure against vibration. In addition, this information can be used to check that the method used has located all the eigenvalues. The number of eigenvalues less than a specified value of λ can be determined using a Sturm sequence.

Consider the solution of equation (11.20) with

$$\mathbf{A} = \begin{bmatrix} a_1 & b_2 & & & \\ b_2 & a_2 & b_3 & & \mathbf{0} \\ & \ddots & \ddots & \ddots & \\ \mathbf{0} & & b_{n-1} & a_{n-1} & b_n \\ & & & b_n & a_n \end{bmatrix} \tag{11.31}$$

which is referred to as a tri-diagonal matrix.

The eigenvalues of (11.20) are given by the solution of

$$\begin{vmatrix} (a_1-\lambda) & b_2 & & \mathbf{0} & \\ b_2 & (a_2-\lambda) & b_3 & & \\ & \ddots & \ddots & \ddots & \\ \mathbf{0} & & b_{n-1} & (a_{n-1}-\lambda) & b_n \\ & & & b_n & (a_n-\lambda) \end{vmatrix} = 0 \tag{11.32}$$

This determinant can be expanded using a recurrence relationship. If $f_r(\lambda)$ denotes the determinant of the leading principal minor of order r, then

$$f_1(\lambda) = (a_1 - \lambda) \tag{11.33}$$

$$\begin{aligned} f_2(\lambda) &= (a_1 - \lambda)(a_2 - \lambda) - b_2^2 \\ &= (a_2 - \lambda) f_1(\lambda) - b_2^2 f_0(\lambda) \end{aligned} \tag{11.34}$$

with

$$f_0(\lambda) = 1. \tag{11.35}$$

Continuing this process gives the recurrence relationship

$$f_{r+1}(\lambda) = (a_{r+1} - \lambda) f_r(\lambda) - b_{r+1}^2 f_{r-1}(\lambda) \tag{11.36}$$

Finally

$$f_n(\lambda) = (a_n - \lambda) f_{n-1}(\lambda) - b_n^2 f_{n-2}(\lambda) \tag{11.37}$$

Equation (11.32) now becomes

$$f_n(\lambda) = 0 \tag{11.38}$$

The roots of this equation are the eigenvalues of $\mathbf{A}$.

The sequence of polynomial functions

$$f_0(\lambda),\ f_1(\lambda), \ldots,\ f_r(\lambda), \ldots,\ f_n(\lambda) \tag{11.39}$$

has certain important properties which can be used to locate the eigenvalues of $\mathbf{A}$. In this connection define a sign count function, $S(\lambda)$, which gives the number of changes in sign in the sequence of polynomial functions (11.39) when evaluated for a given value of λ. $S(\lambda)$ has the following properties:

(1) $S(\lambda)$ only changes when λ passes through a root (unique or multiple) of equation (11.38).
(2) In passing through a root, λ increasing, $S(\lambda)$ will always increase.
(3) If λ passes through a unique root, $S(\lambda)$ will increase by one. When λ passes through a root of multiplicity m, then $S(\lambda)$ will increase by m.
(4) $S(\lambda)$ equals the number of roots of equation (11.38) which are less than or equal to the value of λ being considered.
(5) The number of roots of equation (11.38) between λ_1 and λ_2 is, therefore, $\{S(\lambda_2) - S(\lambda_1)\}$.

A sequence of functions, $f_r(\lambda)$, for which $S(\lambda)$ has the above properties is said to form a Sturm sequence.

EXAMPLE 11.8 Investigate the Sturm sequence properties of the eigenproblem given by the system shown in Figure P1.1 with $k_1 = k_4 = 2$, $k_2 = k_3 = 1$ and $m_1 = m_2 = m_3 = 1$.

The stiffness and inertia matrices are

$$\mathbf{K} = \begin{bmatrix} 3 & -1 & 0 \\ -1 & 2 & -1 \\ 0 & -1 & 3 \end{bmatrix}, \qquad \mathbf{M} = \begin{bmatrix} 1 & 0 & 0 \\ 0 & 1 & 0 \\ 0 & 0 & 1 \end{bmatrix}$$

and so $\mathbf{A} = \mathbf{K}$.

The sequence of polynomials (11.39) are, therefore,

$$\begin{aligned} f_0(\lambda) &= 1 \\ f_1(\lambda) &= (3 - \lambda) \\ f_2(\lambda) &= (2 - \lambda) f_1(\lambda) - (-1)^2 f_0(\lambda) \\ &= (2 - \lambda)(3 - \lambda) - 1 \\ &= 5 - 5\lambda + \lambda^2 \\ f_3(\lambda) &= (3 - \lambda) f_2(\lambda) - (-1)^2 f_1(\lambda) \\ &= (3 - \lambda)(5 - 5\lambda + \lambda^2) - (3 - \lambda) \\ &= (3 - \lambda)(4 - 5\lambda + \lambda^2) \\ &= (3 - \lambda)(1 - \lambda)(4 - \lambda) \end{aligned} \tag{11.40}$$

The roots of $f_3(\lambda) = 0$ are, therefore, 1, 3 and 4 which are the eigenvalues of the problem.

The variation of the functions (11.40) with λ is shown in Figure 11.2. Also the values of these functions at discrete values of λ together with the corresponding value of the sign count function $S(\lambda)$ are given in Table 11.1. In this table, wherever a function $f_r(\lambda)$ is zero, it is given the opposite sign to $f_{r-1}(\lambda)$.

From Table 11.1 it can be seen that $S(\lambda)$ changes when λ passes through a root of $f_3(\lambda) = 0$, that is when $\lambda = 1, 3$ and 4. Each time $S(\lambda)$ increases by one. $S(\lambda)$ does not change as λ passes through the roots of $f_2(\lambda) = 0$, that is

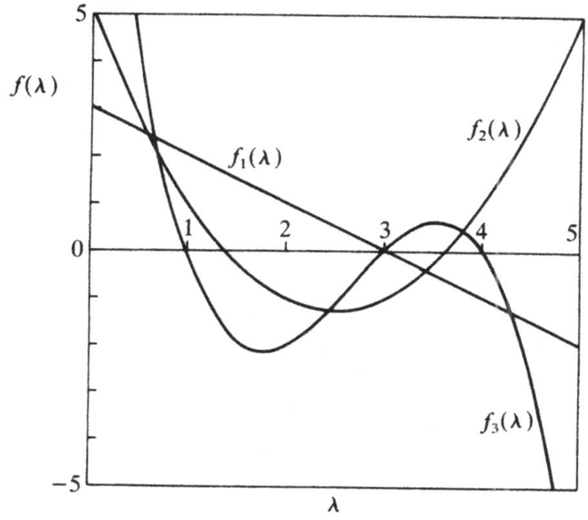

Figure 11.2. Sturm sequence functions (Example 11.8).

$\lambda = 1.382$ and 3.618. $S(\lambda)$ does change as λ passes through the root of $f_1(\lambda) = 0$, that is $\lambda = 3$, because this is identical to one of the roots of $f_3(\lambda) = 0$. Table 11.1 also indicates that $S(\lambda)$ is always equal to the number of roots of $f_3(\lambda) = 0$ which are less than or equal to the value of λ under consideration.

EXAMPLE 11.9 Investigate the Sturm sequence properties of the eigenvalue problem considered in Example 11.3.

In this case

$$\mathbf{A} = \begin{bmatrix} 5 & -2 & 0 \\ -2 & 2 & 0 \\ 0 & 0 & 1 \end{bmatrix}$$

Table 11.1. *Variation of Sturm sequence and sign count function with λ. Example 11.8*

λ	$f_0(\lambda)$	$f_1(\lambda)$	$f_2(\lambda)$	$f_3(\lambda)$	$S(\lambda)$
0.0	1.0	3.0	5.0	12.0	0
0.5	1.0	2.5	2.75	4.375	0
1.0*	1.0	2.0	1.0	−0.0	1
1.5	1.0	1.5	−0.25	−1.875	1
2.0	1.0	1.0	−1.0	−2.0	1
2.5	1.0	0.5	−1.25	−1.125	1
3.0*	1.0	−0.0	−1.0	+0.0	2
3.5	1.0	−0.5	−0.25	0.625	2
3.7	1.0	−0.7	0.19	0.567	2
4.0*	1.0	−1.0	1.0	−0.0	3
4.5	1.0	−1.5	2.75	−2.625	3
5.0	1.0	−2.0	5.0	−8.0	3

* Eigenvalues.

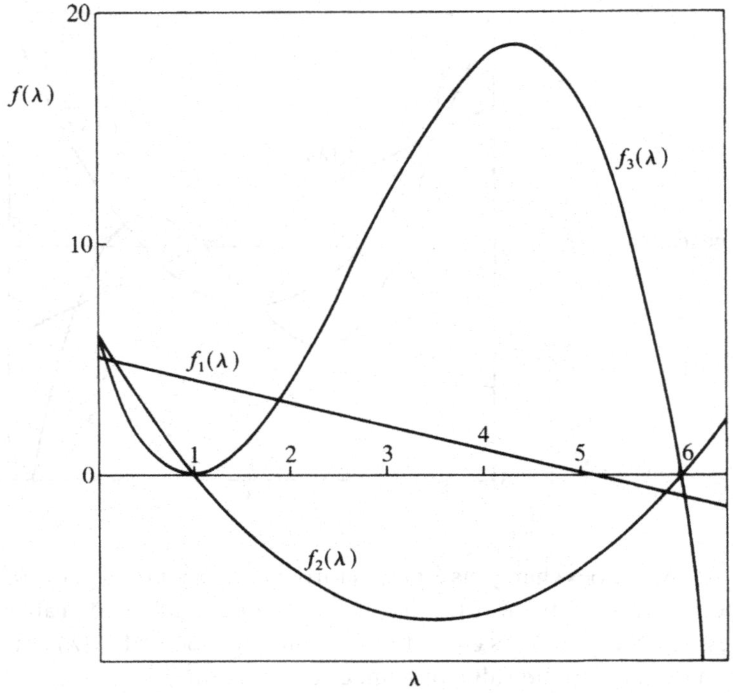

Figure 11.3. Sturm sequence functions (Example 11.9).

The sequence of polynomials (11.39) is, therefore

$$\begin{aligned} f_0(\lambda) &= 1 \\ f_1(\lambda) &= (5-\lambda) \\ f_2(\lambda) &= (2-\lambda)\, f_1(\lambda) - (-2)^2 f_0(\lambda) \\ &= (2-\lambda)(5-\lambda) - 4 \\ &= (1-\lambda)(6-\lambda) \\ f_3(\lambda) &= (1-\lambda)\, f_2(\lambda) - (0)\, f_1(\lambda) \\ &= (1-\lambda)^2(6-\lambda) \end{aligned} \tag{11.41}$$

The roots of $f_3(\lambda) = 0$ are, therefore, 1, 1 and 6. There is a root of multiplicity 2 at $\lambda = 1$.

The values of the polynomials, for a range of λ, are presented in Figure 11.3 and Table 11.2. The table also includes the corresponding sign count function.

As λ passes through the root of $f_1(\lambda) = 0$, that is $\lambda = 5$, the sign count function remains unchanged. In the case of $f_2(\lambda) = 0$ the sign count function does change as λ passes through its roots $\lambda = 1$ and 6. This is due to the fact that these values are also roots of $f_3(\lambda) = 0$. The equation $f_3(\lambda) = 0$ has a double root at $\lambda = 1$ and a single root at $\lambda = 6$. The values of $S(\lambda)$ on each side of the double root are $S(0.5) = 0$ and $S(1.5) = 2$. Therefore, the sign count function has increased by two. For the single root $S(5.5) = 2$ and $S(6.5) = 3$ and so the sign count function has increased by one. Table 11.2 indicates that $S(\lambda)$ is always

Table 11.2. *Variation of Sturm sequence and sign count function with* λ. *Example 11.9*

λ	$f_0(\lambda)$	$f_1(\lambda)$	$f_2(\lambda)$	$f_3(\lambda)$	$S(\lambda)$
0.0	1.0	5.0	6.0	6.0	0
0.5	1.0	4.5	2.75	1.375	0
1.0*	1.0	4.0	−0.0	+0.0	2
1.5	1.0	3.5	−2.25	1.125	2
2.0	1.0	3.0	−4.0	4.0	2
2.5	1.0	2.5	−5.25	7.875	2
3.0	1.0	2.0	−6.0	12.0	2
3.5	1.0	1.5	−6.25	15.625	2
4.0	1.0	1.0	−6.0	18.0	2
4.5	1.0	0.5	−5.25	18.375	2
5.0	1.0	−0.0	−4.0	16.0	2
5.5	1.0	−0.5	−2.25	10.125	2
6.0*	1.0	−1.0	+0.0	−0.0	3
6.5	1.0	−1.5	2.75	−15.125	3

*Eigenvalues.

equal to the number of roots of $f_3(\lambda) = 0$ which are less than or equal to λ even though there is a double root.

When using floating point arithmetic, the values of the $f_r(\lambda)$ can lie outside the permissible range, especially when equation (11.32) has very close or equal roots. This can be overcome by replacing the functions $f_r(\lambda)$ by [11.14]

$$q_r(\lambda) = f_r(\lambda)/f_{r-1}(\lambda) \qquad i = 1, 2, \ldots, n \tag{11.42}$$

In this case the sequence (11.36) is replaced by

$$q_r(\lambda) = (a_r - \lambda) - b_r^2/q_{r-1}(\lambda) \qquad i = 2, 3, \ldots, n \tag{11.43}$$

with

$$q_1(\lambda) = (a_1 - \lambda)$$

If the value of $q_{r-1}(\lambda)$ is zero for any value of r, it is replaced by a suitable small quantity.

The number of negative signs, $S(\lambda)$, of the sequence (11.43) indicates the number of roots of equation (11.32) which are less than or equal to λ.

Now consider any symmetric matrix **A**, rather than a symmetric tridiagonal matrix. Define

$$f_r(\lambda) = |\mathbf{A}_r - \lambda\mathbf{I}| \tag{11.44}$$

where $\mathbf{A}_r$ is the leading principal minor of order r of **A**. Then the sequence of functions defined by (11.44) with $r = 0, 1, 2, \ldots, n$ and $f_0(\lambda) = 1$, also form a Sturm sequence [11.7].

A recurrence relationship is not available for evaluating the sequence defined by (11.44). Instead, Gauss elimination is used to reduce the matrix to an upper triangular matrix. This process involves adding a scalar multiple of one row to another row, an operation which does not affect the determinant of the matrix. The determinant of an upper triangular matrix is the product of its diagonal terms. Depending upon the value of λ used in (11.44), a small or zero pivot could be encountered making the process break down. This is overcome by using row interchanges. In this case the product of the diagonal terms must be multiplied by $(-1)^N$, where N is the number of row interchanges.

EXAMPLE 11.10 Determine how many eigenvalues there are below $\lambda = 0.3$ for the matrix **A** in Example 11.7.

$$\mathbf{A} - 0.3\mathbf{I} = \begin{bmatrix} 0.2 & -0.3873 & 0.1035 & -0.0407 \\ -0.3873 & 0.4 & -0.4543 & 0.1786 \\ 0.1035 & -0.4543 & 0.4286 & -0.6797 \\ -0.0407 & 0.1786 & -0.6797 & 0.6992 \end{bmatrix}$$

It can immediately be seen that

$$f_0(0.3) = 1, \qquad f_1(0.3) = 0.2$$

To determine the values of the other functions in the sequence, a Gauss elimination with row interchanges is carried out. The steps in this process are shown in Table 11.3. It can be seen that the signs of the sequence defined by (11.44) are ++−−+. There are, therefore, two changes of sign indicating two eigenvalues less than or equal to $\lambda = 0.3$.

Finally, consider two symmetric matrices **K** and **M**. Define

$$f_r(\lambda) = |\mathbf{K}_r - \lambda \mathbf{M}_r| \tag{11.45}$$

where $\mathbf{K}_r$ and $\mathbf{M}_r$ are the leading principal minors of order r of **K** and **M** respectively. Then the sequence of functions defined by (11.45) with $r = 0, 1, 2, \ldots, n$ and $f_0(\lambda) = 1$, form a Sturm sequence [11.15]. The determinants in (11.45) are evaluated using Gauss elimination with row interchanges, as described for (11.44).

EXAMPLE 11.11 Determine how many eigenvalues there are below $\lambda = 0.3$ when **K** and **M** are as defined in Example 11.7.

$$\mathbf{K} - 0.3\mathbf{M} = \begin{bmatrix} 0.8 & -1.3 & 0 & 0 \\ -1.3 & 0.8 & -1.3 & 0 \\ 0 & -1.3 & 0.8 & -1.3 \\ 0 & 0 & -1.3 & 0.4 \end{bmatrix}$$

Therefore,

$$f_0(0.3) = 1, \qquad f_1(0.3) = 0.8$$

The Gauss elimination process for evaluating the other functions in the sequence (11.45) is shown in Table 11.4. The signs in the sequence are ++−−+. There are, therefore, two changes of sign, indicating two eigenvalues less than or equal to $\lambda = 0.3$.

Table 11.3. *Gauss elimination and Sturm sequence evaluation. Example 11.10*

Row order	Matrix
2 1 3 4	$\begin{bmatrix} -0.3873 & 0.4 & -0.4543 & 0.1786 \\ 0 & -0.1807 & -0.1311 & 0.0515 \\ 0 & -0.3474 & 0.3072 & -0.6320 \\ 0 & 0.1366 & -0.6320 & 0.6811 \end{bmatrix}$
	$f_2(0.3) = (-1)(-0.3873)(-0.1807) = -0.07$
2 3 1 4	$\begin{bmatrix} -0.3873 & 0.4 & -0.4543 & 0.1786 \\ 0 & -0.3474 & 0.3072 & -0.6320 \\ 0 & 0 & -0.2909 & 0.3802 \\ 0 & 0 & -0.5112 & 0.4326 \end{bmatrix}$
	$f_3(0.3) = (-1)^2(-0.3873)(-0.3474)(-0.2909) = -0.03914$
2 3 4 1	$\begin{bmatrix} -0.3873 & 0.4 & -0.4543 & 0.1786 \\ 0 & -0.3474 & 0.3072 & -0.6320 \\ 0 & 0 & -0.5112 & 0.4326 \\ 0 & 0 & 0 & 0.1340 \end{bmatrix}$
	$f_4(0.3) = (-1)^3(-0.3873)(-0.3474)(-0.5112)(0.1340) = 0.00922$

Table 11.4. *Gauss elimination and Sturm sequence evaluation. Example 11.11*

Row order	Matrix
2 1 3 4	$\begin{bmatrix} -1.3 & 0.8 & -1.3 & 0 \\ 0 & -0.8077 & -0.8 & 0 \\ 0 & -1.3 & 0.8 & -1.3 \\ 0 & 0 & -1.3 & 0.4 \end{bmatrix}$
	$f_2(0.3) = (-1)(-1.3)(-0.8077) = -1.05$
2 3 1 4	$\begin{bmatrix} -1.3 & 0.8 & -1.3 & 0 \\ 0 & -1.3 & 0.8 & -1.3 \\ 0 & 0 & -1.2971 & 0.8077 \\ 0 & 0 & -1.3 & 0.4 \end{bmatrix}$
	$f_3(0.3) = (-1)^2(-1.3)(-1.2971) = -2.1921$
2 3 4 1	$\begin{bmatrix} -1.3 & 0.8 & -1.3 & 0 \\ 0 & -1.3 & 0.8 & -1.3 \\ 0 & 0 & -1.3 & 0.4 \\ 0 & 0 & 0 & 0.4086 \end{bmatrix}$
	$f_4(0.3) = (-1)^3(-1.3)^3(0.4086) = 0.8977$

11.3 Orthogonal Transformation of a Matrix

Consider the eigenproblem (11.20), that is

$$[\mathbf{A} - \lambda \mathbf{I}]\{\boldsymbol{\psi}\} = 0 \tag{11.46}$$

Introducing the transformation

$$\{\boldsymbol{\psi}\} = [\mathbf{P}]\{\boldsymbol{\xi}\} \tag{11.47}$$

where $[\mathbf{P}]$ is a non-singular matrix, and premultiplying by $[\mathbf{P}]^{-1}$ gives

$$[\mathbf{P}]^{-1}[\mathbf{A} - \lambda \mathbf{I}][\mathbf{P}]\{\boldsymbol{\xi}\} = 0 \tag{11.48}$$

or

$$[\mathbf{B} - \lambda \mathbf{I}]\{\boldsymbol{\xi}\} = 0 \tag{11.49}$$

where

$$\mathbf{B} = \mathbf{P}^{-1}\mathbf{A}\mathbf{P} \tag{11.50}$$

The transformation (11.50) is known as a similarity transformation and the matrices **A** and **B** are said to be similar. Both matrices have the same eigenvalues, since

$$\begin{aligned} |\mathbf{B} - \lambda \mathbf{I}| &= |\mathbf{P}^{-1}\mathbf{A}\mathbf{P} - \lambda \mathbf{I}| \\ &= |\mathbf{P}^{-1}(\mathbf{A} - \lambda \mathbf{I})\mathbf{P}| \\ &= |\mathbf{P}^{-1}||\mathbf{A} - \lambda \mathbf{I}||\mathbf{P}| \end{aligned} \tag{11.51}$$

and $|\mathbf{P}|$, $|\mathbf{P}^{-1}|$ are non-zero. Their eignevectors are related by (11.47).

In the case of symmetric matrices, it is convenient to take **P** to be an orthogonal matrix, that is

$$\mathbf{P}\mathbf{P}^{\mathrm{T}} = \mathbf{I} \tag{11.52}$$

and so

$$\mathbf{P}^{-1} = \mathbf{P}^{\mathrm{T}} \tag{11.53}$$

The transformation (11.50) now becomes

$$\mathbf{B} = \mathbf{P}^{\mathrm{T}}\mathbf{A}\mathbf{P} \tag{11.54}$$

which is an orthogonal transformation.

A number of eigenproblem solution methods use a sequence of orthogonal transformations to reduce the matrix **A** to a simpler form. Some of these methods are described in the following section.

11.4 Eigenproblem Solution Methods

The choice of method for solving the eigenproblem (11.2) is mainly guided by the number of degrees of freedom n. For $n \leq 250$ methods such as Jacobi, Givens, Householder, inverse iteration and the LR, QR and QL methods can be used. For $250 \leq n \leq 2500$, the same methods can be used provided full use is made of the band character of the matrices **K** and **M**. In addition, the reduction techniques described in Section 11.5 can also be used. For $n \geq 2500$ the most effective methods are subspace iteration, simultaneous iteration, Lanczos' method, Arnoldi's method, Davidson's method and the Jacobi–Davidson method. These reduce the number of degrees of freedom by means of a transformation. The smaller eigenproblem is then solved by one of the methods referred to previously. Consider the eigenproblem in the form (11.20). The *Jacobi method* applies a sequence of orthogonal transformations to the matrix **A** until it is a diagonal matrix. This process is an iterative one. The eigenvalues of this diagonal matrix are the diagonal elements and the eigenvectors form a unit matrix. The eigenvectors of **A** are given by the products of the orthogonal transformations. The method can be extended to the general eigenproblem (11.2) [11.7].

The *Givens' method* is a modified version of the Jacobi method which reduces **A** to tridiagonal form without iteration.

Householder's method produces a tridiagonal matrix more efficiently than the Givens' method by using an appropriate transformation matrix. Once the matrix **A** has been reduced to tridiagonal form its eigenvalues and eigenvectors can be obtained using the *bisection method* and *inverse iteration* respectively. The bisection method uses Sturm sequences (Section 11.2) to determine the number of eigenvalues within ever decreasing intervals until only one is found within an interval of a given accuracy. This process is repeated until all the required eigenvalues have been found. Inverse iteration is a procedure which converges to the eigenvector corresponding to the eigenvalue having the smallest modulus. The eigenvector corresponding to a particular eigenvalue can be obtained using the method of shifting (Section 11.1).

The *LR*, *QR* and *QL methods* all use a sequence of similarity transformations to transform the matrix **A** to either upper or lower triangular form. The eigenvalues of such a matrix are then equal to the elements of the main diagonal. The method can be applied to a fully populated matrix, but it has been found in practice that it is more efficient to reduce the matrix to tridiagonal form, using either Givens' or Householder's method, before applying one of these techniques. Each method will produce both eigenvalues and eigenvectors, but in practice too much storage is required to calculate the eigenvectors. It is, therefore, usual to determine only the eigenvalues and then use the method of inverse iteration to calculate the eigenvectors one at a time.

Subspace iteration is a very effective method of determining the p lowest eigenvalues and corresponding eigenvectors of equation (11.2) simultaneously by means of an iterative procedure.

Simultaneous iteration is a similar technique to subspace iteration which is applied to equation (11.20). But in this case it converges to the highest eigenvalues of **A**. In order to determine the lowest eigenvalues of (11.2), a Cholesky decomposition is carried out on the matrix **K** instead of **M** (see Section 11.1.2).

The *Lanczos method* produces an orthogonal matrix **P** which can be used to transform the matrix **A** in equation (11.20) to tridiagonal form of smaller order.

The *Arnoldi method* is a generalization of the Lanczos process and reduces to that method when the matrix **A** is symmetric. However, there are a number of numerical difficulties which can be overcome by an implicitly restarted iteration technique [11.16].

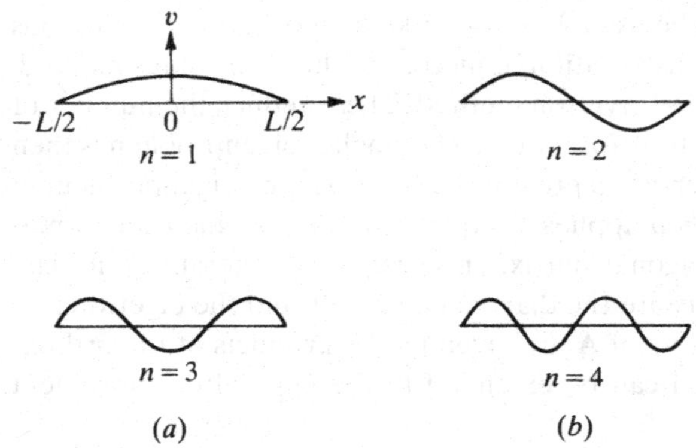

Figure 11.4. Modes of free vibration of a simply supported beam.

The *method of Davidson* is commonly seen as an extension to Lanczos' method, but from the implementation point of view it is more related to Arnoldi's method.

The *Jacobi–Davidson iteration method* is a combination of *Jacobi's orthogonal component correction* and the method of Davidson [11.17].

11.5 Reducing the Number of Degrees of Freedom

The previous section indicates that if the number of degrees of freedom, n, is such that $250 \le n \le 2500$, the methods of solving the eigenproblem for $n \le 250$ can be used provided n is reduced by some means. Several such techniques are presented in this section.

11.5.1 Making Use of Symmetry

If a structure and its boundary conditions exhibit either an axis or plane of symmetry, then the modes of free vibration will be either symmetric or antisymmetric with respect to this axis or plane. For example, consider a uniform, slender beam which is simply supported at both ends. The first four modes of vibration are shown in Figure 11.4. The two modes in Figure 11.4(a) are symmetric with respect to an axis through the mid-point. This means that for $0 \le x \le L/2$

$$v(x) = v(-x) \tag{11.55}$$

and

$$\theta_z(x) = -\theta_z(-x)$$

where $\theta_z = \partial v/\partial x$. This indicates that at $x = 0$, $\theta_z = 0$.

The two modes in Figure 11.4(b) are antisymmetric with respect to an axis through the mid-point. Therefore

$$v(x) = -v(x)$$

and

$$\theta_z(x) = \theta_z(-x). \tag{11.56}$$

Therefore, in this case $v = 0$ at $x = 0$.

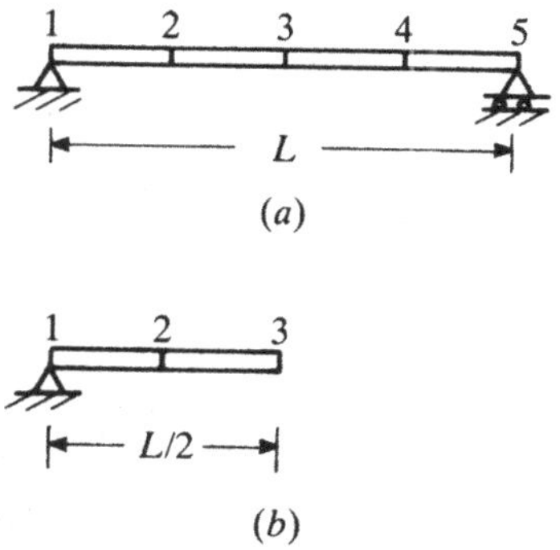

Figure 11.5. Idealisation of a simply supported beam: (*a*) full model, (*b*) half model.

This means that the modes can be calculated using an idealisation of half the beam and applying the condition $\theta_z = 0$ at the right hand boundary for symmetric modes and $v = 0$ for antisymmetric modes. This procedure is illustrated in Figure 11.5. Figure 11.5(*a*) shows the beam represented by four elements. After applying the boundary conditions $v = 0$ at nodes 1 and 5 there are eight degrees of freedom. Figure 11.5(*b*) shows half the beam represented by two elements. Again the boundary condition at node 1 is $v = 0$. Applying the condition $\theta_z = 0$ at node 3 will give the symmetric modes and $v = 0$ at node 3 will give the antisymmetric modes. In both cases there are four degrees of freedom. Therefore, one eigenproblem having eight degrees of freedom has been replaced by two having four degrees of freedom each.

Using expressions (3.132) and (3.135) the assembled matrices for symmetric modes are

$$\mathbf{K} = \frac{EI_z}{2a^3}\begin{bmatrix} 4a^2 & & \text{Sym} & \\ -3a & 6 & & \\ 2a^2 & 0 & 8a^2 & \\ 0 & -3 & -3a & 3 \end{bmatrix} \tag{11.57}$$

$$\mathbf{M} = \frac{\rho Aa}{105}\begin{bmatrix} 8a^2 & & \text{Sym} & \\ 13a & 156 & & \\ -6a^2 & 0 & 16a^2 & \\ 0 & 27 & 13a & 78 \end{bmatrix} \tag{11.58}$$

where $a = L/8$. Similarly, the assembled matrices for the antisymmetric modes are

$$\mathbf{K} = \frac{EI_z}{2a^3}\begin{bmatrix} 4a^2 & & \text{Sym} & \\ -3a & 6 & & \\ 2a^2 & 0 & 8a^2 & \\ 0 & 3a & 2a^2 & 4a^2 \end{bmatrix} \tag{11.59}$$

$$\mathbf{M} = \frac{\rho Aa}{105}\begin{bmatrix} 8a^2 & & \text{Sym} & \\ 13a & 156 & & \\ -6a^2 & 0 & 16a^2 & \\ 0 & -13a & -6a^2 & 8a^2 \end{bmatrix} \tag{11.60}$$

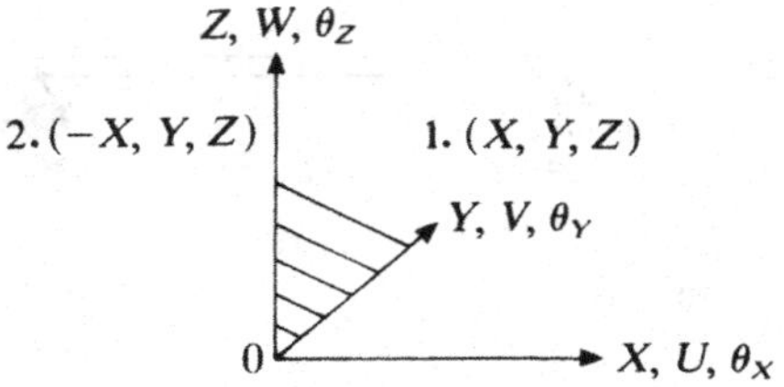

Figure 11.6. Coordinate system for a three-dimensional structure.

The boundary conditions on a plane of symmetry of a three-dimensional structure, can be obtained by considering the displacements of two points which are symmetrically placed with respect to the plane of symmetry. This situation is illustrated in Figure 11.6 where $X = 0$ is the plane of symmetry. Consider the displacements at points 1 and 2 with coordinates (X, Y, Z) and $(-X, Y, Z)$ respectively. For symmetric motion

$$(V, W, \theta_X)_2 = (V, W, \theta_X)_1$$

and

$$(U, \theta_Y, \theta_Z)_2 = -(U, \theta_Y, \theta_Z)_1 \tag{11.61}$$

Therefore, on $X = 0$

$$(U, \theta_Y, \theta_Z) = 0 \tag{11.62}$$

For antisymmetric motion

$$(V, W, \theta_X)_2 = -(V, W, \theta_X)_1$$

and

$$(U, \theta_Y, \theta_Z)_2 = (U, \theta_Y, \theta_Z)_1 \tag{11.63}$$

In this case on $X = 0$

$$(V, W, \theta_X) = 0 \tag{11.64}$$

If the structure has two planes of symmetry, then it is only necessary to idealise one quarter of it and apply four combinations of symmetric and antisymmetric boundary conditions on the two planes. This situation is illustrated in Example 3.9. Similarly, if the structure has three planes of symmetry, then the modes can be calculated by idealising one-eighth of it and applying eight combinations of symmetric and antisymmetric boundary conditions (see Figure 7.3).

11.5.2 Rotationally Periodic Structures

A rotationally periodic structure consists of a finite number of identical components which form a closed ring. There are many examples of such structures including bladed disc assemblies as used in turbines, cooling towers on column supports and antennae for space communications. The natural frequencies and modes of such structures can be calculated using a finite element idealisation of just one component.

Consider first an infinite one-dimensional periodic structure as shown schematically in Figure 11.7. This consists of a number of identical components linked

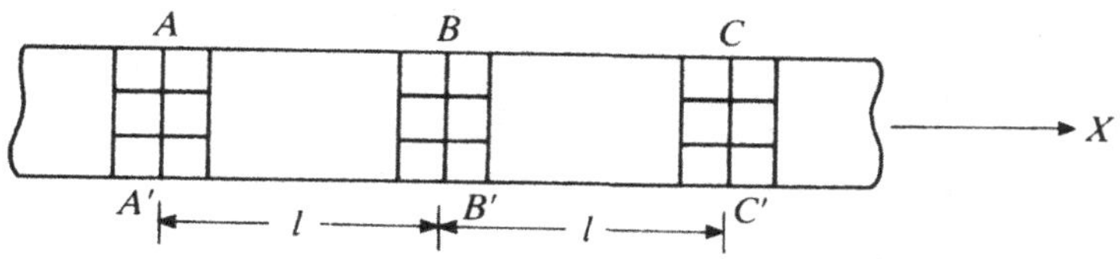

Figure 11.7. Schematic representation of a one-dimensional periodic structure.

together in identical ways. If one of the components, r, is represented by a finite element model, then its equation of motion, when vibrating harmonically with angular frequency ω, is of the form

$$[\mathbf{K}^r - \omega^2 \mathbf{M}^r]\{\mathbf{u}\}^r = \{\mathbf{f}\}^r \tag{11.65}$$

where $\{\mathbf{u}\}^r$ are the nodal degrees of freedom and $\{\mathbf{f}\}^r$ the equivalent nodal forces.

If a free harmonic wave of frequency ω propagates in the X-direction, then a physical quantity ϕ can be expressed in the form

$$\phi = \bar{\phi} \exp\{\mathrm{i}(\omega t - kX)\} \tag{11.66}$$

The wave will propagate without attenuation provided k is real.

Now consider the motion at two points in identical positions, x, within adjacent components r and $(r+1)$, where x is a local coordinate. Then

$$\phi^r = \bar{\phi} \exp\{\mathrm{i}(\omega t - kx)\} \tag{11.67a}$$

$$\phi^{r+1} = \bar{\phi} \exp\{\mathrm{i}(\omega t - kx - kl)\} \tag{11.67b}$$

where l is the dimension of the component in the X-direction. Comparing (11.67a) and (11.67b) shows that

$$\phi^{r+1} = \phi^r \exp(-\mathrm{i}\mu) \tag{11.68}$$

where $\mu = kl$.

The column matrix $\{\mathbf{u}\}^r$ in equation (11.65) will contain degrees of freedom corresponding to nodes on the left hand boundary AA', $\{\mathbf{u}_\mathrm{L}\}^r$, the right hand boundary BB', $\{\mathbf{u}_\mathrm{R}\}^r$, and all other nodes $\{\mathbf{u}_\mathrm{I}\}^r$. Therefore

$$\{\mathbf{u}\}^r = \begin{bmatrix} \mathbf{u}_\mathrm{L} \\ \mathbf{u}_\mathrm{I} \\ \mathbf{u}_\mathrm{R} \end{bmatrix}^r \tag{11.69}$$

Equation (11.68) shows that

$$\{\mathbf{u}_\mathrm{R}\}^r = \{\mathbf{u}_\mathrm{L}\}^{r+1} = \exp(-\mathrm{i}\mu)\,\{\mathbf{u}_\mathrm{L}\}^r \tag{11.70}$$

The forces $\{\mathbf{f}\}^r$ can also be partitioned in a similar way to (11.69) giving

$$\{\mathbf{f}\}^r = \begin{bmatrix} \mathbf{f}_\mathrm{L} \\ \mathbf{f}_\mathrm{I} \\ \mathbf{f}_\mathrm{R} \end{bmatrix}^r \tag{11.71}$$

Since a free wave is propagating $\{\mathbf{f}_\mathrm{I}\}$ will, in fact, be zero. The nodal forces at the boundary nodes, $\{\mathbf{f}_\mathrm{L}\}$ and $\{\mathbf{f}_\mathrm{R}\}$, are not zero, since these forces are responsible for transmitting the wave motion from one component to the next.

For equilibrium at the boundary BB'

$$\{\mathbf{f}_\mathrm{R}\}^r + \{\mathbf{f}_\mathrm{L}\}^{r+1} = 0 \tag{11.72}$$

Therefore

$$\{\mathbf{f}_\mathrm{R}\}^r = -\{\mathbf{f}_\mathrm{L}\}^{r+1} = -\exp(-\mathrm{i}\mu)\{\mathbf{f}_\mathrm{L}\}^r \tag{11.73}$$

The relationships (11.70) and (11.73) give rise to the following transformations

$$\{\mathbf{u}\}^r = \begin{bmatrix} \mathbf{u}_\mathrm{L} \\ \mathbf{u}_\mathrm{I} \\ \mathbf{u}_\mathrm{R} \end{bmatrix}^r = \begin{bmatrix} \mathbf{I} & \mathbf{0} \\ \mathbf{0} & \mathbf{I} \\ \exp(-\mathrm{i}\mu)\mathbf{I} & \mathbf{0} \end{bmatrix} \begin{bmatrix} \mathbf{u}_\mathrm{L} \\ \mathbf{u}_\mathrm{I} \end{bmatrix}^r = \mathbf{W} \begin{bmatrix} \mathbf{u}_\mathrm{L} \\ \mathbf{u}_\mathrm{I} \end{bmatrix}^r \tag{11.74}$$

and

$$\{\mathbf{f}\}^r = \begin{bmatrix} \mathbf{f}_\mathrm{L} \\ \mathbf{f}_\mathrm{I} \\ \mathbf{f}_\mathrm{R} \end{bmatrix}^r = \begin{bmatrix} \mathbf{I} & \mathbf{0} \\ \mathbf{0} & \mathbf{I} \\ -\exp(-\mathrm{i}\mu)\mathbf{I} & \mathbf{0} \end{bmatrix} \begin{bmatrix} \mathbf{f}_\mathrm{L} \\ \mathbf{f}_\mathrm{I} \end{bmatrix}^r \tag{11.75}$$

When the transformations (11.74) and (11.75) are substituted into (11.65), the unknown boundary forces can be eliminated by premultiplying by the matrix $\mathbf{W}^\mathrm{H}$. **H** is used to denote taking the complex conjugate of the matrix and then transposing it. Therefore

$$\mathbf{W}^\mathrm{H} = \begin{bmatrix} \mathbf{I} & \mathbf{0} & \exp(-\mathrm{i}\mu)\mathbf{I} \\ \mathbf{0} & \mathbf{I} & \mathbf{0} \end{bmatrix} \tag{11.76}$$

Taking $\{\mathbf{f}_\mathrm{I}\}^r = 0$ results in the equation

$$[\mathbf{K}^r(\mu) - \omega^2\mathbf{M}^r(\mu)] \begin{bmatrix} \mathbf{u}_\mathrm{L} \\ \mathbf{u}_\mathrm{I} \end{bmatrix}^r = 0 \tag{11.77}$$

where

$$\mathbf{K}^r(\mu) = \mathbf{W}^\mathrm{H}\mathbf{K}^r\mathbf{W}$$

and

$$\mathbf{M}^r(\mu) = \mathbf{W}^\mathrm{H}\mathbf{M}^r\mathbf{W} \tag{11.78}$$

Note that both $\mathbf{K}^r(\mu)$ and $\mathbf{M}^r(\mu)$ are Hermitian matrices, since

$$(\mathbf{W}^\mathrm{H}\mathbf{K}^r\mathbf{W})^\mathrm{H} = \mathbf{W}^\mathrm{H}\mathbf{K}^r\mathbf{W} \tag{11.79}$$

and

$$(\mathbf{W}^\mathrm{H}\mathbf{M}^r\mathbf{W})^\mathrm{H} = \mathbf{W}^\mathrm{H}\mathbf{M}^r\mathbf{W}$$

(Note: a matrix **A** is Hermitian if $\mathbf{A}^\mathrm{H} = \mathbf{A}$).

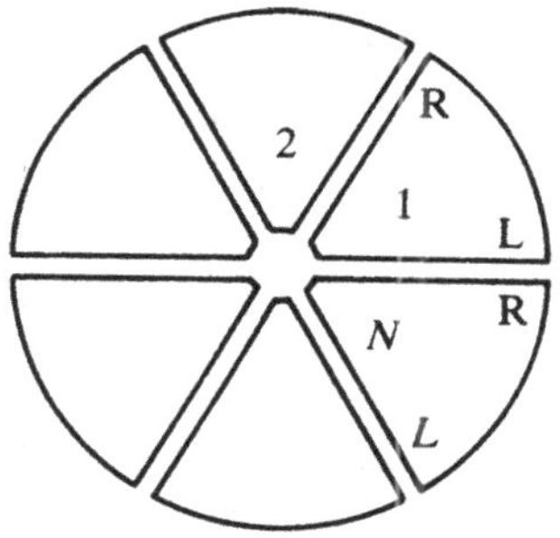

Figure 11.8. Schematic representation of a rotationally periodic structure.

Dropping the superscript r in equation (11.77) and separating the matrices into their real and imaginary parts, that is

$$\begin{aligned} \mathbf{K}(\mu) &= \mathbf{K}^{\mathrm{R}} + \mathrm{i}\mathbf{K}^{\mathrm{I}} \\ \mathbf{M}(\mu) &= \mathbf{M}^{\mathrm{R}} + \mathrm{i}\mathbf{M}^{\mathrm{I}} \end{aligned} \tag{11.80}$$

and

$$\begin{bmatrix} \mathbf{u}_{\mathrm{L}} \\ \mathbf{u}_{\mathrm{I}} \end{bmatrix} = \{\mathbf{u}^{\mathrm{R}}\} + \mathrm{i}\{\mathbf{u}^{\mathrm{I}}\} \tag{11.81}$$

gives

$$[(\mathbf{K}^{\mathrm{R}} + \mathrm{i}\mathbf{K}^{\mathrm{I}}) - \omega^2(\mathbf{M}^{\mathrm{R}} + \mathrm{i}\mathbf{M}^{\mathrm{I}})](\{\mathbf{u}^{\mathrm{R}}\} + \mathrm{i}\{\mathbf{u}^{\mathrm{I}}\}) = 0 \tag{11.82}$$

Separating out the real and imaginary parts and combining the resulting two equations gives

$$\left[\begin{bmatrix} \mathbf{K}^{\mathrm{R}} & -\mathbf{K}^{\mathrm{I}} \\ \mathbf{K}^{\mathrm{I}} & \mathbf{K}^{\mathrm{R}} \end{bmatrix} - \omega^2 \begin{bmatrix} \mathbf{M}^{\mathrm{R}} & -\mathbf{M}^{\mathrm{I}} \\ \mathbf{M}^{\mathrm{I}} & \mathbf{M}^{\mathrm{R}} \end{bmatrix}\right] \begin{bmatrix} \mathbf{u}^{\mathrm{R}} \\ \mathbf{u}^{\mathrm{I}} \end{bmatrix} = 0 \tag{11.83}$$

Since the matrices $\mathbf{K}(\mu)$ and $\mathbf{M}(\mu)$ are Hermitian then $\mathbf{K}^{\mathrm{R}}$ and $\mathbf{M}^{\mathrm{R}}$ are symmetric and

$$-(\mathbf{K}^{\mathrm{I}})^{\mathrm{T}} = \mathbf{K}^{\mathrm{I}}, \qquad -(\mathbf{M}^{\mathrm{I}})^{\mathrm{T}} = \mathbf{M}^{\mathrm{I}} \tag{11.84}$$

Therefore, (11.83) represents a real symmetric eigenproblem. For any given value of μ, which is real, this equation can be solved by one of the techniques presented in the previous sections.

Consider now a rotationally periodic structure as shown schematically in Figure 11.8. The modes of free vibration of such a structure are standing waves. For a free wave, which is propagating round the structure, to be a standing wave, the amplitude and phase of the displacements on the right hand boundary of component N should be equal to the amplitude and phase of the displacements on the left hand boundary of component 1.

Using (11.70) repeatedly shows that

$$\{\mathbf{u}_{\mathrm{R}}\}^{N} = \exp(-\mathrm{i}N\mu)\,\{\mathbf{u}_{\mathrm{L}}\}^{\mathrm{I}} \tag{11.85}$$

Therefore

$$\{\mathbf{u}_{\mathrm{R}}\}^{N} = \{\mathbf{u}_{\mathrm{L}}\}^{\mathrm{I}} \tag{11.86}$$

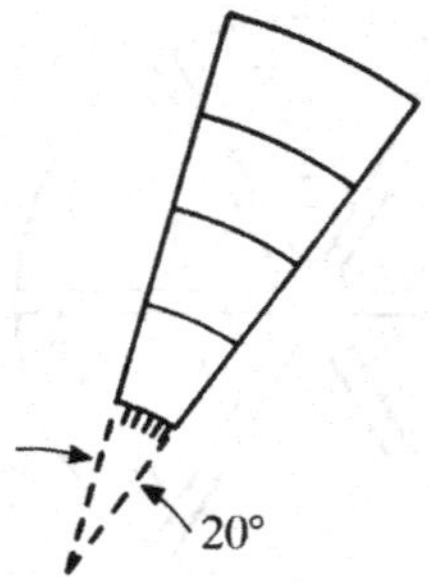

Figure 11.9. Idealisation of a 20° sector of a clamped-free circular annulus.

provided

$$\exp(-\mathrm{i}N\mu) = 1 \tag{11.87}$$

The solutions of this equation are

$$\mu = 2r\pi/N \tag{11.88a}$$

where

$$r = \begin{cases} 0, 1, 2, \ldots, N/2 & N \text{ even} \\ 0, 1, 2, \ldots, (\mathrm{N}-1)/2 & N \text{ odd} \end{cases} \tag{11.88b}$$

The natural frequencies of a rotationally periodic structure are, therefore, given by the solutions of equation (11.83) corresponding to the values of μ given by (11.88). The mode shapes are obtained by substituting the eigenvectors of (11.83) into (11.81) and

$$\begin{bmatrix} \mathbf{u}_{\mathrm{L}} \\ \mathbf{u}_{\mathrm{I}} \end{bmatrix}^{r+1} = \exp(-\mathrm{i}\mu) \begin{bmatrix} \mathbf{u}_{\mathrm{L}} \\ \mathbf{u}_{\mathrm{I}} \end{bmatrix}^{r} \tag{11.89}$$

A discussion of the modes of a rotationally periodic structure can be found in reference (11.18).

EXAMPLE 11.12 Use the theory of rotationally periodic structures to calculate the frequency parameters $\omega a^2(\rho/D)^{1/2}$, corresponding to modes having up to 3 nodal diameters and 1 nodal circle, of an annular plate, whose internal radius b is clamped and outer radius a free. Take $b/a = 0.3$ and $\nu = 0.3$.

The annulus was divided up into eighteen equal sectors, each of which subtended an angle of 20° at the centre. One sector was represented by four elements in the radial direction and one element in the circumferential direction as shown in Figure 11.9. The element used was a modified version of the element RH described in Section 6.7. The element was formulated in polar coordinates instead of Cartesian. This meant that the nodal degrees of freedom consisted of the transverse displacement, the radial rotation and the rotation in the circumferential direction. Details of the element are given in [11.20].

The values of r in (11.89) indicate the number of nodal diameters. The two lowest eigenvalues of equation (11.83) correspond to modes having 0 and 1 nodal circle. The frequency parameters obtained are given in Table 11.5 where they are compared with the analytical frequencies.

Table 11.5. *Frequency parameters, $\omega a^2(\rho/D)^{1/2}$, of a clamped-free circular annulus.* $b/a = 0.3$, $\nu = 0.3$

n	s	FEM [11.23]	Analytical [11.22]	% Difference
1	0	6.56	6.33	3.6
0	0	6.66	6.66	0.0
2	0	8.10	7.96	1.8
3	0	13.6	13.3	2.3
0	1	42.7	42.6	0.2
1	1	44.8	44.6	0.4
2	1	51.8	50.9	1.8
3	1	65.3	62.1	5.2

n = number of nodal diameters; s = number of nodal circles.

Examples of the analysis of bladed disc assemblies are given in references [11.18, 11.20, 11.21, 11.23]. Analyses of cooling towers with column supports are described in references [11.18, 11.24].

11.5.3 Elimination of Unwanted Degrees of Freedom

Very often some of the degrees of freedom are only of secondary importance. These may be, for example, in-plane displacements in a folded plate structure. The analysis may, therefore, be carried out more efficiently if these unwanted degrees of freedom can be eliminated by some procedure before solving the eigenproblem. Such a method was proposed simultaneously in references [11.25, 11.26].

Consider the equation of free vibration

$$\mathbf{M}\ddot{\mathbf{u}} + \mathbf{K}\mathbf{u} = 0 \tag{11.90}$$

The degrees of freedom $\mathbf{u}$ are partitioned into a set $\mathbf{u}_m$ termed master degrees of freedom, which are to be retained, and a set $\mathbf{u}_s$ termed slave degrees of freedom, which are to be eliminated.

Partitioning $\mathbf{M}$ and $\mathbf{K}$ in a compatible manner, equation (11.90) becomes

$$\begin{bmatrix} \mathbf{M}_{mm} & \mathbf{M}_{ms} \\ \mathbf{M}_{sm} & \mathbf{M}_{ss} \end{bmatrix} \begin{bmatrix} \ddot{\mathbf{u}}_m \\ \ddot{\mathbf{u}}_s \end{bmatrix} + \begin{bmatrix} \mathbf{K}_{mm} & \mathbf{K}_{ms} \\ \mathbf{K}_{sm} & \mathbf{K}_{ss} \end{bmatrix} \begin{bmatrix} \mathbf{u}_m \\ \mathbf{m}_s \end{bmatrix} = 0 \tag{11.91}$$

The second of the two matrix equations in (11.91) is

$$\mathbf{M}_{sm}\ddot{\mathbf{u}}_m + \mathbf{M}_{ss}\ddot{\mathbf{u}}_s + \mathbf{K}_{sm}\mathbf{u}_m + \mathbf{K}_{ss}\mathbf{u}_s = 0 \tag{11.92}$$

The assumption is now made that the relationship between $\mathbf{u}_s$ and $\mathbf{u}_m$ is not affected by the inertia terms in this equation. This is an approximation and so the selection of the slave degrees of freedom is of vital importance. Equation (11.92) now reduces to

$$\mathbf{K}_{sm}\mathbf{u}_m + \mathbf{K}_{ss}\mathbf{u}_s = 0 \tag{11.93}$$

Solving for $\mathbf{u}_s$ gives

$$\mathbf{u}_s = -\mathbf{K}_{ss}^{-1}\mathbf{K}_{sm}\mathbf{u}_m \tag{11.94}$$

Therefore

$$\mathbf{u} = \begin{bmatrix} \mathbf{u}_\mathrm{m} \\ \mathbf{u}_\mathrm{s} \end{bmatrix} = \begin{bmatrix} \mathbf{I} \\ -\mathbf{K}_\mathrm{ss}^{-1}\mathbf{K}_\mathrm{sm} \end{bmatrix} \mathbf{u}_\mathrm{m} = \mathbf{R}\mathbf{u}_\mathrm{m} \tag{11.95}$$

The kinetic and strain energy of the system are

$$\begin{aligned} T &= \tfrac{1}{2}\dot{\mathbf{u}}^\mathrm{T}\mathbf{M}\dot{\mathbf{u}} \\ U &= \tfrac{1}{2}\mathbf{u}^\mathrm{T}\mathbf{K}\mathbf{u} \end{aligned} \tag{11.96}$$

Substituting (11.95) into (11.96) gives

$$\begin{aligned} T &= \tfrac{1}{2}\dot{\mathbf{u}}_\mathrm{m}^\mathrm{T}\mathbf{M}^\mathrm{R}\dot{\mathbf{u}}_\mathrm{m} \\ U &= \tfrac{1}{2}\mathbf{u}_\mathrm{m}^\mathrm{T}\mathbf{K}^\mathrm{R}\mathbf{u}_\mathrm{m} \end{aligned} \tag{11.97}$$

where

$$\mathbf{M}^\mathrm{R} = \mathbf{R}^\mathrm{T}\mathbf{M}\mathbf{R}$$

and

$$\mathbf{K}^\mathrm{R} = \mathbf{R}^\mathrm{T}\mathbf{K}\mathbf{R} \tag{11.98}$$

Substituting for **R** from (11.95) into (11.98) gives

$$\mathbf{M}^\mathrm{R} = \mathbf{M}_\mathrm{mm} - \mathbf{M}_\mathrm{ms}\mathbf{K}_\mathrm{ss}^{-1}\mathbf{K}_\mathrm{sm} - \mathbf{K}_\mathrm{ms}\mathbf{K}_\mathrm{ss}^{-1}\mathbf{M}_\mathrm{sm} + \mathbf{K}_\mathrm{ms}\mathbf{K}_\mathrm{ss}^{-1}\mathbf{M}_\mathrm{ss}\mathbf{K}_\mathrm{ss}^{-1}\mathbf{K}_\mathrm{sm} \tag{11.99}$$

$$\mathbf{K}^\mathrm{R} = \mathbf{K}_\mathrm{mm} - \mathbf{K}_\mathrm{ms}\mathbf{K}_\mathrm{ss}^{-1}\mathbf{K}_\mathrm{sm} \tag{11.100}$$

Substituting (11.97) into Lagrange's equations gives

$$\mathbf{M}^\mathrm{R}\ddot{\mathbf{u}}_\mathrm{m} + \mathbf{K}^\mathrm{R}\mathbf{u}_\mathrm{m} = 0 \tag{11.101}$$

as the equation of motion. Comparing (11.101) with (11.90) indicates that the order of the inertia and stiffness matrices has been reduced by the number of slave degrees of freedom. The method is often referred to as the reduction technique or Guyan reduction.

Reference [11.26] indicates that the slave degrees of freedom can be eliminated sequentially rather than simultaneously. This means that all the slave degrees of freedom associated with a particular node can be eliminated as soon as all the elements connected to that node have been assembled. The slave degrees of freedom are, therefore, eliminated during the assembly process resulting in a considerable reduction in computer storage.

If the system has a number of massless degrees of freedom and these are chosen as slaves, then (11.93) is exact and (11.99) reduces to

$$\mathbf{M}^\mathrm{R} = \mathbf{M}_\mathrm{mm} \tag{11.102}$$

Equation (11.100) is unchanged. This result agrees with equation (11.12).

EXAMPLE 11.13 Use the reduction technique to eliminate u_2 from the equations of motion of the system shown in Figure P1.1.

The equations of motion, for free vibration, are

$$\begin{bmatrix} m_1 & 0 & 0 \\ 0 & m_2 & 0 \\ 0 & 0 & m_3 \end{bmatrix} \begin{bmatrix} \ddot{u}_1 \\ \ddot{u}_2 \\ \ddot{u}_3 \end{bmatrix} + \begin{bmatrix} (k_1 + k_2) & -k_2 & 0 \\ -k_2 & (k_2 + k_3) & -k_3 \\ 0 & -k_3 & (k_3 + k_4) \end{bmatrix} \begin{bmatrix} u_1 \\ u_2 \\ u_3 \end{bmatrix} = 0$$

Neglecting the inertia term in the second equation of motion gives

$$-k_2 u_1 + (k_2 + k_3) u_2 - k_3 u_3 = 0$$

Solving for u_2 gives

$$u_2 = \frac{1}{(k_2 + k_3)} (k_2 u_1 + k_3 u_3)$$

Therefore

$$\begin{bmatrix} u_1 \\ u_2 \\ u_3 \end{bmatrix} = \begin{bmatrix} 1 & 0 \\ k_2/(k_2 + k_3) & k_3/(k_2 + k_3) \\ 0 & 1 \end{bmatrix} \begin{bmatrix} u_1 \\ u_3 \end{bmatrix}$$

and so

$$\mathbf{M}^{\mathrm{R}} = \mathbf{R}^{\mathrm{T}}\mathbf{M}\mathbf{R} = \begin{bmatrix} (m_1 + k_2^2 m_2/(k_2 + k_3)^2) & k_2 k_3 m_2/(k_2 + k_3)^2 \\ k_2 k_3 m_2/(k_2 + k_3)^2 & (m_3 + k_3^2 m_2/(k_2 + k_3)^2) \end{bmatrix}$$

and

$$\mathbf{K}^{\mathrm{R}} = \mathbf{R}^{\mathrm{T}}\mathbf{K}\mathbf{R} = \begin{bmatrix} (k_1 + k_2 - k_2^2/(k_2 + k_3)) & -k_2 k_3/(k_2 + k_3) \\ -k_2 k_3/(k_2 + k_3) & (k_3 + k_4 - k_3^2/(k_2 + k_3)) \end{bmatrix}$$

EXAMPLE 11.14 Determine the error in the eigenvalues and eigenvectors caused by the reduction technique for the system defined in Example 11.13 when $m_1 = m_2 = m_3 = m$ and $k_1 = k_2 = k_3 = k_4 = k$.

Putting $\mathbf{u} = \boldsymbol{\phi} \exp(\mathrm{i}\omega t)$ and $\omega^2 m/k = \lambda$ then, without reduction, the eigenproblem is

$$\begin{bmatrix} (2 - \lambda) & -1 & 0 \\ -1 & (2 - \lambda) & -1 \\ 0 & -1 & (2 - \lambda) \end{bmatrix} \begin{bmatrix} \phi_1 \\ \phi_2 \\ \phi_3 \end{bmatrix} = 0$$

The solution of this equation is

$$\lambda = 0.5858, 2.0, 3.4142$$

$$\boldsymbol{\phi} = \begin{bmatrix} 0.7071 & 1.0 & -0.7071 \\ 1.0 & 0.0 & 1.0 \\ 0.7071 & -1.0 & -0.7071 \end{bmatrix}$$

With reduction the eigenproblem is

$$\begin{bmatrix} (6 - 5\lambda) & -(2 + \lambda) \\ -(2 + \lambda) & (6 - 5\lambda) \end{bmatrix} \begin{bmatrix} \phi_1 \\ \phi_3 \end{bmatrix} = 0$$

The solution of this is

$$\lambda = 0.6667, 2.0$$

$$\begin{bmatrix} \phi_1 \\ \phi_3 \end{bmatrix} = \begin{bmatrix} 1 & 1 \\ 1 & -1 \end{bmatrix}$$

Transforming to the complete set of degrees of freedom gives

$$\boldsymbol{\phi} = \begin{bmatrix} 1 & 0 \\ 0.5 & 0.5 \\ 0 & 1 \end{bmatrix} \begin{bmatrix} 1 & 1 \\ 1 & -1 \end{bmatrix} \begin{bmatrix} 1.0 & 1.0 \\ 1.0 & 0.0 \\ 1.0 & -1.0 \end{bmatrix}$$

Comparing the two sets of results shows that the first eigenvalue is in error by 13.8%. Multiplying the first eigenvector by 0.7071 shows that the master degrees of freedom are exact but the slave degree of freedom is in error by −29.3%. The reason for these large errors is that the slave degree of freedom has the largest amplitude in this mode. On the other hand, the second eigenvalue and eigenvector is exact. In this case the slave degree of freedom has zero amplitude in this mode. This example suggests that the smaller the amplitude of the slave degree of freedom is, in relation to the master degree of the freedom, the more accurate are the results.

EXAMPLE 11.15 Repeat Example 11.14 using $m_1 = m_3 = m$, $m_2 = 0.1\ m$ and $k_1 = k_2 = k_3 = k_4 = k$.

Without reduction the eigenproblem is

$$\begin{bmatrix} (2-\lambda) & -1 & 0 \\ -1 & (2-0.1\lambda) & -1 \\ 0 & -1 & (2-\lambda) \end{bmatrix} \begin{bmatrix} \phi_1 \\ \phi_2 \\ \phi_3 \end{bmatrix} = 0$$

The solution of this equation is

$$\lambda = 0.950, 2.0, 21.05$$

$$\boldsymbol{\phi} = \begin{bmatrix} 0.9524 & 1.0 & -0.0525 \\ 1.0 & 0.0 & 1.0 \\ 0.9524 & -1.0 & -0.0525 \end{bmatrix}$$

With reduction the eigenproblem is

$$\begin{bmatrix} (6-4.1\lambda) & -(2+0.1\lambda) \\ -(2+0.1\lambda) & (6-4.1\lambda) \end{bmatrix} \begin{bmatrix} \phi_1 \\ \phi_3 \end{bmatrix} = 0$$

The solution of this is

$$\lambda = 0.9524,\ 2.0$$

$$\begin{bmatrix} \phi_1 \\ \phi_3 \end{bmatrix} = \begin{bmatrix} 1 & 1 \\ 1 & -1 \end{bmatrix}$$

Table 11.6. *Automatic selection of slave degrees of freedom*

	$(K_{ii}/M_{ii})(m/k)$	
i	Example 11.14	Example 11.15
1	2	2
2	2	20
3	2	2

Transforming to the complete set of degrees of freedom gives

$$\boldsymbol{\phi} = \begin{bmatrix} 1 & 0 \\ 0.5 & 0.5 \\ 0 & 1 \end{bmatrix} \begin{bmatrix} 1 & 1 \\ 1 & -1 \end{bmatrix} = \begin{bmatrix} 1.0 & 1.0 \\ 1.0 & 0.0 \\ 1.0 & -1.0 \end{bmatrix}$$

This time the first eigenvalue is in error by 0.25%. Multiplying the eigenvector by 0.9524 shows that the master degrees of freedom are exact, whilst the slave degree of freedom is in error by −4.8%. Again the second eigenvalue and eigenvector are exact. This example indicates that the smaller the inertia associated with a slave degree of freedom, the more accurate the solution is.

The previous two examples indicate that the slave degrees of freedom should contribute very little to the kinetic energy of the system. References [11.27, 11.28] have investigated the accuracy of the method and suggest that the slave degrees of freedom should be chosen in regions of high stiffness and the master degrees of freedom in regions of high flexibility. This is to satisfy the criterion that the lowest eigenvalue of the equation

$$[\mathbf{K}_{ss} - \lambda \mathbf{M}_{ss}]\{\boldsymbol{\phi}_s\} = 0 \tag{11.103}$$

has a maximum value. This will be so if the terms of $\mathbf{M}_{ss}$ are small and/or the terms of $\mathbf{K}_{ss}$ are large. Reference [11.29] therefore suggests that the master and slave degrees of freedom can be selected on the basis of the ratio of the diagonal terms in the stiffness and inertia matrices **K** and **M** in equation (11.90). Those degrees of freedom which yield the largest values of the ratio K_{ii}/M_{ii} are selected as slave degrees of freedom. References [11.29, 11.30] describe automatic procedures for this. Frequencies below $0.3\lambda_s^{1/2}$, obtained in this way, are normally satisfactory. Here, λ_s is the lowest eigenvalue of equation (11.103) [11.31].

EXAMPLE 11.16 Use the automatic selection procedure described above to select one slave degree of freedom for Examples 11.14 and 11.15.

The ratios of the diagonal terms for both examples are given in Table 11.6. In the case of Example 11.15 the largest ratio is 20 for degree of freedom number 2. This will then be selected as slave degree of freedom. However, for Example 11.14 all the ratios are the same. The procedure described in reference [11.29] will tend to select the degrees of freedom assembled first, as master degrees of freedom and those assembled last, as slave degrees of freedom. In such a case it is better to use manual selection rather than automatic selection, to ensure that the master degrees of freedom are evenly distributed.

Experience has shown that the number of master degrees of freedom should be between two and three times the number of frequencies of interest. Examples of the use of the method can be found in references [11.27–11.35].

The accuracy of the recovered slave degrees of freedom can be increased by retaining the inertia terms in equation (11.91). This is illustrated by means of the following example.

EXAMPLE 11.17 Include the inertia terms in recovering the slave degree of freedom u_2 in Example 11.14.

Retaining the inertia terms in the second equation of motion and assuming harmonic motion gives

$$-\omega^2 m u_2 + k(-u_1 + 2u_2 - u_3) = 0$$

Putting $\omega^2 m/k = \lambda$, then

$$u_2 = \frac{1}{(2-\lambda)}(u_1 + u_3)$$

When $\lambda = 0.6667$, $u_2 = 0.75(u_1 + u_3)$.

Transforming to the complete set of degrees of freedom gives

$$\boldsymbol{\varphi} = \begin{bmatrix} 1 & 0 \\ 0.75 & 0.75 \\ 0 & 1 \end{bmatrix} \begin{bmatrix} 1 & 1 \\ 1 & -1 \end{bmatrix} = \begin{bmatrix} 1 & 1 \\ 1.5 & 0 \\ 1 & -1 \end{bmatrix}$$

Multiplying the first eigenvector by 0.7071 shows that the slave degree of freedom u_2 is in error by 6.1% which is an improvement on the previous error of −29.3%.

The transformation (11.94) is an approximate one since the inertia terms in equation (11.91) have been neglected. Increased accuracy can be obtained by retaining them. Assuming harmonic motion (11.91) becomes

$$[\mathbf{K}_{ss} - \omega^2 \mathbf{M}_{ss}]\mathbf{u}_s = -[\mathbf{K}_{sm} - \omega^2 \mathbf{M}_{sm}]\mathbf{u}_m \tag{11.104}$$

Thus

$$\mathbf{u}_s = -[\mathbf{K}_{ss} - \omega^2 \mathbf{M}_{ss}]^{-1}[\mathbf{K}_{sm} - \omega^2 \mathbf{M}_{sm}]\mathbf{u}_m = \mathbf{R}_s \mathbf{u}_m \tag{11.105}$$

Therefore

$$\mathbf{u} = \begin{bmatrix} \mathbf{u}_m \\ \mathbf{u}_s \end{bmatrix} = \begin{bmatrix} \mathbf{I} \\ \mathbf{R}_s \end{bmatrix} \mathbf{u}_m = \mathbf{R}\,\mathbf{u}_m \tag{11.106}$$

Substituting equation (11.106) into equation (11.97) produces reduced inertia and stiffness matrices which are frequency dependent. The resulting eigenproblem is, therefore, a nonlinear one. Several methods have been suggested for solving such a problem [11.36]. Further details of reduction techniques are presented in reference [11.37].

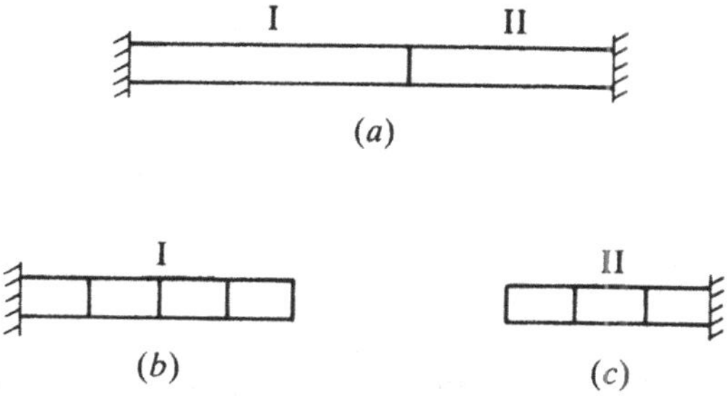

Figure 11.10. Division of a clamped-clamped beam into two components and their idealisation.

11.5.4 Component Mode Synthesis

Another technique for reducing the number of degrees of freedom is known alternatively as component mode synthesis, substructure analysis or building block approach. Such methods involve dividing the structure up into a number of separate components or substructures. Each component is then represented by a finite element model. The next step is concerned with reducing the number of degrees of freedom for each component by modal substitution. All the components are then assembled together and the complete structure analysed. In this way one large eigenproblem is replaced by several smaller ones.

Such a technique has several advantages:

(1) It is more efficient to confirm a large quantity of input data via subsets.
(2) The input data for each component/substructure can be prepared by separate analysts almost independently.
(3) Long computer runs are avoided.
(4) Numerous restart points are automatically provided.
(5) Re-analysis time is minimised when localised modifications are investigated.

There are various methods of component mode synthesis which are reviewed in references [11.38–11.44]. Only the two major ones, which are referred to as fixed and free interface methods respectively, will be considered.

11.5.4.1 Fixed Interface Method

The first step in the analysis, as noted above, is to divide the complete structure into a number of substructures. This is illustrated in Figure 11.10(a) where a clamped–clamped beam has been divided into two. Each substructure is then represented by a finite element idealisation (Figures 11.10(b) and (c)). The energy expressions for a single substructure take the form

$$\begin{aligned} T_s &= \tfrac{1}{2}\{\dot{\mathbf{u}}\}_s^T[\mathbf{M}]_s\{\dot{\mathbf{u}}\}_s \\ U_s &= \tfrac{1}{2}\{\mathbf{u}\}_s^T[\mathbf{K}]_s\{\mathbf{u}\}_s \end{aligned} \qquad (11.107)$$

where the subscript s denotes a substructure. In order to reduce the number of degrees of freedom, the following transformation is applied

$$\{\mathbf{u}\}_s = \begin{bmatrix} \mathbf{u}_I \\ \mathbf{u}_B \end{bmatrix} = \begin{bmatrix} \boldsymbol{\phi}_N & \boldsymbol{\phi}_c \\ \mathbf{0} & \mathbf{I} \end{bmatrix}_s \begin{bmatrix} \mathbf{q}_N \\ \mathbf{u}_B \end{bmatrix}_s = [\mathbf{T}_F]_s\{\mathbf{q}\}_s \qquad (11.108)$$

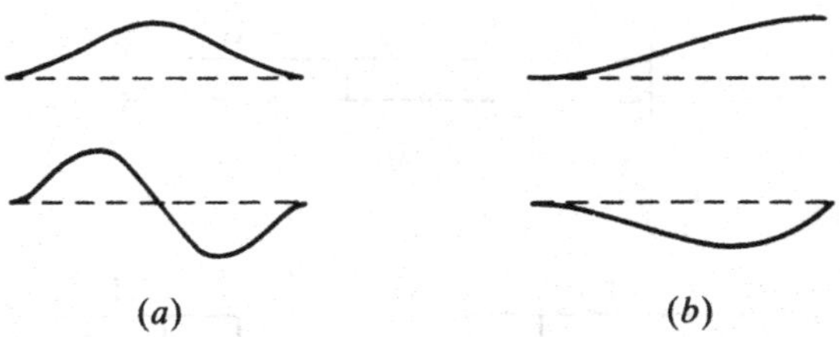

Figure 11.11. (*a*) Fixed interface modes, (*b*) constraint modes.

Here suffix I refers to internal node points and suffix B to node points on boundaries common to two or more substructures.

The columns of the matrix $\boldsymbol{\phi}_N$ are the natural modes of the substructure with interface boundaries fixed. They are obtained by solving the eigen-problem

$$[\mathbf{K}_{II} - \omega^2 \mathbf{M}_{II}]\{\boldsymbol{\phi}_I\} = 0 \tag{11.109}$$

where $\mathbf{K}_{II}$, $\mathbf{M}_{II}$ are the appropriate partitions of $[\mathbf{K}]_s$ and $[\mathbf{M}]_s$. In general, $\boldsymbol{\phi}_N$ is a rectangular matrix with fewer columns than rows. This is because it is assumed that only the first few lower frequency substructure modes contribute significantly to the modes of the complete structure. The degrees of freedom $\mathbf{q}_N$ are generalised coordinates related to the natural modes of the substructure.

The matrix $\boldsymbol{\phi}_c$ is a matrix of constraint modes of the substructure. Each column represents the values assumed by the degrees of freedom at the internal nodes for a unit value of one of the degrees of freedom at an interface boundary node. These are given by the solution of

$$[\mathbf{K}_{II}]\{\mathbf{u}_I\} + [\mathbf{K}_{IB}]\{\mathbf{u}_B\} = 0 \tag{11.110}$$

namely

$$\begin{aligned}\{\mathbf{u}_I\} &= -[\mathbf{K}_{II}]^{-I}[\mathbf{K}_{IB}]\{\mathbf{u}_B\} \\ &= \boldsymbol{\phi}_c\{\mathbf{u}_B\}\end{aligned} \tag{11.111}$$

Both the fixed interface and constraint modes of substructure I in Figure 11.10 are illustrated in Figure 11.11.

Substituting (11.108) into (11.107) gives

$$\begin{aligned}T_s &= \tfrac{1}{2}\{\dot{\mathbf{q}}\}_s^T[\bar{\mathbf{M}}]_s\{\dot{\mathbf{q}}\}_s \\ U_s &= \tfrac{1}{2}\{\mathbf{q}\}_s^T[\bar{\mathbf{K}}]_s\{\mathbf{q}\}_s\end{aligned} \tag{11.112}$$

where

$$[\bar{\mathbf{M}}]_s = [\mathbf{T}_F]_s^T[\mathbf{M}]_s[\mathbf{T}_F]_s = \begin{bmatrix} \bar{\mathbf{M}}_{NN} & \bar{\mathbf{M}}_{NB} \\ \bar{\mathbf{M}}_{BN} & \bar{\mathbf{M}}_{BB} \end{bmatrix}_s \tag{11.113}$$

and

$$[\bar{\mathbf{K}}]_s = [\mathbf{T}_F]_s^T[\mathbf{K}]_s[\mathbf{T}_F]_s = \begin{bmatrix} \bar{\mathbf{K}}_{NN} & \mathbf{0} \\ \mathbf{0} & \bar{\mathbf{K}}_{BB} \end{bmatrix}_s \tag{11.114}$$

Now

$$\begin{aligned}\bar{\mathbf{M}}_{NN} &= \boldsymbol{\phi}_N^T \mathbf{M}_{II} \boldsymbol{\phi}_N \\ \bar{\mathbf{K}}_{NN} &= \boldsymbol{\phi}_N^T \mathbf{K}_{II} \boldsymbol{\phi}_N\end{aligned} \tag{11.115}$$

and so will be diagonal matrices. Also

$$\bar{\mathbf{K}}_{BB} = \mathbf{K}_{BB} - \mathbf{K}_{BI}\mathbf{K}_{II}^{-I}\mathbf{K}_{IB} \tag{11.116}$$

where $\mathbf{K}_{BB}$, $\mathbf{K}_{BI}$, $\mathbf{K}_{IB}$, $\mathbf{K}_{II}$ are partitions of $\mathbf{K}$, which is the stiffness matrix of the substructure in terms of the interface boundary degrees of freedom. Note that the interior node point degrees of freedom have been eliminated by static condensation.

Adding the contributions from the two substructures in Figure 11.10 together gives

$$\begin{aligned}T &= \tfrac{1}{2}\{\dot{\mathbf{q}}\}^T[\mathbf{M}]\{\dot{\mathbf{q}}\} \\ U &= \tfrac{1}{2}\{\mathbf{q}\}^T[\mathbf{K}]\{\mathbf{q}\}\end{aligned} \tag{11.117}$$

where

$$\{\mathbf{q}\} = \begin{bmatrix} \mathbf{q}_N^I \\ \mathbf{q}_N^{II} \\ \mathbf{u}_B \end{bmatrix} \tag{11.118}$$

$$[\mathbf{M}] = \begin{bmatrix} \bar{\mathbf{M}}_{NN}^I & \mathbf{0} & \bar{\mathbf{M}}_{NB}^I \\ \mathbf{0} & \bar{\mathbf{M}}_{NN}^{II} & \bar{\mathbf{M}}_{NB}^{II} \\ \bar{\mathbf{M}}_{BN}^I & \bar{\mathbf{M}}_{BN}^{II} & \bar{\mathbf{M}}_{BB}^I + \bar{\mathbf{M}}_{BB}^{II} \end{bmatrix} \tag{11.119}$$

and

$$[\mathbf{K}] = \begin{bmatrix} \bar{\mathbf{K}}_{NN}^I & \mathbf{0} & \mathbf{0} \\ \mathbf{0} & \bar{\mathbf{K}}_{NN}^{II} & \mathbf{0} \\ \mathbf{0} & \mathbf{0} & \bar{\mathbf{K}}_{BB}^I + \bar{\mathbf{K}}_{BB}^{II} \end{bmatrix} \tag{11.120}$$

where superscripts I and II indicate the substructure.

The equation of motion of the complete structure is

$$[\mathbf{M}] = \{\ddot{\mathbf{q}}\} + [\mathbf{K}]\{\mathbf{q}\} = 0 \tag{11.121}$$

After solving this equation, the displacements for each substructure are calculated using (11.108).

Reference [11.45] presents the original derivation of this method and gives an example of a cantilever plate divided into two substructures.

The method has very good convergence properties as the number of component modes are increased, but can result in a large number of interface degrees of freedom if too many substructures are used.

References [11.40, 11.46] present methods of reducing the number of fixed interface normal modes. These consist of using higher order approximations to the omitted modes. References [11.47, 11.48] increase the efficiency of the analysis by generating a set of vectors to be used in place of the boundary fixed normal modes.

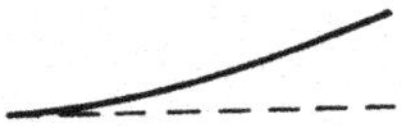

Figure 11.12. Free interface modes.

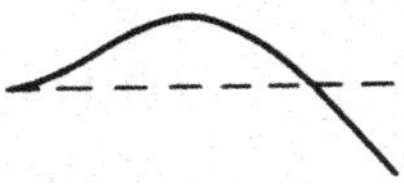

11.5.4.2 Free Interface Method

In the free interface method the number of degrees of freedom for a substructure are reduced using the transformation

$$\{\mathbf{u}\}_s = [\boldsymbol{\phi}_N]\{\mathbf{q}_N\}_s \tag{11.122}$$

where the columns of the matrix $\boldsymbol{\phi}_N$ are the natural modes of the substructure with interface boundaries free. They are obtained by solving the eigenproblem

$$[\mathbf{K} - \omega^2\mathbf{M}]_s\{\boldsymbol{\phi}\} = 0 \tag{11.123}$$

Again $\boldsymbol{\phi}_N$ has fewer columns than rows. Free interface modes for the substructure I in Figure 11.10 are illustrated in Figure 11.12.

If a substructure is completely free, which would be the case for substructure II in Figure 11.10 if the right hand boundary was free, then any rigid body modes are treated as modes with zero frequency.

Substituting (11.122) into (11.107) gives

$$\begin{aligned} T_s &= \tfrac{1}{2}\{\dot{\mathbf{q}}_N\}_s^T[\bar{\mathbf{M}}]_s\{\dot{\mathbf{q}}_N\}_s \\ U_s &= \tfrac{1}{2}\{\mathbf{q}_N\}_s^T[\bar{\mathbf{K}}]_s\{\mathbf{q}_N\}_s \end{aligned} \tag{11.124}$$

where

$$[\bar{\mathbf{M}}]_s = [\boldsymbol{\phi}_N]^T[\mathbf{M}]_s[\boldsymbol{\phi}_N]$$

and

$$[\bar{\mathbf{K}}]_s = [\boldsymbol{\phi}_N]^T[\mathbf{K}]_s[\boldsymbol{\phi}_N] \tag{11.125}$$

which are both diagonal matrices. If the columns of $[\boldsymbol{\phi}_N]$ are mass normalised (see equation (11.4)) then

$$[\bar{\mathbf{M}}]_s = \mathbf{I} \tag{11.126}$$

Putting $s = r$ in (11.15) and introducing (11.126) shows that

$$[\bar{\mathbf{K}}]_s = \boldsymbol{\Lambda}_s \tag{11.127}$$

which is a diagonal matrix containing the eigenvalues on the diagonal.

Adding the contributions from the two substructures in Figure 11.10 together gives expressions of the form (11.117) where

$$\{\mathbf{q}\} = \begin{bmatrix} \mathbf{q}_N^I \\ \mathbf{q}_N^{II} \end{bmatrix} \tag{11.128}$$

$$[\mathbf{M}] = \begin{bmatrix} \mathbf{I} & \mathbf{0} \\ \mathbf{0} & \mathbf{I} \end{bmatrix} \tag{11.129}$$

and

$$[\mathbf{K}] = \begin{bmatrix} \boldsymbol{\Lambda}^I & \mathbf{0} \\ \mathbf{0} & \boldsymbol{\Lambda}^{II} \end{bmatrix} \tag{11.130}$$

The next step is to apply the constraints that the two substructures have the same displacements at their interface. This can be expressed by

$$\{\mathbf{u}_B^I\} = \{\mathbf{u}_B^{II}\} \tag{11.131}$$

where subscript B denotes displacements on the interface boundary.

Substituting (11.122) into (11.131) gives

$$[\boldsymbol{\phi}_B^I]\{\mathbf{q}_N^I\} = [\boldsymbol{\phi}_B^{II}]\{\mathbf{q}_N^{II}\} \tag{11.132}$$

where $\boldsymbol{\phi}_B$ are the rows of $\boldsymbol{\phi}_N$ which relate to the interface boundary degrees of freedom. Equation (11.132) defines a set of linear constraint relations between the generalised coordinates of the two substructures.

Let there be n_I, n_{II} modes representing substructures I and II respectively and n_B degrees of freedom on the interface boundary ($n_B = 2$ for the system shown in Figure 11.10). If $n_I > n_B$, then $[\boldsymbol{\phi}_B^I]$ can be partitioned as follows

$$[\boldsymbol{\phi}_B^I] = [\boldsymbol{\phi}_{B1}^I \quad \boldsymbol{\phi}_{B2}^I] \tag{11.133}$$

$$(n_B, n_I)(n_B, n_B)(n_B, (n_I - n_B))$$

Equation (11.132) becomes

$$[\boldsymbol{\phi}_{B1}^I \boldsymbol{\phi}_{B2}^I]\begin{bmatrix} \mathbf{q}_{N1}^I \\ \mathbf{q}_{N2}^I \end{bmatrix} = [\boldsymbol{\phi}_B^{II}]\{\mathbf{q}_N^{II}\} \tag{11.134}$$

Solving for $\{\mathbf{q}_{N1}^I\}$ gives

$$\{\mathbf{q}_{N1}^I\} = -[\boldsymbol{\phi}_{B1}^I]^{-1}[\boldsymbol{\phi}_{B2}^I]\{\mathbf{q}_{N2}^I\} + [\boldsymbol{\phi}_{B1}^I]^{-1}[\boldsymbol{\phi}_B^{II}]\{\mathbf{q}_N^{II}\} \tag{11.135}$$

Therefore

$$\{\mathbf{q}\} = \begin{bmatrix} \mathbf{q}_N^I \\ \mathbf{q}_N^{II} \end{bmatrix} = \begin{bmatrix} \mathbf{q}_{N1}^I \\ \mathbf{q}_{N2}^I \\ \mathbf{q}_N^{II} \end{bmatrix} = [\mathbf{T}_c]\{\mathbf{r}\} \tag{11.136}$$

where

$$\{\mathbf{r}\} = \begin{bmatrix} \mathbf{q}_{N2}^I \\ \mathbf{q}_N^{II} \end{bmatrix} \tag{11.137}$$

and

$$[\mathbf{T}_c] = \begin{bmatrix} -[\boldsymbol{\phi}_{B1}^{I}]^{-1} - [\boldsymbol{\phi}_{B2}^{I}] & [\boldsymbol{\phi}_{B1}^{I}]^{-1} - [\boldsymbol{\phi}_{B}^{II}] \\ \mathbf{I} & \mathbf{0} \\ \mathbf{0} & \mathbf{I} \end{bmatrix} \tag{11.138}$$

Substituting the transformation (11.136) into the energy expression (11.116) and using (11.129) and (11.130) gives

$$\begin{aligned} T &= \tfrac{1}{2}\{\dot{\mathbf{r}}\}^{T}[\mathbf{M}_R]\{\dot{\mathbf{r}}\} \\ U &= \tfrac{1}{2}\{\mathbf{r}\}^{T}[\mathbf{K}_R]\{\mathbf{r}\} \end{aligned} \tag{11.139}$$

where

$$[\mathbf{M}_R] = [\mathbf{T}_c]^{T}[\mathbf{T}_c] \tag{11.140}$$

and

$$[\mathbf{K}_R] = [\mathbf{T}_c]^{T}[\mathbf{K}][\mathbf{T}_c] \tag{11.141}$$

The equation of motion of the complete structure is

$$[\mathbf{M}_R]\{\ddot{\mathbf{r}}\} + [\mathbf{K}_R]\{\mathbf{r}\} = 0 \tag{11.142}$$

After solving this equation, the displacements for the complete structure are calculated using

$$\begin{bmatrix} \mathbf{u}^{I} \\ \mathbf{u}^{II} \end{bmatrix} = \begin{bmatrix} \boldsymbol{\phi}_{N}^{I} & \mathbf{0} \\ \mathbf{0} & \boldsymbol{\phi}_{N}^{II} \end{bmatrix} [\mathbf{T}_c]\{\mathbf{r}\} \tag{11.143}$$

This method was first presented in reference [11.49]. Its main disadvantage is that the convergence is weak. This can be overcome by including a low-frequency approximation for the contribution of the neglected high frequency modes (see Chapter 13). This is usually termed residual flexibility. Details of this modification are given in references [11.50–11.52]. The efficiency of the analysis can be increased by generating a set of vectors to be used in place of the boundary free normal modes [11.47, 11.48, 11.53]. References [11.40, 11.54] extend the methods to higher frequencies without increasing the number of degrees of freedom. This is achieved by approximating the modes outside the frequency band of interest (i.e. both above and below).

References [11.55–11.57] present an alternative formulation which is particularly useful when applied to machinery. In this method it is assumed that the substructures are connected together via springs, as illustrated in Figure 11.13.

The kinetic and strain energy of the two substructures, in terms of modal coordinates, are again given by (11.116) with $\{\mathbf{q}\}$, $[\mathbf{M}]$ and $[\mathbf{K}]$ defined by (11.128) to (11.130). The strain energy of the connectors is

$$U_c = \frac{1}{2}\begin{bmatrix} \mathbf{u}_{B}^{I} \\ \mathbf{u}_{B}^{II} \end{bmatrix}^{T} [\mathbf{K}_c] \begin{bmatrix} \mathbf{u}_{B}^{I} \\ \mathbf{u}_{B}^{II} \end{bmatrix} \tag{11.144}$$

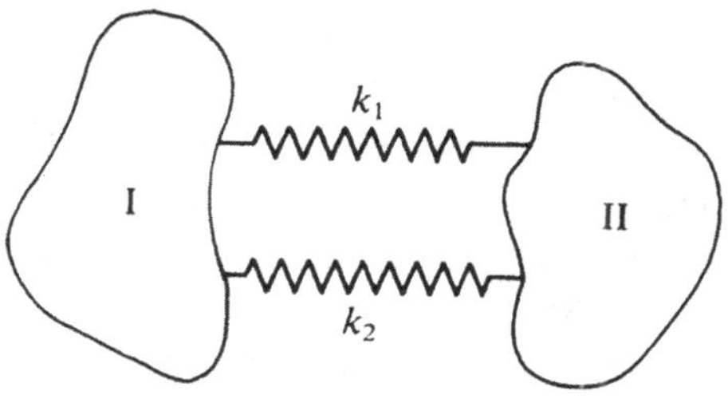

Figure 11.13. Two substructures connected by springs.

Now from (11.122)

$$\begin{bmatrix} \mathbf{u}_B^I \\ \mathbf{u}_B^{II} \end{bmatrix} = \begin{bmatrix} \boldsymbol{\phi}_B^I & \mathbf{0} \\ \mathbf{0} & \boldsymbol{\phi}_B^{II} \end{bmatrix} \begin{bmatrix} \mathbf{q}_N^I \\ \mathbf{q}_N^{II} \end{bmatrix} = [\boldsymbol{\Phi}_B]\{\mathbf{q}\} \tag{11.145}$$

Substituting (11.145) into (11.144) gives

$$U_c = \tfrac{1}{2}\{\mathbf{q}\}^T[\boldsymbol{\Phi}_B]^T[\mathbf{K}_c][\boldsymbol{\Phi}_B]\{\mathbf{q}\} \tag{11.146}$$

The total strain energy of the system is, therefore,

$$U_T = \tfrac{1}{2}\{\mathbf{q}\}^T([\mathbf{K}]^T + [\boldsymbol{\Phi}_B]^T[\mathbf{K}_c][\boldsymbol{\Phi}_B])\{\mathbf{q}\} \tag{11.147}$$

The equation of motion of the complete structure is

$$\{\ddot{\mathbf{q}}\} + ([\mathbf{K}] + [\boldsymbol{\Phi}_B]^T[\mathbf{K}_c][\boldsymbol{\Phi}_B])\{\mathbf{q}\} = 0 \tag{11.148}$$

After solving this equation, the displacements for each substructure are calculated using (11.122).

Reference [11.56] analyses two beams connected by springs, whilst reference [11.55] considers a packet of turbine blades connected by means of a shroud.

The obvious advantage of free interface methods is that the interface degrees of freedom do not appear in the final equations of motion. In the two methods presented here the total number of degrees of freedom are $(n_I + n_{II} + n_B)$ and $(n_I + n_{II})$ respectively for two substructures.

Another advantage is that if a particular component is difficult to model using finite element techniques, its modal representation can be determined experimentally [11.57] and included in the analysis. In this case it is simpler to test the component with interface boundaries free rather than fixed. If joint stiffnesses are to be included they can also be measured [11.58, 11.59].

Problems

11.1 Calculate the natural frequencies and modes of the system shown in Figure P1.1 with $k_1 = k_3 = 9$, $k_2 = k_4 = 3$, $m_1 = m_3 = 3$ and $m_2 = 2$.

11.2 Investigate the Sturm sequence properties of the eigenproblem given by the system shown in Figure P1.1 with $k_1 = 3$, $k_2 = k_3 = 2$, $k_4 = 1$ and $m_1 = m_2 = m_3 = 1$.

11.3 Determine how many eigenvalues there are below $\lambda = 0.5$ when $\mathbf{K}$ and $\mathbf{M}$ are as defined in Example 11.7.

11.4 A two-dimensional structure has $X = 0$ as an axis of symmetry. It is to be analysed using an idealisation of half the structure. Derive the boundary conditions to be applied on the axis of symmetry for the symmetric and antisymmetric modes.

11.5 Figure P11.5 shows a simply supported uniform beam which is represented by two elements. Calculate the two lowest natural frequencies by the following methods:

(1) use a half-model and apply the appropriate conditions of symmetry and antisymmetry.
(2) use a full model and apply the reduction technique to eliminate two degrees of freedom.

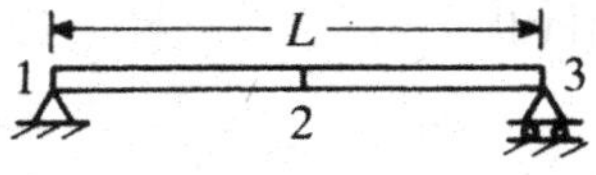

Figure P11.5

Compare the results with the analytical solution $\omega_1 = \pi^2(EI_z/\rho AL^4)^{1/2}$ and $\omega_2 = 4\pi^2(EI_z/\rho AL^4)^{1/2}$.

12 Forced Response I

In Chapters 1 and 3 it is shown that the equation of motion of a structure is

$$\mathbf{M}\ddot{\mathbf{u}} + \mathbf{C}\dot{\mathbf{u}} + \mathbf{K}\mathbf{u} = \mathbf{f} \tag{12.1}$$

where

$\mathbf{u}$ = colum matrix of nodal displacements
$\mathbf{M}$ = inertia matrix
$\mathbf{C}$ = damping matrix
$\mathbf{K}$ = stiffness matrix
$\mathbf{f}$ = column matrix of equivalent nodal forces

Chapters 3 to 7 give details of how to derive the matrices **M**, **K** and **f**. The derivation of the damping matrix **C** is treated in Section 12.2.

The method of solving equation (12.1) depends upon whether the applied forces are harmonic, periodic, transient or random. This and the following chapter present solution techniques for each of these cases.

12.1 Modal Analysis

Whatever the nature of the applied forces, the solution of equation (12.1) can be obtained either directly or by first transforming it into a simpler form. This can be achieved by means of the expansion theorem [12.1].

Any vector in n-dimensional space can be expressed as a linear combination of n linearly independent vectors (see Section 3.1). The eigenvectors $\boldsymbol{\phi}_r$ of the eigenproblem (11.2) are orthogonal (see Section (11.1.1) and so are linearly independent. This means that the solution $\mathbf{u}$ of equation (12.1) can be expressed in the form

$$\mathbf{u} = \boldsymbol{\Phi}\mathbf{q}(t) \tag{12.2}$$

where the columns of $\boldsymbol{\Phi}$ consist of the eigenvectors $\boldsymbol{\phi}_r$.

Substituting (12.2) into the energy expressions

$$\begin{aligned} T &= \tfrac{1}{2}\dot{\mathbf{u}}^{\mathrm{T}}\mathbf{M}\dot{\mathbf{u}}, \qquad & D &= \tfrac{1}{2}\dot{\mathbf{u}}^{\mathrm{T}}\mathbf{C}\dot{\mathbf{u}} \\ U &= \tfrac{1}{2}\mathbf{u}^{\mathrm{T}}\mathbf{K}\mathbf{u}, \qquad & \delta W &= \delta\mathbf{u}^{\mathrm{T}}\mathbf{f} \end{aligned} \tag{12.3}$$

and using Lagrange's equations gives

$$\bar{\mathbf{M}}\ddot{\mathbf{q}} + \bar{\mathbf{C}}\dot{\mathbf{q}} + \bar{\mathbf{K}}\mathbf{q} = \mathbf{Q} \tag{12.4}$$

where

$$\begin{aligned} \bar{\mathbf{M}} &= \mathbf{\Phi}^{\mathrm{T}}\mathbf{M}\mathbf{\Phi} \\ \bar{\mathbf{C}} &= \mathbf{\Phi}^{\mathrm{T}}\mathbf{C}\mathbf{\Phi} \\ \bar{\mathbf{K}} &= \mathbf{\Phi}^{\mathrm{T}}\mathbf{K}\mathbf{\Phi} \\ \mathbf{Q} &= \mathbf{\Phi}^{\mathrm{T}}\mathbf{f} \end{aligned} \tag{12.5}$$

Both $\bar{\mathbf{M}}$ and $\bar{\mathbf{K}}$ are diagonal matrices by virtue of equations (11.18) and (11.19). In addition, if the colummns of $\mathbf{\Phi}$ satisfy (11.4) then

$$\bar{\mathbf{M}} = \mathbf{I} \quad \text{and} \quad \bar{\mathbf{K}} = \mathbf{\Lambda} \tag{12.6}$$

where

$$\mathbf{\Lambda} = \begin{bmatrix} \omega_1^2 & & & \\ & \omega_2^2 & & \\ & & \ddots & \\ & & & \omega_n^2 \end{bmatrix} \tag{12.7}$$

and ω_r is the rth natural frequency. In general $\bar{\mathbf{C}}$ will not be diagonal (but see Section 12.2). Equation (12.4) now reduces to

$$\ddot{\mathbf{q}} + \bar{\mathbf{C}}\dot{\mathbf{q}} + \mathbf{\Lambda}\mathbf{q} = \mathbf{Q} \tag{12.8}$$

This equation is solved for $\mathbf{q}$ and the result substituted into equation (12.2).

12.2 Representation of Damping

Section 2.10 indicates that it is difficult to formulate explicit expressions for the damping forces in a structure. Instead, simplified models, based more on mathematical convenience than physical representation, are used. Two types are considered: structural (sometimes referred to as hysteretic) and viscous.

12.2.1 Structural Damping

Generalising the treatment of structural damping presented in Section 2.10 to a multi-degree of freedom system indicates that equation (12.1) should be replaced by

$$\mathbf{M}\ddot{\mathbf{u}} + [\mathbf{K} + \mathrm{i}\mathbf{H}]\mathbf{u} = \mathbf{f} \tag{12.9}$$

This form of damping can only be used when the excitation is harmonic [12.2]. The complex matrix $[\mathbf{K} + \mathrm{i}\mathbf{H}]$ is obtained by replacing Young's modulus E by a complex one $E(1 + \mathrm{i}\eta)$, where η is the material loss factor, when deriving the element stiffness matrices. Reference [12.3] indicates that η can vary from 2×10^{-5} for pure aluminium to 1.0 for hard rubber.

Equation (12.9) can be simplified using modal analysis (see Section 12.1) to give

$$\ddot{\mathbf{q}} + [\mathbf{\Lambda} + \mathrm{i}\bar{\mathbf{H}}]\mathbf{q} = \mathbf{Q} \tag{12.10}$$

where $\mathbf{\Lambda}$, $\mathbf{Q}$ are defined by (12.7), (12.5) and

$$\bar{\mathbf{H}} = \mathbf{\Phi}^{\mathrm{T}}\mathbf{H}\mathbf{\Phi} \tag{12.11}$$

In general, $\bar{\mathbf{H}}$ will be a fully populated matrix, unless every element has the same loss factor. In this case

$$\mathbf{H} = \eta\mathbf{K} \tag{12.12}$$

and

$$\bar{\mathbf{H}} = \eta\mathbf{\Lambda} \tag{12.13}$$

Equation (12.10) now becomes

$$\ddot{\mathbf{q}} + (1 + \mathrm{i}\eta)\mathbf{\Lambda}\mathbf{q} = \mathbf{Q} \tag{12.14}$$

All the equations in (12.14) are now uncoupled and each one is of the form of a single degree of freedom system.

12.2.2 Viscous Damping

Viscous type damping can be used whatever the form of the excitation. The most common form of such damping is the so-called Rayleigh-type damping [12.4] given by

$$\mathbf{C} = a_1\mathbf{M} + a_2\mathbf{K} \tag{12.15}$$

The advantage of this representation is that the matrix $\bar{\mathbf{C}}$ (equation (12.5)) becomes

$$\bar{\mathbf{C}} = a_1\mathbf{I} + a_2\mathbf{\Lambda} \tag{12.16}$$

which is diagonal. So once again the equations in (12.8) are uncoupled. Each one is of the form

$$\ddot{q}_r + 2\gamma_r\omega_r\dot{q}_r + \omega_r^2 q_r = Q_r \tag{12.17}$$

where γ_r is the modal damping ratio and

$$2\gamma_r\omega_r = a_1 + \omega_r^2 a_2 \tag{12.18}$$

The two factors a_1 and a_2 can be determined by specifying the damping ratio for two modes, 1 and 2, say. Substituting into (12.18) gives

$$\begin{aligned} a_1 + \omega_1^2 a_2 &= 2\gamma_1\omega_1 \\ a_1 + \omega_2^2 a_2 &= 2\gamma_2\omega_2 \end{aligned} \tag{12.19}$$

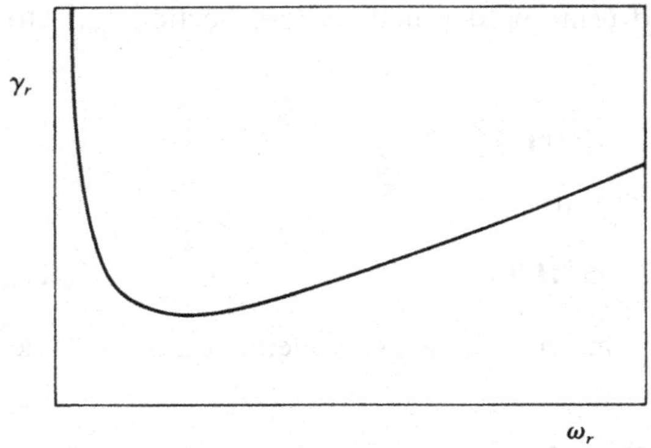

Figure 12.1. Rayleigh damping.

The solution of these two equations gives

$$\begin{aligned} a_1 &= 2\omega_1\omega_2(\omega_2\gamma_1 - \omega_1\gamma_2)/(\omega_2^2 - \omega_1^2) \\ a_2 &= 2(\omega_2\gamma_2 - \omega_1\gamma_1)/(\omega_2^2 - \omega_1^2) \end{aligned} \tag{12.20}$$

The damping ratios in the other modes are then given by (12.18), that is

$$\gamma_r = \frac{a_1}{2\omega_r} + \frac{a_2\omega_r}{2} \tag{12.21}$$

A typical variation of γ_r with ω_r is shown in Figure 12.1.

If a direct solution of equation (12.1) is preferred, the values of a_1 and a_2 given by (12.20) can be substituted into (12.15), to give the required matrix **C**.

Mass-proportional and stiffness-proportional damping can be used separately. For mass-proportional damping $a_2 = 0$ in (12.15) and (12.18). Specifying the damping ratio for mode 1 gives

$$a_1 = 2\gamma_1\omega_1 \tag{12.22}$$

Therefore

$$\gamma_r = \frac{\gamma_1\omega_1}{\omega_r} \tag{12.23}$$

This means that the damping ratios decrease with increase in mode frequency.

For stiffness-proportional damping $a_1 = 0$ in (12.15) and (12.18). Again specifying the damping ratio for mode 1 gives

$$a_2 = \frac{2\gamma_1}{\omega_1} \tag{12.24}$$

Therefore

$$\gamma_r = \frac{\gamma_1\omega_r}{\omega_1} \tag{12.25}$$

In this case the damping ratios increase with increase in mode frequency. Typical variations of mass- and stiffness-proportional damping ratios are shown in Figure 12.2. In practice it has been found that mass-proportional damping can represent friction damping, whilst stiffness-proportional damping can represent internal material damping.

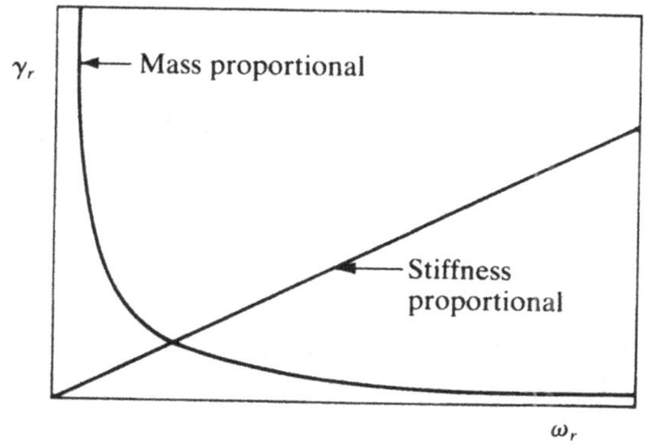

Figure 12.2. Mass and stiffness-proportional damping.

Accepted values of modal damping ratios for typical forms of construction vary from 0.01 for small diameter piping systems, to 0.07 for bolted joint and reinforced concrete structures. If all the modal damping ratios can be estimated, it is not necessary to form the damping matrix. The values of γ_r are substituted into equation (12.17).

In cases where the damping varies considerably in different parts of the structure the above techniques cannot be used directly. An example of this is the analysis of soil-structure interaction problems, where there is significantly more damping in the soil than in the structure. This type of problem should be analysed using component mode synthesis techniques (see Section 13.5.4). In some cases the damping in a structure varies with frequency. Possible methods of deriving such models are given in references [12.5–12.8]. Further information on damping can be found in references [12.9–12.14].

12.3 Harmonic Response

If the nodal forces are harmonic, all with the same frequency, ω, but having different amplitudes and phases, then

$$\mathbf{f}(t) = \mathbf{f}\exp(\mathrm{i}\omega t) \tag{12.26}$$

In general the elements of $\mathbf{f}$ will be complex since

$$f_k = |f_k| \exp(\mathrm{i}\theta_k) \tag{12.27}$$

where θ_k is the phase of force f_k relative to a reference force.

12.3.1 Modal Analysis

Equations (12.5) and (12.26) indicate that the generalised forces take the form

$$\mathbf{Q}(\mathrm{t}) = \mathbf{\Phi}^{\mathrm{T}}\mathbf{f}\exp(\mathrm{i}\omega t) = \mathbf{Q}\exp(\mathrm{i}\omega t) \tag{12.28}$$

Equation (12.8) now becomes

$$\ddot{\mathbf{q}} + \bar{\mathbf{C}}\dot{\mathbf{q}} + \mathbf{\Lambda}\mathbf{q} = \mathbf{Q}\exp(\mathrm{i}\omega t) \tag{12.29}$$

The steady state response is obtained by assuming that the response is harmonic with frequency ω. This gives

$$[\mathbf{\Lambda} - \omega^2\mathbf{I} + \mathrm{i}\omega\,\bar{\mathbf{C}}]\mathbf{q} = \mathbf{Q}\exp(\mathrm{i}\omega t) \tag{12.30}$$

Solving for $\mathbf{q}$ gives

$$\mathbf{q} = [\mathbf{\Lambda} - \omega^2\mathbf{I} + \mathrm{i}\omega\,\bar{\mathbf{C}}]^{-1}\mathbf{Q}\exp(\mathrm{i}\omega t) \tag{12.31}$$

If either structural or proportional damping is used, the matrix to be inverted is diagonal. Its inverse is also a diagonal matrix with diagonal elements

(1) $(\omega_r^2 - \omega^2 + \mathrm{i}\eta\omega_r^2)^{-1}$ for structural damping
or
(2) $(\omega_r^2 - \omega^2 + \mathrm{i}2\gamma_r\omega_r\omega)^{-1}$ for proportional damping

Substituting (12.31) and (12.28) into (12.2) gives

$$\mathbf{u} = \mathbf{\Phi}[\mathbf{\Lambda} - \omega^2\mathbf{I} + \mathrm{i}\omega\,\bar{\mathbf{C}}]^{-1}\mathbf{\Phi}^{\mathrm{T}}\mathbf{f}\exp(\mathrm{i}\omega t) \tag{12.32}$$

Equation (12.32) is often written in the form

$$\mathbf{u} = [\boldsymbol{\alpha}(\omega)]\mathbf{f}\exp(\mathrm{i}\omega t) \tag{12.33}$$

where

$$[\boldsymbol{\alpha}(\omega)] = \mathbf{\Phi}[\mathbf{\Lambda} - \omega^2\mathbf{I} + \mathrm{i}\omega\,\bar{\mathbf{C}}]^{-1}\mathbf{\Phi}^{\mathrm{T}} \tag{12.34}$$

is a matrix of receptances. $\alpha_{jk}(\omega)$ is a transfer receptance which represents the response in degree of freedom j due to a harmonic force of unit magnitude and frequency ω applied in degree of freedom k. Likewise $\alpha_{jj}(\omega)$ represents a point receptance. From (12.34) it can be seen that

$$\alpha_{jk}(\omega) = \sum_{r=1}^{n} \frac{\phi_{jr}\phi_{kr}}{(\omega_r^2 - \omega^2 + \mathrm{i}\eta\omega_r^2)} \tag{12.35}$$

for structural damping, and

$$\alpha_{jk}(\omega) = \sum_{r=1}^{n} \frac{\phi_{jr}\phi_{kr}}{(\omega_r^2 - \omega^2 + \mathrm{i}2\gamma_r\omega_r\omega)} \tag{12.36}$$

for proportional damping. The term $\phi_{jr}\phi_{kr}$ will be referred to as a modal constant [12.15].

If the damping in mode s is small and its frequency ω_s well separated from the frequencies ω_{s-1} and ω_{s+1} of the neighbouring modes, then term s in the series will dominate the response whenever the exciting frequency ω is close to ω_s.

EXAMPLE 12.1 Calculate the receptances $\alpha_{j1}(\omega)$, $j = 1, 2, 3$ for $0 < \omega < 94.25$ rad/s (i.e., 15 Hz) of the system shown in Figure P1.1 with $k_1 = k_4 = 3000$ N/m, $k_2 = k_3 = 1000$ N/m, $m_1 = m_3 = 2$ kg and $m_2 = 1$ kg. Assume no damping.

From Example 11.1 it can be seen that

$$\mathbf{\Lambda} = \begin{bmatrix} 1000 & & \\ & 2000 & \\ & & 3000 \end{bmatrix}$$

$$\mathbf{\Phi} = \frac{1}{2(2)^{1/2}} \begin{bmatrix} 1 & 2^{1/2} & 1 \\ 2 & 0 & -2 \\ 1 & -2^{1/2} & 1 \end{bmatrix}$$

$$\mathbf{Q} = \mathbf{\Phi}^{\mathrm{T}}\mathbf{f} = \frac{1}{2(2)^{1/2}} \begin{bmatrix} 1 & 2 & 1 \\ 2^{1/2} & 0 & -2^{1/2} \\ 1 & -2 & 1 \end{bmatrix} \begin{bmatrix} f_1 \\ 0 \\ 0 \end{bmatrix}$$

$$= \frac{1}{2(2)^{1/2}} \begin{bmatrix} 1 \\ 2^{1/2} \\ 1 \end{bmatrix} f_1$$

Therefore

$$\alpha_{11}(\omega) = \frac{1}{8}\left\{\frac{1(1)}{(1000-\omega^2)} + \frac{(2^{1/2})(2^{1/2})}{(2000-\omega^2)} + \frac{1(1)}{(3000-\omega^2)}\right\}$$

$$\alpha_{21}(\omega) = \frac{1}{8}\left\{\frac{2(1)}{(1000-\omega^2)} + \frac{0(2^{1/2})}{(2000-\omega^2)} + \frac{(-2)(1)}{(3000-\omega^2)}\right\}$$

$$\alpha_{31}(\omega) = \frac{1}{8}\left\{\frac{1(1)}{(1000-\omega^2)} + \frac{(-2^{1/2})(2^{1/2})}{(2000-\omega^2)} + \frac{1(1)}{(3000-\omega^2)}\right\}$$

The variations of the modulus of α_{11}, α_{21} and α_{31} with frequency are shown in Figures 12.3(a) to 12.3(c).

The point receptance α_{11} shown in Figure 12.3(a) has three resonant peaks at the natural frequencies of the system 5.03, 7.12 and 8.72 Hz. Each resonant peak is separated by an anti-resonance, one at 5.72 and the other at 8.28 Hz. The reason for this can be seen by inspecting the series expression for α_{11}. All three modal constants are positive. At a frequency between the first two resonant frequencies the first term is negative whilst the second one is positive. The contribution from the third term is significantly smaller. Therefore, it is possible to find a frequency for which the receptance is zero. Similarly, at a frequency between the second and third resonant frequencies the second term is negative whilst the third is positive, giving rise to an anti-resonance in this range.

The anti-resonant frequencies can be found by equating α_{11} to zero. This gives

$$\begin{aligned}&(2000-\omega^2)(3000-\omega^2) + 2(1000-\omega^2)(3000-\omega^2)\\ &\quad + (1000-\omega^2)(2000-\omega^2) = 0\end{aligned}$$

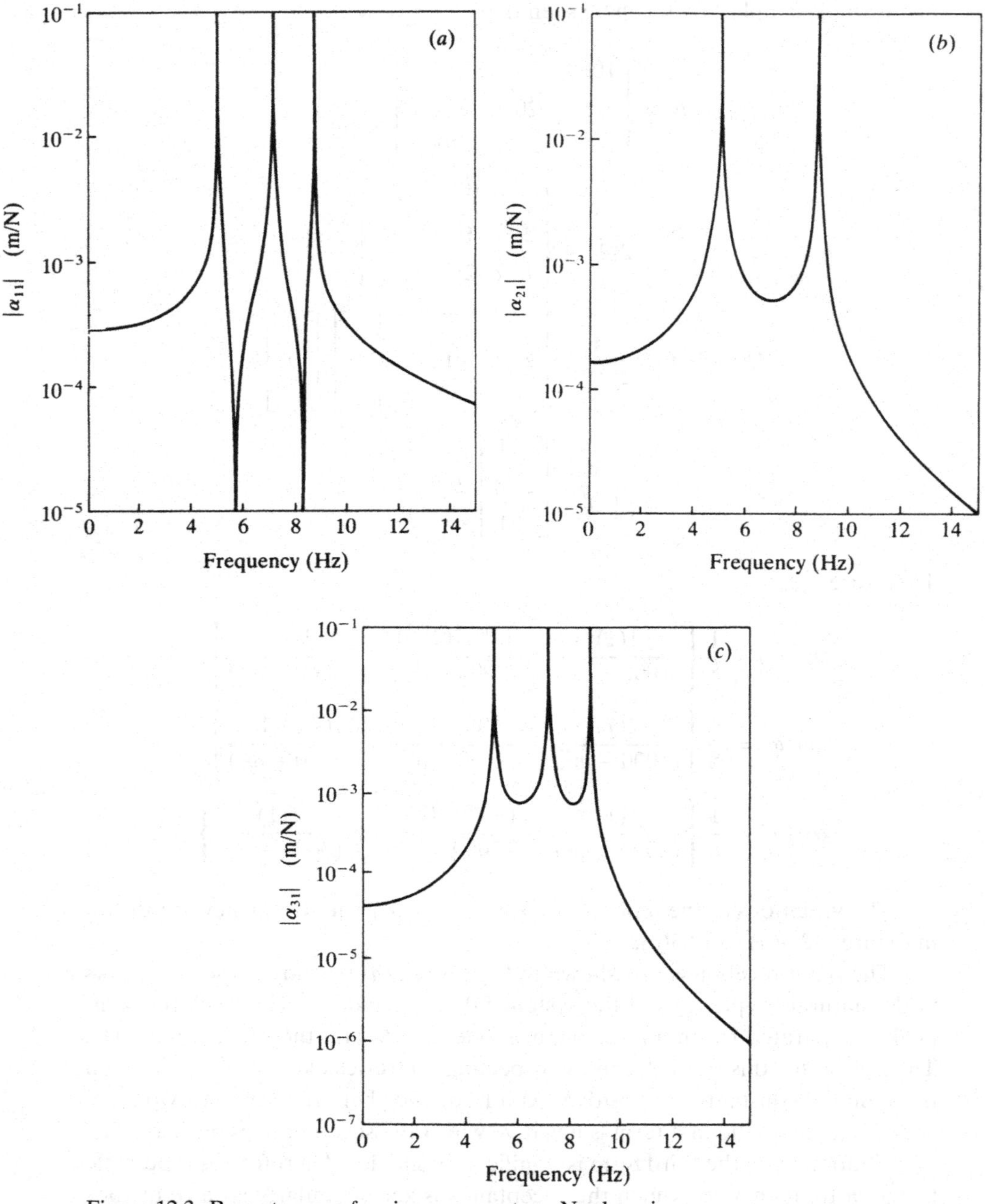

Figure 12.3. Receptances of spring-mass system. No damping.

That is

$$4\omega^4 - 16 \times 10^3\omega^2 + 14 \times 10^6 = 0$$

giving

$$\omega^2 = 1.2929 \times 10^3 \qquad \text{or} \qquad 2.7071 \times 10^3$$

Table 12.1. *Receptances for three degree of freedom system with stiffness proportional damping*

	(Modulus of receptance) $\times 10^3$ m/N		
Frequency (Hz)	α_{11}	α_{21}	α_{31}
5.03	3.17	6.25	3.12
5.72	0.115	0.986	0.694
7.12	2.24	0.493	2.18
8.28	0.291	0.830	0.579
8.72	0.708	1.20	0.598

and

$$\omega = 35.96 \quad \text{or} \quad 52.03\,\text{rad/s}$$

The transfer receptance α_{21} has only two resonant peaks at 5.03 and 8.72 Hz. This is because the modal constant for the second mode is zero, due to the fact that the displacement u_2 is zero in this mode. An anti-resonance does not occur between the two resonances, since the two non-zero terms are both negative in this range. This is because the modal constant for the third mode is negative.

The transfer receptance α_{31} has three resonant peaks at the three natural frequencies. Anti-resonances do not occur between either pair of resonances. In between the first two resonances the first two terms are both negative, whilst in between the second two resonances the second and third terms are both positive. Both these situations are due to the fact that the modal constant in the second mode is negative.

EXAMPLE 12.2 Calculate the receptances $\alpha_{j1}(\omega)$, $j = 1, 2, 3$ for the frequencies 5.03, 5.72, 7.12, 8.28 and 8.72 Hz of the system considered in Example 12.1 when the damping is proportional to the stiffness matrix with $\gamma_1 = 0.02$.

From (12.25) $\gamma_r = (\gamma_1/\omega_1)\omega_r$. Now $\gamma_1 = 0.02$ and $\omega_1 = 31.623$. This gives $\gamma_2 = 0.02828$ and $\gamma_3 = 0.03464$. Including this damping in the expressions in Example 12.1 gives

$$\alpha_{11}(\omega) = \tfrac{1}{8}\{(1000 - \omega^2 + \mathrm{i}1.2649\omega)^{-1} + 2(2000 - \omega^2 + \mathrm{i}2.5294\omega)^{-1} + (3000 - \omega^2 + \mathrm{i}3.7946\omega)^{-1}\}$$

$$\alpha_{21}(\omega) = \tfrac{1}{8}\{2(1000 - \omega^2 + \mathrm{i}1.2649\omega)^{-1} - 2(3000 - \omega^2 + \mathrm{i}3.7946\omega)^{-1}\}$$

$$\alpha_{31}(\omega) = \tfrac{1}{8}\{(1000 - \omega^2 + \mathrm{i}1.2649\omega)^{-1} - 2(2000 - \omega^2 + \mathrm{i}2.5294\omega)^{-1} + (3000 - \omega^2 + \mathrm{i}3.7946\omega)^{-1}\}$$

The values of the moduli of the receptances at the required frequencies are given in Table 12.1. The undamped natural frequencies of the system are 5.03, 7.12 and 8.72 Hz. The receptance $\alpha_{11}(\omega)$ has finite values at these frequencies in contrast to the infinite values obtained for the undamped case. Without damping $\alpha_{11}(\omega)$ exhibited anti-resonances at 5.72 and 8.28 Hz. With damping

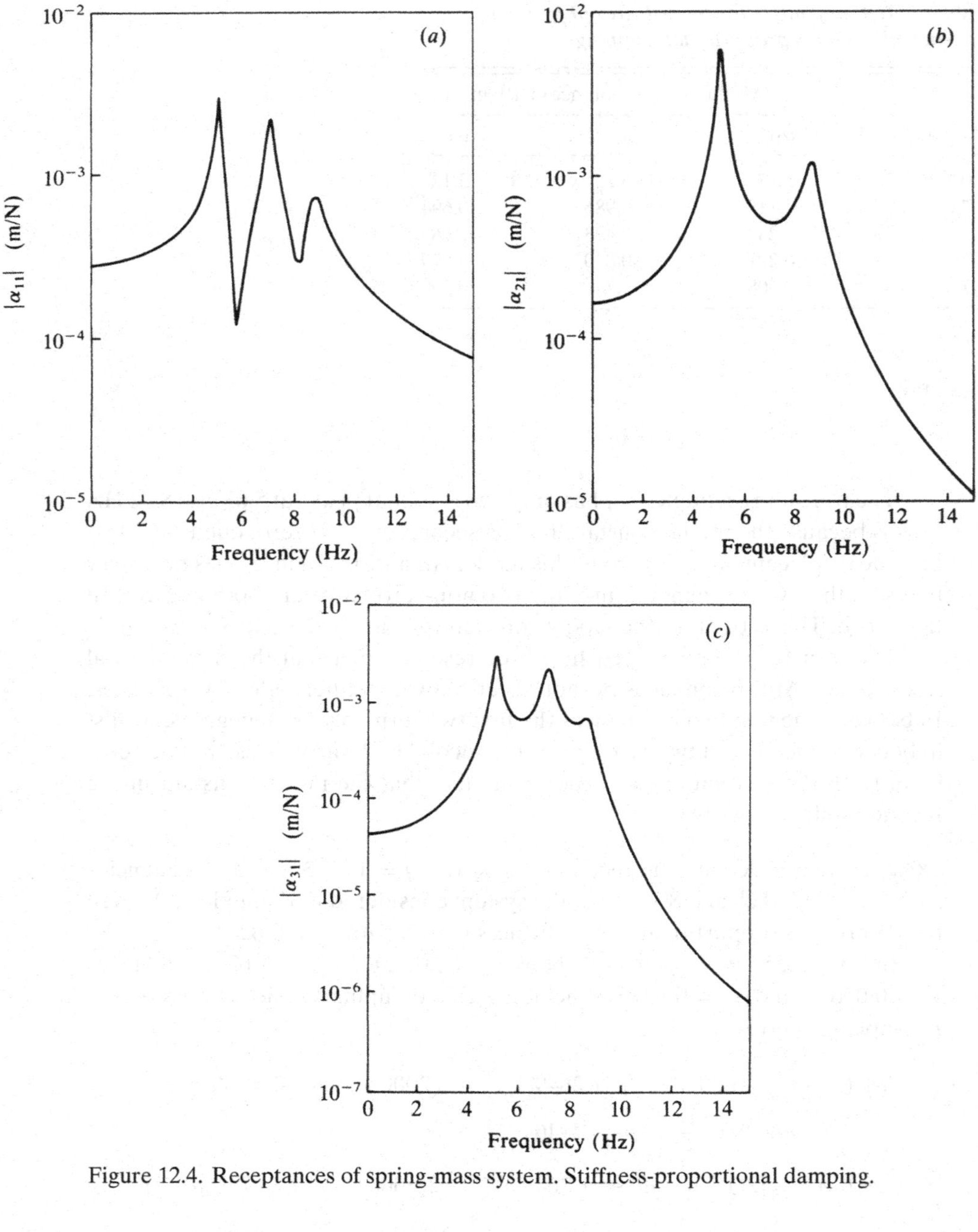

Figure 12.4. Receptances of spring-mass system. Stiffness-proportional damping.

$\alpha_{11}(\omega)$ is no longer zero at these frequencies. Neither $\alpha_{21}(\omega)$ nor $\alpha_{31}(\omega)$ exhibited anti-resonances in the undamped case. $\alpha_{21}(\omega)$ has two infinite peaks at 5.03 and 8.72 Hz which become finite when damping is added. Similarly $\alpha_{31}(\omega)$ had three infinite peaks which are now finite.

The variations of the modulus of α_{11}, α_{21} and α_{31} with frequency are shown in Figure 12.4(a) to 12.4(c). Comparison with Figures 12.3(a) to 12.3(c) clearly indicates the overall effect of including damping.

Table 12.2. *Receptances for three degree of freedom system with structural damping*

	(Modulus of receptance) × 10^3 m/N		
Frequency (Hz)	α_{11}	α_{21}	α_{31}
5.03	3.16	6.24	3.11
7.12	3.14	0.496	3.11
8.72	1.11	2.08	1.04

EXAMPLE 12.3 Calculate the receptances $\alpha_{j1}(\omega)$, $j = 1, 2, 3$ for the frequencies 5.03, 7.12 and 8.72 Hz of the system considered in Example 12.1 when the damping is structural with $\eta = 0.04$.

Including this type of damping in the expressions in Example 12.1 gives

$$\alpha_{11}(\omega) = \tfrac{1}{8}\{(1000 - \omega^2 + \mathrm{i}40)^{-1} + 2(2000 - \omega^2 + \mathrm{i}80)^{-1} + (3000 - \omega^2 + \mathrm{i}120)^{-1}\}$$

$$\alpha_{21}(\omega) = \tfrac{1}{8}\{2(1000 - \omega^2 + \mathrm{i}40)^{-1} - 2(3000 - \omega^2 + \mathrm{i}120)^{-1}\}$$

$$\alpha_{31}(\omega) = \tfrac{1}{8}\{(1000 - \omega^2 + \mathrm{i}40)^{-1} - 2(2000 - \omega^2 + \mathrm{i}80)^{-1} + (3000 - \omega^2 + \mathrm{i}120)^{-1}\}$$

The values of the moduli of the receptances at the required frequencies are given in Table 12.2. The values at 5.03 Hz are similar to the values given in Table 12.1 for stiffness proportional damping. However, the values at 7.12 and 8.72 Hz are larger than the corresponding values in Table 12.1. This is because the effective damping ratios at the natural frequencies are 0.02 compared with the values 0.02, 0.02828 and 0.03464 used in Example 12.2. The values quoted in these tables are not necessarily peak values, since these occur at frequencies which are slightly lower than the undamped natural frequencies. For a single degree of freedom system the maximum response occurs at $\omega_n(1 - 2\gamma^2)^{1/2}$, where ω_n is the undamped natural frequency.

EXAMPLE 12.4 Calculate the receptances $\alpha_{j1}(\omega)$, $j = 1, 2, 3$ for frequencies 5.03, 7.12 and 8.72 Hz of the system considered in Example 12.1 when the damping is of the Rayleigh-type with $\gamma_1 = 0.02$ and $\gamma_3 = 0.03$.

Equations (12.20) and (12.21) give $\gamma_2 = 0.02544$. Including this type of damping in the expressions in Example 12.1 gives

$$\alpha_{11}(\omega) = \tfrac{1}{8}\{2(1000 - \omega^2 + \mathrm{i}1.2649\omega)^{-1} + 2(2000 - \omega^2 + \mathrm{i}2.2754\omega)^{-1} + (3000 - \omega^2 + \mathrm{i}3.2863\omega)^{-1}\}$$

$$\alpha_{21}(\omega) = \tfrac{1}{8}\{2(1000 - \omega^2 + \mathrm{i}1.2649\omega)^{-1} - 2(3000 - \omega^2 + \mathrm{i}3.2863\omega)^{-1}\}$$

$$\alpha_{31}(\omega) = \tfrac{1}{8}\{(1000 - \omega^2 + \mathrm{i}1.2649\omega)^{-1} - 2(2000 - \omega^2 + \mathrm{i}2.2754\omega)^{-1} + (3000 - \omega^2 + \mathrm{i}3.2863\omega)^{-1}\}$$

Table 12.3. *Receptances for three degree of freedom system with Rayleigh damping*

	(Modulus of receptance) × 10^3 m/N		
Frequency (Hz)	α_{11}	α_{21}	α_{31}
5.03	3.16	6.25	3.12
7.12	2.48	0.494	2.43
8.72	0.790	1.39	0.691

The values of the moduli of the receptances at the required frequencies are given in Table 12.3. The values at 5.03 are similar to the values given in Table 12.1. The values at 7.12 and 8.72 Hz are slightly larger than the corresponding values in Table 12.1. This is because the damping ratios in the three modes are 0.02, 0.02544 and 0.03 compared with the values 0.02, 0.02828 and 0.03464 used in Example 12.2.

Section 11.1 indicates that if a system is unsupported, it is capable of moving as a rigid body with one or more zero frequencies. In this case the response to externally applied forces (12.2) can be written in the form

$$\mathbf{u} = [\mathbf{\Phi}_R \quad \mathbf{\Phi}_E]\begin{bmatrix} \mathbf{q}_R \\ \mathbf{q}_E \end{bmatrix} \tag{12.37}$$

where $\mathbf{\Phi}_R$ are the rigid body modes and $\mathbf{\Phi}_E$ the elastic modes. Equation (12.4) can be partitioned in a similar manner. The equation of motion for the rigid body motion is

$$\ddot{\mathbf{q}}_R = \mathbf{Q}_R \tag{12.38}$$

where

$$\mathbf{Q}_R = \mathbf{\Phi}_R^T \mathbf{f} \tag{12.39}$$

The solution of equation (12.38), when the motion starts from rest, is

$$\mathbf{q}_R(t) = \int_0^t \int_0^{\tau_2} \mathbf{Q}_R(\tau_1)\, d\tau_1\, d\tau_2 \tag{12.40}$$

EXAMPLE 12.5 Derive expressions for the response of the three masses of the system shown in Figure 12.5. The units of mass are kg and the stiffnesses in N/m.

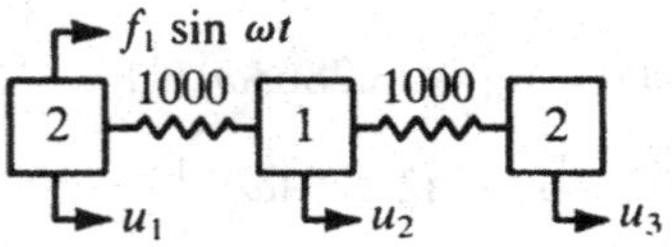

Figure 12.5. Unrestrained system subject to a harmonic force.

From Example 11.2 it can be seen that

$$\mathbf{\Lambda} = \begin{bmatrix} 0 & & \\ & 500 & \\ & & 2500 \end{bmatrix}$$

$$\mathbf{\Phi} = \frac{1}{2(5)^{1/2}} \begin{bmatrix} 2 & -5^{1/2} & -1 \\ 2 & 0 & 4 \\ 2 & 5^{1/2} & -1 \end{bmatrix}$$

$$\mathbf{Q} = \mathbf{\Phi}^{\mathrm{T}}\mathbf{f} = \frac{1}{2(5)^{1/2}} \begin{bmatrix} 2 & 2 & 2 \\ -5^{1/2} & 0 & 5^{1/2} \\ -1 & 4 & -1 \end{bmatrix} \begin{bmatrix} f_1 \\ 0 \\ 0 \end{bmatrix} = \frac{1}{2(5)^{1/2}} \begin{bmatrix} 2 \\ -5^{1/2} \\ -1 \end{bmatrix} f_1$$

Equation of motion for the rigid body mode is

$$\ddot{q}_1 = \frac{1}{5^{1/2}} f_1 \exp(\mathrm{i}\omega t)$$

Integrating gives

$$\dot{q}_1 = \frac{1}{5^{1/2}} f_1 \frac{1}{\mathrm{i}\omega} \{\exp(\mathrm{i}\omega t) - 1\}$$

and

$$q_1 = \frac{1}{5^{1/2}} \frac{f_1}{\omega^2} \{1 + \mathrm{i}\omega t - \exp(\mathrm{i}\omega t)\}$$

Taking the imaginary part gives

$$q_1 = \frac{1}{5^{1/2}} \frac{f_1}{\omega^2} (\omega t - \sin \omega t)$$

Equations of motion for the elastic modes are

$$\begin{bmatrix} \ddot{q}_2 \\ \ddot{q}_3 \end{bmatrix} + \begin{bmatrix} 500 & 0 \\ 0 & 2500 \end{bmatrix} \begin{bmatrix} q_2 \\ q_3 \end{bmatrix} = \frac{2}{2(5)^{1/2}} \begin{bmatrix} -5^{1/2} \\ -1 \end{bmatrix} f_1 \exp(\mathrm{i}\omega t)$$

The solution is

$$\begin{bmatrix} q_2 \\ q_3 \end{bmatrix} = \begin{bmatrix} (500 - \omega^2)^{-1} & 0 \\ 0 & (2500 - \omega^2)^{-1} \end{bmatrix} \begin{bmatrix} -5^{1/2} \\ -1 \end{bmatrix} \frac{f}{2(5)^{1/2}} \exp(\mathrm{i}\omega t)$$

Taking the imaginary part gives

$$q_2 = -\tfrac{1}{2}(500 - \omega^2)^{-1} f_1 \sin \omega t$$

$$q_3 = -\frac{1}{2(5)^{1/2}} (2500 - \omega^2)^{-1} f_1 \sin \omega t$$

The total response is given by

$$u_1 = \frac{1}{20}\left(\frac{4}{\omega^2}(\omega t - \sin\omega t) + \frac{5\sin\omega t}{(500-\omega^2)} + \frac{\sin\omega t}{(2500-\omega^2)}\right) f_1$$

$$u_2 = \frac{1}{20}\left(\frac{4}{\omega^2}(\omega t - \sin\omega t) - \frac{4\sin\omega t}{(2500-\omega^2)}\right) f_1$$

$$u_3 = \frac{1}{20}\left(\frac{4}{\omega^2}(\omega t - \sin\omega t) - \frac{5\sin\omega t}{(500-\omega^2)} + \frac{\sin\omega t}{(2500-\omega^2)}\right) f_1$$

A vibrating system with viscous damping is said to be classically damped if the damping matrix $\mathbf{C}$ transforms into a modal damping matrix $\bar{\mathbf{C}}$ that is diagonal. Thus it can be seen that proportional damping is a special case of classical damping. Reference [12.16] shows that a necessary and sufficient condition under which a system is classically damped is

$$\mathbf{C}\mathbf{M}^{-1}\mathbf{K} = \mathbf{K}\mathbf{M}^{-1}\mathbf{C} \tag{12.41}$$

In the case of non-classical damping, the modal damping matrix can be made diagonal by using the modes of free vibration of the damped system instead of those of the undamped system. The modes of the damped system are complex which means that the phase difference between two points in the system can differ by amounts other than 0° or 180°.

If the damping is assumed to be structural, the matrix $\bar{\mathbf{H}}$ (12.11) can be made diagonal by using a set of forced damped modes which are again complex. The forcing to be used is proportional to the inertia loading.

Further details of both methods are given in references [12.15, 12.17–12.20].

12.3.2 Direct Analysis

The steady state response can also be obtained by solving the equation

$$\mathbf{M}\ddot{\mathbf{u}} + \mathbf{C}\dot{\mathbf{u}} + \mathbf{K}\mathbf{u} = \mathbf{f}\exp(\mathrm{i}\omega t) \tag{12.42}$$

directly. This has the advantage that the frequencies and modes of free vibration of the undamped system do not have to be calculated prior to the response analysis. Assuming that the steady state response is harmonic with frequency ω gives

$$[\mathbf{K} - \omega^2\mathbf{M} + \mathrm{i}\omega\,\mathbf{C}]\mathbf{u} = \mathbf{f}\exp(\mathrm{i}\omega t) \tag{12.43}$$

The solution of this equation is

$$\mathbf{u} = [\mathbf{K} - \omega^2\mathbf{M} + \mathrm{i}\omega\,\mathbf{C}]^{-1}\mathbf{f}\exp(\mathrm{i}\omega t) \tag{12.44}$$

which can be evaluated in various ways. In the following only supported structures are considered. Putting

$$\mathbf{K} - \omega^2\mathbf{M} = \mathbf{A}_\mathrm{R}, \qquad \omega\mathbf{C} = \mathbf{A}_\mathrm{I} \tag{12.45}$$

and

$$[\mathbf{A}_\mathrm{R} + \mathrm{i}\mathbf{A}_\mathrm{I}]^{-1} = [\mathbf{B}_\mathrm{R} + \mathrm{i}\mathbf{B}_\mathrm{I}] \tag{12.46}$$

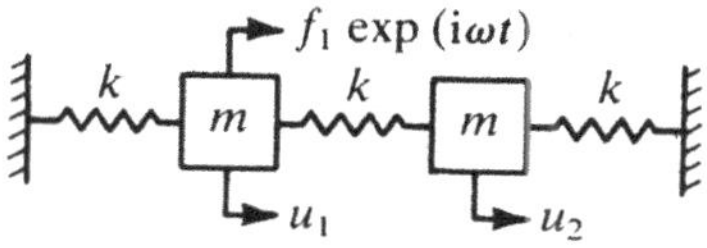

Figure 12.6. Two degree of freedom system subjected to a harmonic force. $m = 1$ kg, $k = 1000$ N/m.

then

$$[\mathbf{A}_\mathrm{R} + \mathrm{i}\mathbf{A}_\mathrm{I}][\mathbf{B}_\mathrm{R} + \mathrm{i}\mathbf{B}_\mathrm{I}] = \mathbf{I} \tag{12.47}$$

Equating real and imaginary parts gives

$$\mathbf{A}_\mathrm{R}\mathbf{B}_\mathrm{R} - \mathbf{A}_\mathrm{I}\mathbf{B}_\mathrm{I} = \mathbf{I} \tag{12.48a}$$

$$\mathbf{A}_\mathrm{I}\mathbf{B}_\mathrm{R} + \mathbf{A}_\mathrm{R}\mathbf{B}_\mathrm{I} = 0 \tag{12.48b}$$

From (12.48b)

$$\mathbf{B}_\mathrm{R} = -\mathbf{A}_\mathrm{I}^{-1}\mathbf{A}_\mathrm{R}\mathbf{B}_\mathrm{I} \tag{12.49}$$

Substituting (12.49) into (12.48a) gives

$$\mathbf{B}_\mathrm{I} = -\left[\mathbf{A}_\mathrm{I} + \mathbf{A}_\mathrm{R}\mathbf{A}_\mathrm{I}^{-1}\mathbf{A}_\mathrm{R}\right]^{-1} \tag{12.50}$$

Finally, substituting (12.50) into (12.49) gives

$$\mathbf{B}_\mathrm{R} = \mathbf{A}_\mathrm{I}^{-1}\mathbf{A}_\mathrm{R}\left[\mathbf{A}_\mathrm{I} + \mathbf{A}_\mathrm{R}\mathbf{A}_\mathrm{I}^{-1}\mathbf{A}_\mathrm{R}\right]^{-1} \tag{12.51}$$

EXAMPLE 12.6 Calculate the response of the system shown in Figure 12.6 at the frequency $5(10)^{1/2}/\pi$ Hz when the damping is structural with $\eta = 0.04$.

The inertia and stiffness matrices for the system shown in Figure 12.6 are

$$\mathbf{M} = \begin{bmatrix} 1 & 0 \\ 0 & 1 \end{bmatrix}, \qquad \mathbf{K} = 10^3 \begin{bmatrix} 2 & -1 \\ -1 & 2 \end{bmatrix}$$

When $\omega = 10(10)^{1/2}$

$$\mathbf{A}_\mathrm{R} = \mathbf{K} - \omega^2\mathbf{M} = 10^3 \begin{bmatrix} 1 & -1 \\ -1 & 1 \end{bmatrix}$$

$$\mathbf{A}_\mathrm{I} = \omega\mathbf{C} = \eta\mathbf{K} = 40 \begin{bmatrix} 2 & -1 \\ -1 & 2 \end{bmatrix}$$

Substituting for $\mathbf{A}_\mathrm{R}$ and $\mathbf{A}_\mathrm{I}$ in (12.50) and (12.51) gives

$$\mathbf{B}_\mathrm{I} = 10^{-3} \begin{bmatrix} -12.515 & -12.485 \\ -12.485 & -12.515 \end{bmatrix}$$

and

$$\mathbf{B}_\mathrm{R} = 10^{-3} \begin{bmatrix} 0.2491 & -0.2491 \\ -0.2491 & 0.2491 \end{bmatrix}$$

Now

$$\mathbf{f} = \begin{bmatrix} 1 \\ 0 \end{bmatrix} f_1 \exp(\mathrm{i}\omega t)$$

and so

$$u_1 = 10^{-3}(0.2491 - \mathrm{i}12.515)\, f_1 \exp(\mathrm{i}\omega t)$$
$$u_2 = 10^{-3}(-0.2491 - \mathrm{i}12.485)\, f_1 \exp(\mathrm{i}\omega t)$$

Note that expressions (12.50) and (12.51) do not involve inverting the matrix $\mathbf{A}_\mathrm{R}$, which is singular whenever the exciting frequency is equal to an undamped natural frequency. The matrix $\mathbf{A}_\mathrm{I}$ does have to be inverted and so the form of damping assumed should ensure that $\mathbf{C}$ and hence $\mathbf{A}_\mathrm{I}$ is not singular. Provided $\mathbf{A}_\mathrm{I}$ is symmetric, the only situation considered here, the product $\mathbf{A}_\mathrm{I}^{-1}\mathbf{A}_\mathrm{R}$ can be obtained by expressing $\mathbf{A}_\mathrm{I}$ in the form

$$\mathbf{A}_\mathrm{I} = \mathbf{L}\mathbf{L}^\mathrm{T} \tag{12.52}$$

by means of a Cholesky decomposition (see Section 11.1.2) and solving the equations

$$\mathbf{L}\mathbf{A} = \mathbf{A}_\mathrm{R}, \qquad \mathbf{L}^\mathrm{T}\mathbf{B} = \mathbf{A} \tag{12.53}$$

by forward and backward substitution for $\mathbf{A}$ and $\mathbf{B}$ respectively. This gives

$$\mathbf{B} = \mathbf{A}_\mathrm{I}^{-1}\mathbf{A}_\mathrm{R} \tag{12.54}$$

The matrix $[\mathbf{A}_1 + \mathbf{A}_\mathrm{R}\mathbf{A}_\mathrm{I}^{-1}\mathbf{A}_\mathrm{R}]^{-1}$ can be obtained by putting

$$\left[\mathbf{A}_\mathrm{I} + \mathbf{A}_\mathrm{R}\mathbf{A}_\mathrm{I}^{-1}\mathbf{A}_\mathrm{R}\right] = \mathbf{L}\mathbf{L}^\mathrm{T} \tag{12.55}$$

and solving

$$\mathbf{L}\mathbf{A} = \mathbf{I}, \qquad \mathbf{L}^\mathrm{T}\mathbf{B} = \mathbf{A} \tag{12.56}$$

for $\mathbf{A}$ and $\mathbf{B}$.

An alternative way of evaluating (12.44) is to write

$$\mathbf{u} = (\mathbf{u}_\mathrm{R} + \mathrm{i}\mathbf{u}_\mathrm{I}) \exp(\mathrm{i}\omega t) \tag{12.57}$$

and

$$\mathbf{f} = (\mathbf{f}_\mathrm{R} + \mathrm{i}\mathbf{f}_\mathrm{I}) \tag{12.58}$$

Substituting (12.57) and (12.58) into (12.43) gives

$$[\mathbf{K} - \omega^2\mathbf{M} + \mathrm{i}\omega\mathbf{C}](\mathbf{u}_\mathrm{R} + \mathrm{i}\mathbf{u}_\mathrm{I}) = (\mathbf{f}_\mathrm{R} + \mathrm{i}\mathbf{f}_\mathrm{I}) \tag{12.59}$$

Separating out the real and imaginary parts results in

$$\begin{bmatrix} \mathbf{K} - \omega^2\mathbf{M} & -\omega\mathbf{C} \\ \omega\mathbf{C} & \mathbf{K} - \omega^2\mathbf{M} \end{bmatrix} \begin{bmatrix} \mathbf{u}_\mathrm{R} \\ \mathbf{u}_\mathrm{I} \end{bmatrix} = \begin{bmatrix} \mathbf{f}_\mathrm{R} \\ \mathbf{f}_\mathrm{I} \end{bmatrix} \tag{12.60}$$

This set of equations has the disadvantage that the number of equations is equal to twice the number of degrees of freedom. The matrix of coefficients is non-symmetric. Equations (12.60) can be solved using Gauss elimination (see Section 11.2). However, since the sub-matrix $[\mathrm{K} - \omega^2\mathrm{M}]$ is singular whenever ω is equal to an undamped natural frequency, it is necessary to use row interchanges (Section 11.2).

Table 12.4. *Gauss elimination with row interchanges. Example 12.7*

Row order	Matrix × 10^{-3}				
1	1	−1	−0.08	0.04	0.001
2	−1	1	0.04	−0.08	0
3	0.08	−0.04	1	−1	0
4	−0.04	0.08	−1	1	0
1	1	−1	−0.08	0.04	0.001
2	0	0	−0.04	−0.04	0.001
3	0	0.04	1.0064	−1.0032	−0.00008
4	0	0.04	−1.0032	1.0016	0.00004
1	1	−1	−0.08	0.04	0.001
3	0	0.04	1.0064	−1.0032	−0.00008
2	0	0	−0.04	−0.04	0.001
4	0	0	−2.0096	2.0048	0.00012
1	1	−1	−0.08	0.04	0.001
3	0	0.04	1.0064	−1.0032	−0.00008
4	0	0	−2.0096	2.0048	0.00012
2	0	0	0	−0.0799044	0.9976115 × 10^{-3}

EXAMPLE 12.7 Obtain the solution to Example 12.6 by solving equation (12.60).

The steps in the solution of equation (12.60) by means of Gauss elimination with row interchanges are shown in Table 12.4. The matrix of coefficients has been augmented by the column on the right hand side of equation (12.60). A process of back substitution using the final set of equations gives

$$\begin{bmatrix} u_{1R} \\ u_{2R} \\ u_{1I} \\ u_{2I} \end{bmatrix} = 10^{-3} \begin{bmatrix} 0.2491 \\ -0.2491 \\ -12.515 \\ -12.485 \end{bmatrix} f_1$$

This gives

$$u_1 = 10^{-3}(0.2491 - \mathrm{i}12.515)\, f_1 \exp(\mathrm{i}\omega t)$$

$$u_2 = 10^{-3}(-0.2491 - \mathrm{i}12.485)\, f_1 \exp(\mathrm{i}\omega t)$$

as before.

An alternative but equivalent method of solving equation (12.60) is to use Doolittle–Crout factorisation [12.21]. In this method the matrix of coefficients on the left hand side of equation (12.60) is expresssed as the product **LU**, where **L** is a lower triangular matrix and **U** an upper triangular matrix. Either **L** or **U** is defined to have unit values on the main diagonal. As in the Gauss elimination process it is necessary to use partial pivoting. This process is then followed by both a forward and backward substitution.

If **U** has unit values on the main diagonal the method is known as Crout factorisation. If the matrix has n rows and columns, then there are n steps. Each step produces one column of **L** and one row of **U**. These can overwrite the corresponding elements of the original matrix **A**, say. At the beginning of the rth step the stored array with $n = 4$, $r = 3$ will be of the form

$$\begin{bmatrix} l_{11} & u_{12} & u_{13} & u_{14} \\ l_{21} & l_{22} & u_{23} & u_{24} \\ l_{31} & l_{32} & a_{33} & a_{34} \\ l_{41} & l_{42} & a_{43} & a_{44} \end{bmatrix}$$

where l_{ij}, u_{ij}, a_{ij} are the elements of **L**, **U** and **A** respectively. Note that the diagonal elements of **U**, which are unity, are not stored. The rth step consists of the following:

(1) Calculate

$$l_{ir} = a_{ir} - \sum_{k=1}^{r-1} l_{ik} u_{kr} \tag{12.61}$$

and overwrite a_{ir} $(i = r, \ldots, n)$.

(2) If int is the smallest integer for which

$$|l_{\text{int},r}| = \max_{i \geq r} |l_{ir}| \tag{12.62}$$

then interchange the whole of rows r and int in the current array.

(3) Calculate

$$u_{ri} = \left(a_{ri} - \sum_{k=1}^{r-1} l_{rk} u_{ki} \right) \Big/ l_{rr} \tag{12.63}$$

and overwrite $a_{ri} (i = r + 1, \ldots, n)$.

After n steps, **A** is replaced by **L** and **U** and the product **LU** gives a matrix $\tilde{\mathbf{A}}$, which is **A** with the row interchanges. It is necessary to apply the same row interchanges to the right hand side of the equation to be solved.

EXAMPLE 12.8 Use Crout factorisation to solve the equations in Example 12.7.

The steps in the Crout factorisation of the matrix of coefficients are shown in Table 12.5. The first step produces the first column of **L** and the first row of **U**. These are identical to the first column and row of **A**. The second step produces the second column of **L** and the second row of **U** after interchanging rows 2 and 3. During the third step rows 3 and 4 are interchanged. This means that the rows of **A** end up in the order 1, 3, 4 and 2.

Solving the equations

$$\begin{bmatrix} 1 & 0 & 0 & 0 \\ 0.08 & 0.04 & 0 & 0 \\ -0.04 & 0.04 & -2.0096 & 0 \\ -1 & 0 & -0.04 & -0.0799044 \end{bmatrix} \begin{bmatrix} x_1 \\ x_2 \\ x_3 \\ x_4 \end{bmatrix} = \begin{bmatrix} 0.001 \\ 0 \\ 0 \\ 0 \end{bmatrix} f_1$$

Table 12.5. *Crout factorisation with row interchanges. Example 12.8*

Step no.	Row order	Matrix × 10^{-3}			
	1	1	−1	−0.08	0.04
	2	−1	1	0.04	−0.08
	3	0.08	−0.04	1	−1
	4	−0.04	0.08	−1	1
1	1	1	−1	−0.08	0.04
	2	−1	1	0.04	−0.08
	3	0.08	−0.04	1	−1
	4	−0.04	0.08	−1	1
2	1	1	−1	−0.08	0.04
	3	0.08	0.04	25.16	−25.08
	2	−1	0	0.04	−0.08
	4	−0.04	0.04	−1	−1
3	1	1	−1	−0.08	0.04
	3	0.08	0.04	25.16	−25.08
	4	−0.04	0.04	−2.0096	−0.9976114
	2	−1	0	−0.04	−0.08
4	1	1	−1	−0.08	0.04
	3	0.08	0.04	25.16	−25.08
	4	−0.04	0.04	−2.0096	−0.9976114
	2	−1	0	−0.04	−0.0799044

by forward substitution gives

$$\begin{bmatrix} x_1 \\ x_2 \\ x_3 \\ x_4 \end{bmatrix} = \begin{bmatrix} 0.001 \\ -0.002 \\ -5.971338 \times 10^{-5} \\ -12.48506 \times 10^{-3} \end{bmatrix} f_1$$

Solving the equations

$$\begin{bmatrix} 1 & -1 & -0.08 & 0.04 \\ 0 & 1 & 25.16 & -25.08 \\ 0 & 0 & 1 & -0.9976114 \\ 0 & 0 & 0 & 1 \end{bmatrix} \begin{bmatrix} u_{1R} \\ u_{2R} \\ u_{1I} \\ u_{2I} \end{bmatrix} = \begin{bmatrix} 0.001 \\ -0.002 \\ -5.971338 \times 10^{-5} \\ -12.48506 \times 10^{-3} \end{bmatrix} f_1$$

for the real and imaginary parts of the displacements by backward substitution gives

$$\begin{bmatrix} u_{1R} \\ u_{2R} \\ u_{1I} \\ u_{2I} \end{bmatrix} = 10^{-3} \begin{bmatrix} 0.2491 \\ -0.2491 \\ -12.515 \\ -12.485 \end{bmatrix} f_1$$

Crout factorisation with row interchanges can also be used to solve equation (12.43) directly if complex arithmetic is used. This procedure is illustrated in the next example.

Table 12.6. *Crout factorisation of a complex matrix with row interchanges. Example 12.9*

Step no.	Row order	Matrix × 10^{-3}		
	1	(2 + i0.16)	(−1 − i0.04)	0
	2	(−1 − i0.04)	(1 + i0.08)	(−1 − i0.04)
	3	0	(−1 − i0.04)	(2 + i0.16)
1	1	(2 + i0.16)	(−0.4984101 + i0.0198728)	0
	2	(−1 − i0.04)	(1 + i0.08)	(−1 − i0.04)
	3	0	(−1 − i0.04)	(2 + i0.16)
2	1	(2 + i0.16)	(−0.4984101 + i0.0198728)	0
	3	0	(−1 − i0.04)	(−2.0031948 − i0.0798722)
	2	(−1 − i0.04)	(0.500795 + i0.0799364)	(−1 − i0.04)
3	1	(2 + i0.16)	(−0.4984101 + i0.0198728)	0
	3	0	(−1 − i0.04)	(−2.0031948 − i0.0798722)
	2	(−1 − i0.04)	(0.500795 + i0.0799364)	(−0.0031069 + i0.1601278)

EXAMPLE 12.9 Use Crout factorisation to solve equation (12.43) for the system defined in Example 12.3 when $\omega^2 = 1000(\text{rad/s})^2$.

The stiffness, inertia and damping matrices are

$$\mathbf{K} = 10^3 \begin{bmatrix} 4 & -1 & 0 \\ -1 & 2 & -1 \\ 0 & -1 & 4 \end{bmatrix}$$

$$\mathbf{M} = \begin{bmatrix} 2 & 0 & 0 \\ 0 & 1 & 0 \\ 0 & 0 & 2 \end{bmatrix}$$

$$\omega\mathbf{C} = \eta\mathbf{K} = 40 \begin{bmatrix} 4 & -1 & 0 \\ -1 & 2 & -1 \\ 0 & -1 & 4 \end{bmatrix}$$

The steps in the Crout factorisation of the matrix $[\mathbf{K} - \omega^2\mathbf{M} + \mathrm{i}\eta\mathbf{K}]$ are shown in Table 12.6. Solving the equations

$$\begin{bmatrix} (2+\mathrm{i}0.16) & 0 & 0 \\ 0 & (-1-\mathrm{i}0.04) & 0 \\ (-1-\mathrm{i}0.04) & (0.500795+\mathrm{i}0.0799364) & (-0.0031069+\mathrm{i}0.1601278) \end{bmatrix} \begin{bmatrix} x_1 \\ x_2 \\ x_3 \end{bmatrix}$$

$$= \begin{bmatrix} 0.001 \\ 0 \\ 0 \end{bmatrix}$$

by forward substitution gives

$$\begin{bmatrix} x_1 \\ x_2 \\ x_3 \end{bmatrix} = 10^{-3} \begin{bmatrix} 0.4968203 - i0.0397456 \\ 0 \\ -0.1844289 - i3.109008 \end{bmatrix}$$

Solving the equations

$$\begin{bmatrix} 1 & (-0.498410 + i0.0198728) & 0 \\ 0 & 1 & (-2.0031948 - i0.0798722) \\ 0 & 0 & 1 \end{bmatrix} \begin{bmatrix} u_1 \\ u_2 \\ u_3 \end{bmatrix}$$

$$= 10^{-3} \begin{bmatrix} (0.4968203) - i0.0397456 \\ 0 \\ (-0.1844289 - i3.109008) \end{bmatrix}$$

for the components of displacement by backward substitution gives

$$\begin{bmatrix} u_1 \\ u_2 \\ u_3 \end{bmatrix} = 10^{-3} \begin{bmatrix} (0.3124 - i3.149) \\ (-0.1211 - i6.243) \\ (-0.1844 - i3.109) \end{bmatrix}$$

The moduli of u_1, u_2 and u_3 agree with the values given in Table 12.2.

12.4 Response to Periodic Excitation

Periodic forces, such as those that arise during the operation of machinery, can be represented by means of a Fourier series, which is a series of harmonically varying quantities of the form

$$f(t) = \frac{1}{2}a_0 + \sum_{r=1}^{\infty} (a_r \cos \omega_r t + b_r \sin \omega_r t) \tag{12.64}$$

where

$$\omega_r = r(2\pi / T) \tag{12.65}$$

$$a_r = \frac{2}{T} \int_0^T f(t) \cos \omega_r t \, dt \tag{12.66a}$$

$$b_r = \frac{2}{T} \int_0^T f(t) \sin \omega_r t \, dt \tag{12.66b}$$

In these expressions, T denotes the period of the force.

Sufficient conditions for the convergence of Fourier series are known as Dirichlet conditions. They state that if a periodic function is piecewise continuous in the interval $0 < t < T$ and has left and right hand derivatives at each point in the

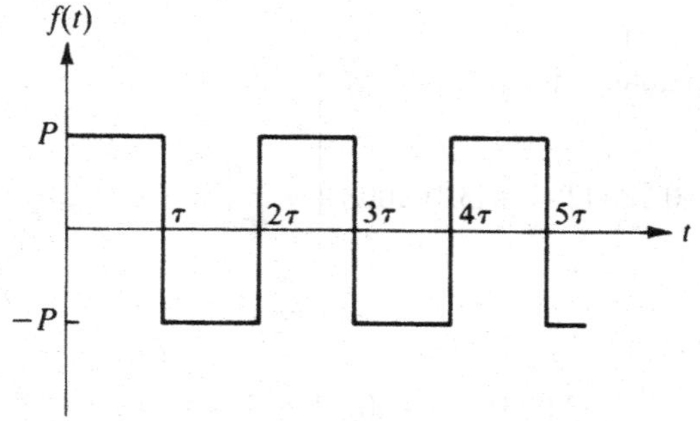

Figure 12.7. Periodic forcing function.

interval, then its Fourier series converges and the sum is $f(t)$, if the function is continuous at t. If the function is not continuous at t, then the sum is the average of the left and right hand limits of f at t.

EXAMPLE 12.10 Find the Fourier series expansion of the forcing function illustrated in Figure 12.7.

The relationships (12.66) give

$$a_0 = \frac{1}{\tau}\int_0^{\tau} P\,\mathrm{d}t - \frac{1}{\tau}\int_{\tau}^{2\tau} P\,\mathrm{d}t = 0$$

For $r \geq 1$

$$a_r = \frac{P}{\tau}\int_0^{\tau} \cos\omega_r t\,\mathrm{d}t - \frac{P}{\tau}\int_{\tau}^{2\tau} \cos\omega_r t\,\mathrm{d}t$$

$$= \frac{P}{\omega_r\tau}(2\sin\omega_r\tau - \sin 2\omega_r\tau) = 0$$

since $\omega_r\tau = r\pi$.

$$b_r = \frac{P}{\tau}\int_0^{\tau} \sin\omega_r t\,\mathrm{d}t - \frac{P}{\tau}\int_{\tau}^{2\tau} \sin\omega_r t\,\mathrm{d}t$$

$$= \frac{P}{\omega_r\tau}(1 - 2\cos\omega_r\tau + \cos 2\omega_r\tau)$$

$$= \frac{2P}{r\pi}\{1 - (-1)^r\}.$$

Therefore

$$b_r = \begin{cases} \dfrac{4P}{r\pi} & r \text{ odd} \\ 0 & r \text{ even} \end{cases}$$

The Fourier series expansion is, therefore

$$f(t) = \frac{4P}{\pi}\left(\sin\omega_1 t + \tfrac{1}{3}\sin\omega_3 t + \tfrac{1}{5}\sin\omega_5 t + \cdots\right)$$

where $\omega_r = r\pi/\tau$.

Equation (12.64) can be expressed in complex form by substituting the relationships

$$\cos \omega_r t = \tfrac{1}{2}\{\exp(\mathrm{i}\omega_r t) + \exp(-\mathrm{i}\omega_r t)\}$$
$$\sin \omega_r t = -\tfrac{1}{2}\mathrm{i}\{\exp(\mathrm{i}\omega_r t) - \exp(-\mathrm{i}\omega_r t)\} \tag{12.67}$$

and defining

$$c_0 = \tfrac{1}{2}a_0, \qquad c_r = \tfrac{1}{2}(a_r - \mathrm{i}b_r), \qquad c_{-r} = \tfrac{1}{2}(a_r + \mathrm{i}b_r) \tag{12.68}$$

This gives

$$f(t) = \sum_{r=-\infty}^{+\infty} c_r \exp(\mathrm{i}\omega_r t) \tag{12.69}$$

where

$$c_r = \frac{1}{T}\int_0^T f(t)\exp(-\mathrm{i}\omega_r t)\,\mathrm{d}t \tag{12.70}$$

EXAMPLE 12.11 Repeat Example 12.10 using the complex form of Fourier series.

The relationship (12.70) gives

$$c_r = \frac{P}{2\tau}\int_0^{\tau} \exp(-\mathrm{i}\omega_r t)\,\mathrm{d}t - \frac{P}{2\tau}\int_{\tau}^{2\tau} \exp(-\mathrm{i}\omega_r t)\,\mathrm{d}t$$

Integrating gives

$$c_r = \frac{\mathrm{i}P}{2\omega_r\tau}\{-1 + 2\exp(-\mathrm{i}\omega_r\tau) - \exp(-\mathrm{i}2\omega_r\tau)\}$$

Now $\omega_r\tau = r\pi$ and so

$$c_r = \frac{\mathrm{i}P}{2r\pi}\{-1 + 2\exp(-\mathrm{i}r\pi) - \exp(-\mathrm{i}2r\pi)\}$$
$$= \frac{\mathrm{i}P}{r\pi}\{(-1)^r - 1\}$$

Therefore

$$c_r = \begin{cases} -\dfrac{\mathrm{i}P}{r\pi} & r \text{ odd} \\ 0 & r \text{ even} \end{cases}$$

If the periodic force cannot be expressed as a mathematical function, only as a set of values $f_0, f_1, \ldots, f_N$ at times $t_0 = 0, t_1, \ldots, t_N = T$, as shown in Figure 12.8, then it can be represented by means of a finite Fourier series [12.22]. Taking N to be even and

$$f_0 = 0, \qquad f_j = f(j\Delta t), \qquad j = 1, 2, \ldots, N \tag{12.71}$$

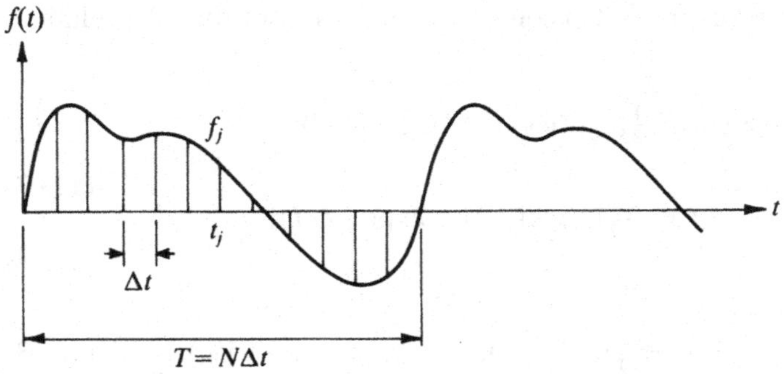

Figure 12.8. Numerical representation of periodic forcing function.

then

$$f(t) = \sum_{r=-N/2}^{N/2} c_r \exp(\mathrm{i}\omega_r t) \tag{12.72}$$

At the points $t = j\Delta t (j = 1, 2, \ldots, N)$

$$f_j = f(j\Delta t) = \sum_{r=-N/2}^{N/2} c_r \exp\{\mathrm{i}(2\pi/N) jr\} \tag{12.73}$$

Multiplying both sides by exp $\{-\mathrm{i}(2\pi/N) js\}$ and summing over j gives

$$c_r = \frac{1}{N} \sum_{j=1}^{N} f_j \exp\{-\mathrm{i}(2\pi/N) jr\} \tag{12.74}$$

since

$$\sum_{j=1}^{N} \exp\{\mathrm{i}(2\pi/N)(r - s) j\} = 0 \qquad \text{for } r \neq s \tag{12.75}$$

The highest frequency present in (12.72) is $\omega_{N/2} = N\pi/T = \pi/\Delta T$ rad/s $= 1/2\Delta t$ Hz. If there are any frequencies higher than this present in the periodic force, they will contribute to frequencies below this maximum. This is referred to as aliasing in the terminology of signal processing. Therefore, it is necessary to choose Δt small enough to ensure aliasing does not occur. The efficiency of computing (12.74) can be increased by using a Fast Fourier Transform algorithm [12.22].

If the complete set of nodal forces acting on a structure is periodic, then

$$\mathbf{f}(t) = \sum_{r=-\infty}^{\infty} \mathbf{c}_r \exp(\mathrm{i}\omega_r t) \tag{12.76}$$

where

$$\mathbf{c}_r = \frac{1}{T} \int_0^T \mathbf{f}(t) \exp(-\mathrm{i}\omega_r t)\, \mathrm{d}t \tag{12.77}$$

The equation of motion (12.1) becomes

$$\mathbf{M}\ddot{\mathbf{u}} + \mathbf{C}\dot{\mathbf{u}} + \mathbf{K}\mathbf{u} = \sum_{r=-\infty}^{+\infty} \mathbf{c}_r \exp(\mathrm{i}\omega_r t) \tag{12.78}$$

Since this equation is linear, the solutions corresponding to each term on the right hand side can be obtained separately and superimposed to give the complete solution. This gives

$$\mathbf{u} = \sum_{r=-\infty}^{+\infty} [\boldsymbol{\alpha}(\omega_r)]\mathbf{c}_r \exp(\mathrm{i}\omega_r t) \tag{12.79}$$

where $[\boldsymbol{\alpha}(\omega_r)]$ is the matrix of receptances evaluated at the frequency ω_r. This matrix is given by (12.34) if the modal method is used and (12.44) if the direct method is used.

12.5 Transient Response

If a structure is excited by a suddenly applied non-periodic excitation, the response is transient since steady state oscillations are not produced. Strictly speaking, the term 'transient' should be applied to the situation when the forces are applied for a short interval of time. Subsequent motion of the structure is free vibration, which will decay due to the damping present. However, it is often applied to a continually changing situation for an indefinite period of time. In this case the column matrix of nodal forces, $\mathbf{f}(t)$, is an arbitrary varying function of time.

12.5.1 Modal Analysis

Assuming viscous damping and that the transformed damping matrix is diagonal, gives equation (12.17) for the rth mode, that is

$$\ddot{q}_r + 2\gamma_r\omega_r\dot{q}_r + \omega_r^2 q_r = Q_r(t) \tag{12.80}$$

The solution of (12.80) at time t is given by the Duhamel integral [12.23]

$$q_r(t) = \int_0^t Q_r(\tau)h_r(t-\tau)\,\mathrm{d}\tau \tag{12.81}$$

if the motion starts from rest, where $h_r(t)$ is the impulse response function which is

$$h_r(t) = \frac{1}{\omega_{\mathrm{d}r}} \exp(-\gamma_r\omega_r t)\sin\omega_{\mathrm{d}r} t \tag{12.82}$$

$\omega_{\mathrm{d}r}$ is the damped natural frequency of mode r which is defined as

$$\omega_{\mathrm{d}r} = \omega_r\left(1-\gamma_r^2\right)^{1/2} \tag{12.83}$$

If the forces are applied for a short time, the maximum response (displacement, stress, etc.) will occur during the first few oscillations. If the damping is small, its

Table 12.7. *Response functions for undamped single degree of freedom systems*

Case no	Forcing function $Q(t)$	Displacement response
1	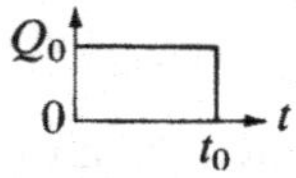	$\dfrac{Q_0}{\omega^2}(1-\cos\omega t) \qquad t<t_0$ $\dfrac{Q_0}{\omega^2}\{\cos\omega(t-t_0)-\cos\omega t\} \qquad t>t_0$
2	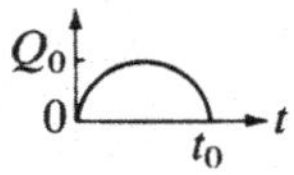	$\dfrac{(Q_0t_0/\omega)}{\{(\omega t_0)^2-\pi^2\}}\{\omega t_0\sin(\pi t/t_0)-\pi\sin\omega t\} \qquad t<t_0$ $\dfrac{-(Q_0\pi t_0/\omega)}{\{(\omega t_0)^2-\pi^2\}}\{\sin\omega(t-t_0)+\sin\omega t\} \qquad t>t_0$
3	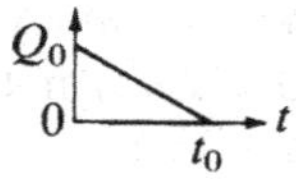	$\dfrac{Q_0}{\omega^2}\left\{1-\cos\omega t-\dfrac{t}{t_0}+\dfrac{\sin\omega t}{\omega t_0}\right\} \qquad t<t_0$ $\dfrac{Q_0}{\omega^2}\left\{-\cos\omega t-\dfrac{\sin\omega(t-t_0)}{\omega t_0}+\dfrac{\sin\omega t}{\omega t_0}\right\} \qquad t>t_0$
4	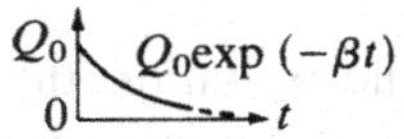	$\dfrac{Q_0}{(\omega^2+\beta^2)}\left\{\exp(-\beta t)-\cos\omega t+\dfrac{\beta\sin\omega t}{\omega}\right\}$

effect on the maximum response will be small. Thus damping is often neglected to simplify the analysis. If, in addition, the applied force can be represented by an analytical function, then equation (12.81) can be evaluated analytically. Reference [12.24] tabulates the results for some typical loading functions. A brief selection is presented in Table 12.7, with subscript r omitted for convenience.

In many practical situations, however, the loading is known only from experimental data. The expression (12.81) must, therefore, be evaluated numerically. Reference [12.25] suggests the following technique. Noting that

$$\sin\omega_{\mathrm{d}r}(t-\tau)=\sin\omega_{\mathrm{d}r}t\cos\omega_{\mathrm{d}r}\tau-\cos\omega_{\mathrm{d}r}t\sin\omega_{\mathrm{d}r}\tau \tag{12.84}$$

then (12.81) can be written in the form

$$q_r(t)=A_r(t)\sin\omega_{\mathrm{d}r}t-B_r(t)\cos\omega_{\mathrm{d}r}t \tag{12.85}$$

where

$$A_r(t)=\frac{\exp(-\gamma_r\omega_rt)}{\omega_{\mathrm{d}r}}\int_0^t Q_r(\tau)\exp(\gamma_r\omega_r\tau)\cos\omega_{\mathrm{d}r}\tau\,\mathrm{d}\tau \tag{12.86a}$$

and

$$B_r(t)=\frac{\exp(-\gamma_r\omega_rt)}{\omega_{\mathrm{d}r}}\int_0^t Q_r(\tau)\exp(\gamma_r\omega_r\tau)\sin\omega_{\mathrm{d}r}\tau\,\mathrm{d}\tau \tag{12.86b}$$

The integrals in the expressions for $A_r(t)$ and $B_r(t)$ are now evaluated numerically using, for example, the trapezium rule or Simpson's rule. Since, however, the time history of the response is required, it is better to evaluate (12.86) in an incremental manner. For example, evaluating $A_r(t)$ and $B_r(t)$ at equal intervals of time

Table 12.8. *Transient response of single degree of freedom system. Example 12.12*

t	Approximate solution	Analytical solution
0	0	0
0.05	0.2786	0.2865
0.1	0.5011	1.036
0.15	1.899	1.964
0.2	2.624	2.714
0.25	2.901	3.0
0.3	2.624	2.714
0.35	1.899	1.964
0.4	0.5011	1.036
0.45	0.2786	0.2865
0.5	0	0
0.55	0.2786	0.2865
0.6	0.5011	1.036

$\Delta\tau$ and using the trapezium rule, $A_r(t)$ can be evaluated as follows

$$\begin{aligned}
A_r(t) &= A_r(t-\Delta\tau) + \frac{\Delta\tau}{2\omega_{dr}}\exp(-\gamma_r\omega_r t) \\
&\quad \times [Q_r(t-\Delta\tau)\exp\{\gamma_r\omega_r(t-\Delta\tau)\}\cos\omega_{dr}(t-\Delta\tau) \\
&\quad + Q_r(t)\exp(\gamma_r\omega_r t)\cos\omega_{dr}t] \qquad (12.87) \\
&= A_r(t-\Delta\tau) + \frac{\Delta\tau}{2\omega_{dr}}[Q_r(t-\Delta\tau)\exp(-\gamma_r\omega_r\Delta\tau)]\cos\omega_{dr}(t-\Delta\tau) \\
&\quad + Q_r(t)\cos\omega_{dr}t]
\end{aligned}$$

$B_r(t)$ is calculated using a similar expression with the cosine functions replaced by sine functions.

The accuracy of the solution given by this procedure will depend upon the choice of $\Delta\tau$. It should be chosen small enough to ensure the loading history and the trigonometric functions are accurately defined.

EXAMPLE 12.12 Calculate the response of the system

$$\ddot{q} + (16\pi^2)q = 24\pi^2$$

with $q = 0$ and $\dot{q} = 0$ at $t = 0$ for $t \leq 0.6$ s, by evaluating the Duhamel integral numerically using $\Delta\tau = 0.05$. Compare the results with the analytical solution [12.24]

$$q = \tfrac{3}{2}(1 - \cos 4\pi t)$$

The values of the response are given in Table 12.8 where they are compared with the analytical solution. The maximum response is underestimated by 3.3%.

An alternative way of calculating the transient response, which is more commonly used, is to solve equation (12.80) numerically by a step-by-step procedure.

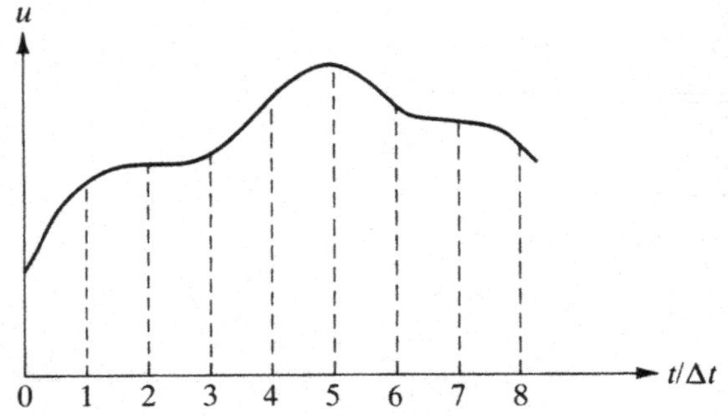

Figure 12.9. Step-by-step solution for transient response.

The method will be presented with reference to the equation

$$m\ddot{u} + c\dot{u} + ku = f \tag{12.88}$$

This will facilitate the extension of the method to the direct analysis of multi-degree of freedom systems in Section 12.5.2. In applying the technique to the modal method put

$$m = 1, \qquad c = 2\gamma_r\omega_r, \qquad k = \omega_r^2, \qquad f = Q_r \tag{12.89}$$

The initial displacement and velocity at time $t = 0$, u_0 and $\dot{u}_0$, are usually known. The acceleration at $t = 0$, $\ddot{u}_0$, can then be calculated using equation (12.88), that is

$$\ddot{u}_0 = (f_0 - c\dot{u}_0 - ku_0)/m \tag{12.90}$$

where f_0 is the value of $f(t)$ at $t = 0$.

In order to evaluate the response at time T, the time interval $(0, T)$ is divided into N equal time intervals $\Delta t = T/N$. The response (u, $\dot{u}$ and $\ddot{u}$) is then calculated at the times Δt, $2\Delta t$, $3\Delta t, \ldots, T$, by an approximate technique. This is illustrated in Figure 12.9. There are many such techniques available, each with its own advantages and disadvantages. A few of the more commonly used ones are described in the following sections.

12.5.1.1 Central Difference Method

The central difference method consists of expressing the velocity and acceleration at time t_j in terms of the displacements at times t_{j-1}, t_j and t_{j+1} using central finite difference formulae. These are obtained by approximating the response curve, shown in Figure 12.10, by a quadratic polynomial within the interval (t_{j-1}, t_{j+1}). That is

$$u = a\tau^2 + b\tau + c \qquad -\Delta t \le \tau \le \Delta\tau \tag{12.91}$$

Evaluating (12.91) at $\tau = -\Delta t,\ 0,\ \Delta t$ gives

$$a(\Delta t)^2 - b\Delta t + c = u_{j-1} \tag{12.92a}$$

$$c = u_j \tag{12.92b}$$

$$a(\Delta t)^2 + b\Delta t + c = u_{j+1} \tag{12.92c}$$

Figure 12.10. Central difference approximation.

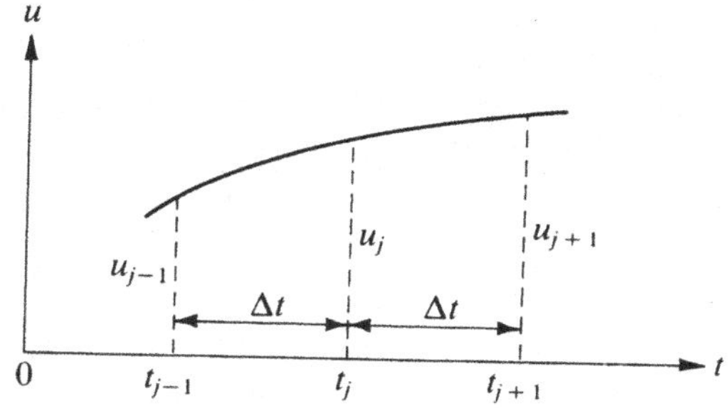

Solving (12.92) for a, b and c gives

$$a = \frac{1}{2(\Delta t)^2}(u_{j+1} - 2u_j + u_{j-1})$$
$$b = \frac{1}{2\Delta t}(u_{j+1} - u_{j-1})$$
$$c = u_j$$

Differentiating (12.91) with respect to τ and evaluating at $t = 0$ gives

$$\dot{u}_j = b = \frac{1}{2\Delta t}(u_{j+1} - u_{j-1}) \tag{12.93}$$

and

$$\ddot{u}_j = 2a = \frac{1}{(\Delta t)^2}(u_{j+1} - 2u_j + u_{j-1}) \tag{12.94}$$

Reference [12.26] shows that the error in these approximations is of the order $(\Delta t)^2$.

The response at time t_{j+1} is obtained by substituting (12.93) and (12.94) into the equation of motion evaluated at time t_j, that is

$$m\ddot{u}_j + c\dot{u}_j + ku_j = f_j \tag{12.95}$$

Methods based upon equations (12.93), (12.94) and (12.95) are called explicit methods. Performing the substitution gives

$$\frac{m}{(\Delta t)^2}(u_{j+1} - 2u_j + u_{j-1}) + \frac{c}{2\Delta t}(u_{j+1} - u_{j-1}) + ku_j = f_j \tag{12.96}$$

Solving for u_{j+1} gives

$$\begin{aligned} &\{m/(\Delta t)^2 + c/2\Delta t\}u_{j+1} \\ &\quad = f_j + \{2m/(\Delta t)^2 - k\}u_j - \{m/(\Delta t)^2 - c/2\Delta t\}u_{j-1} \end{aligned} \tag{12.97}$$

Hence, if the displacements u_{j-1} and u_j are known, then the displacement u_{j+1} can be calculated. The time history of the response can be obtained by taking $j = 1$, $2, \ldots,$ provided u_1 and u_0 are known.

Table 12.9. *Central difference solution of single degree of freedom system. Example 12.13*

	Approximate solutions		
t	$\Delta t = 1$	$\Delta t = 0.5$	Analytical solution
0	1.0	1.0	1.0
0.5		0.8750	0.8776
1.0	0.5	0.5313	0.5403
1.5		0.0547	0.0707
2.0	−0.5	−0.4355	−0.4161
2.5		−0.8169	−0.8011
3.0	−1.0	−0.9940	−0.9900
3.5		−0.9226	−0.9365
4.0	−0.5	−0.6206	−0.6537
4.5		−0.1634	−0.2108
5.0	0.5	0.3346	0.2837

u_1 can be determined by evaluating (12.93) and (12.94) for $j = 0$

$$\dot{u}_0 = \frac{1}{2\Delta t}(u_1 - u_{-1}) \tag{12.98}$$

$$\ddot{u}_0 = \frac{1}{(\Delta t)^2}(u_1 - 2u_0 + u_{-1}) \tag{12.99}$$

and eliminating u_{-1} to give

$$u_1 = u_0 + \Delta t \dot{u}_0 + \tfrac{1}{2}(\Delta t)^2 \ddot{u}_0 \tag{12.100}$$

As previously mentioned, u_0 and $\dot{u}_0$ are given and $\ddot{u}_0$ can be calculated using (12.90). The velocity and acceleration at each time step can be obtained using (12.93) and (12.94).

EXAMPLE 12.13 Calculate the response of the system

$$\ddot{u} + u = 0$$

with $u_0 = 1$, $\dot{u}_0 = 0$ at $t = 0$ using the central difference method with $\Delta t = 0.5$, 1.0 and 3.0. Compare the results with the analytical solution, $u = \cos t$.

When $\Delta t = 1$, $\ddot{u}_0 = -1$ and $u_1 = 0.5$. Repeated application of equation (12.97) gives the values in Table 12.9. When $\Delta t = 0.5$, $\ddot{u}_0 = -1$ and $u_1 = 0.875$. Subsequent values of u are also given in Table 12.9. Both sets of values are compared with the analytical solution. The results given by the smaller time step are closer to the analytical values. However, even the coarse time step gives remarkably good accuracy. But, too large a time step should not be used, as illustrated by the results for $\Delta t = 3$ given in Table 12.10. Not only are the amplitudes in gross error, they increase in magnitude as time increases. In such a situation, the numerical solution is said to be unstable.

The maximum value of Δt for which the solution is numerically stable can be obtained by applying equation (12.97) to free vibration of an undamped system.

Table 12.10. *Central difference solution of single degree of freedom system. Example 12.13*

t	Approximate solution $\Delta t = 3$	Analytical solution
0	1.0	1.0
3	−3.5	−0.9900
6	23.5	0.9602
9	−161.0	−0.9111
12	1103.5	0.8439

That is, $f = 0$ and $c = 0$. This gives

$$\frac{m}{(\Delta t)^2}u_{j+1} = \left\{\frac{2m}{(\Delta t)^2} - k\right\}u_j - \frac{m}{(\Delta t)^2}u_{j-1} \tag{12.101}$$

Multiplying by $(\Delta t)^2/m$ and rearranging gives

$$u_{j+1} + \{(\omega_0 \Delta t)^2 - 2\}u_j + u_{j-1} = 0 \tag{12.102}$$

where $\omega_0 = (k/m)^{1/2}$.

The solution of equation (12.102) is of the form

$$u_j = A\beta^j \tag{12.103}$$

where β is a parameter to be determined and A a constant. Substituting (12.103) into (12.102) gives

$$A\beta^{j+1} + \{(\omega_0 \Delta t)^2 - 2\}A\beta^j + A\beta^{j-1} = 0 \tag{12.104}$$

Dividing by $A\beta^{j-1}$ gives

$$\beta^2 + \{(\omega_0 \Delta t)^2 - 2\}\beta + 1 = 0 \tag{12.105}$$

Equation (12.105) is a quadratic equation in β, the solution of which is

$$\beta_{1,2} = -\tfrac{1}{2}\{(\omega_0 \Delta t)^2 - 2\} \pm \tfrac{1}{2}\{((\omega_0 \Delta t)^2 - 2)^2 - 4\}^{1/2} \tag{12.106}$$

The general solution of equation (12.102) is of the form

$$u_j = A_1\beta_1^j + A_2\beta_2^j \tag{12.107}$$

This will represent an oscillation provided β_1 and β_2 are complex conjugates. This will be the case if

$$\{(\omega_0 \Delta t)^2 - 2\}^2 - 4 < 0 \tag{12.108}$$

That is

$$(\omega_0 \Delta t)^2\{(\omega_0 \Delta t)^2 - 4\} < 0 \tag{12.109}$$

or

$$\omega_0 \Delta t < 2 \tag{12.110}$$

If this is the case then

$$\beta_2 = \beta_1^* \tag{12.111}$$

where

$$\beta_1 = -\tfrac{1}{2}\{(\omega_0\Delta t)^2 - 2\} + \mathrm{i}\frac{(\omega_0\Delta t)}{2}\{4 - (\omega_0\Delta t)^2\}^{1/2} \tag{12.112}$$

and the asterisk denotes the complex conjugate. Putting

$$\beta_1 = \rho\exp(\mathrm{i}\theta) \tag{12.113}$$

then

$$\rho^2 = \left\{\frac{(\omega_0\Delta t)^2}{2} - 1\right\}^2 + (\omega_0\Delta t)^2\left\{1 - \frac{(\omega_0\Delta t)^2}{4}\right\} = 1 \tag{12.114}$$

and so

$$\rho = 1 \tag{12.115}$$

This means that the oscillations will not increase in magnitude and the numerical solution is stable.

Relationship (12.110) is the criterion for numerical stability. In Example 12.13, $\omega_0 = 1$. The results in Tables 12.9 and 12.10 indicate stable solutions when $\Delta t = 0.5$ and 1.0 and an unstable solution for $\Delta t = 3.0$. Since $\omega_0 = 2\pi/\tau_0$, where τ_0 is the period of oscillation, an alternative way of expressing (12.110) is

$$\Delta t/\tau_0 < 1/\pi \simeq 0.318 \tag{12.116}$$

Because of this restriction on step size for numerical stability, the method is said to be conditionally stable.

From equations (12.112) and (12.113)

$$\tan\theta = \frac{(\omega_0\Delta t)\{4 - (\omega_0\Delta t)^2\}^{1/2}}{\{2 - (\omega_0\Delta t)^2\}} \tag{12.117}$$

Equation (12.107) now becomes

$$\begin{aligned} u_j &= A_1\exp\{\mathrm{i}(j\theta)\} + A_2\exp\{-\mathrm{i}(j\theta)\} \\ &= A_3\cos(j\theta) + A_4\sin(j\theta) \end{aligned} \tag{12.118}$$

Now

$$j = t/\Delta t \tag{12.119}$$

and so

$$u_j = A_3\cos\left(\frac{\theta}{\Delta t}\right)t + A_4\sin\left(\frac{\theta}{\Delta t}\right)t \tag{12.120}$$

This equation represents an oscillation of constant amplitude with frequency $(\theta/\Delta t)$. The distortion in frequency caused by the numerical procedure is

$$\frac{\theta}{\omega_0\Delta t} = \frac{1}{\omega_0\Delta t}\tan^{-1}\left[\frac{\omega_0\Delta t\{4 - (\omega_0\Delta t)^2\}^{1/2}}{\{2 - (\omega_0\Delta t)^2\}}\right] \tag{12.121}$$

The variation of $\theta/\omega_0\Delta t$ with $\omega_0\Delta t$ is given in Table 12.11.

Table 12.11. *Variation of the distortion in frequency with time increment for the central difference method*

$\omega_0 \Delta t$	$\theta/\omega_0 \Delta t$
0.00	1.0
0.25	1.0026
0.50	1.0106
0.75	1.0251
1.00	1.0472
1.25	1.0802
1.50	1.1307
1.75	1.2177

Reference [12.27] concludes that for good accuracy $\omega_0 \Delta t = \pi/10$ (i.e., $\Delta t/\tau_0 = 1/20$). This will give a distortion in frequency of 0.4%.

EXAMPLE 12.14 Repeat Example 12.12 using the central difference method and a time step $\Delta t = 0.05$.

Changing to the present notation, the equation of motion is

$$\ddot{u} + (16\pi^2)u = 24\pi^2$$

with $u_0 = 0$ and $\dot{u}_0 = 0$ at $t = 0$. Equation (12.90) gives $\ddot{u}_0 = 24\pi^2$ and (12.100) $u_1 = 0.296$. Repeated application of (12.97) then gives the values in Table 12.12 where they are compared with the analytical solution. The maximum response is underestimated by 0.07%.

12.5.1.2 The Houbolt Method

The Houbolt method [12.28] consists of expressing the velocity and acceleration at time t_{j+1} in terms of the displacements at times t_{j-2} to t_{j+1} using backward difference formulae. These are obtained by approximating the response curve, shown in Figure 12.11, by a cubic polynomial within the interval (t_{j-2}, t_{j+1}). Using Lagrange interpolation functions (Section 3.8) gives

$$\begin{aligned} u = &-\tfrac{1}{6}(2\tau + 3\tau^2 + \tau^3)u_{j-2} + \tfrac{1}{2}(3\tau + 4\tau^2 + \tau^3)u_{j-1} \\ &- \tfrac{1}{2}(6\tau + 5\tau^2 + \tau^3)u_j + \tfrac{1}{6}(6 + 11\tau + 6\tau^2 + \tau^3)u_{j+1} \\ &\text{for } -3 \leq \tau \leq 0 \end{aligned} \qquad (12.122)$$

where $\tau = t/\Delta t$.

Rearranging gives

$$\begin{aligned} u = &\tfrac{1}{6}\{6u_{j+1} + (-2u_{j-2} + 9u_{j-1} - 18u_j + 11u_{j+1})\tau \\ &+ (-3u_{j-2} + 12u_{j-1} - 15u_j + 6u_{j+1})\tau^2 \\ &+ (-u_{j-2} + 3u_{j-1} - 3u_j + u_{j+1})\tau^3\} \end{aligned} \qquad (12.123)$$

Table 12.12. *Transient response of a single degree of freedom system. Example 12.14*

t	Central difference	Analytical solution
0	0	0
0.05	0.2961	0.2865
0.1	1.067	1.036
0.15	2.010	1.964
0.2	2.751	2.714
0.25	2.998	3.0
0.3	2.654	2.714
0.35	1.854	1.964
0.4	0.9148	1.036
0.45	0.2065	0.2865
0.5	0.088	0
0.55	0.3998	0.2865
0.6	1.225	1.036

Now

$$\dot{u}_{j+1} = \frac{1}{\Delta t}\left(\frac{du}{d\tau}\right)_{\tau=0} \tag{12.124}$$

and so

$$\dot{u}_{j+1} = \frac{1}{6\Delta t}(-2u_{j-2} + 9u_{j-1} - 18u_j + 11u_{j+1}) \tag{12.125}$$

Similarly

$$\ddot{u}_{j+1} = \frac{1}{(\Delta t)^2}\left(\frac{d^2u}{d\tau^2}\right)_{\tau=0} \tag{12.126}$$

and so

$$\ddot{u}_{j+1} = \frac{1}{(\Delta t)^2}(-u_{j-2} + 4u_{j-1} - 5u_j + 2u_{j+1}) \tag{12.127}$$

Reference [12.27] indicates that the error in the approximations (12.125) and (12.127) is of the order $(\Delta t)^3$.

The response at time t_{j+1} is obtained by substituting (12.125) and (12.127) into the equation of motion evaluated at time t_{j+1}, that is

$$m\ddot{u}_{j+1} + c\dot{u}_{j+1} + ku_{j+1} = f_{j+1} \tag{12.128}$$

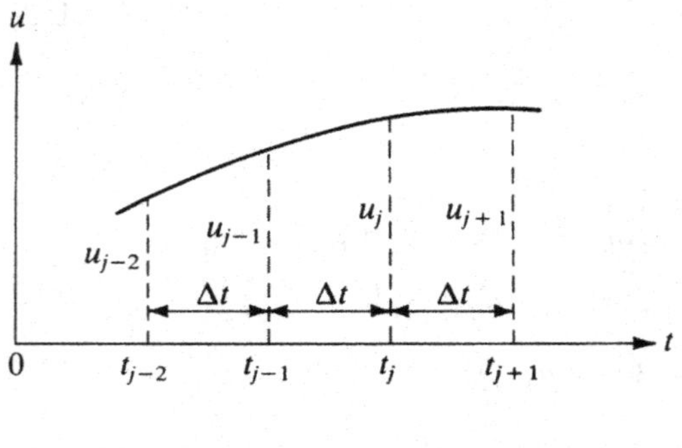

Figure 12.11. Backward difference approximation.

Methods based upon equations (12.125), (12.127) and (12.128) are called implicit methods. Performing the substitution gives

$$\frac{m}{(\Delta t)^2}(-u_{j-2} + 4u_{j-1} - 5u_j + 2u_{j+1})$$
$$+ \frac{c}{6\Delta t}(-2u_{j-2} + 9u_{j-1} - 18u_j + 11u_{j+1}) + ku_{j+1} = f_{j+1} \quad (12.129)$$

Solving for u_{j+1}, gives

$$\{2m/(\Delta t)^2 + 11c/6\Delta t + k\}u_{j+1}$$
$$= f_{j+1} + \{5m/(\Delta t)^2 + 3c/\Delta t\}u_j - \{4m/(\Delta t)^2 + 3c/2\Delta t\}u_{j-1}$$
$$+ \{m/(\Delta t)^2 + c/3\Delta t\}u_{j-2} \quad (12.130)$$

If the displacements u_{j-2}, u_{j-1} and u_j are known, then the displacement u_{j+1} can be calculated. The time history of the response can be obtained by taking $j = 2,3,\ldots,$ provided u_0, u_1 and u_2 are known. u_0 and $\dot{u}_0$ at $t = 0$ are specified. it is usual to obtain u_1 and u_2 using a different numerical procedure, for example, the central difference method. This may involve a smaller time step, Δt, for this initial phase.

Stability of the method is investigated by considering free vibration of an undamped system. Equation (12.129) reduces to

$$\frac{m}{(\Delta t)^2}(-u_{j-2} + 4u_{j-1} - 5u_j + 2u_{j+1}) + ku_{j+1} = 0 \quad (12.131)$$

Multiplying by $(\Delta t^2)/m$ and rearranging gives

$$\{2 + (\omega_0 \Delta t)^2\}u_{j+1} - 5u_j + 4u_{j-1} - u_{j-2} = 0 \quad (12.132)$$

where $\omega_0 = (k/m)^{1/2}$.

The solution of equation (12.132) is of the form

$$u_j = A\beta^j \quad (12.133)$$

Substituting (12.133) into (12.132) gives

$$\{2 + (\omega_0 \Delta t)^2\}A\beta^{j+1} - 5A\beta^j + 4A\beta^{j-1} - A\beta^{j-2} = 0 \quad (12.134)$$

Dividing by $A\beta^{j-2}$ gives

$$\{2 + (\omega_0 \Delta t)^2\}\beta^3 - 5\beta^2 + 4\beta - 1 = 0 \quad (12.135)$$

This is a cubic equation in β. All the coefficients are real and they alternate in sign. Therefore, there will be one real root, which will be positive, and one pair of complex conjugate roots. For stability, the modulus of each root must be less than or equal to unity. The largest modulus is known as the spectral radius. Therefore, for stability the spectral radius must be less than or equal to unity.

EXAMPLE 12.15 Investigate the stability of the Houbolt method for the case $\omega_0 \Delta t = 0.5$.

When $\omega_0 \Delta t = 0.5$ equation (12.135) becomes

$$2.25\beta^3 - 5\beta^2 + 4\beta - 1 = 0 \quad (12.136)$$

Substituting

$$\beta = \frac{1}{2.25}\left(\alpha + \frac{5}{3}\right) \tag{12.137}$$

into (12.136) gives the reduced cubic equation

$$\alpha^3 + a\alpha + b = 0 \tag{12.138}$$

where

$$a = \tfrac{2}{3}, \qquad b = \tfrac{293}{432} \tag{12.139}$$

Now put

$$\alpha = r + s \tag{12.140}$$

where r and s are to be determined. Now

$$\alpha^3 = r^3 + s^3 + 3rs(r+s) \tag{12.141}$$

or

$$\alpha^3 - 3rs\alpha - (r^3 + s^3) = 0 \tag{12.142}$$

Comparing (12.138) and (12.142) gives

$$r^3 + s^3 - b, \qquad rs = -a/3 \tag{12.143}$$

Hence r^3 and s^3 are the roots of the quadratic equation

$$\lambda^2 + b\lambda - a^3/27 = 0 \tag{12.144}$$

and so

$$\begin{aligned} \lambda_1 = r^3 &= \{-b/2 + (b^2/4 + a^3/27)^{1/2}\} \\ \lambda_2 = s^3 &= \{-b/2 + (b^2/4 + a^2/27)^{1/2}\} \end{aligned} \tag{12.145}$$

Substituting for a and b gives

$$r^3 = 0.0158113, \qquad s^3 = -0.6940519 \tag{12.146}$$

The roots of equations (12.146) are

$$\begin{aligned} r &= 0.25099,\ 0.25099\exp(\mathrm{i}2\pi/3),\ 0.25099\exp(\mathrm{i}4\pi/3) \\ s &= -0.885382,\ -0.885382\exp(\mathrm{i}2\pi/3),\ -0.885382\exp(\mathrm{i}4\pi/3) \end{aligned} \tag{12.147}$$

From (12.143) and (12.139)

$$rs = -a/3 = -2/9 \tag{12.148}$$

Therefore the roots r and s can only be taken in the following combinations

$$\begin{aligned} r_1 &= 0.25099, & s_1 &= -0.885382 \\ r_2 &= 0.25099\exp(\mathrm{i}2\pi/3), & s_2 &= -0.885382\exp(\mathrm{i}4\pi/3) \\ r_3 &= 0.25099\exp(\mathrm{i}4\pi/3), & s_3 &= -0.885382\exp(\mathrm{i}2\pi/3) \end{aligned} \tag{12.149}$$

Table 12.13. *Solution of a single degree of freedom system using the Houbolt method. Example 12.16*

t	Approximate solution $\Delta t = 0.5$	Analytical solution
0	1.0	1.0
0.5	0.8750	0.8776
1.0	0.5313	0.5403
1.5	0.0694	0.0707
2.0	−0.4012	−0.4161
2.5	−0.7790	−0.8011
3.0	−0.9869	−0.990
3.5	−0.9866	−0.9365
4.0	−0.7841	−0.6536
4.5	−0.4272	−0.2108
5.0	0.0062	0.2837

Using (12.140), the roots of (12.138) are

$$\begin{aligned} \alpha_1 &= -0.634392 \\ \alpha_2 &= 0.317196 + \mathrm{i}0.984127 \\ \alpha_3 &= 0.317196 - \mathrm{i}0.984127 \end{aligned} \tag{12.150}$$

Finally, substituting the solutions (12.150) into (12.137) gives the roots of equation (12.136), namely

$$\begin{aligned} \beta_1 &= 0.4587887 \\ \beta_2 &= 0.8817167 + \mathrm{i}0.4373897 \\ \beta_3 &= 0.8817167 - \mathrm{i}0.4373897 \end{aligned} \tag{12.151}$$

The spectral radius is, therefore, 0.9842428 and the solution is stable. Reference [12.29] shows that the method is stable however large the time step is. Because of this, the method is said to be unconditionally stable.

EXAMPLE 12.16 Repeat Example 12.13 using the Houbolt method with $\Delta t = 0.5$.

The values u_1 and u_2 were obtained using the central difference method. The same time step was used, as this is within the limit for stability, and so the values are identical to the corresponding ones in Table 12.9. These and subsequent values, obtained by a repeated application of equation (12.130) are given in Table 12.13 where they are compared with the analytical solution. There is a suggestion that the period is increased and the maximum amplitude has decreased. This is confirmed in Figure 12.12 where the response has been plotted for an increased length of time. The figure clearly indicates that the numerical solution has introduced artificial damping or amplitude decay. Reference [12.29] indicates that both period elongation and amplitude decay increase, with an increase in Δt. It is concluded that for good accuracy $\Delta t/\tau_0 = 0.01$ (i.e., $\omega_0 \Delta t = \pi/50$).

Table 12.14. *Transient response of a single degree of freedom system. Example 12.17*

t	Houbolt	Analytical solution
0	0.0	0.0
0.05	0.2961	0.2865
0.1	1.067	1.036
0.15	1.981	1.964
0.2	2.725	2.714
0.25	3.073	3.0
0.3	2.939	2.714
0.35	2.388	1.964
0.4	1.609	1.036
0.45	0.8435	0.2865
0.5	0.3187	0
0.55	0.1755	0.2865
0.6	0.4320	1.036

EXAMPLE 12.17 Repeat Example 12.12 using the Houbolt method and a time step of 0.05.

The values of u_1 and u_2 were obtained using the central difference method using the same time step. These values are given in Table 12.12. The complete set of values are given in Table 12.14, where they are compared with the analytical solution. The maximum response is overestimated by 2.4% and there is period elongation. Comparing Tables 12.12 and 12.14 indicates that the Houbolt method requires a smaller time increment than the central difference method, to give the same accuracy.

12.5.1.3 The Newmark Method

The Newmark method [12.30] is a generalisation of the linear acceleration method. This latter method assumes that the acceleration varies linearly within the interval

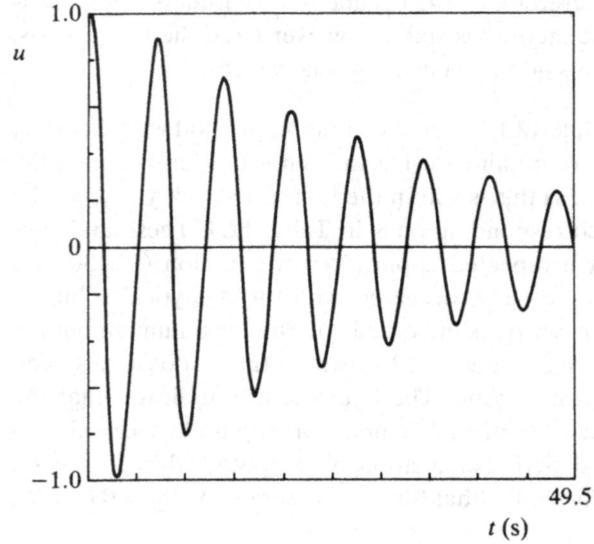

Figure 12.12. Response of a single degree of freedom system using the Houbolt method. $\Delta t = 0.5$.

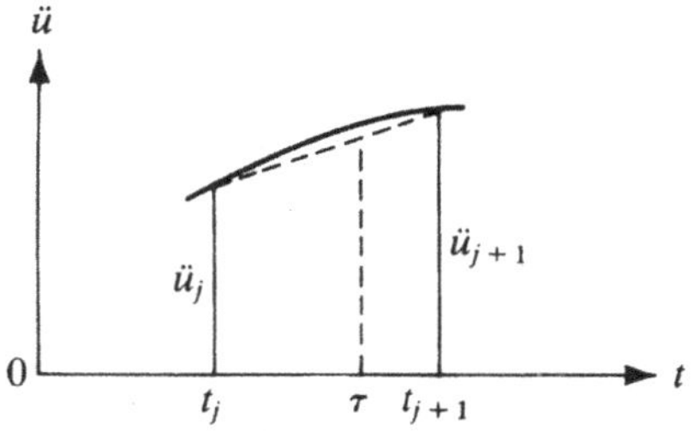

Figure 12.13. Linear acceleration approximation.

(t_j, t_{j+1}) as illustrated in Figure 12.13. This gives

$$\ddot{u} = \ddot{u}_j + \frac{1}{\Delta t}(\ddot{u}_{j+1} - \ddot{u}_j)\tau \qquad \text{for } 0 \leq \tau \leq \Delta t \tag{12.152}$$

Integrating gives

$$\dot{u} = \dot{u}_j + \ddot{u}_j\tau + \frac{1}{2\Delta t}(\ddot{u}_{j+1} - \ddot{u}_j)\tau^2 \tag{12.153}$$

since $\dot{u} = \dot{u}_j$ when $\tau = 0$. Integrating again gives

$$u = u_j + \dot{u}_j\tau + \frac{1}{2}\ddot{u}_j\tau^2 + \frac{1}{6\Delta t}(\ddot{u}_{j+1} - \ddot{u}_j)\tau^3 \tag{12.154}$$

since $u = u_j$ when $\tau = 0$.

Evaluating (12.153) and (12.154) at $\tau = \Delta t$ gives

$$\dot{u}_{j+1} = \dot{u}_j + \frac{\Delta t}{2}(\ddot{u}_j + \ddot{u}_{j+1}) \tag{12.155}$$

and

$$u_{j+1} = u_j + \dot{u}_j\Delta t + \frac{(\Delta t)^2}{6}(2\ddot{u}_j + \ddot{u}_{j+1}) \tag{12.156}$$

In the Newmark method, equations (12.155) and (12.156) are assumed to take the form

$$\dot{u}_{j+1} = \dot{u}_j + \Delta t\{(1-\gamma)\ddot{u}_j + \gamma\ddot{u}_{j+1}\} \tag{12.157}$$

and

$$u_{j+1} = u_j + \dot{u}_j\Delta t + (\Delta t)^2\{(\tfrac{1}{2} - \beta)\ddot{u}_j + \beta\ddot{u}_{j+1}\} \tag{12.158}$$

Taking $\gamma = \frac{1}{2}$ and $\beta = \frac{1}{6}$, equations (12.157) and (12.158) reduce to equations (12.155) and (12.156). The linear acceleration method is, therefore, a special case of the Newmark method. It can be similarly shown that taking $\gamma = \frac{1}{2}$ and $\beta = \frac{1}{4}$ corresponds to assuming that the acceleration is constant and equal to the average value $(\ddot{u}_j + \ddot{u}_{j+1})/2$ within the interval (t_j, t_{j+1}).

The response at time t_{j+1}, is obtained by evaluating the equation of motion at time t_{j+1}, that is

$$m\ddot{u}_{j+1} + c\dot{u}_{j+1} + ku_{j+1} = f_{j+1} \tag{12.159}$$

Equations (12.158) and (12.159) indicate that the method is an implicit one.

In order to get an equation for u_{j+1}, equation (12.158) is solved for $\ddot{u}_{j+1}$ which gives

$$\ddot{u}_{j+1} = \frac{1}{\beta(\Delta t)^2}(u_{j+1} - u_j) - \frac{1}{\beta \Delta t}\dot{u}_j - \left(\frac{1}{2\beta} - 1\right)\ddot{u}_j \tag{12.160}$$

Substituting (12.160) into (12.157) gives

$$\dot{u}_{j+1} = \frac{\gamma}{\beta \Delta t}(u_{j+1} - u_j) + \left(1 - \frac{\gamma}{\beta}\right)\dot{u}_j + \Delta t\left(1 - \frac{\gamma}{2\beta}\right)\ddot{u}_j \tag{12.161}$$

Now substitute (12.160) and (12.161) into (12.159) to give

$$\frac{m}{\beta(\Delta t)^2}(u_{j+1} - u_j) - \frac{m}{\beta \Delta t}\dot{u}_j - m\left(\frac{1}{2\beta} - 1\right)\ddot{u}_j + \frac{\gamma c}{\beta \Delta t}(u_{j+1} - u_j)$$
$$+c\left(1 - \frac{\gamma}{\beta}\right)\dot{u}_j + c\Delta t\left(1 - \frac{\gamma}{2\beta}\right)\ddot{u}_j + ku_{j+1} = f_{j+1} \tag{12.162}$$

Solving for u_{j+1} gives

$$\{m/\beta(\Delta t)^2 + \gamma c/\beta \Delta t + k\}u_{j+1}$$
$$= f_{j+1} + \{m/\beta(\Delta t)^2 + \gamma c/\beta \Delta t\}u_j + \{m/\beta \Delta t - c(1 - \gamma/\beta)\}\dot{u}_j$$
$$+ \left\{m\left(\frac{1}{2\beta} - 1\right) - c\left(1 - \frac{\gamma}{2\beta}\right)\Delta t\right\}\ddot{u}_j \tag{12.163}$$

If u_j, $\dot{u}_j$ and $\ddot{u}_j$ are known, then u_{j+1} can be calculated using (12.163). Equations (12.160) and (12.161) can then be used to calculate $\ddot{u}_{j+1}$ and $\dot{u}_{j+1}$. The time history of the response is obtained by taking $j = 0, 1, 2, \ldots,$ At $t = 0$, u_0, $\dot{u}_0$ are given and $\ddot{u}_0$ can be calculated using equation (12.90) and so no special starting procedure is required.

Reference [12.29] investigates the stability of the method and indicates that it is unconditionally stable provided

$$\gamma \geq \tfrac{1}{2} \quad \text{and} \quad \beta \geq \tfrac{1}{4}\left(\gamma + \tfrac{1}{2}\right)^2 \tag{12.164}$$

Unless γ is taken to be $\frac{1}{2}$, the method introduces artificial damping, which can be negative (when $\gamma < \frac{1}{2}$). This means that the oscillations will increase in amplitude. Note that the constant average acceleration method is unconditionally stable, whilst the linear acceleration method is conditionally stable. Reference [12.29] indicates that for good accuracy, the constant average acceleration method should be used with a time step given by $\Delta t/\tau_0 = 0.01$ (or $\omega_0 \Delta t = \pi/50$).

EXAMPLE 12.18 Repeat Example 12.12 using the Newmark method with $\gamma = \frac{1}{2}$ and $\beta = \frac{1}{4}$ and a time step of 0.05.

Equation (12.90) gives $\ddot{u}_0 = 24\pi^2$. Repeated application of equations (12.163), (12.160) and (12.161) gives the values in Table 12.15 where they are compared with the analytical solution. Comparing Tables 12.14 and 12.15 indicates that the Newmark method, with $\gamma = \frac{1}{2}$ and $\beta = \frac{1}{4}$, produces less period elongation than the Houbolt method.

Table 12.15. *Transient response of a single degree of freedom system. Example 12.18*

t	Newmark $\gamma = \frac{1}{2}, \beta = \frac{1}{4}$	Analytical solution
0	0	0
0.05	0.2695	0.2865
0.10	0.9810	1.036
0.15	1.879	1.964
0.20	2.641	2.714
0.25	2.993	3.0
0.30	2.808	2.714
0.35	2.154	1.964
0.40	1.264	1.036
0.45	0.4596	0.2865
0.50	0.0288	0
0.55	0.1266	0.2865
0.60	0.7179	1.036

Reference [12.23] presents an alternative formulation of the Newmark method. The equation of motion (12.88) is evaluated at times t_{j+1}, t_j and t_{j-1} to give

$$m\ddot{u}_{j+1} + c\dot{u}_{j+1} + ku_{j+1} = f_{j+1} \tag{12.165}$$

$$m\ddot{u}_{j} + c\dot{u}_{j} + ku_{j} = f_{j} \tag{12.166}$$

$$m\ddot{u}_{j-1} + c\dot{u}_{j-1} + ku_{j-1} = f_{j-1} \tag{12.167}$$

Multiplying (12.165) and (12.167) by $(\Delta t)^2\beta$ and (12.166) by $(\Delta t)^2(1 - 2\beta)$ and adding gives

$$\begin{aligned} &(\Delta t)^2 m\big[\{\beta\ddot{u}_{j+1} + (\tfrac{1}{2} - \beta)\ddot{u}_j\} - \{\beta\ddot{u}_j + (\tfrac{1}{2} - \beta)\ddot{u}_{j-1}\} + \tfrac{1}{2}\{\ddot{u}_j + \ddot{u}_{j-1}\}\big] \\ &\quad + (\Delta t)^2 c\big[\tfrac{1}{2}(\dot{u}_j + \dot{u}_{j-1}) + \{\beta(\dot{u}_{j+1} - \dot{u}_j)\} + \{(\tfrac{1}{2} - \beta)(\dot{u}_j - \dot{u}_{j-1})\}\big] \\ &\quad + (\Delta t)^2 k\{\beta u_{j+1} + (1 - 2\beta)u_j + \beta u_{j-1}\} \\ &= (\Delta t)^2\{\beta f_{j+1} + (1 - 2\beta) f_j + \beta f_{j-1}\} \end{aligned} \tag{12.168}$$

Note that the terms multiplied by m and c have been expanded in a form that facilitates simplification. Writing equation (12.157) with $\gamma = \frac{1}{2}$ and (12.158) in the forms

$$(\Delta t)^2\{\beta\ddot{u}_{j+1} + (\tfrac{1}{2} - \beta)\ddot{u}_j\} = u_{j+1} - u_j - \Delta t\dot{u}_j \tag{12.169}$$

$$\frac{\Delta t}{2}(\ddot{u}_j + \ddot{u}_{j-1}) = (\dot{u}_j - \dot{u}_{j-1}) \tag{12.170}$$

$$\Delta t\dot{u}_j = u_{j+1} - u_j - (\Delta t)^2\{(\tfrac{1}{2} - \beta)\ddot{u}_j + \beta\ddot{u}_{j+1}\} \tag{12.171}$$

$$\dot{u}_{j+1} - \dot{u}_j = \frac{\Delta t}{2}(\ddot{u}_{j+1} + \ddot{u}_j) \tag{12.172}$$

and substituting in (12.168) gives, after collecting terms

$$\begin{aligned} m[u_{j+1} - 2u_j + u_{j-1}] + c \cdot \left(\frac{\Delta t}{2}\right)(u_{j+1} - u_{j-1}) \\ + k(\Delta t)^2\{\beta u_{j+1} + (1 - 2\beta)u_j + \beta u_{j-1}\} \\ = (\Delta t)^2\{\beta f_{j+1} + (1 - 2\beta) f_j + \beta f_{j-1}\} \end{aligned} \tag{12.173}$$

Solving for u_{j+1} gives

$$\begin{aligned} \{m + \tfrac{1}{2}c\Delta t + \beta(\Delta t)^2 k\}u_{j+1} \\ = (\Delta t)^2\{\beta f_{j+1} + (1 - 2\beta) f_j + \beta f_{j-1}\} \\ + \{2m - (1 - 2\beta)(\Delta t)^2 k\}u_j - \{m - \tfrac{1}{2}c\Delta t + \beta(\Delta t)^2 k\}u_{j-1} \end{aligned} \tag{12.174}$$

Taking $\beta = 0$, equation (12.174) reduces to equation (12.97). The central difference method is, therefore, a special case of the Newmark method.

This form of the Newmark method requires a special starting procedure as does the central difference method. The displacement u_1 at time Δt is obtained using (12.165), (12.166), (12.157) and (12.158). Taking $j = 0$ in these equations gives

$$m\ddot{u}_1 + c\dot{u}_1 + ku_1 = f_1 \tag{12.175}$$

$$m\ddot{u}_0 + c\dot{u}_0 + ku_0 = f_0 \tag{12.176}$$

$$\dot{u}_1 = \dot{u}_0 + \frac{\Delta t}{2}(\ddot{u}_0 + \ddot{u}_1) \tag{12.177}$$

and

$$u_1 = u_0 + \Delta t\dot{u}_0 + (\Delta t)^2[(\tfrac{1}{2} - \beta)\ddot{u}_0 + \beta\ddot{u}_1] \tag{12.178}$$

Substituting for $\ddot{u}_0$ and $\ddot{u}_1$ in (12.177) and (12.178) from (12.175) and (12.176) gives

$$\begin{aligned} \left(m + \frac{\Delta t}{2}c\right)\dot{u}_1 = -\frac{\Delta t}{2}k(u_0 + u_1) + \left(m - \frac{\Delta t}{2}c\right)\dot{u}_0 \\ + \frac{\Delta t}{2}(f_0 + f_1) \end{aligned} \tag{12.179}$$

and

$$\begin{aligned} \beta(\Delta t)^2 c\dot{u}_1 = \{m - (\tfrac{1}{2} - \beta)(\Delta t)^2 k\}u_0 - \{m + \beta(\Delta t)^2 k\}u_1 \\ + \{m\Delta t - (\tfrac{1}{2} - \beta)(\Delta t)^2 c\}\dot{u}_0 + (\Delta t)^2\{(\tfrac{1}{2} - \beta) f_0 + \beta f_1\} \end{aligned} \tag{12.180}$$

Eliminating $\dot{u}_1$ between (12.179) and (12.180) gives

$$a_1 u_1 = a_2 u_0 + a_3\dot{u}_0 + a_4 f_0 + a_5 f_1 \tag{12.181}$$

where

$$a_1 = m + \frac{\Delta t}{2}c + \beta(\Delta t)^2 k$$
$$a_2 = m + \frac{\Delta t}{2}c - \left(\frac{1}{2} - \beta\right)(\Delta t)^2 k - \left(\frac{1}{4} - \beta\right)(\Delta t)^3 \frac{ck}{m}$$
$$a_3 = m\Delta t - \left(\frac{1}{4} - \beta\right)(\Delta t)^2 \frac{c^2}{m} \tag{12.182}$$
$$a_4 = (\Delta t)^2 \left\{\left(\frac{1}{2} - \beta\right) + \left(\frac{1}{4} - \beta\right)\Delta t \frac{c}{m}\right\}$$
$$a_5 = \beta(\Delta t)^2$$

Stability of the method can be investigated in the same way as the central difference method. Considering free vibration of an undamped system, equation (12.174) becomes

$$\{m + \beta(\Delta t)^2 k\}u_{j+1} = \{2m - (1 - 2\beta)(\Delta t)^2 k\}u_j - \{m + \beta(\Delta t)^2 k\}u_{j-1} \tag{12.183}$$

Dividing by m and rearranging gives

$$\{1 + \beta(\omega_0\Delta t)^2\}u_{j+1} + \{(1 - 2\beta)(\omega_0\Delta t)^2 - 2\}u_j + \{1 + \beta(\omega_0\Delta t)^2\}u_{j-1} = 0 \tag{12.184}$$

where $\omega_0 = (k/m)^{1/2}$.

The solution of (12.184) is of the form

$$u_j = A\delta^j \tag{12.185}$$

Substituting (12.185) into (12.184) and dividing by $A\delta^{j-1}$ gives

$$\{1 + \beta(\omega_0\Delta t)^2\}\delta^2 + \{(1 - 2\beta)(\omega_0\Delta t)^2 - 2\}\delta + \{1 + \beta(\omega_0\Delta t)^2\} = 0 \tag{12.186}$$

The condition that the roots of (12.186) are complex conjugates is

$$\{(1 - 2\beta)(\omega_0\Delta t)^2 - 2\}^2 - 4\{1 + \beta(\omega_0\Delta t)^2\}^2 < 0 \tag{12.187}$$

that is

$$(\omega_0\Delta t)^2\{(1 - 4\beta)(\omega_0\Delta t)^2 - 4\} < 0 \tag{12.188}$$

or

$$(1 - 4\beta) < \left(\frac{2}{\omega_0\Delta t}\right)^2 \tag{12.189}$$

This will be satisfied for any value of $\omega_0\Delta t$, however large, provided

$$\beta \geq \tfrac{1}{4} \tag{12.190}$$

The modulus of the two complex roots is given by

$$|\delta|^2 = \frac{\{1 + \beta(\omega_0\Delta t)^2\}}{\{1 + \beta(\omega_0\Delta t)^2\}} = 1 \tag{12.191}$$

Table 12.16. *Transient response of a single degree of freedom system. Example 12.19*

t	Newmark $\gamma = \frac{1}{2}, \beta = \frac{1}{3}$	Analytical solution
0	0	0
0.05	0.2617	0.2865
0.10	0.9553	1.036
0.15	1.839	1.964
0.20	2.604	2.714
0.25	2.985	3.0
0.30	2.847	2.714
0.35	2.239	1.964
0.40	1.374	1.036
0.45	0.5522	0.2865
0.50	0.0614	0
0.55	0.0726	0.2865
0.60	0.5817	1.036

The method is, therefore, unconditionally stable provided (12.190) holds. This agrees with (12.164) when $\gamma = \frac{1}{2}$. Reference [12.31] recommends using $\beta = \frac{1}{3}$

EXAMPLE 12.19 Repeat Example 12.12 using the alternative formulation of the Newmark method with $\beta = \frac{1}{3}$ and a time step of 0.05.

u_1 is calculated using equation (12.181) whilst subsequent values are calculated using equation (12.174) repeatedly. The values are given in Table 12.16 where they are compared with the analytical solution. Comparing Tables 12.15 and 12.16 indicates that using $\beta = \frac{1}{3}$ produces greater period elongation than when using $\beta = \frac{1}{4}$.

Reference [12.31] indicates that the error in frequency is given by

$$\frac{\omega_f}{\omega_0} = 1 - \tfrac{1}{8}(\omega_0 \Delta t)^2 \tag{12.192}$$

where ω_f is the frequency given by the numerical procedure. This means that the period elongation is

$$\text{PE} = \frac{100(\tau_f - \tau_0)}{\tau_0} = \frac{12.5(\omega_0 \Delta t)^2}{\{1 - (\omega_0 \Delta t)^2/8\}} \tag{12.193}$$

12.5.1.4 The Wilson θ Method

The Wilson θ method [12.32] is an extension of the linear acceleration method. The acceleration is assumed to vary linearly over the extended interval $(t_j, t_{j+\theta})$, where $\theta \geq 1$, as illustrated in Figure 12.14. This gives

$$\ddot{u} = \ddot{u}_j + \frac{1}{\theta \Delta t}(\ddot{u}_{j+\theta} - \ddot{u}_j)\tau \qquad \text{for} \qquad 0 \leq \tau \leq \theta \Delta t \tag{12.194}$$

Integrating gives

$$\dot{u} = \dot{u}_j + \ddot{u}_j \tau + \frac{1}{2\theta \Delta t}(\ddot{u}_{j+\theta} - \ddot{u}_j)\tau^2 \tag{12.195}$$

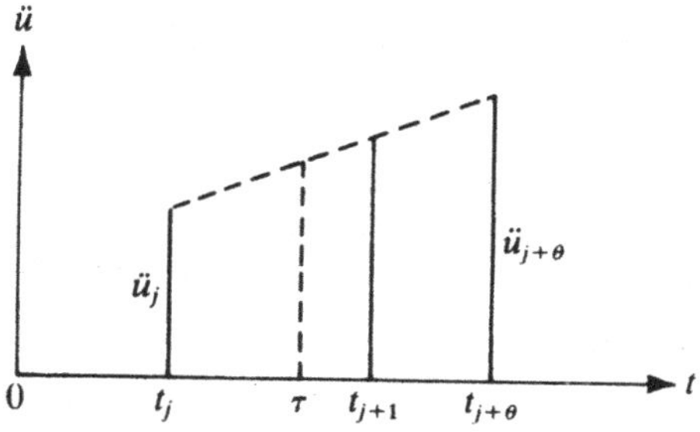

Figure 12.14. Linear acceleration approximation over extended interval.

since $\dot{u} = \dot{u}_j$ when $\tau = 0$. Integrating again gives

$$u = u_j + \dot{u}_j\tau + \frac{1}{2}\ddot{u}_j\tau^2 + \frac{1}{6\theta\,\Delta t}(\ddot{u}_{j+\theta} - \ddot{u}_j)\tau^3 \tag{12.196}$$

since $u = u_j$ when $\tau = 0$.

Evaluating (12.195) and (12.196) at $\tau = \theta\,\Delta t$ gives

$$\dot{u}_{j+\theta} = \dot{u}_j + \frac{\theta\,\Delta t}{2}(\ddot{u}_{j+\theta} + \ddot{u}_j) \tag{12.197}$$

and

$$u_{j+\theta} = u_j + \theta\,\Delta t\dot{u}_j + \frac{(\theta\,\Delta t)^2}{6}(\ddot{u}_{j+\theta} + 2\ddot{u}_j) \tag{12.198}$$

The response at time $t_{j+\theta}$ is obtained by evaluating the equation of motion at time $t_{j+\theta}$, that is

$$m\ddot{u}_{j+\theta} + c\dot{u}_{j+\theta} + ku_{j+\theta} = f_j + \theta(f_{j+1} - f_j) \tag{12.199}$$

where $f_{j+\theta}$ is obtained by linear extrapolation. Equations (12.198) and (12.199) indicate that the method is an implicit one.

In order to get an equation for $u_{j+\theta}$, equation (12.198) is solved for $\ddot{u}_{j+\theta}$ which gives

$$\ddot{u}_{j+\theta} = \frac{6}{(\theta\,\Delta t)^2}(u_{j+\theta} - u_j) - \frac{6}{\theta\,\Delta t}\dot{u}_j - 2\ddot{u}_j \tag{12.200}$$

Substituting (12.200) into (12.197) gives

$$\dot{u}_{j+\theta} = \frac{3}{\theta\,\Delta t}(u_{j+\theta} - u_j) - 2\dot{u}_j - \frac{\theta\,\Delta t}{2}\ddot{u}_j \tag{12.201}$$

Substituting (12.200) and (12.201) into (12.199) and solving for $u_{j+\theta}$ gives

$$\begin{aligned}&\{6m/(\theta\,\Delta t)^2 + 3c/(\theta\,\Delta t) + k\}u_{j+\theta}\\ &\quad = f_j + \theta(f_{j+1} - f_j) + \{6m/(\theta\,\Delta t)^2 + 3c/(\theta\,\Delta t)\}u_j\\ &\quad\quad + \{6m/(\theta\,\Delta t) + 2c\}\dot{u}_j + \{2m + (\theta\,\Delta t)c/2\}\ddot{u}_j\end{aligned} \tag{12.202}$$

If u_j, $\dot{u}_j$ and $\ddot{u}_j$ are known, then $u_{j+\theta}$ can be calculated using (12.202). Putting $\tau = \Delta t$ in (12.194) gives

$$\ddot{u}_{j+1} = \ddot{u}_j + \frac{1}{\theta}(\ddot{u}_{j+\theta} - \ddot{u}_j) \tag{12.203}$$

Substituting for $\ddot{u}_{j+\theta}$ in (12.203) from (12.200) gives

$$\ddot{u}_{j+1} = \frac{6}{\theta^2 \Delta t}\left\{\frac{1}{\theta \Delta t}(u_{j+\theta} - u_j) - \dot{u}_j\right\} + \left(1 - \frac{3}{\theta}\right)\ddot{u}_j \tag{12.204}$$

After calculating $u_{j+\theta}$ from (12.202), $\ddot{u}_{j+1}$ can be calculated using (12.204). $\dot{u}_{j+1}$ and u_j can now be obtained from (12.195) and (12.196) with $\tau = \Delta t$ and $\theta = 1$, that is

$$\dot{u}_{j+1} = \dot{u}_j + \frac{\Delta t}{2}(\ddot{u}_{j+1} + \ddot{u}_j) \tag{12.205}$$

and

$$u_{j+1} = u_j + \Delta t \dot{u}_j + \frac{(\Delta t)^2}{6}(\ddot{u}_{j+1} + 2\ddot{u}_j) \tag{12.206}$$

The time history of the response is obtained by taking $j = 0, 1, 2, \ldots$. At $t = 0$, u_0 and $\dot{u}_0$ are given and $\ddot{u}_0$ can be calculated using equation (12.90) and so no special starting procedure is required.

Reference [12.29] investigates the stability of the method and indicates that it is unconditionally stable provided $\theta \geq 1.37$. It is usual to take $\theta = 1.4$. With $\theta = 1.4$, the method produces less period elongation than the Houbolt method but more than the Newmark method with $\gamma = \frac{1}{2}$ and $\beta = \frac{1}{4}$. The same applies to the amplitude decay. For good accuracy the Wilson θ method should be used with a time step given by $\Delta t/\tau_0 = 0.01$ (or $\omega_0 \Delta t = \pi/50$).

EXAMPLE 12.20 Repeat Example 12.12 using the Wilson θ method with $\theta = 1.4$ and a time step of 0.05.

Equations (12.202), (12.204), (12.205) and (12.206) become

$$u_{j+\theta} = 0.1713469 + 0.8857688 u_j + 0.06200382 \dot{u}_j + 0.001446756 \ddot{u}_j$$

$$\ddot{u}_{j+1} = 874.6355(u_{j+\theta} - u_j) - 61.22449 \dot{u}_j - 1.142857 \ddot{u}_j$$

$$\ddot{u}_{j+1} = \dot{u}_j + 0.025(\ddot{u}_{j+1} + \ddot{u}_j)$$

$$u_{j+1} = u_j + 0.05 \dot{u}_j + 0.0004166667(\ddot{u}_{j+1} + 2\ddot{u}_j)$$

At $t = 0$, $u_0 = 0$, $\dot{u}_0 = 0$ and $\ddot{u}_0 = 24\pi^2$.

Repeated application of these equations gives the values in Table 12.17 where they are compared with the analytical solution. Comparing Tables 12.15 and 12.17 indicates that the Wilson θ method, with $\theta = 1.4$, produces more period elongation than the Newmark method with $\gamma = \frac{1}{2}$ and $\beta = \frac{1}{4}$. Also comparing Tables 12.14 and 12.17 indicates that the Wilson θ method, with $\theta = 1.4$, produces less period elongation than the Houbolt method.

12.5.2 Direct Analysis

The methods presented in the previous sections can be used to solve the equation

$$\mathbf{M}\ddot{\mathbf{u}} + \mathbf{C}\dot{\mathbf{u}} + \mathbf{K}\mathbf{u} = \mathbf{f} \tag{12.207}$$

directly. This has the advantage that the frequencies and modes of free vibration of the undamped system do not have to be calculated prior to the response analysis.

Table 12.17. *Transient response of a single degree of freedom system. Example 12.20*

t	Wilson θ, $\theta = 1.4$	Analytical solution
0	0	0
0.05	0.2719	0.2865
0.10	0.9707	1.036
0.15	1.841	1.964
0.20	2.583	2.714
0.25	2.948	3.0
0.30	2.818	2.714
0.35	2.245	1.964
0.40	1.430	1.036
0.45	0.6526	0.2865
0.50	0.1736	0
0.55	0.1501	0.2865
0.60	0.5827	1.036

The form of the equations as applied to multi-degree of freedom systems are given in the following sections.

12.5.2.1 Central Difference Method

In Section 12.5.1.1 it is shown that for a single degree of freedom system the central difference method consists of

(1) calculating $\ddot{u}_0$ using (12.90),
(2) calculating u_1 using (12.100) and
(3) repeated application of (12.97).

For a multi-degree of freedom system these become

(1) solve the equation

$$\mathbf{M}\ddot{\mathbf{u}}_0 = \mathbf{f}_0 - \mathbf{C}\dot{\mathbf{u}}_0 - \mathbf{K}\mathbf{u}_0 \qquad (12.208)$$

for the column matrix $\ddot{\mathbf{u}}_0$

(2) calculate $\mathbf{u}_1$ using

$$\mathbf{u}_1 = \mathbf{u}_0 + \Delta t\dot{\mathbf{u}}_0 + \{(\Delta t)^2/2\}\ddot{\mathbf{u}}_0 \qquad (12.209)$$

and

(3) repeated solution of the equation

$$[a_1\mathbf{M} + a_2\mathbf{C}]\mathbf{u}_{j+1} = \mathbf{f}_j + [2a_1\mathbf{M} - \mathbf{K}]\mathbf{u}_j - [a_2\mathbf{M} - a_2\mathbf{C}]\mathbf{u}_{j-1} \qquad (12.210)$$

for $\mathbf{u}_{j+1}$ where

$$a_1 = 1/(\Delta t)^2, \qquad a_2 = 1/2\Delta t \qquad (12.211)$$

It is necessary to solve a set of linear equations in order to determine both $\ddot{\mathbf{u}}_0$ and $\mathbf{u}_{j+1}$. These can be solved using the modified Cholesky symmetric decomposition described in Section 11.1.2.

For lightly damped systems the effect of damping on the response is negligible. Putting $\mathbf{C} = 0$ in (12.208) and (12.210) gives

$$\mathbf{M}\ddot{\mathbf{u}}_0 = \mathbf{f}_0 - \mathbf{K}\mathbf{u}_0 \tag{12.212}$$

and

$$a_1\mathbf{M}\mathbf{u}_{j+1} = \mathbf{f}_j + [2a_1\mathbf{M} - \mathbf{K}]\mathbf{u}_j - a_1\mathbf{M}\mathbf{u}_{j-1} \tag{12.213}$$

If the inertia matrix $\mathbf{M}$ is diagonal, then both $\ddot{\mathbf{u}}_0$ and $\mathbf{u}_{j+1}$ can be calculated without solving a set of linear equations.

The inertia matrices for various elements are presented in Chapters 3 to 10. These use the same displacement functions as the derivation of the stiffness matrices. The resulting matrices are termed consistent inertia matrices and are non-diagonal.

One method of obtaining a diagonal inertia matrix is to place discrete masses, that do not have any rotary inertia, at each node. The resulting matrix is termed a lumped mass matrix. The discrete masses are chosen in such a way that the total mass of the element is preserved.

The lumped mass matrix for the rod element presented in Section 3.3 is

$$[\mathbf{m}] = \rho A a \begin{bmatrix} 1 & 0 \\ 0 & 1 \end{bmatrix} \tag{12.214}$$

This result can also be obtained by taking

$$N_1(\xi) = \begin{cases} 1 & -1 \le \xi \le 0 \\ 0 & 0 < \xi \le 1 \end{cases}$$

and (12.215)

$$N_2(\xi) = \begin{cases} 0 & -1 \le \xi < 0 \\ 1 & 0 \le \xi \le 1 \end{cases}$$

in (3.52).

The lumped mass matrix for the beam element presented in Section 3.5 is

$$[\mathbf{m}] = \rho A a \begin{bmatrix} 1 & & & \\ & 0 & & \\ & & 1 & \\ & & & 0 \end{bmatrix} \tag{12.216}$$

This result can be obtained by taking

$$\begin{aligned} N_1(\xi) &= \begin{cases} 1 & -1 \le \xi \le 0 \\ 0 & 0 < \xi \le 1 \end{cases} \\ N_2(\xi) &= 0 \\ N_3(\xi) &= \begin{cases} 0 & -1 \le \xi < 0 \\ 1 & 0 \le \xi \le 1 \end{cases} \\ N_4(\xi) &= 0 \end{aligned} \tag{12.217}$$

in (3.124).

An alternative procedure is to derive the diagonal inertia matrix from the consistent inertia matrix [12.33]. The technique is to compute the diagonal terms of the consistent inertia matrix and then scale them so as to preserve the total mass of the element. The scalar factor is obtained by dividing the total mass, by the sum of the diagonal terms associated with translation.

Applying this technique to the rod element (3.59) gives (12.214). The beam element (3.132) gives the following matrix

$$[\mathbf{m}] = \frac{\rho A a}{78}\begin{bmatrix} 78 & & & \\ & 8a^2 & & \\ & & 78 & \\ & & & 8a^2 \end{bmatrix} \qquad (12.218)$$

EXAMPLE 12.21 Repeat Example 3.3 using the diagonal inertia matrix (12.214).

Example 3.3 analyses a clamped-free rod using one and two element idealisations.

One element solution In this case $a = L/2$ and so the inertia matrix is

$$[\mathbf{m}] = \frac{\rho A L}{2}\begin{bmatrix} 1 & 0 \\ 0 & 1 \end{bmatrix}$$

The equation of motion is

$$\left[\frac{EA}{L} - \omega^2\frac{\rho A L}{2}\right] A_2 = 0$$

the solution of which is

$$\omega_1 = 1.414\left(\frac{E}{\rho L^2}\right)^{1/2}$$

Two element solution

In this case, $a = L/4$ and so

$$[\mathbf{m}] = \frac{\rho A L}{4}\begin{bmatrix} 1 & 0 \\ 0 & 1 \end{bmatrix}$$

The equation of motion is

$$\left[\frac{2EA}{L}\begin{bmatrix} 2 & -1 \\ -1 & 1 \end{bmatrix} - \omega^2\frac{\rho A L}{4}\begin{bmatrix} 2 & 0 \\ 0 & 1 \end{bmatrix}\right]\begin{bmatrix} A_2 \\ A_3 \end{bmatrix} = 0$$

Letting $\omega^2 \rho L^2/8E = \lambda$, this equation simplifies to

$$\begin{bmatrix} (2-2\lambda) & -1 \\ -1 & (1-\lambda) \end{bmatrix}\begin{bmatrix} A_2 \\ A_3 \end{bmatrix} = 0$$

For a non-zero solution

$$(2-2\lambda)(1-\lambda) - 1 = 0$$

that is

$$2\lambda^2 - 4\lambda + 1 = 0$$

Table 12.18. *Comparison of approximate frequencies with the exact solution for a rod*

Mode	FEM solutions 1 element	2 elements	Exact solution
1	1.414	1.531	1.571
2	–	3.695	4.712

The two roots of this equation are

$$\lambda = 0.293 \quad \text{and} \quad 1.707$$

The natural frequencies are therefore

$$\omega_1 = (8\lambda_1)^{1/2}\left(\frac{E}{\rho L^2}\right)^{1/2} = 1.531\left(\frac{E}{\rho L^2}\right)^{1/2}$$

$$\omega_2 = (8\lambda_2)^{1/2}\left(\frac{E}{\rho L^2}\right)^{1/2} = 3.695\left(\frac{E}{\rho L^2}\right)^{1/2}$$

The values of $\omega(\rho L^2/E)^{1/2}$ obtained are compared with exact values in Table 12.18. The approximate frequencies are less than the exact ones and approach them as the number of elements increases. These results should be compared with the ones obtained with consistent inertia matrices in Table 3.3.

EXAMPLE 12.22 Repeat Example 3.7 using the diagonal inertia matrices (12.216) and (12.218).

Example 3.7 analyses a cantilever beam using a one element solution. Since $a = L/2$ matrix (12.216) becomes

$$[\mathbf{m}] = \frac{\rho AL}{2}\begin{bmatrix} 1 & & & \\ & 0 & & \\ & & 1 & \\ & & & 0 \end{bmatrix}$$

The equation of motion is

$$\left[\frac{EI_z}{L^3}\begin{bmatrix} 12 & -6L \\ -6L & 4L^2 \end{bmatrix} - \omega^2\frac{\rho AL}{2}\begin{bmatrix} 1 & 0 \\ 0 & 0 \end{bmatrix}\right]\begin{bmatrix} v_2 \\ \theta_{z2} \end{bmatrix} = 0$$

Letting $\omega^2 \rho AL^4/2EI_z = \lambda$, this equation simplifies to

$$\begin{bmatrix} (12-\lambda) & -6 \\ -6 & 4 \end{bmatrix}\begin{bmatrix} v_2 \\ L\theta_{z2} \end{bmatrix} = 0$$

This equation has non-zero solution provided

$$4(12-\lambda) - 36 = 0$$

that is

$$\lambda = 3$$

and so

$$\omega_1 = (2\lambda)^{1/2}\left(\frac{EI_z}{\rho AL^4}\right)^{1/2} = 2.499\left(\frac{EI_z}{\rho AL^4}\right)^{1/2}$$

Matrix (12.218) becomes

$$[\mathbf{m}] = \frac{\rho AL}{2}\begin{bmatrix} 1 & & & \\ & L^2/39 & & \\ & & 1 & \\ & & & L^2/39 \end{bmatrix}$$

The equation of motion is

$$\left[\frac{EI_z}{L^3}\begin{bmatrix} 12 & -6L \\ -6L & 4L^2 \end{bmatrix} - \omega^2\frac{\rho AL}{2}\begin{bmatrix} 1 & 0 \\ 0 & L^2/39 \end{bmatrix}\right]\begin{bmatrix} v_2 \\ \theta_{z2} \end{bmatrix} = 0$$

Letting $\omega^2\rho AL^4/2EI_z = \lambda$, it becomes

$$\begin{bmatrix} (12-\lambda) & -6 \\ -6 & (4-\lambda/39) \end{bmatrix}\begin{bmatrix} v_2 \\ \theta_{z2} \end{bmatrix} = 0$$

For a non-zero solution

$$(12-\lambda)(4-\lambda/39) - 36 = 0$$

that is

$$\lambda^2 - 168\lambda + 468 = 0$$

the solutions of which are

$$\lambda = 2.834 \quad \text{or} \quad 165.2$$

The natural frequencies are, therefore

$$\omega_1 = (2\lambda_1)^{1/2}\left(\frac{EI_z}{\rho AL^4}\right)^{1/2} = 2.381\left(\frac{EI_z}{\rho AL^4}\right)^{1/2}$$

$$\omega_2 = (2\lambda_2)^{1/2}\left(\frac{EI_z}{\rho AL^4}\right)^{1/2} = 18.18\left(\frac{EI_z}{\rho AL^4}\right)^{1/2}$$

The values of $\omega(\rho AL^4/EI_z)^{1/2}$ obtained are compared with the consistent inertia and analytical solutions in Table 12.19. The frequencies obtained with consistent inertia matrices are greater than the analytical frequencies, whilst those obtained with both types of diagonal matrices are less than the analytical ones.

Reference [12.33] has analysed a simply supported square plate having a span/thickness ratio of 10 and $\nu = 0.3$. Half the plate was represented by a (4×2) mesh of eight node isoparametric elements (RH) as described in Section 6.7. Frequencies are calculated using consistent, lumped and diagonal inertia matrices. The computed frequencies are compared with analytical frequencies obtained using Midlin's thick plate theory in Table 12.20. In this Table *m*, *n* denote the number of

Table 12.19. *Comparision of consistent and diagonal inertia solutions for a cantilever beam using one element*

Inertia matrix	Mode 1	Mode 2
Consistent (3.132)	3.533	34.81
Lumped (12.216)	2.449	–
Diagonal (12.218)	2.381	18.18
Analytical	3.516	22.04

half-waves in the x- and y-directions. The diagonal inertia matrix produces accurate results.

A method of preserving the rotational inertias as well as the masses is given in reference [12.34] and comparisons made with the above two methods.

12.5.2.2 The Houbolt Method

In Section 12.5.1.2, it is shown that for a single degree of freedom system, the time history of displacement can be obtained by repeated application of equation (12.130), after calculating u_1 and u_2 using a different numerical procedure. For a multi-degree of freedom system, equation (12.130) becomes

$$[a_1\mathbf{M} + a_2\mathbf{C} + \mathbf{K}]\mathbf{u}_{j+1} = \mathbf{f}_{j+1} + [a_3\mathbf{M} + a_4\mathbf{C}]\mathbf{u}_j - [2a_1\mathbf{M} + a_5\mathbf{C}]\mathbf{u}_{j-1} + [a_6\mathbf{M} + a_7\mathbf{C}]\mathbf{u}_{j-2} \tag{12.219}$$

where

$$a_1 = 2/(\Delta t)^2, \quad a_2 = 11/6\Delta t, \quad a_3 = 5/(\Delta t)^2$$
$$a_4 = 3/\Delta t, \quad a_5 = 3/2\Delta t, \quad a_6 = 1/(\Delta t)^2, \quad a_7 = 1/3\Delta t \tag{12.220}$$

It is necessary to solve a set of linear equations in order to determine $\mathbf{u}_{j+1}$, even when $\mathbf{C} = 0$ and $\mathbf{M}$ is diagonal. This is because $\mathbf{K}$ is never diagonal.

Table 12.20. *Percentage errors in the natural frequencies of a simply supported square plate when compared with analytical solution [12.23]*

Mode		Type of inertia matrix		
m	n	Consistent	Lumped	Diagonal
1	1	0.11	0.54	0.54
2	1	−0.05	−0.09	0.81
2	2	0.21	−3.6	−2.2
3	1	5.9	−5.1	0.70
3	2	5.5	−9.4	−2.2
3	3	14.9	−18.6	−4.2

12.5.2.3 The Newmark Method

For a single, degree of freedom system the Newmark method consists of repeated application of equations (12.163), (12.160), and (12.161). For a multi-degree of freedom system these become

$$[a_1\mathbf{M} + a_2\mathbf{C} + \mathbf{K}]\mathbf{u}_{j+1} = \mathbf{f}_{j+1} + [a_1\mathbf{M} + a_2\mathbf{C}]\mathbf{u}_j + [a_3\mathbf{M} - a_4\mathbf{C}]\dot{\mathbf{u}}_j + [a_5\mathbf{M} - a_6\mathbf{C}]\ddot{\mathbf{u}}_j \quad (12.221)$$

$$\ddot{\mathbf{u}}_{j+1} = a_1\{\mathbf{u}_{j+1} - \mathbf{u}_j\} - a_3\dot{\mathbf{u}}_j - a_5\ddot{\mathbf{u}}_j \quad (12.222)$$

$$\dot{\mathbf{u}}_{j+1} = a_2\{\mathbf{u}_{j+1} - \mathbf{u}_j\} + a_4\dot{\mathbf{u}}_j + a_6\ddot{\mathbf{u}}_j \quad (12.223)$$

where

$$a_1 = 1/\beta(\Delta t)^2, \quad a_2 = \gamma/\beta\Delta t, \quad a_3 = 1/\beta\Delta t$$
$$a_4 = (1 - \gamma/\beta), \quad a_5 = \{(1/2\beta) - 1\}, \quad a_6 = \{1 - (\gamma/2\beta)\}\Delta t \quad (12.224)$$

It is necessary to solve a set of linear equations in order to determine $\mathbf{u}_{j+1}$.

The modified Newmark method, as applied to a single degree of freedom system consists of

(1) calculating u_1 from (12.181) and
(2) repeated application of (12.174).

For a multi-degree of freedom system these become

$$[\mathbf{M} + a_1\mathbf{C} + a_2\mathbf{K}]\mathbf{u}_1 = [\mathbf{M} + a_1\mathbf{C} - a_3\mathbf{K} - a_4\mathbf{M}^{-1}\mathbf{CK}]\mathbf{u}_0 + [2a_1\mathbf{M} - a_4\mathbf{M}^{-1}\mathbf{C}^2]\dot{\mathbf{u}}_0 + [a_3 + a_4\mathbf{M}^{-1}\mathbf{C}]\mathbf{f}_0 + a_2\mathbf{f}_1 \quad (12.225)$$

and

$$[\mathbf{M} + a_1\mathbf{C} + a_2\mathbf{K}]\mathbf{u}_{j+1} = a_2\mathbf{f}_{j+1} + 2a_3\mathbf{f}_j + a_2\mathbf{f}_{j-1} + [2\mathbf{M} - 2a_3\mathbf{K}]\mathbf{u}_j - [\mathbf{M} - a_1\mathbf{C} + a_2\mathbf{K}]\mathbf{u}_{j-1} \quad (12.226)$$

where

$$a_1 = \Delta t/2, \quad a_2 = \beta(\Delta t)^2$$
$$a_3 = \left(\tfrac{1}{2} - \beta\right)(\Delta t)^2, \quad a_4 = \left(\tfrac{1}{4} - \beta\right)(\Delta t)^3 \quad (12.227)$$

It is necessary to solve a set of linear equations to determine both $\mathbf{u}_1$ and $\mathbf{u}_{j+1}$.

12.5.2.4 The Wilson θ Method

For a single degree of freedom system the Wilson θ method consists of repeated application of equations (12.202), (12.204), (12.205), and (12.206). For a multi-degree of freedom system these become

$$[a_1\mathbf{M} + a_2\mathbf{C} + \mathbf{K}]\mathbf{u}_{j+\theta} = \mathbf{f}_j + \theta(\mathbf{f}_{j+1} - \mathbf{f}_j) + [a_1\mathbf{M} + a_2\mathbf{C}]\mathbf{u}_j + [2a_2\mathbf{M} + 2\mathbf{C}]\dot{\mathbf{u}}_j + [2\mathbf{M} + a_3\mathbf{C}]\ddot{\mathbf{u}}_j \quad (12.228)$$

$$\ddot{\mathbf{u}}_{j+1} = a_4\{\mathbf{u}_{j+\theta} - \mathbf{u}_j\} - a_5\dot{\mathbf{u}}_j + a_6\ddot{\mathbf{u}}_j \quad (12.229)$$

$$\dot{\mathbf{u}}_{j+1} = \dot{\mathbf{u}}_j + a_7\{\ddot{\mathbf{u}}_{j+1} + \ddot{\mathbf{u}}_j\} \quad (12.230)$$

$$\mathbf{u}_{j+1} = \mathbf{u}_j + 2a_7\dot{\mathbf{u}}_j + a_8\{\ddot{\mathbf{u}}_{j+1} + 2\ddot{\mathbf{u}}_j\} \quad (12.231)$$

where

$$\begin{aligned} &a_1 = 6/(\theta\Delta t)^2, \quad a_2 = 3/(\theta\Delta t), \quad a_3 = \theta\Delta t/2 \\ &a_4 = 6/\theta(\theta\Delta t)^2, \quad a_5 = 6/\theta(\theta\Delta t) \\ &a_6 = (1 - 3/\theta), \quad a_7 = \Delta t/2, \quad a_8 = (\Delta t)^2/6 \end{aligned} \quad (12.232)$$

$\mathbf{u}_{j+\theta}$ is obtained by solving a set of linear equations.

12.5.3 Selecting a Time Step

When applying step-by-step integration techniques to multi-degree of freedom systems, the time step Δt is selected on the basis of the shortest period, which corresponds to the highest frequency mode. It should also be small enough to ensure that the time history of the excitation is adequately defined.

For large order systems the required time step will be very small indeed, resulting in many time steps being required to determine the response over the time interval of interest. One way of overcoming this is to use the modal method, retaining only those modes which contribute significantly to the response. These will usually be the lower frequency modes. The number to be included is usually determined from considering the spatial distribution and frequency content of the excitation, as described in Section 13.2. The highest frequency of the modal model will be less than the highest frequency of the complete system, resulting in a larger period on which to base the time increment.

If a direct analysis is used, the time step can be based upon the period corresponding to the highest frequency likely to contribute to the response, if an unconditionally stable method is used. The response in the higher frequency modes will be stable but inaccurate. However, this is of no consequence since their contribution is negligible. Better still, use an unconditionally stable method which gives amplitude decay for large time steps. This will ensure that the unwanted, high frequency components decay rapdily.

The modal method should be used for large order systems or when many time steps are required, especially when the high frequency response is not important. The direct method should be used for small order systems and when only a few time steps are required. It should also be used when the high frequency response is important, such as is the case for shock loading.

12.5.4 Additional Methods

Reference [12.35] gives details and/or references for the following methods

a) collocation methods
b) α-method
c) Bossak's method
d) Bazzi–Anderheggen ρ-method
e) Unified set of single-step methods

Details of other methods are given in references [12.36–12.38].

Reference [12.35] states that one should choose a method which has the following attributes:

1. Unconditional stability when applied to linear problems.
2. No more than one set of implicit equations to be solved at each step.
3. Second order accuracy
4. Controllable algorithmic dissipation in the higher modes.
5. Self starting.

Problems

12.1 Use modal analysis to derive an expression for the transfer receptance α_{21} of the system shown in Figure P1.1 with $k_1 = 3000$, $k_2 = k_3 = 2000$, $k_4 = 1000$ N/m and $m_1 = m_2 = m_3 = 1$ kg. Determine the frequencies at which resonances and anti-resonances occur.

12.2 Use Crout factorisation to solve equation (12.42) for the system defined in Example 12.2 when $\omega^2 = 3000$ (rad/s)^2.

12.3 A clamped-free rod of length 1 m and cross-sectional area 4×10^{-4} m^2 is subject to a harmonic force of magnitude 1000 N at its free end. Calculate the response at the free end at the lowest undamped natural frequency. Take $E = 207 \times 10^9$ N/m^2 and $\rho = 7850$ kg/m^3. Assume structural damping with a loss factor $\eta = 0.04$. Compare the solution with the analytical solution $24.52 \times 10^{-5} \exp\{i(\omega t - 1.5608)\}$.

12.4 Find the Fourier series expansion of the forcing function illustrated in Figure P12.4 where

$$f(t) = P\sin(\pi t/\tau) \qquad 0 \le t \le \tau$$

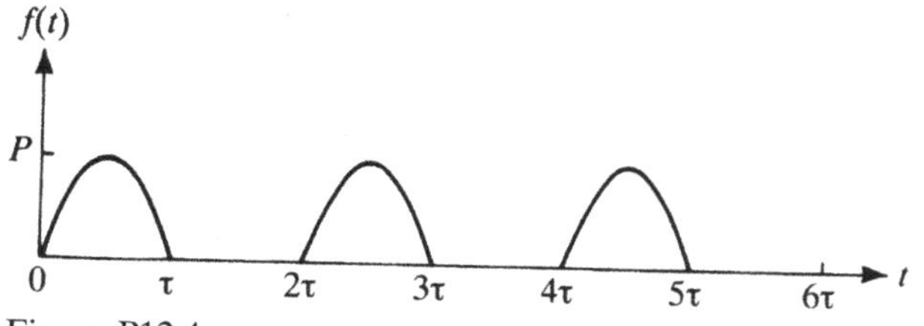

Figure P12.4

12.5 Repeat Example 12.12 using $\Delta\tau = 0.025$.

12.6 Investigate the stability of the central difference method by considering the free vibration of a viscously damped single degree of freedom system.

12.7 Derive expressions for velocity and displacement assuming constant average acceleration within a time increment. Compare them with the corresponding expressions for the Newmark method when $\gamma = \frac{1}{2}$ and $\beta = \frac{1}{4}$.

12.8 Calculate the response of the system defined in Example 12.13 using the linear acceleration method. Show that the method is stable for $\Delta t = 1.0$ but unstable for $\Delta t = 6.0$.

13 Forced Response II

This chapter begins with the solution of equation (12.1) when the applied forces are random. The next section presents methods of improving the convergence and accuracy of the modal method of forced response. This is followed by an analysis of the response of structures to imposed displacements. Finally, the techniques of reducing the number of degrees of freedom presented in Section 11.5 are extended to forced response analysis.

13.1 Response to Random Excitation

Harmonic, periodic and transient forces, which are treated in Chapter 12, are termed deterministic, since their magnitude can be described by explicit mathematical relationships. In the case of random forces, which are caused by gales, confused seas, rough roads, turbulent boundary layers and earthquakes, there is no way of predicting an exact value at a future instant of time. Such forces can only be described by means of statistical techniques.

This section begins by describing how to represent the applied forces statistically. This is followed by an analysis of the response which is also described statistically.

13.1.1 Representation of the Excitation

A typical plot of a randomly varying force, $f(t)$, against t (which represents time) is shown in Figure 13.1. Although it is possible to plot $f(t)$ for a given time interval, if it has been measured during this interval, it is not possible to predict from this the precise value of $f(t)$ at any value of t outside the interval. However, the essential features of the process $f(t)$ can be described by means of statistical concepts. The theory which has been built up to describe these processes is known as random process theory.

Statistical theory is based on the concepts of probability. Defining the probability of realising a value $f(t)$ which is less than some specified value f_0 to be $P(f_0)$ then

$$P(f_0) = \text{Prob}[f(t) < f_0] \tag{13.1}$$

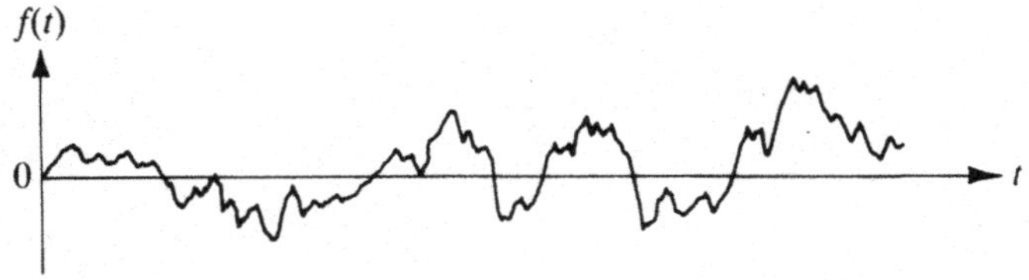

Figure 13.1. Randomly varying force.

The function $P(f)$ is known as a probability distribution function which increases as f increases. The following conditions are satisfied

$$(1) \quad P(-\infty) = 0 \qquad \text{(impossible event)} \tag{13.2}$$

$$(2) \quad P(+\infty) = 1 \qquad \text{(certain event)} \tag{13.3}$$

The probability of the force being between f and $(f + \mathrm{d}f)$ is

$$P(f + \mathrm{d}f) - P(f) = \Delta P(f) \tag{13.4}$$

$\Delta P(f)$ can be considered to be a probability increment. The rate of change of this increment with f is

$$\lim_{\Delta f \to 0} \frac{\Delta P}{\Delta f} = \frac{\mathrm{d}P}{\mathrm{d}f} \tag{13.5}$$

$\mathrm{d}P/\mathrm{d}f$ is called the probability density and is denoted by $p(f)$, thus

$$p(f) = \frac{\mathrm{d}P(f)}{\mathrm{d}f} \tag{13.6}$$

From its definition it is easy to see that

$$(1) \quad P(f + \mathrm{d}f) - P(f) = p(f)\,\mathrm{d}f \tag{13.7}$$

$$(2) \quad P(f) = \int_{-\infty}^{f} p(f)\,\mathrm{d}f \tag{13.8}$$

$$(3) \quad P(f_b) - P(f_a) = \int_{f_a}^{f_b} p(f)\,\mathrm{d}f \tag{13.9}$$

and

$$(4) \quad \int_{-\infty}^{+\infty} p(f)\,\mathrm{d}f = 1 \tag{13.10}$$

Various parameters are used to describe the shape of a probability density curve. The most important one is the expected value or mean which is given by

$$E[f] = \int_{-\infty}^{+\infty} f p(f)\,\mathrm{d}f \tag{13.11}$$

This expression is sometimes called the first moment of f and given the notation μ_f.

The spread, skewness and peakedness of the probability density curve about the mean are given by the second, third and fourth central moments about the mean. The rth central moment about the mean is

$$E[(f - \mu_f)^r] = \int_{-\infty}^{+\infty} (f - \mu_f)^r\, p(f)\,\mathrm{d}f \tag{13.12}$$

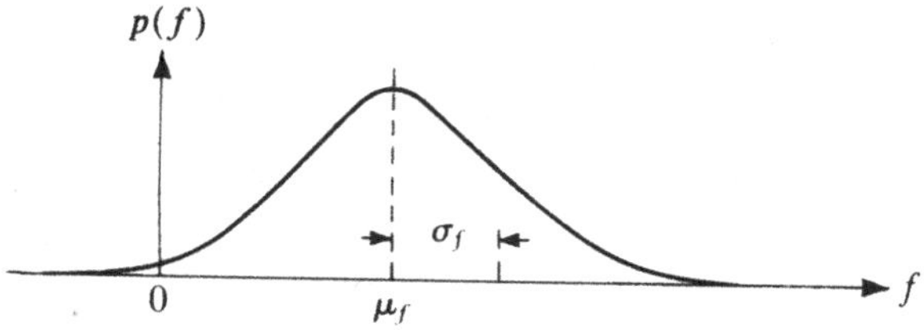

Figure 13.2. Gaussian probability density function.

In particular the second central moment is the variance which is

$$E[(f-\mu_f)^2]=\sigma_f^2 \tag{13.13}$$

where σ_f is the standard deviation of f. It is a measure of the dispersion or spread about the mean. A measure of the departure from symmetry (or skewness) is

$$E[(f-\mu_f)^3]/\sigma_f^3.$$

The peakedness near the mean is measured by the excess of Kurtosis γ_2 where

$$\gamma_2=\frac{E[(f-\mu_f)^4]}{\sigma_f^4}-3 \tag{13.14}$$

A particularly important probability distribution is the Gaussian or normal distribution whose probability density is

$$p(f)=\frac{1}{(2\pi)^{1/2}\sigma_f}\exp\{-(f-\mu_f)^2/2\sigma_f^2\} \tag{13.15}$$

with the shape shown in Figure 13.2.

Note that the Gaussian distribution is completely defined by its mean and standard deviation. It is symmetrical about the mean and so its skewness is zero. The excess of Kurtosis is also zero. The reason for this is that the peakedness of a distribution is measured by its departure from a Gaussian distribution. From (13.15) and (13.9) the probability that

$$|f-\mu_f|<n\sigma_f \tag{13.16}$$

is 0.683, 0.954, 0.997 for $n=1, 2$ and 3 respectively.

The importance of the normal distribution in physical problems may be attributed to the Central Limit Theorem, which essentially states that the sum of a large number of independent variables under fairly general conditions will be approximately normally distributed, regardless of the distribution of the independent variables. Since many physically observed phenomena actually represent the net effect of numerous contributing variables, the normal distribution constitutes a good approximation to commonly occurring distribution functions.

So far the probability distribution of a single random variable has been considered. However, it is often essential to consider the combined probability distribution of two or more random variables which are not completely independent. For the case of two random variables f_1, and f_2, the probability of realising values in the ranges $(f_1, f_1+\mathrm{d}f_1)$ and $(f_2,\ f_2+\mathrm{d}f_2)$ is the joint probability density function. If $P(f_1, f_2)$ is the joint probability distribution function then

$$p(f_1,\ f_2)=\frac{\partial^2 P}{\partial f_1\partial f_2} \tag{13.17}$$

Conversely

$$P(f_1, f_2) = \int_{-\infty}^{f_1}\int_{-\infty}^{f_2} p(f_1, f_2)\,\mathrm{d}f_1\,\mathrm{d}f_2 \tag{13.18}$$

and

$$\int_{-\infty}^{+\infty}\int_{-\infty}^{+\infty} p(f_1, f_2)\,\mathrm{d}f_1\,\mathrm{d}f_2 = 1 \tag{13.19}$$

The probability density function, $p(f_1, f_2)$, can be plotted as a surface above a horizontal plane. Its shape is described by means of the various joint moments $E[f_1^r f_2^s]$ of order $(r+s)$. The quantities of particular interest are

(1) the means

$$\mu_{f_1} = E[f_1] = \int_{-\infty}^{+\infty}\int_{-\infty}^{+\infty} f_1 p(f_1, f_2)\,\mathrm{d}f_1\mathrm{d}f_2 \tag{13.20}$$

$$\mu_{f_2} = E[f_2] = \int_{-\infty}^{+\infty}\int_{-\infty}^{+\infty} f_2 p(f_1, f_2)\,\mathrm{d}f_1\,\mathrm{d}f_2 \tag{13.21}$$

(2) the variances

$$\sigma_{f_1}^2 = E[(f_1-\mu_{f_1})^2] = \int_{-\infty}^{+\infty}\int_{-\infty}^{+\infty} (f_1-\mu_{f_1})^2 p(f_1, f_2)\,\mathrm{d}f_1\,\mathrm{d}f_2 \tag{13.22}$$

$$\sigma_{f_2}^2 = E[(f_2-\mu_{f_2})^2] = \int_{-\infty}^{+\infty}\int_{-\infty}^{+\infty} (f_2-\mu_{f_2})^2 p(f_1, f_2)\,\mathrm{d}f_1\,\mathrm{d}f_2 \tag{13.23}$$

and

(3) the covariance

$$\begin{aligned}\sigma_{f_1 f_2} &= E[(f_1-\mu_{f_1})(f_2-\mu_{f_2})] \\ &= \int_{-\infty}^{+\infty}\int_{-\infty}^{+\infty} (f_1-\mu_{f_1})(f_2-\mu_{f_2}) p(f_1, f_2)\,\mathrm{d}f_1\,\mathrm{d}f_2\end{aligned} \tag{13.24}$$

The normalised covariance

$$\rho_{f_1 f_2} = \frac{\sigma_{f_1 f_2}}{\sigma_{f_1}\sigma_{f_2}} \tag{13.25}$$

is known as the correlation coefficient. It is a measure of the degree of linear dependence between f_1 and f_2. If $\rho_{f_1 f_2} = \pm 1$, then the two variables are perfectly correlated and there is a linear dependence between them. On the other hand, if $\rho_{f_1 f_2} = 0$ the variables are said to be uncorrelated.

If the force, $f(t)$, on n supposedly identical structures subject to the same conditions is measured, then the signals $f^k(t)$ $(k = 1, 2, \ldots, n)$ might look something like the ones shown in Figure 13.3. They are not identical due to influencing factors which cannot be controlled. The ensemble of randomly varying quantities $f^1(t)$, $f^2(t), \ldots$, is called a random process and denoted by $\{f(t)\}$. It is also referred to as a stochastic process or time series.

To characterise the process completely in a probabilistic sense, it is necessary to determine the multivariate probability density function $p(f_1, f_2, \ldots, f_m)$, where

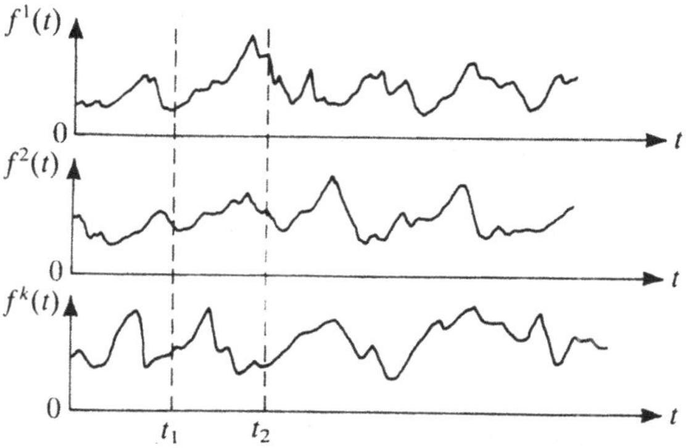

Figure 13.3. Ensemble of random signals.

$f_i = f(t_i)$, for $m = 1, 2, \ldots$. Usually, in engineering applications, it is sufficient to determine only the first two, namely, $p(f_1)$ and $p(f_1, f_2)$. The main parameters which describe these probability densities are:

(1) the means

$$\mu_{f_1} = \mu_f(t_1) \quad \text{and} \quad \mu_{f_2} = \mu_f(t_2) \tag{13.26}$$

where

$$\mu_f = \lim_{n\to\infty} \frac{1}{n} \sum_{k=1}^{n} f^k(t) \tag{13.27}$$

(2) the variances

$$\sigma_{f_1}^2 = \sigma_f^2(t_1) \quad \text{and} \quad \sigma_{f_2}^2 = \sigma_f^2(t_2) \tag{13.28}$$

where

$$\sigma_f^2 = \lim_{n\to\infty} \frac{1}{n} \sum_{k=1}^{n} \{f^k(t) - \mu_f(t)\}^2 \tag{13.29}$$

and

(3) the covariance where

$$\sigma_{f_1 f_2} = \lim_{n\to\infty} \frac{1}{n} \sum_{k=1}^{n} \{f^k(t_1) - \mu_{f_1}\}\{f^k(t_2) - \mu_{f_2}\} \tag{13.30}$$

Expressions (13.27), (13.29) and (13.30) are known as ensemble averages and denoted by the angular brackets $\langle\ \rangle$.

In general, the properties of a random process are time dependent (see equations (13.27), (13.29) and (13.30)). However, in many practical situations the probability density functions are independent of time. A random process is said to be weakly stationary if the probability density functions $p(f_1)$ and $p(f_1, f_2)$ are independent of the time origin. This implies that $p(f_1)$ is independent of the choice of t_1 and $p(f_1, f_2)$ is dependent only on $(t_2 - t_1)$. Thus, for a weakly stationary process, the mean and variance is constant, that is

$$\mu_f = \langle f^k(t) \rangle = \text{constant} \tag{13.31}$$

$$\sigma_f^2 = \langle \{f^k(t) - \mu_f(t)\}^2 \rangle = \text{constant} \tag{13.32}$$

and the covariance is a function of $\tau = (t_2 - t_1)$ only, that is

$$\sigma_{f_1 f_2} = \langle\{f^k(t_1) - \mu_f\}\{f^k(t_1+\tau) - \mu_f\}\rangle \tag{13.33}$$

is independent of t_1. The covariance, as defined in (13.33), can also be written in the form

$$\sigma_{f_1 f_2} = R_f(\tau) - \mu_f^2 \tag{13.34}$$

where

$$R_f(\tau) = \langle f^k(t_1) f^k(t_1+\tau)\rangle \tag{13.35}$$

and is known as the autocorrelation function. If the mean is zero, then the covariance and autocorrelation functions are identical. In vibration analysis a constant mean value represents a static state superimposed upon a dynamic one. Therefore, in what follows, a zero mean will be assumed.

The autocorrelation function has the following properties:

$$(1) \quad R_f(0) = \sigma_f^2 \tag{13.36}$$

$$(2) \quad R_f(-\tau) = R_f(\tau) \text{ and} \tag{13.37}$$

$$(3) \quad |R_f(\tau)| \leq R_f(0) \tag{13.38}$$

Property (1) follows from the definitions (13.32) with $\mu_f = 0$ and (13.35), whilst Property (2) follows from (13.35) since R_f is independent of t_1, thus

$$\begin{aligned} R_f(-\tau) &= \langle f^k(t_1-\tau) f^k(t_1)\rangle \\ &= \langle f^k(t_1) f^k(t_1+\tau)\rangle = R_f(\tau) \end{aligned} \tag{13.39}$$

Therefore, the autocorrelation function is an even function of τ. Next consider

$$\begin{aligned} \langle\{f^k(t_1) \pm f^k(t_1+\tau)\}^2\rangle &= \langle\{f^k(t_1)\}^2\rangle \pm 2\langle f^k(t_1) f^k(t_1+\tau)\rangle + \langle\{f^k(t_1+\tau)\}^2\rangle \\ &= 2\sigma_f^2 \pm 2R_f(\tau) \end{aligned} \tag{13.40}$$

since σ_f^2 is constant. The above expression must be larger than zero and so

$$|R_f(\tau)| \leq \sigma_f^2 = R_f(0) \tag{13.41}$$

The autocorrelation function gives a direct measure of the statistical dependence of the variables $f(t_1+\tau)$ and $f(t_1)$ upon each other. For most stationary processes, the autocorrelation function decays rapidly with increasing value of τ.

All averages introduced so far have been ensemble averages. It is also possible to describe the properties of a stationary random process by means of averages along a single record, that is time averages. For the record $f^k(t)$, the mean and autocorrelation function are:

$$\mu_f^k = \lim_{T\to\infty} \frac{1}{2T} \int_{-T}^{T} f^k(t)\,\mathrm{d}t \tag{13.42}$$

and

$$R_f^k(\tau) = \lim_{T\to\infty} \frac{1}{2T} \int_{-T}^{T} f^k(t) f^k(t+\tau)\,\mathrm{d}t \tag{13.43}$$

The variance is given by $R_f^k(0)$.

Stationary random processes are said to be ergodic if the time averages are equal to the equivalent ensemble averages. This assumption allows the use of only one sample function from a random process in calculating averages instead of the entire ensemble. This implies that the chosen sample function is representative of the complete random process. Since any of the functions, $f^k(t)$, can be taken as representative, the superscript k is omitted from expressions (13.42) and (13.43). Time averages are denoted by a bar and so

$$\mu_f = \bar{f} \tag{13.44}$$

$$\sigma_f^2 = \bar{f}^2 \tag{13.45}$$

$$R_f(\tau) = \overline{f(t)f(t+\tau)} \tag{13.46}$$

Equation (13.45) indicates that the variance is the mean square value.

Equation (13.15) indicates that for a zero mean, the Gaussian probability density function is completely defined by the standard deviation, σ_f, which is also the root mean square value. Its square, the variance, is given by $R_f(0)$ (see (13.45) and (13.46)). A spectral decomposition of the mean square value can also be obtained.

A random signal is essentially a nor-periodic function. Such a function can be represented by means of a Fourier integral, namely,

$$f(t) = \int_{-\infty}^{+\infty} F(\mathrm{i}\omega)\exp(\mathrm{i}\omega t)\,\mathrm{d}\omega \tag{13.47}$$

where

$$F(\mathrm{i}\omega) = \frac{1}{2\pi}\int_{-\infty}^{+\infty} f(t)\exp(-\mathrm{i}\omega t)\,\mathrm{d}t \tag{13.48}$$

ω denoting frequency.

The function $F(\mathrm{i}\omega)$ is the Fourier transform of $f(t)$ and the two quantities $f(t)$ and $F(\mathrm{i}\omega)$ are said to be a Fourier transform pair. A necessary condition for the existence of the Fourier transform is that the integral $\int_{-\infty}^{+\infty}|f(t)|\mathrm{d}t$ be finite. This will be the case if $f(t)$ tends to zero as $t \to \pm\infty$.

A random signal cannot be represented directly by means of Fourier transforms. To have stationary properties a random signal must be assumed to continue over an infinite period of time. In this case the condition that $\int_{-\infty}^{+\infty}|f(t)|\mathrm{d}t$ be finite is not satisfied. But there is no difficulty in determining the Fourier transform of the signal $f_T(t)$ which is identical to $f(t)$ within the interval $-T \le t \le +T$ and zero at all other times, as shown in Figure 13.4. In this case

$$f_T(t) = \int_{-\infty}^{+\infty} F_T(\mathrm{i}\omega)\exp(\mathrm{i}\omega t)\,\mathrm{d}\omega \tag{13.49}$$

and

$$F_T(\mathrm{i}\omega) = \frac{1}{2\pi}\int_{-\infty}^{+\infty} f_T(t)\exp(-\mathrm{i}\omega t)\,\mathrm{d}t \tag{13.50}$$

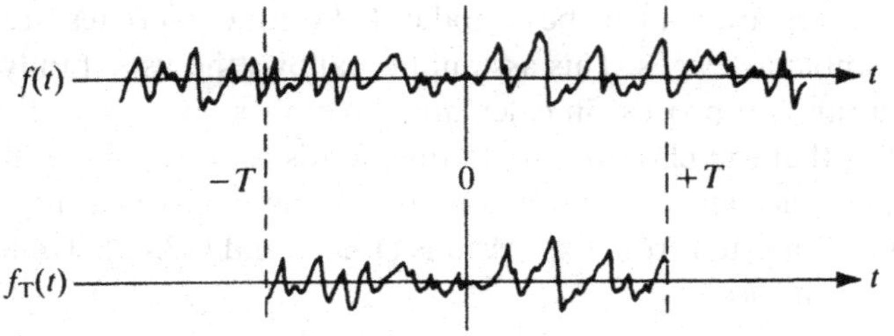

Figure 13.4. Truncation of a random signal.

The mean square or variance of $f_T(t)$ is

$$
\begin{aligned}
\overline{f_T^2(t)} &= \frac{1}{2T}\int_{-\infty}^{+\infty} f_T^2(t)\,\mathrm{d}t \\
&= \frac{1}{2T}\int_{-\infty}^{+\infty} f_T(t)\int_{-\infty}^{+\infty} F_T(\mathrm{i}\omega)\exp(\mathrm{i}\omega t)\,\mathrm{d}\omega\,\mathrm{d}t \\
&= \frac{1}{2T}\int_{-\infty}^{+\infty} F_T(\mathrm{i}\omega)\int_{-\infty}^{+\infty} f_T(t)\exp(\mathrm{i}\omega t)\,\mathrm{d}t\,\mathrm{d}\omega \\
&= \frac{\pi}{T}\int_{-\infty}^{+\infty} F_T(\mathrm{i}\omega)F_T^*(\mathrm{i}\omega)\,\mathrm{d}\omega \\
&= \int_{-\infty}^{+\infty} \frac{\pi}{T}|F_T(\mathrm{i}\omega)|^2\,\mathrm{d}\omega
\end{aligned}
\tag{13.51}
$$

where * denotes a complex conjugate quantity.

The mean square of $f(t)$ is, therefore

$$
\begin{aligned}
\overline{f^2(t)} &= \lim_{T\to\infty}\overline{f_T^2(t)} \\
&= \int_{-\infty}^{+\infty} \lim_{T\to\infty}\frac{\pi}{T}|F_T(\mathrm{i}\omega)|^2\,\mathrm{d}\omega \\
&= \int_{-\infty}^{+\infty} S_f(\omega)\,\mathrm{d}\omega
\end{aligned}
\tag{13.52}
$$

where

$$
S_f(\omega) = \lim_{T\to\infty}\frac{\pi}{T}|F_T(\mathrm{i}\omega)|^2 \tag{13.53}
$$

is called the power spectral density.

The power spectral density function, $S_f(\omega)$, is an even function which is defined over the range $-\infty \leq \omega \leq +\infty$. When making practical measurements it is more convenient to deal with positive frequencies only. In this case a one-sided spectral density function, $G_f(\omega)$, is introduced such that

$$
G_f(\omega) = 2S_f(\omega) \qquad \text{for } \omega > 0 \tag{13.54}
$$

and

$$
\overline{f^2(t)} = \int_0^{\infty} G_f(\omega)\,\mathrm{d}\omega \tag{13.55}
$$

If a random variable has a constant power spectral density over all frequencies it is often referred to as white noise, from analogy with white light which contains waves of all frequencies. However, a white spectrum cannot occur in practice because this would imply an infinite mean square value (see (13.55)). It is useful, though, to consider the power spectral density to be constant over a limited frequency range. In this case it is referred to as band limited white noise.

Both the autocorrelation and power spectral density functions can be used to determine the mean square value. This infers a relationship between them. The autocorrelation function of the signal $f_T(t)$ is

$$R_{f_T}(\tau) = \frac{1}{2T}\int_{-\infty}^{+\infty} f_T(t)\, f_T(t+\tau)\, \mathrm{d}t \tag{13.56}$$

The Fourier transform of this function is

$$\begin{aligned}
&\frac{1}{2\pi}\int_{-\infty}^{+\infty} R_f(\tau)\exp(-\mathrm{i}\omega\tau)\,\mathrm{d}\tau \\
&= \frac{1}{2\pi}\int_{-\infty}^{+\infty}\frac{1}{2T}\int_{-\infty}^{+\infty} f_T(t)\, f_T(t+\tau)\exp(-\mathrm{i}\omega\tau)\,\mathrm{d}t\,\mathrm{d}\tau \\
&= \frac{1}{4\pi T}\int_{-\infty}^{+\infty}\int_{-\infty}^{+\infty} f_T(t)\exp(\mathrm{i}\omega t)\, f_T(t+\tau)\exp\{-\mathrm{i}\omega(t+\tau)\}\,\mathrm{d}t\,\mathrm{d}\tau \\
&= \frac{1}{4\pi T}\int_{-\infty}^{+\infty}\int_{-\infty}^{+\infty} f_T(t)\exp(\mathrm{i}\omega t)\, f_T(s)\exp\{-\mathrm{i}\omega s\}\,\mathrm{d}t\,\mathrm{d}s \qquad (13.57)\\
&= \frac{\pi}{T}\frac{1}{2\pi}\int_{-\infty}^{+\infty} f_T(t)\exp(\mathrm{i}\omega t)\,\mathrm{d}t\,\frac{1}{2\pi}\int_{-\infty}^{+\infty} f_T(s)\exp(-\mathrm{i}\omega s)\,\mathrm{d}s \\
&= \frac{\pi}{T}F_T^*(\mathrm{i}\omega)F_T(\mathrm{i}\omega) \\
&= \frac{\pi}{T}|F_T(\mathrm{i}\omega)|^2
\end{aligned}$$

Taking the limit as $T \to \infty$ of (13.57) gives

$$S_f(\omega) = \frac{1}{2\pi}\int_{-\infty}^{+\infty} R_f(\tau)\exp(-\mathrm{i}\omega\tau)\,\mathrm{d}\tau \tag{13.58}$$

Therefore, the power spectral density is the Fourier transform of the autocorrelation function. This implies that

$$R_f(\tau) = \int_{-\infty}^{+\infty} S_f(\omega)\exp(\mathrm{i}\omega\tau)\,\mathrm{d}\omega \tag{13.59}$$

If a structure is subjected to two random forces there is a possibility that they are related in some way. Assuming that $f_1(t)$ and $f_2(t)$ are stationary, ergodic processes with zero mean values, then the essential features of the process $(f_1(t) + f_2(t))$ are described by its autocorrelation function

$$\begin{aligned}
R_c(\tau) &= \overline{\{f_1(t) + f_2(t)\}\{f_1(t+\tau) + f_2(t+\tau)\}} \\
&= \overline{f_1(t)f_1(t+\tau)} + \overline{f_1(t)f_2(t+\tau)} + \overline{f_2(t)f_1(t+\tau)} + \overline{f_2(t)f_2(t+\tau)} \qquad (13.60)\\
&= R_{f_1}(\tau) + R_{f_1f_2}(\tau) + R_{f_2f_1}(\tau) + R_{f_2}(\tau)
\end{aligned}$$

where

$$R_{f_1 f_2}(\tau) = \overline{f_1(t) f_2(t+\tau)} \tag{13.61}$$

$$R_{f_2 f_1}(\tau) = \overline{f_2(t) f_1(t+\tau)} \tag{13.62}$$

are referred to as cross-correlation functions. They will be zero if $f_1(t)$ and $f_2(t)$ are completely unrelated. If the two processes are in any way related, the cross-correlation functions will not be zero. Since $f_1(t)$ and $f_2(t)$ are stationary, then

$$\begin{aligned} R_{f_1 f_2}(\tau) &= \overline{f_1(t) f_2(t+\tau)} \\ &= \overline{f_1(t-\tau) f_2(t)} \\ &= \overline{f_2(t) f_1(t-\tau)} = R_{f_2 f_1}(-\tau) \end{aligned} \tag{13.63}$$

By taking the Fourier transform of the cross-correlation function of the truncated signals $f_{1T}(t)$ and $f_{2T}(t)$ and letting $T \to \infty$, it can be shown that

$$\frac{1}{2\pi} \int_{-\infty}^{+\infty} R_{f_1 f_2}(\tau) \exp(-\mathrm{i}\omega\tau)\, \mathrm{d}\tau = \lim_{T\to\infty} \frac{\pi}{T} F_{1T}^*(\mathrm{i}\omega) F_{2T}(\mathrm{i}\omega) \tag{13.64}$$

The right-hand side of (13.64) is defined as the cross-spectral density function $S_{f_1 f_2}(\omega)$ and so

$$S_{f_1 f_2}(\omega) = \frac{1}{2\pi} \int_{-\infty}^{+\infty} R_{f_1 f_2}(\tau) \exp(-\mathrm{i}\omega\tau)\, \mathrm{d}\tau \tag{13.65}$$

The inverse relationship is

$$R_{f_1 f_2}(\tau) = \int_{-\infty}^{+\infty} S_{f_1 f_2}(\omega) \exp(\mathrm{i}\omega\tau)\, \mathrm{d}\omega \tag{13.66}$$

The cross-spectral density function is a complex function. Using (13.63) it can be shown that

$$S_{f_1 f_2}(\omega) = S_{f_2 f_1}^*(\omega) \tag{13.67}$$

This analysis can be extended to the case where a structure is subjected to several discrete forces.

When a structure is subjected to distributed forces the random process describing the forcing will be a function of position as well as time. For example, in the case of a one-dimensional structure $f = f(x, t)$.

If the process is stationary and ergodic with a zero mean value, then the essential features of the probability density are described by the cross-correlation between the forces at two points x_1 and x_2, that is

$$R_f(x_1, x_2, \tau) = \overline{f(x_1, t) f(x_2, t+\tau)} \tag{13.68}$$

This function is sometimes known as the space-time correlation function.

The cross-spectral density of the forces at x_1 and x_2 is the Fourier transform of the cross-correlation function and these two functions form a Fourier transform pair and so

$$S_f(x_1, x_2, \omega) = \frac{1}{2\pi} \int_{-\infty}^{+\infty} R_f(x_1, x_2, \tau) \exp(-\mathrm{i}\omega\tau)\, \mathrm{d}\tau \tag{13.69}$$

and

$$R_f(x_1, x_2, \tau) = \int_{-\infty}^{+\infty} S_f(x_1, x_2, \omega) \exp(\mathrm{i}\omega\tau)\, \mathrm{d}\omega \tag{13.70}$$

If the force distribution is weakly homogeneous in space, then the cross-correlation and cross-spectral density functions depend only on the separation of the points and not on their absolute positions, that is

$$R_f(x_1, x_2, \tau) = R_f(x_2 - x_1, \tau) \tag{13.71}$$

and

$$S_f(x_1, x_2, \omega) = S_f(x_2 - x_1, \omega) \tag{13.72}$$

More details of random process analysis can be obtained from references [13.1–13.5].

13.1.2 Response of a Single Degree of Freedom System

The equation of motion of a single degree of freedom system is

$$m\ddot{u} + c\dot{u} + ku = f \tag{13.73}$$

where m, c and k are the mass, damping and stiffness respectively and u the displacement. f is the applied force which is assumed to be a weakly stationary, ergodic process having a Gaussian probability density distribution with a zero mean. The probability density function is, therefore

$$p(f) = \frac{1}{(2\pi)^{1/2}\sigma_f} \exp\left(-f^2/2\sigma_f^2\right) \tag{13.74}$$

The variance or mean square is given by

$$\sigma_f^2 = \int_{-\infty}^{+\infty} S_f(\omega)\, \mathrm{d}\omega \tag{13.75}$$

where $S_f(\omega)$ is the power spectral density function which is given by

$$S_f(\omega) = \lim_{T\to\infty} \frac{\pi}{T} |F_T(\mathrm{i}\omega)|^2 \tag{13.76}$$

where $F_T(\mathrm{i}\omega)$ is the Fourier transform of the truncated function $f_T(t)$.

The response of a linear structure to such an excitation is also weakly stationary, ergodic and has a Gaussian probability density distribution with zero mean [13.6]. This means that the response can be described in a similar manner to the excitation. The first requirement is to calculate the power spectral density of the response which is given by

$$S_u(\omega) = \lim_{T\to\infty} \frac{\pi}{T} |U_T(\mathrm{i}\omega)|^2 \tag{13.77}$$

where $U_T(\mathrm{i}\omega)$ is the Fourier transform of the truncated response function $u_T(t)$.

If both the functions $u(t)$ and $f(t)$ in equation (13.73) are truncated so that they are zero outside the interval $(-T, T)$, then they can be expressed in terms of their

Fourier transforms, namely,

$$u_T(t) = \int_{-\infty}^{+\infty} U_T(\mathrm{i}\omega) \exp(\mathrm{i}\omega t)\, \mathrm{d}\omega \tag{13.78}$$

and

$$f_T(t) = \int_{-\infty}^{+\infty} F_T(\mathrm{i}\omega) \exp(\mathrm{i}\omega t)\, \mathrm{d}\omega \tag{13.79}$$

Substituting (13.78) and (13.79) into (13.73) gives

$$\int_{-\infty}^{+\infty} (-\omega^2 m + \mathrm{i}\omega c + k)\, U_T(\mathrm{i}\omega) \exp(\mathrm{i}\omega t)\, \mathrm{d}\omega = \int_{-\infty}^{+\infty} F_T(\mathrm{i}\omega) \exp(\mathrm{i}\omega t)\, \mathrm{d}\omega \tag{13.80}$$

Since the equality must hold for all values of t, the integrands must be equal so that

$$(k - \omega^2 m + \mathrm{i}\omega c) U_T(\mathrm{i}\omega) = F_T(\mathrm{i}\omega) \tag{13.81}$$

Solving for $U_T(\mathrm{i}\omega)$ gives

$$U_T(\mathrm{i}\omega) = \frac{F_T(\mathrm{i}\omega)}{(k - \omega^2 m + \mathrm{i}\omega c)} = \alpha(\mathrm{i}\omega) F_T(\mathrm{i}\omega) \tag{13.82}$$

where $\alpha(\mathrm{i}\omega)$ is the receptance of the system.

Substituting (13.82) into (13.77) gives

$$S_u(\omega) = |\alpha(\mathrm{i}\omega)|^2 \lim_{T\to\infty} \frac{\pi}{T} |F_T(\mathrm{i}\omega)|^2 \tag{13.83}$$

Equations (13.76) and (13.83) together show that

$$S_u(\omega) = |\alpha(\mathrm{i}\omega)|^2 S_f(\omega) \tag{13.84}$$

The power spectral density of the response is, therefore, equal to the square of the modulus of the receptance multiplied by the power spectral density of the excitation. The mean square response is given by

$$\sigma_u^2 = \int_{-\infty}^{+\infty} S_u(\omega)\, \mathrm{d}\omega \tag{13.85}$$

Substituting for $S_u(\omega)$ from (13.84) gives

$$\sigma_u^2 = \int_{-\infty}^{+\infty} |\alpha(\mathrm{i}\omega)|^2 S_f(\omega)\, \mathrm{d}\omega \tag{13.86}$$

Finally, the probability density function for the response is

$$p(u) = \frac{1}{(2\pi)^{1/2}\sigma_u} \exp\left(-u^2/2\sigma_u^2\right) \tag{13.87}$$

If the excitation is white, then

$$S_f(\omega) = S_0 = \text{constant} \tag{13.88}$$

Figure 13.5. Power spectral density of the response to white noise.

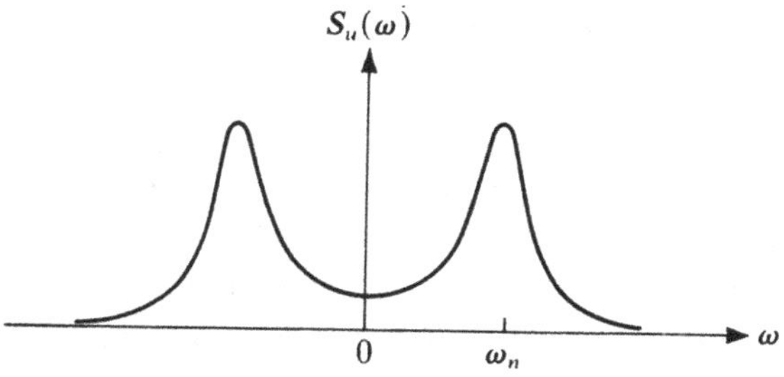

The mean square response is, therefore

$$\sigma_u^2 = S_0 \int_{-\infty}^{+\infty} |\alpha(\mathrm{i}\omega)|^2 \,\mathrm{d}\omega \tag{13.89}$$

Now according to reference [13.5] an integral of the form

$$I = \int_{-\infty}^{+\infty} |H(\omega)|^2 \,\mathrm{d}\omega \tag{13.90}$$

with

$$H(\omega) = \frac{(B_0 + \mathrm{i}\omega B_1)}{(A_0 + \mathrm{i}\omega A_1 - \omega^2 A_2)} \tag{13.91}$$

has the following value

$$I = \frac{\pi \left(A_0 B_1^2 + A_2 B_0^2\right)}{A_0 A_1 A_2} \tag{13.92}$$

Comparing (13.82) and (13.91) it can be seen that

$$\sigma_u^2 = \frac{\pi S_0}{ck} \tag{13.93}$$

The integral (13.89) represents the area under the curve in Figure 13.5. For lightly damped systems the greatest contribution to the area occurs in a relatively narrow frequency band centred on the undamped natural frequency ω_n. This indicates that the predominant frequency components in a sample response, $u(t)$, will be contained in this narrow band. The response envelope of a narrow band system can be expected to show beat characteristics similar to that associated with two harmonics whose frequencies are close together. However, since the predominant frequencies are spread over a narrow band, the beat behaviour is random in character, as shown in Figure 13.6. The response appears as a slightly distorted sine function with a frequency near the natural frequency of the system and with amplitudes that vary slowly in a random fashion.

Figure 13.6. Sample function of a narrow band process.

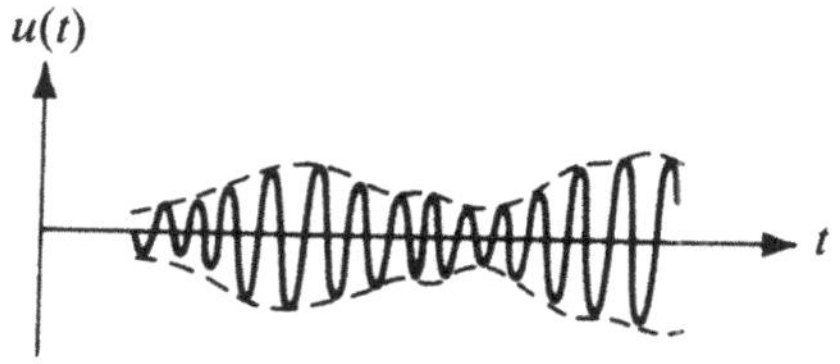

If the excitation is band limited white noise, then the mean square response is

$$\sigma_u^2 = S_0 \int_{-\omega_c}^{\omega_c} |\alpha(\mathrm{i}\omega)|^2 \,\mathrm{d}\omega \tag{13.94}$$

where ω_c is the cut-off frequency. References [13.2, 13.7] give expressions for (13.94), but indicate that its value differs from (13.93) by less than 1%, even for values of ω_c/ω_n as low as 2 and damping ratios as high as 10%.

If $S_f(\omega)$ varies slowly in the vicinity of ω_n, the mean square value for the response is often approximated by

$$\sigma_u^2 = \frac{\pi S_f(\omega_\mathrm{n})}{ck} \tag{13.95}$$

where $S_f(\omega_\mathrm{n})$ is the power spectral density of the excitation at the undamped natural frequency.

13.1.3 Direct Response of a Multi-Degree of Freedom System

The equation of motion of a multi-degree of freedom system is

$$\mathbf{M}\ddot{\mathbf{u}} + \mathbf{C}\dot{\mathbf{u}} + \mathbf{K}\mathbf{u} = \mathbf{f} \tag{13.96}$$

where $\mathbf{M}$, $\mathbf{C}$, $\mathbf{K}$ are the mass, damping and stiffness matrices, $\mathbf{u}$ the displacements and $\mathbf{f}$ the applied forces.

Truncating the time histories of the excitation and response, taking the Fourier transform of (13.96) and solving gives

$$\mathbf{U}_T(\mathrm{i}\omega) = \boldsymbol{\alpha}(\mathrm{i}\omega)\mathbf{F}_T(\mathrm{i}\omega) \tag{13.97}$$

where $\mathbf{U}_T(\mathrm{i}\omega)$ and $\mathbf{F}_T(\mathrm{i}\omega)$ are the Fourier transforms of $\mathbf{u}_T(t)$ and $\mathbf{f}_T(t)$ respectively and

$$\boldsymbol{\alpha}(\mathrm{i}\omega) = [\mathbf{K} - \omega^2\mathbf{M} + \mathrm{i}\omega\mathbf{C}]^{-1} \tag{13.98}$$

is the receptance matrix.

In what fallows the subscript T will be omitted, but it should be remembered that it is always inferred.

The Fourier transform of the response in degree of freedom r, is therefore

$$U_r(\mathrm{i}\omega) = \lfloor\boldsymbol{\alpha}(\mathrm{i}\omega)\rfloor_r \mathbf{F}(\mathrm{i}\omega) \tag{13.99}$$

where $\lfloor\boldsymbol{\alpha}(\mathrm{i}\omega)\rfloor_r$ indicates row r of $\boldsymbol{\alpha}$. Note that degree of freedom r will represent one of the degrees of freedom at a particular node. The power spectral density of the response in degree of freedom r is

$$S_{u_r}(\omega) = \lim_{T\to\infty} \frac{\pi}{T} |U_r(\mathrm{i}\omega)|^2 \tag{13.100}$$

Substituting (13.99) into (13.100) gives

$$S_{u_r}(\omega) = \lfloor\boldsymbol{\alpha}^*(\mathrm{i}\omega)\rfloor_r \lim_{T\to\infty} \frac{\pi}{T} \mathbf{F}^*(\mathrm{i}\omega)\mathbf{F}^\mathrm{T}(\mathrm{i}\omega)\lfloor\boldsymbol{\alpha}(\mathrm{i}\omega)\rfloor_r^\mathrm{T} \tag{13.101}$$

Now

$$\lim_{T\to\infty} \frac{\pi}{T} \mathbf{F}^*(\mathrm{i}\omega)\mathbf{F}^\mathrm{T}(\mathrm{i}\omega) = \mathbf{S}_f(\mathrm{i}\omega) \tag{13.102}$$

is the cross-spectral density matrix of the applied forces. The diagonal terms are the power spectral densities of the individual forces, whilst the off-diagonal terms are the cross-spectral densities of pairs of forces. Substituting (13.102) into (13.101) gives

$$S_{u_r}(\omega) = \lfloor \boldsymbol{\alpha}^*(\mathrm{i}\omega) \rfloor_r \mathbf{S}_f(\mathrm{i}\omega) \lfloor \boldsymbol{\alpha}(\mathrm{i}\omega) \rfloor_r^{\mathrm{T}} \tag{13.103}$$

This is a generalisation of (13.84) to multi-degree of freedom systems.

In the case of a two degree-of-freedom system equation (13.103) gives

$$S_{u_1}(\omega) = \lfloor \boldsymbol{\alpha}_{11}^*(\mathrm{i}\omega) \quad \alpha_{12}^*(\mathrm{i}\omega) \rfloor \mathbf{S}_f(\mathrm{i}\omega) \begin{bmatrix} \alpha_{11}(\mathrm{i}\omega) \\ \alpha_{12}(\mathrm{i}\omega) \end{bmatrix} \tag{13.104}$$

and

$$S_{u_2}(\omega) = \lfloor \boldsymbol{\alpha}_{21}^*(\mathrm{i}\omega) \quad \alpha_{22}^*(\mathrm{i}\omega) \rfloor \mathbf{S}_f(\mathrm{i}\omega) \begin{bmatrix} \alpha_{21}(\mathrm{i}\omega) \\ \alpha_{22}(\mathrm{i}\omega) \end{bmatrix} \tag{13.105}$$

with

$$\mathbf{S}_f(\mathrm{i}\omega) = \begin{bmatrix} S_{f_1 f_1}(\omega) & S_{f_1 f_2}(\mathrm{i}\omega) \\ S_{f_2 f_1}(\mathrm{i}\omega) & S_{f_2 f_2}(\omega) \end{bmatrix} \tag{13.106}$$

If the two forces arise from independent sources, then they will be uncorrelated. This means that

$$R_{f_1 f_2}(\tau) = 0 = R_{f_2 f_1}(\tau) \tag{13.107}$$

Substituting into equation (13.65) gives

$$S_{f_1 f_2}(\mathrm{i}\omega) = 0 = S_{f_2 f_1}(\mathrm{i}\omega) \tag{13.108}$$

Equations (13.104) and (13.105) therefore reduce to

$$S_{u_1}(\omega) = |\alpha_{11}(\mathrm{i}\omega)|^2 S_{f_1 f_1}(\omega) + |\alpha_{12}(\mathrm{i}\omega)|^2 S_{f_2 f_2}(\omega) \tag{13.109}$$

and

$$S_{u_2}(\omega) = |\alpha_{21}(\mathrm{i}\omega)|^2 S_{f_1 f_1}(\omega) + |\alpha_{22}(\mathrm{i}\omega)|^2 S_{f_2 f_2}(\omega) \tag{13.110}$$

The power spectral density of the response is, therefore, equal to the sum of the power spectral densities obtained with the forces acting separately.

If the two forces $f_1(t)$ and $f_2(t)$ are directly related, then

$$f_2(t) = \lambda f_1(t) \tag{13.111}$$

where λ is a constant. This means that

$$R_{f_1 f_2}(\tau) = \lambda R_{f_1 f_1}(\tau) = R_{f_2 f_1}(\tau) \tag{13.112}$$

and

$$R_{f_2 f_2}(\tau) = \lambda^2 R_{f_1 f_1}(\tau) \tag{13.113}$$

Also, from (13.65)

$$S_{f_1 f_2}(\mathrm{i}\omega) = \lambda S_{f_1 f_1}(\omega) = S_{f_2 f_1}(\mathrm{i}\omega) \tag{13.114}$$

and

$$S_{f_2 f_2}(\omega) = \lambda^2 S_{f_1 f_2}(\omega) \tag{13.115}$$

Therefore, equation (13.106) becomes

$$\begin{aligned} \mathbf{S}_f(\mathrm{i}\omega) &= \begin{bmatrix} 1 & \lambda \\ \lambda & \lambda^2 \end{bmatrix} S_{f_1 f_1}(\omega) \\ &= \begin{bmatrix} 1 \\ \lambda \end{bmatrix} \lfloor 1 \quad \lambda \rfloor S_{f_1 f_2}(\omega) \end{aligned} \tag{13.116}$$

Substituting (13.116) into (13.104) and (13.105) gives

$$S_{u_1}(\omega) = |\alpha_{11}(\mathrm{i}\omega) + \alpha_{12}(\mathrm{i}\omega)|^2 S_{f_1 f_1}(\omega) \tag{13.117}$$

and

$$S_{u_2}(\omega) = |\alpha_{21}(\mathrm{i}\omega) + \alpha_{22}(\mathrm{i}\omega)|^2 S_{f_1 f_2}(\omega) \tag{13.118}$$

The power spectral density of the response, in this case, depends upon the vector sum of two receptances.

If a structure is subjected to a distributed pressure distribution, the equivalent nodal force matrix is given by (see Chapter 3)

$$\boldsymbol{f} = \sum_i \mathbf{a}_i^{\mathrm{T}} \mathbf{R}_i^{\mathrm{T}} \int_{A_i} \lfloor \mathbf{N}(\vec{\mathbf{s}}_i) \rfloor^{\mathrm{T}} p(\vec{\mathbf{s}}_i, t)\, \mathrm{d}A_i \tag{13.119}$$

where the summation is over the number of elements. Taking the Fourier transform gives

$$\mathbf{F}(\mathrm{i}\omega) = \sum_i \mathbf{a}_i^{\mathrm{T}} \mathbf{R}_i^{\mathrm{T}} \int_{A_i} \lfloor \mathbf{N}(\vec{\mathbf{s}}_i) \rfloor^{\mathrm{T}} P(\vec{\mathbf{s}}_i, \mathrm{i}\omega)\, \mathrm{d}A_i \tag{13.120}$$

Substituting (13.120) into expression (13.102) for the cross-spectral density matrix of the applied forces gives

$$\mathbf{S}_f(\mathrm{i}\omega) = \sum_i \sum_j \mathbf{a}_i^{\mathrm{T}} \mathbf{R}_i^{\mathrm{T}} \int_{A_i} \int_{A_i} \lfloor \mathbf{N}(\vec{\mathbf{s}}_i) \rfloor^{\mathrm{T}} S_p(\vec{\mathbf{s}}_i, \vec{\mathbf{s}}_j, \mathrm{i}\omega) \lfloor \mathbf{N}(\vec{\mathbf{s}}_j) \rfloor\, \mathrm{d}A_i\, \mathrm{d}A_j\, \mathbf{R}_j \mathbf{a}_j \tag{13.121}$$

where

$$S_p(\vec{\mathbf{s}}_i, \vec{\mathbf{s}}_j, \mathrm{i}\omega) = \lim_{T \to \infty} \frac{\pi}{T} P^*(\vec{\mathbf{s}}_i, \mathrm{i}\omega) P(\vec{\mathbf{s}}_j, \mathrm{i}\omega) \tag{13.122}$$

is the cross-spectral density of the pressure field. In equation (13.121) A_i and A_j represent the surface areas of elements i and j. The power spectral density of the response at a node is again given by (13.103).

Expression (13.121) involves double integrals over pairs of elements and also a double summation over all elements. It has to be evaluated for a range of frequencies. In order to simplify this procedure, reference [13.8] assumes that the cross-spectral density is constant over each pair of elements (which are triangles), the actual value being calculated at the centroids. An improvement in accuracy can be obtained by assuming the cross-spectral density over a pair of elements varies linearly. For example, in the case of a uniform beam [13.9]

$$S_p(x_i, x_j, \mathrm{i}\omega) = e_1 + e_2 x_i + e_3 x_j + e_4 x_i x_j \tag{13.123}$$

The parameters e_1 to e_4 are evaluated in terms of S_p for the four node points of the two elements i and j, taking them two at a time. A similar procedure for triangular elements is given in references [13.10, 13.11]. References [13.8–13.11] are concerned with predicting the response of aircraft structures to jet noise and boundary layer turbulence.

In some engineering applications, the pressure distribution can be assumed to be weakly homogeneous in space. In this case the cross-spectral density of the pressures for two points depends only on the separation of the points and not on their absolute positions, that is

$$S_p(\vec{s}_i, \vec{s}_j, \mathrm{i}\omega) = S_p(\vec{s}_i - \vec{s}_j, \mathrm{i}\omega) \tag{13.124}$$

Examples of such a situation are as follows:

(1) If a randomly distributed pressure field is convected in the x-direction with constant speed U_c, then different points experience the same randomly varying force but with time lags corresponding to their positions. This situation is known as frozen convection and can be considered to be a first approximation for boundary layer turbulence. The cross-spectral density is given by

$$S_p(x_1 - x_2, \mathrm{i}\omega) = S_p(\omega) \exp\{-\mathrm{i}\omega(x_1 - x_2)/U_c\} \tag{13.125}$$

where $S_p(\omega)$ is the power spectral density at all points in the field.

(2) A better approximation to boundary layer turbulence is a convected field with statistical decay. In this case the cross-spectral density is given by

$$S_p(x_1 - x_2, \mathrm{i}\omega) = S_p(\omega) \exp\left\{-\alpha \frac{\omega}{U_c} |x_1 - x_2| - \frac{\mathrm{i}\omega}{U_c}(x_1 - x_2)\right\} \tag{13.126}$$

where α is a boundary layer decay parameter. When $\alpha = 0$ (13.126) reduces to (13.125).

The function (13.124) can be represented in the spatial domain by means of its Fourier transform, namely

$$S_p(\vec{s}_i - \vec{s}_j, \mathrm{i}\omega) = \int S_p(\vec{k}, \omega) \exp\{\mathrm{i}\vec{k} \cdot (\vec{s}_i - \vec{s}_j)\} \, \mathrm{d}\vec{k} \tag{13.127}$$

where the integration is over the entire range of $\vec{k}$. The vector $\vec{k}$ is known as a wavenumber vector. The function $S_p(\vec{k}, \omega)$ is the wavenumber/frequency spectrum of the pressure field. This function has been derived for boundary layer pressure fluctuations in reference [13.12]. A survey of various models of atmospheric turbulence is given in reference [13.13].

Substituting (13.124) and (13.127) into (13.121) gives, after a little manipulation

$$\mathbf{S}_f(\mathrm{i}\omega) = \int \tilde{\mathbf{f}}^*(\vec{k}) \mathbf{S}_p(\vec{k}, \omega) \tilde{\mathbf{f}}^{\mathrm{T}}(\vec{k}) \, \mathrm{d}\vec{k} \tag{13.128}$$

where

$$\tilde{\mathbf{f}}(\vec{k}) = \sum_i \mathbf{a}_i^{\mathrm{T}} \mathbf{R}_i^{\mathrm{T}} \int_{A_i} \lfloor \mathbf{N}(\vec{s}_i) \rfloor^{\mathrm{T}} \exp(-\mathrm{i}\vec{k} \cdot \vec{s}_i) \,\mathrm{d}A_i \tag{13.129}$$

This last expression represents the equivalent nodal forces due to an harmonic pressure wave of unit amplitude, frequency ω and wavenumber k travelling in the direction of $\vec{k}$. Although equation (13.128) is much simpler to evaluate than equation (13.121), it is necessary to perform the calculations for a range of wavenumbers in order to perform the integration.

Substituting (13.128) into (13.103) gives the power spectral density of the response in the form

$$\begin{aligned} S_{u_r}(\omega) &= \int \alpha_r^*(\vec{k}, \omega) S_p(\vec{k}, \omega) \alpha_r(\vec{k}, \omega) \,\mathrm{d}\vec{k} \\ &= \int |\alpha_r(\vec{k}, \omega)|^2 S_p(\vec{k}, \omega) \,\mathrm{d}\vec{k} \end{aligned} \tag{13.130}$$

where

$$\alpha_r(\vec{k}, \omega) = \lfloor \boldsymbol{\alpha}(\mathrm{i}\omega) \rfloor_r \tilde{\mathbf{f}}(\vec{k}) \tag{13.131}$$

This function represents the response of the structure in degree of freedom r due to the harmonic pressure wave

$$p(\vec{s}, t) = \exp\{\mathbf{i}(\omega t - \vec{k} \cdot \vec{s})\} \tag{13.132}$$

and is known as a wave receptance function.

13.1.4 Modal Response of a Multi-Degree of Freedom System

In Chapter 12 it is shown that when the modal method is used the receptance matrix is given by (see equation (12.34))

$$\boldsymbol{\alpha}(\mathrm{i}\omega) = \boldsymbol{\Phi}[\boldsymbol{\Lambda} - \omega^2 \mathbf{I} + \mathrm{i}\omega \bar{\mathbf{C}}]^{-1} \boldsymbol{\Phi}^{\mathrm{T}} \tag{13.133}$$

where the columns of $\boldsymbol{\Phi}$ are the modes of free vibration of the undamped system, $\boldsymbol{\Lambda}$ is a diagonal matrix containing the squares of the natural frequencies and $\bar{\mathbf{C}}$ the modal damping matrix. Substituting (13.133) into (13.103) gives the following expression, for the power spectral density of the response in degree of freedom r

$$S_{u_r} = \lfloor \boldsymbol{\Phi} \rfloor_r [\boldsymbol{\Lambda} - \omega^2 \mathbf{I} - \mathrm{i}\omega \bar{\mathbf{C}}]^{-1} \mathbf{J} [\boldsymbol{\Lambda} - \omega^2 \mathbf{I} + \mathrm{i}\omega \bar{\mathbf{C}}]^{-1} \lfloor \boldsymbol{\Phi} \rfloor_r^{\mathrm{T}} \tag{13.134}$$

where

$$\mathbf{J} = \boldsymbol{\Phi} \mathbf{S}_f \boldsymbol{\Phi}^{\mathrm{T}} \tag{13.135}$$

is the cross-spectral density matrix of the generalised forces in the various modes. In some texts [13.3] non-dimensional forms of the elements of $\mathbf{J}$ are referred to as joint acceptances.

If the natural frequencies are well separated and the damping is small, then terms of the form

$$(\Lambda_i - \omega^2 - \mathrm{i}\omega \bar{c}_{ii})^{-1} J_{ij} (\Lambda_j - \omega^2 + \mathrm{i}\omega \bar{c}_{jj})^{-1}$$

for $j \neq i$, are negligible in comparison with terms of the form

$$(\Lambda_i - \omega^2 - i\omega\bar{c}_{ii})^{-1} J_{ii} (\Lambda_i - \omega^2 + i\omega\bar{c}_{ii})^{-1}$$

These two expressions assume the matrix $\bar{\mathbf{C}}$ is diagonal. Using this approximation (13.134) is reduced to

$$S_{u_r} = \sum_{i=1}^{N} \phi_{ri}^2 \{(\Lambda_i - \omega^2)^2 + (\omega\bar{c}_{ii})^2\}^{-2} J_{ii} \tag{13.136}$$

where N is the number of modes. This expression can also be obtained by omitting the off-diagonal terms of the matrix $\mathbf{J}$ in (13.134).

Now the matrix $\mathbf{S}_f$ is Hermitian. This means that

$$S_f^* = S_f^{\mathrm{T}} \tag{13.137}$$

This can easily be seen from the definition of $\mathbf{S}_f$ in (13.102). Substituting (13.137) into (13.135) shows that the matrix $\mathbf{J}$ is also Hermitian. This means that the diagonal terms J_{ii} are real. Also equation (13.136) gives a real value for $\mathbf{S}_{u_r}$, which is to be expected. Because the J_{ii} are real they can be calculated from

$$J_{ii} = \lfloor \boldsymbol{\phi} \rfloor_i \mathbf{C}_f \lfloor \boldsymbol{\phi} \rfloor_i^{\mathrm{T}} \tag{13.138}$$

where $\mathbf{C}_f$ is the real part of $\mathbf{S}_f$ and is known as the co-spectrum matrix.

Equation (13.136) is a good approximation in regions close to the natural frequencies. Away from these regions some inaccuracy can be permitted since the magnitudes are much smaller.

The question of frequency separation has been investigated in reference [13.14]. The results presented indicate that in any analysis, it would be better to retain the coupling between all modes, whose frequencies are within a certain band, centred around the frequency being considered. Therefore, in equation (13.134) the actual modes retained would vary with frequency. This approach has been used in references [13.8, 13.10, 13.11].

In the case of a weakly homogeneous pressure field, the power spectral density of the response is given by equation (13.130). Substituting (13.133) into (13.131) gives the following expressions for the wave receptance function

$$\alpha_r(\vec{k}, \omega) = \lfloor \boldsymbol{\Phi} \rfloor_r [\boldsymbol{\Lambda} - \omega^2 \mathbf{I} + i\omega\bar{\mathbf{C}}]^{-1} \boldsymbol{\Phi}^{\mathrm{T}} \tilde{\mathbf{f}}(\vec{k}) \tag{13.139}$$

Substituting (13.139) into (13.130) gives an expression of the form (13.134) with $\mathbf{J}$ given by (13.135) where

$$\mathbf{S}_f = \int \tilde{\mathbf{f}}^*(\vec{k})\, S_p(\vec{k}, \omega) \tilde{\mathbf{f}}^{\mathrm{T}}(\vec{k})\, \mathrm{d}\vec{k} \tag{13.140}$$

13.2 Truncation of the Modal Solution

Section 12.1 shows that when the modal method of solution is used the displacements, $\mathbf{u}$, are given by

$$\mathbf{u} = \boldsymbol{\Phi} \mathbf{q}(t) \tag{13.141}$$

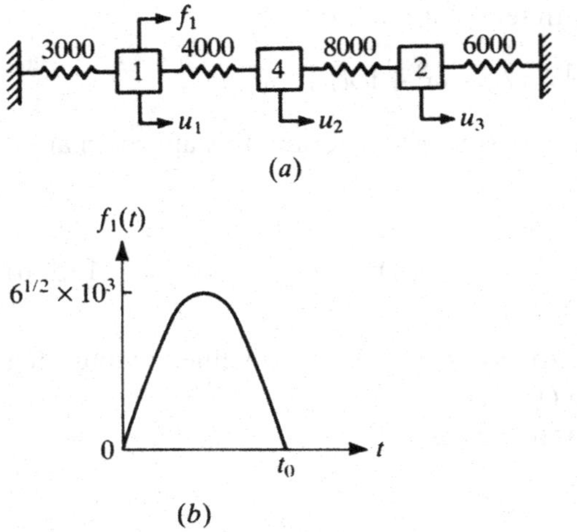

Figure 13.7. (a) Three degree of freedom system, (b) time history of applied force.

where $\mathbf{q}$ is the solution of

$$\ddot{\mathbf{q}} + \bar{\mathbf{C}}\dot{\mathbf{q}} + \boldsymbol{\Lambda}\mathbf{q} = \mathbf{Q} \tag{13.142}$$

The columns of $\boldsymbol{\Phi}$ are the modes of free vibration of the undamped system, whilst $\boldsymbol{\Lambda}$ is a diagonal matrix whose elements are the squares of the natural frequencies. $\mathbf{Q}$ is a column matrix of generalised modal forces and $\bar{\mathbf{C}}$ the modal damping matrix.

When using this method to calculate the maximum response whether it be displacement, velocity, acceleration or stress, it will be found that only some of the modes will contribute significantly. It is, therefore, more efficient to retain only these significant modes in the solution process. This will reduce the computational cost of determining the frequencies and modes of free vibration as well as that of the forced response calculations.

The modes which are to be retained in the analysis are determined by considering

(1) the spatial distribution and time history of the excitation,
(2) the characteristics of the structure, that is, the mass, damping and stiffness, and
(3) the response quantity of interest, that is, displacement, velocity, acceleration or stress.

Some of these factors will be illustrated by means of a simple example.

Figure 13.7 shows a three degree of freedom system which is subject to a half-sine pulse applied to one of the masses. Solution of the equations of free vibration gives

$$\boldsymbol{\Lambda} = \begin{bmatrix} 1000 & & \\ & 7000 & \\ & & 9000 \end{bmatrix} \tag{13.143}$$

and

$$\mathbf{\Phi} = \begin{bmatrix} 3^{1/2}/6 & 6^{1/2}/3 & 2^{1/2}/2 \\ 3^{1/2}/4 & 0 & -2^{1/2}/4 \\ 3^{1/2}/6 & -6^{1/2}/6 & 2^{1/2}/2 \end{bmatrix} \tag{13.144}$$

The generalised modal forces are

$$\mathbf{Q} = \mathbf{\Phi}^{\mathrm{T}} \begin{bmatrix} 6^{1/2} \times 10^3 \\ 0 \\ 0 \end{bmatrix} \sin(\pi t/t_0)$$

$$= \begin{bmatrix} (2^{1/2}/2) \times 10^3 \\ 2 \times 10^3 \\ 3^{1/2} \times 10^3 \end{bmatrix} \sin(\pi t/t_0) \tag{13.145}$$

Therefore, equations (13.142) are

$$\begin{aligned} \ddot{q}_1 + 1000\, q_1 &= (2^{1/2}/2) \times 10^3 \sin(\pi t/t_0) \\ \ddot{q}_2 + 7000\, q_2 &= 2 \times 10^3 \sin(\pi t/t_0) \\ \ddot{q}_3 + 9000\, q_3 &= 3^{1/2} \times 10^3 \sin(\pi t/t_0) \end{aligned} \tag{13.146}$$

If the motion is assumed to start from rest, then at $t = 0$

$$\mathbf{u} = 0, \qquad \dot{\mathbf{u}} = 0 \tag{13.147}$$

Now

$$\mathbf{u} = \mathbf{\Phi q} \tag{13.148}$$

and

$$\mathbf{\Phi}^{\mathrm{T}}\mathbf{M\Phi} = \mathbf{I} \tag{13.149}$$

Premultiplying (13.148) by $\mathbf{\Phi}^{\mathrm{T}}\mathbf{M}$ gives

$$\mathbf{q} = \mathbf{\Phi}^{\mathrm{T}}\mathbf{Mu} \tag{13.150}$$

Therefore, at $t = 0$

$$\mathbf{q} = 0, \qquad \dot{\mathbf{q}} = 0 \tag{13.151}$$

The solutions of equations (13.146) subject to the boundary conditions (13.151) can be obtained using Table 12.7. The solution for the displacements, **u**, are then obtained by substituting into equation (13.141). The response will be investigated for $t_0 = 0.12$ and $t_0 = 0.07$.

When $t_0 = 0.12$ the maximum displacement response in the excitation era and free vibration era are of similar magnitude. This is also true for the displacement u_1 when $t_0 = 0.07$. The maximum values of the displacements u_2 and u_3 occur in the free vibration era when $t_0 = 0.07$. Therefore, for brevity, the expressions for the q_r will only be given for the free vibration era, $t \geq t_0$.

When $t_0 = 0.12$, the solution for $t \geq 0.12$ is

$$\begin{aligned} q_1 &= -1.8607(\sin\omega_1(t-t_0) + \sin\omega_1 t) \\ q_2 &= -0.099107(\sin\omega_2(t-t_0) + \sin\omega_2 t) \\ q_3 &= -0.057487(\sin\omega_3(t-t_0) + \sin\omega_3 t) \end{aligned} \tag{13.152}$$

where

$$\begin{aligned} \omega_1^2 &= 1000 \\ \omega_2^2 &= 7000 \\ \omega_3^2 &= 9000 \end{aligned} \tag{13.153}$$

Equations (13.152) may be written in the alternative form

$$\begin{aligned} q_1 &= 1.1938\sin(\omega_1 t - \alpha_1) \\ q_2 &= -0.060007\sin(\omega_2 t - \alpha_2) \\ q_3 &= -0.095467\sin(\omega_3 t - \alpha_3) \end{aligned} \tag{13.154}$$

where $\alpha_r = \omega_r t_0/2$.

The response is dominated by the first mode and so the contributions from modes 2 and 3 can be neglected. If, however, acceleration is the response quantity of interest, then

$$\begin{aligned} \ddot{q}_1 &= -1193.8\sin(\omega_1 t - \alpha_1) \\ \ddot{q}_2 &= 420.05\sin(\omega_2 t - \alpha_2) \\ \ddot{q}_3 &= 859.20\sin(\omega_3 t - \alpha_3) \end{aligned} \tag{13.155}$$

and the contributions from modes 2 and 3 cannot be neglected.

When $t_0 = 0.07$, the solution for $t \geq 0.07$ is

$$\begin{aligned} q_1 &= 0.88565\sin(\omega_1 t - \alpha_1) \\ q_2 &= 0.42060\sin(\omega_2 t - \alpha_2) \\ q_3 &= 0.23085\sin(\omega_3 t - \alpha_3) \end{aligned} \tag{13.156}$$

In this case the contributions from all three modes are of comparable magnitude.

The reason why one mode is dominant when $t_0 = 0.12$ and why all three contribute to the response when $t_0 = 0.07$ can be seen by examining the frequency content of the excitation. If

$$f(t) = \sin(\pi t/t_0) \tag{13.157}$$

then the modulus of its Fourier transform is [13.15]

$$|F(\omega)| = \frac{t_0}{\pi^2}\left|\frac{\cos(\omega t_0/2)}{1 - 4(\omega t_0/2\pi)^2}\right| \qquad \text{for } \omega \neq \pi/t_0$$

and

$$|F(\omega)| = t_0/4\pi \qquad \text{for } \omega = \pi/t_0 \tag{13.158}$$

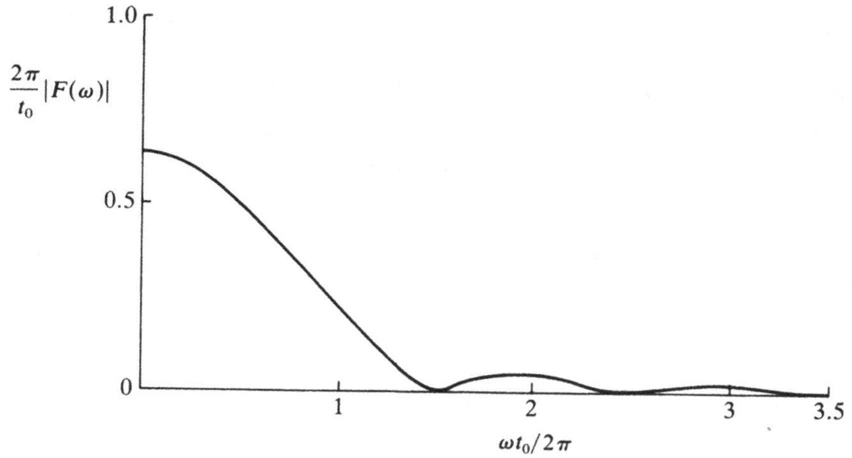

Figure 13.8. Modulus of the Fourier transform of a half-sine pulse.

The variation of $2\pi|F(\omega)|/t_0$ with $\omega t_0/2\pi$ is shown in Figure 13.8. Evaluating $2\pi|F(\omega)|$ at the natural frequencies of the system gives 0.05, 0.003, 0.005 for $t_0 = 0.12$ and 0.04, 0.02, 0.01 for $t_0 = 0.07$. The second and third values are less than one tenth of the first value when $t_0 = 0.12$. However, when $t_0 = 0.07$ all three values are of similar magnitude.

Inspection of the modal matrix $\mathbf{\Phi}$ (equation (13.144)) indicates that the second mode does not contribute to the response of the central mass, since its displacement is zero in this mode. Also, if the excitation is applied to the central mass only, the second mode does not contribute to the response of any of the masses, since the generalised force in this mode is zero.

13.2.1 Mode Acceleration Method

When the response is calculated using the first N_R low frequency modes, where N_R is less than the total number of modes N, then equation (13.141) becomes

$$\mathbf{u} = \mathbf{\Phi}_R \mathbf{q}_R \tag{13.159}$$

where $\mathbf{\Phi}_R$ and $\mathbf{q}_R$ are of order $(N \times N_R)$ and $(N_R \times 1)$ respectively. $\mathbf{q}_R$ is given by the solution of

$$\ddot{\mathbf{q}}_R + \bar{\mathbf{C}}_R \dot{\mathbf{q}}_R + \mathbf{\Lambda}_R \mathbf{q}_R = \mathbf{Q}_R \tag{13.160}$$

where $\bar{\mathbf{C}}_R$ and $\mathbf{\Lambda}_R$ are both of order $(N_R \times N_R)$ and $\mathbf{Q}_R$ is of order $(N_R \times 1)$.

Rearranging (13.160) gives

$$\mathbf{q}_R + \mathbf{\Lambda}_R^{-1} \mathbf{Q}_R - \mathbf{\Lambda}_R^{-1} [\bar{\mathbf{C}}_R \dot{\mathbf{q}}_R + \ddot{\mathbf{q}}_R] \tag{13.161}$$

Substituting (13.161) into (13.159) gives

$$\mathbf{u} = \mathbf{\Phi}_R \mathbf{\Lambda}_R^{-1} \mathbf{Q}_R - \mathbf{\Phi}_R \mathbf{\Lambda}_R^{-1} [\bar{\mathbf{C}}_R \dot{\mathbf{q}}_R + \ddot{\mathbf{q}}_R] \tag{13.162}$$

If the external forces are static instead of dynamic, the second and third terms on the right-hand side of equation (13.162) will be zero. In order to get an accurate solution for the displacements, all the modes will have to be retained in the first term which becomes

$$\boldsymbol{\Phi}\boldsymbol{\Lambda}^{-1}\mathbf{Q} = \boldsymbol{\Phi}\boldsymbol{\Lambda}^{-1}\boldsymbol{\Phi}^{\mathrm{T}}\mathbf{f} \tag{13.163}$$

where $\mathbf{f}$ is the column matrix of externally applied forces.

The product $\boldsymbol{\Phi}\boldsymbol{\Lambda}^{-1}\boldsymbol{\Phi}^{\mathrm{T}}$ can be expressed in an alternative form. From Section 12.1

$$\boldsymbol{\Phi}^{\mathrm{T}}\mathbf{K}\boldsymbol{\Phi} = \boldsymbol{\Lambda} \tag{13.164}$$

Premultiplying by $\boldsymbol{\Lambda}^{-1}$ gives

$$\boldsymbol{\Lambda}^{-1}\boldsymbol{\Phi}^{\mathrm{T}}\mathbf{K}\boldsymbol{\Phi} = \mathbf{I} \tag{13.165}$$

Postmultiplying by $\mathbf{M}\boldsymbol{\Phi}^{\mathrm{T}}$ gives

$$\boldsymbol{\Lambda}^{-1}\boldsymbol{\Phi}^{\mathrm{T}}\mathbf{K}\boldsymbol{\Phi}\mathbf{M}\boldsymbol{\Phi}^{\mathrm{T}} = \mathbf{M}\boldsymbol{\Phi}^{\mathrm{T}} \tag{13.166}$$

Now

$$\boldsymbol{\Phi}\mathbf{M}\boldsymbol{\Phi}^{\mathrm{T}} = \mathbf{I} \tag{13.167}$$

and so (13.166) becomes

$$\boldsymbol{\Lambda}^{-1}\boldsymbol{\Phi}^{\mathrm{T}}\mathbf{K} = \mathbf{M}\boldsymbol{\Phi}^{\mathrm{T}} \tag{13.168}$$

Premultiplying by $\boldsymbol{\Phi}$ gives

$$\boldsymbol{\Phi}\boldsymbol{\Lambda}^{-1}\boldsymbol{\Phi}^{\mathrm{T}}\mathbf{K} = \mathbf{I} \tag{13.169}$$

Finally, postmultiplying by $\mathbf{K}^{-1}$ gives

$$\boldsymbol{\Phi}\boldsymbol{\Lambda}^{-1}\boldsymbol{\Phi}^{\mathrm{T}} = \mathbf{K}^{-1} \tag{13.170}$$

Substituting (13.170) into (13.163) gives

$$\boldsymbol{\Phi}\boldsymbol{\Lambda}^{-1}\mathbf{Q} = \mathbf{K}^{-1}\mathbf{f} \tag{13.171}$$

as would be expected for a static force.

A more accurate expression for the dynamic response is obtained by including all the modes in the first term on the right-hand side of equation (13.162). Using the form (13.171) in (13.162) gives

$$\mathbf{u} = \mathbf{K}^{-1}\mathbf{f} - \boldsymbol{\phi}_{\mathrm{R}}\boldsymbol{\Lambda}_{\mathrm{R}}^{-1}[\bar{\mathbf{C}}_{\mathrm{R}}\dot{\mathbf{q}}_{\mathrm{R}} + \ddot{\mathbf{q}}_{\mathrm{R}}] \tag{13.172}$$

This is the expression used in the mode acceleration method. Increased convergence is obtained because of the presence of $\boldsymbol{\Lambda}_{\mathbf{R}}^{-1}$ in the second and third terms. This means that fewer modes are required for a given accuracy. This is particularly important when calculating stresses, since accurate stress solutions usually require more modes than do accurate displacement solutions. The convergence of the mode acceleration method is investigated in reference [13.16].

To illustrate the use of the method, consider the determination of the response of the third mass within the interval $t \leq t_0$, when $t_0 = 0.12$, for the example in Figure 13.7. Using three modes gives

$$u_3 = 0.66681 \sin \omega t - 0.53714 \sin \omega_1 t + 0.04046 \sin \omega_2 t - 0.04065 \sin \omega_3 t \qquad (13.173)$$

where $\omega = \pi / t_0$.

Using only one mode gives

$$u_3 = 0.64881 \sin \omega t - 0.53714 \sin \omega_1 t \qquad (13.174)$$

Now if one mode is used in the mode acceleration method, the result is

$$u_3 = 0.66825 \sin \omega t - 0.53714 \sin \omega_1 t \qquad (13.175)$$

The first term in (13.174) is in error by -2.7% whilst the same term in (13.175) is in error by $+0.22\%$. Further applications may be found in reference [13.16].

13.2.2 Residual Flexibility

Equation (13.172) can be expressed in an alternative form as follows. From equation (13.160)

$$[\bar{\mathbf{C}}_R \dot{\mathbf{q}}_R + \ddot{\mathbf{q}}_R] = \mathbf{\Phi}_R^T \mathbf{f} - \mathbf{\Lambda}_R \mathbf{q}_R \qquad (13.176)$$

Premultiply (13.176) by $\mathbf{\Phi}_R \mathbf{\Lambda}_R^{-1}$ gives

$$\mathbf{\Phi}_R \mathbf{\Lambda}_R^{-1} [\bar{\mathbf{C}}_R \dot{\mathbf{q}}_R + \ddot{\mathbf{q}}_R] = \mathbf{\Phi}_R \mathbf{\Lambda}_R^{-1} \mathbf{\Phi}_R^T \mathbf{f} - \mathbf{\Phi}_R \mathbf{q}_R \qquad (13.177)$$

Substituting (13.177) into (13.172) gives

$$\mathbf{u} = \mathbf{\Phi}_R \mathbf{q}_R + \mathbf{K}^{-1} \mathbf{f} - \mathbf{\Phi}_R \mathbf{\Lambda}_R^{-1} \mathbf{\Phi}_R^T \mathbf{f} \qquad (13.178)$$

The second term has been shown to represent the static contributions from all the modes. Similarly, the third term represents the static contributions from the retained modes. Combining these two terms gives the static contributions from the omitted modes and is referred to as the residual flexibility. The inclusion of the residual flexibility of the omitted modes increases the accuracy of the solution. Note that equation (13.178) only involves the retained modes. The static effect of the omitted modes is obtained without the need to determine them. An alternative way of deriving equation (13.178) is to include all the modes in the solution and then approximate the contributions from the high frequency modes, which are to be omitted, by their values at zero frequency [13.17].

Reference [13.18] presents an analysis of the response of a simple off-shore platform model to harmonic wave loading. Three analyses are carried out, namely:

(1) a solution using all the modes,
(2) a solution using a reduced number of modes and
(3) a solution using the same number of modes as in (2), but including the residual flexibility of the omitted modes.

It is shown that while solution (2) gives good results for displacements, bending moments are not as well predicted. However, when residual flexibility effects are included, there is a marked improvement in the accuracy of the bending moments.

13.3 Ritz Vector Analysis

Another method of reducing the number of degrees of freedom of a system is to use Ritz vectors. This technique has the advantage that it does not require the solution of a linear eigenproblem to determine the modes of free vibration of the undamped system. Instead, the columns of the matrix $\mathbf{\Phi}$ in equation (12.2) are generated by means of a recurrence relationship.

Separating out the time variation $g(t)$ of the equivalent nodal forces, equation (12.1) becomes

$$\mathbf{M\ddot{u}} + \mathbf{C\dot{u}} + \mathbf{Ku} = \mathbf{f}g(t) \tag{13.179}$$

The solution of this equation is expressed in the form

$$\mathbf{u} = \mathbf{\Phi q}(t) \tag{13.180}$$

where $\mathbf{\Phi}$ is a matrix of Ritz vectors. The first column of $\mathbf{\Phi}$ is obtained by solving the equation

$$\mathbf{K}\bar{\mathbf{\Phi}}_1 = \mathbf{f} \tag{13.181}$$

for $\bar{\mathbf{\Phi}}_1$. This amounts to finding the static response to the load vector $\mathbf{f}$. The next step is to mass-orthogonalise $\bar{\mathbf{\Phi}}_1$ that is

$$\mathbf{\Phi}_1 = \bar{\mathbf{\Phi}}_1/(\bar{\mathbf{\Phi}}_1^{\mathrm{T}}\mathbf{M}\bar{\mathbf{\Phi}}_1)^{1/2} \tag{13.182}$$

so that

$$\mathbf{\Phi}_{11}^{\mathrm{T}}\mathbf{M}\mathbf{\Phi}_1 = 1 \tag{13.183}$$

Additional vectors are generated as follows. Solve

$$\mathbf{K}\bar{\Phi}_i = \mathbf{M}\mathbf{\Phi}_{i-1} \tag{13.184}$$

for $\bar{\mathbf{\Phi}}_i$ (i = 2,..., N_R) where N_R is the number of required vectors. To ensure that $\bar{\mathbf{\Phi}}_i$ is mass-orthogonal to $\mathbf{\Phi}_1, \ldots, \mathbf{\Phi}_{i-1}$, replace it with $\bar{\bar{\mathbf{\Phi}}}_i$, where

$$\bar{\bar{\mathbf{\Phi}}}_i = \bar{\mathbf{\Phi}}_i - \sum_{j=1}^{i-1} \alpha_j \mathbf{\Phi}_j \tag{13.185}$$

Premultiplying by $\mathbf{\Phi}_k^{\mathrm{T}}\mathbf{M}$ gives

$$\begin{aligned} \mathbf{\Phi}_k^{\mathrm{T}}\mathbf{M}\bar{\bar{\mathbf{\Phi}}}_i &= \mathbf{\Phi}_k^{\mathrm{T}}\mathbf{M}\bar{\mathbf{\Phi}}_i - \sum_{j=1}^{i-1} \alpha_j \mathbf{\Phi}_k^{\mathrm{T}}\mathbf{M}\mathbf{\Phi}_j \\ &= \mathbf{\Phi}_k^{\mathrm{T}}\mathbf{M}\bar{\mathbf{\Phi}}_i - \alpha_k \end{aligned} \tag{13.186}$$

since

$$\mathbf{\Phi}_k^{\mathrm{T}}\mathbf{M}\mathbf{\Phi}_j = \begin{cases} 0 & j \neq k \\ 1 & j = k \end{cases} \tag{13.187}$$

and so

$$\mathbf{\Phi}_k^{\mathrm{T}}\mathbf{M}\bar{\bar{\mathbf{\Phi}}}_j = 0 \tag{13.188}$$

provided

$$\alpha_k = \mathbf{\Phi}_k^{\mathrm{T}}\mathbf{M}\bar{\mathbf{\Phi}}_i \tag{13.189}$$

Substituting equation (13.189) into equation (13.185) gives

$$\bar{\bar{\mathbf{\Phi}}}_i = \bar{\mathbf{\Phi}}_i - \sum_{j=1}^{i-1}\left(\mathbf{\Phi}_j^{\mathrm{T}}\mathbf{M}\bar{\mathbf{\Phi}}_i\right)\mathbf{\Phi}_j \tag{13.190}$$

Finally, $\bar{\bar{\mathbf{\Phi}}}_i$ is mass-orthogonalised to give $\mathbf{\Phi}_i$, where

$$\mathbf{\Phi}_i = \bar{\bar{\mathbf{\Phi}}}_i/\left(\bar{\bar{\mathbf{\Phi}}}_i^{\mathrm{T}}\mathbf{M}\bar{\bar{\mathbf{\Phi}}}_i\right)^{1/2} \tag{13.191}$$

Equation (13.181) can be solved using the decomposition $\mathbf{K} = \mathbf{L}\,\mathbf{L}^{\mathrm{T}}$, where $\mathbf{L}$ is a lower triangular matrix. The elements of $\mathbf{L}$ can be determined using Cholesky's symmetric decomposition (see Section 11.1.2). The elements of $\bar{\mathbf{\Phi}}_1$ can then be obtained by a forward substitution followed by a backward substitution. Once $\mathbf{K}$ has been decomposed in this manner it can also be used to solve equation (13.184) at each step. Alternatively, the decomposition $\mathbf{K} = \mathbf{L}\,\mathbf{D}\,\mathbf{L}^{\mathrm{T}}$ can be used where $\mathbf{L}$ is now a lower triangular matrix with unit values on the diagonal and $\mathbf{D}$ is a diagonal matrix [13.19]. This method of generating Ritz vectors was first presented in reference [13.20]. Alternative methods using Krylov and Lanczos vectors are described in reference [13.22].

Following Section 12.1, the transformation (13.180) is substituted into the energy expressions (12.3) and use made of Lagrange's equations to give the following reduced set of equations

$$\bar{\mathbf{M}}\ddot{\mathbf{q}} + \bar{\mathbf{C}}\dot{\mathbf{q}} + \bar{\mathbf{K}}\mathbf{q} = \bar{\mathbf{Q}}g(t) \tag{13.192}$$

where $\bar{\mathbf{M}}$, $\bar{\mathbf{C}}$, $\bar{\mathbf{K}}$ and $\bar{\mathbf{Q}}$ are given by equation (12.5). However, in this case $\bar{\mathbf{M}}$ is the only matrix which is diagonal because of the way the columns of $\mathbf{\Phi}$ are generated, and is in fact a unit diagonal matrix.

Equation (13.192) can now be solved by either direct or modal methods as previously described. Reference [13.20] illustrates the fact that the superposition of Ritz vectors yields more accurate solutions with fewer vectors than the modal method of solution. A proof of the convergence of the Ritz vector method is presented in reference [13.23].

Among the advantages of the Ritz vector method are:

(1) They can be automatically generated with a fraction of the numerical effort required for the calculation of the exact mode shapes.
(2) The number of vectors is less than the number of eigenvectors required in a conventional mode superposition analysis.
(3) The vectors are generated by taking into account the spatial distribution of the dynamic loading whereas the direct use of the mode shapes neglects this very important information.

13.4 Response to Imposed Displacements

Previous sections deal with the response of structures to externally applied forces. In some cases the excitation is caused by prescribed displacements at the supports or boundaries of the structure. Examples of this are the response of buildings to earthquakes and the response of vehicles travelling over rough ground surfaces. Methods of analysing this type of problem are presented in this section.

13.4.1 Direct Response

The equation of motion of a structure is given by equation (12.1), namely

$$\mathbf{M}\ddot{\mathbf{u}} + \mathbf{C}\dot{\mathbf{u}} + \mathbf{K}\mathbf{u} = \mathbf{f} \tag{13.193}$$

The column matrix of displacements, $\mathbf{u}$, can be partitioned into prescribed boundary displacements, $\mathbf{u}_\mathrm{B}$, and the remaining internal displacements, $\mathbf{u}_\mathrm{I}$, giving

$$\mathbf{u} = \begin{bmatrix} \mathbf{u}_\mathrm{I} \\ \mathbf{u}_\mathrm{B} \end{bmatrix} \tag{13.194}$$

Partitioning equation (13.193) in a similar manner gives

$$\begin{bmatrix} \mathbf{M}_\mathrm{II} & \mathbf{M}_\mathrm{IB} \\ \mathbf{M}_\mathrm{BI} & \mathbf{M}_\mathrm{BB} \end{bmatrix} \begin{bmatrix} \ddot{\mathbf{u}}_\mathrm{I} \\ \ddot{\mathbf{u}}_\mathrm{B} \end{bmatrix} + \begin{bmatrix} \mathbf{C}_\mathrm{II} & \mathbf{C}_\mathrm{IB} \\ \mathbf{C}_\mathrm{BI} & \mathbf{C}_\mathrm{BB} \end{bmatrix} \begin{bmatrix} \dot{\mathbf{u}}_\mathrm{I} \\ \dot{\mathbf{u}}_\mathrm{B} \end{bmatrix} + \begin{bmatrix} \mathbf{K}_\mathrm{II} & \mathbf{K}_\mathrm{IB} \\ \mathbf{K}_\mathrm{BI} & \mathbf{K}_\mathrm{BB} \end{bmatrix} \begin{bmatrix} \mathbf{u}_\mathrm{I} \\ \mathbf{u}_\mathrm{B} \end{bmatrix} = \begin{bmatrix} \mathbf{O} \\ \mathbf{f}_\mathrm{B} \end{bmatrix} \tag{13.195}$$

Note that the applied forces corresponding to internal node points are zero. The forces $\mathbf{f}_\mathrm{B}$ are unknown support forces at the nodes where the motion is prescribed.

Separating out the equations of motion corresponding to the internal node points gives

$$\begin{bmatrix} \mathbf{M}_\mathrm{II} & \mathbf{M}_\mathrm{IB} \end{bmatrix} \begin{bmatrix} \ddot{\mathbf{u}}_\mathrm{I} \\ \ddot{\mathbf{u}}_\mathrm{B} \end{bmatrix} + \begin{bmatrix} \mathbf{C}_\mathrm{II} & \mathbf{C}_\mathrm{IB} \end{bmatrix} \begin{bmatrix} \dot{\mathbf{u}}_\mathrm{I} \\ \dot{\mathbf{u}}_\mathrm{B} \end{bmatrix} + \begin{bmatrix} \mathbf{K}_\mathrm{II} & \mathbf{K}_\mathrm{IB} \end{bmatrix} \begin{bmatrix} \mathbf{u}_\mathrm{I} \\ \mathbf{u}_\mathrm{B} \end{bmatrix} = [\mathbf{O}] \tag{13.196}$$

Now the total displacement at the internal nodes can be considered to be made up of two separate parts. The first, $\mathbf{u}_\mathrm{I}^\mathrm{s}$, is due to the prescribed boundary motions and can be calculated using static relationships. Static equilibrium at the internal nodes is given by

$$\mathbf{K}_\mathrm{II}\mathbf{u}_\mathrm{I}^\mathrm{s} + \mathbf{K}_\mathrm{IB}\mathbf{u}_\mathrm{B} = 0 \tag{13.197}$$

Solving this equation gives

$$\mathbf{u}_\mathrm{I}^\mathrm{s} = -\mathbf{K}_\mathrm{II}^{-1}\mathbf{K}_\mathrm{IB}\mathbf{u}_\mathrm{B} \tag{13.198}$$

$\mathbf{u}_\mathrm{I}^\mathrm{s}$ are referred to as quasi-static displacements. The second part consists of dynamic displacements, $\mathbf{u}_\mathrm{I}^\mathrm{d}$ relative to fixed boundary nodes. Therefore

$$\mathbf{u}_\mathrm{I} = \mathbf{u}_\mathrm{I}^\mathrm{s} + \mathbf{u}_\mathrm{I}^\mathrm{d} \tag{13.199}$$

Substituting (13.199) into (13.196) gives, after some manipulation

$$\mathbf{M}_\mathrm{II}\ddot{\mathbf{u}}_\mathrm{I}^\mathrm{d} + \mathbf{C}_\mathrm{II}\dot{\mathbf{u}}_\mathrm{I}^\mathrm{d} + \mathbf{K}_\mathrm{II}\mathbf{u}_\mathrm{I}^\mathrm{d} = \mathbf{f}_\mathrm{eff} \tag{13.200}$$

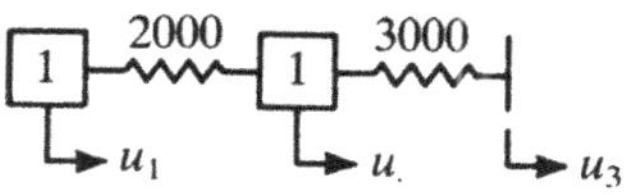

Figure 13.9. Mass-spring system subject to imposed displacements.

where $\mathbf{f}_{\text{eff}}$ is an effective force which is given by

$$\mathbf{f}_{\text{eff}} = -\mathbf{M}_{\text{II}}\ddot{\mathbf{u}}_{\text{I}}^{\text{s}} - \mathbf{M}_{\text{IB}}\ddot{\mathbf{u}}_{\text{B}} - \mathbf{C}_{\text{II}}\dot{\mathbf{u}}_{\text{I}}^{\text{s}} - \mathbf{C}_{\text{IB}}\dot{\mathbf{u}}_{\text{B}} \tag{13.201}$$

Substituting (13.198) into (13.201) gives

$$\mathbf{f}_{\text{eff}} = [\mathbf{M}_{\text{II}}\mathbf{K}_{\text{II}}^{-1}\mathbf{K}_{\text{IB}} - \mathbf{M}_{\text{IB}}]\ddot{\mathbf{u}}_{\text{B}} + [\mathbf{C}_{\text{II}}\mathbf{K}_{\text{II}}^{-1}\mathbf{K}_{\text{IB}} - \mathbf{C}_{\text{IB}}]\dot{\mathbf{u}}_{\text{B}} \tag{13.202}$$

Since $\ddot{\mathbf{u}}_{\text{B}}$ and $\dot{\mathbf{u}}_{\text{B}}$ are given, $\mathbf{f}_{\text{eff}}$ can be calculated. Once the time history of the effective force has been calculated using (13.202), the dynamic response relative to the fixed boundary nodes can be determined using equation (13.200). The total displacement is then given by equation (13.199).

In general, the damping terms in (13.202) are much smaller than the inertia terms and can be neglected. Also, if a lumped mass representation is used $\mathbf{M}_{\text{IB}}$ is zero. The expression for the effective force reduces to

$$\mathbf{f}_{\text{eff}} = \mathbf{M}_{\text{II}}\mathbf{K}_{\text{II}}^{-1}\mathbf{K}_{\text{IB}}\ddot{\mathbf{u}}_{\text{B}} \tag{13.203}$$

Consider the system shown in Figure 13.9 which is subject to the prescribed displacement u_3. The inertia and stiffness matrices are

$$\mathbf{M} = \begin{bmatrix} 1 & 0 & 0 \\ 0 & 1 & 0 \\ 0 & 0 & 0 \end{bmatrix}, \qquad \mathbf{K} = 10^3 \begin{bmatrix} 2 & -2 & 0 \\ -2 & 5 & -3 \\ 0 & -3 & 3 \end{bmatrix} \tag{13.204}$$

The required partitions are

$$\mathbf{M}_{\text{II}} = \begin{bmatrix} 1 & 0 \\ 0 & 1 \end{bmatrix}, \qquad \mathbf{K}_{\text{II}} = 10^3 = \begin{bmatrix} 2 & -2 \\ -2 & 5 \end{bmatrix} \tag{13.205}$$

$$\mathbf{K}_{\text{IB}} = 10^3 \begin{bmatrix} 0 \\ -3 \end{bmatrix} \tag{13.206}$$

Substituting into (13.203) gives the effective force

$$\mathbf{f}_{\text{eff}} = \begin{bmatrix} 1 & 0 \\ 0 & 1 \end{bmatrix} \frac{1}{6} \begin{bmatrix} 5 & 2 \\ 2 & 2 \end{bmatrix} \begin{bmatrix} 0 \\ -3 \end{bmatrix} \ddot{u}_3 \tag{13.207}$$

That is

$$\mathbf{f}_{\text{eff}} = \begin{bmatrix} -1 \\ -1 \end{bmatrix} \ddot{u}_3 \tag{13.208}$$

Equation (13.200) is, therefore,

$$\begin{bmatrix} \ddot{u}_1^{\text{d}} \\ \ddot{u}_2^{\text{d}} \end{bmatrix} + 10^3 \begin{bmatrix} 2 & -2 \\ -2 & 5 \end{bmatrix} \begin{bmatrix} u_1^{\text{d}} \\ u_2^{\text{d}} \end{bmatrix} = \begin{bmatrix} -1 \\ -1 \end{bmatrix} \ddot{u}_3 \tag{13.209}$$

The quasi-static displacements are given by

$$\begin{bmatrix} u_1^s \\ u_2^s \end{bmatrix} = \frac{-1}{6}\begin{bmatrix} 5 & 2 \\ 2 & 2 \end{bmatrix}\begin{bmatrix} 0 \\ -3 \end{bmatrix} u_3 = \begin{bmatrix} 1 \\ 1 \end{bmatrix} u_3 \tag{13.210}$$

If the time variation of u_3 is harmonic with frequency ω, then equation (13.209) becomes

$$\begin{bmatrix} (2000-\omega^2) & -2000 \\ -2000 & (5000-\omega^2) \end{bmatrix}\begin{bmatrix} u_1^d \\ u_2^d \end{bmatrix} = \begin{bmatrix} 1 \\ 1 \end{bmatrix}\omega^2 u_3 \tag{13.211}$$

The solution of this equation is

$$\begin{bmatrix} u_1^d \\ u_2^d \end{bmatrix} = (1000-\omega^2)^{-1}(6000-\omega^2)^{-1}\begin{bmatrix} (7000-\omega^2) \\ (4000-\omega^2) \end{bmatrix}\omega^2 u_3 \tag{13.212}$$

Substituting (13.210) and (13.212) into (13.199) gives

$$u_1 = \frac{6\times 10^6}{(1000-\omega^2)(6000-\omega^2)}u_3$$

$$u_2 = \frac{3000(2000-\omega^2)}{(1000-\omega^2)(6000-\omega^2)}u_3 \tag{13.213}$$

13.4.2 Modal Response

A modal solution can be carried out using the modes of free vibration of the structure with the constraints $\mathbf{u}_B = 0$ applied. These are obtained by solving the equation

$$[\mathbf{K}_{II} - \omega^2\mathbf{M}_{II}]\mathbf{u}_I = 0 \tag{13.214}$$

to give the matrix of eigenvalues, $\mathbf{\Lambda}_c$, and eigenvectors $\mathbf{\Phi}_c$ such that

$$\mathbf{\Phi}_c^T\mathbf{M}_{II}\mathbf{\Phi}_c = \mathbf{I}, \qquad \mathbf{\Phi}_c^T\mathbf{K}_{II}\mathbf{\Phi}_c = \mathbf{\Lambda}_c \tag{13.215}$$

These modes can be used to transform equation (13.200) using the relationship

$$\mathbf{u}_I^d = \mathbf{\Phi}_c\mathbf{q} \tag{13.216}$$

Substituting (13.216) into (13.200) and premultiplying by $\mathbf{\Phi}_c^T$ gives

$$\ddot{\mathbf{q}} + \mathbf{\Phi}_c^T\mathbf{C}_{II}\mathbf{\Phi}_c\dot{\mathbf{q}} + \mathbf{\Lambda}_c\mathbf{q} = \mathbf{\Phi}_c^T\mathbf{f}_{eff} \tag{13.217}$$

Substituting (13.202) into (13.217) gives

$$\begin{aligned}\ddot{\mathbf{q}} + \mathbf{\Phi}_c^T\mathbf{C}_{II}\mathbf{\Phi}_c\dot{\mathbf{q}} + \mathbf{\Lambda}_c\mathbf{q} &= \mathbf{\Phi}_c^T[\mathbf{M}_{II}\mathbf{K}_{II}^{-1}\mathbf{K}_{IB} - \mathbf{M}_{IB}]\ddot{\mathbf{u}}_B \\ &\quad + \mathbf{\Phi}_c^T[\mathbf{C}_{II}\mathbf{K}_{II}^{-1}\mathbf{K}_{IB} - \mathbf{C}_{IB}]\dot{\mathbf{u}}_B\end{aligned} \tag{13.218}$$

When applied to the above example, this equation gives

$$\begin{bmatrix} \ddot{q}_1 \\ \ddot{q}_2 \end{bmatrix} + \begin{bmatrix} 1000 & 0 \\ 0 & 6000 \end{bmatrix}\begin{bmatrix} q_1 \\ q_2 \end{bmatrix} = \frac{1}{5^{1/2}}\begin{bmatrix} -3 \\ 1 \end{bmatrix}\ddot{u}_3 \tag{13.219}$$

Figure 13.10. Idealisation of a simply supported beam subject to a point force.

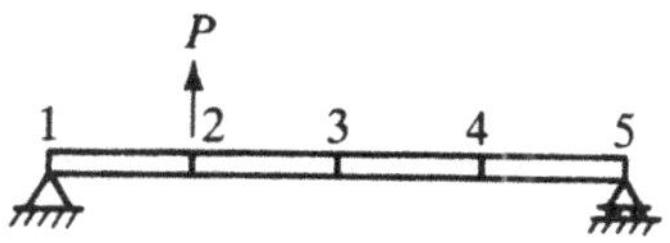

If the time variation of u_3 is harmonic with frequency ω, then equation (13.219) becomes

$$\begin{bmatrix} (1000-\omega^2) & 0 \\ 0 & (6000-\omega^2) \end{bmatrix} \begin{bmatrix} q_1 \\ q_2 \end{bmatrix} = \frac{1}{5^{1/2}} \begin{bmatrix} +3 \\ -1 \end{bmatrix} \omega^2 u_3 \qquad (13.220)$$

The solution of this equation is

$$\begin{bmatrix} q_1 \\ q_2 \end{bmatrix} = \frac{1}{5^{1/2}} (1000-\omega^2)^{-1} (6000-\omega^2)^{-1} \begin{bmatrix} 3(6000-\omega^2) \\ -(1000-\omega^2) \end{bmatrix} \omega^2 u_3 \qquad (13.221)$$

Substituting (13.221) into (13.216) gives

$$\begin{bmatrix} u_1^{\mathrm{d}} \\ u_2^{\mathrm{d}} \end{bmatrix} = (1000-\omega^2)^{-1} (6000-\omega^2)^{-1} \begin{bmatrix} (7000-\omega^2) \\ (4000-\omega^2) \end{bmatrix} \omega^2 u_3 \qquad (13.222)$$

which agrees with (13.212). This has then to be substituted into (13.199) together with (13.210) to give the complete solution (13.213).

13.5 Reducing the Number of Degrees of Freedom

Section 11.5 describes various techniques for reducing the number of degrees of freedom in the case of free vibration. This section indicates how these same techniques can be applied when determining the forced response of the structure.

13.5.1 Making Use of Symmetry

If a structure and its boundary conditions exhibit either an axis or plane of symmetry, then the response can be calculated by idealising only half the structure, even when the applied loads are non-symmetric. To illustrate this, consider the simply supported beam shown in Figure 11.5 with a single force at node 2. This configuration is shown in Figure 13.10.

Figure 13.11 shows how the load P at node 2 in Figure 13.10 can be represented by the sum of symmetric and antisymmetric loads $P/2$ at nodes 2 and 4. The solution of configuration 13.11(a) can be obtained by considering half the structure (nodes 1 to 3) subject to the load $P/2$ at node 2 with symmetric boundary conditions applied at node 3. Similarly the solution of configuration 13.11(b) can be obtained by repeating the analysis with antisymmetric boundary conditions applied at node 3.

If $\mathbf{u}_{\mathrm{S}}$ and $\mathbf{u}_{\mathrm{A}}$ are the responses for configurations 13.11(a) and (b) respectively, then the total response for the half structure 1 to 3 will be

$$\mathbf{u} = \mathbf{u}_{\mathrm{S}} + \mathbf{u}_{\mathrm{A}} \qquad (13.223)$$

To obtain the total response for the half structure 3 to 5 the responses $\mathbf{u}_{\mathrm{S}}$ and $\mathbf{u}_{\mathrm{A}}$ are first reflected in the axis of symmetry through node 3, to give $\mathbf{u}_{\mathrm{S}}^{\mathrm{R}}$ and $\mathbf{u}_{\mathrm{A}}^{\mathrm{R}}$. If the

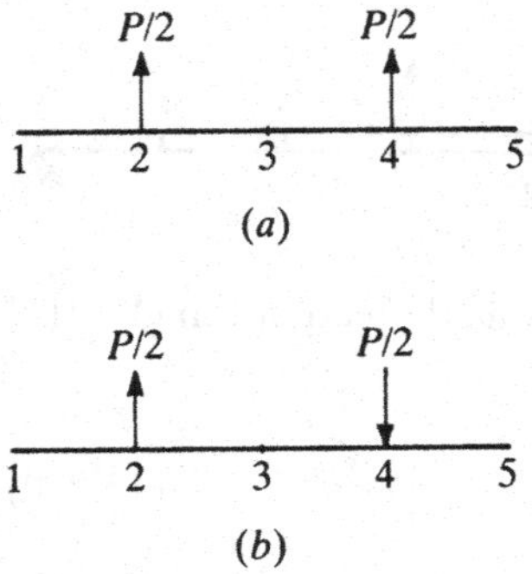

Figure 13.11. Representation of a non-symmetric load by the sum of (a) a symmetric loading and (b) an antisymmetric loading.

nodal degrees of freedom are $(v,\ \theta_z)$ then this operation consists of putting

$$\begin{aligned} v_{(6-i)} &= v_i \\ \theta_{z(6-i)} &= -\theta_{zi} \end{aligned} \qquad i = 1, 2 \tag{13.224}$$

The total response is given by

$$\mathbf{u} = \mathbf{u}_{\mathrm{S}}^{\mathrm{R}} - \mathbf{u}_{\mathrm{A}}^{\mathrm{R}} \tag{13.225}$$

If the loading extends across the axis of symmetry, then the loading over each half structure is treated in the above manner and the responses added. If the modal method of solution is used then equation (12.8) is solved twice, once with the symmetric modes and once with the antisymmetric modes. The techniques to be used when there are two and three planes of symmetry are presented in reference [13.23].

13.5.2 Rotationally Periodic Structures

The response of a rotationally periodic structure to a distribution of harmonic forces can be obtained by carrying out a series of response calculations for just one component. To this end the force distribution is expressed in terms of a number of propagating waves. Thus the equivalent nodal forces due to the external loading on the component r are expressed in the form

$$\{\mathbf{f}^{\mathrm{e}}\}^r \exp(\mathrm{i}\omega t) = \sum_{p=1}^{N} \{\mathbf{A}\}_p \exp[\mathrm{i}\{\omega t - 2\pi(r-1)p/N\}] \tag{13.226}$$

where N is the number of components. This expression indicates that the forces on the first component are

$$\{\mathbf{f}^{\mathrm{e}}\}^r \exp(\mathrm{i}\omega t) = \sum_{p=1}^{N} \{\mathbf{A}\}_p \exp(\mathrm{i}\omega t) \tag{13.227}$$

Also, the forces on adjacent components have the same magnitude but with a phase difference of $2\pi p/N$.

In order to determine the $\{\mathbf{A}\}_p$, equation (13.226) is written for all the components in a single matrix expression

$$[\mathbf{f}^{\mathrm{e}}] = [\mathbf{A}][\boldsymbol{\phi}] \tag{13.228}$$

where the rth column of $[\mathbf{f}^e]$ is $\{\mathbf{f}\}^r$, the pth column of $[\mathbf{A}]$ is $\{\mathbf{A}\}_p$ and element (p, r) of $[\boldsymbol{\phi}]$ is $\exp[-i2\pi(r-1)p/N]$. $[\boldsymbol{\phi}]$ is a square matrix of order N and so (13.228) can be solved for $[\mathbf{A}]$ giving

$$[\mathbf{A}] = [\mathbf{f}^e][\boldsymbol{\phi}]^{-1} \tag{13.229}$$

Reference (13.24) shows that

$$[\boldsymbol{\phi}]^{-1} = [\boldsymbol{\phi}]^H \tag{13.230}$$

and so element (p, r) of $[\boldsymbol{\phi}]^{-1}$ is $\exp\{i2\pi r(p-1)/N\}$.

The response of the structure to each of the propagating waves in (13.226) is determined separately. As one of the force waves propagates round the structure, it will induce a similar displacement wave in the structure having the same frequency and phase variation.

The equation of motion for component r is, on omitting the factor $\exp(i\omega t)$

$$[\mathbf{K}^r - \omega^2\mathbf{M}^r + i\omega\mathbf{C}^r]\{\mathbf{u}\}^r = \{\mathbf{f}\}^r \tag{13.231}$$

The nodal displacements and forces are partitioned into those corresponding to nodes on the left- and right-hand boundaries and all other nodes as in (11.69) and (11.71). The nodal forces consist of forces due to the externally applied loading and also boundary forces due to the motion of adjacent components. They may, therefore, be written in the form

$$\{\mathbf{f}\}^r = \begin{bmatrix} \mathbf{f}_L^e \\ \mathbf{f}_I^e \\ \mathbf{f}_R^e \end{bmatrix}^r + \begin{bmatrix} \mathbf{f}_L^b \\ 0 \\ \mathbf{f}_R^b \end{bmatrix}^r \tag{13.232}$$

where the superscripts e and b denotes external and boundary loading.

Equation (13.226) indicates that

$$\begin{bmatrix} \mathbf{f}_L^e \\ \mathbf{f}_I^e \\ \mathbf{f}_R^e \end{bmatrix}^r = \begin{bmatrix} \mathbf{A}_L \\ \mathbf{A}_I \\ \exp(-i\varepsilon)\mathbf{A}_L \end{bmatrix}_p^r \tag{13.233}$$

where $\varepsilon = 2\pi p/N$. Following the analysis given in Section 11.5.2 shows that

$$\begin{bmatrix} \mathbf{f}_L^b \\ 0 \\ \mathbf{f}_R^b \end{bmatrix}^r = \begin{bmatrix} \mathbf{f}_L^b \\ 0 \\ -\exp(-i\varepsilon)\mathbf{f}_L^b \end{bmatrix}^r \tag{13.234}$$

and

$$\{\mathbf{u}\}^r = \mathbf{w}\begin{bmatrix} \mathbf{u}_L \\ \mathbf{u}_I \end{bmatrix}^r \tag{13.235}$$

where $\mathbf{w}$ is defined by (11.74) with μ replaced by ε.

When expressions (13.232) to (13.235) are substituted into (13.231), the unknown boundary forces $\{\mathbf{f}_L^b\}^r$ can be eliminated by premultiplying by $\mathbf{w}^H$ (see equation (11.76)). This results in the equation

$$[\mathbf{K}^r(\varepsilon) - \omega^2\mathbf{M}^r(\varepsilon) + i\omega\mathbf{C}^r(\varepsilon)]\begin{bmatrix}\mathbf{u}_L\\ \mathbf{u}_I\end{bmatrix}^r = \begin{bmatrix}2\mathbf{A}_L\\ \mathbf{A}_I\end{bmatrix}_p^r \tag{13.236}$$

where $\mathbf{K}^r(\varepsilon)$ and $\mathbf{M}^r(\varepsilon)$ are defined in (11.78). Also

$$\mathbf{C}^r(\varepsilon) = \mathbf{w}^H\mathbf{C}^r\mathbf{w} \tag{13.237}$$

Equation (13.236) is solved N times corresponding to the N values of $\varepsilon = 2\pi p/N$ and forces $\{\mathbf{A}\}_p$ $(p = 1, 2, \ldots, N)$. Each time the complete set of displacements on the component is obtained using (13.235). If this analysis is carried out for the first component and the N solutions assembled into a matrix $[\mathbf{u}]$ having N columns, then the displacements on the whole structure due to the complete loading are given by

$$[\mathbf{U}] = [\mathbf{u}][\boldsymbol{\phi}] \tag{13.238}$$

where the rth column of $[\mathbf{U}]$ gives the displacements of component r. Further details can be found in reference [13.24].

13.5.3 Elimination of Unwanted Degrees of Freedom

In Section 11.5.3 it is shown that the full set of degrees of freedom can be related to a reduced set of master degrees of freedom by means of the relation

$$\mathbf{u} = \mathbf{R}\mathbf{u}_m \tag{13.239}$$

where $\mathbf{R}$ is defined by (11.95).

Substitituting (13.239) into the energy expressions (12.3) and using Lagrange's equations gives

$$\mathbf{M}^R\ddot{\mathbf{u}}_m + \mathbf{C}^R\dot{\mathbf{u}}_m + \mathbf{K}^R\mathbf{u}_m = \mathbf{f}^R \tag{13.240}$$

where $\mathbf{M}^R$ and $\mathbf{K}^R$ are defined in (11.98) and

$$\mathbf{C}^R = \mathbf{R}^T\mathbf{C}\mathbf{R}, \qquad \mathbf{f}^R = \mathbf{R}^T\mathbf{f} \tag{13.241}$$

Reference [13.25] investigates the effect of the use of this technique on the response of plates. Fuller details on reduction techniques for forced response calculations are given in reference [13.27].

13.5.4 Component Mode Synthesis

In Section 11.5.4 three methods of component mode synthesis for free vibration are presented. The first uses fixed interface modes, whilst the other two use free interface modes. The application of these techniques to forced response is illustrated in this section by considering the second of the two methods which use free interface modes. The configuration considered is illustrated in Figure 11.13.

The kinetic and strain energies of the two substructures are given by equations (11.117), (11.128), (11.129) and (11.130). The strain energy of the connectors is given by (11.146). The dissipation function for a single substructure is of the form

$$D_s = \tfrac{1}{2}\{\dot{\mathbf{u}}\}_s^T[\mathbf{C}]_s\{\dot{\mathbf{u}}\}_s \tag{13.242}$$

where $[\mathbf{C}]_s$ is the damping matrix for the substructure. Introducing the transformation (11.122) gives

$$D_s = \tfrac{1}{2}\{\dot{\mathbf{q}}_N\}_s^T[\bar{\mathbf{C}}]_s\{\dot{\mathbf{q}}_N\}_s \tag{13.243}$$

where

$$[\bar{\mathbf{C}}]_s = [\boldsymbol{\phi}_N]^T[\mathbf{C}]_s[\boldsymbol{\phi}_N] \tag{13.244}$$

Adding the contributions from the two substructures gives

$$D = \tfrac{1}{2}\{\dot{\mathbf{q}}\}^T[\mathbf{C}]\{\dot{\mathbf{q}}\} \tag{13.245}$$

where

$$[\mathbf{C}] = \begin{bmatrix} \bar{\mathbf{C}}_I & \mathbf{0} \\ \mathbf{0} & \bar{\mathbf{C}}_{II} \end{bmatrix} \tag{13.246}$$

and $\{\mathbf{q}\}$ is defined by (11.128).

The dissipation function for the connectors is

$$D_c = \frac{1}{2}\begin{bmatrix} \dot{\mathbf{u}}_B^I \\ \dot{\mathbf{u}}_B^{II} \end{bmatrix}^T [\mathbf{C}_c] \begin{bmatrix} \dot{\mathbf{u}}_B^I \\ \dot{\mathbf{u}}_B^{II} \end{bmatrix} \tag{13.247}$$

Introducing the transformation (13.128) into (13.247) gives

$$D_c = \tfrac{1}{2}\{\dot{\mathbf{q}}\}^T[\boldsymbol{\Phi}_B]^T[\mathbf{C}_c][\boldsymbol{\Phi}_B]\{\mathbf{q}\} \tag{13.248}$$

The dissipation function for the complete system is, therefore,

$$D_T = \tfrac{1}{2}\{\dot{\mathbf{q}}\}^T[[\mathbf{C}] + [\boldsymbol{\Phi}_B]^T[\mathbf{C}_c][\boldsymbol{\Phi}_B]]\{\dot{\mathbf{q}}\} \tag{13.249}$$

The Virtual work done by the forces, $\{\mathbf{f}\}_s$, applied to a single substructure is

$$\delta W_s = \{\delta\mathbf{u}\}_s^T\{\mathbf{f}\}_s \tag{13.250}$$

Introducing the transformation (11.122) gives

$$\delta W_s = \{\delta\mathbf{q}_N\}_s^T\{\mathbf{Q}\}_s \tag{13.251}$$

where

$$\{\mathbf{Q}\}_s = [\boldsymbol{\phi}_N]^T\{\mathbf{f}\}_s \tag{13.252}$$

The virtual work done by the applied forces for the complete system is

$$\delta W_T = \{\delta\mathbf{q}\}^T\begin{bmatrix} \mathbf{Q}_I \\ \mathbf{Q}_{II} \end{bmatrix} \tag{2.253}$$

The equation of motion of the complete structure is, therefore

$$\{\ddot{\mathbf{q}}\} + [[\mathbf{C}] + [\boldsymbol{\Phi}_B]^T[\mathbf{C}_c][\boldsymbol{\Phi}_B]]\{\dot{\mathbf{q}}\} + [[\mathbf{K}] + [\boldsymbol{\Phi}_B]^T[\mathbf{K}_c][\boldsymbol{\Phi}_B]]\{\mathbf{q}\} = \begin{bmatrix} \mathbf{Q}_I \\ \mathbf{Q}_{II} \end{bmatrix} \tag{13.254}$$

Although the matrix [**C**] can be made to be a diagonal matrix by an appropriate choice for the matrices $[\mathbf{C}]_s$, as described in Section 12.2, the product $[\boldsymbol{\Phi}_B]^T[\mathbf{C}_c][\boldsymbol{\Phi}_B]$ will not be diagonal.

As well as reducing the number of degrees of freedom, this method has the advantage that components having widely differing damping characteristics can easily be represented.

Problems

13.1 Calculate the Ritz vectors for the three degree of freedom system shown in Figure 13.10.

14 Computer Analysis Techniques

A new user of finite element analysis is unlikely to start writing a computer program. The reason for this is that there is a large number of general purpose finite element programs which can be obtained commercially. All are available on a wide range of powerful desktop computers. There is also an increasing number available for running on personal computers. These tend to be a subset of the desktop version. They can be used to analyse small scale structures and also prepare the input data for large scale structures which are to be analysed on a powerful desktop.

This chapter assumes that the reader intends to use one of these commercial programs. Details of programming aspects can be found in references [14.1–14.6]. Some of the problems at the ends of earlier chapters require the use of a finite element program as indicated. Those readers who do not have such a program available can use the program MATLAB [14.7].

Different programs can give different solutions to the same problem. This is illustrated in Figure 14.1, which shows the variation of the frequency of the first bending mode of a twisted cantilever plate, as a function of the angle of twist. Analyses were carried out by sixteen different establishments using both finite element and analytical methods. Figure 14.1 represents a subset of the results presented in reference [14.8]. Plots 1 and 2 were obtained using a triangular facet-shell element, the first with a consistent mass matrix and the second with a lumped mass matrix. Plots 3 and 4 were both computed with quadrilateral elements, the first being a facet shell and the second a doubly curved shell. Both used a lumped mass representation. Plot 5 was obtained using a super-parametric thick shell element with a consistent mass matrix. Plots 6 and 7 were computed using eight and sixteen node isoparametric solid elements, having lumped and consistent mass matrices respectively.

The various analyses are in good agreement for zero angle of twist and the predicted non-dimensional frequency is close to the measured one.

However, as the angle of twist increases, the predicted results diverge and only three of them are reasonably close to the measured values.

References [14.9, 14.10] present the results of a separate survey. In this case the structure consisted of a cylinder and an I-section beam with variable cross-section, which were connected by two relatively soft springs. The twelve participants were

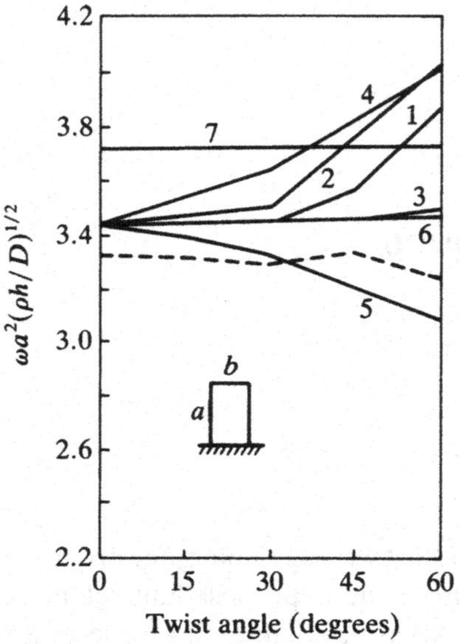

Figure 14.1. Frequencies of a twisted plate, as a function of twist angle for first bending mode $a/b = 3$, $b/h = 20$. —— finite element, - - - experimental.

asked to predict various quantities which included all natural frequencies below 2500 Hz, direct and transfer frequency response functions within the frequency range 1–2500 Hz, and the transient response to two separate impulses.

The number of modes predicted having frequencies less than 2500 Hz varied from 8 to 42. However, comparison of the lowest nine non-zero frequencies showed that the mean predicted values compared reasonably closely with measured values.

There was a certain amount of agreement between the frequency response functions in the range 1–150 Hz. However, there was considerable disagreement at higher frequencies. Some of this is due to the disagreement in the number of modes in the range considered. The transient response calculations produced even greater scatter. Not only did the time histories disagree, the maximum responses differed by several orders of magnitude.

Both surveys showed that not only did different programs produce different results, but also different analysts using the same program for the same problem can produce different results. Possible causes of this are different choices of idealisation and/or element types and incorrect data. The accuracy of computed results is, therefore, a function of the experience of the user as well as the accuracy of the program. It is essential that users of finite element programs be well trained in both the finite element method and the use of the program to be used.

14.1 Program Format

A typical finite element analysis consists of three phases, as indicated in Figure 14.2. The pre-processing phase consists of specifying and checking the input data. This is followed by the solution phase in which the analysis is carried out. The final phase, which is known as the post-processing phase, is concerned with the interpretation and presentation of the results of the analysis.

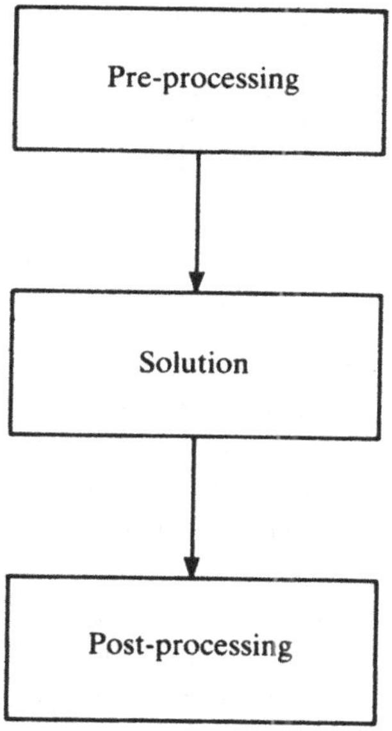

Figure 14.2. The three phases of analysis.

14.1.1 Pre-processing

In the pre-processing phase the following input data is prepared for a free vibration analysis:

(1) Element types
(2) Element geometric properties
(3) Element material properties
(4) Nodal coordinates
(5) Element definition
(6) Boundary conditions
(7) Multi-point constraints
(8) Master degrees of freedom
(9) Analysis options

Items (1) to (5) are required to define the idealisation. This information is used to calculate the element inertia and stiffness matrices and the assembly of these into the inertia and stiffness matrices for the complete structure.

The specification of the element types consists of defining which elements in the program element library are to be used in the idealisation. These may be beams, plates, shells or solids or a combination of them. The element definition consists of specifying the node numbers for each element in the mesh. This information, together with the nodal coordinates, is used to calculate the area or volume of each element and the orientation of its local axes, if required. Certain geometric properties of elements, such as the area and second moment of area of the cross-section of a beam and the thickness of a plate, cannot be calculated from the nodal coordinates and so have to be input separately. These are referred to as element geometric properties. The material properties, such as Young's modulus and density, are also

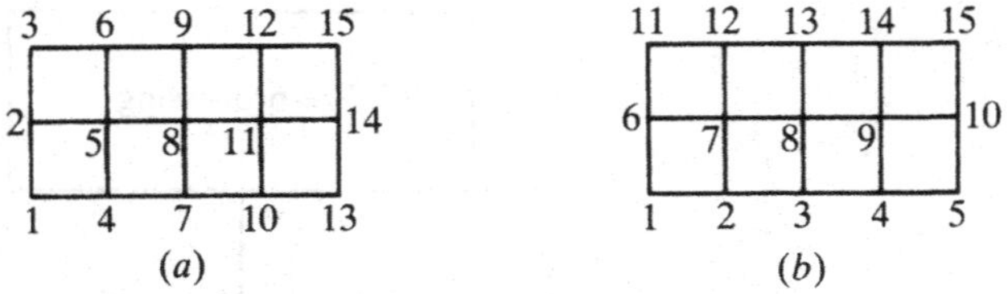

Figure 14.3. Node numbering schemes.

required for each element. Sufficient information has now been defined for the calculation of the element inertia and stiffness matrices and their assembly into the inertia and stiffness matrices for the complete structure.

There are various ways of inputting the nodal coordinates and element definition. The first is by direct definition. This consists of specifying the coordinates, in an appropriate coordinate system, of each node in turn and then the node numbers for each element in the mesh. This process can be quite tedious. Alternatively, the geometry of the structure can be defined by means of a solid modeller program and the finite element mesh created automatically. It may be that the geometry has been defined using a CAD (computer aided design) system. In this case the model may contain more detail than is required for finite element idealisation (see Section 14.2). These should be removed retaining only the detail which would have been produced by a solid modeller.

The numbering of the nodes and elements depends upon the solution method incorporated into the program to be used. In some, the node numbering is important and the element numbering unimportant, whilst in others the reverse situation holds.

Node numbering is important if the symmetric half of the inertia and stiffness matrices are stored in one-dimensional arrays. In order to minimise storage only the non-zero terms beneath a skyline are included in these arrays [14.11]. It is, therefore, important that the nodes be numbered such that the non-zero terms are as close as possible to the main diagonal. Some programs contain node renumbering facilities [14.3]. If these are not available, then the nodes should be numbered along the topologically shortest path. For example, the numbering scheme shown in Figure 14.3(*a*) is preferred to that in Figure 14.3(*b*).

Element numbering is important if a front solution method is used [14.12]. In this method the slave degrees of freedom at a node are eliminated as soon as all the elements connected to that node have been assembled. The order of assembling the elements is, therefore, important. Some programs contain element reordering facilities.

Having defined the nodes and elements, it is advisable to check these to ensure that they are correct, before proceeding with the analysis. This checking is best done graphically. This means that a high resolution graphics terminal, possibly capable of colour shading, and a hard copy facility are required.

Two-dimensional idealisations can be checked relatively easily, as only two-dimensional plots are required. The program should be capable of plotting nodes or elements and including node and/or element numbers on request. Outline drawing will emphasise the overall shape of the idealisation without internal details. Also element shrinking will indicate if any element is missing from the idealisation.

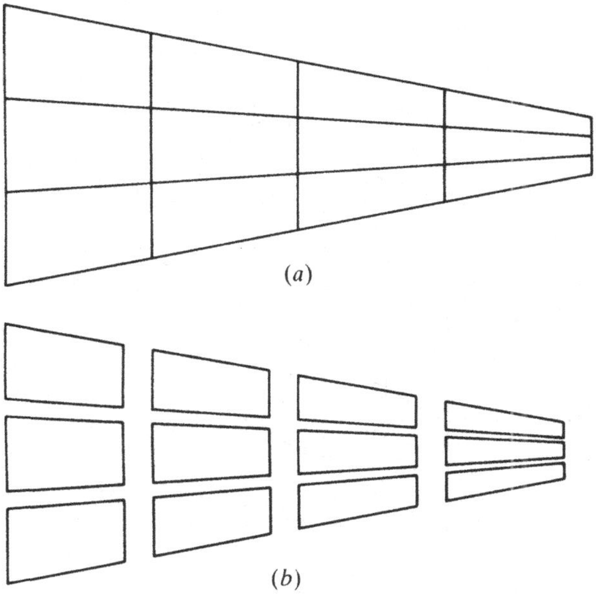

Figure 14.4. Idealisation of a tapered cantilever. (*a*) Without shrinking, (*b*) with shrinking.

Figure 14.4 shows an idealisation without and with shrinking. It will also be useful if the distribution of element geometric or material properties can be indicated on the element plot. This can be done by means of colour shading.

Additional features required for three-dimensional idealisations are plotting of the mesh from any viewpoint, selective plotting of elements and hidden line plots. Figure 14.5 to 14.9 show various plots of an open box with two internal partitions. Figure 14.5 is a see-through plot of the idealisation. With such a plot it is difficult to check all the details. Figure 14.6 shows a hidden line plot of the same structure. It is now easy to check all the elements which are immediately in view. Other elements can be checked by hidden line plots from different viewpoints. This is illustrated in Figure 14.7 which shows a view from underneath. Another useful technique is to

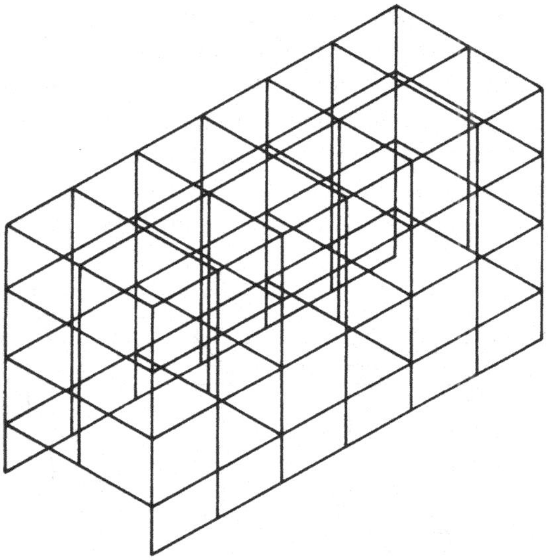

Figure 14.5. Idealisation of an open box with two internal webs.

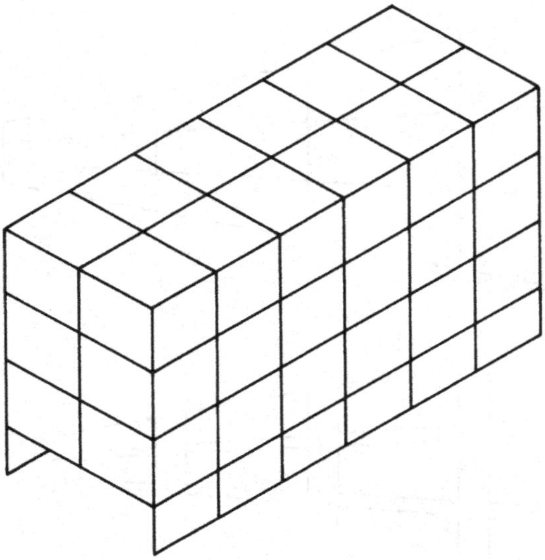

Figure 14.6. Hidden line plot of structure shown in Figure 14.5.

plot sub-components of the structure, as shown in Figure 14.8. Figure 14.9 shows section plots through the two vertical planes of symmetry of the structure.

Boundary conditions are of two types: those that occur on a true boundary and the symmetric or antisymmetric conditions at a plane of symmetry. It is helpful if these conditions can be indicated on an element plot. In Figure 14.10 zero values of two components of in-plane displacements at each node point on the left hand boundary are indicated by means of arrows.

Multi-point constraints are used to represent rigid off-sets, which are discussed in more detail in Section 3.11. This type of constraint is difficult to indicate graphically.

Master degrees of freedom can be chosen manually, automatically or by a combination of the two. An indication of manually selected masters on an element plot is helpful. This is illustrated in Figure 14.11. The component of displacement normal to the plane of the cantilever at each node has been selected as a master degree of freedom and is indicated by means of an arrow. Automatically selected masters cannot be indicated as they are selected during the element assembly procedure.

Most finite element programs will give the user a choice of more than one eigenproblem solution technique. It is, therefore, necessary to specify which one is to be used and any information the chosen method requires. This may be the number of frequencies and modes required to be calculated.

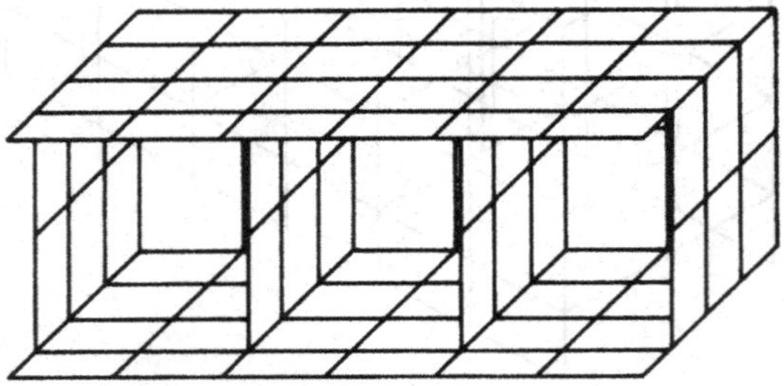

Figure 14.7. Internal view of structure shown in Figure 14.5 with hidden lines.

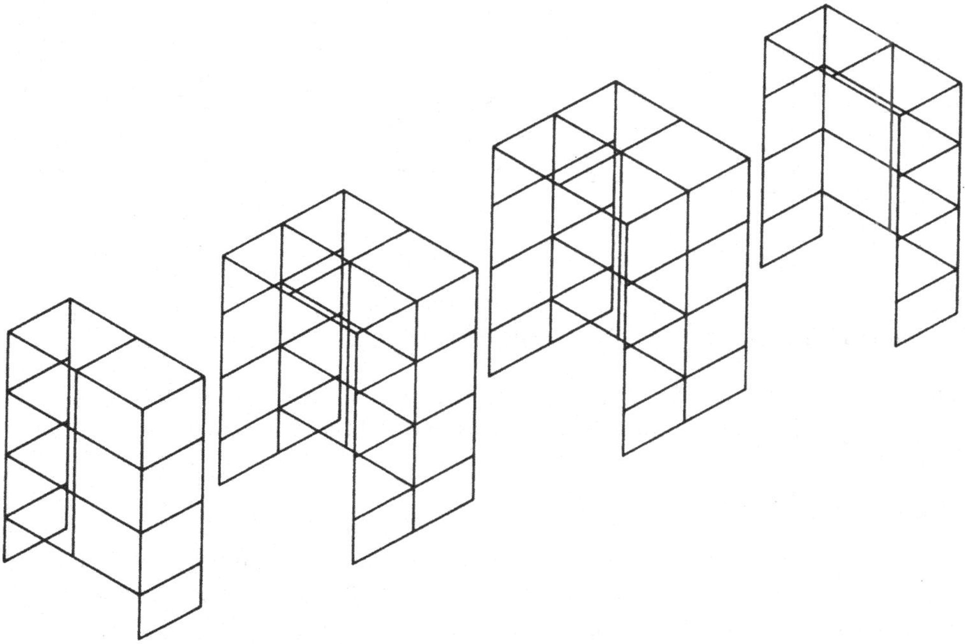

Figure 14.8. Four sub-components of structure shown in Figure 14.5.

The additional input data required for forced response analysis consists of specifying the applied forces or imposed displacements and the damping. The excitation will be either in the frequency or time domain. Precise details of the form of the input and the additional information required for the solution procedure to be used, can be obtained from Chapters 12 and 13. The spatial distribution should be

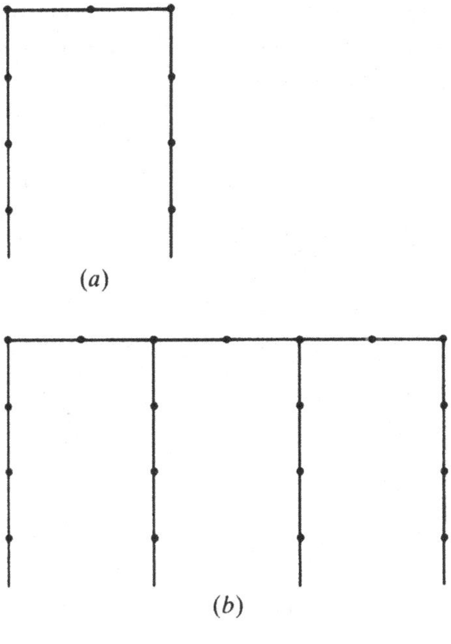

Figure 14.9. Section plots of the structure shown in Figure 14.5.

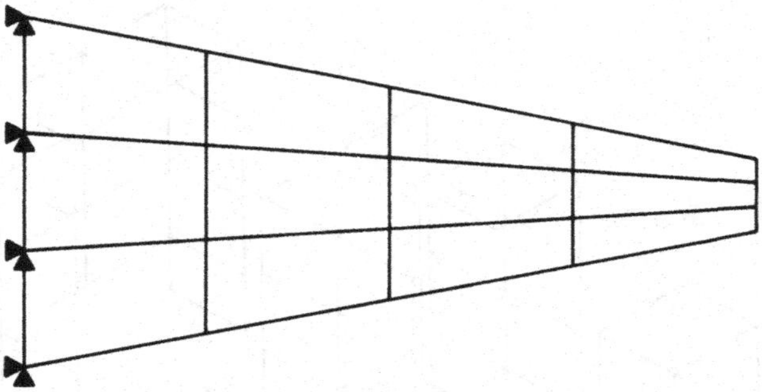

Figure 14.10. Indication of boundary conditions using arrows.

checked by indications on an element plot. The time history and/or the frequency distribution of the excitations should also be plotted for visual checking.

14.1.2 Solution Phase

Once the input data has been prepared and checked, the solution can be carried out. Figure 14.12 shows the various types of analysis and the order in which they are carried out. For forced response the first choice is between a modal solution (M) or a direct solution (D). For a modal solution the frequencies and modes of free vibration have to be calculated first. In either case, the next choice is between a frequency or time domain solution. In the frequency domain, harmonic response is a prerequisite to both periodic (P) and random response (R) as well as being an important solution procedure itself. In the time domain the choice is between a time history of the response (TH) or predicting the peak response using the response spectrum method (RS) [14.13].

The solution can be carried out in either interactive or batch mode, depending on the estimated run time. Good finite element programs will carry out further checks on the data during the solution phase. Fatal errors will cause the execution to be terminated. Non-fatal errors will be indicated by warning messages, but execution will continue.

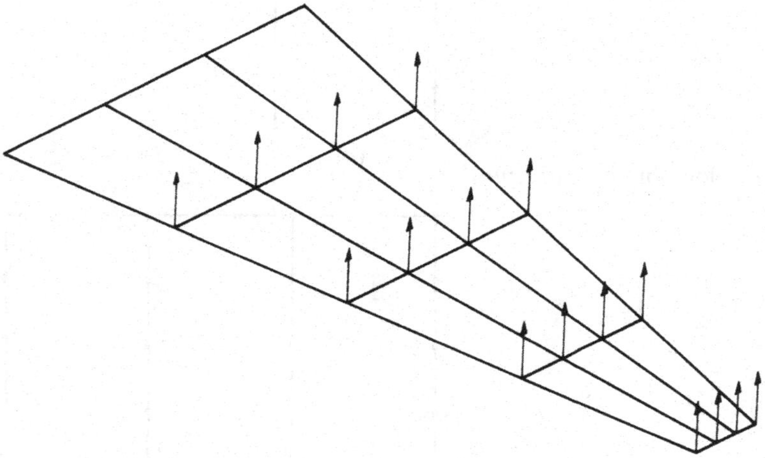

Figure 14.11. Indication of master degrees of freedom.

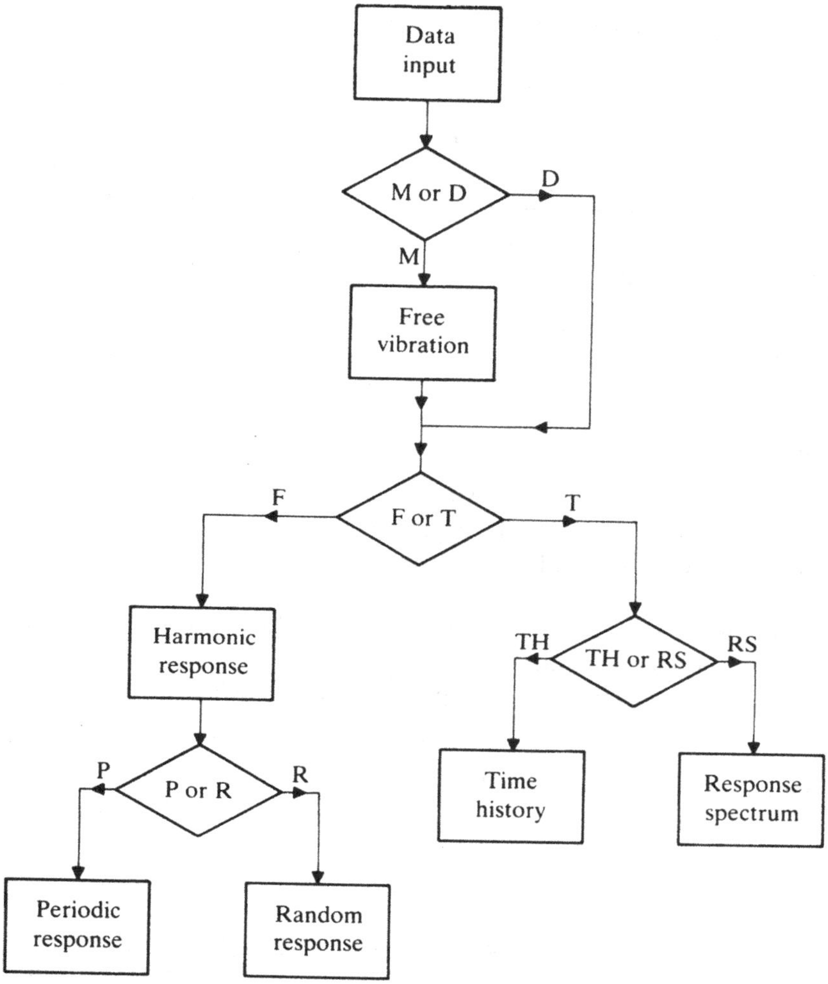

Figure 14.12. Flow chart of the solution process.

14.1.3 Post-processing

The post-processing phase is concerned with interrogating the results of the analysis. In the case of free vibration this will be the natural frequencies and modes of free vibration. The distorted mode shapes should be plotted from a specified viewpoint. It is useful if the undistorted idealisation, either in full or in outline, can be superimposed for comparison. This is illustrated in Figure 14.13 which shows a distorted mode shape of a square cantilever plate, superimposed upon an undistorted outline. A better understanding of the motion can be obtained if the mode shape can be animated.

When analysing response in the frequency domain, response quantities are plotted against frequency of excitation. A choice of linear or logarithmic scales on both axes is required. For time domain solutions response quantities are plotted against time.

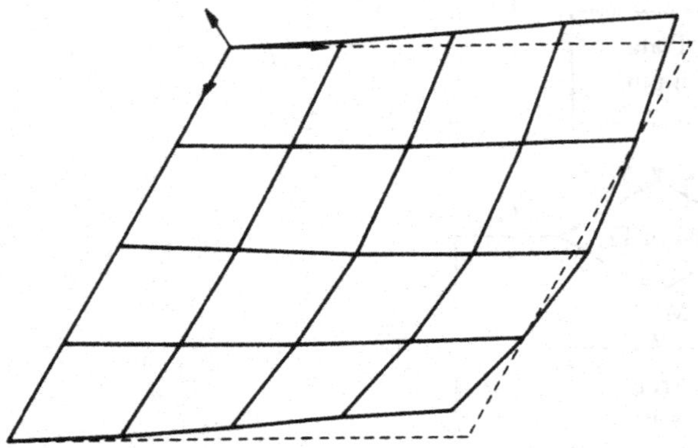

Figure 14.13. Free vibration mode of a square cantilever plate.

14.2 Modelling

Modelling is a two stage process. To start with, the actual structure is replaced by a simplified one which retains the essential features to be investigated. The reason for this is that it would be too costly and time-consuming to try and model every small detail of the structure. Expertise in carrying out this simplification is usually gained by working in conjunction with experienced engineers and designers. The second stage is to represent the simplified model by a finite element model.

Before a finite element mesh can be specified, the following information is required:

(1) The geometry of the simplified model
(2) The boundary conditions
(3) The applied loads
(4) The required results

Note that both the boundary conditions and applied loadings need to be adequately represented and, therefore, will affect the choice of mesh as well as the geometry. Previous chapters have indicated that the size of mesh affects the accuracy of the results. This means that the size of the mesh should be chosen to ensure that the quantities of interest are predicted accurately.

It is a simple matter to represent uniform regions of a structure by means of a uniform mesh. Several examples are given in earlier chapters. When using triangular elements, the arrangement of the triangles influences the results. If the region to be modelled has two axes of symmetry, then arrangements (*a*) and (*b*) of Figure 14.14 are preferable to (*c*) and (*d*), which cannot predict symmetric responses. Arrangement (*c*) can, however, predict a symmetric response in one direction. If the boundaries are all fully fixed, then arrangements (*b*) to (*d*) have elements which are completely inactive (**I**), since all the degrees of freedom at all three nodes will be zero.

Quadrilateral elements should be used in preference to triangles as they tend to give more accurate results for the same arrangement of node points. Triangular elements should only be used where the structural shape requires it.

When modelling irregular geometries, the shape of the element should be controlled. The basic shape of a triangle is an equilateral triangle, and that of a quadrilateral is a square. Accuracy tends to deteriorate as elements are distorted from their

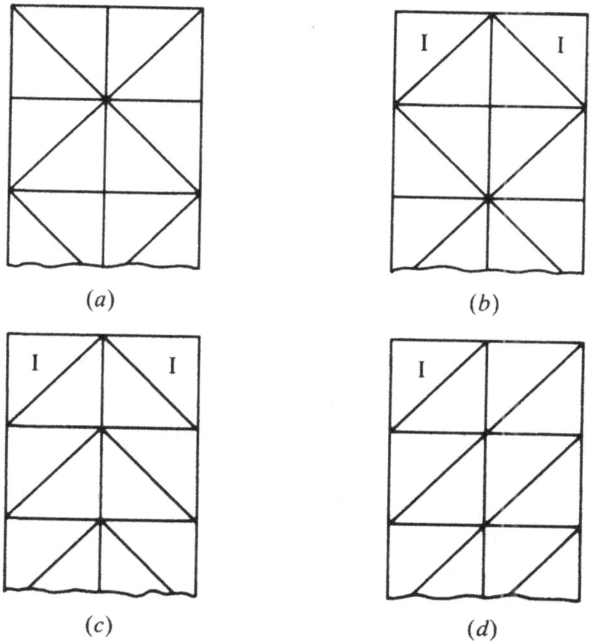

Figure 14.14. Arrangements of triangular elements

basic shape. Various types of distortion for a quadrilateral are shown in Figure 14.15. The first (*a*) indicates a change in aspect ratio, whilst (*b*) and (*c*) show two types of angular distortion. Both of these can be controlled by requiring the included angle between two adjacent sides to be close to 90°.

Elements with mid-side nodes tend to give more accurate solutions than those without. However, several precautions should be observed when using such elements. They should have straight sides, except when modelling curved boundaries. In such cases the curve should not be excessive. Also a 'mid-side' node should be

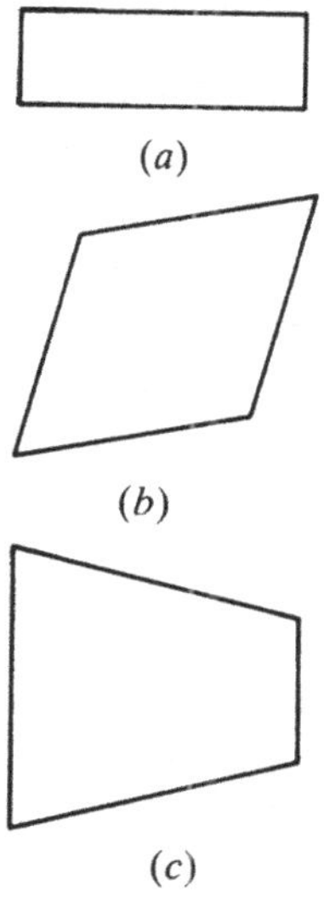

Figure 14.15. Distortion of quadrilaterals: (*a*) aspect ratio, (*b*) angular (skew), (*c*) angular (taper).

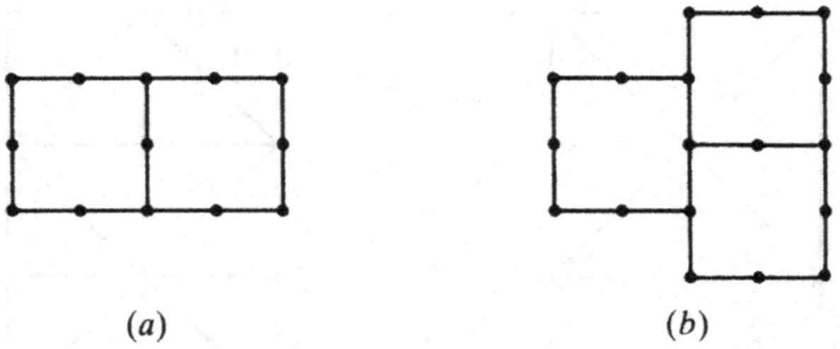

Figure 14.16. Connecting elements with mid-side nodes.

located close to the point lying midway between the corner nodes. Adjacent elements should have the same number of nodes along the common side. A corner node should only be connected to a corner node of the adjacent element and not a mid-side node. Therefore arrangement (*a*) in Figure 14.16 is preferable to (*b*). When mixing element types it may be necessary to remove the mid-side node using constraint equations. This is illustrated in Figure 14.17 where node 7 should be removed from element *A* before connecting to element *B* to ensure they have common interpolations along the interface. Distributed edge loads and surface pressures are not always allocated to the element nodes according to commonsense. Figure 14.18 indicates the distribution of unit loads [14.14]. Reaction forces tend to be similarly distributed. The mass at mid-side nodes is also greater than at the corner nodes. This should be kept in mind when selecting master degrees of freedom.

Difficulties arise when two neighbouring regions, which are to be represented by different mesh densities, are connected together. Two ways of doing this are illustrated in Figure 14.19. Both methods use triangular elements to connect meshes of rectangular elements. This is acceptable provided they have common interpolations along the interfaces. Figure 14.20 illustrates a case where dissimilar element types do not have common interpolations along the interface. Element *A* has a quadratic variation of displacement, whilst each of elements *B*1 and *B*2 has a linear variation. Multi-point constraints can be used to enforce compatibility in the case illustrated in Figure 14.21. The displacements at nodes 7 and 9 of elements *B*1 and *B*2 are constrained to be defined by the displacements at nodes 6, 8 and 10 using the interpolation function for element *A*.

Problems often arise in modelling physical discontinuities in structures such as joints and other connections. Such a problem occurs with bolted joints or joints that are spot welded. To model such details will require a very fine mesh. Such refinement is rarely used. Instead, a coarse mesh is defined, which means that the details of the connection are smeared out. This fact should be borne in mind when interpreting the results. The joining of beams in frameworks and plates in folded plate structures can also cause modelling problems. The reason for this is that very often beam and plate elements are assumed to be thin. But in practice the thickness

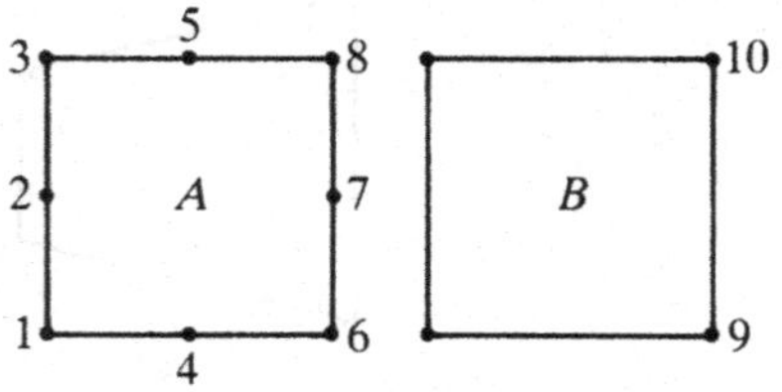

Figure 14.17. Connection of different element types.

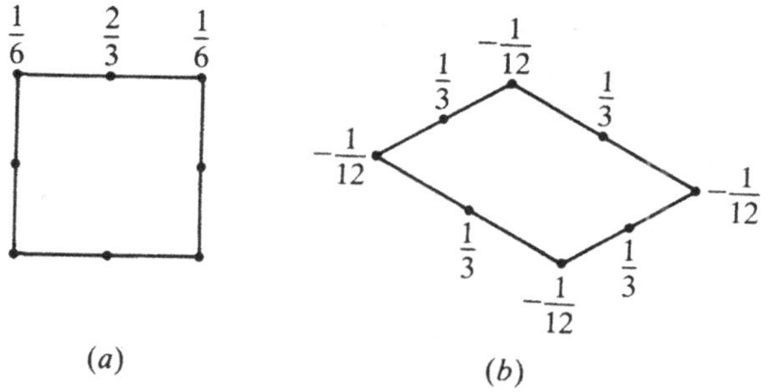

Figure 14.18. Distribution of unit loads: (*a*) along an edge, (*b*) on a surface.

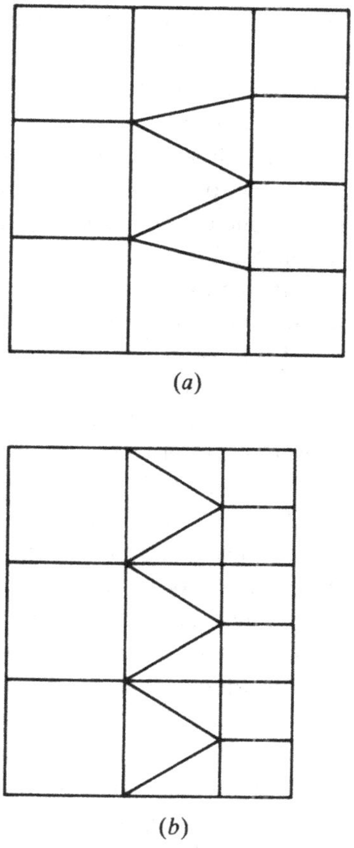

Figure 14.19. Mesh grading using compatible elements.

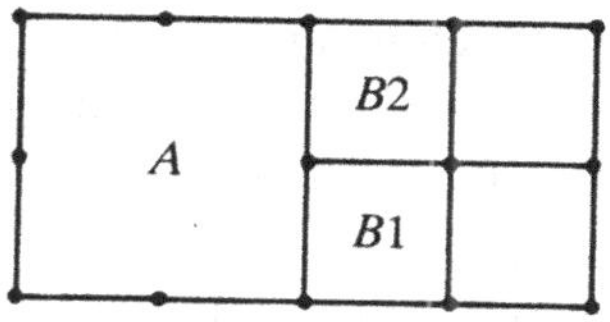

Figure 14.20. Mesh grading using non-compatible elements.

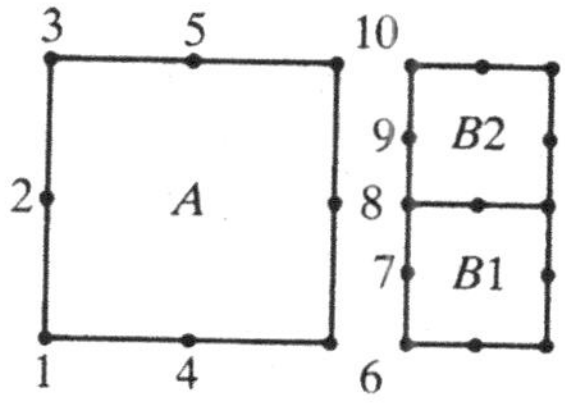

Figure 14.21. Mesh grading using the same element type.

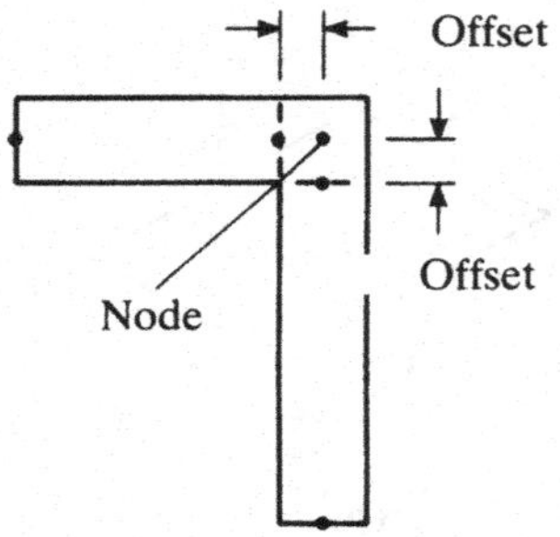

Figure 14.22. Off-set connection of beam elements.

of the structure can be significant. This is illustrated in Figure 14.22 where two beam elements are connected at right angles. If the off-sets of the node from the ends of the beam elements are significant, then rigid off-sets, as described in Section 3.11, should be used.

Care should be taken in modelling support conditions, as often there are alternate ways of defining them. Figure 14.23 shows a beam which is modelled using membrane elements. Two methods of defining fixed boundary conditions are illustrated. Similarly, Figure 14.24 shows two ways of defining a simple support. In each case the analyst should decide which is the more relevant one for the particular case under consideration.

There are many occasions when the structure has one or more planes of symmetry. These can be used to reduce the cost and simplify the analysis. In the case of free vibration, the mode shapes are either symmetric or antisymmetric about a plane of symmetry. In practice the symmetric and antisymmetric modes are calculated separately using an idealisation of a portion of the structure and appropriate boundary conditions along the plane of symmetry. Details are given in Section 11.5.1. Symmetry can also be exploited in forced response analysis as described in Section 13.5.1. Two other types of symmetry exist, axisymmetry and cyclic symmetry. Analysis of axisymmetric structures is presented in Sections 5.1 to 5.6. Cyclic symmetric, or rotationally periodic structures, are considered in Sections 11.5.2 and 13.5.2.

Problems arise when modelling semi-infinite regions. Such a situation occurs when considering soil–structure interaction problems. It is usual to model the structure and a finite portion of the semi-infinite medium adjacent to the structure using finite elements. The remaining portion of the semi-infinite region is then represented by (a) applying transmitting boundary conditions, (b) semi-infinite elements or (c) boundary elements. Reference [14.15] discusses these and then proposes three finite element-based procedures for the semi-infinite region to overcome the limitations of the aforementioned methods.

Guidelines for the use of solid elements follow a similar pattern to those for two-dimensional elements. Hexahedra should be used in preference to pentahedra and tetrahedra, the latter two being only used where the structural shape requires

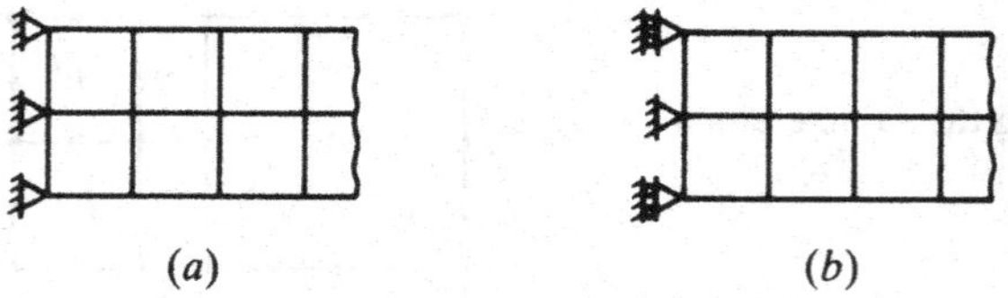

Figure 14.23. Types of fixed support: (*a*) fully fixed, (*b*) Engineer's theory of bending.

Figure 14.24. Modelling of a simple support.

it. Elements behave best when they are least distorted. The basic shape of a hexahedra is a square cube. The performance of higher order elements deteriorates as the curvature of the edges increases. Mid-edge nodes should be as near to the half-way point as possible.

Further hints on modelling can be found in reference [14.16]. These and the ones presented here, should be considered to be suggestions rather than rules, as important exceptions may exist. There is no substitute for knowledge and experience.

14.3 Using Commercial Codes

The choice of which commercial program to use depends upon many factors. If several types of analyses are required, then the choice is between the general purpose programs. However, if only one type is required, then a special purpose program might prove more efficient computationally. The available programs tend to have particular strengths depending upon the interests of the developers, so it would be wise to determine these by talking to experienced users.

The chosen program should be capable of providing accurate solutions. More and more developers are subjecting their programs to stringent quality assurance tests [14.17, 14.18]. However, it is very difficult for a developer to envisage every possible use a customer might make of his software. The quality assurance tests may, therefore, concentrate more on checking the accuracy of the code.

Having chosen the program, the next step is familiarisation with its use. Most program developers provide Verification and/or Examples Manuals with their product. Running some of these examples will provide some experience of its use.

The next stage is to build up confidence in the accuracy of the elements selected for the idealisation and the solution procedures to be used. The accuracy of the elements depends upon the representation of the stiffness, inertia and applied forces. The accuracy of stiffness and spatial distribution of the applied forces can be checked by applying a set of static validation tests. A number of papers have been published on this topic, many of which are listed in reference [14.19], which represents the current state of the art for membrane and plate bending elements. Reference [14.20] discusses the problems involved in setting up a similar procedure for shell elements.

Reference [14.19] proposes the following set of tests:

(1) Single element completeness tests
(2) Completeness tests for a patch of elements
(3) False zero energy mode tests
(4) Invariance tests
(5) Single element, shape sensitivity tests
(6) Benchmark tests

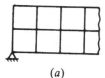

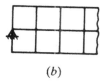

The single element tests have been applied to the elements of a number of commercial finite element systems in references [14.21–14.23]. Benchmark tests for membrane, plate bending, shell and folded plate structures are proposed in references [14.24, 14.25], whilst composites are dealt with in references [14.26, 14.27].

The single element completeness tests consist of showing that the element displacement assumptions contain zero strain (rigid body) modes and constant strain modes. If each element in a model passes these tests, and the model is a conforming one, the results will converge monotonically as the number of elements is increased. If the model is a non-conforming one, then constant strain states should be applied to an assemblage, or patch, of elements. All elements should contain the same constant strains. If this is so, then the results will converge, though not necessarily monotonically.

The false zero energy mode tests consist of determining whether an element contains deformation modes, other than rigid body modes, which have zero strain energy. This information can be determined by calculating the eigenvalues of the element stiffness matrix. The number of false zero energy modes is then equal to the number of zero eigenvalues minus the number of rigid body modes. If an element does contain false modes, then the test should be applied to a patch of elements. In many cases it will be found that the patch does not exhibit false modes.

In developing a finite element model, the assumed displacement functions are very often related to a local set of axes. These are defined by the order of the element specifying nodes. The invariance test consists of taking a finite element model and analysing it for a given set of element input nodes. The problem is re-analysed after specifying the element nodes differently, by taking another corner node as the first specifying node for each element. If the results are the same in both cases, then the element is invariant.

Single element, shape sensitivity tests consist of determining the sensitivity of elements to variations in aspect ratio, skew, taper and a combination of these. The element to be examined is cut out of a rectangular continuum and loaded using specified nodal displacements and forces and analysed. The resulting displacements and/or stresses are compared with the exact ones. This process is repeated after introducing further distortions. This will indicate limits on the allowable variations. As an illustration of this, Figure 14.25 shows the effect of aspect ratio on the solution for displacement, for various elements which are subject to a twisting moment. As these results were published in 1978, the commercial codes containing these elements are not indicated. It does illustrate though, that the aspect ratio of many elements should be limited in order to preserve accuracy.

Benchmarks are fully specified standard problems which are used for evaluating the performance of element assemblies. They resemble instances found in industrial applications wherever possible. Reference values for the assessment of benchmarks are, as far as possible, obtained from known analytical results.

Benchmarks can also be used for convergence studies and checking elements for locking effects. Beams, membranes and plates, which are based upon theories requiring continuity of displacements only (often referred to as C^0 theory), suffer from shear locking. This has been discussed in Sections 3.10, 4.2 and 6.3. Curved

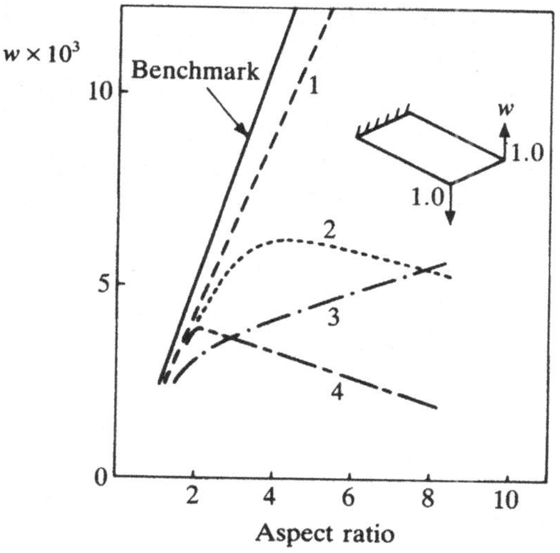

Figure 14.25. Effect of aspect ratio on accuracy of plate bending elements [14.28].

shells which are based upon C^0 theory suffer from both shear and membrane locking [14.11]. Three-dimensional elements are subject to volumetric locking when applied to incompressible, or nearly incompressible materials [14.11].

The accuracy of element inertia matrices can be checked by carrying out benchmark tests for free vibration. Such a set has been proposed in reference [14.29]. These tests are also designed to test the following aspects of eigenproblem solution techniques:

(1) The prediction of zero frequency rigid body modes
(2) The calculation of closely spaced and coincident eigenvalues and associated eigenvectors
(3) The extraction of all eigenvalues in a given frequency range (i.e., none are missed)
(4) The ability to choose universally satisfactory starting vectors for iterative schemes
(5) The ability to calculate eigenvalues and eigenvectors irrespective of the conditioning of the stiffness and inertia matrices
(6) The reliability and robustness of automatic master selection schemes when reduction techniques are employed

Benchmarks for the response of beams and plates subject to harmonic, periodic, transient and random forces are presented in reference [14.30].

At this stage the user should have sufficient confidence to analyse a practical structure. As a final check it may be advantageous to represent the structure initially by means of a very coarse mesh. Analysing this will highlight any potential problems in running the analysis before the fine mesh is used. By this means, long wasted computer runs are avoided. Finally, always disbelieve the output of a computer run until you can prove that it is satisfactory. Comprehensive checks should always be made to verify the validity of the results.

Another method of validating a finite element model is to compare it with an experimental model. This may be a spatial, modal or response model, the most popular being a modal one. The objective is to locate the errors in the finite element model so that they may be updated to achieve correlation between theoretical predictions and experimental results. Details of such methods can be found in references [14.31–14.33].

APPENDIX 1

Equations of Motion of Multi-Degree of Freedom Systems

In Chapter 1 the equations of motion of single degree of freedom systems were derived in various forms. In this appendix the analogous derivations for multi-degree of freedom systems are presented.

A1.1 Hamilton's Principle

Consider a system of N masses, m_j, which undergo displacements $\vec{u}_j$, when subject to forces $\vec{f}_j$. Applying the principle of virtual displacements gives

$$\sum_{j=1}^{N}(\vec{f}_j \cdot \delta\vec{u}_j - m_j\ddot{\vec{u}}_j \cdot \delta\vec{u}_j) = 0 \tag{A1.1}$$

where the $\delta\vec{u}_j$ are virtual displacements.

Now

$$\sum_{j=1}^{N}(\vec{f}_j \cdot \delta\vec{u}_j) = \delta W \tag{A1.2}$$

where δW is the virtual work done by the forces; also

$$\begin{aligned}\ddot{\vec{u}}_j \cdot \delta\vec{u}_j &= \frac{\mathrm{d}}{\mathrm{d}t}(\dot{\vec{u}}_j \cdot \delta\vec{u}_j) - \dot{\vec{u}}_j \cdot \delta\dot{\vec{u}}_j \\ &= \frac{\mathrm{d}}{\mathrm{d}t}(\dot{\vec{u}}_j \cdot \delta\vec{u}_j) - \delta\left(\tfrac{1}{2}\dot{\vec{u}}_j \cdot \dot{\vec{u}}_j\right)\end{aligned} \tag{A1.3}$$

Therefore

$$\begin{aligned}\sum_{j=1}^{N} m_j\ddot{\vec{u}}_j \cdot \delta\vec{u}_j &= \sum_{j=1}^{N} m_j\frac{\mathrm{d}}{\mathrm{d}t}(\dot{\vec{u}}_j \cdot \delta\vec{u}_j) - \delta\sum_{j=1}^{N}\frac{1}{2}m_j\dot{\vec{u}}_j \cdot \dot{\vec{u}}_j \\ &= \sum_{j=1}^{N} m_j\frac{\mathrm{d}}{\mathrm{d}t}(\dot{\vec{u}}_j \cdot \delta\vec{u}_j) - \delta T\end{aligned} \tag{A1.4}$$

where T is the kinetic energy of the system.

Substituting equations (A1.2) and (A1.4) into equation (A1.1) gives

$$\delta W - \sum_{j=1}^{N} m_j \frac{\mathrm{d}}{\mathrm{d}t}(\dot{\vec{u}}_j \cdot \delta \vec{u}_j) + \delta T = 0 \tag{A1.5}$$

or, on rearranging,

$$\delta T + \delta W = \sum_{j=1}^{N} m_j \frac{\mathrm{d}}{\mathrm{d}t}(\dot{\vec{u}}_j \cdot \delta \vec{u}_j) \tag{A1.6}$$

To determine the true path between two instants of time t_1 and t_2 where $\delta \vec{u}_j = 0$, equation (A1.6) is integrated with respect to time between these limits, giving

$$\begin{aligned}\int_{t_1}^{t_2} (\delta T + \delta W)\,\mathrm{d}t &= \int_{t_1}^{t_2} \sum_{j=1}^{N} m_j \frac{\mathrm{d}}{\mathrm{d}t}(\dot{\vec{u}}_j \cdot \delta \vec{u}_j)\,\mathrm{d}t \\ &= \sum_{j=1}^{N} \int_{t_1}^{t_2} m_j \frac{\mathrm{d}}{\mathrm{d}t}(\dot{\vec{u}}_j \cdot \delta \vec{u}_j)\mathrm{d}t \\ &= \sum_{j=1}^{N} \left[m_j \dot{\vec{u}}_j \cdot \delta \vec{u}_j\right]_{t_1}^{t_2} = 0\end{aligned} \tag{A1.7}$$

This gives

$$\int_{t_1}^{t_2} (\delta T + \delta W)\,\mathrm{d}t = 0 \tag{A1.8}$$

Separating the forces into conservative and non-conservative forces gives

$$\delta W = -\delta V + \delta W_{\mathrm{nc}} \tag{A1.9}$$

where δV is the change in potential energy of the conservative forces (see Chapter 1). Substituting equation (A1.9) into equation (A1.8) gives

$$\int_{t_1}^{t_2} \{\delta(T - V) + \delta W_{\mathrm{nc}}\}\,\mathrm{d}t = 0 \tag{A1.10}$$

Taking the potential energy V to be strain energy U gives

$$\int_{t_1}^{t_2} \{\delta(T - U) + \delta W_{\mathrm{nc}}\}\,\mathrm{d}t = 0 \tag{A1.11}$$

A1.2 Lagrange's Equations

For a system of N masses which are free to move in three dimensions the kinetic energy is

$$T = \frac{1}{2}\sum_{j=1}^{N} m_j \dot{\vec{u}}_j \cdot \dot{\vec{u}}_j \tag{A1.12}$$

Since each vector displacement $\vec{u}_j$ can be resolved into three scalar components, then the kinetic energy can be expressed as follows

$$T = T(\dot{q}_1, \dot{q}_2, \ldots, \dot{q}_n) \tag{A1.13}$$

where the q_j are $n = 3N$ independent displacements. Therefore

$$\delta T = \sum_{j=1}^{n} \frac{\partial T}{\partial \dot{q}_j} \delta \dot{q}_j \tag{A1.14}$$

Similarly the strain energy of the system can be expressed as

$$U = U(q_1, q_2, \ldots, q_n) \tag{A1.15}$$

and so

$$\delta U = \sum_{j=1}^{n} \frac{\partial U}{\partial q_j} \delta q_j \tag{A1.16}$$

Also, the virtual work done by the non-conservative forces is

$$\delta W_{\mathrm{nc}} = \sum_{k=1}^{N} \vec{\boldsymbol{f}}_k \cdot \delta \vec{\boldsymbol{u}}_k - \sum_{k=1}^{N} c_k \dot{\vec{\boldsymbol{u}}}_k \cdot \delta \vec{\boldsymbol{u}}_k \tag{A1.17}$$

where $\vec{\boldsymbol{f}}_k$ and $c_k \dot{\vec{\boldsymbol{u}}}_k$ are the applied force and viscous damping force acting on the mass m_k.

Since $\vec{\boldsymbol{u}}_k = \vec{\boldsymbol{u}}_k(q_1, q_2, \ldots, q_n)$, the first term of equation (A1.17) becomes

$$\begin{aligned}
\sum_{k=1}^{N} \vec{\boldsymbol{f}}_k \cdot \delta \vec{\boldsymbol{u}}_k &= \sum_{k=1}^{N} \vec{\boldsymbol{f}}_k \cdot \sum_{j=1}^{N} \frac{\partial \vec{\boldsymbol{u}}_k}{\partial q_j} \delta q_j \\
&= \sum_{j=1}^{n} \left(\sum_{k=1}^{N} \vec{\boldsymbol{f}}_k \cdot \frac{\partial \vec{\boldsymbol{u}}_k}{\partial q_j} \right) \delta q_j \\
&= \sum_{j=1}^{n} Q_j \delta q_j
\end{aligned} \tag{A1.18}$$

where

$$Q_j = \sum_{k=1}^{N} \vec{\boldsymbol{f}}_k \cdot \frac{\partial \vec{\boldsymbol{u}}_k}{\partial q_j} \tag{A1.19}$$

The second term of equation (A1.17) becomes

$$\begin{aligned}
\sum_{k=1}^{N} c_k \dot{\vec{\boldsymbol{u}}}_k \cdot \delta \vec{\boldsymbol{u}}_k &= \sum_{k=1}^{N} c_k \dot{\vec{\boldsymbol{u}}}_k \cdot \sum_{j=1}^{n} \frac{\partial \vec{\boldsymbol{u}}_k}{\partial q_j} \delta q_j \\
&= \sum_{j=1}^{n} \left(\sum_{k=1}^{N} c_k \dot{\vec{\boldsymbol{u}}}_k \cdot \frac{\partial \vec{\boldsymbol{u}}_k}{\partial q_j} \right) \delta q_j
\end{aligned} \tag{A1.20}$$

Now

$$\dot{\vec{\boldsymbol{u}}}_k = \sum_{j=1}^{N} \frac{\partial \vec{\boldsymbol{u}}_k}{\partial q_j} \dot{q}_j$$

Therefore

$$\frac{\partial \dot{\vec{\boldsymbol{u}}}_k}{\partial \dot{q}_j} = \frac{\partial \vec{\boldsymbol{u}}_k}{\partial q_j} \tag{A1.21}$$

Using equation (A1.21) in (A1.20) gives

$$\sum_{k=1}^{N} c_k \dot{\vec{u}}_k \cdot \delta \vec{u}_k = \sum_{j=1}^{n} \left(\sum_{k=1}^{N} c_k \dot{\vec{u}}_k \cdot \frac{\partial \dot{\vec{u}}_k}{\partial \dot{q}_j} \right) \delta q_j$$
$$= \sum_{j=1}^{n} \frac{\partial}{\partial \dot{q}_j} \left(\sum_{k=1}^{N} \frac{1}{2} c_k \dot{\vec{u}}_k \cdot \dot{\vec{u}}_k \right) \delta q_j \tag{A1.22}$$

Defining

$$D = \frac{1}{2} \sum_{k=1}^{N} c_k \dot{\vec{u}}_k \cdot \dot{\vec{u}}_k \tag{A1.23}$$

equation (A1.22) becomes

$$\sum_{k=1}^{N} c_k \dot{\vec{u}}_k \cdot \delta \vec{u}_k = \sum_{j=1}^{n} \frac{\partial D}{\partial \dot{q}_j} \delta q_j \tag{A1.24}$$

Substituting equations (A1.18) and (A1.24) into (A1.17) gives

$$\delta W_{\text{nc}} = \sum_{j=1}^{n} \left(Q_j - \frac{\partial D}{\partial \dot{q}_j} \right) \delta q_j \tag{A1.25}$$

Applying Hamilton's principle (equation (A1.11)) and using equations (A1.14), (A1.16) and (A1.25) gives

$$\sum_{j=1}^{n} \int_{t_1}^{t_2} \left(\frac{\partial T}{\partial \dot{q}_j} \delta \dot{q}_j - \frac{\partial U}{\partial q_j} \delta q_j + Q_j \delta q_j - \frac{\partial D}{\partial \dot{q}_j} \right) \mathrm{d}t = 0 \tag{A1.26}$$

Now

$$\delta \dot{q}_j = \delta \left(\frac{\mathrm{d}q_j}{\mathrm{d}t} \right) = \frac{\mathrm{d}}{\mathrm{d}t} (\delta q_j) \tag{A1.27}$$

Hence, integrating the first term by parts gives

$$\int_{t_1}^{t_2} \frac{\partial T}{\partial \dot{q}_j} \delta \dot{q}_j \, \mathrm{d}t = \left[\frac{\partial T}{\partial \dot{q}_j} \delta q_j \right]_{t_1}^{t_2} - \int_{t_1}^{t_2} \frac{\mathrm{d}}{\mathrm{d}t} \left(\frac{\partial T}{\partial \dot{q}_j} \right) \delta q_j \, \mathrm{d}t$$
$$= - \int_{t_1}^{t_2} \frac{\mathrm{d}}{\mathrm{d}t} \left(\frac{\partial T}{\partial \dot{q}_j} \right) \delta q_j \, \mathrm{d}t \tag{A1.28}$$

since the $\delta q_j = 0$ at $t = t_1$ and t_2.

Substituting (A1.28) into (A1.26) gives

$$\sum_{j=1}^{n} \int_{t_1}^{t_2} \left(-\frac{\mathrm{d}}{\mathrm{d}t} \left(\frac{\partial T}{\partial \dot{q}_j} \right) - \frac{\partial D}{\partial \dot{q}_j} - \frac{\partial U}{\partial q_j} + Q_j \right) \delta q_j \, \mathrm{d}t = 0 \tag{A1.29}$$

Since the virtual displacements δq_j are arbitrary and independent, then

$$\frac{\mathrm{d}}{\mathrm{d}t} \left(\frac{\partial T}{\partial \dot{q}_j} \right) + \frac{\partial D}{\partial \dot{q}_j} + \frac{\partial U}{\partial q_j} = Q_j, \qquad j = 1, 2, \ldots, n \tag{A1.30}$$

APPENDIX 2

Transformation of Strain Components

From equation (3.195) the relationship between displacements in local coordinates and those in global coordinates is

$$\begin{bmatrix} \bar{u} \\ \bar{v} \\ \bar{w} \end{bmatrix} = \begin{bmatrix} l_1 & m_1 & n_1 \\ l_2 & m_2 & n_2 \\ l_3 & m_3 & n_3 \end{bmatrix} \begin{bmatrix} u \\ v \\ w \end{bmatrix} \tag{A2.1}$$

where l_i, m_i and n_i are the direction cosines between local and global axes.

With respect to local axes

$$\frac{\partial \bar{u}}{\partial \bar{x}} = l_1 \frac{\partial u}{\partial \bar{x}} + m_1 \frac{\partial v}{\partial \bar{x}} + n_1 \frac{\partial w}{\partial \bar{x}} \tag{A2.2}$$

Now

$$\begin{aligned} \frac{\partial u}{\partial x} &= \frac{\partial u}{\partial x}\frac{\partial x}{\partial \bar{x}} + \frac{\partial u}{\partial y}\frac{\partial y}{\partial \bar{x}} + \frac{\partial u}{\partial z}\frac{\partial z}{\partial \bar{x}} \\ &= l_1 \frac{\partial u}{\partial x} + m_1 \frac{\partial u}{\partial y} + n_1 \frac{\partial u}{\partial z} \end{aligned} \tag{A2.3}$$

Similarly, for $\partial v / \partial \bar{x}$ and $\partial w / \partial \bar{x}$

Substituting into equation (A2.2) gives

$$\begin{aligned} \frac{\partial \bar{u}}{\partial \bar{x}} &= l_1 \left(l_1 \frac{\partial u}{\partial x} + m_1 \frac{\partial u}{\partial y} + n_1 \frac{\partial u}{\partial z} \right) + m_1 \left(l_1 \frac{\partial v}{\partial x} + m_1 \frac{\partial v}{\partial y} + n_1 \frac{\partial v}{\partial z} \right) \\ &\quad + n_1 \left(l_1 \frac{\partial w}{\partial x} + m_1 \frac{\partial w}{\partial y} + n_1 \frac{\partial w}{\partial z} \right) \end{aligned}$$

$$= \lfloor l_1^2 \quad l_1 m_1 \quad l_1 n_1 \quad m_1 l_1 \quad m_1^2 \quad m_1 n_1 \quad n_1 l_1 \quad n_1 m_1 \quad n_1^2 \rfloor \begin{bmatrix} \frac{\partial u}{\partial x} \\ \frac{\partial u}{\partial y} \\ \frac{\partial u}{\partial z} \\ \frac{\partial v}{\partial x} \\ \vdots \\ \frac{\partial w}{\partial z} \end{bmatrix} \tag{A2.4}$$

Therefore,

$$\bar{\varepsilon} = \frac{\partial \bar{u}}{\partial \bar{x}} = \lfloor l_1^2 \quad m_1^2 \quad n_1^2 \quad l_1 m_1 \quad n_1 l_1 \quad m_1 n_1 \rfloor \begin{bmatrix} \varepsilon_x \\ \varepsilon_y \\ \varepsilon_z \\ \gamma_{xy} \\ \gamma_{xz} \\ \gamma_{yz} \end{bmatrix} \tag{A2.5}$$

Similarly for the other components of strain.

The transformation matrix in equation (8.22) is, therefore

$$[\mathbf{R}_\varepsilon] = \begin{bmatrix} l_1^2 & m_1^2 & n_1^2 & l_1 m_1 & n_1 l_1 & m_1 n_1 \\ l_2^2 & m_2^2 & n_2^2 & l_2 m_2 & n_2 l_2 & m_2 n_2 \\ l_3^2 & m_3^2 & n_3^2 & l_3 m_3 & n_3 l_3 & m_3 n_3 \\ 2l_1 l_2 & 2m_1 m_2 & 2n_1 n_2 & (l_1 m_2 + l_2 m_1) & (n_1 l_2 + n_2 l_1) & (m_1 n_2 + m_2 n_1) \\ 2l_3 l_1 & 2m_3 m_1 & 2n_3 n_1 & (l_3 m_1 + l_1 m_3) & (n_3 l_1 + n_1 l_3) & (m_3 n_1 + m_1 n_3) \\ 2l_2 l_3 & 2m_2 m_3 & 2n_2 n_3 & (l_2 m_3 + l_3 m_2) & (n_2 l_3 + n_3 l_2) & (m_2 n_3 + m_3 n_2) \end{bmatrix} \tag{A2.6}$$

Answers to Problems

Chapter 1

1.1 $$\begin{bmatrix} m_1 & 0 & 0 \\ 0 & m_2 & 0 \\ 0 & 0 & m_3 \end{bmatrix}\begin{bmatrix} \ddot{u}_1 \\ \ddot{u}_2 \\ \ddot{u}_3 \end{bmatrix} + \begin{bmatrix} (k_1+k_2) & -k_2 & 0 \\ -k_2 & (k_2+k_3) & -k_3 \\ 0 & -k_3 & (k_3+k_4) \end{bmatrix}\begin{bmatrix} u_1 \\ u_2 \\ u_3 \end{bmatrix} = \begin{bmatrix} f_1 \\ f_2 \\ f_3 \end{bmatrix}$$

1.2 $$\begin{bmatrix} m_2 & 0 \\ 0 & m_2 \end{bmatrix}\begin{bmatrix} \ddot{u}_1 \\ \ddot{u}_2 \end{bmatrix} + \begin{bmatrix} 2k_1 & -2k_1 \\ -2k_1 & (2k_1+2k_2) \end{bmatrix}\begin{bmatrix} u_1 \\ u_2 \end{bmatrix} = 0$$

1.3 $$\begin{bmatrix} m_1 & 0 & 0 \\ 0 & m_2 & 0 \\ 0 & 0 & m_3 \end{bmatrix}\begin{bmatrix} \ddot{u}_1 \\ \ddot{u}_2 \\ \ddot{u}_3 \end{bmatrix} + \begin{bmatrix} (c_1+c_2) & -c_2 & 0 \\ -c_2 & (c_2+c_3) & -c_3 \\ 0 & -c_3 & c_3 \end{bmatrix}\begin{bmatrix} \dot{u}_1 \\ \dot{u}_2 \\ \dot{u}_3 \end{bmatrix}$$
$$+ \begin{bmatrix} (k_1+k_2) & -k_2 & 0 \\ -k_2 & (k_2+k_3) & -k_3 \\ 0 & -k_3 & k_3 \end{bmatrix}\begin{bmatrix} u_1 \\ u_2 \\ u_3 \end{bmatrix} = 0$$

1.4 $$\begin{bmatrix} m & 0 \\ 0 & I_p \end{bmatrix}\begin{bmatrix} \ddot{u} \\ \ddot{\theta} \end{bmatrix} + \begin{bmatrix} (c_1+c_2) & (-c_1L_1+c_2L_2) \\ (-c_1L_1+c_2L_2) & (c_1L_1^2+c_2L_2^2) \end{bmatrix}\begin{bmatrix} \dot{u} \\ \dot{\theta} \end{bmatrix}$$
$$+ \begin{bmatrix} (k_1+k_2) & (-k_1L_1+k_2L_2) \\ (-k_1L_1+k_2L_2) & (k_1L_1^2+k_2L_2^2) \end{bmatrix}\begin{bmatrix} u \\ \theta \end{bmatrix} = 0$$

1.5 $[\mathbf{M}]\{\ddot{\mathbf{q}}\} + [\mathbf{C}]\{\dot{\mathbf{q}}\} + [\mathbf{K}]\{\mathbf{q}\} = 0$

where

$$\{\mathbf{q}\}^{\mathrm{T}} = \lfloor w \quad \theta_x \quad \theta_y \rfloor, \qquad [\mathbf{M}] = \begin{bmatrix} m & 0 & 0 \\ 0 & I_x & 0 \\ 0 & 0 & I_y \end{bmatrix}$$

$$[\mathbf{C}] = \begin{bmatrix} 4c & 2(b_1 - b_2)c & -2(a_1 - a_2)c \\ 2(b_1 - b_2)c & 2(b_1^2 + b_2^2)c & (-a_1b_1 + a_2b_1 - a_2b_2 + a_1b_2)c \\ -2(a_1 - a_2)c & (-a_1b_1 + a_2b_1 - a_2b_2 + a_1b_2)c & 2(a_1^2 + a_2^2)c \end{bmatrix}$$

$$[\mathbf{K}] = \begin{bmatrix} 4k & 2(b_1 - b_2)k & -2(a_1 - a_2)k \\ 2(b_1 - b_2)k & 2(b_1^2 + b_2^2)k & (-a_1b_1 + a_2b_1 - a_2b_2 + a_1b_2)k \\ -2(a_1 - a_2)k & (-a_1b_1 + a_2b_1 - a_2b_2 + a_1b_2)k & 2(a_1^2 + a_2^2)k \end{bmatrix}$$

1.6 $$\begin{bmatrix} m_1 & 0 & 0 \\ 0 & m_2 & 0 \\ 0 & 0 & m_3 \end{bmatrix} \begin{bmatrix} \ddot{u}_1 \\ \ddot{u}_2 \\ \ddot{u}_3 \end{bmatrix} + \begin{bmatrix} (k_1 + k_2) & -k_2 & 0 \\ -k_2 & (k_2 + k_3) & -k_3 \\ 0 & -k_3 & k_3 \end{bmatrix} \begin{bmatrix} u_1 \\ u_2 \\ u_3 \end{bmatrix} = \begin{bmatrix} f_1 \\ f_2 \\ f_3 \end{bmatrix}$$

1.7 $(I_1 + 2n^2 I_2)\,\ddot{\theta}_1 + (k_1 + 2n^2 k_2)\,\theta_1 = 0$
where $n = R_1/R_2$

1.8 $$\begin{bmatrix} I_1 & 0 & 0 & 0 \\ 0 & (I_2 + 4I_3 + 9I_4) & 0 & 0 \\ 0 & 0 & I_5 & 0 \\ 0 & 0 & 0 & I_6 \end{bmatrix} \begin{bmatrix} \ddot{\theta}_1 \\ \ddot{\theta}_2 \\ \ddot{\theta}_3 \\ \ddot{\theta}_4 \end{bmatrix}$$

$$+ \begin{bmatrix} k_1 & -k_1 & 0 & 0 \\ -k_1 & (k_1 + 4k_2 + 9k_3) & 2k_3 & 3k_3 \\ 0 & 2k_2 & k_2 & 0 \\ 0 & 3k_3 & 0 & k_3 \end{bmatrix} \begin{bmatrix} \theta_1 \\ \theta_2 \\ \theta_3 \\ \theta_4 \end{bmatrix} = 0$$

Chapter 3

In Problems 3.1–3.9, 3.20 the percentage difference between the approximate solution and analytical solution is given.

3.1 0.64, 3.12

3.2 0.55, 18.6

3.3 11.0, 27.2

3.4 0.34, 1.61

3.5 0.88, 5.76

3.6 4.58, 16.9

3.7 10.2. Rigid body mode

3.8 11.0, 27.2

3.9 1.63, 32.4

3.11 (1) $T_e = \frac{1}{2}(\rho A \times 2a)\dot{v}^2$, $U_e = 0$
(2) $T_e = \frac{1}{2}(\frac{2}{3}\rho A a^3)\dot{\theta}^2$, $U_e = 0$
$\rho A \times 2a =$ mass, $\frac{2}{3}\rho A a^3 =$ moment of inertia

3.15 16.06, 44.26 Hz
Modes 1T, 2T

3.16 Eigenvector $\lfloor 0 \quad 1 \quad 0.7068 \rfloor^T$
Stress varies linearly from 1.827 E/L to 0.173 E/L
Gauss point stresses in error by −0.50 and +2.07%

3.20 2.4, 18.6

Chapter 4

4.1 Gives mass due to rigid body translation in the x-direction

4.2 No forces are required for a rigid body translation in the x-direction

4.9 (2 × 2) for both inertia and stiffness matrices

4.10 (3 × 3), (2 × 2)

4.13 $(bp_x/3, 0)$, $(bp_x/3, 0)$, $(4\,bp_x/3, 0)$ at nodes 2, 3, 6 respectively.

Chapter 5

5.1 $$P_r = \frac{p\phi}{\pi} + \sum_{n=1}^{\infty} \frac{2p}{\pi} \frac{\sin n\phi}{n} \cos n\theta$$

5.2 $$p_r = \frac{p}{2\pi a} + \sum_{n=1}^{\infty} \frac{p}{\pi a} \cos \theta$$

5.10 Inertia (4 × 3), stiffness (3 × 2)

5.11 $(-bcp_x/3, 0, 0)$ at nodes 2, 3, 6, 7 and $(+4bcp_x/3, 0, 0)$ at nodes 10, 14, 18, 19.

Chapter 6

6.4 (4 × 4), (4 × 4)

Chapter 11

11.1 1, 2, 3 rad/s: $\left(\frac{1}{3}, 1, 1\right)$, $\left(1, 0, -\frac{1}{3}\right)$, $\left(-\frac{1}{5}, 1, -\frac{3}{5}\right)$

11.3 2

11.5 $\begin{bmatrix} 4 & -3^{1/2} & 0 & 0 \\ & 4/3 & 2^{1/2}/3 & 0 \\ & & 14/3 & 0 \\ \text{Sym} & & & 4 \end{bmatrix}$

11.7 1.5550

11.8 $\lfloor 1.0 \quad 0.445 \quad -0.802 \rfloor$

11.9 0.2087, 4, 4, 4.7913

11.10 Symmetric $(u, \theta_z) = 0$
Antisymmetric $v = 0$

11.11 (1) 0.40, 11.0; (2) 0.72, 27.2

Chapter 12

12.1 $\alpha_{21}(\omega) = \frac{1}{9}[2(1000 - \omega^2)^{-1} + 2(4000 - \omega^2)^{-1} - 4(7000 - \omega^2)^{-1}]$
Resonances at 5.03, 10.07, 13.32 Hz
Anti-resonance at 8.72 Hz

12.2 $u_1 = 10^{-3}(-0.4290 - \text{i}0.6630)$
$u_2 = 10^{-3}(0.1249 + \text{i}1.198)$
$u_3 = 10^{-3}(0.0615 - \text{i}0.5950)$

12.3 $25.78 \times 10^{-5} \exp \text{i}(wt - 1.5633)$

12.4 $f(t) = \frac{p}{\pi}\left[1 + \frac{\pi}{2}\sin\frac{\pi t}{\tau} - \frac{2}{3}\cos\frac{2\pi t}{\tau} - \frac{2}{15}\cos\frac{4\pi t}{\tau} - \frac{2}{35}\cos\frac{6\pi t}{\tau} - \cdots\right]$

12.5 Maximum response underestimated by 0.82%

12.6 $w_0 \Delta t < 2(1 - \zeta^2)^{1/2}$

12.7 $\dot{u}_{j+1} = \dot{u}_j + \frac{\Delta t}{2}(\ddot{u}_j + \ddot{u}_{j+1})$

$$u_{j+1} = u_j + \Delta t \dot{u}_j + \frac{(\Delta t)^2}{4}(\ddot{u}_j + \ddot{u}_{j+1})$$

Bibliography

CHAPTER 1

A. R. Collar and A. Simpson (1987) *Matrices and Engineering Dynamics*. Chichester: Ellis Horwood.

R. W. Clough and J. Penzien (1975) *Dynamics of Structures*. New York: McGraw-Hill.

C. L. Dym and J. H. Shames (1973) *Solid Mechanics: A Variational Approach*. New York: McGraw-Hill.

M. J. Forray (1968) *Variational Calculus in Science and Engineering*. New York: McGraw-Hill.

C. Fox (1950) *Calculus of Variations*. Oxford University Press.

L. Meirovitch (1967) *Analytical Methods in Vibrations*. New York: Macmillan.

L. Meitrovitch (1975) *Elements of Vibration Analysis*. New York: McGraw-Hill.

W. T. Thomson (1972) *Theory of Vibrations with Applications*. New Jersey: Prentice-Hall.

G. B. Warburton (1976) *The Dynamical Behaviour of Structures*, 2nd edn. Oxford: Pergamon.

CHAPTER 2

A. D. Barr (1962) Torsional waves in uniform rods of non-circular cross-section. *J. Mech. Eng. Sci.* **4**, 127–35.

C. L. Dym and J. H. Shames (1973) *Solid Mechanics: A Variational Approach*. New York: McGraw-Hill.

R. F. Hearmon (1961) *An Introduction to Applied Anisotropic Elasticity*. Oxford: Oxford University Press.

H. Kolsky (1963) *Stress Waves in Solids*. New York: Dover Publications.

S. G. Lekhnitskii (1963) *Theory of Elasticity of an Anisotropic Elastic Body*. Translation from Russian by P. Fern. San Francisco: Holden Day.

R. D. Mindlin (1951) Influence of rotatory inertia and shear on flexural motions of isotropic, elastic plates. *J. Appl. Mech., Trans. ASME* **18**, 31–8.

S. Timoshenko and J. N. Goodier (1970) *Theory of Elasticity*, 3rd edn. New York: McGraw-Hill.

S. Timoshenko and S. Woinowsky-Krieger (1959) *Theory of Plates and Shells*, 2nd edn. New York: McGraw-Hill.

G. B. Warburton (1976) *The Dynamical Behavior of Structures*, 2nd edn. Oxford: Pergamon.

CHAPTER 5

J. H. Argyris, K. E. Buck, I. Grieger and G. Maraczek (1970) Application of the matrix displacement method to the analysis of pressure vessels. *J. Eng. Ind., Trans. ASME* **92**, 317–29.

M. A. J. Bossak and O. C. Zienkiewicz (1973) Free vibration of initially stressed solids, with particular reference to centrifugal-force effects in rotating machinery. *J. Strain Anal.* **8**, 245–52.

H. E. Ermutlu (1968) Dynamic Analysis of Arch Dams Subject to Seismic Disturbances. Ph.D. thesis, University of Southampton.

S. Ghosh and E. L. Wilson (1969) Dynamic stress analysis of axisymmetric structures under arbitrary loading. *Report EERC* 69–10, College of Engineering, University of California, Berkeley.

P. Kelen (1985) A Finite Element Analysis of the Vibration Characteristics of Rotating Turbine Blade Assemblies. Ph.D. thesis, University of Surrey.

M. Langballe, E. Aasen and T. Mellem (1974) Application of the finite element method to machinery. *Computers and Structures* **4**, 149–92.

F. E. Sagendorph (1976) Natural frequencies of mid-span shrouded fan blades. In *Structural Dynamic Aspects of Bladed Disc Assemblies*, ed. A. V. Srinivasan, 93–9. New York: The American Society of Mechanical Engineers.

G. Waas (1972) Linear Two-dimensional Analysis of Soil Dynamics Problems in Semi-infinite Layered Media. Ph.D. thesis, University of California, Berkeley.

E. L. Wilson (1965) Structural analysis of axisymmetric solids. *AIAA J.* **3**, 2269–74.

E. L. Wilson, R. L. Taylor, W. P. Doherty and J. Ghaboussi (1973) Incompatible displacement models. In *Numerical and Computer Methods in Structural Mechanics*, ed. S. J. Fenves, N. Perrone, A. R. Robinson and W. C. Schnbrich, 43–57. New York: Academic.

CHAPTER 6

D. J. Dawe (1965) A finite element approach to plate vibration problems. *J. Mech. Eng. Sci.* **7**, 28–32.

R. J. Guyan (1965) Distributed mass matrix for plate element bending. *AIAA J.* **3**, 567–8.

J. S. Przemieniecki (1966) Equivalent mass matrices for rectangular plates in bending. *AIAA J.* **4**, 949–50.

B. M. Irons and K. J. Draper (1965) Inadequacy of nodal connections in a stiffness solution for plate bending. *AIAA J.* **3**, 961.

J. L. Tocher and K. K. Kapur (1965) Comment on 'Basis for derivation of matrices for the direct stiffness'. *AIAA J.* **3**, 1215.

References

CHAPTER 2

[2.1] J. T. Oden (1967) *Mechanics of Elastic Structures*, New York: McGraw-Hill.
[2.2] G. R. Cowper (1966) The shear coefficient in Timoshenko's beam theory. *J. Appl. Mech. Trans. ASME* **88**, 335–40.

CHAPTER 3

[3.1] E. Kreysziz (1972) *Advanced Engineering Mathematics*, 3rd edn, New York: Wiley.
[3.2] R. Courant and D. Hilbert (1953) *Methods of Mathematical Physics*, Vol. I. New York: Interscience.
[3.3] L. V. Kanotorovich and V. I. Krylov (1958) *Approximate Methods of Higher Analysis*, Groningen: Noordhoff.
[3.4] S. G. Mikhlin (1964) *Variational Methods in Mathematical Physics*, New York: Macmillan.
[3.5] S. G. Mikhlin (1965) *The Problem of the Minimum of a Quadratic Functional*, San Francisco: Holden-Day.
[3.6] R. E. D. Bishop, G. M. L. Gladwell and S. Michaelson (1965) *The Matrix Analysis of Vibration*, Cambridge: Cambridge University Press.
[3.7] C. L. Dym and I. H. Shames (1973) *Solid Mechanics: A Variational Approach*, New York: McGraw-Hill.
[3.8] G. B. Warburton (1976) *The Dynamical Behaviour of Structures*, 2nd edn, Oxford: Pergamon.
[3.9] J. T. Oden (1972) *Finite Elements of Nonlinear Continua*, New York: McGraw-Hill.
[3.10] J. S. Przemieniecki (1968) *Theory of Matrix Structural Analysis*, New York: McGraw-Hill.
[3.11] S. Timoshenko and J. N. Goodier (1970) *Theory of Elasticity*, 3rd edn, New York: McGraw-Hill.
[3.12] A. E. H. Love (1944) *Mathematical Theory of Elasticity*, New York: Dover.
[3.13] R. Courant (1943) Variational methods for the solution of problems of equilibrium and vibrations. *Bull. Amer. Math. Soc.* **49**, 1–29.
[3.14] F. A. Leckie and G. M. Lindberg (1963) The effect of lumped parameters on beam frequencies. *Aeronaut. Quart.* **14**, 224–40.
[3.15] F. F. Rudder (1970) Effect of stringer eccentricity on the normal mode stress response of stiffened flat panel arrays. Conference on Current Developments in Sonic Fatigue, Southampton University. Paper J. Available from ISVR, University of Southampton.
[3.16] J. T. Oden (1967) *Mechanics of Elastic Structures*, New York: McGraw-Hill.

[3.17] T. Y. Yang and C. T. Sun (1973) Axial-flexural vibration of frameworks using finite element approach. *J. Acoust. Soc. Amer.* **53**, 137–46.

[3.18] G. M. L. Gladwell (1964) The vibration of frames. *J. Sound Vibration* **1**, 402–25.

[3.19] W. Carnegie, J. Thomas and E. Dokumaci (1969) An improved method of matrix displacement analysis in vibration problems. *Aeronaut. Quart.* **20**, 321–32.

[3.20] Z. Kopal (1961) *Numerical Analysis*. London: Chapman and Hall.

[3.21] F. Scheid (1968) *Numerical Analysis*. Schaum's Outline Series. New York: McGraw-Hill.

[3.22] D. L. Thomas, J. M. Wilson and R. R. Wilson (1973) Timoshenko beam finite elements. *J. Sound Vibration* **31**, 315–30.

[3.23] A. W. Lees and D. L. Thomas (1982) Unified Timoshenko beam finite element. *J. Sound Vibration* **80**, 355–66.

[3.24] S. Corn, N. Bouthaddi and J. Piranda (1997) Transverse vibrations of short beams: finite element models obtained by a condensation method. *J. Sound Vibration* **201**, 353–63.

[3.25] T. C. Huang (1961) The effect of rotary inertia and of shear deformation on the frequency and normal mode equations of uniform beams with simple end conditions. *J. Appl. Mech. Trans. ASME* **28**, 579–84.

[3.26] T. C. Huang and C. S. Kung (1963) New tables of eigenfunctions representing normal modes of vibration of Timoshenko beams. *Developments in Theoretical and Applied Mechanics* I, New York: Plenum Press, 59–71.

[3.27] D. J. Dawe (1978) A finite element for the vibration analysis of Timoshenko beams. *J. Sound Vibration* **60**, 11–20.

[3.28] T. Moan (1973) On the local distribution of errors by finite element approximations. In *Theory and Practice in Finite Element Structural Analysis*, ed. Y. Yamada and R. H. Gallagher. Tokyo: University of Tokyo Press, 43–60.

[3.29] E. Hinton and J. S. Campbell (1974) Local and global smoothing of discontinuous finite element functions using a least squares method. *Int. J. Num. Meth. Eng.* **8**, 461–80.

[3.30] E. Hinton, F. C. Scott and R. E. Ricketts (1975) Least squares stress smoothing for parabolic isoparametric elements. *Int. J. Num. Meth. Eng.* **9**, 235–56.

[3.31] J. Barlow (1976) Optimal stress locations in finite element models. *Int. J. Num. Meth. Eng.* **10**, 243–51.

[3.32] T. J. R. Hughes (1977) A simple and efficient finite element for plate bending. *Int. J. Num. Meth. Eng.* **11**, 1529–43.

[3.33] G. R. Bhashyam and G. Prathap (1981) The second frequency spectrum of Timoshenko beams. *J. Sound Vibration* **76**, 407–20.

[3.34] A. H. Vermeulen and G. R. Heppler (1998) Predicting and avoiding shear locking in beam vibration problems using the β-spline field approximation method. *Comp. Methods Appl. Mech. Eng* **158**, 311–27.

[3.35] M. D. Olson (1975) Compatibility of finite elements in structural mechanics. Proc. World Cong. on Finite Element Methods in Structural Mechanics, Bournemouth, England. Okehampton: Robinson and Associates. H1–H33.

[3.36] V. Z. Vlasov (1961) *Thin Walled Elastic Beams*, 2nd edn, Washington, DC: National Science Foundation.

[3.37] M. Tanaka and A. N. Bercin (1997) Finite element modelling of the coupled bending and torsional free vibration of uniform beams with an arbitrary cross-section. *Appl. Math. Modell.* **21**, 339–44.

[3.38] C. A. Mota Soares and J. E. Barradas Cardoso (1979) Finite element dynamic analysis of structures based on the Vlasov beam theory. In *Numerical Analysis of the Dynamics of Ship Structures, Euromech* 122, Paris.

[3.39] Y. Hu, X. Jin and B. Chen (1996) A finite element model for the static and dynamic analysis of thin-walled beams with asymmetric cross-sections. *Comput. Struct.* **61**, 897–908.

[3.40] Q.-F. Wang (1997) Spline finite member element method for vibration of thin-walled members with shear lag. *J. Sound Vibration* **206**, 339–52.

[3.41] J. H. Kim and Y. Y. Kim (2000) One-dimensional analysis of thin-walled closed beams having general cross-sections. *Int. J. Num. Math. Eng.* **49**, 653–68.

[3.42] G. M. Lindberg (1963) Vibration of non-uniform beams. *Aeronaut. Quart.* **14**, 387–95.

[3.43] W. L. Cleghorn and B. Tabarrock (1992) Finite element formulation of a tapered Timoshenko beam for free lateral vibration analysis. *J. Sound Vibration* **152**, 461–70.

[3.44] J. Thomas and E. Dokumaci (1974) Simple finite elements for pre-twisted blading vibration. *Aeronaut. Quart.* **25**, 109–18.

[3.45] B. Yardimoglu and T. Yildirim (2004) Finite element model for vibration analysis of pre-twisted Timoshenko beam, *J. Sound Vibration* **273**, 741–54.

[3.46] R. S. Gupta and S. S. Rao (1978) Finite element eigenvalue analysis of tapered and twisted Timoshenko beams. *J. Sound Vibration* **56**, 187–200.

[3.47] P. Raveendranath, G. Singh and B. Pradhan (2000) Free vibration of arches using a curved beam element based on a coupled polynomial displacement field. *Comput. Struct.* **78**, 583–90.

[3.48] P. Litewka and J. Rakowski (2001) Free vibrations of shear-flexible and compressible arches by FEM. *Int. J. Num. Math. Eng.* **52**, 273–86.

[3.49] J.-S. Wu and L.-K. Chiang (2004) A new approach for free vibration analysis of arches with effects of shear deformation and rotary inertia considered. *J. Sound Vibration* **277**, 49–71.

[3.50] R. R. Rossi (1989) In-plane vibrations of circular rings of non-uniform cross-section with account taken of shear and rotatory inertia effects. *J. Sound Vibration* **135**, 443–52.

[3.51] J.-S. Wu and L.-K. Chiang (2004) Free vibration of a circularly curved Timoshenko beam normal to its initial plane using finite curved beam elements. *Comput. Struct.* **82**, 2525–40.

[3.52] N.-I. Kim, K.-J. Seo and M.-Y. Kim (2003) Free vibration and spatial stability of non-symmetric thin-walled curved beams with variable curvatures. *Int. J. Solids Structures* **40**, 3107–28.

[3.53] D. L. Thomas and R. R. Wilson (1973) The use of straight beam finite elements for analysis of vibrations of curved beams. *J. Sound Vibration* **26**, 155–8.

CHAPTER 4

[4.1] P. C. Dunne (1968) Complete polynomial displacement fields for finite element method. *Aeronaut. J.* **72**, 245–6.

[4.2] B. M. Irons, J. Ergatoudis and O. C. Zienkiewicz (1968) Comment on ‘Complete polynomial displacement fields for finite element method’. *Aeronaut. J.* **72**, 709–11.

[4.3] J. B. Carr (1970) The effect of shear flexibility and rotatory inertia on the natural frequencies of uniform beams. *Aeronaut. Quart.* **21**, 79–90.

[4.4] J. Barlow (1976) Optimal stress locations in finite element models. *Int. J. Num. Meth. Eng.* **10**, 243–51.

[4.5] M. A. Eisenberg and L. E. Malvern (1973) On finite element integration in natural co-ordinates. *Int. J. Num. Meth. Eng.* **7**, 574–5.

[4.6] K. J. Bathe (1996) *Finite Element Procedures*. Upper Saddle River, NJ: Prentice Hall.

[4.7] O. C. Zienkiewicz and R. L. Taylor (1989) *The Finite Element Method. Vol. 1: Basic Formulation and Linear Problems*, 4th edn, London: McGraw-Hill Book Company.

[4.8] T. K. Hellen (1976) Numerical Integration considerations in two and three dimensional isoparametric finite elements. In *The Mathematics of Finite Elements and Applications II* ed. J. R. Whiteman, London: Academic Press, 511–24.

[4.9] T. Krauthammer (1979) Accuracy of the finite element method near a curved boundary. *Computers and Structures* **10**, 921–9.

[4.10] R. D. Cook, D. S. Malkus, M. E. Plesha and R. J. Witt (2002) *Concepts and Applications of Finite Element Analysis*, 4th edn, New York: John Wiley & Sons.

[4.11] R. D. Cook (1975) Avoidance of parasitic shear in plane element. *J. Struct. Div. Proc. ASCE* **101**, 1239–53.

[4.12] R. W. Clough and A. K. Chopra (1966) Earthquake stress analysis in earth dams. *J. Eng. Mech. Proc. ASCE* **92**, 197–211.

CHAPTER 5

[5.1] P. C. Hammer, O. J. Marlowe and A. H. Stroud (1956) Numerical integration over simplexes and cones. *Mathematical Tables and other Aids to Computation* **10**, 130–6.

[5.2] M. E. Laursen and M. Gellert (1978) Some criteria for numerically integrated matrices and quadrature formulas for triangles. *Int. J. Num. Meth. Eng.* **12**, 67–76.

[5.3] T. Belytschko (1972) Finite elements for axisymmetric solids under arbitrary loadings with nodes on origin. *AIAA J.* **10**, 1532–3.

[5.4] K. E. Buck (1973) Comment on 'Finite elements for axisymmetric solids under arbitrary loadings with nodes on origin'. *AIAA J.* **11**, 1357–8.

[5.5] T. Belytschko (1973) Reply by Author to K. E. Buck. *AIAA J.* **11**, 1358.

[5.6] H. Deresiewicz and R. D. Mindlin (1955) Axially symmetric flexural vibrations of a circular disc. *J. Appl. Mech., Trans. ASME* **22**, 86–8.

[5.7] D. C. Gazis and R. D. Mindlin (1960) Extensional vibrations and waves in a circular disc and a semi-infinite plate. *J. Appl. Mech. Trans. ASME* **27**, 541–7.

[5.8] W. E. Baker and J. M. Daly (1967) Dynamic analysis of continuum bodies by direct stiffness method. *Shock and Vibration Bull.* **36**, Part 5, 55–68.

[5.9] K. K. Gupta (1984) STARS – A general purpose finite element computer program for analysis of engineering structures. *NASA Reference Publication* 1129.

[5.10] B. M. Irons (1971) Quadrature rules for brick based finite elements. *Int. J. Num. Meth. Eng.* **3**, 293–4.

[5.11] M. A. Eisenberg and L. E. Malvern (1973) On finite element integration in natural coordinates. *Int. J. Num. Meth. Eng.* **7**, 574–5.

[5.12] D. P. Gao and M. Petyt (1983) Prediction of frequencies of a practical turbine disc. *ISVR Memorandum* No. 636, University of Southampton.

[5.13] S. S. Rao and A. S. Prasad (1975) Vibrations of annular plates including the effects of rotary inertia and transverse shear deformation. *J. Sound Vibration* **42**, 305–24.

[5.14] K. J. Bathe (1996) *Finite Element Procedures*. Upper Saddle River, NJ: Prentice Hall.

[5.15] O. C. Zienkiewicz and R. L. Taylor (1989) *The Finite Element Method. Vol. 1: Basic Formulation and Linear Problems*, 4th edn, London: McGraw-Hill Book. Company.

[5.16] S. E. Johnson and E. I. Field (1973) Three isoparametric solid elements for NASTRAN. NASA TM X-2893 NASTRAN: Users' Experiences, 423–37.

[5.17] C. W. S. To (1982) Application of the finite element method for the evaluation of velocity response of anvils. *J. Sound Vibration* **84**, 529–48.

[5.18] T. K. Hellen (1976) Numerical integration considerations in two and three dimensional isoparametric finite elements. In *The Mathematics of Finite Elements and Applications II* ed. J. R. Whiteman, London: Academic Press, 511–24.

[5.19] T. K. Hellen (1972) Effective quadrature rules for quadratic solid isoparametric finite elements. *Int. J. Num. Meth. Eng.* **4**, 597–600.

[5.20] A. M. Salama (1976) Finite Element Dynamic Analysis of Blade Packets and Bladed Disc Assemblies. Ph.D. Thesis, University of Southampton.

[5.21] A. E. Armenakas, D. C. Gazis and G. Herrmann (1969) *Free Vibrations of Circular Cylinder Shells*, Oxford: Pergamon Press.

CHAPTER 6

[6.1] A. Adini and R. W. Clough (1961) Analysis of Plate Bending by the Finite Element Method. Report submitted to the National Science Foundation, G7337.

[6.2] R. J. Melosh (1963) Basis for derivation of matrices for the direct stiffness method. *AIAA J.* **1**, 1631–7.

[6.3] C. V. Smith (1970) Finite Element Model, with Applications to Buildings and K-33 and K-31. Union Carbide Corporation Report Number CTC-29.

[6.4] R. A. Tinawi (1972) Anisotropic tapered elements using displacement models. *Int. J. Num. Meth. Eng.* **4**, 475–89.

[6.5] G. M. Lindberg, M. D. Olson and H. A. Tulloch (1969) Closed Form Finite Element Solutions for Plate Vibrations. National Research Council of Canada, Aeronautical Report, LR-518.

[6.6] G. M. Lindberg and M. D. Olson (1970) Convergence studies of eigenvalue solutions using two finite plate bending elements. *Int. J. Num. Meth. Eng.* **2**, 99–116.

[6.7] R. E. Reid (1965) Comparison of Methods in Calculating Frequencies of Corner Supported Rectangular Plates. NASA TN D-3030.

[6.8] D. J. Gorman (1981) An analytical solution for the free vibration analysis of rectangular plates resting on symmetrically distributed point supports. *J. Sound Vibration* **79**, 561–74.

[6.9] M. Petyt and W. H. Mirza (1972) Vibration of column supported floor slabs. *J. Sound Vibration* **21**, 355–64.

[6.10] T. Wah (1964) Vibration of stiffened plates. *Aeronaut. Quart.* **15**, 285–98.

[6.11] N. J. Huffington (1956) Theoretical determination of rigidity properties of orthogonally stiffened plates. *J. Appl. Mech. Trans. ASME* **23**, 15–20.

[6.12] F. K. Bogner, R. L. Fox and L. A. Schmit (1966) The generation of inter-element-compatible stiffness and mass matrices by the use of interpolation formulas. In Matrix Methods in Structural Mechanics, AFFDL-TR-66-80, 397–443.

[6.13] G. A. Butlin and F. A. Leckie (1966) A study of finite elements applied to plate flexure. Symposium on Numerical Methods for Vibration Problems, University of Southampton, 3, 26–37. Availabe from ISVR, Southampton.

[6.14] R. E. Rossi (1997) A note on a finite element for vibrating thin, orthotropic rectangular plates. *J. Sound Vibration* **208**, 864–68.

[6.15] V. Mason (1967) On the use of rectangular finite elements. *ISVR Report* No. 161, University of Southampton.

[6.16] V. Mason (1968) Rectangular finite elements for analysis of plate vibrations. *J. Sound Vibration* **7**, 437–48.

[6.17] R. R. Wilson and C. A. Brebbia (1971) Dynamic behaviour of steel foundations for turbo-alternators. *J. Sound Vibration* **18**, 405–16.

[6.18] T. J. R. Hughes, R. L. Taylor and W. Kanoknukulcha (1977) A simple efficient finite element for plate bending. *Int. J. Num. Meth. Eng.* **11**, 1529–43.

[6.19] E. D. L. Pugh, E. Hinton and O. C. Zienkiewicz (1978) A study of quadrilateral plate bending elements with reduced integration. *Int. J. Num. Meth. Eng.* **12**, 1059–79.

[6.20] E. Hinton and N. Bicanic (1979) A comparison of Lagrangian and Serendipity Mindlin plate elements for free vibration analysis. *Computers and Structures* **10**, 483–93.

[6.21] S. Srinivas, C. V. Joga Rao and A. K. Rao (1970) An exact analysis for vibration of simply supported homogeneous and laminated thick rectangular plates. *J. Sound Vibration* **12**, 187–99.

[6.22] J. Robinson (1978) Element evaluation – a set of assessment points and standard tests. In *Finite Element Methods in the Commercial Environment*, ed. J. Robinson. Okehampton: Robinson and Associates, 218–47.

[6.23] J. L. Tocher (1962) Analysis of Plate Bending using Triangular Elements. Ph.D. dissertation, California University, Berkeley.

[6.24] M. Petyt (1967) Finite Element Vibration Analysis of Cracked Plates in Tension. Ph.D. thesis, University of Southampton.

[6.25] M. Petyt (1966) Structural vibration analysis using triangular finite elements. Symposium on Numerical Methods for Vibration Problems, Southampton University, **3**, 55–64. Available from ISVR, Southampton.

[6.26] P. N. Gustafson, W. F. Stokey and C. F. Zorowski (1953) An experimental study of natural vibrations of cantilevered triangular plates. *J. Aeronaut. Sci.* **20**, 331–7.

[6.27] R. W. Clough and J. L. Tocher (1966) Finite element stiffness matrices for plate bending. In Matrix Methods in Stuctural Mechanics, AFFDL-TR-66-80, 515–45.

[6.28] R. W. Clough and C. A. Felippa (1968) A refined quadrilateral element for analysis of plate bending. In Matrix Methods in Structural Mechanics, AFFDL-TR-68-150, 399–440.

[6.29] S. M. Dickinson and R. D. Henshell (1969) Clough–Tocher triangular plate bending element in vibration. *AIAA J.* **7**, 560–1.

[6.30] J.-L. Batoz, K.-J. Bathe and L.-W. Ho (1980) A study of three-node triangular plate bending elements. *Int. J. Num. Meth. Eng.* **15**, 1771–812.

[6.31] P. P. Lynn and B. S. Dhillon (1971) Triangular thick plate bending elements. Proc. 1st Int. Conf. on Structural Mechanics in Reactor Technology, Berlin, **6**, 365–89.

[6.32] A. L. Deak and T. H. H. Pian (1967) Application of the smooth surface interpolation to the finite element analysis. *AIAA J.* **5**, 187–9.

[6.33] G. Birkhoff and H. L. Garabedian (1960) Smooth surface interpolation. *J. Math. Phys.* **39**, 258–68.

[6.34] B. Fraeijs De Veubeke (1968) A conforming finite element for plate bending. *Int. J. Solids Structures* **4**, 95–108.

[6.35] R. M. Orris and M. Petyt (1973) A finite element study of the vibration of trapezoidal plates. *J. Sound Vibration* **27**, 325–44.

[6.36] T. Rock and E. Hinton (1974) Free vibration and transient response of thick and thin plates using the finite element method. *Int. J. Earthquake Eng. Struct. Dyn.* **3**, 51–63.

[6.37] A. Razzaque (1984) On the four noded discrete Kirchhoff shell elements. In *Accuracy, Reliability and Training in FEM Technology*, ed. J. Robinson. Okehampton: Robinson and Associates, 473–83.

[6.38] A.-K. Soh and C. Ling (2000) An improved discrete Kirchhoff triangular element for bending, vibration and buckling analysis. *Eur. J. Mech. A. Solids* **19**, 891–910.

[6.39] K.-J. Bathe (1996) *Finite Element Procedures*. Upper Saddle River, NJ: Prentice Hall.

[6.40] E. Hernández, L. Hervella-Nieto and R. Rodriguez (2003) Computation of the vibration modes of plates and shells by low-order MITC quadrilateral finite elements. *Comput. Struct.* **81**, 615–28.

[6.41] B. S. Al Janabi and E. Hinton (1987) A study of the free vibrations of square plates with various edge conditions. In E. Hinton (ed.) *Numerical Methods and Software for Dynamic Analysis of Plates and Shells*, Swansea: Pineridge Press, 167–204.

[6.42] A. W. Leissa (1969) *Vibration of Plates*, Washington DC: US Government Printing Office. (Reprinted 1993 by The Acoustical Society of America.)

[6.43] G. R. Cowper, E. Kosko, G. M. Lindberg and M. D. Olson (1968) A High Precision Triangular Plate Bending Element. National Research Council of Canada, Aeronautical Report LR-514.

[6.44] S. Ghazzi, F. A. Barki and H. M. Safwat (1997) Free vibration analysis of penta, hepta-gonal shaped plates. *Comp. Struct.* **62**, 395–407.

[6.45] N. Popplewell and D. McDonald (1971) Conforming rectangular and triangular plate bending elements. *J. Sound Vibration* **19**, 333–47.

[6.46] R. W. Claassen and C. J. Thorne (1961) Vibrations of thin rectangular isotropic plates. *J. Appl. Mech. Trans ASME* **28**, 304–5.

[6.47] G. M. Lindberg (1967) The Vibration of Stepped Cantilever Plates. National Research Council of Canada, Aeronautical Report, LR-494.

[6.48] R. Plunkett (1963) Natural frequencies of uniform and non-uniform rectangular cantilever plates. *J. Mech. Eng. Sci.* **5**, 146–56.

CHAPTER 7

[7.1] M. D. Olson and G. M. Lindberg (1970) Free Vibrations and Random Response of an Integrally Stiffened Panel. National Research Council of Canada, Aeronautical Report, LR-544.

[7.2] R. N. Yurkovich, J. H. Schmidt and A. R. Zak (1971) Dynamic analysis of stiffened panel structures. *J. Aircraft* **8**, 149–55.

[7.3] R. P. McBean (1968) Analysis of Stiffened Plates by the Finite Element Method. Ph.D. Thesis, Stanford University.

[7.4] M. D. Olson and C. R. Hazell (1977) Vibration studies on some integral rib-stiffened plates. *J. Sound Vibration* **50**, 43–61.

[7.5] M. Petyt (1977) Finite strip analysis of flat skin-stringer structures. *J. Sound Vibration* **54**, 537–47.

[7.6] M. N. Bapu Rao, P. Guruswamy, M. Venkateshwara Rao and S. Pavithran (1978) Studies on vibration of some rib-stiffened cantilever plates. *J. Sound Vibration* **57**, 389–402.

[7.7] C. K. Ramesh and R. M. Belkune (1973) Free vibrations of plate-beam systems. In *Theory and Practice in Finite Element Structural Analysis*, ed. Y. Yamada and R. H. Gallagher. Tokyo: University of Tokyo Press, 357–70.

[7.8] R. E. Miller (1980) Dynamic aspects of the error in eccentric beam modelling. *Int. J. Num. Meth. Eng.* **15**, 1447–55.

[7.9] R. E. Grandle and C. E. Rucker (1971) Modal analysis of a nine-bay skin-stringer panel. In NASA TM X-2378 NASTRAN: Users' Experiences, 343–61.

[7.10] E. A. Thornton (1972) A NASTRAN Correlation Study for Vibrations of a Cross-stiffened Ship's Deck. NASA TM X-2637 NASTRAN: Users' Experiences, 145–59.

[7.11] R. R. Wilson and C. A. Brebbia (1971) Dynamic behaviour of steel foundations for turbo-alternators. *J. Sound Vibration* **18**, 405–16.

[7.12] P. S. Nair and M. S. Rao (1984) On vibration of plates with varying stiffener length. *J. Sound Vibration* **95**, 19–29.

[7.13] M. S. Rao, P. S. Nair and S. Durvasula (1985) On vibration of eccentrically stiffened plates with varying stiffener length. *J. Sound Vibration* **99**, 568–71.

[7.14] A. Mukherjee and M. Mukhopadhyay (1988) Finite element free vibration of eccentrically stiffened plates. *Comp. Struct.* **30**, 1303–17.

[7.15] G. S. Palani, N. R. Iyer and T. V. S. R. Appa Rao (1992) An efficient finite element model for static and vibration analysis of eccentrically stiffened plates/shells. *Comp. Struct.* **43**, 651–61.

[7.16] G. S. Palami, N. R. Iyer and T. V. S. R. Appa Rao (1993) An efficient finite element model for static and vibration analysis of plates with arbitrary located eccentric stiffeners. *J. Sound Vibration* **166**, 409–27.

[7.17] Y. Y. Lee and C.-F. Ng (1998) Sound insertion loss of stiffened enclosure plates using the finite element method and the classical approach. *J. Sound Vibration* **217**, 239–60.

[7.18] N. Popplewell (1971) The vibration of a box-type structure I. Natural frequencies and normal modes. *J. Sound Vibration* **14**, 357–65.

[7.19] S. M. Dickinson and G. B. Warburton (1967) Vibration of box-type structures. *J. Mech. Eng. Sci.* **9**, 325–35.

[7.20] N. Popplewell, N. A. N. Youssef and D. McDonald (1976) Economical evaluation of the vibration characteristics of rectangular structures with sloping roofs. *J. Sound Vibration* **44**, 493–7.

[7.21] T. C. Huang and C. S. Kung (1963) New tables of eigenfunctions representing normal modes of vibration of Timoshenko beams. *Developments in Theoretical and Applied Mechanics* I. New York: Plenum Press, 59–71.

[7.22] A. W. Lees, D. L. Thomas and R. R. Wilson (1976) Analysis of the vibration of box beams. *J. Sound Vibration* **45**, 559–68.

[7.23] L. A. Schmit, F. K. Bogner and R. L. Fox (1968) Finite deflection structural analysis using plate and shell discrete elements. *AIAA J.* **6**, 781–91.

[7.24] R. Hickling and M. M. Kamal (Eds) (1982) *Engine Noise: Excitation, Vibration and Radiation*, New York: Plenum Press.

[7.25] N. Lalor and M. Petyt (1982) Noise assessment of engine structure designs by finite element techniques. In Reference [7.24], 211–44.

[7.26] O. C. Zienkiewicz and R. L. Taylor (1991) *The Finite Element Method. Vol. 2: Solid and Fluid Mechanics, Dynamics and Non-linearity*, 4th edn, London: McGraw-Hill Book Company.

[7.27] R. D. Cook, D. S. Malkus, M. E. Plesha and R. J. Witt (2002) *Concepts and Applications of Finite Element Analysis*, 4th edn, New York: John Wiley & Sons.

[7.28] R. H. MacNeal and R. L. Harder (1988) A refined four-noded membrane element with rotational degrees of freedom. *Comp. Struct.* **28**, 75–84.

[7.29] R. D. Cook (1994) Four-node flat shell element: drilling degrees of freedom, membrane-bending coupling, warped geometry, and behaviour. *Comp. Struct.* **50**, 549–55.

[7.30] A. G. Razaqpur, O. Aziz and M. Nofal (1990) A new quadrilateral facet shell element. In *FEM in the Design Process*, Okehampton: Rabinson and Associates, 502–10.

[7.31] C. T. F. Ross (1975) Free vibration of thin shells. *J. Sound Vibration* **39**, 337–44.

[7.32] J. J. Webster (1968) Free vibration of rectangular curved panels. *Int. J. Mech. Sci.* **10**, 571–82.

[7.33] M. Petyt (1971) Vibration of curved plates. *J. Sound Vibration* **15**, 381–95.

[7.34] E. Hernández, L. Hervella-Nieto and R. Rodríguez (2003) Computation of the vibration modes of plates and shells by low-order MITC quadrilateral finite elements. *Comp. Struct.* **81**, 615–28.

[7.35] T. Irie, G. Yamada and Y. Kobayashi (1984) Free vibration of a cantilever folded plate. *J. Acoust. Soc. Amer.* **76**, 1743–8.

CHAPTER 8

[8.1] G. Wemper (1989) Mechanics and finite elements of shells. *ASME Appl. Mechanics Rev.* **42**, 129–42.

[8.2] R. H. MacNeal (1989) The evolution of lower order plate and shell elements in MSC/NASTRAN. *Finite Elements Anal. Design* **5**, 197–222.

[8.3] A. K. Noor (1990) Bibliography on monographs and surveys on shells. *ASME Appl. Mech. Rev.* **43**, 223–34.

[8.4] W. Gilewski and M. Radwanska (1991) A survey of finite element models for the analysis of moderately thick shells. *Finite Elements Anal. Design* **9**, 1–21.

[8.5] M. Bernadou (1996) *Finite Element Methods for Thin Shell Problems*, New York: John Wiley & Sons.

[8.6] M. L. Bucalem and K. J. Bathe (1997) Finite element analysis of shell structures. *Arch. Comp. Meth. Eng.* **4**, 3–61.

[8.7] G. H. Lim (1999) Vibration of plates and shells using finite elements (1996–1997). *Finite Elements Anal. Design* **31**, 223–30.

[8.8] J. Mackerle (1999) Finite element vibration analysis of beams, plates and shells. *Shock and Vibration* **6**, 97–109.

[8.9] H. Y. T. Yang, S. Saigal, A. Masud and R. K. Kapania (2000) A survey of recent shell finite elements. *Int. J. Num. Meth. Eng.* **47**, 101–27.

[8.10] M. S. Qatu (2002) Recent research advances in the dynamic behaviour of shells 1989–2000. Part 2: homogeneous shells. *ASME Appl. Mech. Rev.* **55**, 415–34.

[8.11] H. Kraus (1967) *Thin Elastic Shells*, New York: John Wiley & Sons.

[8.12] A. W. Leissa (1993) *Vibration of Shells*, Sewickley, PA: Acoustical Society of America.

[8.13] G. B. Warburton (1970) Dynamics of shells. In *Symposium on Structural Dynamics*, Loughborough University. Paper A1.

[8.14] V. V. Novozhilov (1964) *Thin Shell Theory*, Groeningen: Noordhoff.

[8.15] M. Petyt and C. C. Fleischer (1973) Vibration of curved structures using quadrilateral finite elements. In *The Mathematics of Finite Elements and Applications*, London: Academic Press, 367–78.

[8.16] J. J. Webster (1968) Free vibration of rectangular curved panels. *Int. J. Mech. Sci.* **10**, 571–82.

[8.17] M. D. Olson and G. M. Lindberg (1971) Dynamic analysis of shallow shells with a doubly-curved triangular finite element. *J. Sound Vib.* **19**, 229–318.

[8.18] O. C. Zienkiewicz and R. L. Taylor (1991) *The Finite Element Method*, 4th edn, Vol. 2, London: McGraw-Hill.

[8.19] C. T. F. Ross (1984) *Finite Element Programs for Axisymmetric Problems in Engineering*, Chichester: Ellis Horwood.
[8.20] J. J. Webster (1967) Free vibrations of shells of revolution using ring finite elements. *Int. J. Mech. Sci.* **9**, 559–70.
[8.21] S. K. Sen and P. L. Gould (1974) Free vibration of shells of revolution using FEM. *ASCE J. Eng. Mech. Div.* **100**, EM2, 283–303.
[8.22] S. C. Fan and M. H. Luah (1989) Spline finite element for axisymmetric free vibrations of shells of revolution. *J. Sound Vib.* **132**, 61–72.
[8.23] K. S. Surana (1980) Transition finite elements for three-dimensional stress analysis. *Int. J. Num. Meth. Eng.* **15**, 991–1020
[8.24] S. Ahmed, B. M. Irons and O. C. Zienkiewicz (1970) Analysis of thick and thin shell structures by curved finite elements. *Int. J. Num. Meth. Eng.* **2**, 419–51.
[8.25] O. C. Zienkiewicz, J. Too and R. L. Taylor (1971) Reduced integration technique in general analysis of plates and shells. *Int. J. Num. Meth. Eng.* **3**, 275–90.
[8.26] W. Weaver Jr. and P. R. Johnston (1987) *Structural Dynamics by Finite Elements*, Englewood Cliffs, NJ: Prentice-Hall.
[8.27] K.-J. Bathe (1996) *Finite Element Procedures*. Upper Saddle River, NJ: Prentice-Hall.
[8.28] R. D. Cook, D. S. Malkus, M. E. Plesha and R. J. Witt (2002) *Concepts and Applications of Finite Element Analysis*, 4th edn, New York: John Wiley.
[8.29] L. D. Hofmeister and D. A. Evensen (1972) Vibration problems using isoparametric shell elements. *Int. J. Num. Meth. Eng.* **5**, 142–45.
[8.30] M. D. Olson and G. M. Lindberg (1969) Vibration analysis of cantilevered curved plates using a new cylindrical shell finite element. *Proceedings of the Second Conference on Matrix Methods in Structural Mechanics*, Wright-Patterson Air Force Base, Ohio, AFFDL-TR-150, 247–70.
[8.31] C. A. Mota Scares and M. Petyt (1978) Finite element dynamic analysis of practical bladed discs. *J. Sound Vib.* **61**, 561–70.
[8.32] K. S. Surana (1980) Transition finite elements for axisymmetric stress analysis. *Int. J. Num. Meth. Eng.* **15**, 809–32.
[8.33] M. Özakça and E. Hinton (1994) Free vibration analysis and optimisation of axisymmetric plates and shells – 1. finite element formulation. *Comp. Struct.* **6**, 1181–1197.
[8.34] H. Kunieda (1984) Flexural axisymmetric free vibrations of a spherical dome: exact results and approximate solutions. *J. Sound Vib.* **92**, 1–10.

CHAPTER 9

[9.1] J. N. Reddy (2003) *Mechanics of Laminated Composite Plates and Shells: Theory and Analysis*, 2nd edn, Boca Raton, FI: CRC Press.
[9.2] J. N. Reddy (1985) A review of the literature on finite element modelling of laminated composite plates. *Shock Vib. Dig.* **17**(4), 3–8.
[9.3] A. K. Noor and W. S. Burton (1989) Assessment of shear deformation theories for multi-layered composite plates. *Appl. Mech. Rev.* **42**, 1–13.
[9.4] J. N. Reddy (1989) Refined computational models of composite laminates. *Int. J. Num. Meth. Eng.* **27**, 361–82.
[9.5] A. K. Noor and W. S. Burton (1990) Assessment of computational models for multi-layered anisotropic plates. *Composite Struct.* **14**, 233–65.
[9.6] J. N. Reddy (1990) A review of refined theories of laminated plates. *Shock Vib. Dig.* **22**(7), 3–17.
[9.7] A. K. Noor (1992) Mechanics of anisotropic plates and shells – a new look at an old subject. *Comput. Struct.* **44**, 499–514.
[9.8] J. N. Reddy and D. H. Robbins (1994) Theories and computational models for composite laminates. *Appl. Mech. Rev.* **47**, 147–69.
[9.9] E. Carrera (2002) Theories and finite elements for multi-layered, anisotropic, composite plates and shells. *Arch. Comput. Math. Eng.* **9**, 87–140.
[9.10] L. P. Kollar and G. P. Springer (2003) *Mechanics of Composite Structures*, Cambridge: Cambridge University Press.

[9.11] M. S. Qatu (2004) *Vibration of Laminated Shells and Plates*, Kidlington: Elsevier.

[9.12] C. Mei and C. B. Prasad (1989) Effects of large deflection and transverse shear on responses of rectangular symmetrical composite laminates subjected to acoustic excitation. *J. Comp. Materials* **23**, 606–39.

[9.13] E. A. Thornton (1977) Free vibrations of unsymmetrically laminated cantilevered composite panels. *Shock Vibration Bull.* **47** (Part 2), 79–88.

[9.14] T. S. Chow (1971) On the propagation of flexural waves in an orthotropic laminated plate and its response to an impulsive load. *J. Comp. Materials* **5**, 306–319.

[9.15] J. M. Whitney (1973) Shear correction factors for orthotropic laminates under static load. *J. Appl. Mech. Trans. ASME* **40**, 302–304.

[9.16] W. H. Wittrick (1987) Analytical three-dimensional elasticity solutions to some plate problems and some observations on Mindlin plate theory. *Int. J. Solids Struct.* **23**, 441–64.

[9.17] J. N. Reddy (1979) Free vibration of antisymmetric angle-ply laminated plates including transverse shear deformation by the finite element method. *J. Sound Vibration* **66**, 565–76.

[9.18] C. W. Bert and T. L. C. Chen (1978) Effect of shear deformation on vibration of antisymmetric angle-ply laminated rectangular plates. *Int. J. Solids Struct.* **14**, 465–73.

[9.19] J. N. Reddy (1984) A simple higher-order theory for laminated composite plates. *J. Appl. Mech. Trans. ASME* **51**, 745–52.

[9.20] J. N. Reddy (1984) A refined nonlinear theory of plates with transverse shear deformation. *Int. J. Solids Struct.* **20**, 881–906.

[9.21] J. N. Reddy (1990) A general non-linear third-order theory of plates with moderate thickness. *Int. J. Nonlinear Mech.* **25**, 677–86.

[9.22] S.-Y. Lee and S.-C. Wooh (2004) Finite element vibration analysis of composite box structures using the high order plate theory. *J. Sound Vibration* **277**, 801–14.

[9.23] S. Liu (1991) A vibration analysis of composite laminated plates. *Finite Elements Anal. Design* **9**, 295–307.

[9.24] A. K. Ghosh and S. S. Dey (1994) Free vibration of laminated composite plates – a simple finite element based on higher order theory. *Comput. Struct.* **52**, 397–404.

[9.25] M. Soula, R. Nasri, A. Ghazel and Y. Chevalier (2006) The effects of kinematic model approximations on natural frequencies and modal damping of laminated composite plates. *J. Sound Vibration* **297**, 315–28.

[9.26] D. A. Saravanos, P. R. Heyliger and D. A. Hopkins (1997) Layerwise mechanics and finite elements for the dynamic analysis of piezoelectric composite plates. *Int. J. Solids Struct.* **34**, 359–78.

[9.27] D. A. Saravanos and P. R. Heyliger (1999) Mechanics and computational models for laminated piezoelectric beams, plates and shells. *Appl. Mech. Rev.* **52**, 305–19.

[9.28] A. Benjeddo (2000) Advances in piezoelectric finite element modelling of adaptive structural elements. *Comput. Struct.* **76**, 347–63.

[9.29] M. S. Qatu (2002) Recent research advances in the dynamic behaviour of shells: 1989–2000, part 1: Laminated composite shells. *Appl. Mech. Rev.* **55**, 325–49.

[9.30] H. Matsunaga (2002) Assessment of a global higher-order deformation theory for laminated composite and sandwich plates. *Composite Struct.* **56**, 279–291.

[9.31] F. J. Plantema (1966) *Sandwich Construction: The Bending and Buckling of Sandwich Beams. Plates and Shells*, New York: John Wiley & Son.

[9.32] H. G. Allen (1969) *Analysis and Design of Structural Sandwich Panels*, Oxford: Pergamon Press.

[9.33] R. A. Shenoi and J. F. Wellicome (eds) (1993) *Composite Materials in Maritime Structures. Vol. 1: Fundamental Aspects. Vol. 2: Practical Considerations*, Cambridge: Cambridge University Press.

[9.34] Mallikarjura and T. Kant (1993) A critical review and some results of recently developed refined theories of fibre-reinforced laminated composites and sandwiches. *Composite Struct.* **23**, 293–312.

[9.35] J. Mackerle (2002) Finite element analysis of sandwich structures: a bibliography (1980–2001). *Eng. Comput.* **19**, 206–45.

[9.36] A. K. Nayak, S. S. J. Moy and R. A. Shenoi (2002) Free vibration analysis of composite sandwich plates based on Reddy's higher-order theory. *Composites: Part B* **33**, 505–19.

[9.37] M. E. Raville and C. E. S. Ueng (1967) Determination of natural frequencies of vibration of a sandwich plate. *Exp. Mech.* **7**, 490–93.

CHAPTER 10

[10.1] L. Meirovitch and H. Baruh (1983) On the inclusion principle for the hierarchical finite element method. *Int. J. Num. Meth. Eng.* **19**, 281–91.

[10.2] A. Peano (1976) Hierarchies of conforming finite elements for plane elasticity and plate bending. *Comput. Math. Appl.* **2**, 211–24.

[10.3] D. C. Zhu (1986) Development of hierarchical finite element methods at BIAA. *Proceedings of the International Conference on Computational Mechanics*, Tokyo.

[10.4] S. I. Hayek (2001) *Advanced Mathematical Methods in Science and Engineering*, New York: Marcel Dekker.

[10.5] O. Beslin and J. Nicolas (1997) A hierarchical functions set for predicting very high order plate bending modes with any boundary conditions. *J. Sound Vibration* **202**, 633–55.

[10.6] N. S. Bardell (1989) The application of symbolic computing to the hierarchical finite element method. *Int. J. Num. Meth. Eng.* **28**, 1181–1204.

[10.7] N. S. Bardell (1991) Free vibration analysis of a flat plate using the hierarchical finite element method. *J. Sound Vibration* **151**, 263–89.

[10.8] A. W. Leissa (1969) *Vibration of Plates*, Washington DC: US Government Printing Office. (Reprinted 1993 by the Acoustical Society of America.)

[10.9] N. S. Bardell (1992) The free vibration of skew plates using the hierarchical finite element method. *Comput. Struct.* **45**, 841–74.

[10.10] K. M. Liew and K. Y. Lam (1990) Applications of two-dimensional orthogonal plate functions to flexural vibration of skew plates. *J. Sound Vibration* **139**, 241–252.

[10.11] W. Han and M. Petyt (1996) Linear vibration analysis of laminated rectangular plates using the hierarchical finite element method – I. Free vibration analysis. *Comput. Struct.* **61**, 705–12.

[10.12] S. T. Chow, K. M. Liew and K. Y. Lam (1992) Transverse vibration of symmetrically laminated rectangular composite plate. *Compos. Struct.* **20**, 213–26.

[10.13] N. S. Bardell, J. M. Dunsdon and R. S. Langley (1995) Free vibration analysis of thin rectangular laminated plate assemblies using the h-p version of the finite element method. *Compos. Struct.* **32**, 237–46.

[10.14] D. W. Jensen and E. F. Crawley (1984) Frequency determination techniques for cantilevered plates with bending-torsion coupling. *AIAA J* **22**, 415–20.

[10.15] N. S. Bardell (1991) The free vibration of cylindrically-curved rectangular panels. In *Structural Dynamics: Recent Advances*, London: Elsevier, 254–63.

[10.16] A. W. Leissa (1993) *Vibration of Shells*, Sewickley, PA: Acoustical Society of America.

[10.17] L. J. West, N. S. Bardell, J. M. Dunsdon and P. M. Loasby (1997) Some limitations associated with the use of K-orthogonal polynomials in hierarchical versions of the finite element method. In *Structural Dynamics: Recent Advances*, Southampton ISVR: University of Southampton, 217–31.

[10.18] A. Y. T. Leung and J. K. W. Chan (1998) Fourier p-element for the analysis of beams and plates. *J. Sound Vibration* **212**, 179–85.

[10.19] A. Y. T. Leung, B. Zhu, J. Zheng and H. Yang (2004) Analytical trapezoidal Fourier p-element for vibrating plane problems. *J. Sound Vibration* **271**, 67–81.

[10.20] A. Y. T. Leung and B. Zhu (2004) Transverse vibration of thick polygonal plates using analytically integrated trapezoidal Fourier p-element. *Comput. Struct.* **82**, 109–19.

[10.21] C. M. Wang, J. N. Reddy and K. H. Lee (2000) *Shear Deformable Beams and Plates*, Oxford: Elsevier.
[10.22] N. S. Bardell, J. M. Dunsdon and R. S. Langley (1997) Free vibration analysis of coplanar sandwich panels. *Comput. Struct.* **38**, 463–75.
[10.23] M. E. Raville and C. E. S. Veng (1967) Determination of natural frequencies of vibration of a sandwich plate. *Ex. Mech.* **7**, 490–93.
[10.24] N. S. Bardell, J. M. Dunsdon and R. S. Langley (1998) Free vibration of thin, isotropic, open, conical panels. *J. Sound Vibration* **217**, 297–320.
[10.25] N. S. Bardell, R. S. Langley, J. M. Dunsdon and G. S. Aglietti (1999) An h-p finite element vibration analysis of open conical sandwich panels and conical sandwich frusta. *J. Sound Vibration* **226**, 345–77.
[10.26] D. J. Wilkins, C. W. Bert and D. M. Egle (1970) Free vibration of orthotropic sandwich conical shells with various boundary conditions. *J. Sound Vibration* **13**, 211–28.

CHAPTER 11

[11.1] J. H. Wilkinson (1965) *The Algebraic Eigenvalue Problem*, Oxford: Clarendon Press.
[11.2] R. E. D. Bishop, G. M. L. Gladwell and S. Michaelson (1965) *The Matrix Analysis is of Vibration*, Cambridge: Cambridge University Press.
[11.3] A. R. Gourlay and G. A. Watson (1973) *Computational Methods for Matrix Eigenproblems*, Chichester: Wiley.
[11.4] T. J. R. Hughes (1987) *The Finite Element Method: Linear Static and Dynamic Finite Element Analysis*, Englewood Cliffs, NJ: Prentice Hall.
[11.5] H. Kardestuncer and D. H. Norrie (eds) (1988) *Finite Element Handbook*, New York: McGraw-Hill.
[11.6] N. S. Sehmi (1989) *Large Order Structural Eigenanalysis Techniques*, Chichester: Ellis Horwood Ltd.
[11.7] A. Jennings and J. J. McKeown (1992) *Matrix Computation*, 2nd edn, Chichester: John Wiley & Sons.
[11.8] K. J. Bathe (1996) *Finite Element Procedures*, Upper Saddle River, NJ: Prentice Hall.
[11.9] M. Geradin and D. Rixen (1997) *Mechanical Vibrations: Theory and Applications to Structural Dynamics*, 2nd edn, Chichester: John Wiley & Sons.
[11.10] M. Petyt (1998) *Introduction to Finite Element Vibration Analysis*, 1st edn, Cambridge: Cambridge University Press.
[11.11] G. W. Stewart (2001) *Matrix Algorithms Vol. II: Eigensystems*, Philadelphia: SIAM.
[11.12] R. D. Cook, D. S. Malkus, M. E. Plesha and R. J. Witt (2002) *Concepts and Applications of Finite Element Analysis*, 4th edn, New York: John Wiley & Sons.
[11.13] G. M. L. Gladwell (1961) Vibrating systems with equal natural frequencies. *J. Mech. Eng. Sci.* **3**, 178–81.
[11.14] W. Barth, R. S. Martin and J. H. Wilkinson (1967) Calculation of the eigenvalues of a symmetric tridiagonal matrix by the method of bisection. *Numerische Mathematik* **9**, 386–93.
[11.15] G. Peters and J. H. Wilkinson (1969) Eigenvalues of $Ax = \lambda Bx$ with band symmetric A and B. *Computer J.* **12**, 398–404.
[11.16] R. B. Lehoucq and D. C. Sorensen (1996) Deflation techniques for an implicitly restarted Arnoldi iteration. *SIAM J. Matrix Anal. Appl.* **17**, 789–821.
[11.17] G. L. G. Sleijpen and H. A. van der Vorst (1996) A Jacobi-Davidson iteration method for linear eigenvalue problems. *SIAM J. Matrix Anal. Appl.* **17**, 401–25.
[11.18] D. L. Thomas (1979) Dynamics of rotationally periodic structures. *Int. J. Num. Meth. Eng.* **14**, 81–102.
[11.19] A. V. Srinivasan (ed.) (1976) *Structural Dynamic Aspects of Bladed Disk Assemblies*, New York: The American Society of Mechanical Engineers.
[11.20] C. A. Mota Soares, M. Petyt and A. M. Salama (1976) Finite element analysis of bladed discs. In Reference [11.19], 73–91.

[11.21] A. M. Salama, M. Petyt and C. A. Mota Soares (1976) Dynamic analysis of bladed disks by wave propagation and matrix difference techniques. In Reference [11.19], 45–56.

[11.22] A. W. Leissa (1969) Vibration of Plates. NASA SP-160.

[11.23] C. A. Mota Soares and M. Petyt (1978) Finite element analysis of practical bladed discs. *J. Sound Vibration* **61**, 561–70.

[11.24] R. L. Nelson and D. L. Thomas (1978) Free vibration analysis of cooling towers with column supports. *J. Sound Vibration* **57**, 149–53.

[11.25] R. J. Guyan (1965) Reduction of stiffness and mass matrices. *AIAA J.* **3**, 380.

[11.26] B. M. Irons (1965) Structural eigenvalue problems: elimination of unwanted variables. *AIAA J.* **3**, 961–2.

[11.27] G. C. Wright and G. A. Miles (1971) An economical method for determining the smallest eigenvalues of large linear systems. *Int. J. Num. Meth. Eng.* **3**, 25–34.

[11.28] M. Geradin (1971) Error bounds for eigenvalue analysis by elimination of variables. *J. Sound Vibration* **19**, 111–32.

[11.29] R. D. Henshall and J. H. Ong (1975) Automatic masters for eigenvalue economization. *Int. J. Earthquake Engineering and Structural Dynamics* **3**, 375–83.

[11.30] V. N. Shah and M. Raymund (1982) Analytical selection of masters for the reduced eigenvalue problem. *Int. J. Num. Meth. Eng.* **18**, 89–98.

[11.31] N. Bouhaddi (1992) A method for selecting master DOF in dynamic substructuring using the Guyan condensation method. *Comput. Struct.* **45**, 941–46.

[11.32] R. G. Anderson, B. M. Irons and O. C. Zienkiewicz (1968) Vibration and stability of plates using finite elements. *Int. J. Solids Structures* **4**, 1031–55.

[11.33] R. Levy (1971) Guyan reduction solutions recycled for improved accuracy. In NASTRAN: Users' Experiences, NASA TM X-2378, 201–20.

[11.34] N. Popplewell, A. W. M. Bertels and B. Arya (1973) A critical appraisal of the elimination technique. *J. Sound Vibration* **31**, 213–33.

[11.35] D. L. Thomas (1982) Errors in natural frequency calculations using eigenvalue economization. *Int. J. Num. Meth. Eng.* **18**, 1521–27.

[11.36] R. Lin and Y. Xia (2003) A new eigensolution of structures via dynamic condensation. *J. Sound Vibration* **266**, 93–106.

[11.37] Z.-Q. Qu (2004) *Model Order Reduction Techniques with Applications in Finite Element Analysis*, Berlin: Springer.

[11.38] L. Meirovitch (1980) *Computational Methods in Structural Dynamics*, Rijn, The Netherlands: Sijthoff & Noordhoff.

[11.39] R. R. Craig Jr. and A. J. Kurdila (2006) *Fundamentals of Structural Dynamics*, New York: John Wiley.

[11.40] K. Kubomura (1982) A theory of substructure modal synthesis. *Trans. AMSE, J. Appl. Mech.* **49**, 903–9.

[11.41] R. R. Craig Jr. (1987) A review of time-domain and frequency-domain component mode synthesis. *J. Modal Anal.* **2**, 59–72.

[11.42] R. R. Craig Jr. (1995) Substructure methods in vibration. *Trans. ASME* **117**, 207–13.

[11.43] W.-H. Shyu, Z.-D. Ma and G. M. Hulbort (1997) A new component mode synthesis method: quasi-static mode compensation. *Finite Elements Anal. Design* **24**, 271–81.

[11.44] D.-M. Tran (2001) Component mode synthesis methods using interface modes: application to structures with cyclic symmetry. *Comput. Struct.* **79**, 209–22.

[11.45] R. R. Craig and M. C. C. Bampton (1968) Coupling of substructures for dynamic analysis. *AIAA J.* **6**, 1313–19.

[11.46] L. E. Suarez and M. P. Singh (1992) Improved fixed interface method for modal synthesis. *AIAA J.* **30**, 2952–58.

[11.47] R. R. Craig Jr. and A. L. Hale (1988) Block-Krylov component synthesis method for structural model reduction. *AIAA J. Guidance, Control Dynamics* **11**, 562–70.

[11.48] J. H. Wang and H. R. Chen (1990) A substructure modal synthesis method with high computational efficiency. *Comput. Methods Appl. Mech. Eng.* **79**, 203–17.

[11.49] S.-N. Hou (1969) *Review of modal synthesis techniques and a new approach. Shock and Vibration Bull.* **40**(4), 25–39.

[11.50] L. Jezequel and S. T. Tchere (1991) A procedure for improving component-mode representation in structural dynamic analysis. *J. Sound Vibration* **144**, 409–19.

[11.51] N. Bhouhaddi and J. P. Lombard (2000) Improved free-interface substructures representation method. *Comput. Struct.* **77**, 269–83.

[11.52] M. A. Tournour, N. Atalla, O. Chiello and F. Sgard (2001) Validation, performance, convergence and application of free interface component mode synthesis. *Comput. Struct.* **79**, 1861–76.

[11.53] B. Haggblad and L. Eriksson (1993) Model reduction methods for dynamic analyses of large structures. *Comput. Struct.* **43**, 735–49.

[11.54] J. H. Kuang and Y. G. Tsuei (1983) A more general method of substructure mode synthesis for dynamic analysis. *AIAA J.* **23**, 618–23.

[11.55] A. L. Klosterman (1976) Modal survey of weakly coupled systems. SAE Paper 760876.

[11.56] K. H. Ghlaim and K. F. Martin (1984) Reduced component modes in damped systems. In *Proc. Int. Conf. on Modal Analysis*. Schenectady, NY: Union College, 683–9.

[11.57] D. J. Ewins (2000) *Modal Testing: Theory, Practice and Application*, 2nd edn, Letchworth: Research Studies Press.

[11.58] T. Yang, S.-H. Fan and C.-S. Lin (2003) Joint stiffness identification using FRF measurements. *Comput. Struct.* **81**, 2549–56.

[11.59] J. H. Wang and S. C. Chuang (2004) Reducing errors in the identification of structural joint parameters using error functions. *J. Sound Vibration* **273**, 295–316.

CHAPTER 12

[12.1] L. Meritovitch (1967) *Analytical Methods in Vibration*, New York: Macmillan.

[12.2] S. H. Crandall (1970) The role of damping in vibration theory. *J. Sound Vibration* **11**, 3–18.

[12.3] C. F. Beards (1983) *Structural Vibration Analysis*, Chichester: Ellis Horwood.

[12.4] Lord Rayleigh (1945) *The Theory of Sound*, vol. I. New York: Dover.

[12.5] S. Adhirkari (2006) Damping modelling using generalized proportional damping. *J. Sound Vibration* **293**, 156–70.

[12.6] D. I. G. Jones (2001) *Handbook of Viscoelastic Vibration Damping*, Chichester: John Wiley & Sons.

[12.7] S. Sorrentino and A. Fasana (2007) Finite element analysis of vibrating linear systems with fractional derivative viscoelastic models. *J. Sound Vibration* **299**, 839–853.

[12.8] K. Dovstam (1995) Augmented Hooke's law in frequency domain. A three-dimensional, material damping formulation. *Int. J. Solids Struct.* **32**, 2835–52.

[12.9] J. C. Snowdon (1963) Representation of the mechanical damping possessed by rubberlike materials and structures. *J. Acoust. Soc. Amer.* **35**, 821–9.

[12.10] J. C. Snowdon (1968) *Vibration and Shock in Damped Mechanical Systems*, New York: Wiley.

[12.11] B. Lazan (1968) *Damping of Materials and Members in Structural Mechanics*, New York: Pergamon.

[12.12] E. E. Ungar (1973) The status of engineering knowledge concerning the damping of built-up structures. *J. Sound Vibration* **26**, 141–54.

[12.13] C. W. Bert (1973) Material damping: an introductory review of mathematical models, measures and experimental techniques. *J. Sound Vibration* **29**, 129–53.

[12.14] A. D. Nashif, D. I. G. Jones and J. P. Henderson (1985) *Vibration Damping*, New York: John Wiley & Sons.

[12.15] D. J. Ewins (2000) *Modal Testing: Theory, Practice and Application*, 2nd edn, Letchworth: Research Studies Press.

[12.16] T. K. Caughey and M. E. J. O'Kelly (1965) Classical normal modes in damped linear dynamic systems. *ASME J. Appl. Mech.* **32**, 583–88.

[12.17] K. A. Foss (1958) Coordinates which uncouple the equations of motion of damped linear dynamic systems. *ASME J. Appl. Mech.* **25**, 361–64.

[12.18] D. J. Mead (1970) The existence of normal modes of linear systems with arbitrary damping. *Proceedings of the Symposium on Structural Dynamics*, Paper C5, Loughborough University of Technology.

[12.19] M. Geradin and D. Rixen (1997) *Mechanical Vibrations: Theory and Applications to Structural Dynamics*, 2nd edn, Chichester: John Wiley & Sons.

[12.20] M.-C. Kim and I.-W. Lee (1999) A computationally efficient algorithm for the solution of eigenproblems for large structures with non-proportional damping using Lanczos method. *Int. J. Earthquake Eng. Struct. Dynamics* **28**, 157–172.

[12.21] H. J. Bowdler, R. S. Martin, G. Peters and J. H. Wilkinson (1966) Solution of real and complex systems of linear equations. *Numerische Mathematik* **8**, 217–34.

[12.22] J. S. Bendat and A. G. Piersol (1971) *Random Data: Analysis and Measurement Procedures*, New York: Wiley-Interscience.

[12.23] G. B. Warburton (1976) *The Dynamical Behaviour of Structures*, 2nd edn, Oxford: Pergamon Press.

[12.24] J. S. Przemieniecki (1968) *Theory of Matrix Structural Analysis*, New York: McGraw-Hill.

[12.25] R. W. Clough and J. Penzien (1995) *Dynamics of Structures*, New York: McGraw-Hill.

[12.26] J. E. Grant (1971) Response computation using truncated Taylor series. *J. Eng. Mech. Proc. ASME* **97** EM2, 295–304.

[12.27] S. Levy and J. P. D. Wilkinson (1976) *The Component Element Method in Dynamics*, New York: McGraw-Hill.

[12.28] J. C. Houbolt (1950) A recurrence matrix solution for the dynamic response of elastic aircraft. *J. Aeronaut. Sci.* **17**, 540–50.

[12.29] K.-J. Bathe (1996) *Finite Element Procedures*, Upper Saddle River, NJ: Prentice Hall.

[12.30] N. M. Newmark (1959) A method of computation for structural dynamics. *J. Eng. Mech. Proc. ASCE* **85**, 67–94.

[12.31] R. H. MacNeal and C W. McCormick (1971) The NASTRAN computer program for structural analysis. *Comput. Struct.* **1**, 389–412.

[12.32] E. L. Wilson, I. Farhoomand and K. J. Bathe (1973) Nonlinear dynamic analysis of complex structures. *Int. J. Earthquake Engineering and Structural Dynamics* **1**, 241–52.

[12.33] E. Hinton, T. Rock and O. C. Zienkiewicz (1976) A note on mass lumping and related processses in the finite element method. *Int. J. Earthquake Engineering and Structural Dynamics* **4**, 245–9.

[12.34] G. C. Archer and T. M. Whalen (2005) Development of rotationally consistent diagonal mass matrices for plate and beam elements. *Comput. Methods Appl. Mech. Eng.* **194**, 675–89.

[12.35] T. J. R. Hughes (1987) *The Finite Element Method. Linear Static and Dynamic Finite Element Analysis*, Englewood Cliffs, NJ: Prentice Hall.

[12.36] H. Kardestuncer and D. H. Norrie (eds) (1988) *Finite Element Handbook*, New York: McGraw-Hill.

[12.37] A.-V. Idesman, M. Schmidt and R. L. Sierakowski (2008) A new explicit predictor-multicorrector high-order accurate method for linear elastodynamics. *J. Sound Vibration* **310**, 217–29.

[12.38] W. L. Wood (1990) *Practical Time-Stepping Schemes*, Oxford: Clarendon Press.

CHAPTER 13

[13.1] J. D. Robson (1963) *An Introduction to Random Vibration*, Edinburgh: Edinburgh University Press.

[13.2] S. H. Crandall and W. D. Mark (1963) *Random Vibration in Mechanical Systems*, New York: Academic Press.

[13.3] Y. K. Lin and G. Q. Cai (2004) *Probabilistic Structural Dynamics*, New York: McGraw-Hill.

[13.4] J. S. Bendat and A. G. Piersol (2000) *Random Data: Analysis and Measurement Procedures*, 3rd edn, New York: John Wiley.

[13.5] D. E. Neweland (1975) *An Introduction to Random Vibrations and Spectral Analysis*, London: Longman.

[13.6] R. W. Clough and J. Penzien (1993) *Dynamics of Structures*, 2nd edn, New York: McGraw-Hill.

[13.7] G. B. Warburton (1976) *The Dynamical Behaviour of Structures*, 2nd edn, Oxford: Pergamon Press.

[13.8] M. D. Olson and G. M. Lindberg (1970) Free vibrations and random response of an integrally stiffened panel. Proc. Conf. Current Developments in Sonic Fatigue, Southampton University. Available from ISVR, Southampton University.

[13.9] M. D. Olson (1972) A consistent finite element method for random response problems. *Comput. Struct.* **2**, 163–80.

[13.10] M. D. Olson and G. M. Lindberg (1970) Free Vibrations and Random Response of an Integrally Stiffened Panel. National Research Council of Canada Aeronautics Report LR-544.

[13.11] M. D. Olson and G. M. Lindberg (1971) Jet noise excitation of an integrally stiffened panel. *J. Aircraft* **8**, 847–55.

[13.12] D. J. Mead and K. K. Pujara (1971) Space harmonic analysis of periodically supported beams: response to converted random loading. *J. Sound Vibration* **14**, 525–41.

[13.13] B. Etkin (1972) *Dynamics of Atmospheric Flight*, New York: Wiley.

[13.14] R. E. Davis (1966) Statistical dependence effect of normal mode response. *AIAA J.* **4**, 2033–4.

[13.15] C. M. Harris and A. G. Piersol (eds) (2001) *Shock and Vibration Handbook*, 5th edn, New York: McGraw-Hill.

[13.16] R. E. Cornwell, R. R. Craig and C. P. Johnson (1983) On the application of the mode-acceleration method to structural engineering problems. *Int. J. Earthquake Engineering and Structural Dynamics* **11**, 679–88.

[13.17] D. J. Ewins (2000) *Modal Testing: Theory, Practice and Application*, 2nd edn, Letchworth: Research Press.

[13.18] O. E. Hansteen and K. Bell (1979) On the accuracy of mode superposition analysis in structural dynamics. *Int. J. Earthquake Engineering and Structural Dynamics* **7**, 405–11.

[13.19] K.-J. Bathe (1996) *Finite Element Procedures*, Upper Saddle River, NJ: Prentice Hall.

[13.20] E. L. Wilson, M.-Y. Yuun and J. M. Dickens (1982) Dynamic analysis by direct superposition of Ritz vectors. *Int. J. Earthquake Eng. Struct. Dynamics* **10**, 813–82.

[13.21] R. R. Craig Jr. (2002) Krylov-Lanczos methods. In *Encyclopedia of Vibration*, Vol. 2, London: Academic Press, 691–98.

[13.22] R. R. Arnold (1985) Application of Ritz vectors for dynamic analysis of large structures. *Comput. Struct.* **21**, 901–8.

[13.23] T. G. Butler (1982) Using NASTRAN to solve symmetric structures with nonsymmetric loads. *Tenth NASTRAN Users' Colloquium, NASA Conference Publication* **2249**, 216–32.

[13.24] D. L. Thomas (1979) Dynamics of rotationally periodic structures. *Int. J. Num. Meth. Eng.* **14**, 81–102.

[13.25] N. Popplewell, A. W. M. Bertels and B. Arya (1973) A critical appraisal of the elimination technique. *J. Sound Vibration* **31**, 213–33.

[13.26] Z.-Q. Qu (2004) *Model Order Reduction Techniques with Applications in Finite Element Analysis*, Berlin: Springer.

CHAPTER 14

[14.1] I. M. Smith and D. V. Griffiths (eds) (1997) *Programming the Finite Element Method*, Chichester: John Wiley & Sons.

[14.2] E. Hinton and D. R. J. Owen (1979) *An Introduction to Finite Element Computations*, Swansea: Pineridge Press.

[14.3] R. D. Cook, D. S. Malkus, M. E. Plesha and R. J. Witt (2002) *Concepts and Applications of Finite Element Analysis*, 4th edn, New York: John Wiley & Sons.

[14.4] E. Hinton and D. R. J. Owen (1984) *Finite Element Software for Plates and Shells*, Swansea: Pineridge Press.

[14.5] E. Hinton (ed.) (1988) *Numerical Methods and Software for Dynamic Analysis of Plates and Shells*, Swansea: Pineridge Press.

[14.6] S. S. Rao (2005) *The Finite Element Method in Engineering*, 4th edn, Oxford: Elsevier.

[14.7] Y. W. Kwon and H. Bang (2000) *The Finite Element Method using* MATLAB, 2nd edn, Boca Raton FL: CRC Press.

[14.8] R. E. Kielb and A. W. Leissa (1985) Vibrations of twisted cantilever plates – a comparison of theoretical results. *Int. J. Num. Meth. Eng.* **21**, 1365–80.

[14.9] D. J. Ewins and M. Imregun (1986) State-of-the-art assessment of structural dynamic response analysis methods (DYNAS). *Shock and Vibration Bull.* **56**(1), 59–90.

[14.10] D. J. Ewins and M. Imregun (1987) A survey to assess structural dynamic response prediction capabilities: DYNAS. In *Quality Assurance in FEM Technology* (ed. J. Robinson). Okehampton: Robinson and Associates, 604–15.

[14.11] T. J. R. Hughes (1987) *The Finite Element Method: Linear Static and Dynamic Finite Element Analysis*. Englewood Cliffs: Prentice-Hall.

[14.12] B. Irons and S. Ahmad (1980) *Techniques of Finite Elements*, Chichester: Ellis Horwood.

[14.13] M. Petyt (1990) *Introduction to Finite Element Vibration Analysis*, 1st edn, Cambridge: Cambridge University Press.

[14.14] O. C. Zienkiewicz (1977) *The Finite Element Method*, 3rd edn, London: McGraw-Hill.

[14.15] J. P. Wolfe and C. Song (1996) *Finite Element Modelling of Unbounded Media*, Chichester: John Wiley & Sons.

[14.16] (1986) *A Finite Element Primer*, Glasgow: National Engineering Laboratory.

[14.17] (1987) *Proc. Int. Conf. on Quality Assurance and Standards in Finite Element Analysis*, Glasgow: National Engineering Laboratory.

[14.18] J. Robinson (ed.) (1987) *Quality Assurance in FEM Technology*, Okehampton: Robinson and Associates.

[14.19] J. Robinson (1985) Basic and shape sensitivity tests for membrane and plate bending finite elements. NAFEMS C2.

[14.20] A. J. Morris (1985) Shell finite element evaluation tests. NAFEMS C4.

[14.21] J. Robinson (1979) An Evaluation of Lower Order Membranes as contained in the MSC/NASTRAN, ASAS and PAFEC systems. Report to Royal Aircraft Establishment, Farnborough. MoD Contract No. A93b/494.

[14.22] J. Robinson and S. Blackham (1981) An evaluation of lower order membranes as contained in the ANSYS and SAP4 FEM systems. *Finite Element News*, Issue No. 2.

[14.23] J. Robinson and S. Blackham (1981) An evaluation of plate bending elements – MSC/NASTRAN, ASAS, PAFEC, ANSYS, SAP4. Robinson and Associates report. ISBN 0 9507649 0 6. Okehampton: Robinson and Associates.

[14.24] D. Hitchings, A. Kamoulakos and G. A. O. Davies (1987) Linear static benchmarks – Vol. 1. NAFEMS P07.

[14.25] D. Hitchings, A. Kamoulakos and G. A. O. Davies (1987) Linear static benchmarks – Vol. 2. National Engineering Laboratory Report, NAFEMS P08.

[14.26] I. C. Taig (1992) *Finite Element Analysis of Composite Materials*, East Kilbride: NAFEMS R0003.

[14.27] S. Hardy (2001) *Composite Benchmarks*, East Kilbride: NAFEMS R0031.

[14.28] J. Robinson (1978) Element evaluation – a set of assessment points and standard tests. *Finite Element Methods in the Commercial Environment*, 217–47. Okehampton: Robinson and Associates.

[14.29] F. Abbassian, F. Dawswell and N. C. Knowles (1987) *Selected Benchmarks for Natural Frequency Analysis*, East Kilbride: NAFEMS R0015.

[14.30] A. Rahman and M. Petyt (1995) *Fundamental Tests for Forced Vibrations of Linear Elastic Structures*, East Kilbride: NAFEMS R0034.

[14.31] M. J. Friswell and J. E. Mottershead (1995) *Finite Element Model Updating in Structural Dynamics*. Dordrecht: Kluwer.

[14.32] D. J. Ewins (2000) *Modal Testing. Theory, Practice and Applications*, 2nd edn, Baldock: Research Studies Press.

[14.33] S. G. Braun, D. J. Ewins and S. S. Rao (eds) (2002) *Encyclopedia of Vibration*, London: Academic Press.

Index

amplitude decay, 393
anisotropic material, 19
anti-resonance, 363
area coordinates, 137
Arnoldi method, 335
artificial damping, 393
aspect ratio of elements, 459
assembly of element matrices, 60, 75, 82, 91, 124
autocorrelation function, 418
automatic mesh generation, 452
axial symmetry, 33, 148
axisymmetric elements, 153, 160
axisymmetric shell elements, 277

band limited white noise, 421
beam elements, 72, 77, 84, 95, 108, 112
benchmark tests, 463
bisection method, 335
boundary conditions
  antisymmetric, 91, 336
  geometric, 40
  imposition of, 36
  natural, 40
  skew, 82
  symmetric, 91, 336
building block approach, 349

central difference method, 384, 403
central limit theorem, 415
central moments, 414
Cholesky decomposition, 323
classical plate theory, 285
complete polynomial, 47
completeness tests, 463
complex stiffness, 36
component mode synthesis, 349, 446
composites, 284
condensation
  massless degrees of freedom, 344
  static, 140, 183, 321
conditional stability, 388
conservative force, 4
consistent inertia matrix, 404
constant acceleration method, 395
constraint conditions, 12, 82
constraint modes, 350
coordinate systems, 78, 85
core elements, 160
correlation coefficient, 416
co-spectrum matrix, 431
covariance, 416
cross-correlation function, 422
cross-spectral density, 422
Crout factorisation, 373

d'Alembert's principle, 2
damping, 2, 35, 358, 370
damping matrix, 11
Davidson method, 336
deterministic forces, 413
direct analysis, 370, 402, 426
direction cosine array, 80
Dirichlet conditions, 377
discrete Kirchhoff shear elements, 240
dissipation function, 9, 35
Doolittle-Crout factorisation, 373
Duhamel integral, 381
dynamic equilibrium, 1, 62

effective force, 440
eigenproblem, 49, 316
eigenvalues, 48, 316
eigenvalue shift, 320
eigenvectors, 49, 316
element displacement functions, 55
element distortion, 458
element geometric properties, 451
element material properties, 451
element reordering, 452
element shrinking, 452
ensemble averages, 418
equations of motion, 1, 8–12, 36
ergodic process, 419
excess of Kurtosis, 415

expected value, 414
explicit integration, 385

facet shell elements, 252
false zero energy mode tests, 462
Fast Fourier Transform algorithm, 380
finite element, 55
finite element method, 53
first moment, 414
first-order shear deformation plate theory, 288
flat plate bending elements, 192
folded plates, 257
forced vibration, 357, 413
forward substitution, 323
Fourier integral, 419
Fourier series, 149, 377
Fourier transform, 419
free vibration, 48, 316
front solution method, 452

Gauss elimination, 331
Gaussian probability distribution, 415
Gauss–Legendre integration, 102
generalised coordinates, 140
generalised forces, 10
geometric invariance, 119
Givens' method, 335
graphics, 452
Guyan reduction, 344

Hamilton's principle, 4, 7
harmonic response, 361
Hermitian interpolation functions, 95
Hermitian matrix, 340, 431
hexahedral elements, 165, 170, 183
hidden line plot, 451
Hierarchical finite element method, 297
Houbolt method, 389, 408
Householder method, 325

implicit integration, 395
imposed displacements, 440
impulse response function, 381
integration formulae, 101, 129, 156, 168
inertia force, 1
inertia matrix
  Axisymmetric shell, 282
  Axisymmetric solid, 154, 159, 160, 162
  beam, 74, 98, 111, 302
  facet shell, 254
  membrane, 123, 128, 134, 138
  middle surface shell, 276
  plate, 196, 210, 217, 220, 227, 231, 234
  plate stiffener, 250, 255
  rod, 58, 93, 300, 308
  shaft, 71
  solid, 166, 172, 176, 181
  three-dimensional framework, 87
  two-dimensional framework, 79, 81
invariance, 119
invariance tests, 463
isoparametric elements, 133, 143, 170, 183, 189, 239
isotropic material, 19

Jacobi-Davidson iterative method, 336
Jacobian matrix, 134, 172
Jacobi method, 335
joining unlike elements, 459
joint probability density, 415

kinetic energy, 6, 21, 23, 24, 26, 28, 30, 31, 33, 34, 267, 287, 290, 294
Kirchhoff hypothesis, 240

Lagrange's equations, 8
Lagrange interpolation functions, 94
laminated plates, 285
laminated shells, 295
Lanczos method, 335
least squares, 106
Legendre polynomials, 104, 206, 297
linear acceleration method, 394
linear dependency, 46
linear elements
  axisymmetrical solid, 153
  beam, 108
  membrane, 120, 127, 132, 138
  plate, 209, 234
  rod, 56
  shaft, 70
  solid, 165, 170, 174, 180
load matrix
  axisymmetric shell, 283
  axisymmetric solid, 159, 160, 163
  beam, 75, 98
  membrane, 124, 131, 136
  middle surface shell, 277
  plate, 199, 213, 218, 220, 228, 236
  rod, 60, 93
  shaft, 71
  solid, 169, 174, 177, 182
  three-dimensional framework, 88
  two-dimensional framework, 80, 81
locking of a mesh, 464
loss factor, 36
lower triangular matrix, 323
LR method, 335
lumped mass matrix, 404

massless degrees of freedom, 344
mass matrix, *see* inertia matrix, lumped mass matrix
mass-proportional damping, 360
master degrees of freedom, 343
mean, 414
membrane elements, 119
mid-side nodes, 93, 140
middle-surface shell elements, 268, 269
modal analysis, 357, 361, 381, 430, 442

modal constant, 362
modal damping ratio, 359
mode acceleration method, 435
modelling, 458
mode shape, 49
multiple eigenvalue, 316
multi-point constraints, 451
multivariate probability density, 416

narrow band process, 425
natural coordinates, 139, 159
natural frequencies, 48
Newmark method, 394, 409
Newton's second law, 1
nodal degrees of freedom, 53
nodeless degrees of freedom, 140, 183
node point, 53
node renumbering, 452
non-axisymmetric loads, 149
non-conservative forces, 4
normal distribution, 415
normalised eigenvectors, 316
numerical integration, 101, 129, 156, 168

orthogonality of eigenvectors, 322
orthogonal matrix, 334
orthogonal similarity transformation, 334
orthotropic material, 19

parasitic shear, 130
patch test, 463
peakedness, 415
pentahedral elements, 174, 188
period elongation, 393
periodic excitation, 377
periodic structures, 338
plane strain element, 147
plane stress element, 26
plate elements, 192
Poisson's ratio, 28
positive definite matrix, 12
positive semi-definite matrix, 12
potential energy, 5
power spectral density, 420
pre- and post-processors, 450
principle of virtual displacements, 3
probability, 413
probability density, 414
probability distribution function, 414
proportional damping, 360

QL method, 335
QR method, 335
quadratic elements, 139, 183, 239
quadrilateral elements, 132
quasi-static displacements, 440

random excitation, 413
random process, 416
random process theory, 413
Rayleigh–Ritz method, 45
Rayleigh-type damping, 359
receptance, 362
rectangular elements, 127
reduced energy expressions, 152
reduced integration, 109
reduction technique, 344
residual flexibility, 437
resonance, 363
response time history, 381
rigid body modes, 319
rigid links, 113
Ritz vector analysis, 438
rod elements, 56, 93
rotary inertia, 25, 95
rotationally periodic structures, 338, 444
rotation of axes, 80

Saint-Venant theory of torsion, 21
sandwich plates and shells, 295
semi-infinite regions, 462
shaft elements, 70
shape functions, 55
shape sensitivity tests, 463
shear deformation, 25
shell elements, 266
shifting of eigenvalues, 320
shrink plot, 452
sign count function, 327
similarity transformation, 334
Simpson's rule, 382
simultaneous iteration, 335
single element test, 463
singular matrix, 217
skew boundary conditions, 82
skewness, 415
skyline, 452
slave degrees of freedom, 343
soil-structure interaction, 462
solid elements, 148
solids of revolution, 33, 148
space-time correlation function, 422
spectral radius, 391
spread, 414
stability
  conditional, 388
  unconditional, 393
standard deviation, 415
static condensation, 140, 183, 321
stationary process, 417
step-by-step integration methods, 383
stiffened plates, 248
stiffness matrix
  axisymmetric shell, 282
  axisymmetric solid, 155, 159, 160, 162
  beam, 75, 98, 110, 301
  facet shell, 254
  membrane, 124, 129, 135, 138
  middle surface shell, 275
  plate, 197, 210, 218, 220, 228, 232, 235

stiffness matrix (*cont.*)
  plate stiffener, 251, 256
  rod, 59, 93, 299, 308
  shaft, 71
  solid, 168, 173, 176, 181
  three-dimensional framework, 88
  two-dimensional framework, 79, 81
stiffness-proportional damping, 366
stochastic process, 416
strain components, 19, 34, 266, 286
strain-displacement relationships, 19, 34, 267, 286, 288, 292
strain energy, 5, 21, 23, 24, 26, 28, 30, 31, 32, 34, 266, 273, 286, 289, 293
stress components, 19, 160
stress computation, 66, 71, 77, 106, 160, 169, 199
stress-strain relationships, 27, 32
structural damping, 35, 358
Sturm sequence, 326
sub-parametric element, 142
subspace iteration, 335
substructure analysis, 349
symmetry, 336, 443

tetrahedral elements, 180, 189
third-order shear deformation plate theory, 290
three-dimensional frameworks, 84
time averages, 418
time series, 416
torsion, 21
transformation of generalised eigenproblem, 322
transient response, 381
trapezium rule, 382
triangular decomposition, 322
triangular elements, 120, 138, 241
tri-diagonal matrix, 327
truncation of modes, 431
two-dimensional frameworks, 77

unconditional stability, 393
upper triangular matrix, 373

variance, 415
virtual displacement, 3
virtual work, 21, 25, 28, 30, 33, 34
viscous damping, 2, 35, 359
volume coordinates, 178

warping, 21
wavenumber, 429
wavenumber/frequency spectrum, 429
wavenumber vector, 429
wave propagation, 339
wave receptance function, 430
weakly stationary process, 417
Wilson θ method, 400, 409

zero energy modes, 463

Printed in the United States
By Bookmasters